The Western Woodlands of Ethiopia

A study of the woody vegetation and flora between the Ethiopian Highlands and the lowlands of the Nile Valley in the Sudan and South Sudan

Abstract

The western woodlands of Ethiopia have been less studied than the highlands, with a first description published in 1940. Based on field work and herbarium studies we intend to increase the knowledge of these woodlands. Some authors describe them as one continuous zone from south to north, spanning ca. 10° latitude and dominated by species of *Combretum* and *Terminalia* but sometimes with various names applied to the vegetation at its limits. The *Atlas of the Potential Vegetation of Ethiopia* (2010) followed the continuous model. In the Sudan and South Sudan, the same vegetation is described as distinct types defined on soil and rainfall. Reviewing species composition, environment, research-history and human influence, we present our own observations of the woody species and map their distribution in Ethiopia, review their ecological adaptations and model the total distribution in Africa of twelve characteristic species. We use our data for clustering and principal component analyses to study continuity and discontinuity of the vegetation and the drivers of variation. The appearance of the western woodlands is variable, and cluster analyses indicate that a high number of small and narrowly defined units might be identified. A significant number of trees are widespread from north to south, but some are restricted to the north, a few to the centre and south, and many just enter the woodlands with a main distribution elsewhere. At a high level, our analysis defines two or three weakly defined clusters, (1) the northern woodlands to just south of the Abay River and continuing far south as a brim along the western escarpment of Ethiopia, (2) woodlands south of the Abay at altitudes above cluster no. (1), and (3) dry woodlands of the upper Tacazze and Abay Rivers. Many environmental factors explain small but important parts of the variation. The clusters relate to variables such as latitude, altitude, climate and soil types, while slope, fire frequency and other parameters are less important. Variation in the ecological adaptation of the woody species is limited, and there are no sharp discontinuities in species diversity. The Omo Valley, previously included with the western woodlands, is probably best treated as a transition zone to the Somalia-Masai *Acacia-Commiphora* bushland.

Ib Friis
Natural History Museum of Denmark
University of Copenhagen
Denmark

Paulo van Breugel
HAS University of Applied Sciences
's-Hertogenbosh
The Netherlands

Odile Weber
Botanical Section
National Museum of Natural History
Luxembourg

Sebsebe Demissew
Dept. of Plant Biology and Biodiversity Management
Addis Ababa University
Ethiopia

The Western Woodlands of Ethiopia

A study of the woody vegetation and flora between the Ethiopian Highlands and the lowlands of the Nile Valley in the Sudan and South Sudan

by Ib Friis, Paulo van Breugel, Odile Weber and Sebsebe Demissew

Scientia Danica. Series B, Biologica · vol. 9

DET KONGELIGE DANSKE VIDENSKABERNES SELSKAB

Printed in Denmark by Narayana Press
Graphic design by Dan Eggers
ISSN 1904-5484 · ISBN 978-87-7304-440-7

Submitted to the Academy December 2020
Published January 2022

Table of Contents

1
Introduction

This work is a continuation of ideas and research presented in *Forests and Forest Trees of Northeast Tropical Africa* (Friis 1992) and *Atlas of the Potential Vegetation of Ethiopia* (Friis et al. 2010). Here we focus on the vegetation of the western escarpment of the Ethiopian Highlands and in the lowlands between the escarpment and the border with the Sudan and South Sudan. This is an up to 200 km wide zone stretching from the southernmost point at the border with Kenya at ca. 4° 25' N to the northernmost point of the border with Eritrea at ca. 14° 50' N, with a latitudinal span of 10° 25', and covering ca. 1170 km between the northern and southernmost points. An equivalent distance at the same latitudes in America would be from Bogota in Columbia to the border between Guatemala and Mexico, or in Asia from the southern point of Sri Lanka to the southern border of the state of Goa in India, or in Europe, at higher latitudes, from Rome in Italy to Hamburg in northern Germany. We describe and analyse the woody flora and vegetation of those areas in a broad context with relation to the Ethiopian Highlands, to the vegetation in Sudan and South Sudan and to the belt of woodlands across Africa to the Atlantic Ocean, as well as generally from south to north.

In previous works, the vegetation on the western escarpment of the Ethiopian Highlands and most of the western lowlands of Ethiopia has been described either as several zones along the western escarpment of the Ethiopian Highlands (Pichi Sermolli 1957) or as one continuous zone ending some distance north of Lake Turkana (White 1983). The entire woodland vegetation on the western Ethiopian escarpment and in the western lowlands between the Ethiopian borders with Eritrea and Kenya is what we in this work refer to as the western woodlands of Ethiopia. Although the vegetation in the largest part of the western woodlands is dominated by species of *Combretum* and *Terminalia*, there is also vegetation in which the genera *Acacia* (in the traditional meaning of this generic name, including *Senegalia* and *Vachellia*), *Balanites*, *Boswellia*, *Anogeissus*, *Pterocarpus* and other genera are important.

The vegetation on just the other side of the border with Sudan and South Sudan has been described as a sequence of zones with alternating vegetation types and floristic patterns along a south-north rainfall gradient, ranging from relatively high rainfall in the south to very low rainfall in the north (Harrison & Jackson 1958), and some of the main research questions in this work have dealt with the coherence of this vegetation, how it is delimitated towards the vegetation types further to the north and further to the south, and if there are more vegetation types involved than the extensive woodlands with species of *Combretum*, *Terminalia*, and *Boswellia*, *Pterocarpus lucens*, *Anogeissus leiocarpa* and *Stereospermum kunthianum*.

In a study of the forests and forest flora of the Horn of Africa (Ethiopia, Eritrea, Djibouti, Somalia) the first author of this work (Friis 1992) reviewed the ecological range and distribution of the trees of the Horn of Africa and proposed an initial classification of the forest types, as well as a phytogeographical classification of the species. That work (Friis 1992) included only vegetation that could be characterized as closed forest, not woodlands, but many of the questions relating to vegetation and phytogeography also touched upon in this work were already approached in 1992.

In the *Atlas of the Potential Vegetation of Ethiopia* (Friis et al. 2010), we described the highlands of Ethiopia as a sequence of vegetation types from *Moist evergreen Afromontane Forest* (MAF) in the south to *Dry Afromontane Forest* (DAF) in the north, but in the western lowlands we adhered to a continuous model, according to which, with one exception, the whole range of woodlands was classified as one vegetation type, the *Combretum-Terminalia woodland and wooded grassland* (CTW), extending from the border with Kenya and the Omo Valley in the south to Eritrea, the Tacazze Valley (sometimes spelt Tekeze Valley) and the Mareb (sometimes spelt the Mereb Valley) in the north. The exception was the *Wooded grassland of the western Gambela region* (WGG), which is restricted to the low-lying parts of the Gambela Region west of the town of Gambela and is in fact a continuation into Ethiopia of the 'Flood region' in South Sudan, an area of flooded grasslands and swamps in the Nile Basin. The few previous studies of the western woodlands of Ethiopia have followed different classifications of the *Combretum-Terminalia woodland and wooded grassland* (CTW). Pichi Sermolli (1957), White (1983) and Sebsebe Demissew et al. (2004) have in the north indicated vegetation with other names that mainly described the physiognomy, such as 'shrub steppe,' 'grassland mosaic with *Acacia*' or 'desert and semi-desert scrubland.' And in the south and south-west the same authors indicated 'various types of savanna,' 'Somalia-Masai *Acacia-Commiphora* deciduous bushland and thicket,' '*Acacia-Commiphora* woodland' and 'desert and semi-desert scrubland.' Thus, there seemed to be a need for further studies of the continuity of the vegetation and the floristic patterns along the western escarpment of Ethiopia and more attention given to the northern and southern limits of the *Combretum-Terminalia* woodland.

Note that although we are aware of the distinction introduced by Kyalangalilwa et al. (2013) between *Senegalia* and *Vachellia* for species previously referred to *Acacia,* we have maintained the nomenclature with all the species in these genera as in *Acacia* (and cited the species names in *Senegalia* and *Vachellia* in the relevant places in the systematic Section 6), finding that we wanted to maintain such terms for vegetation as '*Acacia-Commiphora* deciduous bushland' and '*Acacia* wooded grassland', rather than change to '*Senegalia* bushland' or '*Vachellia* wooded grassland', dependent on the genus to which the dominant species previously in *Acacia* now belongs to.

We also felt that a reconsideration of the more central part of the *Combretum-Terminalia* woodland would be relevant, as it was documented by the study of the fire regimes and their influence on the Ethiopian vegetation by Breugel et al. (2016a) that there were strikingly different fire frequencies in various parts of the western Ethiopian lowlands and that there were virtually no fires in the extreme north and extreme south, which yet had floristic similarity with the central part. However, Breugel et al. (2016a) did not connect these findings to specific differences in the composition or structure of the vegetation, except for the abundance of grass that functioned as inflammable biomass in the dry season. The present work expands the knowledge of the woodlands in this part of Ethiopia, based on field work mainly carried out during the Ethiopian Flora Project, which was active during the years 1980-2009, and on field work focussed on our recording of data from 2014 to 2018 in relevés along almost the whole western border of Ethiopia.

The western escarpment of the Ethiopian Highlands and the western lowlands have traditionally been far less studied than the highlands, and they were also less known than the eastern lowlands in the Afar Region and the south-eastern lowlands towards Somalia. Apart from a few early publications in Italian, some of our own work (for example Jensen & Friis 2001) and the work by the Southwest Ethiopia Forest Inventory Project (Chaffey 1978a, 1978b, 1978c, 1978d, 1979), very little floristic and ecological work has been published on the vegetation of the western lowlands of Ethiopia.

The parts of this work are of varying length and of different contents; we have therefore termed them Sections rather than Chapters, as the latter term would imply a more uniform sequence of parts. In Section 2, we have chosen to give a comprehensive introduction to the topography, climate, geology and soil, history, political division and protection of the western Ethiopian lowlands, and we try to explain the potential coverage of the huge reservoir behind the GERD Dam on the Abay River, which will inundate some of our relevés. We felt that we missed detailed information about these points during our field work, and there are few reviews of these subjects available, even on the ultimate size of the flooding behind the GERD Dam.

During the field work, we also asked ourselves questions about why and how the western boundary of Ethiopia was drawn as it has, and how this border related to ecological patterns and ethnic boundaries. We traced the original documents of the boundary commission, and a review of the available information on how the boundary was drawn is given in Section 2.7. We have also in other parts of Section 2 tried to explain the floristic regions of the *Flora of Ethiopia and Eritrea,* which we use throughout this work for easy geographical reference. We expect that the information on the wide range of subjects regarding the western lowlands of Ethiopia summarized in Section 2 will be new to many readers of this book, and that they may find it useful in their further studies.

In Section 3 we give a review of previous works dealing with the vegetation and phytogeography of that part of Ethiopia. The earliest works, Pichi Sermolli (1938, 1940, 1957), are in Italian and have here been introduced so that they are intelligible to non-Italian readers. Another important work, Harrison & Jackson (1958), is from the other side of the border between Ethiopia, Sudan and South Sudan, and the printed text of this document is rare and not available on-line. A paper by Breugel et al. (2016b) pointed to the need for further studies of the *Combretum-Terminalia* woodland; this was an analysis of the transition between the vegetation of the eastern part of the Ethiopian Highlands and the *Acacia-Commiphora* deciduous bushland covering large areas of south-eastern Ethiopia towards Somalia, demonstrating that a number of species, including species of *Combretum* and *Terminalia,* prolong their distribution from the western woodlands into a more scattered distribution in the semi-evergreen bushland on and at the foot of the eastern escarpment of the Ethiopian Highlands. Breugel et al. (2016b) suggested that this vegetation should be given status as a distinct vegetation type termed *Transitional semi-evergreen bushland* (TSEB), and that its relation to the western woodlands should be studied further. We have explored these questions in this work, combined with a comparison of two maps of the ecosystems identified for Ethiopia by Sayer et al. (2013, 2020). Later works by others, including works on remote sensing of fires, attempting at analysing the biodiversity based on environmental parameters and at computerized phytogeographical analyses, have also been reviewed and commented on.

In Section 4 we outline in considerable detail the materials, methods and terms used in our own work, explaining how we have studied the vegetation and our ideas about the phytogeographical concepts and terms, we ask research questions to be answered in the concluding part of the book, and compare the woody flora of the *Combretum-Terminalia* woodlands with the description of the vegetation in Friis et al. (2010).

In Section 5 we present our field observations, made in the years 2014-2018 and represented by 151 relevés in 17 profiles (during 2014-2018, we could not go south of Maji to visit what must be the extreme southernmost border-region of the *Combretum-Terminalia* woodland); these observations are illustrated with images of landscapes and species.

In Section 6 we provide ecological and distributional information for all the 169 trees and larger shrubs that we observed in our relevés, including detailed dot-maps for their distribution in Ethiopia; we also assign the species to chorological types, 'elements', and to various ecological categories. This information is based on a combination of information from herbaria in Ethiopia, the UK, and elsewhere, from our field observations and from the literature. We hope that the distribution maps of Section 6 will be useful for the interpretation of our field observations and the observations of future researchers.

In Sections 7 and 8 we seek for patterns in the phytogeography and ecological properties of the 169 species of woody plants.

Section 9 presents species distribution models of the entire African distribution of 12 woodland species that occur in western Ethiopia and are characteristic of the Sudanian Centre of Endemism according to White (1983). The models suggest that the species have a wider distribution than documented by the recorded specimens, both in Ethiopia and in the area towards the Atlantic Ocean. Despite these species being all restricted to the Sudanian region, they have slightly varying distribution patterns, both in Ethiopia and in western Africa, and these patterns are not always clearly related.

In Section 10 we analyse the floristic composition and environmental parameters of the 151 relevés with clustering and principal component analysis. We use clustering and principal component analysis of our field observations in the 151 relevés to explore and test this possible division into separated clusters representing vegetation types.

In Section 11 we try to synthesize the findings of Sections 5 to 10 and see to what extent we can answer our research questions from Section 4.

In short, we hope that this book may be a useful step forward in the understanding of the vegetation and flora of western Ethiopia, but we do not expect that we will have the final word on the many questions that can be asked. Our work should, however, support a widening of the scope of botanical studied that can be undertaken in the vast areas of western Ethiopia.

The original field observations of the relevés are summarized in Appendix 1. In Appendix 2 we have listed the species seen in the relevés, the species codes used in the analyses, diagrams and graphs in Section 10, the number of specimens studied for each species, the number of times the species have been observed in relevés, and a tabular comparison with our present results and the results we found in Friis et al. (2010), which is discussed in Section 4.3.

The environmental data for each relevé, both what was observed in the field and what was later derived from environmental databases, are listed in Appendix 3. A summary of the distributional, phytogeographical and ecological categories of the species in the relevés are given in Appendix 4 and 5.

Details about the selection of clustering methods and the number of clusters are presented in Appendix 6. The distribution of all species, excluding singletons, and of CTW-species on three clusters according to *f.UPGMA* (unweighted pair group method with arithmetic mean) and *f.WPGMA* (weighted pair group method with arithmetic mean) are listed in Appendix 7.

We hope, as with the above-mentioned *Forests and Forest Trees of Northeast Tropical Africa* (Friis 1992) and *Atlas of the Potential Vegetation of Ethiopia* (Friis et al. 2010), that this book may become a widely used source of reference and a foundation on which others will be able to build further research, or at least a framework into which they can attempt to fit in their results, or maybe find that our frame is not suitable to use and need to be considerably improved or reshaped. Many Ethiopian scientists now produce interesting papers on the vegetation and ecology of specific areas, often located near the universities where they work, but only few of these new papers cover large areas, such as major parts of or the whole of Ethiopia, Eritrea, Sudan, South Sudan and other countries in the Horn of Africa, and relatively few new papers try to draw on the information available in the literature from before the digital age, written in other languages than English, or from observations made in the Sudan and South Sudan, but relevant for Ethiopia.

A recent subdivision of the Afromontane forests into three vegetation types (Abiyot Berhanu et al. 2018), that is *Moist evergreen Afromontane Forest* (MAF), *Intermediate Afromontane Forest* (IMF) and *Dry Afromontane Forest* (DAF) may have limited consequences for our woodland studies, but is also taken into account here. A recent review of the current knowledge of the vegetation and plant diversity of Ethiopia (Mengesha Asefa et al. 2020) points out that most detailed vegetation studies in Ethiopia have focussed on forests, and how also other Ethiopian vegetation types do provide highly relevant opportunities to investigate how complex biotic interactions, physical environment, ecological, and evolutionary processes determine vegetation dynamics at regional scales and beyond. We hope that this work to some extent will meet the wishes of Mengesha Asefa et al.

In this work we use acronyms drawn from a wide range of sources, including terms for vegetation types, biogeographical categories, standard abbreviations for books, statistical terms, etc. These are defined in the text where it is most relevant, but in order to facilitate the reading of other parts of the text we have also produced an accumulated glossary of acronyms before the List of References.

The authors, introduced in 'About authors and assistants' at the end of the work, have contributed to this work in various ways. Ib Friis, Sebsebe Demissew and Paulo van Breugel discussed the basic research-questions and ideas for field work. In collaboration with Sebsebe Demissew, Ib Friis organized field work in 2014-2018 with a range of people mentioned in the acknowledgements, including the co-authors Sebsebe Demissew and Odile Weber. The information from the herbaria ETH, FT and K was gathered by Ib Friis, Sebsebe Demissew and Odile Weber, and the on-line information from L [WAG] and BR was extracted by Odile Weber, who also organized the many spreadsheets used to analyse the data from the herbarium collections and wrote parts of the text. The general writing of the manuscript was done by Ib Friis, with inputs from the three other authors: from Paulo van Breugel the whole of Section 10, from Odile Weber large parts of Section 7, 8, and 9, and many scattered inputs in the whole text were provided by Sebsebe Demissew. All authors have read the whole manuscript.

1.1. Acknowledgements

This work involves many scientific activities by the authors in and about Ethiopia during the last 50 years, and we want to acknowledge the help we have received. Ib Friis' first visit to Ethiopia in October-December 1970 was supported by the Carlsberg Foundation (through a grant to Professor Thorvald Sørensen, then head of the Institute of Systematic Botany, University of Copenhagen); this visit took Friis, senior lecturer Knud Jakobsen and the botany student Asfaw Hunde to the limits of the western woodlands in the provinces of Kefa and Ilubabor, as these were then named. The trip, which lasted for three months, was followed by another trip of five months' duration (October 1972 to February 1973), also supported by the Carlsberg Foundation. The 1972-1973-trip was made with the later Kew botanist Kaj Vollesen and the later senior lecturer at the Botanical Institute, University of Copenhagen, Finn N. Rasmussen, as well as with two lecturers at the Addis Ababa University, Getachew Aweke and Michael G. Gilbert. The trip in 1972-1973 revisited most of the areas from the first trip in 1970 and went further into the western woodlands, but also to Sidamo and parts of Shewa.

A major revolt took place in Ethiopia soon after Ib Friis' second trip, overthrowing the rule of Emperor Haile Selassie I, and during the following years field work in Ethiopia was not possible. However, by 1980 the situation was moderately settled, and the international Ethiopian Flora Project was initiated due to a bilateral agreement between the Ethiopian and Swedish governments and under the leadership of Professor Tewolde Berhan Gebre Egziabher, Addis Ababa University, on the Ethiopian side, and Professor Olov Hedberg and Professor Inga Hedberg, Uppsala University, on the Swedish side. The core activities of the Flora Project were implemented by grants from Swedish Agency for Research Cooperation (SIDA/SAREC) through Addis Ababa and Uppsala Universities (Hedberg 2009). The activities in connection with the Flora Project were also supported by scientists employed at many other institutions, of which the major ones were the Royal Botanical Gardens, Kew, UK, and the Universities of Copenhagen in Denmark and in Oslo in Norway (Sebsebe Demissew et al. 2011).

From the beginning of the Flora project to its conclusion in 2009, Ib Friis took part in the field work in little-studied parts of Ethiopia, which provided a total of ca. 15,000 collections of plant specimens to the National Herbarium of Ethiopia. During the duration of the Ethiopian Flora Project, the National Herbarium of Ethiopia grew from less than 20,000 specimens in 1980 to over 80,000 in 2009, many new collections provided by Ethiopian participants in the Ethiopian Flora Project, particularly Sebsebe Demissew and Mesfin Tadesse.

Most of Ib Friis' Ethiopian field trips were supported by the Carlsberg Foundation, and during these trips he worked together with a range of scientists, students and technicians from Ethiopia, Denmark, United Kingdom and United States of America: Knud Jakobsen, Asfaw Hunde, Kaj Vollesen, Finn N. Rasmussen,

Getachew Aweke, Michael G. Gilbert, Sue Edwards, Zerihun Woldu, Sebsebe Demissew, Mesfin Tadesse, Damtew Teferra, Jonas Lawesson, Lisanework Nigatu, Anders Michelsen, Sally Bidgood, Peter Høst, Melaku Wandafrash, Shigulte Kebede, Malaku Legesse, Michael Lock, Dessalegn Desissa, Fantahun Semon, Michael Jensen, Menassie Gashaw, Lemessa Keneei, Tesfaye Awash, Gashaw Gebre-Hiwot, Victoria C. Friis, Nigist Asfaw, Amsalu Ayana, Getu Tefera, Gregory McKee, Berhanu Yitbarek, Muligeta Gichite, Assefa Hailu, Ermias Getachew, Wege Abebe, Abubaker Atem, Elias Tadesse, Timothy Harris, Frances Crawford, Odile Weber and Ergua Atinafe. All these people are warmly thanked for their contributions.

An important spin-off of the Ethiopian Flora Project, planned for, but not finalized during the project-period, was the book about Ethiopian vegetation by Ib Friis, Sebsebe Demissew and Paulo van Breugel, 'Atlas of the Potential Vegetation of Ethiopia' (Friis et al. 2010), which was generously published by the Royal Danish Academy of Sciences and Letters, skilfully edited by Marita Akhøj Nielsen and designed by Mette Mourier. The Academy and its editor and book-designer are thanked for publishing this handsome and internationally much cited publication.

During many field trips to the western lowlands of Ethiopia (in the years 1970, 1972-1973, 1982, 1984, 1986, 1988, 1990, 1995, 1996, 1997, 1998, 2000, 2001, 2003, 2004, 2005, 2006, 2007, 2008, 2009, 2010, 2011 and 2013) it had been realized how poorly known the vegetation and flora of that part of Ethiopia was in relation to other parts of the country, and in 2013 it was decided that a special study should be dedicated to the flora and vegetation of the western lowlands. For the success of this activity, the authors of the present work want to thank the leaders and staff of the Ethiopian National Herbarium and the Gullele Botanic Garden, both in Addis Ababa, and the leaders and staff at the College of Natural and Computational Science, Addis Ababa University, for helping us to organize our field trips and lending us one of their cars.

We are particularly appreciative of the help we have received from Ermias Getachew, driver for the Faculty of Science (now College of Natural and Computational Sciences), Addis Ababa University, and recently retired, and from Wege Abebe, still actively employed with the National Herbarium of Ethiopia, Addis Ababa University. They are both further introduced in 'About authors and assistants' at the end of the work. Ermias is almost certainly the person (apart from Ib Friis) that has taken part in most field trips behind this work, and he looked after and drove the various vehicles we used in the field, managing repairs, finding fuel and oil when it was scarce, changing punctured tyres, getting us along poor roads and out of mud, and even saving us when our car collided with a cow that suddenly jumped out from the dense woodland. Wege Ababa has taken part in nearly as many field trips and has always been responsible for the bureaucratic arrangements, communication with the local authorities, finding us accommodation and food while in the field, managing negotiations with local farmers, helping with the collection of plants, pressing them and sending the material for identification in Europe. Without the always cheerful and efficient help from Ermias Getachew and Wege Abebe this work would not have been possible.

It is sad that another long-time participant in the trips to the western lowlands, Sally Bidgood, Royal Botanic Gardens, Kew, who took part in field trips from 1995 to 2007, is no longer with us.

We thank the Herbarium, Royal Botanic Gardens, Kew, UK, and the *Centro Studi Erbario Tropicale*, University of Florence, and their staff for providing access to the rich collections kept there and helping us in many other ways. Michael G. Gilbert, in his retirement associated with the Royal Botanic Gardens, Kew, has kindly shared his extensive knowledge of Ethiopia with us since 1970, and he has continued to do so while we have been preparing this work.

Two emeriti geologists, Lotte Melchior Larsen, GEUS, and Asger Ken Pedersen, Natural History Museum of Denmark, have kindly read and commented on the text on geology in the western Ethiopian lowlands.

Victoria C. Friis, who is a native English speaker, is thanked for kindly having looked through most of the text, weeding out what grammatical and other errors she has spotted.

Most of Ib Friis' field work for the Ethiopian Flora Project and this project has been supported by the Carlsberg Foundation, but support was also received from DANIDA and the Danish National Research Council for several previous projects, the results of which have gone into this work. As always, we are grateful to the Carlsberg Foundation, Copenhagen, for the financial support to our fieldwork and herbarium studies relating to the Ethiopian flora and vegetation. Directly related to the field work and herbarium studies connected with this project are the following grants: 2012_01_0035, 2013_01_0051, CF14-0047, CF15-074, CF16-0040, CF17-0165, and CF18-0093.

2
The western lowlands of Ethiopia: topography, geology, soils, climate, fires, history, and population

In this Section, we place the western Ethiopian woodlands in an African setting (Fig. 2-1) and outline the geographical, historical and ethnic set-ups of the western lowlands so that readers are provided with difficult-to-access background information that may help to follow the remaining parts of the text and back-up future research. The general topography of Ethiopia is shown in Fig. 2-2, the names of the surrounding countries are indicated on Fig. 2-28. The floristic regions, based on the administrative division of Ethiopia during the reign of Emperor Haile Selassie I and slightly modified for use in the *Flora of Ethiopia and Eritrea* and in this work, are shown in Fig. 2-28 and described in Table 2-1. The regions or regional states of Ethiopia, which are the current (2021) administrative divisions of the country at the highest level, are indicated on Fig. 2-15 (right map) and Fig. 2-29. The sizeable part of the western woodlands that, according to the currently available information, will be covered by the reservoir on the Abay, the Dabus and the Didessa Rivers behind the GERD Dam is outlined on Fig. 2-15. The geographical variation of major climatic parameters are shown in Fig. 2-5 to Fig. 2-10.

2.1. Ethiopia and the zonation of the vegetation and flora across Africa

The scholarly studies of the African vegetation and its latitudinal zonation go back to the second half of the 19th century (Friis 1998). In Africa north of the Equator and west of the Ethiopian Highlands, the vegetation zones and phytochoria (see definition in Section 4.2) have been perceived as forming a sequence of parallel bands across the continent (Fig. 2-1), with the Guineo-Congolian rainforest closest to the Equator, followed towards the north by various types of woodlands and wooded grasslands, then by the open bushlands of the Sahel and the various zones of desert in the Sahara, and finally the Mediterranean vegetation at the northern shores of Africa. A pioneer work in the 19th century was that of Engler (1882), who recognized the rainforests, the woodlands, the desert and the Mediterranean vegetation, and in the 20th century the works by Lebrun (1947) and Monod (1957), who studied the zonation of the woodlands and their flora in more detail than Engler, and finally several works by Frank White which will be referred to in more detail in the following. The works by Lebrun and Monod focussed on the zonation in the areas to the west of the Nile; limited attention was given to the vegetation between the Nile and the Ethiopian Highlands and to the Ethiopian Highlands themselves before the works by White.

Although the vegetation zones from the Atlantic Ocean to the Nile have had changing definitions and terminology since Engler, the zones from the Equator to the Tropic of Cancer and the Sahara were generally characterized as being of increasing aridity and the rainfall being increasingly restricted to a short period in the summer, and then again from the Sahara to the Mediterranean by increasing humidity and a more pronounced winter rainfall. This view has not changed substantially for the vegetation to the west of the Nile. To the east of the Nile, where the latitudinally arranged zones of West Africa would meet the Ethiopian highlands, the highland floras and vegetation were initially considered subsets of those of the lowlands, but that view changed significantly afterwards, particularly due to the work by White, who reviewed the vegetation and phytogeography of Africa by returning to and updating classic phytogeographical concepts and terminology (see also our review of phytogeographical terminology in Section 4.2).

White's view of the situation to the east of the Nile

differed from the outset from the views of Lebrun and Monod, as can be seen from his first chorological classification of the Horn of Africa (White 1965; reproduced here as Fig. 2-1A; see also Section 4 and 7). In his 1965-publication, White first of all noted the distinctive features of the Afromontane vegetation and flora, which separated them from their counterparts in the lowlands, but he also gave particular attention to the lowland flora to the east of the Ethiopian Highlands and its similarity with parts of the driest part of the Sudanian vegetation, the Sahel, writing: "In distinctness of floristic composition, that part of East Africa [parts of Kenya and northern Tanzania] and Ethiopia occupied by wooded steppe with abundant *Acacia* and *Commiphora* seems to be comparable to the Sudanian and Zambezian Domains and should be recognized as the true Oriental Domain [a term that Lebrun had used for the evergreen forests along the eastern coast of Africa, mainly in Kenya and Tanzania]. It is the home of many endemic species, particularly in *Acacia, Commiphora* and *Terminalia*." And: "In phytogeographic relationships the Sahel appears to be best treated as an impoverished western extension of the floristically rich Oriental Domain." However, on his map White (1965) indicated the Oriental Domain as surrounding and enclosing the entire Ethiopian Highlands on the eastern as well as on the western side and, in agreement with his statement, indicated the flora and vegetation of the Sahel as forming an extension of that domain. This agrees with the way the lowland vegetation surrounding the whole of Ethiopia was represented on the AETFAT map in 1:10,000,000 (Aubréville et al. 1958) by the vegetation type '25. Wooded steppe with abundant *Acacia* and *Commiphora*.' A phytogeographical classification rather similar to that was presented by Clayton & Hepper (1974: 223, map in their Fig. 1), adopted and modified by Wickens (1976: 40-42, map in his Fig. 18), and adopted and modified by Brenan (1978: 443, map in his Fig. 2). In these works, the name of the lowlands surrounding the Ethiopian Highlands on all sides, including Somalia and eastern Kenya, was for clarity changed from 'Oriental Domain' to the 'Afroriental Domain' of the Sudano-Zambesian Region.

In his later works, following from his phytogeographical analyses in Chapman & White (1970) and onwards, White removed the part of the Oriental or Afroriental Domain that covered the western lowlands of Ethiopia and followed the idea of Lebrun and Monod that the west African zones of flora and vegetation met directly with the western escarpment of the Ethiopian highlands (Fig. 2-1B). In fact, all White's later phytogeographical maps show a widening-out in a north-south direction of the broadest zone across Africa from the Atlantic to Ethiopia, the 'Sudanian regional centre of endemism,' before it meets the western escarpment of the Ethiopian Highlands (Fig. 2-1B and Fig. 2-1C). This is in agreement with the distribution of major climatic parameters across Africa as seen on UNEP (2008: 8-10), and it also reflects the pattern of grass fires from the Atlantic coast to the coast of the Indian Ocean (Fig. 2-1D). Almost coinciding with the belt across Africa of strongly contrasting summer rain and winter drought, there is a belt of intense grass fires, recorded from Senegal on the Atlantic Coast across Africa to Nile in the Sudan and South Sudan, to the east of which it widens out in a north-south direction, reaching into the western parts of Eritrea, the deep river valleys of Ethiopia and into large parts of Uganda and parts of Kenya (Barbosa et al.1999: Plate 1). As described by Jensen & Friis (2001) for the Gambela Region of Ethiopia, the grass fires are correlated with high biomass production of grasses and herbs, a biomass that regenerates annually, while several the woody plants have structures, particularly thick bark, which protects them from fire, a feature that we will consider in Section 8.4. However, not all species in the 'Sudanian regional centre of endemism' follow the same pattern from west to east, and may demonstrate local variation in the western woodlands of Ethiopia, as will be discussed later in Section 9.

The striking local variation, contrasting with the large-scale similarity of the vegetation over large parts of the belt of woodlands across Africa in the 'Sudanian regional centre of endemism' has been demonstrated by Kershaw (1968). In his study of woodlands of northern Nigeria he listed, from an area spanning over approximately 300 km (about

1/3 of the latitudinal span of the area studied in this work), many species that also occur in our study area in the western Ethiopian woodland, including *Annona senegalensis, Anogeissus leiocarpus, Carissa edulis, Senna (Cassia) singueana, Combretum collinum* (synonym *C. binderianum*), *Combretum molle, Commiphora africana, Crossopteryx febrifuga, Dichrostachys cinerea, Diospyros mespiliformis, Entada africana, Faurea rochetiana* (synonym *F. speciosa*), *Ficus platyphylla, Gardenia ternifolia, Grewia mollis, Lannea schimperi, Lonchocarpus laxiflorus, Piliostigma thonningii, Pseudocedrela kotschyi, Psorospermum corymbiferum* (West African near relative of *P. febrifugum*), *Pterocarpus erinaceus* (relative of *P. lucens*), *Sarcocephalus (Nauclea) latifolia, Sclerocarya birrea, Securidaca longipedunculata, Steganotaenia araliacea, Stereospermum kunthianum, Strychnos innocua, Tamarindus indica, Terminalia laxiflora, Terminalia schimperiana* (*T. glaucescens*), *Vitex doniana*, and *Ximenia americana*.

However, Kershaw (1968) also demonstrated the striking local variation in the floristic compositions of the woodlands of Northern Nigeria and recognized eighteen local plant associations over the distance of ca. 300 km. White (1983) rejected to recognize these eighteen plant associations in his mapping of the vegetation of Africa, as they were too poorly defined and not covering sufficiently extensive areas to permit the recognition of well-defined vegetation types on a broad scale. White further pointed out that any well-defined floristic pattern, which may formerly have existed in the belt, had been greatly obscured because of the human influence and degradation, which most Sudanian vegetation has suffered.

The current state of knowledge therefore emphasizes a floristic connection between the western Ethiopian lowlands and the belt of Sudanian vegetation across Africa, rather than a connection with the lowlands to the east of the Ethiopian Highlands, and in the following parts of this work the floristic and vegetational connection between the western Ethiopian lowlands and the belt of the Sudanian vegetation across Africa will be discussed wherever it is relevant, particularly in Section 9. However, our maps in Section 6 also show that there is a certain floristic connection between the woodlands in the western lowlands of Ethiopia and the vegetation on the eastern side of the Ethiopian Highlands, either in the form of species mainly distributed in the *Acacia-Commiphora woodland and bushland* (ACB) in south-eastern lowlands, but penetrating into the northern and southern parts of the western woodlands, or in the form of species with main distribution in the western woodlands extending to the Rift Valley or the south-eastern slopes of the highland, particularly in the vegetation type termed *Transitional semi-evergreen bushland* (TSEB) by Breugel et al. (2016b). The floristic relations between the western woodland and the general image of the Ethiopian flora and vegetation are discussed in Section 7.

White (1983; see in this work in Section 3.4) proposed a special subtype of Sudanian woodland, referred to as 'Undifferentiated [Sudanian] woodland – Ethiopian,' in the border region between Ethiopia and the Sudan and South Sudan, and he distinguished this both by the absence of any of the five species of *Isoberlinia* or the eight species of *Julbernardia* (both in the Leguminosae subfam. Caesalpinioideae), which are genera otherwise well represented in the Sudanian and Zambezian woodlands, and by the presence of an endemic species of *Combretum*, *C. hartmannianum*; this proposal is further presented in Section 3.4, and is in part the background for the use in this work of a separate local phytochorion for the western woodlands of Ethiopia.

A

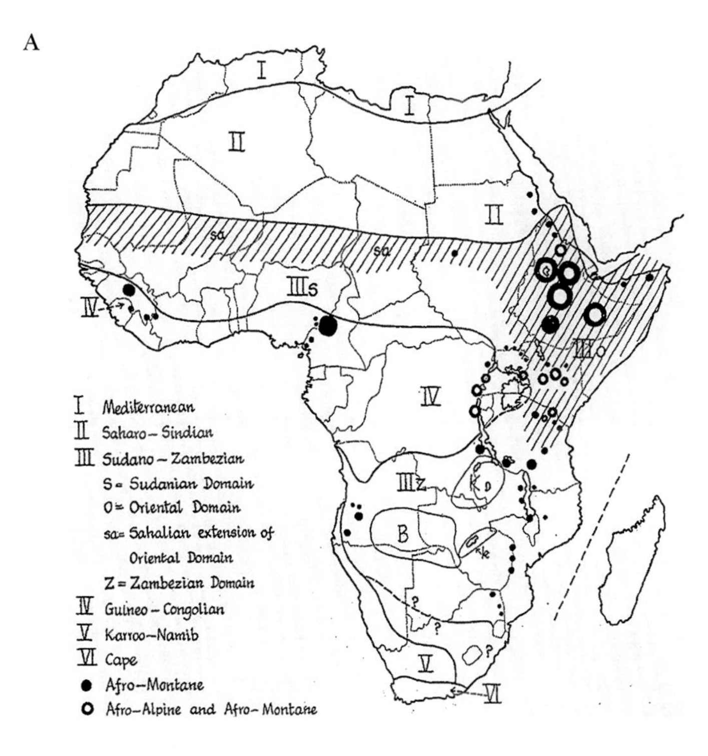

B

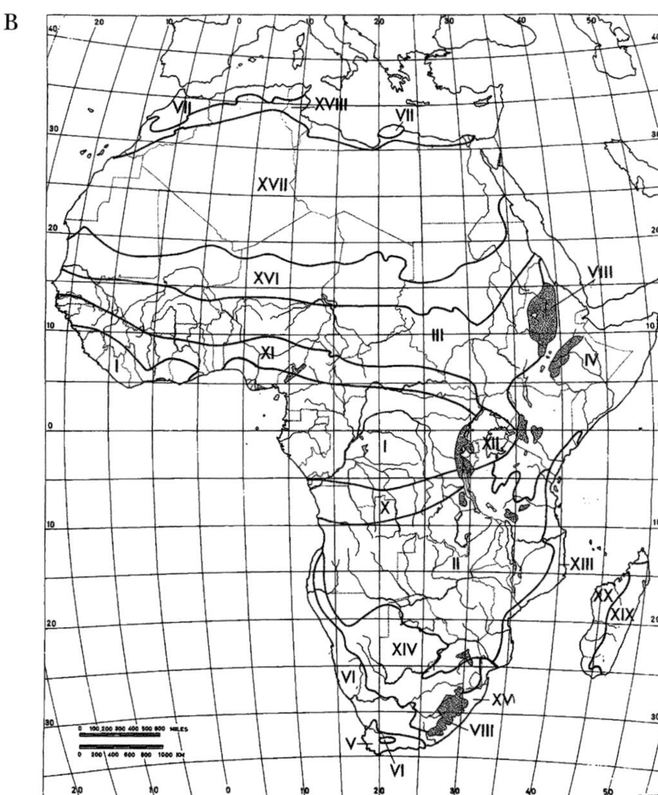

Fig. 2-1. The western Ethiopia lowlands from a continental African view.

A. White's first map of African vegetation and centres of endemism, showing all lowlands around Ethiopia as the 'Oriental Domain' (in later versions renamed the 'Afroriental Domain' or the 'Somalia-Masai regional centre of endemism'). Reproduced with permission from White (1965, Fig 1).

B. White's latest map of African vegetation and centres of endemism, showing all the western lowlands of Ethiopia as 'III. Sudanian regional centre of endemism.' Reproduced with permission from White (1993: Fig. 1).

Fig. 2-1 continues on next page.

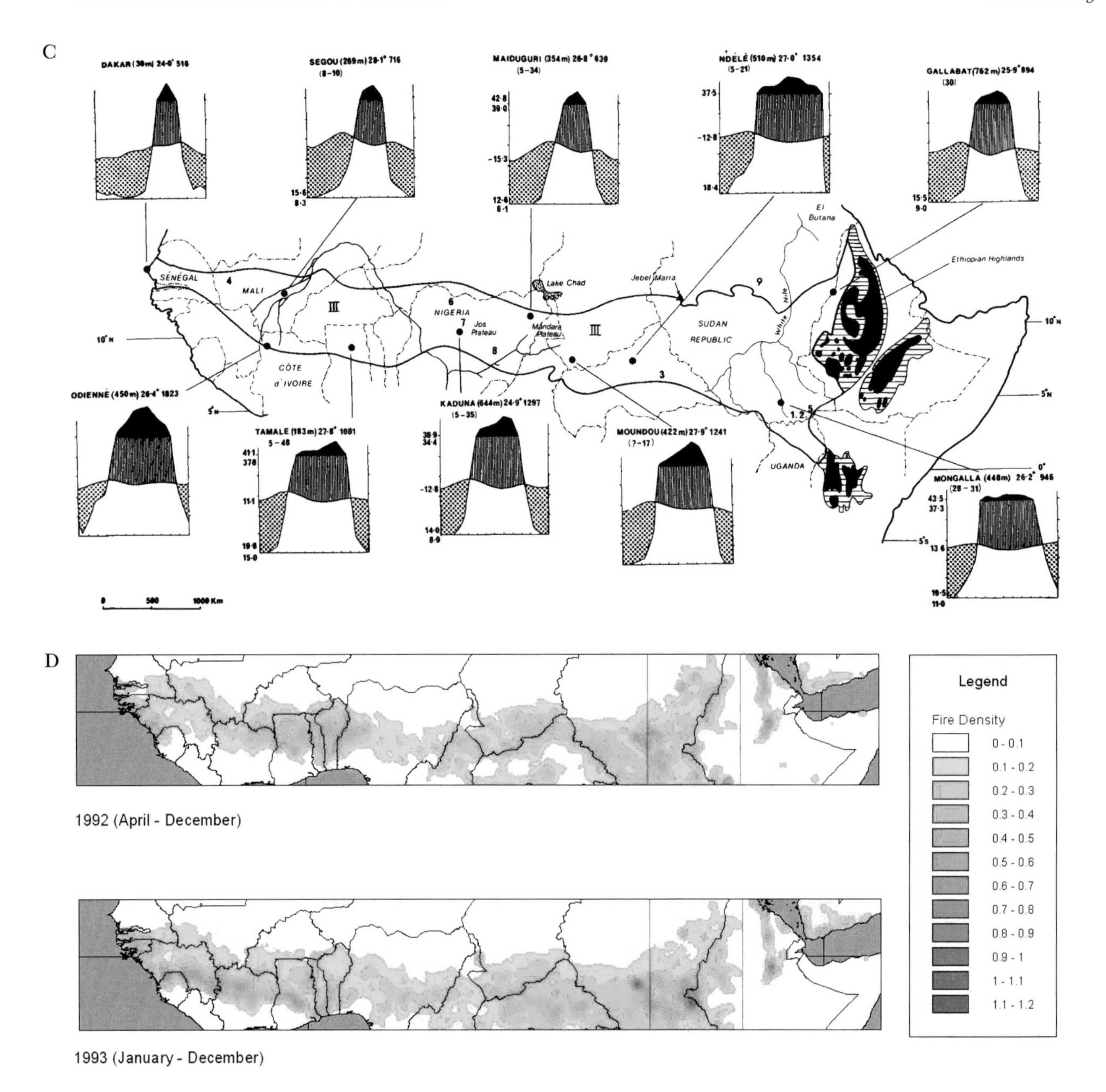

Fig. 2-1 (continued).

C. The Sudanian region with approximately the same delimitation of the 'Sudanian regional centre of endemism', with climatic diagrams added from Walter & Lieth (1960-1967), indicating that the entire Sudanian region has summer rains of varying intensity and length. Reproduced with permission from White (1983: Fig. 7).

D. Fire density across Africa in April-December 1992 and January-December 1993. Fires were recorded by satellite within square area of 50 × 50 km and marked on the map with a grid size of 0.1 degree. The map is based on data from archive files of the Global Vegetation Monitoring Unit's FIRE project at Space Application Institute, Ispra, and was produced by Thomas Theis Nielsen for the Danish-Ethiopian FITES project. Reproduced with permission from Jensen & Friis (2001, Fig. 3).

2.2. Topography of Ethiopia

In Section 2.1 we have presented Ethiopia mainly as a block of mountains, seen in relation to the surrounding lowlands. In the following we present more detailed outlines of the topography, climate, geology, soils, and fire regimes. Since Ethiopia is also a political entity of ca. 1,105,000 sqkm, slightly more than 1/10 the size of Europe including the European part of Russia, we find it also necessary to outline its political borders and the borders of the major administrative subunits; this will be summarized in Section 2.7.

Ethiopia is located on the Horn of Africa, bordered to the north and north-east by Eritrea, to the east by Djibouti and Somalia, to the south by Kenya and to the west by Sudan and South Sudan (Fig. 2-2 and Fig. 2-28). Since the independence of Eritrea, the latitudinal extension of Ethiopia is from ca. 2° 30' to ca. 14° 50' N and the longitudinal extension from ca. 33° to ca. 48° E, and the country measures ca. 1280 km from north to south and 1650 km from east to west. The surrounding countries (indicated on Fig. 2-28) consist mainly of lowland below 1500 m a.s.l., while more than one third of Ethiopia is made up by the Ethiopian Highlands, mostly above ca. 1800 m a.s.l. and reaching a maximum height of 4530 m a.s.l. at the top of Ras Dashen in the Semien Mountains in the north and 4377 m a.s.l. at the top of Tulu Dimtu in the Bale Mountains in the south. The Ethiopian Highlands divide the low-lying parts of the country into the western and the eastern lowlands, and the Highlands are in themselves divided by the Rift Valley, the bottom of which has altitudes mostly between 1500 and 1800 m a.s.l. (but down to below 1200 m at the level of the southern Rift Valley lakes, Lake Abaya and Lake Chamo). To the north, the Rift Valley is prolonged into the much widened Afar Depression. The lowest part of the Afar Depression, the 40 × 10 km Danakil Depression is between 150 and 100 m below sea level, and the lowest point is at about 125 m below sea level (however, Lake Assal on the Djibouti side of the border in the Afar triangle is the lowest point on the Horn of Africa, with the water level at 155 m below sea level). To the south the Rift continues in Lake Turkana (level of the lake at ca. 360 m a.s.l.) and Lake Chew Bahir (level at ca. 400 m a.s.l.). (this data from Google Earth Pro v. 7.3.3.7699 (64-bit)).

Both from the west and east, river valleys cut deeply into the Ethiopian Highlands, and prolong narrow strips of the lowland vegetation continue deeply into the highlands. The western and southern rivers (Fig. 2-3), with woodland vegetation as dealt with in this work, are mainly the Tacazze, the Abay (Blue Nile) and the Baro Rivers, which all flow to the Nile, and the Omo River, which flows to Lake Turkana.

A very large dam, the GERD Dam, is recently (2021) being constructed on the Abay River near the border with the Sudan; the filling of the reservoir behind this dam will considerably change the landscape and inundate significant areas described in this work (see more in Section 2.4).

Among the largest eastern rivers is the Awash River, which does not reach the sea, but flows into a number of lakes in the Afar Depression; other eastern rivers are the Dawa, Ganale and Web, a group of rivers which meet at the border between Ethiopia and Somalia to form the Juba River, which flows into the Indian Ocean, and finally the easternmost river, the Webe Shebele, which almost reaches the sea before seeping out in the sand-dunes of southern Somalia.

In this work, the western escarpment and the western lowlands of Ethiopia are defined to the east by the western border of the Ethiopian Plateau, at approximately 1500-1800 m a.s.l., and to the west by the present political borders towards Eritrea, Sudan, South Sudan and Kenya, which borders were only demarcated in the beginning of the 20th century (see Section 2.7 and Fig. 2-25). While the delimitation of the western lowlands and the western escarpment towards the Ethiopian Plateau is a natural altitudinal limit, although not always well defined (see discussion in Section 7), there is no natural western limit between the woodlands inside Ethiopia and those in Sudan and South Sudan, and the vegetation of the western Ethiopian lowlands continue into the neighbouring

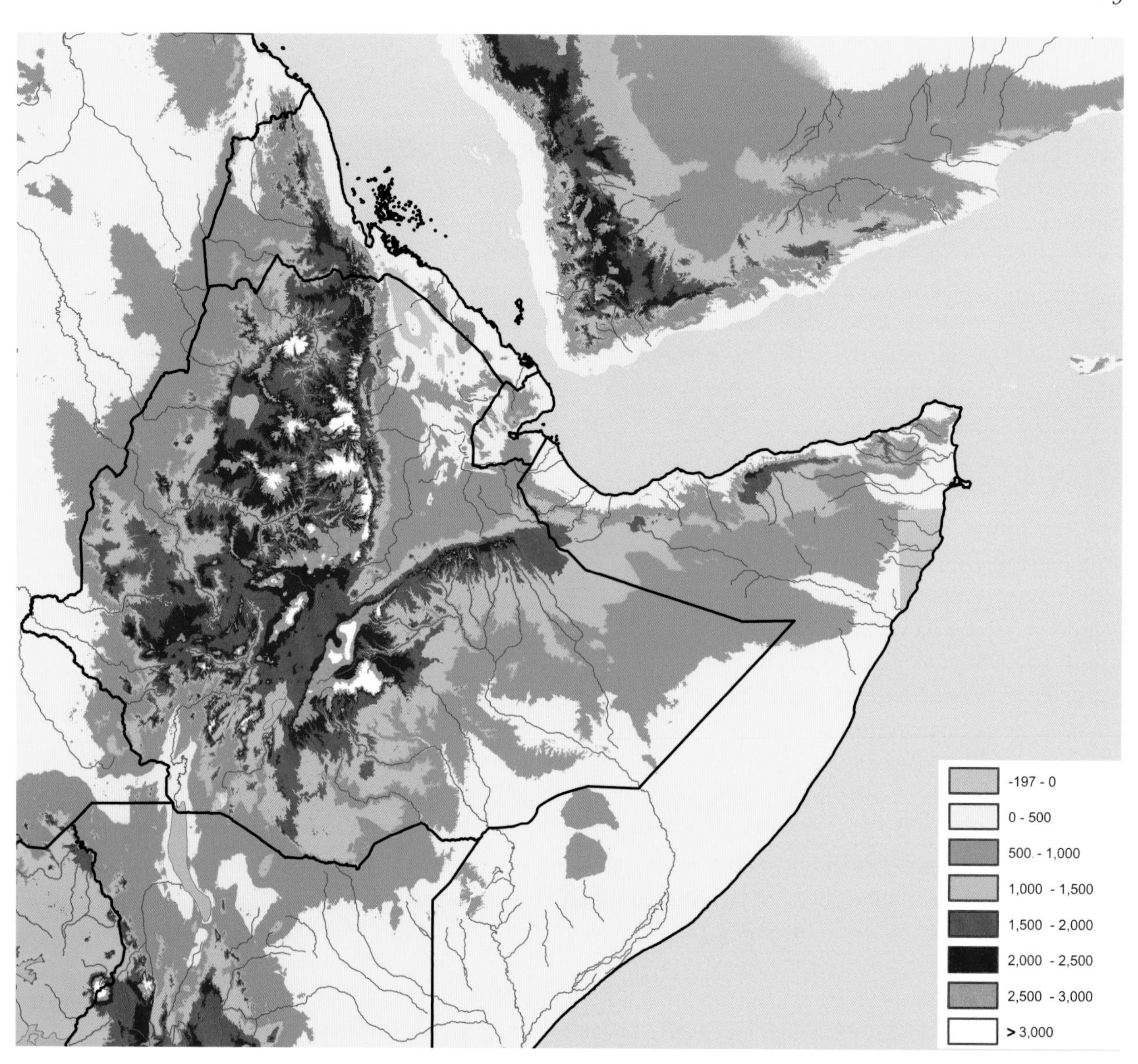

Fig. 2-2. Topography, major rivers, lakes, and borders of Ethiopia. The map shows the topography with altitudinal intervals of 500 m. The names of the surrounding countries are specified on the map in Fig. 2-28. The high-resolution digital elevation model is based on data from https://bigdata.cgiar.org/srtm-90m-digital-elevation-database/. The shapefile with the rivers is based on https://www.naturalearthdata.com/downloads/10m-physical-vectors/10m-rivers-lake-centerlines/. The borders of Ethiopia with Eritrea, Sudan, South Sudan, Kenya and Somalia are based on a shapefile at http://www.maplibrary.org/library/stacks/Africa/Ethiopia/index.htm. The border between the Sudan and South Sudan was not included in that shapefile.

countries. The political boundaries also cut through ethnic groups living to either side of them (see Section 2.7). Because of the entirely political background of the boundary, some features of the vegetation of the neighbouring parts of Eritrea, Sudan, South Sudan, and Kenya will also be accounted for when appropriate.

An important feature of the topography of the Ethiopian Highlands is that they are very rugged, and in the highlands only the alluvial valley-bottoms, filled with deposits from erosion from the slopes or from areas higher up along the valley, are flat. The lowlands have more extensive areas of flat ground. No previous analysis of level ground in Ethiopia has been published, and therefore results of one of our previous unpublished studies of areas with impeded drainage has been included here (Fig. 2-4); the study was made for the VECEA vegetation map (see Section 3.7; Lillesø et al. 2011). Slope in % and run-off, the latter expressed by the Terrain Wetness Index (TWI), were calculated using the data in CGIAR-CSI (2008) as input. Different GIS-layers were created; the topographical layer showed raster cells classified as 1 if they were calculated to have slopes < 2% and a calculated Topographic Wetness Index > 8. All others were assigned a value of 0. A soil layer showed areas with a clay content in the topsoil > 50% from the Harmonized World Soil Database (2008), which excluded the extensive flat areas with very sandy soils in the eastern lowlands. The results (Fig. 2-4) includes areas with vertisol ('black cotton soil'), edaphic grasslands, salt pans and wetlands. These flat areas, both in the lowlands and in the highlands, but saline areas excepted, have typically vertisols, which will be described and their extension discussed further in the text on soils (Section 2.5). The flat areas of the western lowlands are primarily in the Gambela Region, in the Omo Valley and in the north-western lowlands.

2.3. Ethiopian climate

With a latitudinal extension between ca. 2° 30' and ca. 14° 50' N, Ethiopia is entirely located between the Equator and the Tropic of Cancer (which is currently at ca. 23° 26' 11.8" N). The climate of Ethiopia is therefore tropical, and as usual in tropical climates, the yearly variation in rainfall per month is more prominent than the yearly variation in average temperature per month. However, the variation in both temperature and rainfall is strongly modified by the complexity of the relief (Liljequist 1986), as this has been coarsely outlined in Section 2.2. The characteristics and variation of the climate of the western woodlands are illustrated by four images of bioclimatic variable from the worldwide Bioclim dataset (http://www.diva-gis.org/climate) (Fig. 2-5 to Fig. 2-8) and two additional environmental parameters, the moisture index (Fig. 2-9) and the number of dry months (Fig. 2-10) (data derived from Africlim, as presented by Platts et al. 2014), all maps with a resolution of 30 degree seconds and covering the period between ca. 1950 and 2000. The values per pixel of the worldwide bioclimatic variables are not based on direct observations, but are calculated by interpolation from the monthly temperatures and rainfall values measured at meteorological stations, often near towns or airports. These variables represent annual trends (e.g., mean annual temperature, annual precipitation), seasonality (e.g., annual range in temperature and precipitation) and diurnal ranges in temperature. The data sets are used to provide general information here, but are also used in the modelling in Section 10 (data in Appendix 3).

The average annual temperature is mainly a function of the altitude; the north-south extension of Ethiopia of ca. 10° latitude does not cause significant variation in the average annual temperature at sea level, and the average annual temperature is estimated to decrease with altitude by 0.5-0.7° C per 100 metres increase in altitude (Hahn & Knoch 1932; Liljequist 1986; Friis et al. 2010). However, the average annual temperature is also controlled by the cloud cover, for which reason the rainy summer months are generally the coldest despite the sun being north of the Equator

A

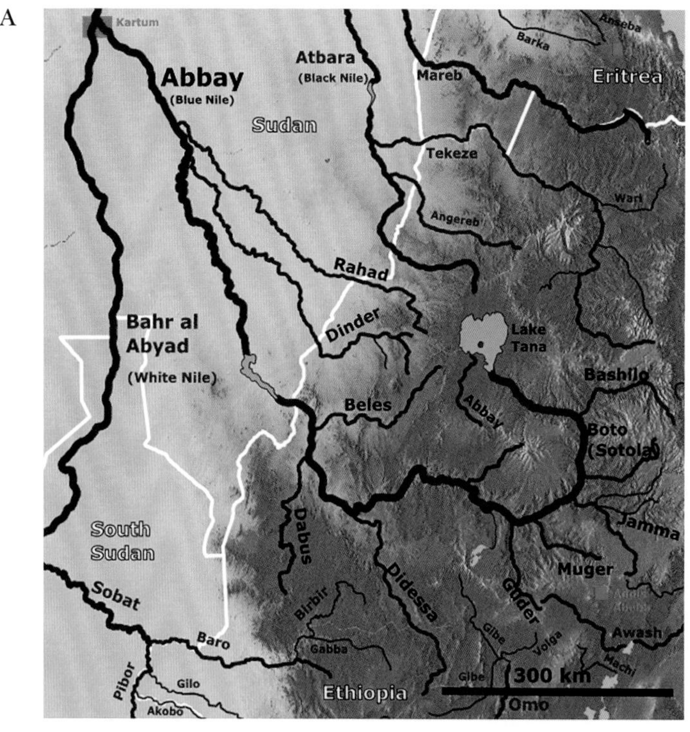

B

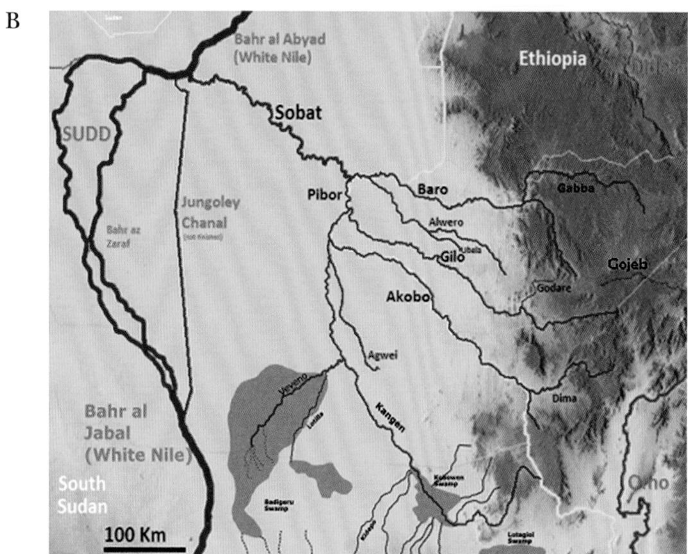

Fig. 2-3. Rivers running through the western Ethiopian lowlands to the Nile or Lake Turkana. A. Rivers in the Atbara and the Abay basins (Blue Nile Basin). B. Rivers in the Sobat-Baro-Akobo and the Omo basins. (Maps by Hans Braxmeier https://maps-for-free.com/, reproduced with licence CC BY-SA 2.0 https://commons.wikimedia.org/w/index.php?curid=75273409). Note that on this map the river, which we refer to as the 'Tacazze River' is indicated with an alternative spelling, 'Tekeze'.

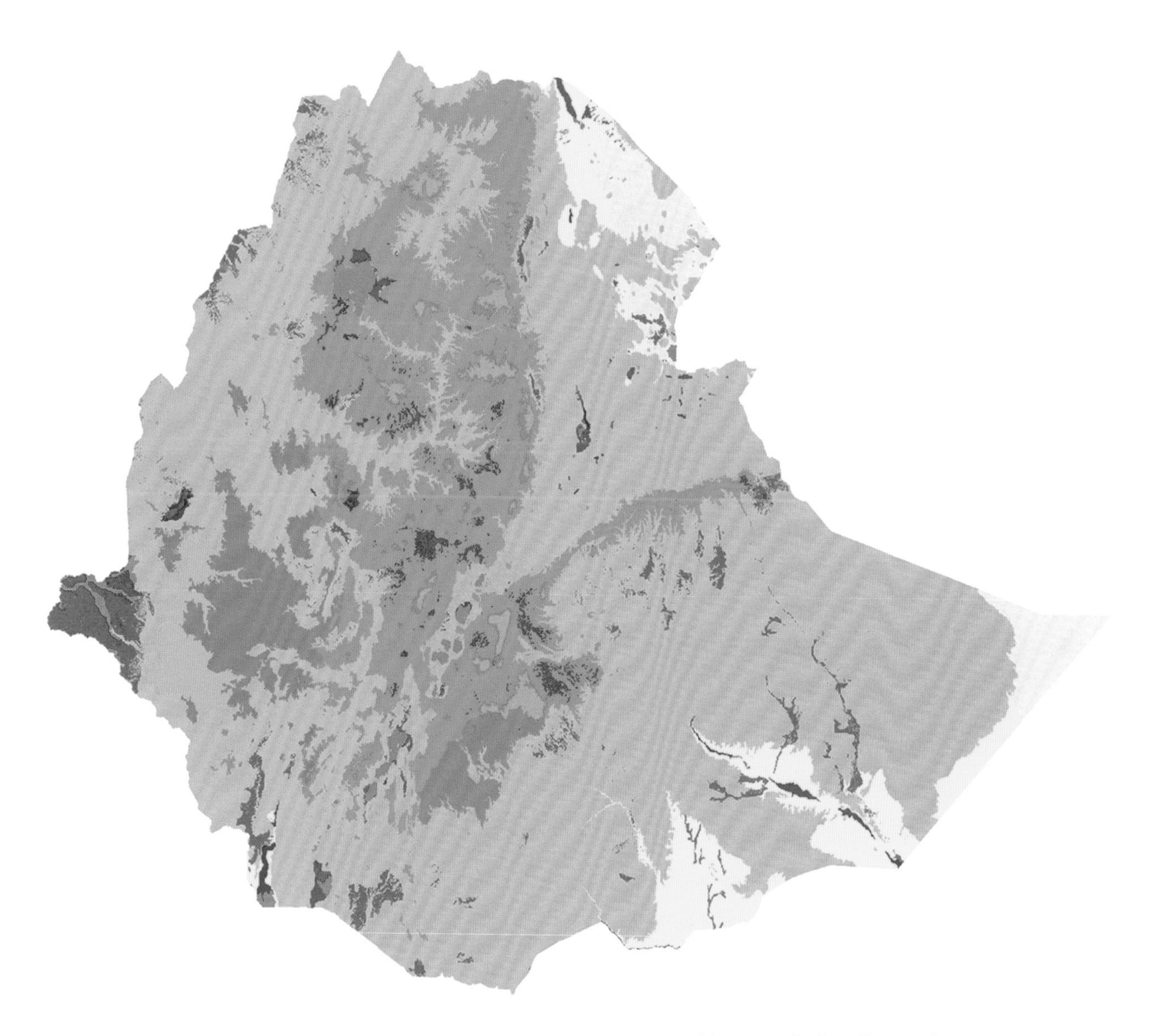

Fig. 2-4. Flat areas and areas with impeded drainage in Ethiopia, excluding flat, sandy plains in eastern Ethiopia. Areas marked with red have slopes < 2%, TWI values > 8 and a clay content > 50 %; these areas agree largely with the larger vertisol areas indicated on the soil map of western Ethiopia in Jones et al. (2013). The data set for this map was produced by P. van Breugel for the VECEA atlas (Lillesø et al. 2011).

at that time of the year. Fig. 2-5 shows a map of this environmental parameter. The rugged topography of Ethiopia is clearly mirrored in the temperature patterns, and the figure shows the high temperatures in the western and eastern lowlands and in the deep river valleys in relation to the lower temperatures in the highlands. The limit of the *Combretum-Terminalia* woodlands as defined in this paper coincides well with the delimitation with highland areas with average annual temperatures below 20° C. Generally, the average annual temperature in the western lowlands of Ethiopia is in the interval between 25 and 30° C near the borders with Sudan and South Sudan and in the Omo Valley, while it is in the interval between 20 and 25° C near the border with the highlands. There is no visible north-south gradient in the western Ethiopian lowlands in this environmental parameter, which agrees with the striking similarity in the vegetation along the same gradient, as analysed in Section 10 and discussed in Section 11, although there is a significant variation on the local scale.

Although the tropical climates are characterized by relatively constant monthly average and rather constant maximum and minimum temperatures during the year, the diurnal variation may be considerable, with hot days and cold nights. Fig. 2-6 shows a map of the annual mean diurnal temperature range, indicating the variation of the interval between the daily minimum and daily maximum temperatures (but this climatic parameter has been calculated from minimum and maximum values averaged over longer periods). The hottest areas, along the Red Sea and the Gulf of Aden with average temperatures above 30° C, have the lowest variation between day and night (2.5-5.0° C). The lowest diurnal variation in the western Ethiopian lowlands is slightly higher, in the interval 10.0-12.5° C, and this is typically encountered in some areas in the southern part of the western lowlands. The diurnal temperature range increases towards the north and reaches higher values in the western lowlands of the GJ and GD floristic regions (for outline of the floristic regions, see Fig. 2-28), locally in the valleys of the Abay, Gibe and Didessa Rivers and in the upper reaches of the Tacazze River (15.0-17.5° C), and the diurnal temperature range reaches an absolute maximum in the lower part of the Tacazze Valley (17.5-20.0° C). Thus, despite the rather constant average annual temperature in the western lowlands of Ethiopia, not all environmental parameters relating to temperature are constant in that part of the country, even at the same altitudes. So, despite the similarity in the vegetation along the north-south gradient mentioned above, we may in Section 10 and 11 expect to find an influence of latitude.

The rainfall regimes of Ethiopia are generally controlled by the movement of the intertropical convergence zone (ITCZ) and thus the prevailing winds. The peak rainfall is related to the passage of the ITCZ in northern and southern directions, respectively, and the associated directions of the wind (Liljequist 1986). Two contrasting wind systems exist, depending upon the position of the ITCZ. In the summer months on the northern hemisphere, from May to October, the ITCZ is north of the Equator, and during this period the prevailing wind over most of Ethiopia is south-westerly. These moisture-laden wind-systems produce rain over large areas of the western escarpment and the Ethiopian Highlands. The highest rainfall is in the south-western part of the western highlands, where the humidity of the air is highest and the ascent of the wind is steepest. In the winter months on the northern hemisphere, from November to April, the ITCZ is to the south of Ethiopia and the prevailing winds are north-easterly, bringing only little moisture from the Red Sea, and very little rain falls on the western escarpment and the western lowlands (but some winter rain falls on the eastern of Eritrea). However, due to local conditions, rain may occur throughout the year in the south-western part of the western escarpment and the adjacent highlands and lowlands, so this area has not only the highest rainfall in the country but also the most evenly distributed precipitation. Our analyses and discussions in Section 10 and 11 do show an influence of the rainfall patterns on the vegetation.

In the highlands there is normally a dry period between the two rainy periods, between the rains in May-June and the rains in August-September. In July the ITCZ is in its most northern position (Liljequist

1986), when the rain falls mainly, but not exclusively, in Eritrea and northernmost Ethiopia, while the maximum rainfall in the northern and central highlands of Ethiopia follows in August-September. The northern and western parts of the western highlands and the lowlands towards Sudan, extending from the western part of the TU floristic region through the GD, GJ, WG and partly the IL floristic regions, receive a basically unimodal rainfall of varying length and intensity during the summer, typically from June-July to September. The length and the intensity of the rainy period decrease markedly from the south towards the north, but throughout this part of the country varying months during the summer represent the wettest period. In Fig. 2-7 it can be seen that the lowest rainfall in the western lowlands is 500-700 mm /year in the northernmost parts of the GD and TU floristic regions, apart from the deviating pattern in the Omo Valley (see below). Similar or slightly lower values are indicated for the upper reaches of the Tacazze Valley and parts of the Abay Gorge. The lowlands in the southern part of GD and most of GJ receive an average annual rainfall of 750-1000 mm, and the values gradually increase to 1250-1500 mm in WG near the border with the highlands. Further south, the rainfall decreases again and there is a slight tendency to bimodal rainfall regimes with slightly less rain in July-August, as can be seen from the studies of the vegetation on the Gambela Region by Jensen & Friis (2001, Fig. 2). Further south, the southern part of the KF and GG floristic regions, this tendency towards a bimodal rainfall becomes more pronounced (Friis et al. 2010, Fig. 7), but the average annual rainfall declines further, until there is a very low rainfall around and to the east of the delta of the Omo River and Lake Chew Bahir, with average annual rainfall of 250-750 mm.

It must be assumed that not only the total average annual rainfall and its distribution throughout the year may be of importance for the flora and vegetation, but also the rainfall seasonality, the variation between months with maximum and months with minimum rainfall. This variation has been calculated and expressed in Fig. 2-8. It is striking that the seasonality in the lowlands, both to the east and to the west of the Ethiopian Highlands, is lowest in the areas with low rainfall, that is in southern Eritrea along the Red Sea Coast and in the AF region, as well into the Omo Valley and in the southernmost lowlands in the KF floristic region (but there is also small variation in the high rainfall areas in eastern IL floristic region). Apart from these patterns, there is a general increase in rainfall seasonality throughout the western lowlands, starting from low seasonality in the Gambela lowlands in the IL floristic regions and on the escarpment in the WG floristic region (75-100%), increasing through the GJ and southern GD floristic regions (100-125%) to the highest figures in the northern GD and TU floristic regions (125-150%).

To sum up, there are strong differences between the average annual temperature of the Ethiopian Highlands and the western lowlands, but not a marked and well defined temperature gradient between north and south in this parameter. Less general temperature-parameters, such as diurnal variation, may show a slight north-south gradient. In the western lowlands, parameters relating to rainfall do show a north-south gradient, but for the average annual rainfall the maximum values are in the WG, IL and KF floristic regions and with decreasing values to the north and south. Rainfall sesonality has a maximum in the north and decreasing values towards the south. With these patterns, it is to be expected that the climatic parameters of the western lowlands will influence the flora and vegetation.

In many applications for agriculture, forestry and ecology the above-mentioned individual parameters outlined above are combined to indices known to be relevant to the growth of plants. Important indices for the availability of sufficient water are the climactic moisture indexes, which are measurement of the surplus or deficit of moisture available for the vegetation. Fig. 2-9 shows the distribution in Ethiopia and surrounding areas of one of the simplest forms according to which a climatic moisture index can be calculated, the annual rainfall divided by the potential evapotranspiration, calculated according to the 1985 formula of Hargreaves (Hargreaves & Allen 2003). The former variable is derived from Bio12, the latter

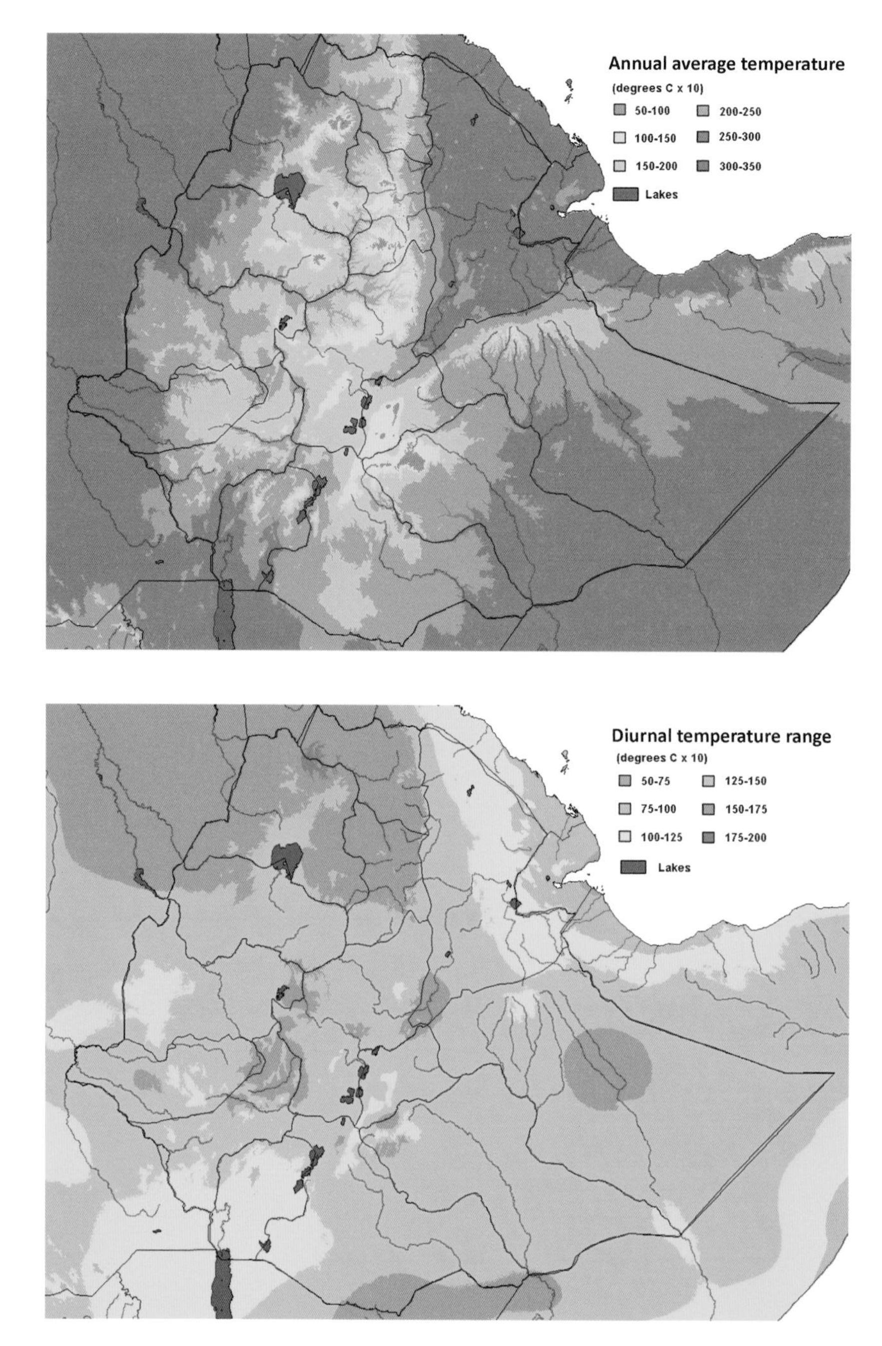

Fig. 2-5. Annual average temperature. The units are degrees Celsius × 10. Mapped from data in BIOCIM, bio1. Added to the climatic data are the borders of the floristic regions of the *Flora of Ethiopia and Eritrea*.

Fig. 2-6. Annual mean diurnal temperature range. The units are degrees Celsius × 10. Map constructed from data in BIOCIM, bio2. Because of the way the climatic data for BIOCLIM has been gathered, it is not possible to record the diurnal temperature range directly; it has to be calculated as the mean of the monthly temperature ranges (monthly maximum minus monthly minimum). The climate data inputs are monthly or averaged monthly records from multiple years. Using monthly averages in this manner is equivalent to calculating the temperature range for each day in a month and averaging these values for the month. Added to the climatic data are the borders of the floristic regions of the *Flora of Ethiopia and Eritrea*.

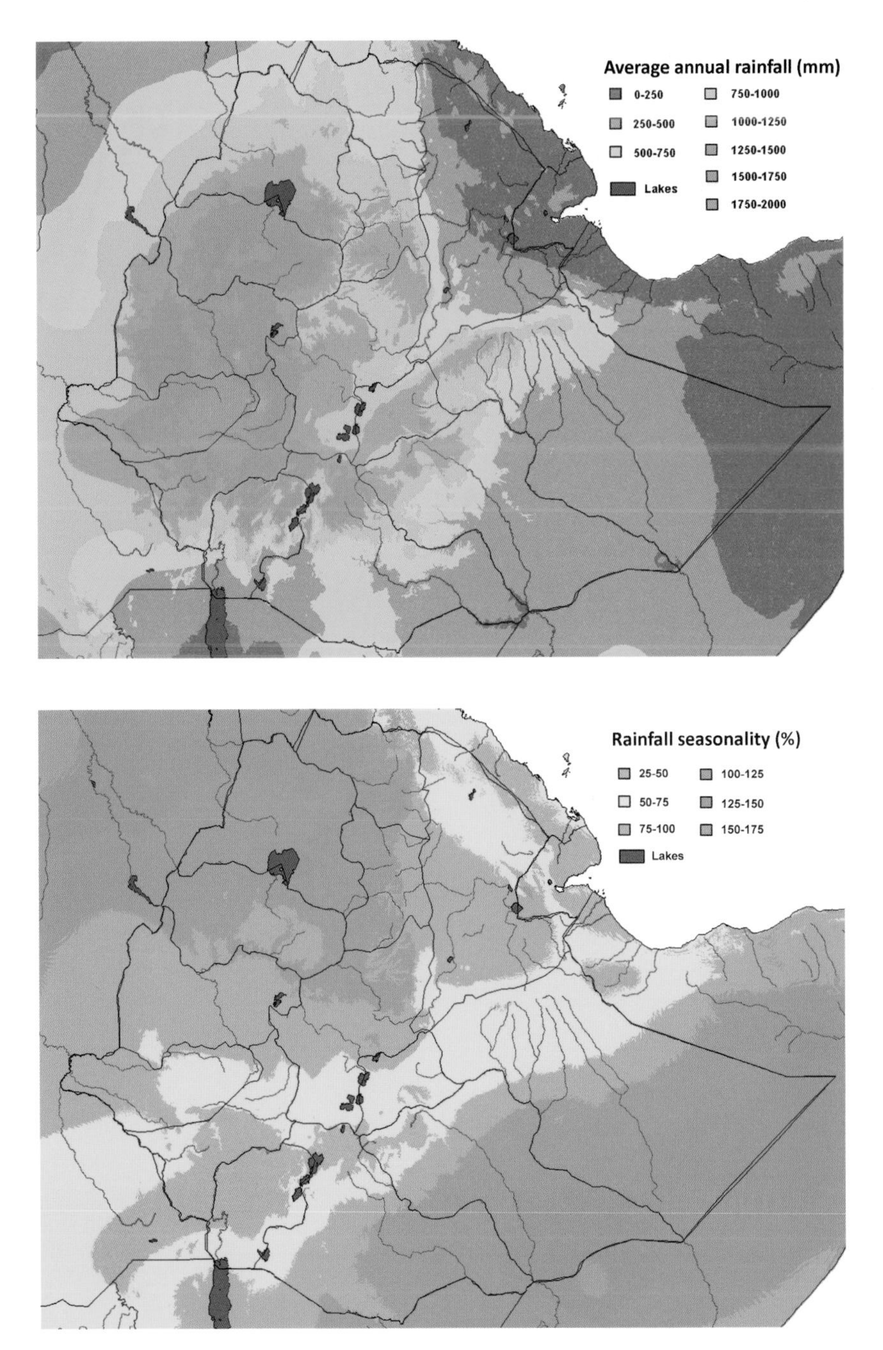

Fig. 2-7. Average annual rainfall. The units are millimetres of precipitation. The map is constructed from data in Bio 12. The values of the annual precipitation have been calculated as the sum of all total monthly precipitation values. Added to the climatic data are the borders of the floristic regions of the *Flora of Ethiopia and Eritrea*.

Fig. 2-8. Rainfall seasonality (coefficient of variation). The coefficient is expressed in percentage. The map is constructed from data in Bio 15. The rainfall seasonality is a measure of the variation in monthly precipitation totals over the course of the year. The index is expressed as the ratio of the standard deviation of the monthly total precipitation to the mean monthly total precipitation. Added to the climatic data are the borders of the floristic regions of the *Flora of Ethiopia and Eritrea*.

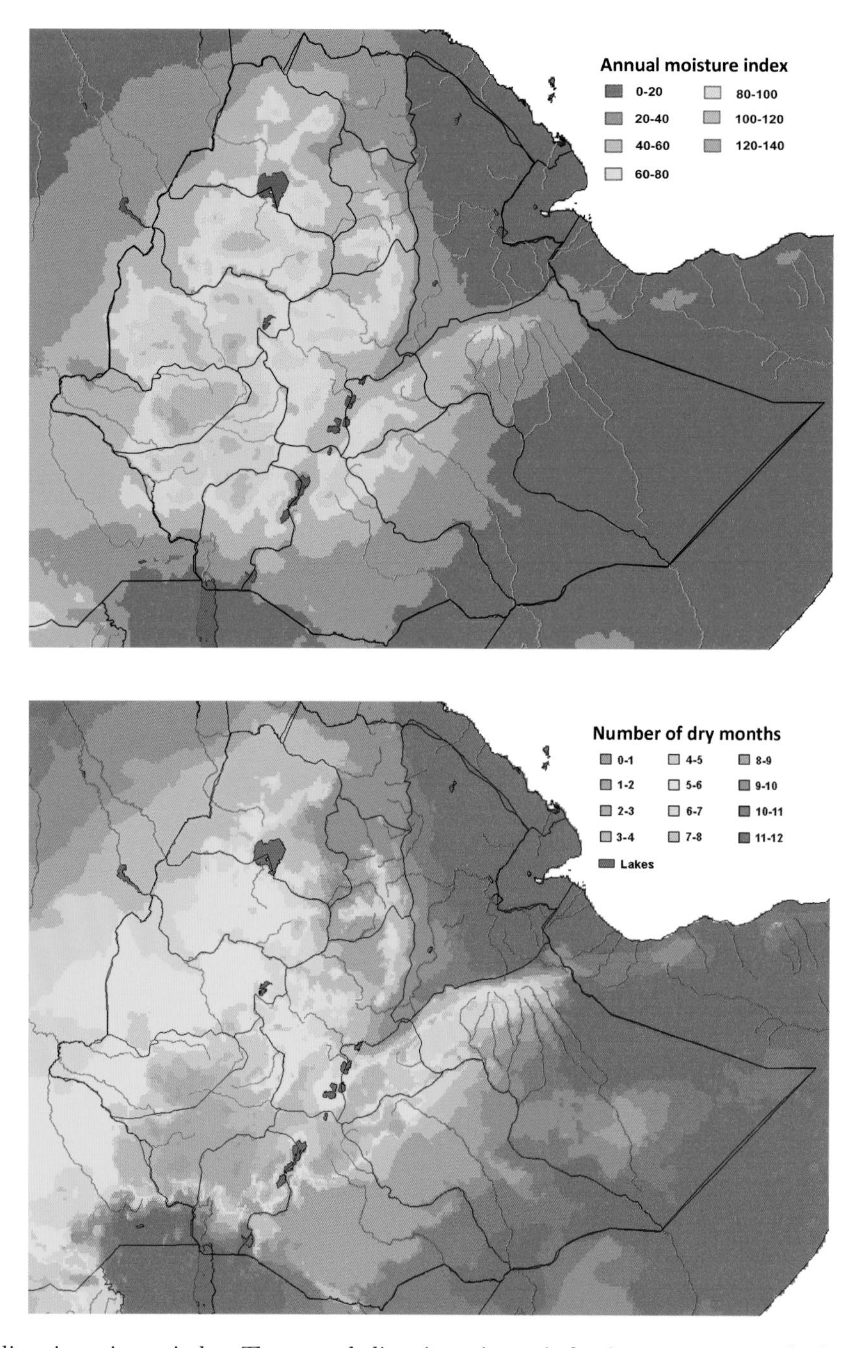

Fig. 2-9. Annual climatic moisture index. The annual climatic moisture index is a percentage calculated as 100 × Bio12/PE, where PE is the potential evapotranspiration, calculated according to the Hargreaves 1985-method. Both Bio12 and PE are in mm. Due to the method used, the index may exceed 100%. Modelled by Platts et al. (2014) and made available in the dataset Africlim (https://www.york.ac.uk/environment/research/kite/resources/). Added to the climatic data are the borders of the floristic regions of the *Flora of Ethiopia and Eritrea*.

Fig. 2-10. Number of dry months. According to UNEP (1997), months are recorded as dry if the moisture index for that month is less than 50%. Calculated and modelled by Platts et al. (2014) and made available in the dataset Africlim. Added to the climatic data are the borders of the floristic regions of the *Flora of Ethiopia and Eritrea*.

calculated according to Platts et al. (2014) and made available in the data set Africlim (https://www.york.ac.uk/environment/research/kite/resources/). Most frequently, the Annual moisture index is calculated by the formula 100 × (S-D)/PE, where S is the water surplus, D is the water deficit, and PE is the potential evapotranspiration (all measurements in mm or cm). The geographical distribution of a high moisture index shows good agreement with the area where the woodlands of the second cluster (blue) is found, as seen on the map in Fig. 10-6.

Also, the distribution of the moisture during the year is a parameter known to be highly relevant to the growth of plants. A highly derived data-set, indicating the number of dry months, has been presented by Platts et al. (2014) in Africlim, and this data has been used for our Fig. 2-10. A dry month is defined as one in which the climatic moisture index is below 50%. The number of dry months in the western woodlands are 8-10 in the extreme north and the upper reaches of the deep valleys of the Tacazze and the Abay Rivers (profiles C, D and H in Section 5), 9-11 in the extreme south in the lower Omo Valley and around Lake Turkana and Lake Chew Bahir (profile R in Section 5), and a large area with 6-9 dry months in the western woodlands from the Ethiopian border with Eritrea to just south of the Abay River, south of which an area with 6-7 dry months stretch approximately from the Abay River to south of the Gambela lowlands. The lowland areas from the Gambela lowlands to the Omo Valley show a surprisingly low number of dry months, 2-5 (but not with very high rainfall during the rains), followed by a very steep gradient to the above mentioned 9-11 months in the extreme south. From our field observations we would assume that the areas from the Gambela lowlands to the Omo Valley would have a higher number of dry months than the 2-5 indicated by the data set, but the very few meteorological stations in that part of Ethiopia may suggest uncertainty for such derived data as the number of dry months.

2.4. Ethiopian geology; the GERD Dam and the reservoir

The soil types in Ethiopia are largely dependent on the rocks from which the soil has been derived, and therefore a short overview of the geology of western Ethiopia is relevant here. The rocks of western Ethiopia are mainly basaltic, granitic or limestone. Generally, basaltic rocks erode to fertile soils with high concentrations of Ca and Mg, while soils derived from granites are poor in these elements and less fertile. Limestones also erode to fertile soils with high concentrations of cations, especially Ca, while quartzite and quartzo-feldspathic rocks erode to less fertile soils.

A new edition of a detailed geological map of Ethiopia at the scale of 1:2,000,000 was last published by Mengesha Tefera et al. (1996). A short introduction to the geological history of the whole of Ethiopia was written by Tarekegn Tadesse (2005) and later reviewed and expanded by Tadesse Alemu (2015), whose simplified map of the geology of Ethiopia is reproduced here as Fig. 2-16. A brief review was also presented as part of an introduction to the vegetation atlas of Ethiopia by Friis et al. (2010). The following information is a compilation from these sources, referring primarily to the map by Tadesse Alemu (2015), but restricted to the western part of the country.

The northernmost part of the western lowlands is mapped as being of early Tertiary (Palaeogene) volcanic origin, from the highlands and right down to the Sudan border, and beginning from close to the Ethiopian border with Eritrea and southwards to the GJ floristic region at the level of Lake Tana.

During the early Tertiary (Palaeogene) there was a dramatic uplift of the highlands, accompanied by massive outpouring of lava from numerous volcanoes, resulting in the often more than 1000 m thick, mainly basaltic shield (the Trap Series of lava deposits), which covers most of the highlands and the northwestern part of the highly eroded western escarpment (Fig. 2-12). On the eroded escarpment slopes, for example along the profiles we have studied and named A and B in Fig. 4-3., exposures show horizontal layers of lava flows alternating with intravolcanic sedimentary deposits, from fossilized plants in which lignite is commonly mined. The volcanic rocks include dominantly basalts but also rhyolites, trachytes, tuffs, ignimbrites, and agglomerates.

Apart from the Trap lava, there are small, flat areas of Quaternary superficial deposits, which agrees with our observations of flat areas with black vertisols ('black cotton soil') in this area near the Sudan border. Heavy rainfall and fracturing of the uplifted highlands during and after the Palaeogene have initiated the development of the continuously expanding river gorges in the highlands and described above in the text on topography (Section 2.2).

In the northernmost part of Tigray, mainly removed from the Sudan border and extending into Eritrea, the rocks consist partly of what Tadesse Alemu calls 'Low grade volcano-sedimentary rocks and associated intrusions' (Fig. 2-16). Thick layers of Mesozoic sediments are exposed in the big river valleys and elsewhere in Tigray, due to heavy erosion of the upper strata. The Mesozoic rocks consist of sandstone and limestone, among which the Enticho Sandstone, Adigrat Sandstone, Angula Shale and Antalo Limestone are particularly prominent (Fig. 2-11).

Approximately from just north of the Abay River and southwards nearly to Assosa, in the Beni Shangul-Gumuz Regional State, the geology of the western lowlands is very complex, with both older and newer rocks, and the rocks of the region are strongly faulted. There are areas with granites and other Precambrian metamorphic rocks, alternating with Proterozoic sedimentary rocks such as schists and sandstones. In the Abay Gorge, a long geological succession is exposed, including crystalline metamorphic rocks, Mesozoic sedimentary rocks, including limestone, and above these strata the Trap lava deposits. Other rocks of the western escarpment and the lowlands in this area are, as in parts of Tigray, made up of the above mentioned 'Low-grade volcano-sedimentary rocks and associated intrusives' and small areas of 'High grade gneisses and magmatites with associated intrusives'.

At this point in history (2021), while describing the Abay Gorge, it is relevant to mention the extent of the nearly finished GERD Dam on the Abay River (Wossenu Abtew & Shimelis Behailu Dessu 2019: Fig. 1.4), which, if the planned reservoir is filled completely, will flood a considerable part of the lower Abay Valley in the strongly faulted part of the Beni Shangul-Gumuz Regional State, as well as the lowermost parts of the Dabus and Didessa Rivers. See in Fig. 2-13 and Fig. 2-14 the situation as it looked when we made our field work well before the construction of the dam, and see the expected extent of the reservoir in Fig. 2-15. The bottom of the dam is at ca. 500 m a.s.l., and the crest is at ca. 655 m a.s.l., which will mark the water level of the filled reservoir. The storage capacity of the planned reservoir is expected to be ca. 74 cubic kilometres and it is expected to cover a surface area of 1,874 sqkm at a level of level of 640 m a.s.l. If filled to this level, the reservoir will be up to 30 km wide at the widest point and more than 100 km long from the dam to the highest flooded point, which is beyond the points where the Beles and the Dabus Rivers join the Abay, and probably reaching just above the point where the Didessa River joins the Abay. The filling of the reservoir behind the GERD Dam on the Abay River (Wossenu Abtew & Shimelis Behailu Dessu 2019: Fig. 1.4), began in July 2020. The planned completely filled reservoir will have considerable influence on the infrastructure and the population of the entire lower Abay Valley in Ethiopia, possibly also on the local climate, and the water will cover several of the relevés studied in this work (profile J.1, described in Section 5).

To the south of Assosa, the Precambrian rocks are interrupted by a belt of higher ground that stretches as a broad volcanic ridge across the Sudan border and reaching as far south as to beyond Dembidolo; this belt is a broad, elongated dome or a low ridge of early Tertiary (Palaeogene) volcanic origin, and again marked in Fig. 2-16 as 'Trap (plateau) volcanics.' Moreover, between Gidami. Dembidolo is the westernmost point of the Yerer-Tullu Wellel Volcano Tectonic Lineament, a tectonic zone stretching across the Ethiopia from Mt. Yerer at the edge of the Rift Valley to Mt. Tullu Wellel in western WG floristic region.

South of this volcanic ridge between Assosa and Dembidolo, and along the western escarpment to

Fig. 2-11. Upper reaches of the Tacazze River, with almost bare rocks of limestone. Near Abi Adi (TU floristic region). Image 0827 by Ib Friis. Oct. 2005.

Fig. 2-12. Rocky slopes south of Humera towards Gondar (GD floristic region). The grass fires have exposed the view to thick horizontal layers of Trap basaltic lava. Early dry season. Image 8397 by Ib Friis. Oct. 2014.

Fig. 2-13. Precambrian rocks in the riverbed of the Abay River near Sudan border. The river is surrounded by mixed open woodland (border between GJ and WG floristic regions). This area will be covered by the reservoir behind the GERD Dam on the Abay River. Early dry season. Image 5724 by Ib Friis. Nov. 2006.

Fig. 2-14. Lower reach of the Abay River. The river is surrounded by mixed open woodland and to the left there is a clump of the palm *Hyphaene thebaica* near the river (border between GJ and WG floristic region). The site was ca. 35 km upstream from the present site of the GERD Dam, and the area will be flooded by the reservoir. Early dry season. Image 6232 by Ib Friis. Nov. 2012.

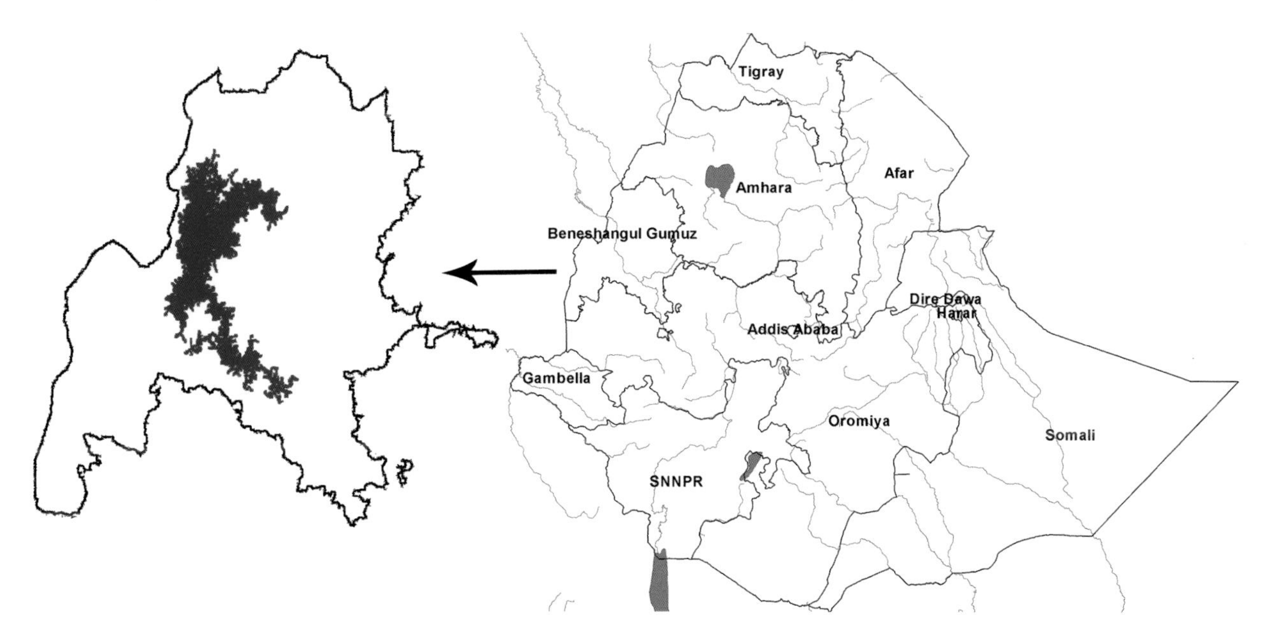

Fig. 2-15. Predicted future outline of the reservoir above the GERD Dam on the Abay River. To the left, the extent of the filled reservoir, marked in blue and imposed on an outline map of the Beni Shangul-Gumuz Regional State. To the right, the position of the Beni Shangul-Gumuz Regional State and the present flow of the major rivers in Ethiopia. If completely filled, the reservoir will flood ca. 1,874 sqkm of *Combretum-Terminalia* woodland up to a level of ca. 640 m a.s.l., covering the lower parts of the Abay River, the lower parts of the Beles and the Dabus Rivers, reaching probably beyond the point where the Didessa River joins the Abay. Based on information from Mohamed & Elmahdy (2017).

south of the Gambela Region, part of the western escarpment again consists of 'Low-grade volcano-sedimentary rocks and associated intrusives,' but also 'High grade gneisses and magmatites with associated intrusives'. These Precambrian rocks are supposed to underlie all other rocks in Ethiopia, although their exact extension is not known. They form a peneplaned basement of extremely folded, metamorphosed sediments and old igneous intrusions, including metamorphosed sandstone, schist, amphibolite, quartzite, and quartzo-feldspathic rocks, as well as occasional intrusions of granites. These rocks are sometimes also visible on other steep slopes of the escarpments of the highlands, and they form the rock near the bottom of many of the deep river valleys.

The majority of the Gambela lowland below the escarpment is made up of Quaternary superficial deposits, which agrees with our observation of this area as partly covered with black vertisols (profile N in Section 5).

South of the Gambela lowlands, the western escarpment loses its well-defined character and consists of low ridges of Tertiary volcanic rocks of the Trap series, alternating with gneisses and magmatites with associated intrusives, between which there are lowland areas, mostly with the great rivers running to Lake Turkana and Lake Chew Bahir, and the valleys of these rivers are filled with Quaternary superficial deposits (see image of the vegetation in Fig. 5-99).

For several reasons, particularly the lack of detailed field observations, the geology has not been included in the environmental parameters used for modelling and analyses of the relevés in this work. It was difficult for us as non-geologists to establish the geology of the relevés in the field, and it was considered too unreliable to attempt a subsequent identification of the relevés with areas on geological maps.

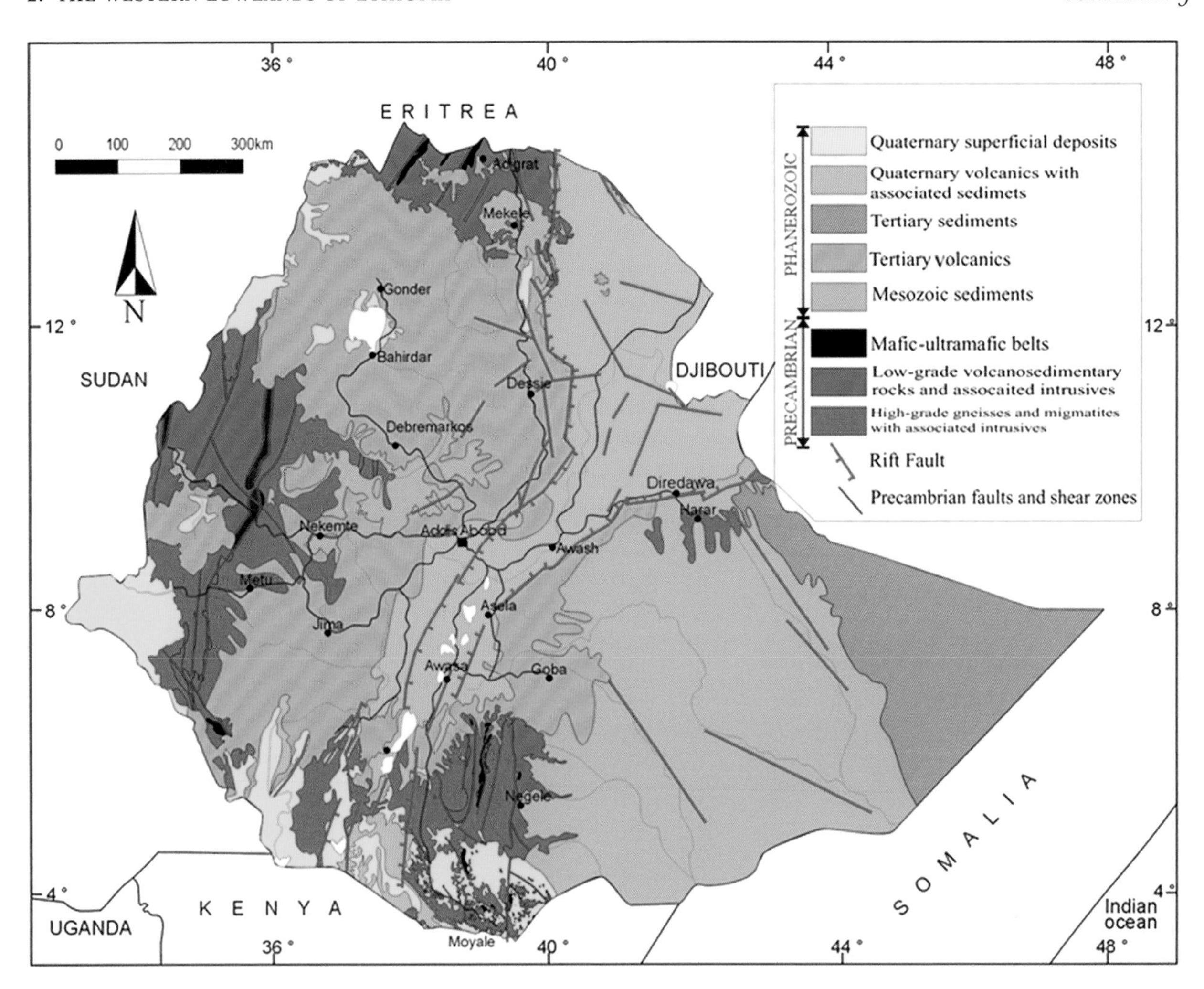

Fig. 2-16. Simplified geological map of Ethiopia. Based on the *Geological Map of Ethiopia* in 1:2,000,000 (Mengesha Tefera et al. 1996), simplified by Tadesse Alemu (2015) and reproduced with permission from Tadesse Alemu.

2.5. Ethiopian soils

The mosaic of soil types in Ethiopia is complex and dependent on the complicated topography and geology, as well as on the climate, but according to a harmonized continental-scale atlas of the soil types of Africa (Dewitte et al. 2013; Jones et al. 2013: Map plates 103 & 112), the soils on the western escarpment and in the lowlands consist of relatively few types (Fig. 2-18). On the Sudanese side of the border there are huge expanses of haplic vertisols (typical vertisols; the characteristic East African 'black cotton soil'), which stretches in Sudan and South Sudan all along western Ethiopia from the Eritrean border right to the border with Kenya, and this soil type is also found in flat areas of the western lowlands of Ethiopia, both in the lowlands in the Nile Valley and in flat areas that form 'shelves' on the western escarpment and in flat areas of the highlands (Fig. 2-17 and Fig. 2-18A). This soil type, which is derived from the weathering of rocks of the surrounding higher ground, has a high content of expansive clay minerals that form deep cracks in the dry season and alternate shrinking and swelling causes 'self-ploughing,' where the soil material consistently mixes itself by top-soil falling into the deep cracks. The areas of vertisol in Ethiopia, Sudan and South Sudan are by far the largest in Africa; further to the west in the Sudanian belt, vertisols only occur on the big floodplains.

Although scattered, haplic luvisols, which are dark brown soils with a well-developed clay horizon, are common in the north, in the GD and GJ floristic regions (Fig. 2-18B), where luvisols are the typical deep soils in the western lowlands between Tigray and Guba. Dark brown luvisols are also common on and below the escarpment from Dembidolo and southwards in the KF floristic region. Luvisols are scattered across Africa in the Sudanian belt.

According to Jones et al. (2013), the most common soil type of large areas of the western escarpment and the highlands above are the nitisols, which are deep, red, well-drained soils with a clay content of more than 30%, mostly rich in organic matter. They are found mostly above the areas with vertisols from approximately around the town of Guba in the GJ floristic region and southwards to Maji in the KF floristic region, where one finds large areas of umbric nitisols (Fig. 2-18C). West of Ethiopia, these soils tend to occur at the most humid southern edge and south of the Sudanian belt

Again, according to the same sources, on the western escarpment and deep river valleys in the TU floristic region, the soils are typically leptosols (Fig. 2-18D), very shallow soils over hard rock, or deeper soils that are extremely gravelly or stony. This type of soil is very common on the western escarpment of Ethiopia. In some of the upper reaches of the Tacazze River, the underlying rocks may be limestone or sandstone, but mostly the rocks are basaltic, and where the layers of soil are thick enough, these soils can be fertile, but have low water retention. These leptosols are characterized as either eutric (shallow soils over hard rocks) or lithic (soils with exposed rocks). Extensive areas with eutric leptosols are also marked on the western escarpment between Lake Tana and the Sudan border. West of Ethiopia, the leptosols are widely distributed in rocky areas.

Lake Tana itself is surrounded by areas with vertisols (mainly in flat areas along the lake shores), but also areas with nitisols and luvisols.

Where the run-off from the Gambela escarpment meets the extensive lowland plains large areas of fluvisols have developed (Fig. 2-18E); these are recently deposited, young soils and are predominantly brown (aerated soils) or grey (waterlogged soils). West of Ethiopia, fluvisols are common along the big river systems.

Fluvisols and vertisols are very common in the river valleys in the southernmost part of the western lowlands, often surrounded by areas with cambisols (soils in the beginning of soil formation or older soils with comparatively slow soil formation).

Another broad soil classification, consisting of six regional groups for the whole of Ethiopia, was outlined by Last (2009). The soil groups were classified according to their geographical position and their parental rocks; only the ones relevant to the western escarpment and lowlands are mentioned here.

Fig. 2-17. Vertisol in fallow land south-west of Shehedi (GD floristic region). Height of dry season. Image 9313 by Ib Friis. Feb. 2015.

The volcanic plateau soil region: This region, which covers the highlands and reaches the western escarpment. The soils are derived from basaltic and other volcanic rocks, and the rainfall is high or relatively high. The soils are mainly red lateritic soils but can also in river valleys be black vertisols ('black cotton soils'). They are usually fertile nitisols.

The crystalline highlands: These are highland soils derived from metamorphosed, crystalline rocks of the Precambrian origin. These soils mainly occur at the edges of the highlands and are acid, thin, grey-brown, or brown. They are usually not fertile. They may be leptosols or other types.

The alluvial plains: These are the plains of the large rivers. The soils are derived from material transported from the drainage area of the river, including topsoil from the highlands and may therefore be fertile. The extensive Gambela lowlands, which are crossed by the Baro and the Akobo Rivers, and the soils of the plains in the lower Omo Valley are also classified as alluvial plains, and their soils will typically be vertisols and fluvisols.

Neither of these two summaries gives a fully adequate description of the Ethiopian soils. Apart from the broad distinctions between lowlands and highlands and some patterns related to geology and rainfall, it has not been possible for us to relate the soil types on a continental scale to the vegetation types used in this work, and we have not included soil types from Jones et al. (2013) in the analyses of our relevés in Section 10. Instead, we have used our subjective observations of soils from the field (distinguishing between vertisol and other types), soil colour (dark brown, brown or red, but we were not able to distinguish between luvisol and nitisol), amount of stones, and presence of rocks (but not stipulating a soil as

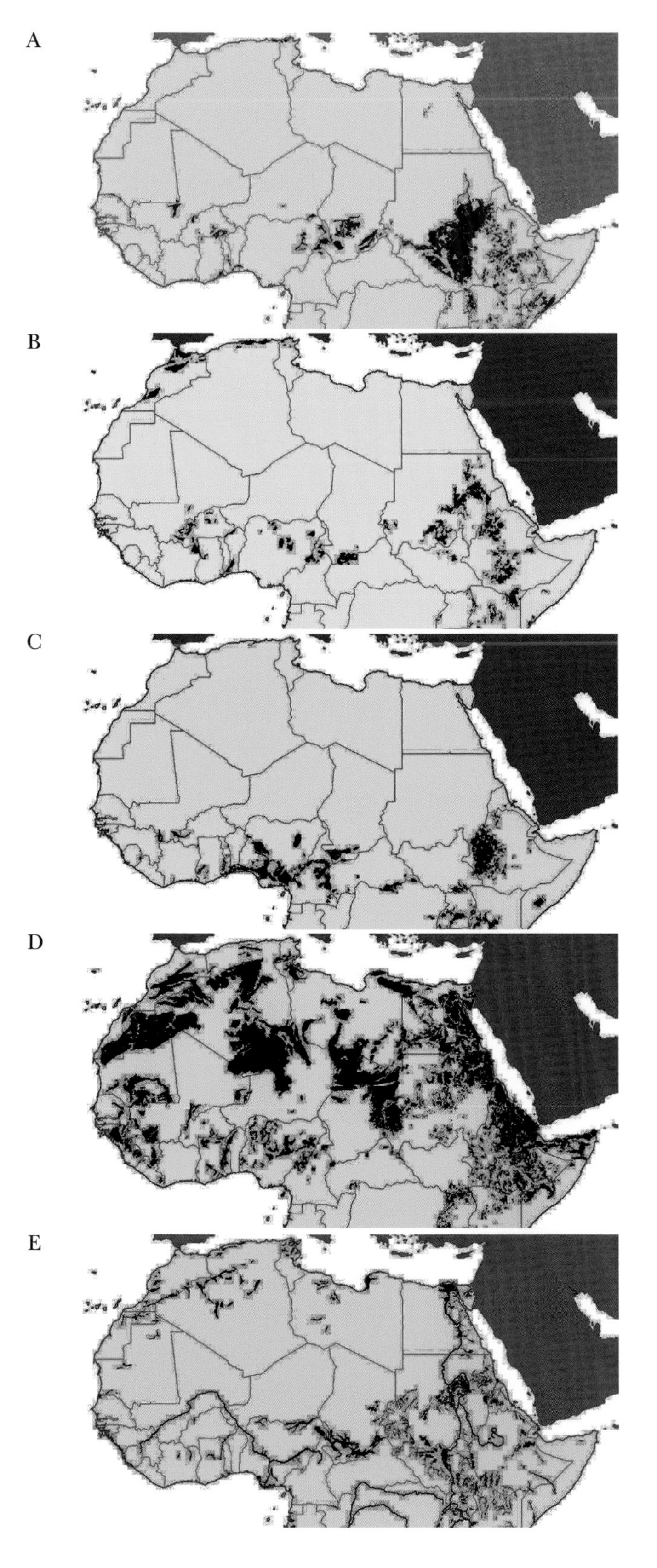

Fig. 2-18. Soil types found on the western Ethiopian escarpment and in the lowlands, as distributed in Africa north of Equator. A. Vertisols. B. Luvisols. C. Nitisols. D. Leptosols. E. Fluvisols. Redrawn from data in various unnumbered figures in Jones et al. (2013).

leptosol). The most distinctive differences were seen between areas with vertisols, areas with deeper red and brown soils and areas with thin or no soil over rocks. The soil categories used in Section 10, are indicated for each relevé in Appendix 1 and are listed in Appendix 3 and illustrated in Fig. 2-19.

2.6. Fires in Ethiopian vegetation

Grass fires during the dry season (Fig. 2-20), often several times in a year in the months between September and May, is an important environmental factor in the western lowlands and on the western escarpment (Fig. 2-24). The fires may burn over large areas or in relatively restricted patches (Fig. 2-21), mainly depending on the topography and biomass available for burning. Fires have been recognized as a major factor shaping the balance between woodlands and forests at the continental or landscape scale (Barbosa et al. 1999), where areas with the same topography and climate may represent either forest (little or no burning) or woodland (frequent and intensive burning) as alternative stable or semi-stable states with different floristic composition (but see also the exception below in the woodland of *Syzygium guineense* subsp. *macrocarpum*).

More specifically, the balance between lowland forest and woodland has been studied at the Imatong Mountains in the south-eastern South Sudan. The shifts of the boundaries between woodlands and the closed forests at Talanga, Lotti and Laboni were found to be dependent on annual grass fires (Friis & Vollesen 2005; Jackson 1956), and similar processes can be seen in Ethiopia along the edges of the *Transitional rain forest* (TRF), the changing boundaries of which has been studied by Breugel et al. (2016a,

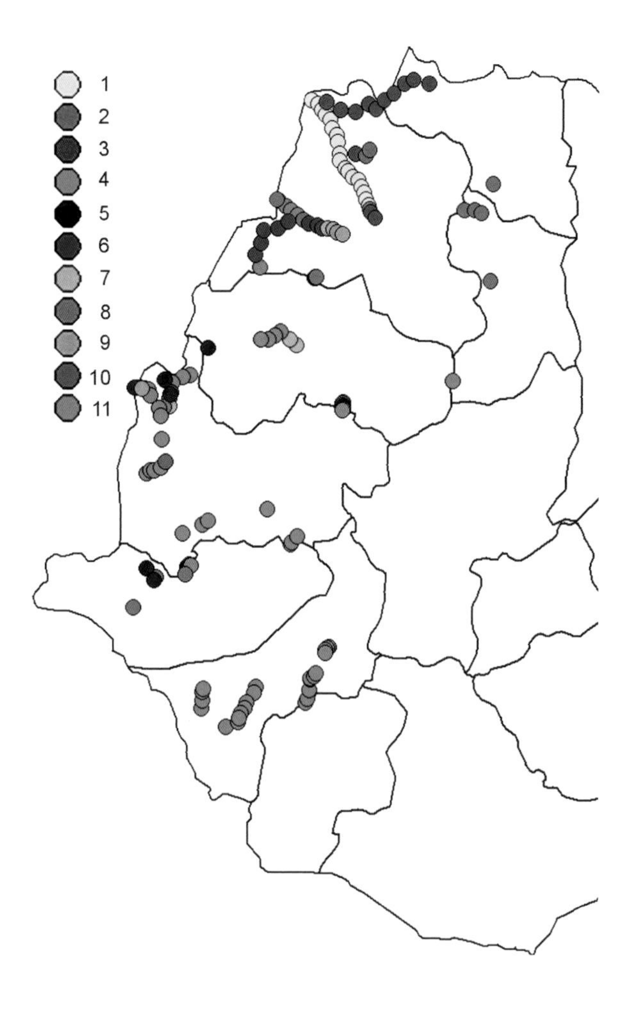

Fig. 2-19. Conversion of field observation of soils into numbered categories. Soil types in descriptive terms, based on the field observation indicated in Appendix 1, are converted into 11 intuitively numbered categories, according the weathering of the rocks from which the soils are derived, the colour of the soil and the grain size. The sign => indicates range of variation within the category:

1. Black cotton soil.
2. Brown soil => black cotton soil; brown soil => fine sand & black cotton soil; brown, stony soil => black cotton soil; dark brown, cracking soil (=> black cotton soil); stony, brown soil => black cotton soil; mosaic of stony soil and black cotton soil.
3. Brown soil.
4. Brown soil on rocks; brown, stony soil; stony brown soil; stony, brown (red) soil; stony, brown soil; stony, brown soil (sandy, pale reddish => almost black cotton soil); stony, brown soil (with lava blocks).
5. Stony, dark red soil; stony, red soil.
6. Dark red soil; red soil.
7. Grey soil; red soil => sandy soil; rocky, red soil; sandy, reddish soil.
8. Stony, grey (or nearly white) soil; stony, grey soil; grey, rocky soil.
9. Stony, sandy soil.
10. Rocky, and sandy soil; stony soil.
11. None [no soil]; rocks with little soil; rocky ground; steep rock).

See also Appendix 3.

2016b). However, the relationship between fire and vegetation is ultimately based on climatic factors, as sufficient inflammable biomass production is a necessary condition for the fires. The amount of rainfall during the rainy season and the severity of the dry season are thus crucial factors that control the frequency and severity of natural fires (Breugel et al. 2016). Also, we have observed in the Gambela region, where the soil is sufficiently moist, that fires may induce regrowth of grass during the dry season and, with that added fuel, make possible more grass fires during the same dry season (Jensen & Friis 2001; Breugel et al. 2016a); see an example of this in Fig. 5-71.

As seen in Fig. 2-24, fires occur widely in the *Combretum-Terminalia woodland* (CTW) and in the *Wooded Grassland of the western Gambela region* (WGG). In fact, the borderlines between the *Combretum-Terminalia* woodland and other vegetation types (except the border to the WGG) follow rather closely the limits of the frequently burning areas. Fires are almost absent from the low-rainfall areas with low ground-cover productivity, particularly in the northern TU floristic region, the middle and upper Tacazze Valley, and in the upper Abay Valley, and are infrequent in the lower parts of the Omo Valley and its surroundings. This agrees with areas where the grass-stratum is low and thin or areas where grass is nearly absent. The lower limit in average annual rainfall for natural fires is around the 650-700 mm (Breugel et al. 2016), which agrees with a limit in the zone between 500 and 750 mm in our Fig. 2-7. Once the forest has been destroyed and replaced by wooded grasslands, woodland species with bark that protects against fire may become important in the vegetation (Fig. 2-23);

the statistical importance of protection against fire is studied in Section 8.4.

A surprising aberration from this rule of a close agreement between rainfall and fire frequency is seen in the low fire frequency in the highest central part of the WG floristic region, which is dominated by woodland of the evergreen species *Syzygium guineense* subsp. *macrocarpum,* where there is low sparse ground-cover and no sign of burning (described in Section 5, Profile J, and shown in Fig. 5-52). There is no climatic factor in the range in Fig. 2-5 to Fig. 2-8 that show correlation with this area of low fire frequency, and we can only attempt to explain it with our observations in the field, such as predominance of evergreen woody plants, shade and hence the undergrowth consisting of a relatively thin grass stratum with mainly annual grasses, or human influence in the form of grazing and farming. Both the thin grass stratum and the grazing and farming reduce the biomass available for burning in the dry season. According to the long-established and still generally accepted absence paradigm, of fire in Sudanian woodland will lead to forest (Friis & Vollesen 2005: 698-700). Here, however, it has led to a fire-free woodland!

The highest number of fires in the western lowlands, and in Ethiopia as a whole, has been observed in the north-western part of the Gambela Region, where there is ample water supply in the rainy season and moisture in the ground to allow the resprouting of the grasses after grass fires in the dry season, allowing several subsequent grass fires. This applies to the flood plains along the Baro River, where a very tall and dense grass stratum develops (Fig. 5-76). Moreover, this area is relatively densely populated with a population (the Nuers) that are mainly pastoralist (see Section 2.7). The freshly resprouting grass after grass fires is favourable for the grazing animals during the dry season.

There are other regions of Ethiopia, where fires can be seen and have been reported to play an important role in vegetation dynamics (in the Afroalpine and Ericaceous belts in the Ethiopian highlands, in few and limited places in the *Acacia-Commiphora* bushland in the SD, AR, BA and HA floristic regions and in swampy places in the AF floristic region). An extremely high number of records of what looks like fire-incidents can be seen in one spot in northern AF, but they coincide with the place of the crater of Erta Ale, a continuously active basaltic shield volcano, and the melted lava is recorded as highly frequent fires.

Summing up, the influence of fires must to some extent be correlated to climatic parameters, but it shows a characteristic pattern not congruent with any of the patterns shown by the climatic parameters in Fig. 2-5 to Fig. 2-8. Yet fires are most clearly correlated with areas where there is a sufficiently long rainy season and a sharp contrast between the wet and the dry season, which allows a tall grass cover to develop and let it dry out completely. Fires are almost entirely absent from the TU floristic region, where the groundcover is very low and thin, but fires occur frequently throughout the western Ethiopian lowlands from the GD to the GG floristic region, where tall grass grows during the rains and provides abundant biomass for fuel. Areas with frequent fires are relatively fewer or smaller in the GD, the KF and the GG floristic regions than in the GJ, WG and IL floristic regions that have the largest areas with the highest frequencies of fire in the entire country. The tallest grass, more than 2 m tall, providing biomass for fires during the dry season was observed in the IL floristic region (relevés 86-88).

Generally, at least according to our observations in western Ethiopia, the vegetation that produces sufficient biomass for intense fires and large fire-scars (Fig. 2-21 and Fig. 2-22) are dominated by the tall grasses of the tribe Andropogoneae and a few other genera, including the genus *Loudetia*. These fire-scars may vary from less than hundred metres to several kilometres across, and reflect the areas where suitable biomass from the grass stratum has accumulated during the wet season.

Woody plants above a certain size seem mostly to survive woodland grass fires, but in the frequently burning woodlands 50% of the species and more have special adaptations that protect them (Fig. 2-23). A list of trees with what appears to be adaptations to protect them against fire is given in Appendix 5, and the distribution of species with protection against

Fig. 2-20. Grass fire in mixed open woodland between Djemu and Maji (KF floristic region). Early dry season. Image 4602 by Ib Friis. Oct. 2018.

Fig. 2-21. Black fire-scars in woodlands. Image above: fire-scar from recent grass fire on slope with mixed open woodland between Shehedi and Aykel (GD floristic region). Height of dry season. Feb. 2015. Image below: fire-scar from recent burning on hill with mixed open woodland and wooded grassland. South-west of Shehedi. In the foreground a shrub of *Calotropis procera*. Height of dry season. Image 9298 & 9319 by Ib Friis. Feb. 2015.

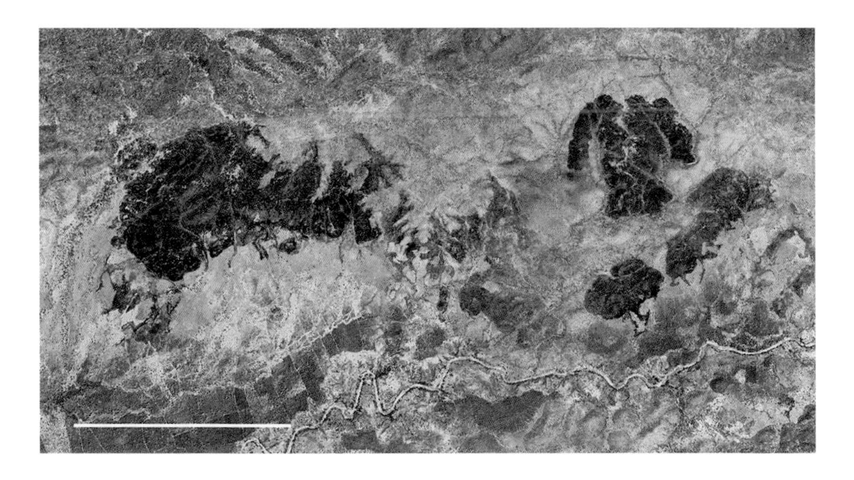

Fig. 2-22. Black fire-scars on northern slope of low hills south of the Setit-Tacazze River, north of the road between Humera and Shire and inside the Kafta-Shiraro National Park (15° 11' N, 36° 55' E, 800 m a.s.l.). This area, ca. 10 km south of the border between Ethiopia and Eritrea, is one of the northernmost localities with intense burning, see Fig. 2-24. Length of the white scale bar bar 1 km. From Google Earth, US Dept. of State Geographer Image © 2020 CNES Airbus Image © 2020 Maxar Technologies; reproduced under general license.

Fig. 2-23. Fire-resistant bark of trees. Left image: trunk of *Boswellia pirottae* with fire-scorched bark in mixed open woodland and wooded grassland above the inner Abay Gorge on the road between Bure and Nekemt (WG floristic region; north-facing side of the gorge). Early dry season. Oct. 2010. Right image: thickened partly fire-resistant bark of *Cussonia ostinii* in mixed open woodland and wooded grassland just above the Abay Gorge on the road between Bure and Nekemt, (WG floristic region; north-facing side of the gorge). Early dry season. Image 1370 & 1408 by Ib Friis. Oct. 2010.

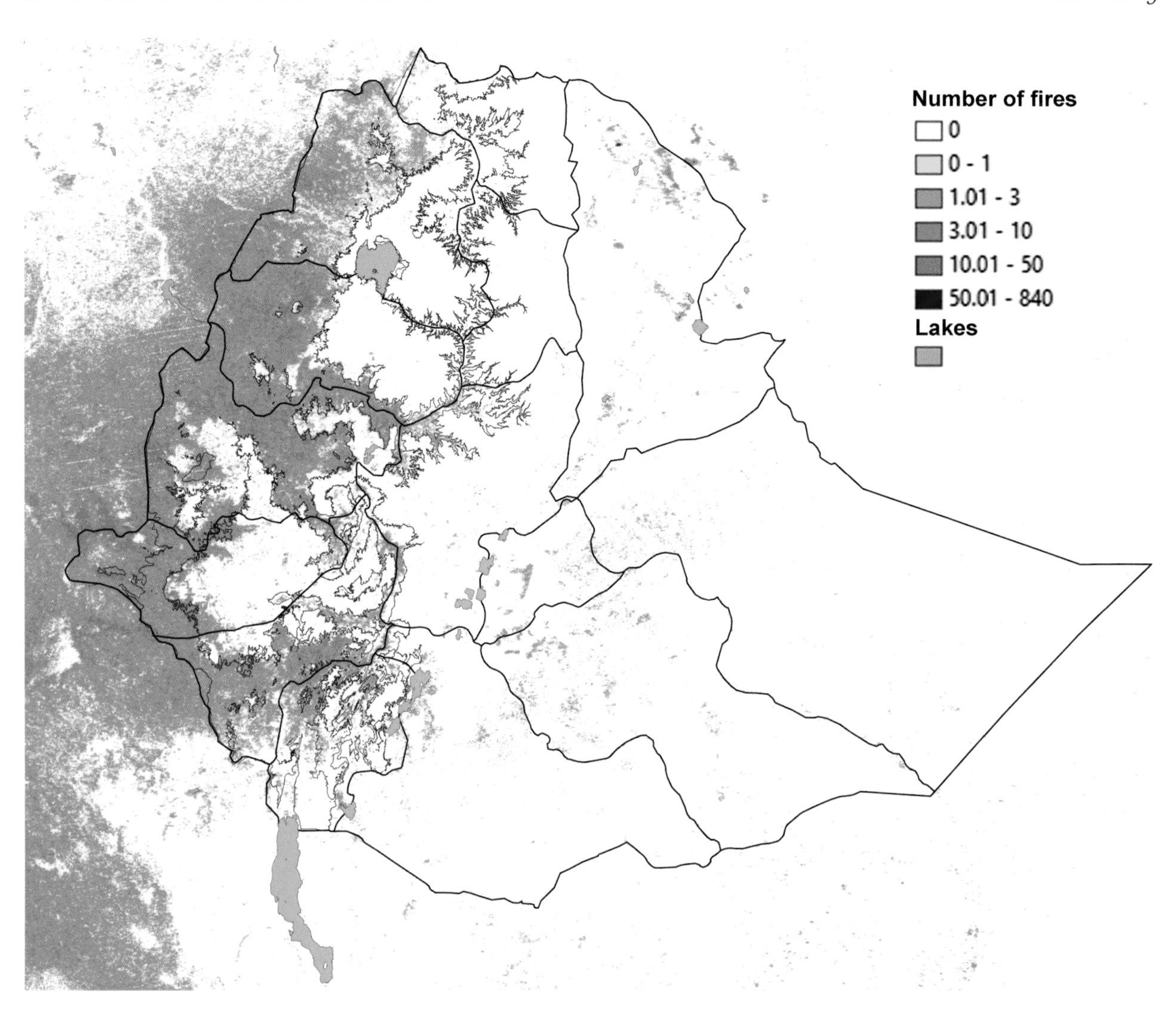

Fig. 2-24. Total number of fires in Ethiopia and adjacent areas recorded on eight day periods. Recording made during the years 2003-2013. Added to the fire frequency data are the borders of the floristic regions of the *Flora of Ethiopia and Eritrea* and a line indicating border of the *Combretum-Terminalia* woodlands as defined in this work. The map is based on MODIS satellite data (https://modis.gsfc.nasa.gov/) for the years 2003-2013. Seaa also Appendix 3 and list of acronyms.

fires is discussed in Section 8.4. The distribution of high frequencies of fires seems to be correlated with the extent of the western woodlands, particularly the *Combretum-Terminalia* woodlands (Fig. 2-24), and in Section 10 it is demonstrated that the variation in the maximum fire frequency (Fig. 10-8, lower right figure) explains 40% of the variance between our relevés in ordination space, with more variation in the northern part of the woodlands than in the southern part (see more in Section 10).

2.7. History of the western Ethiopian lowlands: border-zone, demarcation of a border, historical events of environmental importance

Little or no information about the western Ethiopian lowlands reached outside that area, or indeed outside Ethiopia, before the 17th century. In the north, the boundary area between the Kingdom of Meroë, now in present-day Sudan, and the Axumite Kingdom, now mainly in present-day Ethiopia and Eritrea, was subject to continued rivalry, which ended with the collapse of Meroë in the fourth century AD and a subsequent territorial expansion of the Axumite kingdom into the lowlands. But eventually the borderland seems to have broken up into smaller political entities with local rulers (Bahru Zewde 2003; Zach 2007). It is not known how far the Axumite kingdom extended into the western lowlands, but presumably there was no fixed border, and it is likely that there was a gradual transition to only vaguely influenced or completely independent lowland areas now in the GD and GJ floristic regions.

It was an Egyptian, the Nestorian Christian merchant named Kosmas Indicopleustes, who in the sixth century AD provided the earliest evidence of trade between the Ethiopian Highlands and the western lowlands. He referred to a land called Sasu in the lowlands to the south-west of the Kingdom of Axum, to which expeditions with up to five hundred traders were sent every other year, the traders bartering individually with the local people to obtain their products. In this bartering, the Axumite used rock-salt from the Afar Depression, while the most important local product bartered for was gold in small nuggets, a product which even today is an important produce from the western escarpment (Pankhurst 1997).

Probably identical, at least in part, with the 'Sasu' of Kosmas Indicopleustes is the large areas of the western Ethiopian lowlands now known as Beni Shangul and Gumuz, which stretched from around Metema southwards to near the present-day Assosa in the WG floristic region (and Beni Shangul-Gumuz is now the name of a regional state in federal Ethiopia, see Section 2.8). Traditionally, this area formed a buffer zone between the mainly Muslim Sennar sultanate on the Blue Nile (in present-day Sudan) and the Christian Ethiopian highlands (Wallmark 1986; Bender 2005; Abbink 2005). The Beni Shangul and Gumuz areas were (and are) inhabited by two ethnic groups, the Beni Shangul, also known as the Berta, mainly settled to the south of the Abay River, and the Gumuz, mainly settled to the north of the river (Abbink 2005). From the 16th century onwards, these areas were part of or associated the Fung kingdom of Fazugli, under the Sennar Sultanate in the present-day Sudan (Spaulding 2003).

The western woodlands were an area, where the Sudanese trading network could interact with highland Ethiopia, and from which the Ethiopians and the Arabs in the Nile Valley could obtain not only gold, but also slaves. People from these areas were for millennia captured and traded (Bustorf 2010; Miran 2010; Pankhurst 2010a). A map by Miran (2010) shows the known slave-raiding areas in the Ethiopian-Sudanese lowlands and southern Ethiopia in the 19th century, when the slave-trade was at its highest. Apart from gold and slaves, cattle, horses, iron, civet, coffee, ivory and honey were traded from the western lowlands of Ethiopia to the highlands (Triulzi 2003; Abbink 2005).

By the end of the 19th century, the western borderlands of Ethiopia were still ruled by local rulers, and not from Khartoum or Addis Ababa. But in April 1891, the Emperor Menelik II sent a circular letter to the European powers defining what he considered should be the future western boundary of Ethiopia.

This included an enormous territory from Gedaref, now in the fertile areas in the Sudan north of the Abay (Blue Nile) and near the border between Ethiopia and Eritrea, to the point in South Sudan near the grass swamps in the Sudd where the Sobat River meets the White Nile (Marcus 1963). Due to its influence in Egypt, the British government had strong interests in this part of the Sudan, but the British were deeply involved in the Mahdist War (1881-1899) about the control of the territory. Therefore, negotiations between Emperor Menelik II and the British government about the western boundaries of Ethiopia only began in 1897, when John Lane Harrington, the British consular officer at Zeila (then in British Somaliland), was appointed Her British Majesty's Agent in Ethiopia. Menelik soon realized that the British would only accept to relinquish what they considered Sudanese territory after an Ethiopian *de facto* occupation of the territory that he wanted to become Ethiopian, and in 1898, after having been part of the Fung Kingdom for approximately 300 years, most of the Beni Shangul and the Gumuz areas, east of Mt. Fazugli, were occupied by Menelik's army. This happened during the last British campaigns against the Mahdists, and only the British victory at Khartoum initiated in earnest the British-Ethiopian negotiations. According to Marcus, at an early point in the negotiations, Menelik II drew his finger on a map over all lowlands bordered on the west by the White Nile between 2° and 14° northern latitude, saying, "That all belongs to me" (see Menelik II's territorial claims and changes in the boundaries of Ethiopia during his reign on map in Clapham 2007: 926, and see also map in Fig. 2-25).

Intense negotiations between Ethiopia and Britain began in April 1899, when Menelik II told Harrington that he stood firmly on his claims of 1891, and Harrington pointed out that the British government had not accepted Menelik's 1891-declaration, but was willing to negotiate. In the negotiations, the areas of the Beni Shangul and Gumuz became an important point, and Harrington could report that, except for the major part of Beni Shangul-Gumuz (see present boundaries of the currently Ethiopian part of Beni Shangul-Gumuz in Fig. 2-15 and Fig. 2-29), very little territory that had formerly been under Egyptian influence would have to be given up in order to satisfy Menelik, and that much of what would have to be ceded to Ethiopia "... was formerly a bone of contention between Egypt and Abyssinia as the frontiers were never properly defined between these two countries." With that agreed, the only serious remaining problem was who would control Metema. But since Great Britain only required the fort, which was in the new town, Menelik II could be given the old town east of the Khor Abnakara, and a division of the town into a Sudanese and an Ethiopian part along the Khor was agreed to. Slow communication and Menelik's wish to ponder over the results delayed the final conclusions, but by early April 1901 there was agreement on a draft treaty, and Great Britain, Italy and Ethiopia could sign two treaties at Addis Ababa, settling the borders of Sudan, Ethiopia and Eritrea. Great Britain's final ratification of the treaty was presented to Menelik II on 28 October 1902.

The wording of the relevant articles of the treaty is as follows (Ullendorff 1967): "Article I. The frontier between the Soudan and Ethiopia agreed on between the two Governments shall be: the line which is marked in red on the map annexed to this treaty in duplicate and traced from Khor Um Hagar [a wadi near the present town of Humera] to Gallabat [Metema], to Abay [Blue Nile], Baro, Pibor and Akobo rivers, to Melile [a village now apparently deserted, according to map of Anglo-Egyptian Sudan in 1:250,000, sheet 78H, July 1923, at ca. 6° 45' N, 34° 34' E], thence to the intersection of the 6th degree north latitude with the 35th degree longitude East of Greenwich [This is the point where the present northern boundary of the Equatoria Province in South Sudan meets the present Ethiopian-South Sudan boundary. The remaining part of the boundary to Lake Turkana (then Lake Rudolph) was demarcated in connection with the Ethio-British treaty of the 6th of December, 1907, on the boundary between Ethiopia and the British East Africa Protectorate, later to become the Kenya Colony and present-day Kenya]. – Article II. The boundary as defined in Article I shall be delimited and marked on the ground by a Joint Boundary Commission, which shall be nominated by the two High Contracting Parties, who shall notify the same to their subjects after delimitation."

The following articles III-V stipulate that Ethiopia was not allowed to construct or permit to be constructed any work across the Blue Nile [Abay River], at Lake Tana or on the Sobat River, which would arrest the flow of their waters into the Nile except with the acceptance of the British government and the Government of the Sudan (an article now much debated in connection with the Ethiopian GERD Dam on the Abay River), that the British government and the government of the Sudan could establish an exclave on the Baro River in the Gambela lowlands to be used for a British trading post, and that the British government and the Government of the Soudan should have the right to construct a railway through Ethiopian territory to connect the Sudan with Uganda. Ullendorff (1967) published both the English and the Amharic version of the treaty and concluded: "The English version is a good deal more specific, though, by and large, the Amharic text adequately conveys the gist of the agreement." The British copy of the map with the boundary suggested as a red line is in the National Archives (part of FO 93/2/4/5) and has been reproduced by Mulatu Wubneh (2015: 450 & 466).

In December 1902, Major Charles William Gwynn took up the task of the Sudan Commissioner in the demarcation of the Sudan-Ethiopian border, the nearly final outline of which can be seen in Fig. 2-25. From an earlier visit in 1899, Gwynn had studied Beni Shangul, which was to be Ethiopian territory (Gwynn 1937: 153-155): "Our real start was made from Roseires ... Almost our first task was to ascertain the limits of the Beni Shangul District and the conditions prevailing in it. So far as could be ascertained by short excursions into it, the district was a fine bit of country with perennial streams. Ferns, jasmine, and blackberries were pleasant to see in contrast to the Sudan vegetation. It is separated to the East from the main plateau by the Yabus River [the Yabus and the Dabus Rivers are often confused; the Dabus River is a tributary of the Abay River, while the Yabus River is a smaller stream that flows directly into South Sudan; Gwynn's remarks suggest that he actually meant the Dabus River] and forms a salient pointing and sloping from south to north." ... "As part of this the district had been definitely assigned to Abyssinia the chief problem was to ascertain its limits, especially how far the outlying hills should be included in it, and which could be taken as frontier marks. Some of the hills were inhabited, some not; but there appeared to be no considerable tribal organisation and names which had appeared in the old maps had ceased to have any meaning." ... "I decided therefore to recommend that most of the detached hills should be included in the Sudan ..." The party continued southwards from Beni Shangul: "After several days of difficult marching through belts of bamboo, and without meeting any inhabitants, we suddenly came on a new type of country at the main plateau level. Delightful park-like country with short grass, a pleasure to canter over, neat farmsteads and a considerable amount of cultivation. The people were Gallas [Oromo] ... We had actually entered the western extremity of Welega, one of the most prosperous and fertile districts of Abyssinia, then practically unknown to Europeans." The party soon after reached Gidami, where it was held for a month before it could continue to the Sobat River in the Gambela lowlands and the White Nile. After 1902 (Gwynn 1937: 158), the whole border from Eritrea down to Melile (6° 29' N, 35° 00' E) in the south was demarcated by selecting peaks of named inselbergs (Jebels) as boundary markers, or, in the south, rivers running to the White Nile as natural boundaries.

The 1902-boundary is shown on a British map in Fig. 2-25, published soon after the agreement and showing an alignment, which, apart from the not demarcated or missing parts in the extreme south, is close to the present situation and how the western Ethiopian woodlands have been defined for this work. Most, if not all legal documents on the Ethio-Sudanese boundary up to 1979, including detailed information on Gwynn's demarcation, have been reproduced by Brownlie & Burns (1979: 852-887).

Marcus (1963: 94) wrote about the Ethio-British negotiations, "The care taken in these negotiations

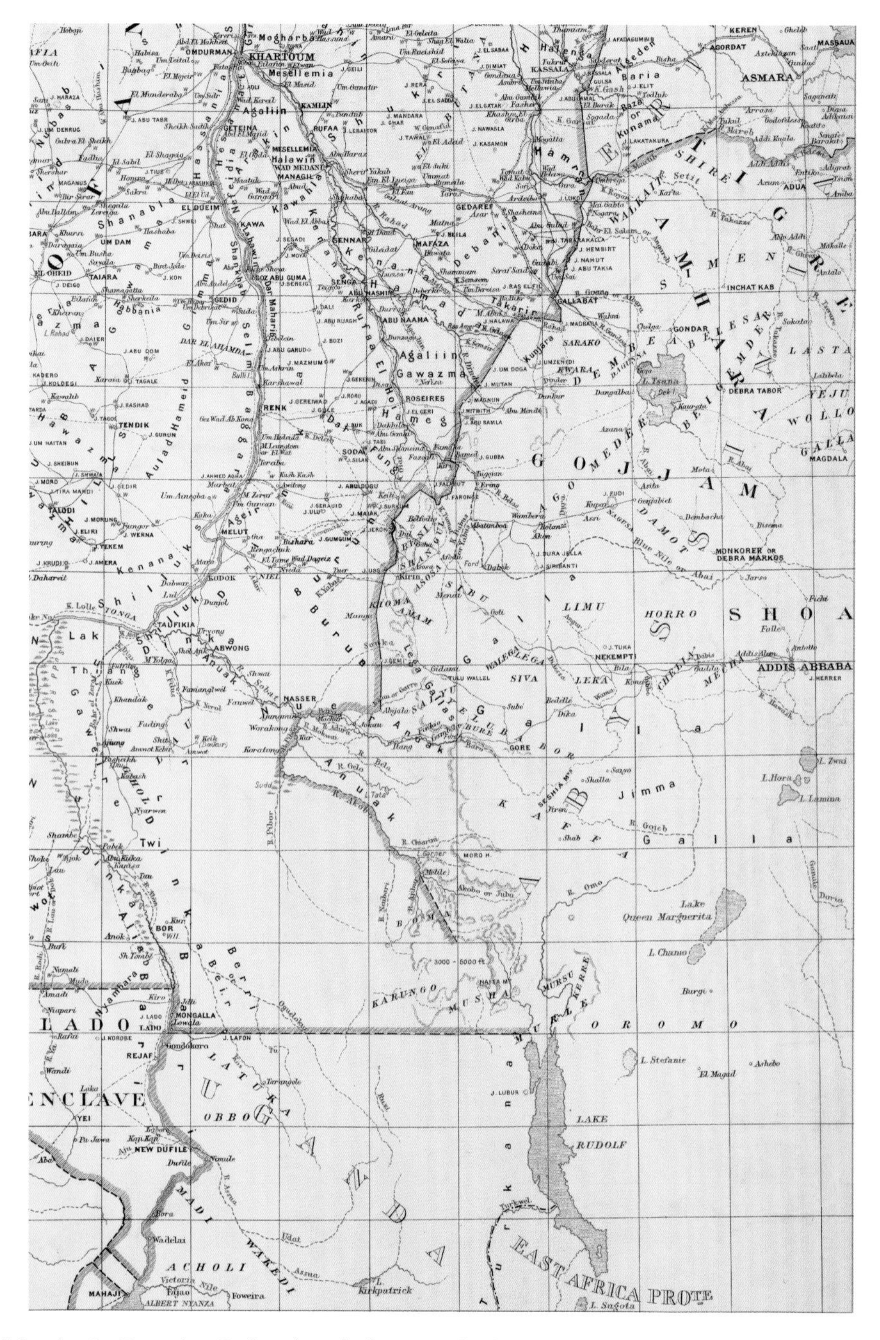

Fig. 2-25. Map of the Anglo-Egyptian Sudan (1904) showing the first agreed boundary with Ethiopia. The boundary was agreed with the Ethio-British treaty of 1902 and demarcated in 1903. Isolated inselbergs away from the western Ethiopian escarpment were selected as boundary markers on the boundary from Eritrea to the Baro River and are here indicated and named ("J" stands for Jebel). To the south of the part of the boundary where rivers were chosen to form natural borders is the locality Melile at ca. 6° 45' N, 34° 34' E. From there, a non-demarcated boundary continued to 5° N, 35° E. The remaining border to Lake Turkana (Lake Rudolf) was settled by treaty in 1907. Note that the boundaries between Uganda and the 'East Africa Protectorate' (Kenya) were significantly altered during the colonial period. Map from Great Britain. [London.] War Office, General Staff. Geographical Section. Intelligence Division. September 1904. Library of Congress. [Public domain]. https://hdl.loc.gov/loc.gmd/g8310.ct011330

Fig. 2-26. An Ethio-Sudanese boundary crossing in 2005. The northern edge of the Ethiopian village of Gizen north of Mengi (WG floristic region) marks a point on the Ethio-Sudanese boundary. The unpaved road is the old trail between Assosa in Ethiopia and Roseires and the villages around Mt. Fazugli in the Sudan. The local ethnic group on both sides of the border is Beni Shangul (Berta). This and similar crossings at Kurmuk and Bumbadi (on the Abay, GJ floristic region, not the same as Bambasi) have seen large influx of refugees, but in 2005, a rope across the road (here lying on the ground) was the only mark of the boundary, guarded only by one armed watchman in a home-built shelter (left side of road). Image 3724 by Ib Friis. Nov. 2005.

Fig. 2-27. Bambasi refugee camp in former mixed open woodland (WG floristic region). Early dry season. Image 6180 by Ib Friis. Nov. 2012.

can still be appreciated in the fact that the border exists today much as it was set in 1902 ..." However, as Bahru Zewde (1976) pointed out, "The peoples of the Borderland, the parties most directly affected by the issue, had little say in the outcome of the negotiations. But some of them did try to influence it to a degree hitherto scarcely recognized. Formal delimitation aside, the frontier peoples pursued a life of virtual independence from both Khartoum and Addis Ababa." This is confirmed by a more resent study (Mulatu Wubneh 2015), which has also uncovered that in 2007, the then government, the Ethiopian People's Revolutionary Democratic Front, entered into a secret agreement with the Sudan to adjust on the boundary.

It is not the task of this work to go into the recent disagreements over the boundary, but during our field work we have experienced how often the border is crossed by local people, who indeed, as described by Bahru Zewde, pursue "a life of virtual independence from both Khartoum and Addis Ababa" (Fig. 2-26). With all the efforts mentioned by Marcus, the boundary is not natural, it divides ethnic groups, and the flora and vegetation are identical on both sides of the line. Nor is it our task here to discuss the eventual complications, ecological and political, caused by the construction of the GERD Dam.

Except for local contacts, the travelling- and trading-routes between the Ethiopian Highland and the western lowlands and further into the Sudan followed rather few and fixed paths. The old northernmost route went from Gondar in the highlands to Metema on the border with Sudan, more southern routes went from the central highlands of Ethiopia to Kurmuk on the Sudan border and between Gore and Dembidolo in the highlands and via Gambela on the Baro River into the Sudan (Pankhurst 2010b, map on p. 977). The 1902 treaty with Emperor Menelik II on the Ethio-Sudan boundary granted the Anglo-Egyptian Sudan special extraterritorial rights over the town of Gambela and the right to trade on the Baro River (Ullendorff 1967; Kurimoto 2005).

The Gambela lowlands, which formed part of what the British called the 'Baro Salient', stretched into the British controlled Sudan and was inhabited by two ethnic groups, who live on either side of the present South Sudan-Ethiopian border: the mainly agricultural Anuak [Anwaa], who in Ethiopia primarily live in the more wooded lowlands of the Gambela Region, and the mainly pastoral Nuer, who in Ethiopia live in the areas with mainly grassland and deciduous woodland to the north and west of the Anuak (Kurimoto 2003; Falge 2007). More modern trade-routes across the boundary went from Gondar to Humera [Om Ager] where the present borders of Ethiopia, Eritrea and Sudan meet, and from Bahr Dar on the southern shore of Lake Tana via Dangla and Chagni to Guba [Mankush], where the Abay River [Blue Nile] enters Sudan from Ethiopia. We cannot go further into detail with the ethnic groups that are divided by the boundary between Ethiopia and the Sudan and South Sudan, but refer to the text and the maps on p. 10 and p. 73 in Bahru Zewde (1976).

With the revolution that followed the disposal of Emperor Haile Selassie I in 1974 came a new interest in the western lowlands of Ethiopia, areas which, in the eyes of the Provisional Military Government of Socialist Ethiopia (the 'derg' that ruled the country from 1974 to 1987) and the People's Democratic Republic of Ethiopia (PDRE) (in power 1987-1991), were sparsely populated and offered new opportunities for settlements (Wolde-Selassie Abute 2010). The lowlands, particularly the extensive lowland areas with vertisols, are fertile and could be used for resettlement of farmers from the Ethiopian Highlands and for the establishment of mechanized government farms.

Yet the population density of the western lowlands, as indeed of all Ethiopian lowlands, has remained much lower than in the highlands, and in 2010 more than 80% of the Ethiopian population lived above 1800-2000 m a.s.l., on 30% of the territory (Wolde-Selassie Abute 2010). An important factor determining this uneven distribution of the Ethiopian population has been the prevalence of malaria in the lowlands, spread by mosquitoes among humans, and the prevalence of trypanosomiasis, spread by tsetse flies among both humans and cattle (Gascon 2010). Both human trypanosomiasis (sleeping sickness) and various kinds of trypanosomiasis in cattle and pack animals are common below ca. 1500 m a.s.l. Especially horses and mules are vulnerable, and these animals are hardly ever met with in the western lowlands. Camels, the traditional pack-animal in the western lowlands, may be attacked by a form of trypanosomiasis, but the disease is not spread by insect vectors (Gascon 2010).

Almost immediately after the major famine in 1973-1974 that triggered the 1974 revolution, many drought-stricken farmers from the northern highlands were resettled on 'under-utilised' land in the western and south-western lowlands. Moreover, the population in the south-west was mainly agro-pastoral or even hunter-gatherers (Abbink 2010), with a traditional high degree of mobility. The initial resettlements expanded during the years 1975-1984, when 80,000 drought-stricken farmers from the northern highlands were resettled to fill 'empty spaces' in the western lowlands, being placed between settlements of the traditional ethnic groups practicing subsistence agriculture. Approximately 600,000 new resettlements were planned in 1984-1985, and many of these took place in the initial period, after which the number of resettlements declined. Conflicts between the original population and the resettled highlanders had existed since the beginning of the scheme, but after the fall of the Marxist government in 1991 more conflicts flared up. The Transitional Government of Ethiopia (TGE; 1991-1995) continued the resettlements, and a scheme to resettle about 2,000,000 was set up for the years 2003-2005 (Wolde-Selassie Abute 2010). Even now, resettlement in the western lowlands is not abandoned, but it is our impression that more voluntary programmes may now be favoured over forced resettlements, and the new administrative structure of Ethiopia based on regional states with relatively homogenous ethnic populations does not agree with resettlement from other groups (see Section 2.8).

The second effect of the 1974 revolution on the western lowlands was the establishment of large, mechanized state farms for the mass production of oil seeds, cotton, sugar cane and other tropical agricultural products, a number of which had not been tried before in Ethiopia. Although state farms were established soon after 1974, both near Humera, south of the Abay River and in the Gambela lowlands, they have generally not been successful. Despite heavy machinery, the productivity remained low and the operational efficiency of the state farms poor. As far as information can be obtained, the state farms never contributed more than 5% of the agricultural production of the country (Mekete Belachew & Metz 2003). It is our impression from travels in the western lowlands of Ethiopia that after ca. 2000 most of the state farms have been privatized or leased to foreign investors, and that they are productive due to a combination of foreign capital and hired, migrant labour, mostly from inside Ethiopia. For example, a Saudi Arabian company, Saudi Star PLC, has acquired the rights to farm large areas of land in the lowlands of Gambela Region according to a 60-year contract with the Ethiopian government (https://www.oaklandinstitute.org/sites/oaklandinstitute.org/files/SaudiStar-Agreement.pdf).

As a result of civil unrest in Sudan and South Sudan, the western lowlands have received a substantial number of refugees, sometimes refugees belonging to a ethnic group living on the Sudanese or South Sudanese side of the border and escaping to areas inhabited by the same ethnic group on the Ethiopian side of the border, as has for example been the case with the Nuer refugees in the Gambela Region (Falge 2007). Refugee camps have been established near the western borders inside Ethiopia; in a recent review of women's health in Ethiopian refugee camps (Tefera Darge Delbiso et al. 2016), two camps were listed in western Tigray, one in Beni Shangul-Gumuz Regional State, and three in the Gambela Region. However, during our field work in the western lowlands we have been working near six refugee camps in Beni Shangul-Gumuz Regional State (Tongo, Bambasi, Sherkole, Tsore, Gure-Shomboda, and Add Damazin), frequently scattered out over considerable areas, and each main camp with several more or permanent outposts. Of necessity, these camps have a considerable impact on the vegetation in the localities where they are situated. In the time of writing (December 2020) a new flow of refugees is crossing the boundary from northern Ethiopia to Sudan.

The history of the changing administrative division of the western lowlands is dealt with in the following Section 2.8.

2.8. Higher administrative divisions of Ethiopia (provinces, regional states, etc.); ethnic groupings and population in the western lowlands

Imperial Ethiopia has had a century-old tradition for regional administrative units under the control of feudal governors. This system was gradually modernized under the long rule of Haile Selassie I (regent 1917-1930, emperor of Ethiopia 1930-1936 and 1941-1974), but completely reorganized during the Italian occupation (1936-1941). The names of the traditional units have been widely used in the literature for practical references; the following outline is based on Mantel-Niećko (2003) and form the basis for many geographical references in this work.

The geographical delimitation of the traditional governorates, usually called provinces, often agreed roughly with boundaries of ethnic groupings, but occasionally also with natural boundaries that conflicted with the traditional areas of ethnic grouping. The boundaries followed mainly the large rivers that cut deeply into the highlands, as for example the province or governorate of Gojam, which has always been largely delimited by the big bend of the Abay River and now largely agrees with the GJ floristic region of the *Flora of Ethiopia and Eritrea*. According to the first written constitution of Ethiopia from 1931 there should be 14 administrative units, each having a main town with two palaces, one for the emperor and one for the provincial governor. Particularly in northern Ethiopia, these units retained their historical names from far back in history, but parts of the western lowlands were only loose and casual attachments to the adjacent highland areas. However, the new division of Ethiopia according to the 1931 constitution was not fully achieved by the time of the Italian occupation in 1935.

After 1935, the Italian colonial administration divided Ethiopia, including Eritrea, into six administrative units, based on the idea that each contained people speaking languages that belonged to very broadly defined language groups. They were: (1) 'Il Governato dell'Eritrea' (capital: Asmara), which included present-day Eritrea and most of Tigray, as the highland population in this whole area would mainly speak Tigrinya, but there were also many lowland minorities, like the Afar. (2) 'Il Governato di Amara [Amhara]' (capital: Gondar), in which the highland population would mainly speak Amharic, but several western lowland minorities were included, including speakers of Gumuz. (3) 'Il Governato di Harar' (capital: Harar), where the highland population would mainly speak Oromiffa, Amharic or Harari, but most of the lowland population would speak Somali. The whole of the Ogaden was transferred from Haile Selassie's Harar Province to (4) 'il Governato della Somalia' (capital: Mogadishu), with predominantly speakers of Somali. (5) 'Il Governato di Galla e Sidama' (capital: Gimma [Jimma]), where a proportion of the population would speak Oromiffa, but the rest, including a significant number of ethnic groups living in the western lowlands from the Abay River to the Omo River would speak a very large number of minority languages. Finally, (6) 'Lo Stato dello Scioa e Addis Abeba' (capital: Addis Ababa) was the capital district of the entire 'Africa Orientale Italiana,' with a relatively small hinterland with speakers of Oromiffa.

After the overturn by Ethio-British forces of the Italian rule in 1941 and the reinstatement of Emperor Haile Selassie I, the former Ethiopian system was largely re-established, and until the inclusion of Eritrea in the Empire in 1952, there were 12 governorate-generals or provinces in Ethiopia. Apart from some redefinition of administrative units, such as the establishment of Bale in 1960, this system remained unaltered until 1974, and the positions of governors were, as for centuries, in the hands of the landed gentry. Although the Ethiopian Flora Project (see Section 3) was initiated during the period after the 1974 revolution, it was decided to define the floristic regions on the traditional system of Administrative units, as this would result in entities of roughly comparable size, delimited to some extent on natural boundaries like river valleys or escarpments (Fig. 2-28). The abbreviations used in the *Flora of Ethiopia and Eritrea* for the floristic regions are listed and explained in Table 2-1. For the same reasons, this system is used to indicate specific areas and distributions of plants in this work.

The period from the revolution in 1974 and to the constitution of the Federal Ethiopian Republic in 1994 saw many major changes. A new system of administrative regions replaced the provinces of imperial Ethiopia in 1992 and was formalized by the new Ethiopian constitution in 1995. Based largely on ethnicity and language, the system had slight resemblance with the system during the Italian occupation, where also a special status for large cities with populations of mixed background was introduced, but no distinct territory for the Oromo had been demarcated. Another significant difference from any previous system was the establishment of administrative units for ethnic groups in the western lowlands, primarily the Beni Shangul-Gumuz Regional State in the west on either side of the Abay River, and the Gambela Regional State in the lowlands of the 'Baro Salient,' which is shared between the mainly nomadic Nuers in the predominantly grassland areas in the north-western part and the Anuak and other minor groups, which are partly farmers in the more densely wooded areas in the southern and eastern parts. The rest of the south-west is even more ethnically heterogeneous. Here, following the first elections of regional councils on 21 June 1992, originally five regional states were merged to form the multi-ethnic Southern Nations, Nationalities, and Peoples' Regional State (SNNPR). The largest ethnic groups in SNNPR are the Sidama, Welayta, Hadiya, Gurage, Gamo, Kafficho, and Silt'e, each with more than 5% of the population, but there are more than forty other, smaller ethnic groups, particularly in the lowlands towards South Sudan and Kenya (for the outline of SNNPR, see Fig. 2-29). In the following, we will for brevity refer to Regional States as Regions.

Table 2-1. The floristic regions of the *Flora of Ethiopia and Eritrea*, also used for geographical references in this work, see Fig. 2-28.

AF	Afar, consisting of the lowland parts of the former Tigray, Welo, Shewa and Hararge regions, below and to the east of the 1000 m contour on the east side of the Ethiopian Highlands, extending to the Eritrean border in the east and the border with HA in the south.
AR	The former Arsi region.
BA	The former Bale region.
EE	The eastern lowlands of Eritrea, below and to the east of the 1000 m contour on the east side of the Eritrean Highlands.
EW	The highlands and western lowlands of Eritrea, above and to the west of the 1000 m contour on the east side of the Eritrean Highlands.
GD	The former Gondar region.
GG	The former Gamo Gofa region.
GJ	The former Gojam region.
HA	Hararge uplands and southern lowlands, part of the former Hararge region, above and to the west of the 1000 m contour on the north side of the Ethiopian Highlands.
IL	The former Ilubabor region.
KF	The former Kefa region.
SD	The former Sidamo region.
SU	Shewa uplands, Part of the former Shewa region, above and to the west of the 1000 m contour on the east side of the Ethiopian Highlands.
TU	Tigray uplands, part of the former Tigray region, above and to the west of the 1000 m contour on the east side of the Ethiopian Highlands.
WG	The former Welega region.
WU	Welo uplands, part of the former Welo region, above and to the west of the 1000 m contour, above and to the west of the 1000 m contour on the east side of the Ethiopian Highlands.

Most of the western woodlands have still a low population density, as seen in Fig. 2-30. The system of new regions has attempted to draw local boundaries around populations, which are as ethnically homogenous as possible. Despite the generally low population density, this may lead to conflicts where one region is significantly more densely populated than the neighbouring regions, as is the case with the western lowlands of Oromia seen in relation to the lowlands of Beni Shangul-Gumuz and Gambela. This again may lead to increasing environmental pressure. We have not been able to document our statements here with scholarly studies of others, but refer them on our own impression from our visits to the field.

2.9. Protected areas in Ethiopia, eastern Sudan and South Sudan

Protection of vegetation has old traditions in the Ethiopian Highlands, where the woody vegetation in the often extensive church compounds was protected (Aerts et al. 2016). During his reign, Menelik II extended protection to various forested areas around Addis Ababa, some of which had been crown land for centuries (Sebsebe Demissew 1988). Similar traditions are not documented for vegetation in the western lowlands. The establishment of protected areas in the lowlands, in agreement with the definitions of the International Union for the Conservation of Nature (IUCN), goes back to 1966-1979 (national parks) and the 1980s (game reserves, wildlife sanctuaries, and the 'national forest priority areas', which are a specific Ethiopian type of protected forests). In the western lowlands the first protected area were in the south the Omo National Park (1966), the Gambela National Park (1974), the Mago National Park (1979), the Tama Wildlife Reserve (foundation date unknown), and in the north the Shire Wildlife Reserve (1968), which are all partly or entirely located in the western woodlands.

The national parks of Ethiopia have never attained the national or international attention of the national parks in East Africa, and their management has often been little more than symbolic, while the local populations have often found that their traditional rights were restricted by the establishment of protected areas. An attempt by the Ethiopian government in 2004 to boost particularly the Omo National Park by a contract with the Dutch-based "African Parks Foundation" (APF) failed, and in 2008 the APF withdrew (Gascon & Abbink 2010).

A review of the database of protected areas (UNEP-WCMC & IUCN 2020), shows that there are relatively few protected areas in the large northern parts of the western woodlands (Fig. 2-31). From the Eritrean border to the Gambela lowlands there are only two national parks, the Kafta-Shiraro National Park, founded in 2007 (in the GD floristic region with a small overlap with the TU floristic region) and bordering the Eritrean Gash-Setit Wildlife Reserve, the Alatish National Park, founded in 2007 in the GD floristic region) and bordering the Dinder National Park in the Sudan, and two wildlife reserves, the Shire Wildlife Reserve in the TU floristic region, founded in 1968, adjacent to Kafta-Shiraro National Park, and the Dabus Valley Wildlife Reserve (in the WG floristic region, foundation unknown). The latter will be affected by the reservoir behind the GERD Dam on the Abay River (Fig. 2-15). Elagib & Basheer (2021) suggest that around the maximum reservoir area of 1904 sqkm surface evaporation and local precipitation and humidity will increase, causing change in the local climate and vegetation, and it is also likely that the reservoir may cause more extreme weather to occur.

Thryambakam and Saini (2014) mention 'Hermi Natural Forest' and 'Hermi Forest Priority Area,' but the areas with natural vegetation at Hermi in Tigray are not yet recorded in the IUCN database as a formally protected area. Hermi is located south-west of Axum in the upper reaches of tributaries to the Tacazze River (TU floristic region) and contains a mosaic of *Combretum-Terminalia woodland* (CTW) and *Dry Afromontane Forest* (DAF), with species that also occur in *Acacia-Commiphora bushland* (ACB); the vegetation has been analysed by Mehari Girmai et al. (2020) under the name of *Hirmi woodlands*. Mehari Girmai et al. (2020) "underscore the immediate need for conservation measures in the Hirmi woodland vegetation ...",

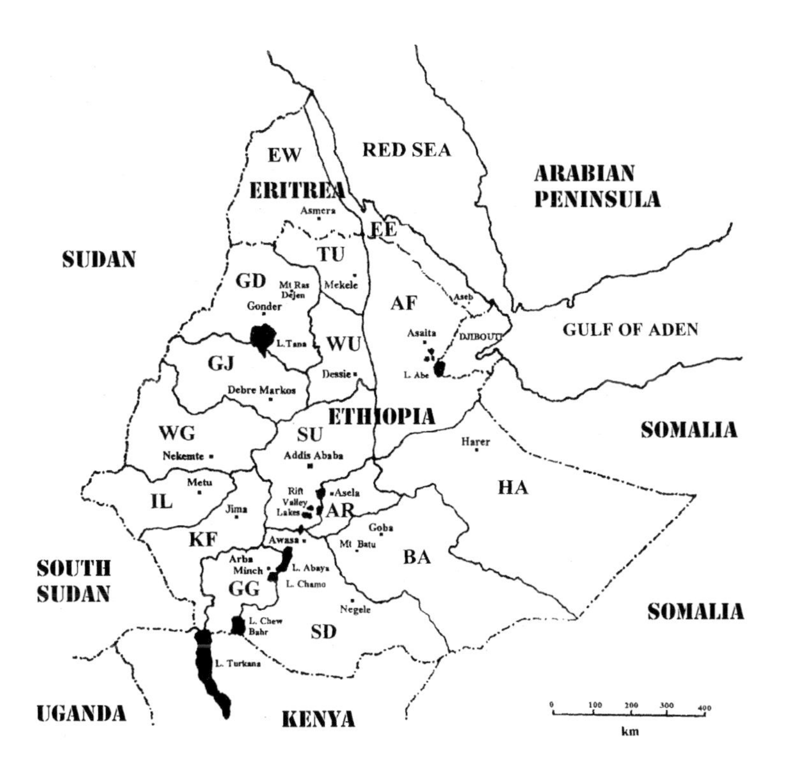

Fig. 2-28. The floristic regions of Ethiopia and Eritrea used in the *Flora of Ethiopia and Eritrea* and in this work. The regions are convenient units when information from the Ethiopian flora is cited or discussed. Their delimitations are slightly modified from the administrative division of Ethiopia used during the reign of Emperor Haile Selassie I after the Italian occupation (1941-1974). For the meaning of the abbreviations and the definitions of the floristic regions, see Table 2-1 and the list of acronyms at the end of this work.

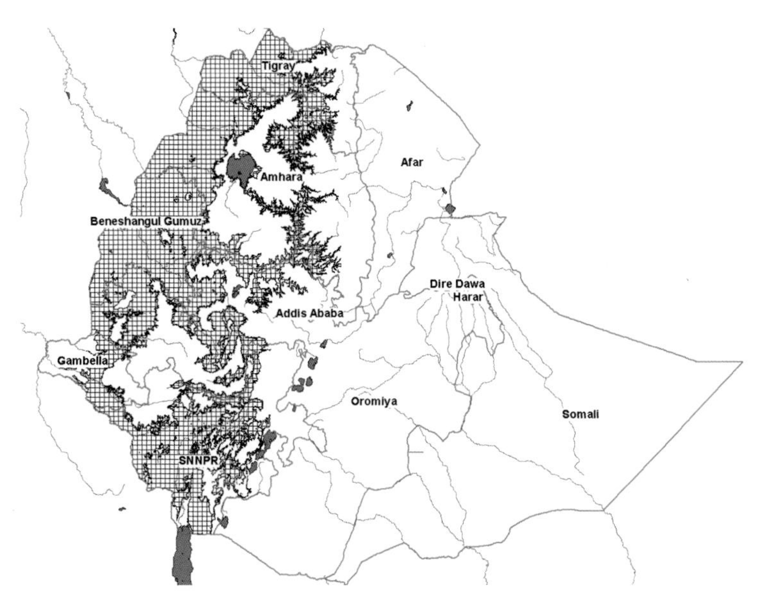

Fig. 2-29. The extent of the western woodland in Ethiopia (cross-hatched), mainly with *Combretum-Terminalia* woodland, in this work referred to as the CTW-area. Shown with red outline are the current highest-level administrative units of Ethiopia, the nine so-called Regional States and the two chartered cities of Addis Ababa (surrounded by Oromia), and Dire Dawa (surrounded by both Oromia and Somali). The small regional state of Harar is surrounded by Oromia. The acronym SNNPR stands for Southern Nations, Nationalities, and Peoples' Regional State. The administrative units are redrawn here, based on shapefile at http://www.maplibrary.org/library/stacks/Africa/Ethiopia/index.htm

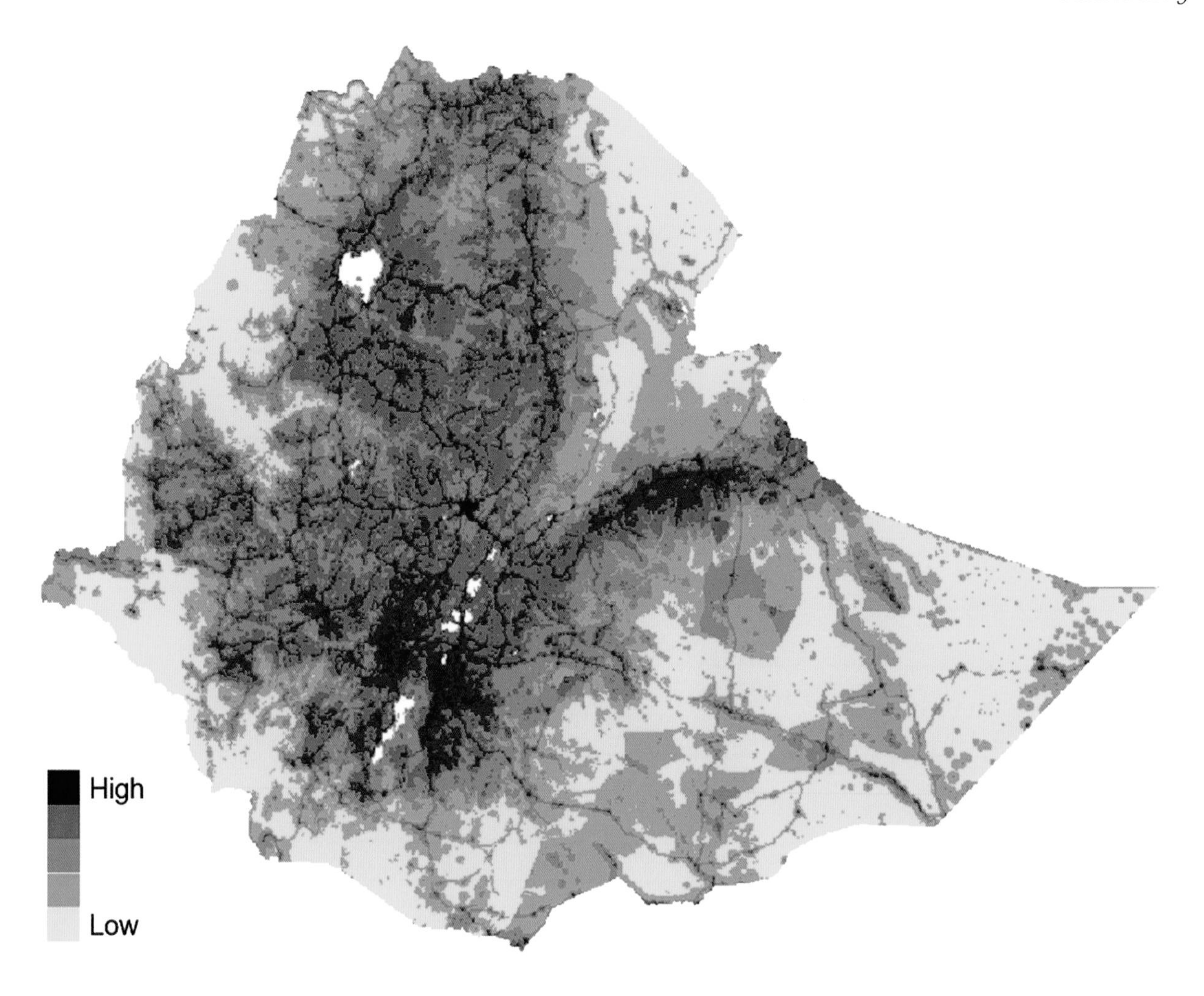

Fig. 2-30. Population density of Ethiopia. Estimated population density in 2019, measured as number of people/sqkm. Lowest (yellow): less than 10 people/sqkm. Highest (dark blue): more than 1000 people/sqkm. Estimate by the WorldPop project, Geography and Environmental Science, University of Southampton; Department of Geography and Geosciences, University of Louisville; Department of Geography, University of Namur) and Center for International Earth Science Information Network (CIESIN), Columbia University. Reproduced under Creative Commons Attribution 4.0 International License from https://www.worldpop.org/geodata/summary?id=41052.

but they do not mention if Hirmi at the time of the writing had status as a potential forest priority area.

In the highlands south-west of Lake Tana near the western escarpment, is the Kahatasa-Guangua National Forest Priority Area (in the GJ floristic region) that protects an important part of the *Intermediate evergreen Afromontane forest* (IAF); it is adjacent to *Combretum-Terminalia* woodland, but it seems that no sizeable woodland vegetation is part of it.

Near the north-eastern point of the Gambela Region (IL floristic region) there are two reserves, one not specified in the IUCN database, the Gargeda-Anfillo National Forest Priority Area (WG floristic region) and another, the Jikao Controlled Hunting Area (partly WG, partly IL floristic regions); both of these are only partly in the western woodlands, and the Gargeda-Anfillo is mainly created to protect the patches of *Transitional rain forest* (TRF) near Dembidolo. To the east, at the Gibe River, the Abelti-Gibe National Forest Priority Area (KF floristic region) is partly created to protect the deciduous woodlands on the western slopes Gibe Valley.

In western Welega (WG floristic region) to the north of Gidami and south of Bambasi, the establishment of a national park to be named the Dati Wolel National Park was initiated in 2006, covering large areas of wetland and riverine forest in the catchment area of the Dabus River. Moist woodlands exist in the peripheral parts of this area, which is almost congruent with the freshwater swamps in western Welega seen on the map in Fig. 4-1. The very local woodland tree *Zanthoxylum gilletii* (Section 6.28. Rutaceae) is restricted to the surroundings of this park, the core area of which has not been studied for this work. The park was not recorded by UNEP-WCMC & IUCN (2020).

The old Gambela National Park (IL floristic region) contains several vegetation types, mainly the *Wooded Grassland of the western Gambela region* (WGG), but also *Combretum-Terminalia woodland and wooded grassland* (CTW) and the lowermost reaches of *Transitional rain forest* (TRF). It is surrounded by a number of protected areas with lower status: to the south the Tedo and Akobo Controlled Hunting Areas (both IL floristic region), which is connected with the Boma National Park and Boma Extension National Park in South Sudan, and to the east the Abobo-Gog National Forest Priority Area (again IL floristic region), also created to protect patches of *Transitional rain forest* (TRF). Most of these protected areas have limited representation of the *Combretum-Terminalia* woodland.

A large area to the east of the Gambela lowlands, in the IL and KF floristic regions, is or has formerly been covered with *Moist evergreen Afromontane forest* (MAF) or *Transitional rain forest* (TRF), as seen on the vegetation map in Fig. 4-1. Many remaining patches of forest are now in areas with some protected status (Fig. 2-31), for example the Shako, Bonga, Mizan Teferi, Gura Ferda, Yayu, Yeki, and Belete Gera National Forest Priority Area (IL and KF floristic regions), but basically outside the *Combretum-Terminalia* woodland as defined in Fig. 4-1, and not created to protect the biodiversity of that vegetation type. At the southern edge of this widely forested area is the Chebera-Churchura National Park, founded in 2006 (KF floristic region), with a range of vegetation types, including *Combretum-Terminalia* woodland, *Moist evergreen Afromontane Forest* (MAF) and *Transitional rain forest* (TRF) (Weber et al. 2020, Fig. 4).

The main protected areas in the south are the long established Omo National Park on the western bank of the Omo River (in KF floristic region) and the Mago National Park on the eastern bank (in GG floristic region). These are surrounded by the Tama Wildlife Reserve in the north (in GG floristic region), and to the south Murle (in GG floristic region) and Omo West Controlled Hunting Areas (in KF floristic region). The entire complex of protected areas in this part of Ethiopia are on the southern border of the *Combretum-Terminalia* woodland (see discussion in Section 5, Profile R, and Section 11.1), and are adjacent to the Loelle National Park in South Sudan. According to our vegetation map in Fig. 4-1, the Chelbi Wildlife Reserve around the Lake Chew Bahir is entirely inside the *Acacia-Commiphora bushland* (ACB). Isolated in the Omo Valley, but inside the area of the *Combretum-Terminalia woodland* (CTW), is the small Maze Controlled Hunting Area, sometimes referred to as the Maze National Park.

While working on this project and noticing the low number of protected areas in the northern part of the western woodlands, we considered the potential importance of these protected areas for the conservation of the woodland flora. However, the facts that far from all the protected areas were yet clearly described, that at least one (the Dati Wolel National Park) was not registered or mapped by the IUCN, and that the distribution of the flora, even of the 169 woody plants dealt with in this work, is only moderately well known, we found that a detailed gap-analysis based on plant-distribution must wait. All we can do here is to point out that, according to our personal impression, the Kafta-Shiraro National Park, the Alatish National Park and the Shire Wildlife Reserve and the Dabus Valley Wildlife Reserve very likely do not offer adequate protection to the flora of the northern and central part of the western woodland, which includes species of restricted range like *Boswellia papyrifera* (not Red-listed by IUCN, but according to e.g., Addisalem Ayele Bekele et al. (2016) in need of protection), *Boswellia pirottae* (Vulnerable, according to IUCN Red List, *Tantani* (IUCN 2020a)) and *Combretum hartmannianum* (Vulnerable, according to the Red List of IUCN, *Combretum hartmannianum* (IUCN 2020b). Our map in Fig. 4-5 shows the number of species (of the 169 species we deal with in this work) in squares of varying size. As mentioned in Section 4.1, the area between Kurmuk, Gizen and the Abay River is species-rich in comparison with most other areas of the western woodlands. Probably this is not only due to the area being well collected, but does reflect high plant diversity due to diverse topography. The area between Kurmuk, Gizen and the Abay River is densely populated (Fig. 2-30) and has received a high number of refugees (Fig. 2-26) and the establishment of large refugee camps of long standing (Section 2.7; Fig. 2-27), so the establishment of a protected area here, possibly in connection with the Dabus Valley Wildlife Reserve, will cause humanitarian complications. Another complication in that area is the future extension of the reservoir behind the GERD Dam on the Abay River. We hope that our data in this work may be of use in further studies and considerations concerning the protection of the western woodland flora.

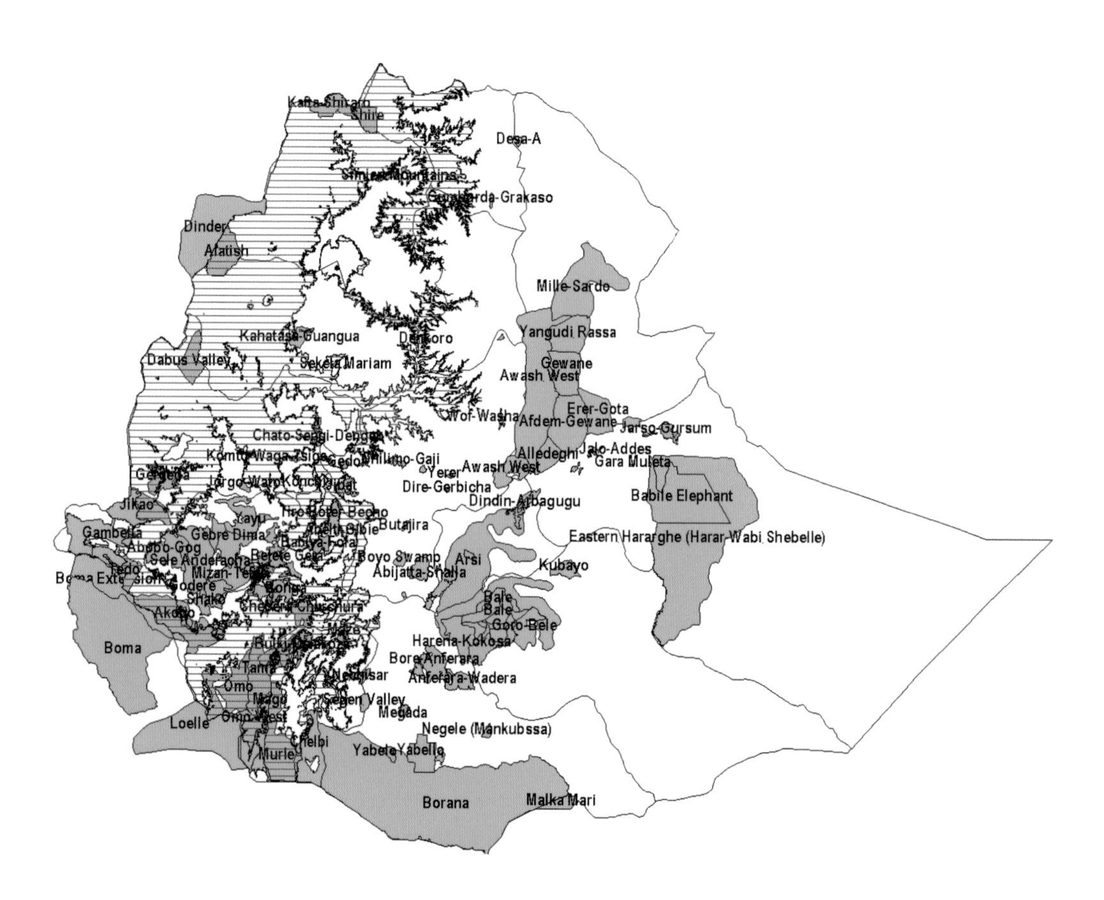

Fig. 2-31. Protected areas of Ethiopia (pale green) according to UNEP-WCMC & IUCN (2020). With the same colour, the map also shows closely adjacent protected areas in the Sudan and South Sudan, including the Dinder, Boma, Boma Extension and Loelle National Parks). The *Combretum-Terminalia* woodlands (Fig. 4-1) and associated vegetation types (the CTW-area) are horizontally hatched, and the floristic regions of the *Flora of Ethiopia and Eritrea* (Fig. 2-28) are outlined. White areas are not protected. The names of the many small National Forest Priority Areas of south-western Ethiopia are too crowded to be shown clearly on a map of the whole country, but they are included here to indicate their number and their gaps in coverage in comparison with the areas with *Combretum-Terminalia* woodland. The names and exact extents of the protected areas can be seen at https://www.protectedplanet.net/en.

3
Previous studies of the western Ethiopian woodlands

The literature about the previous botanical studies of the western woodlands of Ethiopia is scarce and difficult to find. It is partly written in Italian and difficult to access. Another complication is that studies covering wider parts of Africa, or even studies on a continental scale, have sometimes drawn conclusions about the western woodlands of Ethiopia without much data from that actual area. In this Section we have therefore tried to give a review of what has been published on the western woodlands of Ethiopia so far, both including direct studies of the area and studies that try to view the western woodlands of Ethiopia in a wider context without direct studies of the area.

The earliest direct observation of the western Ethiopian woodlands made by European travellers were notes made by the French physician Charles Chaques Poncet, who was called from Cairo to attend to the Ethiopian Emperor Iyasu I and his son in Gondar in 1698-1700; both the Ethiopian royalties suffered from leprosy. Poncet travelled up the western Ethiopian escarpment from Sennar in the Sudan, and in his travel account (Poncet 1709) there are notes of botanical interest from the journey, describing six species of trees, of which two may be identified as *Adansonia digitata* and *Borassus aethiopum*. Close to the Ethiopian border Poncet mentions the high density of the trees in the woodland and the presence of an evergreen tree, probably identifiable as *Tamarindus indica,* which he noticed growing beside the streams. In his description of the journey up the western Ethiopian escarpment to the town of Chelga (Aykel) Poncet also mentions the continued high density of trees on the escarpment, but none of the species can be identified from his descriptions. Poncet's return from Gondar to Europe took place via the Red Sea.

No specific plant products from the western woodlands are mentioned in early sources describing trade between the western lowlands and the Ethiopian Highlands (Pankhurst 1961, 1997, 2010b). One of the potential products, frankincense, that is now extracted from *Boswellia papyrifera*, a characteristic tree of the north-western woodlands, may have come from different species of *Boswellia* growing in other parts of the Ethiopian lowlands. The first specific recording of the use of *Boswellia papyrifera* to produce frankincense is a note in Italian with drawings and descriptions of the plant by James Bruce's artist, Luigi Balugani, who wrote that the Abyssinians used the plant for production of incense, that the tree was called 'Anguah', and that he and Bruce saw it in the Tacazze Valley (Hulton et al. 1991: Fig. 84). After the death of Balugani in Ethiopia, Bruce travelled to the Sudan via the western lowlands of Ethiopia (Friis 2013).

Bruce (1790) translated and edited parts of Balugani's notes on *Boswellia papyrifera* and had two engravings made from Balugani's drawings of the tree. In the account of his descent from the Ethiopian Highlands to the Nile Valley, Bruce gave no information on the floristic composition of the vegetation in the western lowlands, but like Poncet, he mentioned the high density of trees on the western escarpment and in the lowlands, the tall grass and the extensive areas of black soil, stating that this type of soil extended along the western escarpment of Ethiopia from ca. 13° to ca. 16° N. Bruce also recorded that further away from the escarpment the woodlands were replaced by bushland.

The first European to visit Ethiopia after Bruce and Balugani was the British antiquarian scholar and diplomat Henry Salt, who travelled twice to Ethiopia (1804-1806 and 1809-1811), but these journeys started from the Red Sea and took him no closer to the western woodlands than a visit to an outpost of this vegetation in the upper reach of the Tacazze Valley (Friis 2013), near our Profile C in Section 5.

In 1831, the German naturalist Eduard Rüppell on

his way to the Semien Mountains also crossed the Tacazze Valley near the place visited by Salt, and later in the year he visited the western Ethiopian lowlands northwest of Gondar as far as the Angereb River, where he collected animals and plants in the *Combretum-Terminalia* woodlands (Friis 2013). However, his visit to the western woodland took place during the dry season, and he seems only to have collected animals on this part of his journey.

In late 1836 or early 1837 the German naturalist and professional collector of natural history specimens, Georg Wilhelm Schimper, arrived in Ethiopia and made that country his home until his death in 1878 (Friis 2009a). To describe Schimper's life and botanical activities here will lead us far from the western woodlands, the main part of which he never visited, but his collecting activity in the western part of the Tacazze Valley, particularly around the village of Djeladjeranne, led to the discovery of a significant number of plant species that are common in the western lowlands of Ethiopia. Several European scientists visited Schimper at the places where he collected, also at Djeladjeranne, where he met the French geologists and botanists P.V.A. Ferret and J.G. Galinier in 1839-1842 (Friis & Sebsebe Demissew 2020), and probably also the French naturalist expedition-members R. Quartin-Dillon and A. Petit, who travelled in northern Ethiopia in 1839-1843 (Friis 2009b). Schimper's collections are the earliest in the material we have used for mapping the distributions of the woodland species in Section 6.

In 1865 the German botanist and explorer Georg August Schweinfurth spent several months studying the flora and vegetation around the twin border-town of Gallabat (later in the Sudan) and Metema (later in Ethiopia) (Friis 2009b).

3.1. Observations during the Italian colonization of Eritrea and occupation of Ethiopia

The Italian colonisation of Eritrea began in 1869 or 1870, when the Rubattino Shipping Company purchased land surrounding the Bay of Assab on the Red Sea Coast. This Italian territory was soon further expanded by the occupation of the port Masawa (Friis 2009a). The confused situation in Ethiopia caused by an invasion by Muslim fundamentalists, the Mahdists, allowed an Italian army to occupy most of the highlands along the Eritrean coast, and soon after Italy proclaimed the establishment of their new colony, Italian Eritrea. Botanical research activity in western Eritrea soon followed, particularly carried out by the Italian botanist Agostino Pappi during the years 1901-1914, which brought new information on the vegetation in that part of Eritrea (Friis 2009b). Information about the woody plants of Eritrea was summarized by the Italian botanist Adriano Fiori (1909-1912).

The Italian botanist Emilio Chiovenda, the first keeper of the Italian colonial herbarium in Rome (now part of *Centro Studi Erbario Tropicale* (FT) of the University of Florence), travelled in 1909 from Eritrea to Gondar, passing through a part of the western woodlands in the wide western part of the Tacazze Valley; the journey was made in association with the Italian commercial agent in Gondar, Giuseppe Ostini (Friis 2009b), and Chiovenda published brief notes on the vegetation in this area (Chiovenda 1912).

After the Italian army had occupied Ethiopia in 1935, the botanist Rodolfo E.G. Pichi Sermolli took part in a biological expedition, *Missione di Studio al Lago Tana*, organized by the *Reale Accademia d'Italia*. The field work took place in the beginning of 1937, during the dry season, and Pichi Sermolli travelled with a convoy of vehicles from Humera on the border with Eritrea to Gondar in the highlands while studying the vegetation in north-western Ethiopia (Fig. 3-1). His first published description of the vegetation in this area (Pichi Sermolli 1938: 78-80) was very brief, but later replaced by a paper specially dedicated to the lowlands and western escarpment of Ethiopia (Pichi Sermolli 1940). The description of the vegetation on the journey from Humera [then Om Ager or Omager] to Tekeldengy [then Tucur Dinghia] near Gondar gives the first published ideas about the mosaic-like woodland vegetation in that area. As no other similar description has been published since or in any other language, a translation of the Italian text is therefore given here, with some interpretation of the species mentioned.

> From Tessenei [in western Eritrea] to Baker River the road crosses an extensive plain at an altitude of 500-600 m a.s.l., furrowed by scarce watercourses, the main ones being the Setit [the name used for the lower reach of the Tacazze River] and a stream that pours into it and has various names along its course (Royan, Selassil, Scie and Baker). This plain is covered by a homogeneous vegetation, a typical savanna, of which the main tree-components are *Balanites aegyptiaca* or *Acacia seyal* or sometimes both species together. The herbaceous layer, on average about 1.50 m tall, consists mainly of species of *Andropogon* with some other less common grasses (*Setaria* sp.) and rarely shrubs (*Cadaba rotundifolia*, etc.). At some points, but covering very limited areas, one can find places where the savanna assumes the appearance of a spiny bushland of the type observed in the plains between Cheren and Tessenei [both places in Eritrea]. In places where the plain is crossed by watercourses, the vegetation thickens into two bands of woody plants flanking the streams, the appearance of which, however, is no longer similar to the riverine forests that I observed along the Barca and Gasc Rivers in Eritrea, but looks rather like a very dense bush with evergreen leaves, mostly made up of shrubs (*Balanites, Zizyphus, Tamarix*, etc.). Just beyond the confluence between the streams of Scie and Barker the plain ceases and the first hills appear, representing the base of the western slopes of the Abyssinian Highlands. The road, which we followed to Gondar, runs for a long way along this system of low hills, always at altitudes ranging from 800 to 900 m a.s.l. The change in morphology of the terrain is followed by changes in the appearance of the vegetation, and from the savannah we move on to the deciduous woodlands formed by sparse young trees only 5-6 m tall, among which *Boswellia papyrifera* dominates together with a layer of tall grasses and spaced-out shrubs. Plant communities of this type occupy the entire lowermost part of the slopes of the Abyssinian highlands between the rivers of Baker and Sua; the density of the vegetation varies according to the greater or lesser fertility of the substrate, while also the percentage of the components of the vegetation changes. At this point, and beyond the above-mentioned species, the vegetation also includes various Combretaceae. At some points, in fact, I have seen areas clearly dominated by *Combretum hartmannianum,* for example at Mai Agam. As one approaches the stream of Sua, I noticed clumps of the lowland bamboo, *Oxytenanthera abyssinica*; a species, as we shall see later, indeed characterizes the landscape where it grows. In the vicinity of the stream of Sua, which has permanent water, the vegetation becomes denser. *Boswellia* disappears as the main component of the woodland, and the Combretaceae become more abundant, first of all *Terminalia schimperiana*. In this part of the woodland, although it is made up of deciduous species, almost all the trees were in leaf. Along the banks of Sua there is a luxuriant and dense arboreal vegetation, while the pebbly bed of the stream is spotted with bushes of *Kanahia laniflora*. In the immediate vicinity of the stream the indigenous people practice cultivation of various crops, including cotton and durra. Continuing far beyond the boundaries of the drainage area of the Sua, the vegetation consists of woodland dominated by Combretaceae with very few trees of *Boswellia*; but while in the valley of this stream, and despite the dry season (January), the trees all had leaves. However, when we passed into the low hills the woodland was completely bare, although the floristic composition of the vegetation was the same. I note, however, that generally, where the hills are higher and cooler, the woodland becomes denser, the Combretaceae more abundant and *Boswellia* scarcer. Yet physiognomically the landscape varies little along this chain of hills that is uninterruptedly covered by deciduous woodland. In some places, and especially in the valleys, one can also find scrub of *Oxytenanthera*. Thus, we reach the stream of Sorocà, which is also bordered by two wide walls of dense riverine vegetation. Beyond this stream, the road, until now running at an altitude of about 800-850 meters, tends to climb. The growth of *Oxytenanthera* becomes thicker and thicker and penetrates more deeply into the woodland dominated by *Terminalia*; thus, we pass to the *Oxytenanthera* scrub, which covers the whole territory between the Sorocà and the confluence of the streams of Sengià and Tznatè Feteràt. This type of scrub is very dense and, in some places, even impenetrable; the trees that appear in it are in many places highly sporadic. This species of bamboo rises to a height from 6 to 8 m and forms a layer under which mostly nothing, but a few herbaceous species can grow. Here I travelled through vast areas of pure *Oxytenanthera* scrub, alternating in places where other woody species occur, mostly with woodland dominated by species of *Terminalia*. In the stretch between the streams of Bascurà and Zagba

the bamboo does not descend to the bottom of the narrow valley where the stream of Sengià flows, flanked by luxuriant vegetation, but only grows on the upper part of the slopes. The lower part is instead occupied by deciduous woodland, the main components of which are *Terminalia*, *Anogeissus leiocarpa*, *Gardenia lutea* and *Boswellia papyrifera*. After the confluence of Tznatè Feteràt and Sengià we climb slightly and find ourselves on an expanse of flat ground, which we cross along its greatest diameter. Here the appearance of the vegetation is that of a wooded savanna, whose main arboreal cover consists of *Bauhinia thonningii*, *Gardenia lutea*, *Acacia* sp. and *Combretum* sp., while the herbaceous layer is made up of a thin layer of tall grasses, mostly Andropogoneae (*Hyparrhenia*). This area is not very large and soon the road starts to climb again, bringing us to a slightly higher altitude than before, 1100-1300 m a.s.l., rather than 900-1100 m). The vegetation of this area up to the stream of Avellana appears like a woodland that is slightly denser than the ones transversed until now, with a richer shrub layer and trees of different height, among which I have noticed the usual *Terminalia*, *Combretum collinum*, *Anogeissus leiocarpa*, *Gardenia lutea* and *Bauhinia thonningii*. Along some small streams in this territory, I have seen some magnificent examples of *Albizzia*. The vegetation by the stream of Avellana is a dense woodland of *Anogeissus leiocarpa*, which is almost the only representative of the canopy layer, and *Gardenia lutea*, which is abundant in the low tree layer, slightly exceeds in height the tall grasses and shrubs that form the ground-cover. From this region we descend slightly and cross again the river of Sengià, which has its banks wonderfully covered by a fresh and dense shore vegetation; the presence of water also makes its beneficial action felt here. The slope of the valley that we ascend after crossing the river of Sengià is furrowed by torrents that mostly carry permanent water. The presence of these streams has a great influence on the vegetation that is here no longer dry and sparse, but green and dense. The thicker vegetation and the richness of the undergrowth heralds a near change in the vegetation zones. There are still *Anogeissus*, *Combretum* and *Gardenia*, but in the undergrowth, there is an evergreen species: *Carissa edulis*. In the region of Checc at an altitude of about 1700 m a.s.l., the vegetation changes completely and the undergrowth is replaced by evergreen tropical forest. The appearance of this vegetation is so lush and dense that one almost has the impression of being at a rainforest. I was able to make a small collection of that vegetation during a short stop on our trip, among which the more abundant species were *Mimusops kummel*, *Ficus riparia* [*Ficus sur*], *Gymnosporia schimperi* [*Maytenus serrata*], *Strychnos unguacha* [*Strychnos innocua*], *Phoenix reclinata*, etc. In this place the soil is fertile and permanently moist, thanks to the presence of numerous small streams that descend into this narrow valley from all sides. The dense forest covers the whole slope up to Tucur Dinghià, also embracing the basal part of the two mountains above this village. In some places the forest has been destroyed by the local people to be replaced with crops of cereals and other edible plants. Above around 2300 m a.s.l. the vegetation becomes sparser, and *Acacia abyssinica*, with its umbrella-like canopy, makes the landscape particularly beautiful and characteristic.

The route of the Italian mission can be approximately localized by the river-crossings mentioned; because hardly any collections were made between Humera and Gondar there is no further information in the monograph of the collection (Pichi Sermolli 1951). After leaving Om Ager [Humera], they crossed the Baker River at ca. 13° 58' N, the Sorocà River at ca. 13° 15' N, 36° 05' E. the name 'Bascurà' is unidentifiable with that spelling, but there is a 'Buscaràt' at ca. 13° 08' N, 37° 22' E, just south of the large Angereb River, which Pichi Sermolli surprisingly does not mention. The mission crossed the Sengià River twice, the first time at ca. 12° 55' N, 37° 25' E, and the last time just before the mission reached the watershed and the highlands, at ca. 12° 48' N, 37° 25' E.

Many observations in this text agree with our findings along our Profile A (described in Section 5; for route, see Fig. 4-3, particularly Pichi Sermolli's distinction between five plant communities that replace each other with increasing altitudes: (1) the *Balanites aegyptiaca-Acacia* wooded grassland [which is found on black cotton soil], sometimes replaced by thorny *Acacia* bushland, (2) the vegetation at the base of the foothills dominated by *Boswellia papyrifera* and sometimes *Combretum hartmannianum*, (3) the *Combretum-Terminalia* woodland proper, often at higher altitudes than the *Boswellia* woodland, and sometimes leafless during the dry season and sometimes not, (4) the patch-like and sometimes large areas of scrub made up

Fig. 3-1. The convoy of the Italian *Missione di Studio al Lago Tana*, 1937, between Humera and Gondar. The mission, in which the botanist R.E.G. Pichi Sermolli took part, travelled in January 1937, at the height of the dry season, through the western Ethiopian woodlands, following the trail from Om Ager (Humera) to Gondar. Pichi Sermolli's observations of the woodlands, the first ever published, were made during short stops on this trip. Image above: photograph taken during a short stops (archival photograph 501/2537) by L. Cipriani, anthropologist. Image below: photograph (archival photograph 501/1200) taken of the convoy on the move through the almost leafless woodlands by G. Bini, ornithologist. Both images were taken near Caza Jesus, north of Angereb River. Reproduced with licence from the *Società Geografica Italiana*.

Fig. 3-2. Thicket of lowland bamboo (*Oxytenanthera abyssinica*) on the trail from Om Ager (Humera) to Gondar near the river Soroca, a tributary of the Angereb River. Image photographed and published by Pichi Sermolli (1940; 1957, Fig. 43); in public domain.

uniformly by the lowland bamboo (*Oxytenanthera abyssinica*), (5) a mosaic of *Terminalia*, *Anogeissus leiocarpa* and species of *Terminalia*, and at even higher altitudes (6) woodlands of *Bauhinia thonningii*, *Gardenia lutea*, *Acacia* sp. and *Combretum* sp., followed by a transition to the Afromontane highland vegetation with *Acacia abyssinica*, species of *Albizzia* and *Carissa spinarum*.

As documented in Section 5, Profile A, we have observed similar patchy vegetation, not always with the same combinations of the woody species, and patchiness of the vegetation is further discussed in Section 10, where the clustering methods expose the many narrowly-defined combinations of species in various patches of the woodland, making it very difficult to define plant communities, except as clusters on the highest level.

Apart from the patchiness, two further points are striking in Pichi Sermolli's description: The first point is that he did not mention any observations of fire along the profile, which he described as having tall grass (he generally did not notice anything about the importance of fires in the western deciduous woodlands). As he travelled through the woodlands at the height of the dry season, one would expect that he had seen signs of fire. In fact, the ground in the lower image in Fig. 3-1 looks as if the ground-cover has been burnt, as there is no litter. The second point is that he did not mention the notable difference between the soils, the vertisols ['black cotton soil'] in the flat areas and the brown soils with stones on the rocky and stony slopes, and even the slopes with almost bare rocks.

It is also striking that Pichi Sermolli emphasized the extensive areas with lowland bamboo *Oxytenanthera,* as shown in Fig. 3-2, while such large areas of that species no longer seem to exist in that part of Ethiopia (but see in Fig. 5-51 photographs of the extensive bamboo thicket Anbessa Chaka, located much further south near Assosa). We cannot suggest a likely reason for the disappearance of the large areas of bamboo, but it may possibly be due to seed defect after mass-flowering, when large parts of the population die back (Kassahun Embaye Yikuno et al. 2015).

At the same time, in his vegetation map of the Horn of Africa, the first ever published, Negri (1940) defined a long, narrow strip along the western escarpment of Ethiopia as *Boscaglia e bosco tropical caducifolio*, 'deciduous tropical scrub and woodland', Pichi Sermolli's woodlands, with an even narrower strip closer to the highlands defined as 'Bambuseto.'

3.2. The western Ethiopian woodlands on Pichi Sermolli's map of the vegetation on the Horn of Africa (1957)

For his vegetation maps of Africa and the Horn of Africa, Pichi Sermolli (1955, 1957) subsequently compiled all available observations on the vegetation of the Horn of Africa. The western lowlands of Ethiopia, as seen on Pichi Sermolli's map from 1957, are shown in Fig. 3-3. In the north, with a southern limit approximately at the current border between Eritrea and Ethiopia, Pichi Sermolli classified the lowland vegetation as *Steppa arbustata*, 'shrub steppe'. In the areas south of the Eritrean-Ethiopian border he divided the main part of our present study area into two vegetation types *Bosco caducifolio*, 'deciduous woodland' with extensive patches of *Boscaglia a bambù* (*Oxytenanthera*)', 'bamboo thicket (*Oxytenanthera*).' The lowland vegetation penetrating the Ethiopian Highlands along the Abay River (Blue Nile) was classified as *Savanna (vari tipi)*, 'savanna (various types).' The vegetation in the lowlands of the Gambela Region was mapped as *Boscaglia xerophila*, 'xerophilous open woodland', and the lowlands along the Omo River also as *Savanna (vari tipi)*, 'savanna (various types).'

However, only Pichi Sermolli's description of *Boscaglia xerophila,* 'xerophilous open woodland,' was based on his experience from western Eritrea, and only his descriptions of *Bosco caducifolio*, 'deciduous woodland,' *Boscaglia a bambù (Oxytenanthera)*', 'bamboo thicket *(Oxytenanthera)*' were based on his experience from western Ethiopia. The description of the *Boscaglia xerophila,* 'xerophilous open woodland', from 1957 is cited below.

> Starting from the north, the xerophilous woodland covers a good part of the "quolla" [hot lowlands] of Eritrea where it covers very extensive areas especially on the sides of the hills that represent the last outposts of the Ethiopian Plateau. ... The predominant species belong to the genus *Acacia*, among which *Acacia seyal*, *A. nefasia* [*Acacia sieberiana*], *A. senegal*, *A. orfota* [*Acacia oerfota*] and *A. mellifera*, to which we can add a few specimens of *Adansonia digitata* and *Dobera glabra*. In some localities the *Acacia* species are mixed with species of *Combretum* and *Terminalia*. In these consortia of xerophilous bush we often notice *Sterculia tomentosa* [*Sterculia setigera*], *Capparis decidua*, *Boscia senegalensis* and *Boswellia papyrifera*, which develops with much more luxuriance and frequency in the deciduous woodlands.

This agrees with the woodlands and bushlands around Humera in our Profile A and A.2, but not with the vegetation in the Gambela Region.

Pichi Sermolli's 1957 descriptions of the western woodlands further south are based on other sources than his own observations. His definition of *Savanna (vari tipi)*, 'savanna (various types),' is physiognomic, based on the presence of a well-defined herbaceous layer and thus includes treeless grasslands. In this text, he mentions the importance of grass fires for the first time:

> The savannah as a physiognomic unit can be defined as a tropical formation consisting of herbaceous plants, generally xeromorphic, among which grasses and Cyperaceae are the main components, forming an ecologically dominant herbaceous layer, practically continuous, at least 80 cm tall, sometimes with scattered shrubs or trees that may be isolated or in small groups. In Africa, at least, the savannah is usually subject to fire every year. ... I have distinguished various types of savannah; among those represented in East Africa are the treeless savannah, which appears as a homogeneous grassland without large shrubs and trees, the shrubby savannah, scattered with large shrubs with protruding foliage above the herbaceous layer, with their trunk hidden in it, the wooded savannah, with scattered trees or saplings whose trunk is clearly visible above the herbaceous layer, and the clumped-tree savannah with scattered, very dense and almost impenetrable clumps of trees, shrubs and lianas. ... In Eritrea, the savannah grows only in the western lowlands where it represents the continuation of the Sudanese savannahs. They are savannahs with trees (Pichi Sermolli 1940) and with a herbaceous layer of Andropogoneae, to which are added *Setaria* sp. and some small shrubs (*Cadaba rotundifolia*, etc.), about 1.50 m tall; the trees are *Balanites aegyptiaca* or *Acacia seyal* or both these species together. These savannahs continue southwards, penetrating also in the plains where the Blue Nile leaves the Ethiopian mountains. Small patches, too small in extension to be indicated on the map, can also be seen in the flat stretches of the western slopes of the plateau. The vegetation is also here a savanna of Andropogoneae (*Hyparrhenia*) with small trees of *Piliostigma thonningii*, *Gardenia lutea*, *Acacia* sp. and *Combretum* sp. (Pichi Sermolli 1940). Other patches of savannah, not very extensive but very interesting, are present in the plains surrounding Lake Tana (Pichi Sermolli 1938, 1952 [the latter quoted reference refers to the notes on the habitats of the *Usnea* collections cited in Motyka & Pichi Sermolli 1952]). ... Also, in the southern part of Ethiopia there are vast extensions of savannah entering the valleys of the Didessa and Baro Rivers (Dei Gaslini 1940). From the brief hints that Dei Gaslini gives, it seems to be savannah with trees or shrubs where the woody species belong to the genera *Acacia*, *Gardenia*, *Protea* and *Stereospermum*.

According to our observations, this description is physiognomic and cannot be used for a floristic classification, as it mixes species like *Cadaba rotundifolia* from the *Acacia-Commiphora* bushland with *Gardenia*, *Protea* and *Stereospermum* from the *Combretum-Terminalia* woodland. Yet it is an early example of species with their main distribution in *Acacia-Commiphora* bushland, but penetrating into *Combretum-Terminalia* woodland.

It is Pichi Sermolli's own observations on the *Bosco caducifolio*, 'deciduous woodland' and *Boscaglia a bambù (Oxytenanthera)*', 'bamboo thicket (*Oxytenanthera*)' that are his main and lasting contributions to the subject of this book, including his further discussions (Pichi Sermolli 1957: 70-72) and his mapping of these vegetation types on the map in 1:5,000,000, showing how the characterization of the vegetation type is widened in relation to Pichi Sermolli (1940), although sometimes incorrectly so.

> The deciduous woodland is a type of vegetation made up of not very tall (5-12 m) deciduous trees with slender trunks that branch not very high above the ground and whose canopy forms a continuous or almost continuous layer, but always low. Below the canopy is developed a very heterogeneous layer, but always denser than the arboreal layer, including various shrubs or small trees, 2-3 m tall, mostly deciduous, and abundant suffrutices. In the ground-cover grasses and other perennials are always sporadic [this observation of Pichi Sermolli's is not always correct]. This low layer therefore forms a mass of woody species of various heights which, starting from a few centimetres above the ground, and forming a continuous mass of ground-covering vegetation [in our experience the ground-cover is mostly herbaceous; a ground-cover of woody species would mostly be destroyed by fire]. The surface of the soil is mostly rough due to the presence of stones and small boulders. The physiognomy of these woodlands is known to us only through some illustrations among which we can cite those of Schweinfurth (1905, t. 58), Scott (1955, f. 3) and Chiovenda (1936: 462). Fiori (1909-1912) and Chiovenda (1912) have highlighted some biological characteristics of the plants that are part of the deciduous woodlands, including their flowering time and the time of the fall of their leaves. In some species, for example in *Boswellia papyrifera*, which for a long period of time in the year remain bare of foliage, the flowering takes place in the dry season after the plant has been stripped of the leaves, while in other species, for example in some *Combretum* and *Odina schimperi* [*Lannea schimperi*], the flowers develop immediately before or are contemporary with the young leaves. In these latter species, the period in which the plant is without foliage is generally shorter. Another interesting biological characteristic of the tree species of the deciduous woodlands is the structure of the foliage, made up of relatively small or sparse leaves oriented in such a way that they allow a great amount of sunlight to pass through. The deciduous woodlands are present in the western part of the Ethiopian Plateau where they cover the slopes below the evergreen mountain scrub and bushland, and where they are interrupted by stretches of bamboo scrub (*Oxytenanthera*). They are also present in some of the deep inland river valleys such as the Tacazze Valley. The altitudinal limits can be set between 700-800 and 1400-1800 m a.s.l. with considerable variation from region to region. Therefore, on the whole, this type of vegetation is part of what the Ethiopian farmers call "quolla" [hot lowlands]. Even though the deciduous woodlands cover a considerable area, their floristic constitution is not very well known, mainly because it is most typical in places that are not easily accessible. ... Chiovenda (1912) gives us some information about these woodlands but does not mention the constitution of the plant communities. ... it seems to me that I can conclude that the main species in the deciduous woodlands are: *Boswellia papyrifera*, *B. pirottae*, *Anogeissus leiocarpa*, *Terminalia brownii*, *Combretum collinum*, *C. hartmannianum*, *Odina schimperi* [*Lannea schimperi*], *Lonchocarpus laxiflorus*, *Stereospermum kunthianum*, *Commiphora africana*, *C. schimperi*, *Erythrina abyssinica*, *Dalbergia melanoxylon*, *Gardenia lutea*, *Dombeya multiflora* [*Dombeya quinqueseta*], *Balanites aegyptiaca*, *Piliostigma thonningii*, etc. Some other information, but always very sketchy, we may find in Scott (1955). In the Tacazze valley, at about 1000 m a.s.l., the deciduous woodlands contains many trees of *Boswellia papyrifera*. Higher up, on the left bank of the river, the woodland is formed by *Anogeissus schimperi* [*Anogeissus leiocarpa*] to which are added *Stereospermum kunthianum*, *Lonchocarpus laxiflorus* and *Dombeya* sp. On the western side of the Ethiopian Plateau this type of forest is very extensive. According to my own observation (Pichi Sermolli 1940) we know that here the deciduous woodland consists of two types that gradually give way to each other. On the lower

horizon the deciduous woodland has as dominant species *Boswellia papyrifera*, *Combretum hartmannianum* and *Terminalia brownii*, while on the upper horizon *Anogeissus leiocarpa* predominates, in some points dominating over the other species, *Terminalia brownii*, *Combretum collinum*, *Gardenia lutea* and *Piliostigma thonningii*. These two consortia, characterized by *Boswellia papyrifera* and *Anogeissus leiocarpa*, come into contact and interpenetrate in a wide area in which we also find the lowland bamboo (*Oxytenanthera*). The deciduous woodland continues further to the south, but we have no records of the composition there.

Pichi Sermolli has therefore correctly characterized the *Bosco caducifolio* (Pichi Sermolli 1957: 70-72) as a vegetation type dominated by deciduous, broadleaved trees, mostly 5-12 metres tall and forming a clear tree stratum. Among the genera of trees are the ones he mentioned in his description from 1940, but also *Lonchocarpus laxiflorus*, *Stereospermum kunthianum*, *Erythrina abyssinica* and species of *Dombeya,* probably *D. quinqueseta*. He emphasizes various characteristic features of this vegetation: that all strata are deciduous, that the trees flower either when the leaves have fallen, just before or when the new leaves are coming out, and that the soil is dominated by stones and rocks. According to Pichi Sermolli's 1957 map the *Bosco caducifolio* occurs in Ethiopia in a broad fringe along or on the western escarpment of the north-western highlands. Along the southern part of the western escarpment, it is shown as only reaching the Omo River, but not further to the east.

Unlike us, who suggest that the lowland bamboo (*Oxytenanthera*) occurs patchy throughout most of the western Ethiopian woodlands, Pichi Sermolli (1957: 50-51) considered the *Boscaglia a bambù* (*Oxytenanthera*)', 'bamboo thicket (*Oxytenanthera*), as a separate vegetation type and he described it like this, in our translation.

This vegetation appears as large tuft-like clumps of *Oxytenanthera abyssinica* or *Oxytenanthera borzii* with the central stems erect and the peripheral ones slightly bending outwards. While the upper part of the clumps is widened so that the peripheral culms of each clump intertwine abundantly with those of the neighbouring ones to form a continuous layer, the lower part of each clump is compact and distinct from that of its neighbours. However, the space on the ground without vegetation always remains scarce and fallen old culms keep lying on the ground and leaning against each other to form such a dense tangle that the bamboo scrub is almost impenetrable. The culms are up to 10-12 m tall, but only exceptionally they reach that height. Mostly the bamboo scrub is only 6-8 meters tall. According to Giordano (1940), the culms of *Oxytenanthera abyssinica* reach 4-6 cm in diameter at breast-height, and the number of culms in a continuous scrub would be higher than 30,000 per hectare. The bamboo may be present either in pure stands, in stands in which sporadic trees of modest size may protrude from the continuous layer of *Oxytenanthera*, or in stands that allow large shrubs to thrive underneath. The *Oxytenanthera* scrub, both at its lower and upper limits, grades into its nearby vegetation types with a mosaic of transitional associations, in which the bamboo in almost pure colonies are scattered like islands in the middle of the woodland. The *Oxytenanthera* scrub develops only on the western side of the Ethiopian Plateau towards the Sudan and, as far as I know, it has never been reported on the eastern side facing the Great Rift Valley, or on the escarpment facing Dancalia (Afar) and the Red Sea. To the south, the bamboo scrub reaches as far as the last spurs of the Ethiopian Highlands, and here too it remains on the western side and has never been reported on the side facing Somalia. In this southern part, the bamboo scrub makes contact upwards with the woodlands and the evergreen mountain bushland, with which it forms mosaics. There are two species of *Oxytenanthera* that form this scrub: in Eritrea there is a species called *Oxytenanthera borzii* that grows in the valleys of Anseba, Barca and Mareb Rivers (Fiori 1909-1912), while further to the south one finds only *Oxytenanthera abyssinica*. The first species is not very abundant, and forms stands of limited extension and importance (Giordano 1940), always mixed with scrub of other species and deciduous woodland (Senni 1938). This species, again according to Senni (1938), is found between 1800-2100 m above sea level, much higher than where the other species occurs. Unlike *Oxytenanthera borzii*, *Oxytenanthera abyssinica* forms pure or almost pure stands, often of remarkable dimensions, which reach their maximum development between 800 and 1200 m a.s.l. in the northern part and at about 700-1000 m a.s.l. in the south. The bamboo scrub is very well represented in the Tacazze valley,

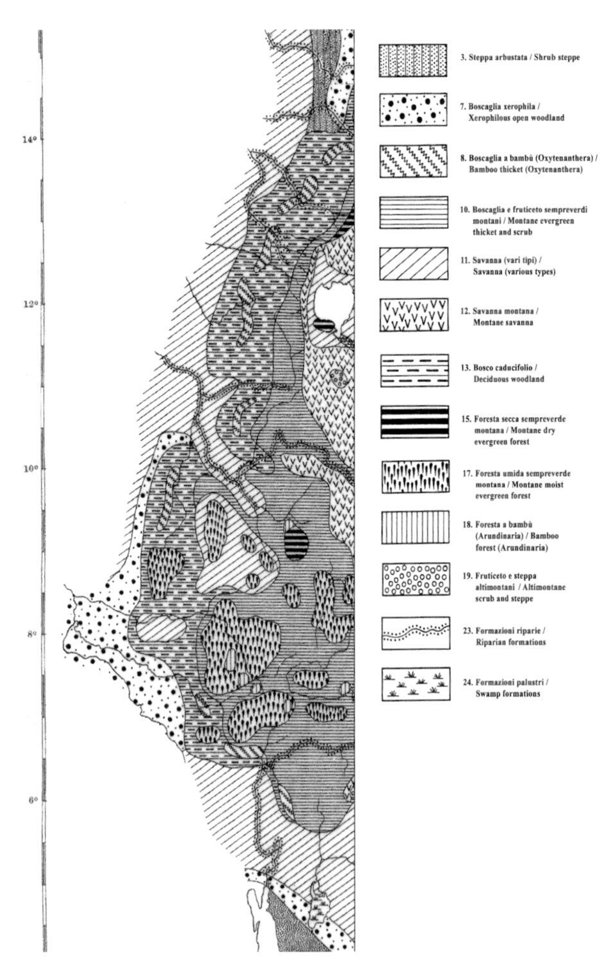

Fig. 3-3. Pichi Sermolli's classification of the vegetation of the western woodlands. Map reproduced from Pichi Sermolli (1957); in public domain.

where it forms extensive stands (Chiovenda 1912), spreading, although not always in pure stands, very far inland (Scott 1954). Further to the south, in the valleys of the Angereb and Sengià Rivers, in the stretch between their respective tributaries Soroca and Tznatè Feteràt, I had the opportunity to observe magnificent pure stands of *Oxytenanthera abyssinica,* of which I have published a photograph showing the physiognomy of this scrub (Pichi Sermolli 1940; our Fig. 3-2). In this region, towards its lower limit, the bamboo forms a mosaic of transitional associations with the deciduous woodland of *Boswellia papyrifera*, *Terminalia* and *Combretum,* while towards the upper limits, the *Oxytenanthera* scrub forms compact colonies in the middle of the deciduous woodland dominated by *Terminalia*, *Combretum*, *Anogeissus* and *Gardenia*. The bamboo scrub is also present in extensive stands on the part of the Ethiopian Plateau which is in western Gojam, the territory of Beni Sciangul [Beni Shangul], Garo [a former independent kingdom inside the area of the KF floristic region], Welega and Kefa, and generally in the vast basin between the Baro and the Akobo Rivers to reach the Bacò [an older name for the town now known as Jinka] (Cei & Pichi Sermolli 1940; Giordano 1940; Dei Gaslini 1940). Towards the south, however, the pure *Oxytenanthera* associations are always of limited extension and are deeply penetrated by various arboreal species, of which unfortunately very little is known.

Oxytenanthera borzii, which Pichi Sermolli considered replaced *Oxytenanthera abyssinica* in Eritrea, was described by Mattei (1909) on material from the upper reaches of the Anseba River, which joins the Barka River in western Eritrea; the species is now, according Philips (1995), synonymous with *Oxytenanthera abyssinica,* and the slight ecological differences mentioned by Pichi Sermolli are apparently consequence of the drier climate in the north. We have no record of *Oxytenanthera* as far south as Jinka, nor have we found evidence for the presence of *Oxytenanthera* in the Gambela lowlands (Pichi Sermolli's 'vast basin between the Baro and the Akobo Rivers'), but it has been observed in the Omo Valley between Kefa and Gemu-Gofa.

3.3. Just across the border in the Sudan and South Sudan: Harrison & Jackson's classification of the vegetation

Harrison & Jackson (1958) published a detailed description and ecological classification of the vegetation of the Sudan, then included the present South Sudan. Like the almost contemporary text by Pichi Sermolli (1957), Harrison & Jackson (1958) is a descriptive account with a vegetation map, interpreting the vegetation of the whole of the Sudan in 1:4,000,000. It was based on the authors' field experience from most of the Sudan. Their classification was primarily based on plant communities correlated with rainfall and soil. Another important feature of their presentation is their recognition of so-called 'hill catenas', a relatively constant vegetation pattern found repeat-

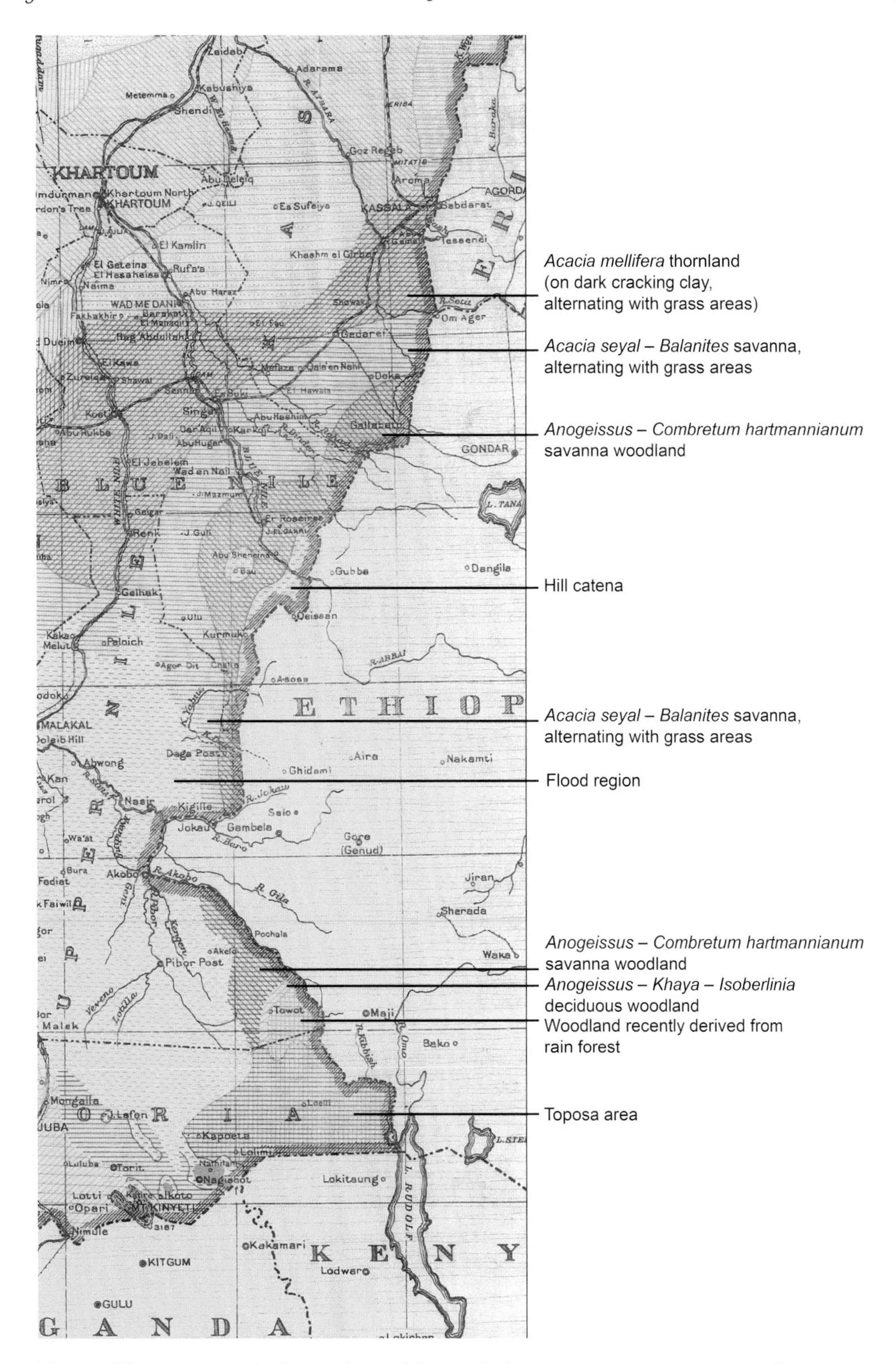

Fig. 3-4. The vegetation in the Sudan and South Sudan opposite the western woodlands of Ethiopia. This map, an eastern strip of the entire map by Harrison and Jackson (1958), shows the part closest to the Ethiopian boundary. Formerly Crown copyright; now in public domain.

edly on rocky hills and slopes, which in the Sudan is mainly found around inselbergs. This text and their map are of significant interest to our study because they represent the vegetation types as they observed across the ecologically arbitrary, political border between Ethiopia, the Sudan, and South Sudan and the changes in the vegetation from north to south. The comparison is outlined here in some detail because such a comparison between the vegetation of the western lowlands of Ethiopia and the vegetation as described by Harrison & Jackson (1958) has not been made before. Moreover, Harrison & Jackson's descriptive memoir and vegetation map are not easily available.

Our observations agree best with those of Harrison & Jackson (1958) regarding the north, while we find notable disagreements between our work and that of Harrison & Jackson in the south. The description of the vegetation types begin from the north, and they can be followed on the reproduction of the relevant part of their vegetation map on Fig. 3-4.

***Acacia mellifera* thornland (on dark cracking clays, alternating with grass areas):** This vegetation type is part of Harrison & Jackson's 'III.A. 1. Low Rainfall Woodland Savannah on Clay.' It occurs on dark, cracking clay [vertisols] at a rainfall range of between 400 and 570 mm approximately. At the lower rainfall limit, the vegetation forms a gradual transition to 'Semi-Desert Grassland on Clay,' and above about 570 mm there is a transition to '*Acacia seyal-Balanites* savannah,' within which patches of this vegetation may also occur. Harrison & Jackson defined the distinction between *Acacia mellifera* thornland and *Acacia seyal-Balanites* savannah as based on the presence or absence of an appreciable numbers of trees of *Acacia seyal* away from water. On their map this vegetation is mostly extended along the border between Eritrea and Sudan and has its southern limit approximately at level with the present border between Eritrea and Ethiopia, but, as they mention, the distinction between this vegetation and the *Acacia seyal-Balanites* savannah is not sharp. In the *Acacia mellifera* thornland, *A. mellifera* forms dense and almost pure thickets which are almost impenetrable and beneath which grasses can hardly grow, so that these thickets do not burn. However, in older stands the closed canopy is often broken, grasses may invade the area and fires may follow. The remaining bushes or trees of *Acacia mellifera* are soon killed by these fires, and the site is occupied by more grassland which persists for a number of years before recolonization by *Acacia mellifera* takes place. This is what Harrison & Jackson call the *Acacia*-grassland cycle. Associated with this vegetation are a number of evergreen species of Capparidaceae (*Cadaba glandulosa, C. rotundifolia*, *Boscia senegalensis*) and *Dichrostachys cinerea*. The presence of *Acacia nubica* [*Acacia oerfota*] is a sign of overgrazing. We observed vegetation like this along the lower parts of our Profile A and Profile A.2 near Humera, and the description agrees with the observations by Pichi Sermolli (1940).

***Acacia seyal-Balanites* savannah, alternating with grass areas:** This vegetation type is also part of Harrison & Jackson's 'III.A. 1. Low Rainfall Woodland Savannah on Clay.' Where the rainfall exceeds 570 mm, '*Acacia mellifera* thornland' passes gradually into '*Acacia seyal-Balanites* savannah,' and on the same type of soils this vegetation passes gradually into '*Anogeissus-Combretum hartmannianum* savannah woodland' with an annual rainfall above 800 mm. On Harrison & Jackson's map this vegetation is extended along the border between Ethiopia and the Sudan and has its southern limit approximately at 13° 30' N, between Humera and Metema. *Acacia seyal* and *Balanites aegyptiaca* are usually more or less equally mixed, but, given the fertile soil (vertisol), the areas with this type of vegetation are often cultivated, and, following abandoned cultivation, stands of pure *Balanites* are frequent. In dry parts of this vegetation almost pure stands of *Acacia mellifera* are found, and in slightly wetter areas, also stands of *Acacia senegal,* though these stands are never as pure or dense as in *A. senegal* savannah on sand. On low-lying areas *Acacia campylacantha* [*Acacia polyacantha* subsp. *campylacantha*] (here mostly reduced to a straggly shrub not more than two metres tall) occurs, and on areas liable to flooding *Acacia fistula* [*Acacia* *seyal* var. *fistula*] and *A. drepanolobium* occur. *A. fistula* [*Acacia seyal* var. *fistula*] is also found on non-flooded areas in dense belts under a slightly higher rainfall than is needed for *Acacia senegal.* Along the Abay River (Blue Nile) and its tributaries in this area woodlands of *Acacia nilotica* may form dense stands, often at the bends of the rivers which are occasionally flooded, and *A. nilotica* depends on flooding for its moisture, rather than on the rainfall. We observed similar vegetation along our Profile A near Humera, although we did not notice pure stands of *Acacia drepanolobium* or *A. nilotica*.

***Anogeissus-Combretum hartmannianum* savanna woodland:** This vegetation type is part of Harrison & Jackson's 'III.A. 1. Low Rainfall Woodland Savannah on Clay.' As for the previous types, it occurs on the dark cracking clays, but at the highest rainfall. Further south on Harrison & Jackson's map this vegetation gives way to the 'Hill catenas' vegetation approximately south of the Dinder River. However, on Harrison & Jackson's map this vegetation is indicated as reappearing approximately south of Assosa and continuing southwards along the border to the northern limit of the Gambela Region. We have not observed this because we find that *Combretum hartmannianum* is restricted to the northern woodlands and does not reappear south of Assosa. Although this vegetation may occur on flat ground, it is most common on slightly sloping ground; it is rarely found far from hills, and rock fragments are frequently found in the soil. There is therefore similarity with some areas of the 'Hill catenas' vegetation. The *Combretum* and *Anogeissus* nearly always have a little *Acacia seyal* among them and large areas of this *Acacia* may alternate with broadleaved trees. *Sterculia setigera* occurs sporadically [we have identified our records of *Sterculia* from this area as *S. africana*, but according to Vollesen (1995b) *S. setigera* also occurs in the north-western lowlands of Ethiopia]. The dominant grasses are tall species of *Hyparrhenia, Andropogon* and *Setaria,* and the annual grasses of the drier clay plains to the north give way to perennial species of the 'High Rainfall Woodland Savannah.'

We observed similar vegetation along the Profile A near Humera, but, as mentioned above, we have not seen it reappear south of Assosa.

Hill catenas: The hill catenas form Harrison & Jackson's category 'III.A. 3. (b). Hill Catenas' of 'III.A. 3. Special Areas of Low Rainfall Woodland Savannah.' On the Ethiopian border the rainfall is 1000-1500 mm. Throughout central Sudan small and large rocky hills (inselbergs, Arabic 'Jebels') protrude through the superficial deposits of the plains. Sometimes the hills are grouped together in large masses, such as the Nuba Mountains. Closer to the Ethiopian border and south of the Abay River, there are many small hills surrounding the larger Mt. Fazugli. The hills have a characteristic vegetation, which is generally moister in character than the surrounding plains and shows erosion with catena development. On Harrison & Jackson's map the 'Hill catena' zone on the Ethiopian border extend from approximately south of the Dinder River at ca. 12° N southwards to somewhat south of the town of Assosa at ca. 9° N. From the top of the rocks to the surrounding plains, Harrison & Jackson have characterized these levels of the 'Hill catenas:'

(i) The rocky summit of the hill, often bare rock, devoid of vegetation except for species of *Ficus,* which root in cracks in the rock.

(ii) the rocky slopes of the hill, with some soil, but often very stony and with boulder outcrops, with a variety of tree species, among which Harrison & Jackson cite these as the most characteristic: *Boswellia papyrifera, Sterculia setigera* [according to us perhaps *S. africana*], *Lonchocarpus laxiflorus, Combretum hartmannianum, Terminalia brownii, Anogeissus schimperi* [*Anogeissus leiocarpa*], *Stereospermum kunthianum, Dichrostachys glomerata* [*Dichrostachys cinerea*] and *Adenium honghel* [*Adenium obesum*].

(iii) Hard-surfaced soils, with little water penetration and forming an apron at the foot of the steep slopes, made up of detritus from above ('qardud'). Trees are often absent and there is only a scanty grass cover, but patches of *Sclerocarya birrea* and the *Lannea humilis* may occur. *Adansonia digitata* often occurs at the foot of the hills just above the hard zone. Where the ground is less hard and more permeable the dom-palm, *Hyphaene thebaica,* is common.

(iv) A zone round the base of the hill in which the soil and vegetation are in general similar to the surrounding plains, but it shows some peculiarities due to the neighbourhood of the hill. The soil of this zone is often transitionary and, though clay, has numerous rock fragments and the rock is often not far below the surface. Here *Anogeissus schimperi* [*Anogeissus leiocarpa*] and *Combretum hartmannianum* are characteristic, gradually merging into the *Acacia seyal-Balanites* savannah of the plain. These areas may be considered as small pockets of *Anogeissus-Combretum hartmannianum* savannah woodland. [According to us, *Anogeissus leiocarpa* does not occur on the Ethiopian border with South Sudan south of Kurmuk, although the species does reappear far from the border at the Bure Escarpment in the Gambela Region, and we have no record of *Combretum hartmannianum* occurring south of Kurmuk and Assosa. Either the distribution of *Anogeissus* and *Combretum hartmannianum* is very different on either side of the border, or the observations of Harrison & Jackson are erroneous on this point. Our maps of the two species on an African scale in Fig. 9-1 and Fig. 9-4 do not support the view of Harrison and Jackson].

(v) Seasonal watercourses, "khors" and "wadis," formed from the run-off of the hill catchments, bordered by belts of fertile sites. The run-off from small inselbergs in the clay plains often collects in depressions where *Acacia fistula* [*Acacia seyal* var. *fistula*] is dominant.

On and below the western escarpment we have observed many places with vegetation similar to that described above, but not as structured as in the description by Harrison & Jackson and, as mentioned, some of the catena-species do not seem to reach as far south as indicated by Harrison & Jackson. Because of the history behind the border delimitation, it is evident that what described as the vegetation of 'hill catenas' in Sudan will have to be slightly reinterpreted for a description of the vegetation on the lower escarpment slopes in Ethiopia. On the lower slopes of the Ethiopian escarpment there are many more extensive areas with well-developed *Combretum-Terminalia* woodland than on Harrison & Jackson's inselbergs. Unfortunately, Harrison & Jackson are not very specific about the ground-cover of the hill catenas: although not mentioned specifically by Harrison & Jackson, the ecological conditions in this area along the Ethiopian border allow perennial grasses grow to a height of more than 2 m and burn frequently during the dry season.

Flood region: There is only one formally recognized vegetation unit in Harrison & Jackson's category 'IV. Flood region.' This vegetation is also termed 'Flood region.' The rainfall here is not particularly high, 750-1250 mm, but large quantities of water are supplied to the region from elsewhere by rivers. Harrison & Jackson divide the 'Flood region' into three informal categories, 'High Land,' 'Intermediate Land' and 'Swamp.'

The 'High Land' is rarely flooded; the soils are mostly sandy or sandy loams, and four sub-types of vegetation are distinguished: (a) Palm vegetation dominant, with *Hyphaene thebaica* generally present, locally also *Borassus aethiopum*; (b) Poorly developed 'Broad-Leaved vegetation type,' similar to that of the 'High Rainfall Woodland Savannah' (see below), but poorer in species; (c) 'Mixed *Acacia*-dominated vegetation type' in which *Acacia sieberiana* is the dominant species and (d) *Acacia seyal-Balanites* occurring on clay soils corresponding to the *A. seyal-Balanites* savannah.

'Intermediate Land' consists of land flooded during the rainy season but dry during the dry season; it is by far the largest constituent of the flood region. There are small areas of '*Acacia seyal-Balanites* savannah' in the 'Intermediate Land' but most of it is tall grassland that regularly burns each year.

'Permanent swamps,' dominated by *Cyperus papyrus* and seasonal swamps with *Echinochloa stagnina,* and *E. pyramidalis*, *Phragmites communis*, *Vetiveria nigritana* [*Chrysopogon nigritanus*], and *Hyparrhenia rufa*.

According to our observations, and what we have presented in Friis et al. (2010), this description agrees better with our vegetation type *Wooded grasslands of the western Gambela region* (WGG) than with the *Combretum-Terminalia woodland and wooded grassland* (CTW).

Anogeissus-Khaya-Isoberlinia **deciduous woodland:** This vegetation type is part of Harrison & Jackson's 'III.B. High Rainfall Woodland Savannah,' which is well represented to the west of the Nile Valley towards D.R. Congo, but to the east of the Nile only indicated as a very small area along the north-north-western slope of the Boma Plateau at 6° 03'-6° 17' N near the Ethiopian border. The type occurs at a rainfall between 900 and 1300 mm and is far from uniform in composition due to catena development. The differences caused by local variations in topography are greater than those caused by differences in rainfall. The three characteristic species of woody plants, *Khaya senegalensis, Isoberlinia doka,* and *Anogeissus schimperi* [*Anogeissus leiocarpa*] do not generally occur mixed together but in separate patches on different soil types. *Khaya senegalensis* has not been recorded from Ethiopia, and as mentioned in Section 2.1, species of the genus *Isoberlinia* do not occur east of the Nile, but *Butyrospermum niloticum* [*Vitellaria paradoxa*], the Shea Butter tree, may become dominant in this vegetation, which may then develop the appearance of parkland, where species other than *Butyrospermum niloticum* have been removed during shifting cultivation.

We find it difficult to support the mapping of this area in the work by Harrison & Jackson. Although the annual rainfall in the woodlands of south-western Ethiopia is as high as required for this type of vegetation, we have no documentation of any plant communities as the *Anogeissus-Khaya-Isoberlinia* deciduous woodland described there, nor do we have evidence of the presence of *Khaya senegalensis* or *Isoberlinia doka* anywhere on the Ethiopian side of the border. We have observed stands of *Vitellaria paradoxa* (*Friis et al.* 2513) in the lowlands of the Gambela Region south of Gambela, and stands of *Anogeissus leiocarpa* on the western escarpments of the Ethiopian Highlands, but neither *Anogeissus leiocarpa,* nor *Vitellaria paradoxa* have been recorded in Ethiopia from south of the Gambela Region, and our field work in areas not far from the Boma Plateau, both in 2018 and earlier, has failed to document any vegetation type similar to the one described here.

Woodland recently derived from rain forest: This vegetation type is also part of Harrison & Jackson's 'III.B. High Rainfall Woodland Savannah.' Its extent in South Sudan near the Ethiopian border approximately agrees with that of the Boma Plateau at 6° 03'-6° 17' N. The annual rainfall is indicated to be ca. 1000-1250 mm. According to Harrison & Jackson, the larger trees are *Terminalia glaucescens* [*Terminalia schimperiana*], *Albizzia zygia* [*Albizia zygia*], *Vitex doniana* and, though not in all areas, *Acacia campylacantha* [*Acacia polyacantha* subsp. *campylacantha*]. Smaller trees include *Combretum binderianum* [*Combretum collinum* subsp. *binderianum*], *Grewia mollis, Annona chrysophylla* [*Annona senegalensis*], *Bridelia scleroneuroides* [*Bridelia scleroneura*] and *Dombeya quinqueseta.*

On the Ethiopian side of the border in southern Ethiopia we have not seen any plant community like the one described above. Here we note that Harrison & Jackson list many of the typical trees recorded from the woodlands further north in Ethiopia, except for *Albizia zygia,* which has not been recorded from Ethiopia. The rainfall seems hardly to be sufficient to have supported a rain forest, with exception of the type that occurs in the area of the *Transitional rain forest* (TRF) (Friis et al. 2010) in Ethiopia, where the rainfall is mostly 1500-2000 mm (Fig. 2-7), and at the base of the Imatong Mountains, the nearest areas of this vegetation type to the west, where the annual rainfall is 1500 mm (Friis & Vollesen 2005). From the information on the floristic composition, we would undoubtedly classify the vegetation as woodland, and we see no convincing evidence that it should have been derived from rain forest, rather than from dense woodland. Presumably, this vegetation only occurs on the South Sudan side of the border.

Toposa area: This is the southernmost vegetation type along the western Ethiopia border and is classified by Harrison & Jackson as the only category in their 'III.A. 3. Special Areas of Low Rainfall Woodland Savannah', listed as 'III.A. 3. (a). Toposa Area.' It does not fit well into Harrison & Jackson's main classification. The rainfall is low, 250-750 mm, particularly towards Lake Turkana [Rudolf], and it is erratically distributed throughout the whole year, unlike in the greater part of the Sudan and South Sudan. On the Ethiopian side of the border, this vegetation corresponds reasonably well with the extensive plains of the lower Omo Valley that have alternatively been considered part of the western deciduous woodlands dealt with here or as the *Acacia-Commiphora bushland* (ACB) (see further in Section 5, Profile R).

On both sides of the border between southernmost Ethiopia and South Sudan, the area is inhabited by many semi-nomadic ethnic groups dependent on cattle grazing for their livelihood and traditional lifestyle. It is an area mainly characterized by flat grasslands with sometimes unpalatable, sometimes palatable perennial grasses, some areas of annual grasses and '*Acacia mellifera* thornland,' which occurs in dense thickets and would be classified with Harrison & Jackson's '*Acacia mellifera* thornland on dark cracking clays, alternating with grass areas.' Harrison & Jackson suggest that the Toposa Area may best be regarded as transitionary from the Sudanese vegetation types '*Acacia mellifera* Thornland' and 'Flood Region' to 'Semi-Desert Grassland' of Kenya.

We have not ourselves studied the vegetation in the Omo Valley to the west of the Omo River, but according to other information (again, see Profile R in Section 5) the vegetation on the Ethiopian side of the border opposite the Toposa region is not entirely dominated by grasslands. The woody flora in this part of Ethiopia seems to be richer than on the South Sudan side, with several species of *Terminalia,* including *T. brownii*, which is used for shade and timber by the Konso ethnic group, and evergreen species of Capparidaceae. As appears from our quotation from Carr (1998) and Jacobs and Schloeder (Schloeder 1999; Jacobs & Schloeder 2002) under Profile R in Section 5, there are a considerable number of species in the Omo Valley that are widespread in the *Acacia-Commiphora bushland* (ACB) towards Somalia and do not penetrate further north and west into the western woodlands of Ethiopia. The consequences of this for the southern delimitation of the western Ethiopian woodlands and the *Combretum-Terminalia* woodlands (CTW) are further discussed in the conclusions in Section 11.1.

3.4. The western Ethiopian woodlands in F. White's 'Vegetation of Africa' (1983)

In 1965, the Vegetation Map Committee of the organisation 'Association pour l'Etude Taxonomique de la Flore d'Afrique Tropicale (AETFAT)' was asked to collaborate with UNESCO in the preparation of a new vegetation map of Africa as part of the latter's programme of mapping the world's vegetation at a scale of 1:5,000,000 (White 1983, introduction). The materials used in compiling this map were exceedingly diverse, but mainly information gathered by the methods of the school of Langdale-Brown et al. (1964), Dobremez (1976) and Trapnell (2001), see further in Section 4.1. For much of tropical Africa, many vegetation maps at various scales had been prepared with these methods for a wide variety of purposes. The Oxford botanist Frank White, expert on African flora and vegetation and curator of the herbaria at the University of Oxford, was made responsible for attempting to standardize the source material and compile it into a coherent whole. The purpose of the map was to indicate the main features of African vegetation, and to provide a continental framework within which more detailed local studies could be compared and fitted. In doing this, White applied updated classic phytogeographical concepts and terminology (see also our discussion and use of phytogeographical terminology in Section 4.2). No new observations on the vegetation of the western woodlands of Ethiopia were made for this vegetation map; for Ethiopia, White drafted the map with additional contributions from R.E.G. Pichi Sermolli.

On the resulting map (White 1983), most of the western lowlands of Ethiopia were characterized as the mapping unit '(29b) Undifferentiated [Sudanian] woodland (B) Ethiopian.' White (1983) defined woodland as open stands of trees, the crowns of which form a canopy from 8 to 20 metres or more in height and with a cover of at least 40 per cent of the surface. The ground-cover might consist of grass or other herbs or dwarf shrubs. The Sudanian woodlands were defined as consisting of a mainly deciduous tree stratum with a rather open canopy and dense herbaceous undergrowth, or various floristically related types with a more open canopy. The particular Ethiopian type of the Sudanian woodlands had the negative characteristic that it almost completely lacked the caesalpiniaceous trees that were characteristic of the Sudanian woodlands to the west of the Nile Valley (particularly *Isoberlinia* and *Julbernardia*), and that it contained an endemic species of *Combretum, C. hartmannianum*.

According to the map of White (1983), these woodlands occur along the Western escarpment and in the Western lowlands from the extreme north near the border with Eritrea to the upper parts of the Omo Valley in the south, where from the west they nearly reach the western slope of the Rift Valley in the east. We have scanned and digitalized the maps by Pichi Sermolli (1957) and White (1983) and these show that White's mapping unit 'Undifferentiated [Sudanian] woodland (B) Ethiopian' largely coincides with a combination of Pichi Sermolli's *Bosco caducifolio* and *Boscaglia a bambù (Oxytenanthera)*, but also covers small parts of several of Pichi Sermolli's other mapping units in the western woodlands of Ethiopia (see also discussion in Section 11.1). As with Pichi Sermolli's accounts, White has no specific mentioning of the importance of fire in these woodlands, but the special soil types, the vertisols, the dark, cracking clays, are mentioned, and White lists the floristic contents of woody species in the 'Undifferentiated [Sudanian] woodland (B) Ethiopian' thus:

> In the extreme east of the Sudan Republic, against the frontier with Ethiopia, a narrow strip of dark cracking clays on sloping ground is dominated by *Anogeissus leiocarpus* and *Combretum hartmannianum* with sporadic *Sterculia setigera* [no mentioning of *S. africana*]. In western Ethiopia the woodland consists chiefly of *Anogeissus leiocarpus, Balanites aegyptiaca, Boswellia papyrifera, Combretum collinum, C. hartmannianum, Commiphora africana, Dalbergia melanoxylon, Erythrina abyssinica, Gardenia ternifolia (G. lutea), Lannea schimperi, Lonchocarpus laxiflorus, Piliostigma thonningii, Stereospermum kunthianum* and *Terminalia brownii*.

To the north of this, most of western Eritrea is marked on White's map as covered by '(43) Sahel *Acacia*

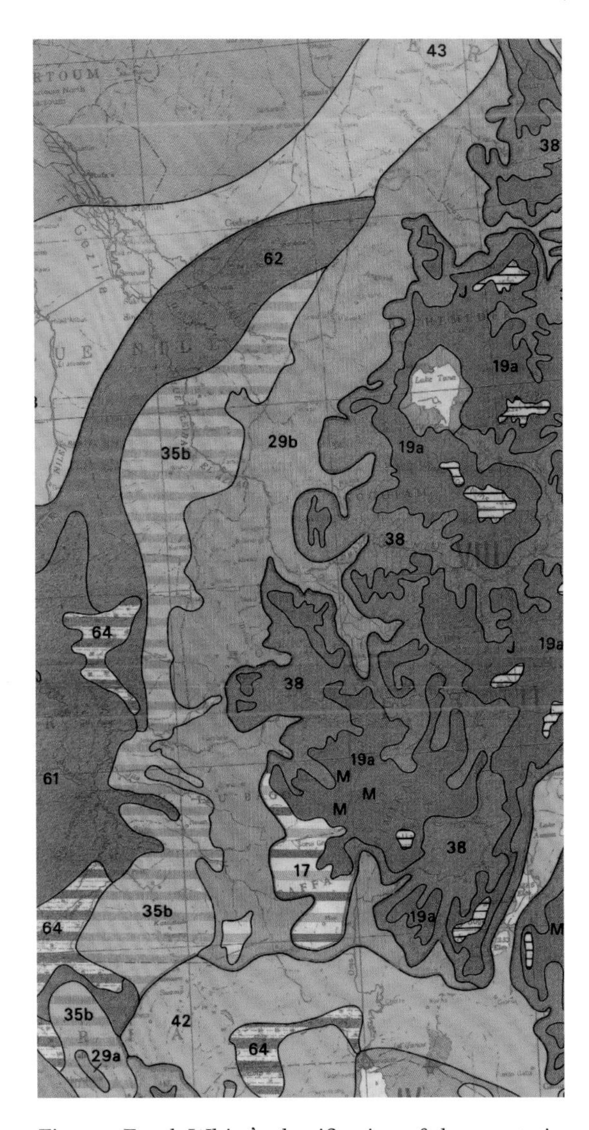

Fig. 3-5. Frank White's classification of the vegetation of the border region between Ethiopia and Sudan (White 1983). Areas hatched with horizontal black lines: Ericaceous bushland and Afroalpine vegetation. 17: Cultivation and secondary grassland replacing upland and montane forest. 19a: Undifferentiated montane vegetation: "J" = *Juniperus procera* forest. "M" = Mixed forest.]. 29b: Undifferentiated [Sudanian] woodland – Ethiopian. 35b: Transition from undifferentiated woodland to *Acacia* deciduous bushland and wooded grassland – Ethiopian. 38: Evergreen and semi-evergreen bushland and thicket – East African. 42: Somalia-Masai *Acacia-Commiphora* deciduous bushland and thicket. 43: Sahel Acacia wooded grassland and deciduous bushland. 61: Edaphic grassland in the Upper Nile basin. 62: Edaphic grassland mosaic with *Acacia* wooded grassland. 64: Edaphic grassland mosaic with semi-aquatic vegetation. Reproduced from Friis et al. (2010).

wooded grassland and deciduous bushland,' which is largely equivalent with Pichi Sermolli's *Steppa arbustata*, 'shrub steppe' and Harrison & Jackson's '*Acacia mellifera* thornland.' In the lowlands inside the extreme north-western Ethiopia a small area is marked as '(62) Edaphic grassland mosaic with *Acacia* wooded grassland', which is not independently characterized in the text. From its geographical position, it seems to represent vegetation types that are somewhat similar to Harrison & Jackson's '*Acacia seyal-Balanites* savannah, alternating with grass areas.'

On White's map, the extensive and sometimes flooded wooded grasslands and grasslands in the western part of Gambela are characterized as '(35) Transition from undifferentiated woodland to *Acacia* deciduous bushland and wooded grassland (B) Ethiopian type' and '(61) Edaphic grassland of the Upper Nile basin.' These areas are part of Harrison & Jackson's 'Flood region' and covers our 'Wooded grassland of the western Gambela region' in Friis et al. (2010). On the same map (White 1983), the lower Omo Valley is indicated as a westward extension of '42: Somalia-Masai *Acacia-Commiphora* deciduous bushland and thicket', which extends into Harrison & Jackson's 'Toposa area' in South Sudan. Thus, White's 'Undifferentiated [Sudanian] woodland (B) Ethiopian' largely agrees with our '*Combretum-Terminalia* woodland' (Friis et al. 2010), except for in the extreme north-west and the part of southernmost Ethiopia marked as *Acacia-Commiphora* bushland. The northern and southern limits of the *Combretum-Terminalia* woodland will be discussed in Section 11.

3.5. 'The Fire Ecology Project'; research on grass fires in woodlands from Senegal to Ethiopia

The 'Fire Ecology Project' was a multidisciplinary project working in several countries in the fire-prone Sudanian zone across northern tropical Africa; it was funded by a joint grant from DANIDA's Council for Development Research as part of a 'Centre for the Study of Fire in Tropical Ecosystems (FITES),' organized at the University of Copenhagen and with participants from a number of African universities in the Sudanian zone across Africa. The field work in Ethiopia originated in part from the Ethiopian Flora Project, in part from a general interest in fire ecology and the little-known vegetation of the western woodlands of Ethiopia. FITES operated from 1996 to 2001 and involved collaboration between the University of Copenhagen and universities in Senegal, Burkina Faso, Ghana, and Ethiopia. The project was coordinated by Kjeld Rasmussen, Department of Geography, Copenhagen, and the ecological and experimental aspects of the Ethiopian part of the project were supervised by Anders Michelsen, Botanical Institute, Copenhagen. The Ethiopian Flora Project, active from ca. 1980 to 2009 was, though a project that should publish a flora of Ethiopia and Eritrea, seen as a facilitator for a range of other projects (Sebsebe Demissew et al. 2011).

The floristic and phytosociological aspects of the field work in Ethiopia were supervised by Sebsebe Demissew and Ib Friis, and included mainly field studies in the Gambela Region and the WG floristic region. Various aspects of the ecology of fire-adapted vegetation were also studied, including the balance between forests and woodlands and the importance of fire in this balance. The project involved training of two Ph.D. students, one from Ethiopia, Minassie Gashaw, and one from Denmark, Michael Jensen, who both successfully defended their Ph.D. theses in respectively 1999 and 2000. This resulted in several publications in international journals (e.g., Jensen et al. 2001; Minassie Gashaw & Michelsen 2001; Minassie Gashaw et al. 2002a, 2002b) and presentations at a *Flora of Ethiopia and Eritrea* symposium at the Carlsberg Academy in Denmark in 1999 and subsequently published in the proceedings of that symposium.

Particularly important for this work is the published vegetational data and the detailed analysis of the data set from five localities in the Gambela lowlands by Jensen & Friis (2001), ranging from dry lowland forest to wooded grassland. The Gambela Region has largely been inaccessible for further studies since the field work as part of the fire project in the 1990s, and data from the woodlands and wooded grasslands gathered during the fire project have therefore been incorporated as relevés in this work.

3.6. The western Ethiopian woodlands in the 'Atlas of the potential vegetation of Ethiopia' (2010)

Large-scale vegetation maps are sometimes produced in connection with flora projects (see for example Wild & Grandvaux Barbosa 1967), and the possibility of producing a vegetation map to accompany the *Flora of Ethiopia and Eritrea* was discussed at the above-mentioned symposium at the Carlsberg Academy in Copenhagen in 1999 (Friis & Sebsebe Demissew 2001), but for several reasons it was not possible to consider production of a detailed vegetation map as part of the Flora Project itself. Instead, a brief Introduction and a small-scale generalized vegetation map was produced and published in Volume 8 of the *Flora of Ethiopia and Eritrea* (Sebsebe Demissew & Friis 2009). Unfortunately, this solution left unresolved many of the problems that had been raised and discussed by Friis and Sebsebe Demissew (2001).

The possibility of producing a vegetation map of Ethiopia appeared when Friis and Sebsebe Demissew in 2008 were introduced to the VECEA-project, which acronym stands for 'Vegetation and Climate Change in Eastern Africa.' The project aimed at producing a high-resolution digital vegetation map for land use planning, natural resource management and conservation of biodiversity in Eastern Africa, and was proposed jointly by the institute Forest & Landscape Denmark, University of Copenhagen, and scientists at the World Agroforestry Centre (ICRAF), Nairobi, Kenya, under CGIAR (formerly the Consultative Group for International Agricultural Research). A first grant was given from the Rockefeller Foundation to the VECEA-project to cover work in the years 2008-2010, and work began in 2008, involving scientists from a range of East African countries (Ethiopia, Kenya, Uganda, Tanzania, Rwanda, Malawi and Zambia). In the work on the production of a vegetation map of Ethiopia, Friis and Sebsebe Demissew were joined by Paulo van Breugel, then working for ICRAF. The VECEA-project attempted to mobilise existing knowledge of the distribution of vegetation types in Africa, both represented in the form of existing vegetation maps, and in the knowledge of botanists familiar with the vegetation of the involved countries.

For most of the countries involved in the VECEA-project, generalized vegetation maps had been produced during the colonial period or in connection with more recent development projects. However, for Ethiopia the situation was different. The two previous large-scaled vegetation maps covering Ethiopia, both in 1:5,000,000 (Pichi Sermolli 1957; White 1983) showed, despite collaboration between Pichi Sermolli and White, considerable differences regarding the vegetation types mapped and their extent. The team behind the VECEA-project realized that, for the inclusion of Ethiopia in the project, it would be unsatisfactory to build only on a compromise between the maps of Pichi Sermolli and White. Sufficiently detailed topographic maps of Ethiopia had not been published to cover the need of vegetation mapping. However, modern satellite images had made available detailed topographical information about even the remotest parts of Ethiopia. It was therefore decided that for Ethiopia the VECEA project should build on a new vegetation map of this topographically complex country.

Based on information from the extensive field experience of Friis and Sebsebe Demissew, previous literature, including 'grey' publications from the British Southwest Ethiopia Forest Inventory Project (Chaffey 1978a, 1978b, 1978c, 1978d, 1979) and the *Flora of Ethiopia and Eritrea*, the survey of forest trees by Friis (1992) and observations made during the Danish-Ethiopian fire-project in the 1990s (Jensen & Friis 2001), the vegetation of Ethiopia was divided into twelve major types, some of these further divided into subtypes. These vegetation types were correlated to environmental parameters, using as simple criteria for the vegetation types as possible, thus relating them to altitude and other topographical features and rainfall patterns, but always checking the results with ground-truthing whenever possible. The Royal Danish Academy of Sciences and Letters kindly took on to publish not only

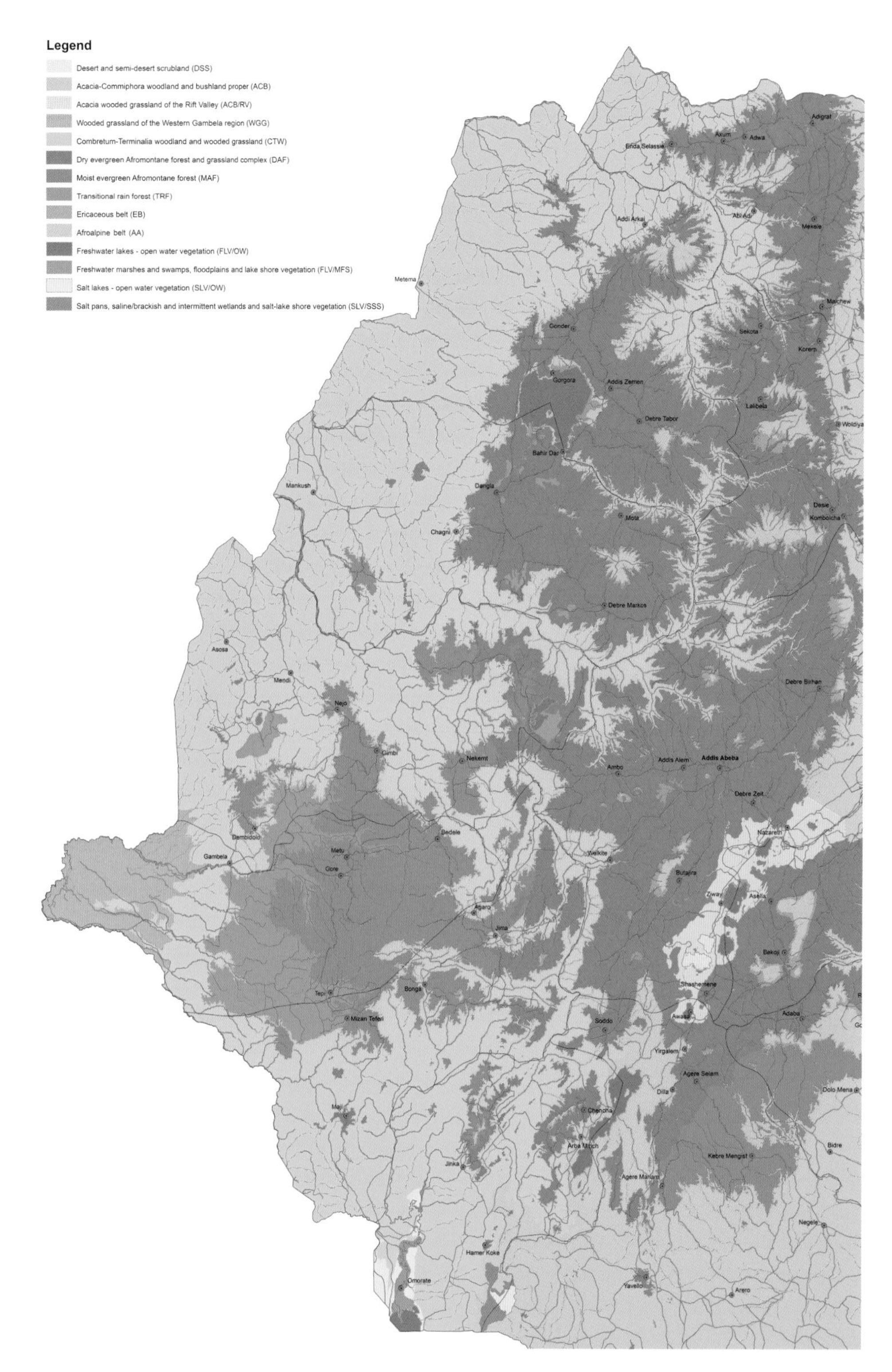

Fig. 3-6. The vegetation of western Ethiopia according to Friis et al. (2010). Reproduced from a digital version of the map distributed on request to users of Friis et al. (2010).

the finalized map in 1:2,000,000 but also a descriptive text (Friis et al. 2010), in time for the information to be integrated in the VECEA project (Fig. 3-6 and 3-7).

On the map published in 2010, we accepted a wide concept of the vegetation in the western Ethiopian woodlands. Almost all the western lowlands were referred to as the *Combretum-Terminalia woodland and wooded grassland* (CTW), which stretched from the border with Eritrea to the border with Kenya, reaching deeply into the large river valleys running to the Nile and Lake Turkana. The *Combretum-Terminalia* woodland was characterized by small to moderate sized trees with fairly large deciduous leaves, particularly species of the genera *Combretum* and *Terminalia* (Combretaceae) were considered characteristic, but also woody species of Fabaceae, both *Acacia* and other species, for example *Lonchocarpus laxiflorus*, *Pterocarpus lucens*, *Dalbergia melanoxylon* and *Piliostigma thonningii*. A review of the information provided by the *Flora of Ethiopia and Eritrea* brought to light a list of taxa of trees, shrubs and lianas that occurred in the area covered by CTW. For a comparison between the species recorded from the western woodlands of Ethiopia and the results of the present studies, see Section 4.3 and Appendix 2.

It was assumed to be characteristic of the CTW-area that the biomass built up during the rainy season would burn in one or several grass fires during the dry season; repeated fires sometimes occur because the perennial grasses sprout again after burning. It was realized in the descriptive text to the vegetation map that there might be significant local variation in the CTW-area, but this was not studied in detail and a subdivision of the vegetation was not attempted. The questions on variation in fire frequency and local variation in other parameters were left unsolved, and have been taken up by Breugel et al. (2016a) and in various parts the present study, particularly Section 10.

The vegetation type referred to as *Combretum-Terminalia woodland and wooded grassland* (CTW) was assumed to occur along the western escarpment of the Ethiopian highlands, from the border region between Ethiopia and Eritrea approximately to the estuary of the Omo River, where it was assumed to occur at 500-1800 metres altitude. At the upper limit the vegetation met in humid areas the *Moist evergreen Afromontane forest* (MAF) and in drier areas the *Dry Afromontane evergreen forest* (DAF). Based on results from his studies around Lake Tana, Pichi Sermolli (1957) had proposed that much of the vegetation in the Lake Tana Basin formed an exclave of the '*Combretum-Terminalia* woodland and wooded grassland.' This was accepted for the vegetation map; the water level of Lake Tana is fluctuating, but always at altitudes slightly below the limit here defined as the upper limit for the *Combretum-Terminalia* woodland.

A few variants of the *Combretum-Terminalia* woodland were described. Woodland on the western side of the highlands include occasionally pure stands of *Acacia seyal*, as observed in the woodlands east of Metema, in the Gibe Valley, and in locations close to the small town of Dima on the Akobo River. In the Tacazze gorge and near Humera on the Sudanian-Ethiopian border field studies have recorded dense stands of *Acacia hecatophylla, Acacia senegal* and *Acacia mellifera* on a mosaic of sand and black cotton soil, as already described by Harrison & Jackson (1958) from the Sudanese side of the border. Another rather distinct variant of *Combretum-Terminalia* woodland and wooded grassland, with very little representation of these two genera, exists south of the Abay River near the upper limit of the *Combretum-Terminalia* woodland and wooded grassland; it is dominated by the evergreen or semi-deciduous tree *Syzygium guineense* subsp. *macrocarpum* (Myrtaceae), which forms a 5-8 metres high and rather homogenous tree stratum. As far as we can see, this vegetation type had not been described in the literature before our mentioning in Friis et al. (2010), but it was briefly mentioned by Solomon Tilahun et al. (1996), with 'Friis (pers. com.)' given as a source, from the Beni Shangul-Gumuz Regional state, where it was called 'Doqma woodland' and classified on level with 'Sudanian woodland' and 'Palms and Bamboo.'

The hitherto underestimated complexity in the vegetation in south-western Ethiopia was realized while we worked on the vegetation map, and two major vegetation types were removed from the general category of deciduous woodlands. One was the partly flooded vegetation in the western part of the Gambela Region, which was considered 'xerophyllous open woodland' by Pichi Sermolli (1957), but was in fact floristically very close to Harrison & Jackson's 'Flood region' and had already been characterized as a temporarily moist vegetation type by White (1983), 'Edaphic grasslands of the upper Nile basin.' On our map (Friis et al. 2010) it was recorded as a temporarily flooded vegetation referred to as 'Wooded grassland of the western Gambela region'. The other vegetation type removed from the western woodlands and reclassified as 'Transititional Rain Forest' was the forest stretching from the lowlands of the Gambela Region at ca. 450 m a.s.l. to about 1500 m a.s.l. in south-western Ethiopia. Pichi Sermolli (1957) had classified this vegetation as part of his other highland and lowland vegetation types, and White (1983), partly in agreement with what was indicated on the map by Harrison & Jackson for the adjacent Boma Plateau in the Sudan, assumed to be mainly cultivated and degraded from forest to secondary woodland or grassland. The previously reported (Pichi Sermolli 1940), almost pure stands of the lowland bamboo (*Oxytenanthera abyssinica*) had been found to be dramatically changing in extent after flowering, when there has been massive die-backs (Kassahun Embaye Yikuno et al. 2015).), and the *Oxytenanthera* stands were therefore included as a variable minor plant association in the *Combretum-Terminalia* woodland and wooded grassland.

3.7. Ethiopian vegetation in the VECEA vegetation map of eastern Africa

The VECEA map of eastern and southern African (Ethiopia, Kenya, Uganda, Rwanda, Tanzania, and Zambia) was, as mentioned above, the product of a project funded by The Rockefeller Foundation and implemented by the now former institute of Forest and Landscape in Denmark, and the World Agroforestry Centre, Nairobi, in collaboration with botanical experts in the seven countries. The project also benefited from previous support to botanists at the relevant departments at the universities of Makerere and Dar es Salaam by an ENRECA programme provided by DANIDA, and previous support to Ethiopian Flora Project provided by SIDA/SAREC and through grants from the Carlsberg Foundation. After the publication of the Ethiopian Vegetation Atlas (Friis et al. 2010), the data set was included in the publications VECEA project with the permission of the Royal Danish Academy of Sciences and Letters, and the publications of the project appeared on-line. The first publication, the 'Potential natural vegetation of eastern Africa. Volume 1: The Atlas' (Lillesø et al. 2011) contained separate vegetation maps for all the participating countries, Ethiopia, Kenya, Malawi, Rwanda, Tanzania, Uganda and Zambia; for Ethiopia, see Fig. 3-7 in this work). Due to the different sources, it was not possible to achieve a complete standardization of the legends, but they were made to agree as much as possible. Apart from the colour scheme, the Ethiopian VECEA vegetation map was much as that in Friis et al. (2010), but the legends for desert and sub desert scrubland were modified and some adaptions of vegetation limits were made along the common border between Ethiopia and Kenya to avoid sudden discontinuities in the distribution patterns of the vegetation on either side of the border.

3.8. Ethiopian vegetation in Sayre et al., 'A New Map of Standardized Terrestrial Ecosystems of Africa' (2013) and later ecosystem maps

An ecosystem map for Africa by Sayre et al. (2013) was based on concepts very different from those of the previously described maps and summarised in the map of White (1983). The values of a suite of environmental parameters (including elevation, landforms, lithology, bioclimate, and regional phytogeography) were correlated with sample locations. The values of the environmental parameters were thus associated with previously defined vegetation types. A panel of thirty-seven experts from eighteen countries had, from their personal experience, developed a system of previously defined vegetation types arranged into a multilevel hierarchical classification of African vegetation

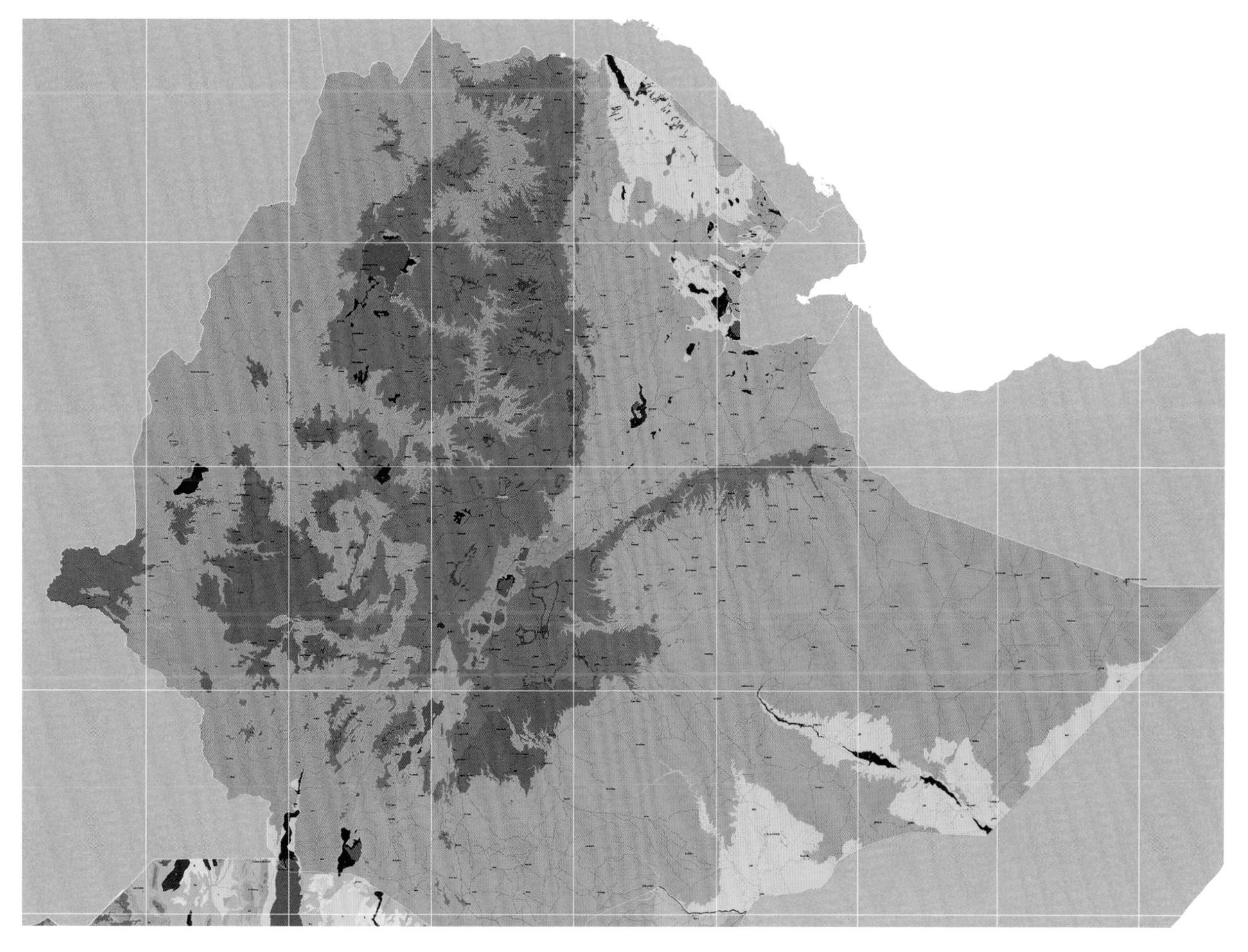

Fig. 3-7. Overview of the vegetation of Ethiopia according to the VECEA vegetation map of countries in eastern Africa (Lillesø et al. 2011). The figure shows the visual index for Ethiopia as presented in Vol. 1. The white lines indicate the sheets of the Ethiopian map, which, apart from some later improvements, follows Friis et al. (2010) rather closely.

categories and provided field-based sample locations where these vegetation types had been observed. The classifications were developed a new, but had taken as their starting point vegetation classification from southern Africa (Mucina & Rutherford 2006). Point samples representing 32,078 known locations of these pre-defined vegetation types were used in the analysis. Regression relationships were established between the environmental variables used as predictors and the dependent variables, which were the vegetation type samples. "The model then recursively partitions all space into a 'most probable vegetation type' for each pixel ..." (Sayre et al. 2013: 5). The units at the lowest level in this hierarchy, equivalent to the 'most probable vegetation type,' were called 'macrogroups,' of which 163 were defined and 126 mapped). A 90 m spatial resolution was used for the pixels.

The result of this study for Ethiopia is shown in Fig. 3-8. It has 21 macrogroups, including open water, while Friis et al. (2010) has 14 vegetation types. Single pixels and groups of pixels indicating these macrogroups occur scattered over the map, often with single pixels detached by a considerable distance from each other or from larger clusters of pixels. This is difficult to see in Fig. 3-8, and the positions of the macrogroups have therefore been studied one by one from the published data, and described in Table 3-1. In many cases, this information disagrees with our field observations, but it will take us too far from the main purpose of this work to go into detailed discussion of all the macrogroups. Here we will only point out one obvious example of discrepancy, Sayre et al.'s 'Moist Evergreen Montane Forest' *versus* our *Moist evergreen Afromontane forest* (MAF). In more detail, we will also discuss the relation of the western woodlands to the vegetation to the east of the Rift Valley, which Sayre et al. have indicated as woodlands with *Combretum* and *Terminalia*.

The upper map in Fig. 3-9 shows Sayre et al.'s 'Moist Evergreen Montane Forest,' overlaid on the *Combretum-Terminalia* woodland *sensu* our concept in the vegetation map in Fig. 4-1. Apart from the 'Eastern African Lowland Semi-Evergreen Forest' and the 'Guineo-Congolian Semi-deciduous Rainforest' in the Gambela lowlands, and a few pixels in south-western Ethiopia south-east of the areas in the Gambela lowlands, the 'Moist Evergreen Montane Forest' is the only forest type recorded by Sayre et al. from Ethiopia. In previous publications we have classified the Ethiopian forests into more categories than the three of Sayre et al. Friis (1992) recognized six forest types. In Friis et al. (2010) this classification was simplified to three types (1) *Transitional rain forest* (TRF); (2) *Moist evergreen Afromontane forest* (MAF); (3) *Dry evergreen Afromontane forest* (DAF). The two latter types are widespread in the highlands, and their distribution only slightly agree with the findings of Sayre et al.

The extent of Sayre et al.'s 'Moist Evergreen Montane Forest' in Fig. 3-9 is much wider than the extent of the two montane forest types in Friis et al. (2010). However, the most surprising discrepancy between our work and that of Sayre et al. is that 'Moist Evergreen Montane Forest', apart from being indicated for areas that we also record as forest, is the macrogroup indicated for large areas of hot and dry lowlands that we consider typical *Combretum-Terminalia* woodland. This is particularly the case in parts of the Tacazze Valley, in the whole of the Abay Valley inside the highland areas, in several lowland areas of the WG floristic region and in other river valleys in the south. These predictions based on model predictions on environmental data cannot be correct, because we have observed deciduous woodlands in these areas and they have different temperature, rainfall and fire regimes from those of the highland forests (see Fig. 2-5, Fig. 2-7, Fig. 2-10, Fig. 2-24).

Sayre et al. (2013) did not recognize a macrogroup similar to White's 'Evergreen and semi-evergreen bushland and thicket – East African.' This vegetation type is no. 38 in White's classification (White 1983), and it occurs on either side of the Ethiopian Highlands. Inside these areas Sayre et al. indicate the presence of *Combretum-Terminalia* woodlands and wooded grasslands (Fig. 3-9, lower map). The relative positions of these, as represented by White and by Sayre et al., were studied by Breugel et al. (2016b, Appendix S5). According to Sayre et al. (2013), the macrogroup '2.A.1.Fg.1. Dry *Combretum* – Mixed Woodland and Savanna' takes up large areas of the western lowlands almost as far south as the border of the Gambela Region, mixed with smaller areas of Sayre et al.'s '2A.1.Fg.2. Dry *Acacia* Woodland & Savanna' and even smaller areas of Sayre et al.' '2.A1. Fg.3. Dry *Acacia-Terminalia-Combretum* Woodland and Savanna.' There is limited overlap with White's 'Evergreen and semi-evergreen bushland and thicket,' but mostly White's vegetation occurs at higher altitudes than the woodlands of Sayre et al. (2013). On the eastern slopes of the Ethiopian Highlands Sayre et al.'s '2.A.1.Fg.1. Dry *Combretum* – Mixed Woodland and Savanna,' '2A.1.Fg.2. Dry *Acacia* Woodland and Savanna' and '2.A1.Fg.3. Dry *Acacia-Terminalia-Combretum* Woodland and Savanna' almost completely coincide with White's s Evergreen bushland, and also with the *Transitional semi-evergreen bushland* (TSEB), as accepted here.

In our opinion, the dominant species of *Combretum* and *Terminalia* are different on the western and eastern side of the Ethiopian Highlands, excluding that the vegetation on either side of the highlands should be classified together. But we have observed, as analysed and documented in Section 7.2 and in Fig. 7-3, that nearly half the number of the species that occur in our CTW-area penetrate eastward to the *Transitional semi-evergreen bushland* (TSEB), a total of ca. 80 taxa, recorded in Section 6 and later as CTW[all+ext], CTW[c+s+ext] and CTW[s+ext], out of the total sample of 169 species. Most of the species classified as CTW[wide+ext], CTW[c+s+ext] and CTW[s+ext] are either widespread in the western woodlands or restricted to the southern part. They are not, as one should expect from the map by Sayre et al. (2013), mainly found in the northern part with their vegetation type '2.A.1.Fg.1. Dry *Combretum* – Mixed Woodland and Savanna.'

In Breugel et al. (2016b), we created predictive distribution models of the natural semi-evergreen bushland in Ethiopia using various environmental distribution models, confirming the existence of the above-mentioned type of semi-evergreen bushland, *Transitional semi-evergreen bushland* (TSEB), forming a transitional zone between the *Acacia-Commiphora* woodland and bushland and the Afromontane forest on the eastern and south-eastern escarpments of Ethiopia. The TSEB has a significant number of species in common with the western woodlands, but is also characterized by several species that are unique to this zone (Breugel et al. 2016b). This vegetation type is seen on the new vegetation map in Fig. 4-1.

From these two examples and the very scattered and fragmented overlap with our floristically based vegetation types seen in Table 3-1, we must conclude that the map by Sayre et al. (2013) must be used with caution for characterisation of vegetation, and that it should be tested against observations made in the field.

Later, Sayre et al. have produced ecological maps covering the entire world, showing "Global Ecological Land Units" (Sayre et al. 2014) and "Ecosystems of the world" (Sayre et al. 2020), also based on global databases but this time without input from local botanists, and with a larger pixel size of 250 × 250 m. These later maps define the documented units as "a delineation of the set of unique physical environments to which biota, and vegetation in particular, respond and distribute." The term 'ecosystems' may sound like a kind of 'vegetation types,' but that's is not what they are. The maps are based on combined data layers representing climate, landforms and land cover, modified with a major phytogeographical classification (Sayer et al. 2020: 4): "The resulting World Ecosystems units therefore represent unique combinations of climate region, landform, and vegetation/land cover. We also stratify the World Ecosystems by biogeographic realm, recognizing that the same combinations of climate, landform, and vegetation on different continents may be compositionally different in terms of biodiversity." The global biogeographic realms used in this classification are broad (Sayer et al. 2020: 6): 'Neotropical, Nearctic, Afrotropical, Palearctic, Australasian, Indomalayan, and Oceanian).'

The layer for 'vegetation/land cover' is derived from data from European Space Agency (ESA 2017) and represents physiognomic vegetation types: "The World Vegetation and Land Cover 2015 layer contains forest, shrubland, grassland, cropland, sparsely or non-vegetated (bare) area, settlements, snow and ice, and water classes, and was derived from the 300 m spatial resolution 2015 global land cover data produced by the European Space Agency (ESA 2017)." For the vegetation of Ethiopia, the product is exceptionally abstract and fragmented, indeed more so than in Sayer et al. (2013). However, the maps may be useful and sufficient for their purposes, to demonstrate the representation in protected areas of combinations of environmental parameters and land forms termed 'ecotypes.'

Table 3-1. Position and extent of "macrogroups" indicated for Ethiopia by Sayre et al. (2013).

Macrogroup name	Macro-group number	Position in Ethiopia	Relation to our vegetation map in Fig. 4-1.
1.A. 2.Fd.3-Guineo-Congolian Semi-Deciduous Rainforest.	3	Only a small area in south-western Ethiopia near border south of Gambela lowlands, ca. at 6° N, 35° 10' E.	Small area in *Combretum-Terminalia woodland and wooded grassland* (CTW) in south-western Ethiopia.
1.A.2.Ff.1-Eastern African Lowland Semi-Evergreen Forest.	6	Forming a belt in north-south direction across the Gambela lowlands around Gambela town.	Overlapping belt on the boundary between *Combretum-Terminalia woodland and wooded grassland* (CTW) and *Wooded grassland of the western Gambela region* (WGG).
1.A.3.Ff.4.-Moist Evergreen Montane Forest.	27	Covering vast areas in the highlands and sometimes descending to lowland areas.	Very largely overlapping the *Dry evergreen Afromontane Forest* (DAF), entirely overlapping the *Moist evergreen Afromontane forest* (MAF), the *Intermediate evergreen Afromontane forest* (IAF), overlapping marginal parts of the *Combretum-Terminalia woodland and wooded grassland* (CTW) and almost entirely the parts in the deep river valleys, and parts of the *Transitional Rain Forest* (TRF).
2.A.1.Ff.2-Western African Mesic Woodland_Grassland.	112	Small areas at the north-western point of the Gambela lowlands and at the base of the Gambela escarpment; also in scattered areas in lowlands south-west of Lake Tana.	Overlapping small areas in the *Combretum-Terminalia woodland and wooded grassland* (CTW), *Transitional Rain Forest* (TRF), and *Wooded grassland of the western Gambela region* (WGG).
2.A.1.Fg.1-Dry *Combretum* – Mixed Woodland_Savanna.	116	Large areas in the north-western and western lowlands as far south as to near the northern border and along the eastern border of Gambela; also in many river valleys and along the south-eastern escarpment to the east of the Rift.	Large areas of the *Combretum-Terminalia woodland and wooded grassland* (CTW), *Transitional Rain Forest* (TRF), in the *Acacia-Commiphora bushland* on higher ground, in *Transitional semi-evergreen bushland* (TSEB), and in small areas of the *Dry evergreen Afromontane Forest* (DAF).
2.A.1.Fg.2-Dry *Acacia* Woodland_Savanna.	117	Small areas, nearly everywhere associated with areas of 2.A.1.Fg.1; also in many river valleys and along the south-eastern escarpment to the east of the Rift.	Small areas of the central *Combretum-Terminalia woodland and wooded grassland* (CTW), in *Transitional semi-evergreen bushland* (TSEB), and in small areas in the *Acacia-Commiphora* bushland on higher ground.
2.A.1.Fg.3-Dry *Acacia – Terminalia – Combretum* Woodland_Savanna.	118	In relatively small areas, mainly in the upper reaches of river valleys, and in many small areas along the south-eastern escarpment to the east of the Rift.	Small and very scattered areas in the *Combretum-Terminalia woodland and wooded grassland* (CTW), in *Transitional semi-evergreen bushland* (TSEB), and in small areas in the *Acacia-Commiphora bushland* (ACB).
2.A.1.Fi.1 -Sudano-Sahelian Herbaceous Savanna.	131	Many small areas in the western lowlands along the border with Sudan and South Sudan, all with relatively few pixels.	Small and scattered areas in the *Combretum-Terminalia woodland and wooded grassland* (CTW), also in the moist parts.
2.A.1.Fi.2-Sudano-Sahelian Shrub Savanna.	132	One large area near the Sudan border around and north of the Rahad River, elsewhere many small areas towards the northern border of the Gambela lowlands.	Scattered areas in the *Combretum-Terminalia woodland and wooded grassland* (CTW), one large area in the north-west, many small areas

Macrogroup name	Macro-group number	Position in Ethiopia	Relation to our vegetation map in Fig. 4-1.
2A.1.Fi.3-Sudano-Sahelian Treed Savanna.	133	In a rather broad belt along the north-eastern edge of the Gambela lowlands; elsewhere scattered in a few small areas.	Scattered areas in the *Combretum-Terminalia woodland and wooded grassland* (CTW), largest areas on the border between the woodland and the *Wooded grassland of the western Gambela region* (WGG), and *Transitional Rain Forest* (TRF).
2.A.1.Fo.1-Moist *Combretum – Terminalia* Woodland_Savanna.	101	Widespread in the Gambela lowlands on both sides of 1.A.2.Ff.1, widespread in the Rift Valley, surprisingly on the dry eastern escarpment of the highlands north of the Rift Valley, and in small, scattered areas elsewhere.	Large part of the *Wooded grassland of the western Gambela region* (WGG), where the lowland is not supposed to be covered by the 1.A.2.Ff.1-Eastern African Lowland Semi-Evergreen Forest (according to Sayre et al.); on the eastern escarpment supposed to be represented by a long north-south strip on the border between the *Dry evergreen Afromontane Forest* (DAF) and *Transitional semi-evergreen bushland* (TSEB), also overlapping with *Acacia wooded grassland in the Rift Valley* (ACB/RV) and in smaller areas with *Acacia-Commiphora bushland* (ACB).
2.A.1.Fo.2-Moist *Acacia – (Combretum)* Woodland_Savanna.	102	A few, small areas in the southwestern Ethiopia south of the Gambela lowlands.	Small patches in the *Combretum-Terminalia woodland and wooded grassland* (CTW) of south-west Ethiopia.
2.A.2.Fe.2-Afro-Alpine Moorland.	142	Nearly everywhere in the high mountains.	Almost completely agreeing with the extent of our combined *Ericaceous belt* and *Afroalpine belt* (EB; AA).
2.A.2.Fe.3-Afromontane Grassland.	143	A few places below 1.A.3.Ff.4 in north-western Ethiopia, a large area below 1.A.3.Ff.4 to the west of Lake Tana, in small and scattered areas elsewhere.	Three very distant patches in the *Combretum-Terminalia woodland and wooded grassland* (CTW), one patch in *Transitional Rain Forest* (TRF).
2.B.7.Fi.1-Eastern African Salt Marsh.	236	Relatively large areas around Lake Chew Bahir and in the Omo Valley, scattered small areas in Afar and in river valleys elsewhere, particularly in the Rift Valley.	Large area around Lake Chew Bahir and many smaller places elsewhere in *Acacia-Commiphora bushland* (ACB), here and there in small patches in the *Combretum-Terminalia woodland and wooded grassland* (CTW). Except for the area around Lake Chew Bahir, this does not overlap with our *Salt lakes with open water* (SLV/OW) or *Salt pans, saline wetlands* (SLV/SSS).
3.A.2.Fe.1 -Eastern African Bushland_Thicket.	176	Extensive, but scattered areas in the lowlands of south-eastern Ethiopia and in the Omo Valley.	Overlap with *Acacia-Commiphora bushland* (ACB), and with *Semi-desert scrubland* (SDS),
3.A.2.Fe.2-Eastern African Semi-Desert Scrub.	177	Vast areas in the lowlands of south-eastern Ethiopia, in the Afar and in the Omo Valley.	Overlapping with large areas in the lower *Acacia-Commiphora bushland* (ACB) and *Semi-desert scrubland* (SDS), in the Omo Valley overlap with *Combretum-Terminalia woodland and wooded grassland* (CTW), also some slight overlap with *Desert* (DES) in Afar.

Macrogroup name	Macro-group number	Position in Ethiopia	Relation to our vegetation map in Fig. 4-1.
3.A.2.Fe.3-Eastern African *Acacia* Woodland.	178	Almost as large areas as 3.A.2.Fe.2 in the lowlands of south-eastern Ethiopia, in the Afar lowlands and in the Omo Valley; very extensive areas in the Rift Valley, many scattered areas in the river valleys in the highlands and around Lake Tana.	Overlapping with a long range of vegetation types, from lowland to highland; the largest overlap is with the *Acacia-Commiphora bushland* (ACB) in Afar, more restricted overlap with *Acacia-Commiphora bushland* (ACB) in the south-eastern lowlands, with *Acacia wooded grassland in the Rift Valley* (ACB/RV), and a very scattered overlap with the *Combretum-Terminalia woodland and wooded grassland* (CTW), particularly around Lake Tana; also extensive overlap in the *Dry evergreen Afromontane Forest* (DAF) up to near the lower limit of the our combined *Ericaceous belt* and *Afroalpine belt* (EB; AA), and a limited overlap with *Transitional semi-evergreen bushland* (TSEB).
3.A.2.Fe.4-Eastern African *Acacia – Commiphora* Woodland.	179	Extensive areas in the lowlands of south-eastern Ethiopia, but more scattered than 3.A.2.Fe.3, on the eastern escarpments of the highlands and in the river valleys, large areas to the north and north-west of Lake Tana.	Extensive areas in the higher parts of the *Acacia-Commiphora bushland* (ACB), small and scattered areas of the *Combretum-Terminalia woodland and wooded grassland* (CTW), particularly in the river valleys, extensive areas in *Dry evergreen Afromontane forest* (DAF), even in small, open areas in the *Moist evergreen Afromontane forest* (MAF).
3.A.2.Pf.3-North Sahel Treed Steppe_ Grassland.	183	Very small area in the northernmost part of the TU floristic region near the border with Eritrea.	Very small area in the northernmost part of the *Combretum-Terminalia woodland and wooded grassland* (CTW)
3.A.2.Pj.3-Saharan Desert Dune_Sand Plain.	208	Areas below sea-level in the Afar depression.	Overlaps areas of *Desert* (DES), *Salt pans, saline wetlands* (SLV/SSS), *Salt lakes with open water* (SLV/OW), and *Semi-desert scrubland* (SDS)

3.9. Further studies of fire ecology in western Ethiopia

Since studies made during the fire ecology project in 1996-2001, it has been realized that fire is an important factor in restricting the *Combretum-Terminalia* woodlands in western Ethiopia. In continuation of the VECEA-project we decided to use the high-resolution map of potential natural vegetation types, in combination with the MODIS fire products, to model and investigate the importance of fire as driver of vegetation distribution patterns in Ethiopia. Products from NASA's Moderate Resolution Imaging Spectroradiometer (MODIS) were used to create decadal fire distribution estimates. These products are daily 30 arc-seconds gridded composites of fire-pixels or other thermal anomalies detected in each grid cell. All layers were summed to get the total count of 8-day periods with observed fires over the period 2003-2013.

For this purpose, Paulo van Breugel employed statistical modelling techniques to estimate the distribution of fire and the vegetation types under current climatic conditions. The results showed congruence between distribution patterns of fire and major vegetation types, as seen in Fig. 2-24 in this work, which has been produced with the data sets by Breugel et al. (2016a, 2016b). Discussion of the importance of fire will be included in the Section 9, 10 and 11.

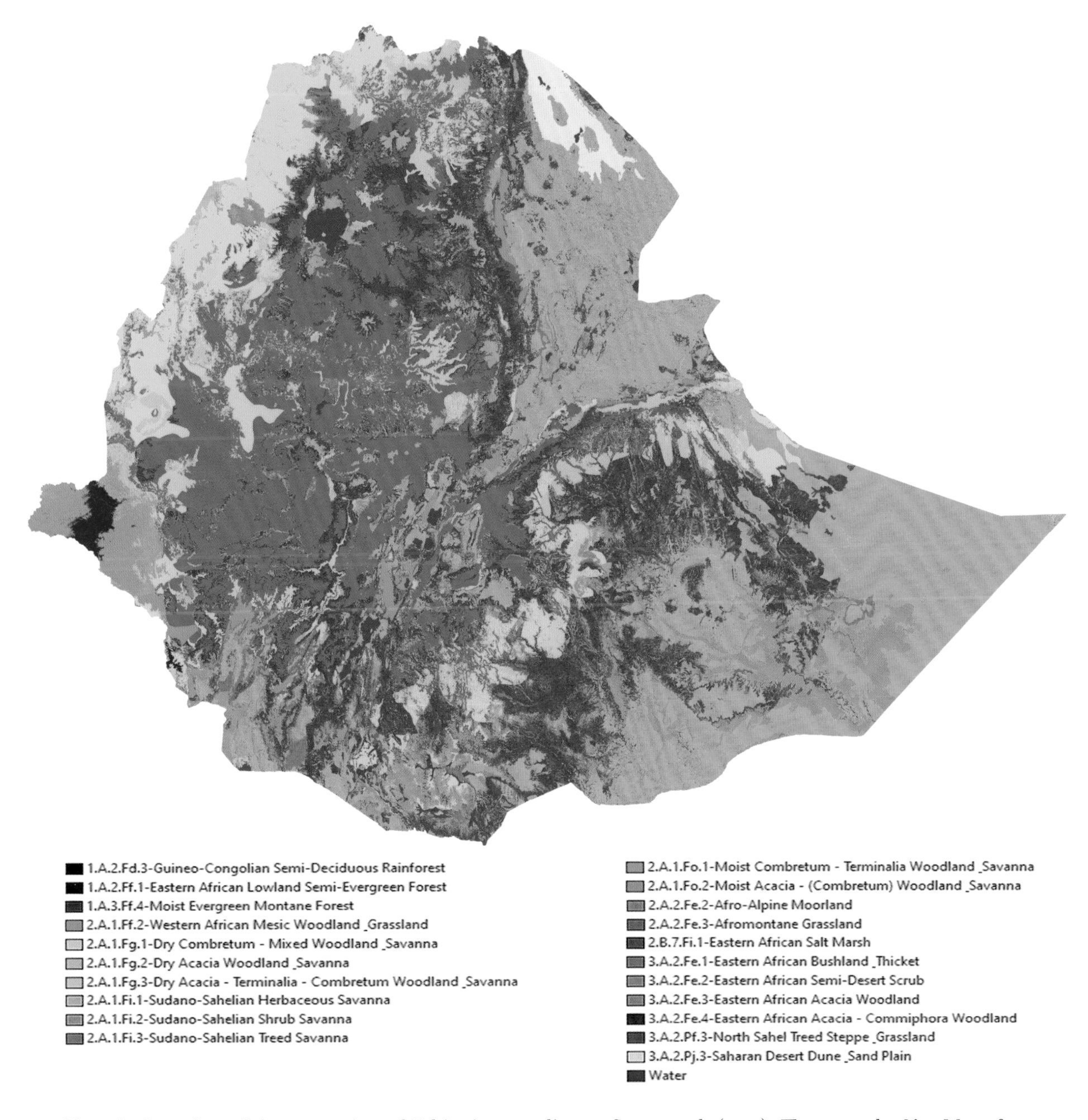

Fig. 3-8. Overview of the vegetation of Ethiopia according to Sayre et al. (2013). The map, the *New Map of Standardized Terrestrial Ecosystems of Africa*, is based on a data set freely available on https://rmgsc.cr.usgs.gov/outgoing/ecosystems/AfricaData/ (the use of the file requires ArcGis 10 or later versions). Due to the small pixels (90 × 90 m), it is difficult to locate and recognize all macrogroups on a map on this scale; guidance for macrogroups found in Ethiopia is provided in Table 3-1. For studies on the pixel-level, the original data must be consulted. Map redrawn from the digital data from https://rmgsc.cr.usgs.gov/.

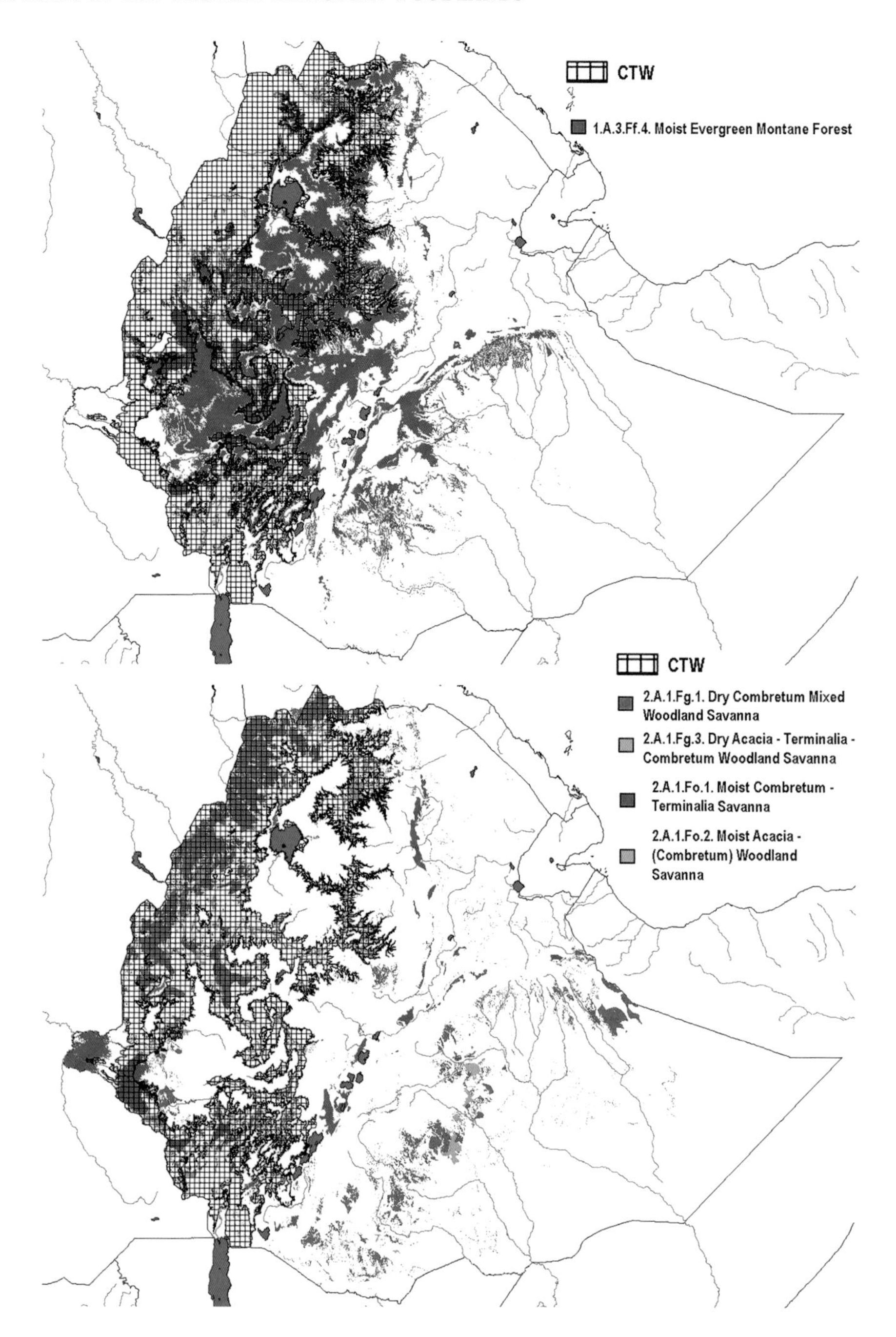

Fig. 3-9. Comparisons between vegetation types in Friis et al. (2010) and Sayre et al. (2013). Image above: Comparison between *Combretum-Terminalia woodland and wooded grassland* (CTW) according to Friis et al. (2010) and 1.A.3.Ff.4. Moist Evergreen Montane Forest in Sayre (2013). The 'Moist Evergreen Montane Forests', according to Sayre et al. (2013), occur in deep valleys, including the Tacazze Valley, in the Abay Valley and the Didessa Valley, and in areas designed by us after field work as CTW in the WG, IL and KF floristic regions. Image below: Comparison between *Combretum-Terminalia woodland and wooded grassland* (CTW) according to Friis et al. (2010) and 2.A.1.Fg.1-Dry *Combretum* – Mixed Woodland Savanna, 2.A.1.Fg.2-Dry *Acacia* Woodland Savanna, 2.A.1.Fg.3-Dry *Acacia* – *Terminalia* – *Combretum* Woodland Savanna, and 2.A.1.Fo.2-Moist *Acacia* – (*Combretum*) Woodland Savanna in Sayre (2913). According to Sayre (2013) the *Combretum-Terminalia* woodlands is more restricted than the CTW-area according to Friis et al. (2010) and Fig. 4-1.

3.10. Studies of *Boswellia papyrifera* in natural populations in Ethiopia

As recorded already in the 18th century (introduction to Section 3), one of the characteristic woody species of the northern part of the *Combretum-Terminalia* woodland, *Boswellia papyrifera*, provided frankincense, a bark resin used in many domestic ceremonies in Ethiopia, as well as in the liturgy of the Ethiopian Orthodox Church (Thulin 2020). Frankincense has been a commodity of domestic and international trade since ancient times and is harvested from natural populations in the *Combretum-Terminalia* woodlands in northern and north-western Ethiopia.

Several recent studies show that in Ethiopia *Boswellia papyrifera* populations have decreased tremendously to smaller and more isolated remnant patches, and that many woodlands in the north-western and north-eastern parts of Ethiopia completely lack recruitment of saplings, which threatens the persistence of the species (Abrham Abiyu et al. 2010). The problem has led to a number of genetic and ecological studies, both of the ecology of the woodlands in which *Boswellia papyrifera* occurs and of the biology and genetics of the plant itself (Addisalem Ayele Bekele et al. 2016; Abeje Eshete Wassie 2011; Mindaye Teshome 2013).

Recently it has been discovered that populations of *Boswellia papyrifera* near Kurmuk, close to the southern limit in Ethiopia of that species, are in fact naturally regenerating (Mindaye Teshome et al. 2017). All these studies of *Boswellia papyrifera* have produced published data of relevés from their work in the western woodlands; the data from these relevés are incorporated in Appendix 1 and the analyses in the present work.

3.11. The *Intermediate evergreen Afromontane forest* (IAF) in north western Ethiopia

Most forests in north-western Ethiopia (in the Amhara Regional State and the GJ floristic region) were considered by Friis et al. (2010) to be part of (5) *Dry evergreen Afromontane forest and grassland complex* (DAF), including the forests around and south-west of Lake Tana. In these areas, at altitudes between 1800 and 2500 m a.s.l., the average annual rainfall varies between 800 mm along the western shore of Lake Tana, rising in many parts of the Lake Tana basin up to 1500 mm or occasionally higher, even higher than 1800 mm in the southern hills around the source of the Little Abay River (Heide 2012: 22). For the first attempt at mapping the potential area of the *Dry evergreen Afromontane forest and grassland complex* (DAF), Friis et al. (2010: 239) used the following criterion for defining potential DAF: "Land between 1800 and 3000 m a.s.l. and average annual rainfall less than 1700 mm." Because, among other places, the Lake Tana Basin and the hills to the south-west of the lake would be divided in a complicated way by this criterion, a list of subsequent criteria were implemented, including this: "in the floristic region GJ all areas between 1800 and 3000 metres altitude that fall within the boundaries of "correction zone 3 (see Friis et al. 2010: Fig 37) were classified as DAF, including areas with rainfall > 1700 millimetres/year." This excluded all areas with forest at and southwest of Lake Tana from the vegetation type *Moist Afromontane Forest* (MAF). Although Friis et al. (2010) divided DAF into four subtypes, they did not further discuss forests around Lake Tana and in Gojam. For a study of these forests by Abiyot Berhanu et al. (2018), primary data was collected from 154 plots or relevés in the forests of Awi Zone in Gojam Floristic region (GJ) and secondary data from Zege Peninsula, the forests on the islands of Lake Tana and an area to the east of Lake Tana on the slope of Mt. Guna. The floristic composition of the forests in the studied area was investigated and compared with other areas and vegetation types. Plant communities were classified, and indicator species were identified for each community in the study area. The indicator species were used for potential distribution modelling and mapping and ground-truthing was carried out for potential areas of a new vegetation type.

In composition, the forests studied were found to be somewhat similar with the *Dry evergreen Afromontane forest* (DAF), but also somewhat similar with the

Moist evergreen Afromontane forest (MAF), except for the absence of important species of the DAF and particularly the MAF. Vegetation distribution modelling showed that highly similar forests were probably distributed in the Gojam (GJ) and Gondar (GD) Floristic Regions, and, as expected, those areas had an intermediate climate type between those of MAF and DAF. It was concluded that these forests were best recognized as a new vegetation type to be named *Intermediate evergreen Afromontane forest* (IAF). Most of this new vegetation type includes forests formerly classified with DAF, while a smaller proportion were formerly included in MAF. The modelled potential distribution of forests related to IMF was shown on Abiyot Berhanu et al. (2018: Fig. 4), and the potential distribution of IAF has now been incorporated in the vegetation map used for this book (Fig. 4-1), except for the small areas overlapping with the *Combretum-Terminalia* woodlands west of the escarpment west of Chagni, where the landscape drops from nearly 2100 to approximately 1400 m a.s.l. According to our field observations, the vegetation at the lower end of this very steep escarpment is clearly *Combretum-Terminalia* woodland and not part of the IAF.

3.12. Western Ethiopia in computerized African chorology

Since the basic synthesis by White (1983), a number of papers on African chorology on a continental scale have employed computerized methods to re-evaluate African phytochoria based on the distribution of species based on samples of the flora of specified areas. Because traditional phytogeographical methods have been criticized for being too subjective (for example by authors referred to in Section 4.2), it is important to assess if the newer computerized methods do really provide better results. An important foundation for better results is of course that the computerized methods used adequate and sufficiently detailed data, and therefore we will discuss the hitherto produced computerized studies in some detail, particularly regarding what they are supposed to state about western Ethiopia.

A very first attempt to produce a phytogeographical map of Africa based entirely on a computerized analysis of the distribution of species was made by Denys (1980), who used 2½ degree squares for recording the occurrence of 494 species and applied factor analysis (Principal Component Analysis) for the calculations. Linder et al. (2005) used the distribution of 5438 species, resulting in 79648 data points recorded on a gridded map with one degree squares and applied cluster analysis and non-metric multidimensional scaling for the calculations. The grid-cells were clustered by UPGMA (unweighted pair group method with arithmetic mean); large groups of cells were searched for and non-metric multidimensional scaling was used to explore regional patterns. The resulting map of the phytochoria from these analyses was shown in Linder et al. (2005: Fig. 1), where the phytochoria were also named. The units were hierarchically arranged at two levels, 'broad phytochoria,' subdivided into 'narrow phytochoria' (Linder et al. 2005: Table 2). Most parts of the Ethiopian Highlands were referred to a 'narrow phytochorion' named the Zambezian-Ethiopian-Kenyan, which stretched from northern Eritrea through Ethiopia and Kenya approximately to the southern limit of the Somalia-Masai Region in Tanzania, as shown in White (1983). The 'narrow phytochorion' Zambezian-Ethiopian-Kenyan was placed in the Zambezian 'broad phytochorion' with other Zambezian 'narrow phytochoria' in what approximately covered White's Sudano-Zambesian Region. The surprising inclusion of the Afromontane areas in Ethiopia, Sudan, Uganda, Kenya and northern Tanzania in the mainly low-altitude Zambezian Region was thus explained (Linder et al. 2005: 241-242): "The grouping of the uplands of Ethiopia, Sudan and East Africa into the Zambezian-Ethiopia-Kenya phytochorion is consistent with the distributions of woody and herbaceous Ethiopian Afromontane species, where the most common distribution pattern is of species restricted to Ethiopia and the mountains of East Africa, and the second most common distribution pattern is of species widespread in the Zambezian woodlands (Friis 1994). More curious though, is the inclusion of those cells with predominantly Somalian species in these clusters. ... [the presence of these Somali species could be a] result from the inclusion of species from the Somalian centre that penetrate along the lowlands and rift valleys of eastern Africa."

Both phenomena, the merging of Afromontane and Zambezian lowland species and the intrusion of Somalian species, can be explained by the complex landscapes of Ethiopia and the Kenyan-Ugandan highlands; in this part of Africa highland areas interdigitated with lowlands having in the west the Sudanian flora and in the east the Somalian flora. With one degree squares as the basal unit, and without taking altitude into consideration, one is almost bound to combine highland and lowland species together in the sampling. In southern Ethiopia, at 5° N, a one degree square is 110.6 × 110.9 km, and in northern Ethiopia, at 14° N, it is 110.6 × 108.0 km. With the highly three-dimensional landscape of Ethiopia, one will invariably record both highland and lowland species in a considerable number of squares. Within a distance of nearly hundred kilometres in the central highlands of Ethiopia there is both the bottom of the Abay (Blue Nile Valley) south of Bure with a typical western lowland flora and vegetation at 810 m a.s.l., and the peak of the Choke Mountains at 4100 m a.s.l. with typical high montane flora and vegetation. The reference by Linder et al. (2005) to the work by Friis (1994) does not help to solve the problem of the complex topography of the mountainous parts of Eastern Africa, because in that study Friis used even larger areas including both lowland and highland, and compared these by clustering without regard to altitude.

Linder et al. (2005) did not refer a number of squares along the western boundary of Ethiopia to any phytochorion, almost certainly because of lack of data, but those that were assigned a category and a name were referred to a Sudanian undifferentiated 'narrow phytochorion', part of the Sudanian 'broad phytochorion.'

Linder et al. (2012) continued and expanded the methods employed by Linder et al. (2005), again using one degree squares, but analysing the distributions of 4142 vertebrate species and 5881 plant species with cluster analyses in order to define a pattern of biochoria. In this analysis a broad Ethiopian region was detected, covering the entire Ethiopian and Eritrean Highlands and the Afar Depression. This Ethiopian region was subdivided into an Ethiopian subregion and a Djibouti subregion, covering Afar plus Djibouti, but with no connection with regions further south in Somalia. Squares along the border between Ethiopia on one side and Sudan and South Sudan on the other were referred to a Sudanian region.

Stropp et al. (2016) and Sosef et al. (2017) used publicly available data sets to explore plant distributions and floristic diversity on a continental scale. Using a dataset of 934,676 collections belonging to 47,238 taxa (species and infraspecific taxa), Stropp et al. (2016) found that with a standard sampling unit of squares measuring 25 × 25 km (approximately our 1/16 degree squares) only 0.6% of the sampling unites could be considered well sampled in the public databases and, as shown in their Fig. 3B, they found a zone of sample units with high data deficiency along the whole western boundary of Ethiopia, particularly from the Gambela Lowlands to Lake Turkana. They concluded that in the increasing application of publicly available species-occurrence data in biogeographical research and biodiversity conservation note should be taken of the fact that the completeness of species-occurrence data in this data is still low and that 'maps of ignorance' should be developed to show the shortcomings of the data. Employing a dataset of 614,022 collections belonging to 25,356 species, Sosef et al. (2017) used standard sampling unit of quarter degree squares, measuring approximately 55 × 55 km, and four or even eight degrees squares for some of their analyses covering poorly sampled areas. Two illustrations in Sosef et al. (2017: Fig. 1a & 1b) show (a) the number of specimens recorded and (b) the observed species-richness on a continental scale, both per quarter degree square, with many blank squares and squares with low values along the border between Ethiopia, Sudan and South Sudan; generally, there is a striking agreement between the records of number of specimens and observed diversity, well known from previous studies Sosef et al. (2017: 12) and probably indicating that the sampling is incomplete nearly everywhere (squares with the highest number of specimens and the highest species-richness also exactly coincide around the towns of Jimma and Harar). Two illustrations in Sosef et al. (2017: Fig. 2a & 2b) show (a) the estimated species diversity based on the Chao1 estimator for each quarter degree sampling unit with more than 100 records and (b) the 'effective number of species' estimated using the Nielsen statistic for each quarter degree sampling unit with more than 100 records (in calculating the 'effective number of species' different weights are given to common and rare species, as in Droissart et al. 2012), but again most squares along the border between Ethiopia, Sudan and South Sudan have too few records to be included in these calculations. Two illustrations in Sosef et al. (2017: Fig. 6a & 6b) show floristic turnover rates across tropical Africa, with (a) meso-scale floristic turnover rate and (b) large-scale turnover rate; in both maps no information was provided for the western Ethiopian woodlands north of the Abay River, moderately high turnover rates were indicated for the areas around Assosa with both calculations and low turnover indicated for from the Gambela Lowlands to Lake Turkana. It is not clear how the low sampling along the entire border between Ethiopia, Sudan and South Sudan influence these results. A last group of illustrations Sosef et al. (2017: Fig. 8a-8e) show the distribution of growth form diversity across Africa as analysed with varying sample unit size. The highest diversity along the border between Ethiopia, Sudan and South Sudan is seen among the herbs, highest north of the Gambela Lowlands, slightly lower from the Gambela Lowlands to Lake Turkana. Lower diversity along the border between Ethiopia, Sudan and South Sudan is found among the shrubs, again higher in the south than in the north, and even lower diversity is seen among the trees, again with higher diversity in the south.

Droissart et al. (2018) analysed African phytochoria based on the distribution of 24719 plant species of different life-forms in tropical Africa with 593861 data points, but as Sosef et al. (2017) they supplemented this general study with analyses of the distribution of different life-forms, here herbs, lianas, shrubs and trees. The units used for recording presence of species were not constant in size, but as for Sosef et al. (2017) they varied between quarter degree squares as the minimum in parts of Africa where the records were densely distributed, and two degree squares as the maximum where the records were laxly distributed. The phytochoria were identified using a bipartite occurrence network first presented by Vilhena & Antonelli (2015). As a result of this study, Droissart et al. (2018) identified 14 phytochoria (one subdivided into three subunits) and 11 transition zones. For western Ethiopia, Droissart et al. (2018) identified two relevant phytochoria, no. 1, the Guineo-Sudanian bioregion, reaching from Senegal to a row of eastern squares along the Ethiopia-Sudan border above the Gambela lowlands, and no. 2, the East African montane bioregion, reaching from the highlands of northern Eritrea to northern Tanzania. Following a subsequent analysis of four plant-growth forms, herbs, lianas, shrubs and trees, Droissart et al. (2018) demonstrated incongruence between the phytochoria revealed by the species of different growth forms, both regarding species diversity and phytogeographical patterns, underlining that more attention should be given to differences in growth form in biogeographical studies.

Fayolle et al. (2018) studied floristic information on woody species for 298 relevés of African woodland (savanna) vegetation across Africa between 18° N and 33° S and 17° W to 48° E. The relevés were selected from published and unpublished sources that could provide lists of woody plant species from savanna areas. By ordination and clustering they identified eight floristic units in their material (Sudanian, Guinean, Ethiopian, Ugandan, Mozambican Zambezian, Namibian and South African), which in turn were grouped into two larger macro-units, the N & W savannas with the Sudanian and the Guinean floristic units, and the S & E Savannas, with the remaining six floristic regions. Of the sampled areas with floristic information listed in Fayolle et al. (2018, Supplementary material S2), only twenty were from inside the boundaries of Ethiopia and only one from Sudan near the western woodlands of Ethiopia and based on observed data, but through information transmitted from pastoralists (Sulieman & Ahmed 2013). In the list in Supplementary material S2, the sites are georeferenced, and from the georeferencing the sites are classified to the ecoregions of Olson et al. (2001, see Fig. 3-10). The sites inside Ethiopia or at the Ethiopian boundary are from the following ecoregions: 'Ethiopian montane forest' (8 sites), 'Ethiopian montane grassland and woodland' (three sites, but in fact only two, as the site on Zegie is clearly from the small patch of forest on the Zegie peninsula), 'Masai xeric grassland and shrubland' (one site) and 'Somali *Acacia-Commiphora* bushland and thicket' (8 sites).

A review of the vegetation of Ethiopia as seen by Olson et al. (2001) is relevant in connection with a discussion of the findings of Fayolle et al. Olson et al. presented a system of ecoregions on global scale; the bioregions of the Ethiopian Highlands and the river valleys are largely mapped as by White (1983), but differently classified. White's 'Undifferentiated montane vegetation' is by Olson et al. (2001) renamed 'Ethiopian montane grassland and woodland,' with a lower altitudinal limit at ca. 1800 m a.s.l. The vegetation zone below this limit, which White (1983) classified as 'Evergreen and semi-evergreen bushland and thicket' is by Olson et al. (2001) referred to as 'Ethiopian montane forest' (see Fig. 3-10 and Table 3-2). In Friis et al. (2010), we have classified only vegetation above 1500-1800 m a.s.l. as montane, and above that limit we reckon with several types of montane forest (mainly dry (DAF) and moist (MAF)), evergreen bushland, montane grasslands and woodlands, all basically depending on topography. Below 1500 m a.s.l. we have accepted a vegetation type named *Combretum-Terminalia* woodland to the west of the highlands and in the valleys of the deep rivers running towards the Nile, and *Acacia-Commiphora* bushland to the south and east of the highlands. From the criteria used to select floristic sample lists by Fayolle et al. (2018: 3), it is not clear that all the Ethiopian samples should be included for the vegetation they have been referred to, for example the above mentioned coffee forest from the Zege peninsula, which Fayolle et al. (2018) list as 'Ethiopian montane grassland and woodland.' None of the sites inside Ethiopia used by Fayolle et al. (2018) are in fact from the western woodlands as defined by us, and the sample from Gadarif is partly based on data from local informants using vernacular names.

Fayolle et al. (2018) analysed their data with unconstrained ordination and several clustering methods. Their analyses of the environmental space showed that Ethiopian and the Zambezian clusters had the highest altitudinal variation in the entire data set, with the samples in the Ethiopian cluster varying from a minimum altitude at ca. 700 m a.s.l., a 1st Quertile at ca. 1000 m a.s.l., a 3rd Quertile at ca. 1500 m a.s.l., and with a maximum altitude at ca. 2200 m a.s.l. An interesting conclusion is this (Fayolle et al. 2018: 9): "Our Ugandan cluster was associated with the Victoria Basin Forest-Savanna Mosaic (n = 4 samples) of Olson et al. (2001) and the East Sudanian [savannas] (n = 3), whereas our Ethiopian cluster included the Ethiopian Montane Forest (n = 7) of Olson et al. (2001), the Northern *Acacia-Commiphora* (n = 2) and the Somali *Acacia-Commiphora* Bushland and Thicket (n = 9). We indeed found that *Commiphora* and *Vachellia* [formerly *Acacia*] were important genera for the Ethiopian cluster." None of the samples in the analyses by Fayolle et al. (2018) came from areas within Ethiopia with Victoria Basin Forest-Savanna Mosaic or East Sudanian savanna according to Olson et al., as these areas have been very little studied in the past. Such studies, including those we have carried out for this work, will illustrate how western Ethiopia forms the boundary between Fayolle et al.'s two major regions, the N&W savannas and the S&E savannas, but Fayolle et al.'s studies of the border region do not bring more evidence on this.

A tentative comparison of the 169 taxa we use in this work for analyses with the species list in Fayolle et al. (2018, Supplementary material S1) gave the result that 65 of our taxa occurred in the N&W savannas, while 117 of our taxa occurred in the S&E savannas. The 13 taxa in western Ethiopia both occurring in the N&W savannas and the S&E savannas are counted for each of these distributions. Since our taxa all occur inside the boundaries of Ethiopia, and the Ethiopian cluster of Fayolle et al. (2018) is classified with the

S&E savannas, it is to be expected that a sample from western Ethiopia would have overweight towards the S&E savannas, which indeed seems to be the case. Another influencing factor is that our 169 taxa are drawn from field work ranging from ca. 500 to 2500 m a.s.l., and therefore the sample includes a number of highland taxa that are only distributed towards the East African highlands in the south-east, while the samples from the N&W savannas are all from altitudes below ca. 1300 m a.s.l. Fayolle et al. (2018: Table 1) present the floristic relationships between all their clusters; according to this, the relationship between the Ethiopian cluster and the Sudanian cluster is expressed by the highest figure (0.174), followed in declining sequence by the Ugandan cluster (0.156), the South African cluster (0.108), the Mozambican cluster (0.084), the Namibian cluster (0.080), the Guinean cluster (0.034) and the Zambezian cluster (0.017). This is surprising, since the Ethiopian cluster was classified with the S&E savannas by of Fayolle et al., presumably due to a numerous overweight of highland and eastern sites from Ethiopia.

Because the data set of Fayolle et al. (2018) includes weak or no sampling from western Ethiopia, a moderate number of samples from the central Ethiopian Highlands, and a relatively large number of samples from the eastern and southern lowlands (our *Acacia-Commiphora* bushland), it is not surprising that *Commiphora* and *Vachellia* appear to be important genera for the Ethiopian cluster. As described in Friis et al. (2010), the genus *Commiphora* has a high number of species in south-eastern and southern Ethiopia (and in Somalia and northern and eastern Kenya), but is virtually absent from the Ethiopian Highlands above 1500 m a.s.l., and the species of *Acacia* (in the broad sense) also differ between the highlands and the south-eastern lowland. It will be interesting to see if a distinct Ethiopian cluster will continue to be prominent with further sampling of sites with genuine highland woodlands and sites in the south-east with *Acacia-Commiphora* bushland.

The studies of Marshall et al. (2020) are, like those of Linder et al. (2005, 2012), based on computerized analyses of the floristic and environmental characterisation of one degree squares. Their data set included 31046 tropical African plant species represented by 531314 data points in 1197 one degree squares. The phytochoria which they define are based on cluster analyses and environmental correlates with non-metric multidimensional scaling. Twelve environmental variables, including mean altitude with a 30 arc second resolution, climatic values from Bio1 to Bio35, surface lithology, and land cover classes, are used to interpolate and increase the resolution across tropical Africa. According to us it introduces a major complication to use the mean altitude in one degree squares as an environmental parameter, as the altitudinal range in one degree squares in Ethiopia may span from 400-500 m a.s.l. to more than 4000 m a.s.l., and many species recorded in the one degree square may not be from anywhere near the mean altitude. Thus, the square may represent a phytochorion that is floristically wrongly characterized for the mean altitude. We agree when Marshall et al. (2020) distance their work from the methods of Sayre et al (2013, 2020), stating: "Our framework has the advantage over physiognomic classifications (Arino et al. 2012; Sayre et al. 2013) that the units are diagnosed by, and fully characterized by, plant species distributions as currently represented by the plant biological record at one degree square resolution. It has the advantage over previous quantitative phytochorological classifications (Droissart et al. 2018; Fayolle et al. 2014, 2018; Linder et al. 2005, 2012) that it is spatially complete across its scope. It has the advantage over qualitatively interpolated biogeographical frameworks (Olson et al. 2001; White 1983) that our unit boundaries are drawn in a reproducible, objective and quantitative manner." But the complication of using the mean altitude in one degree squares as the representative for the entire flora of that square in a topographically highly divers landscape like the Ethiopian remains.

What is the result of the use of this method for western Ethiopia? In Fig. 5 of Marshall et al. (2020), reconstructed here from the data in the supplementary material as Fig. 3-11, the Ethiopian Highlands are categorized as a huge central block, Marshall et al.'s province no. 15.88, Addis Ababa, within their region no. 15, the Ethiopian Highlands Region. What we refer to as the western woodlands consist, according to Marshall et al., of several phytochoria belonging to several regions, districts and provinces. The core of our western woodlands (Marshall et al.'s province no. 15.80, Bahir Dar) is classified in their region 15, district no. 15.88 (Addis Ababa) and the northern highlands (district no. 15.89, Asmara). As seen from our categories, partly based on observation of vegetation in the field, this seems all right for the highlands of 15.80, but it is not clear how much influence it would have had on the other categories if Marshal et al. had distinguished between lowland and highland.

The phytogeographical division by Marshall et al. of the lowlands on the border between Ethiopia and Sudan is also interesting and does not always agree with our field observations of the vegetation. Already in the abstract Marshall et al. (2020) already state that: "We find a novel arrangement of the arid regions." And in chapter "4.2 How do our phytogeographical regions compare with previously defined regions?": "Of note is our reinterpretation of the arid flora affinities in tropical Africa: our results divide White's Somali-Masai into two regions for the first time (18: Kenyan, 19: Somalian), and separate the arid flora of Djibouti, north-eastern Ethiopia and Eritrea away from this region. Instead, the Djibouti flora clusters with cells under the same aridity regime right across to Mauritania (Tagant plateau) in the west of the continent. The new region is named 6: Tagant-Djibouti." In our work we have seen a few particular distribution patterns that agree with that, principally demonstrated by species in the family Capparidaceae, but generally we do not understand why the flora of Djibouti, north-eastern Ethiopia and Eritrea should be removed from the rather similar arid Somalian flora.

Further to the south along the Sudan-Ethiopian border there are representations of 'Region 8. The Congolian Perifery,' with '8.81, Asosa' (which according to us has a typical woodland vegetation). Further south again, there is an vertically elongated phytochorion, that would include the Bebeka and Dima areas in our profile O (see Fig. 4-3); it is referred to '8.29 Kano' (a town in the peripheral forest zone in Nigeria). This could be because of the presence of patches of *Transitional rain forests* (TRF) in this area, which does have Congolian and West African species. But this area has also typical Sudanian woodland and frequent fires, as elsewhere in the western Ethiopian woodlands.

Marshall et al. (2020) classify the western part of the Gambela lowlands as part of 'Region 2. Sudanian,' with the same hatching as the district '2.14, Tamale,' a large area in the West African woodland between Senegal and Togo. The vegetation in the western part of the Gambela lowland is frequently flooded woodlands and grasslands, forming a transition to the swamps along the Upper Nile, the Sudd. Finally, touching but not crossing the Sudan-Ethiopian border is the area which Harrison & Jackson (1958) named the Toposa area and classified as dry grassland. Marshall et al. classify the Toposa area with the district '18.77 Nairobi.' Although we have not seen this area, it is from descriptions very similar to the grasslands in the Omo Valley in Ethiopia, which Marshall et al. classify as district '19.84 Harar.'

Traditionally, plant geography has been seen as a synthesis of vegetation and species distributions. The works referred to in Section 3.12 are partly based on floristically defined vegetation (Fayolle et al. 2018), partly on recorded species- or taxon-distributions in squares defined by geographical coordinates (Linder et al. 2005, 2012; Droissart et al. 2018; Marshall et al. 2020), although environmental parameters are also included in some of the modelling. The detailed results of these studies are somewhat different from each other, and for western Ethiopia they do not in all detail agree with our field observations. It seems that the findings by Linder et al. (2005, 2012), Droissart et al. (2018) and Marshall et al. (2020) are limited by the coarse resolution (one degree squares), restrictions to the sampling, and reliance on the Ohlson map, which is an interpretation of the White (1983) map, whereas our conclusions are based on relevés of such limited extent that they are almost point records. And additionally, one has to note that according to both Stropp et al. (2016) and Sosef et al. (2017) that the data from the international data bases must of necessity be incomplete and contain errors.

A concluding remark on the importance of scale in a topographically diverse area as the western slopes of the Ethiopian Highlands. As repeatedly pointed out here, a relatively coarse-scale study based on species lists from areas of a certain size in Ethiopia will include species representative of different altitudinal zone and different ecology. In Section 10 we find that for the western Ethiopian woodlands our data from the relevés allows many, small clusters, but we also conclude that it is meaningless to go beyond clustering them into more than three major types. Beyond that the clusters become so narrowly defined that they each represent only one or a few relevés. How do those rather few units determined by clustering agree with our statement that large pixel sizes will produce misleading results? We suggest that this is because of our active selection in the field of relevés that are floristically representative of the vegetation. We selected our relevés in such a way that they were relatively homogenous, did not cover a span in altitudinal range of more than 20-50 m a.s.l. and were representative of our broad impression of western Ethiopian woodlands, excluding areas that in our opinion would belong to other vegetation types, for example riverine or riparian vegetation. We do, despite this, find that the relevés may be classified into many small clusters, which to us indicates a high environmental heterogeneity on the small scale, even when we have looked for homogenous relevés.

Table 3-2. Ecoregions and Biomes in Ethiopia and surrounding countries according to Olson et al. (2001). Abbreviations as in Fig. 3-10.

Abbreviation in Fig. 3-10	Ecoregion according to Olson et al. (2001)	Biome according to Olson et al. (2001)
EAmf	East African montane forests	Tropical and Subtropical Moist Broadleaf Forest
ESs	East Sudanian savanna	Tropical and Subtropical Grasslands, Savannas and Shrublands
Ecd	Eritrean coastal desert	Deserts and Xeric Shrublands
Emf	Ethiopian montane forest	Tropical and Subtropical Moist Broadleaf Forest
Emg&w	Ethiopian montane grassland and woodland	Montane Grasslands and Shrublands
Emm	Ethiopian montane moorland	Montane Grasslands and Shrublands
Exg&s	Ethiopian xeric grasslands and shrublands	Deserts and Xeric Shrublands
Hgas	Hobyo grassland and shrubland	Deserts and Xeric Shrublands
Mxg&s	Masai xeric grassland and shrubland	Deserts and Xeric Shrublands
NACb&t	Northern *Acacia-Commiphora* bushlands and thickets	Tropical and Subtropical Grasslands, Savannas and Shrublands
Sfg	Saharan flooded grasslands	Flooded Grasslands and Savannas
SAs	Sahelian *Acacia* savanna	Tropical and Subtropical Grasslands, Savannas and Shrublands
SACb&t	Somali *Acacia-Commiphora* bushlands and thickets	Tropical and Subtropical Grasslands, Savannas and Shrublands
Smxw	Somali montane xeric woodland	Deserts and Xeric Shrublands
Vbfsm	Victoria basin forest-savanna mosaic	Tropical and Subtropical Grasslands, Savannas and Shrublands

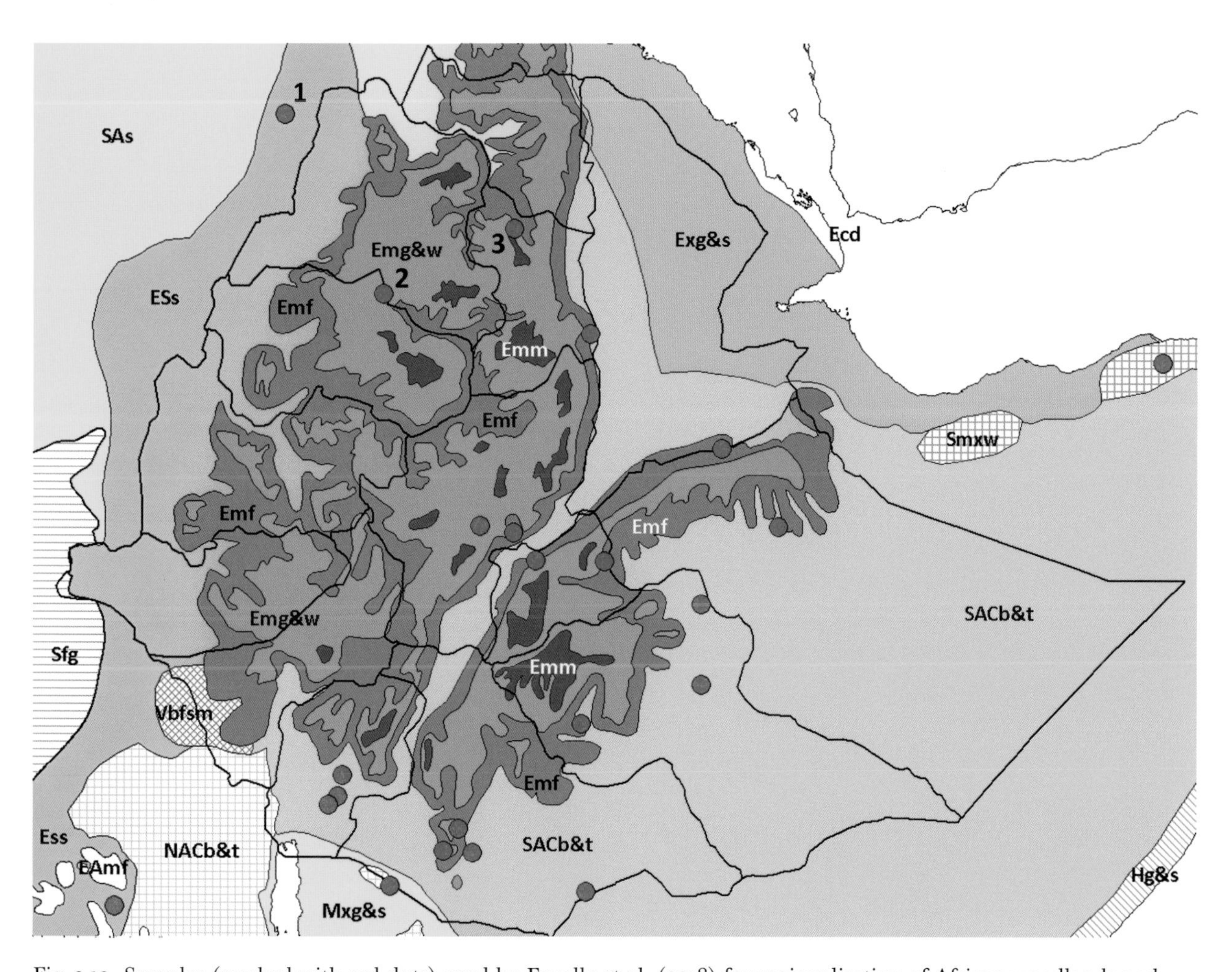

Fig. 3-10. Samples (marked with red dots) used by Fayolle et al. (2018) for regionalisation of African woodlands and wooded grasslands ('savannas') and projected on the map of *Ecoregions of the world* by Olson et al. (2001); the three samples by Fayolle et al. nearest to the western woodlands are: 1. Gadarif (Sulieman & Ahmed 2013; actually in Sudan). 2. Zegie at Lake Tana (Alelign et al. 2007). 3. Tiya in Welo near the eastern edge of the Ethiopian Highlands (Mengistu et al. 2005). For names of the ecoregions, see Table 3-2. Map constructed from data for sites by Fayolle et al. (2018), https://onlinelibrary.wiley.com/doi/abs/10.1111/jbi.13475, in jbi13475-sup-0002-AppendixS2.pdf, and from data for the ecoregions by Olson et al. from https://www.worldwildlife.org/publications/terrestrial-ecoregions-of-the-world.

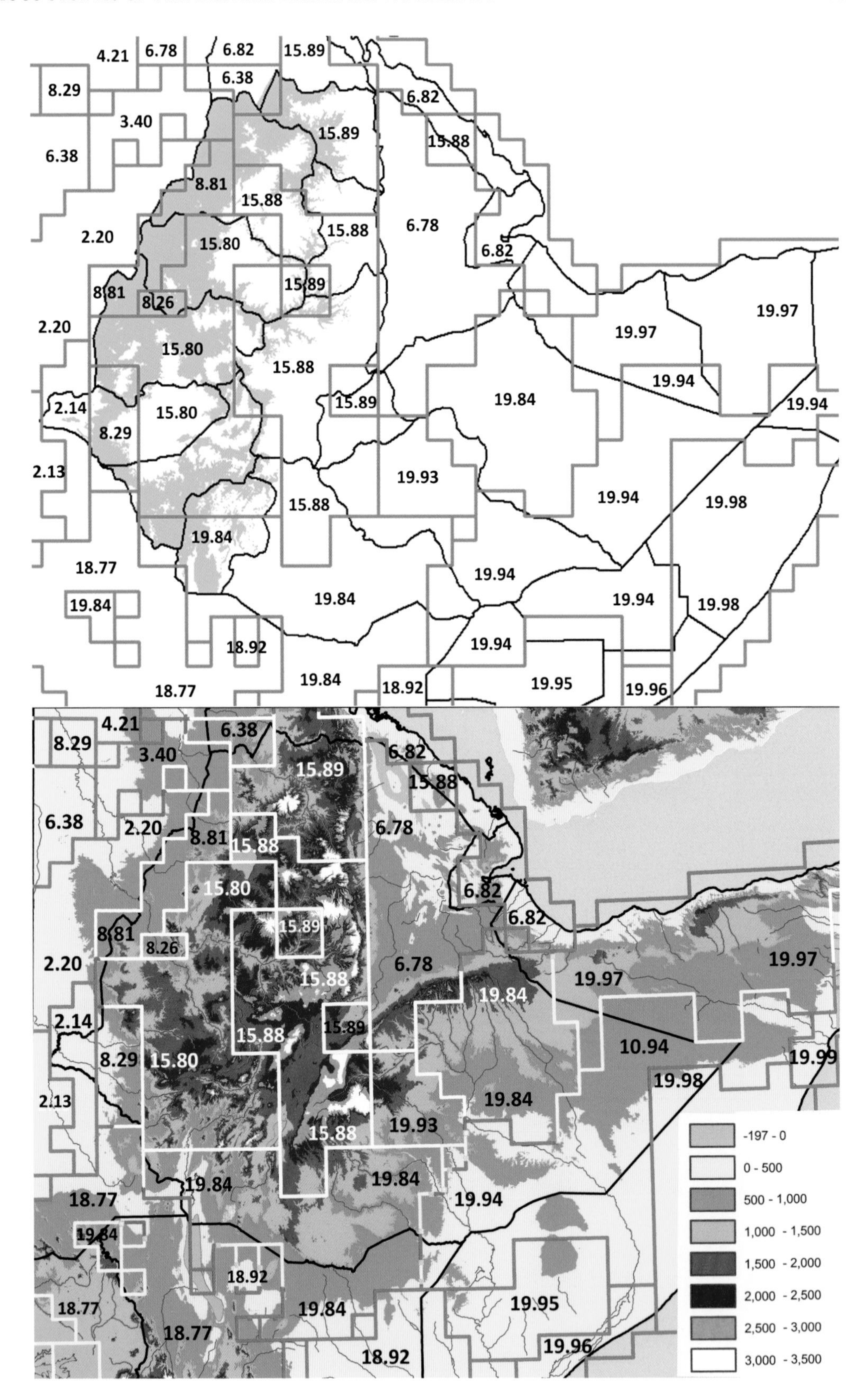
4.21
6.78
6.82
15.89
8.29
6.38
3.40
15.88
8.81
2.20
15.80
6.82
8.26
19.97
19.94
19.84
2.14
2.13
19.93
19.98
18.77
18.92
19.95
19.96
10.94
19.99
-197 - 0
0 - 500
500 - 1,000
1,000 - 1,500
1,500 - 2,000
2,000 - 2,500
2,500 - 3,000
3,000 - 3,500

Fig. 3-11. Phytochoria of Ethiopia and surrounding countries according to Marshall et al. (2020). These maps of Ethiopia and the neighbouring countries show a combination of the districts of Marshall et al. (2020, Fig. 5; with 19 regions (capitals) and 99 districts (italics)) and two of our files. Image above: the districts of Marshall et al. and our shapefile of the *Combretum-Terminalia* woodlands in Ethiopia (orange-brown on the map). Image below: the districts of Marshall et al. and the topography as represented in Fig. 2-2. The districts of Marshall et al. are named after towns, the country of which are indicated here in sharp brackets. 2. SUDANIAN. 2.13: *Accra* [Ghana]. 2.14: *Tamale* [Ghana]. 2.20: *Garoua* [Cameroon]. 3. SAHEL SOUTH. 3.40: *Ndjamena*. [Chad]. 4. SAHEL NORTH. 4.21: *Khartoum* [Sudan]. 6. TAGANT [in Mauritania] -DJIBOUTI. 6.38: *Maiduguri* [Nigeria]. 6.78: *Dire Dawa* [Ethiopia]. 6.82: *Djibouti* [Djibouti]. 8. CONGOLIAN PERIPHERY. 8.26: *Bamenda* [Cameroon]. 8.29: *Kano* [Nigeria]. 8.81: *Asosa* [Assosa; Ethiopia]. 15. ETHIOPIAN HIGHLANDS. 15.80: *Bahir Dar* [Ethiopia]. 15.88: *Addis Ababa* [Ethiopia]. 15.89: *Asmara* [Eritrea]. 18. KENYAN. 18.77: *Nairobi* [Kenya]. 18.92: *Garissa* [Kenya]. 19. SOMALIAN. 19.84: *Harar* [Ethiopia]. 19.93: nr. *Goba* [Ethiopia]. 19.94: *Gode* [Ethiopia]. 19.95: *Baydhabo* [Baidoa; Somalia]. 19.96: *Mogadishu* [Somalia]. 19.97: *Hargeysa* [Somalia]. 19.98: *Gaalkacyo* [Somalia]. 19.99: *Eyl* [Somalia]. Data from Marshall et al. (2020), https://onlinelibrary.wiley.com/doi/full/10.1111/jbi.13976 in two files in supplementary material (jbi13976-sup-0001-Supinfo.docx and jbi13976-sup-0002-Supinfo.rar). Illustration constructed from the data with the authors' permission.

3.13. Why the difference between the Sudanian woodlands to the west of Ethiopia and the Zambezian woodlands to the south?

A review paper by Assédé et al. (2020) has summed up the significant differences between the Sudanian and the Zambezian woodlands by reviewing 141 publications on the subject (but without taking note of Fayolle et al. 2018). Taking into account the geological history of Africa, they have proposed theories on the reasons for the differences between the two phytogeographical entities. They conclude that the Zambezian floristic region is far more diverse than the Sudanian floristic region, counting at least 8500 plant species, of which 54% are endemic, while there are possibly no more than 2750 plant species in the Sudanian floristic region.

Three distinct woodland types are ecologically important and clearly differentiated in the Zambezian woodlands, while the combined effect of wide tolerances of the species and the gradual geographical change in the climate in the Sudanian woodlands makes it difficult at all to recognize any distinct woodland systems in that floristic region. The continental drift, moving Africa gradually northwards from a position in the late Cretaceous (75 my BP) with the Equator across the Sahara to the present position, has caused the development of widespread volcanic activity and great rifts and swells in Zambezian Africa, with intense tectonic activity since Palaeocene (55 my BP) and fragmentation of habitats as a consequence, while such development has hardly taken palace in Sudanian Africa. According to Assédé et al. (2020) these differences in geological history and their environmental consequences may mainly or at least partly explain the difference in the vegetation composition between the Sudanian and Zambezian Africa and the higher diversity and plant endemism in the Zambezian zones.

The western woodlands of Ethiopia are an important part of the previously poorly known border zone between the less diverse Sudanian region in northwestern Africa and the more diverse Zambezian region in eastern and southern Africa. This underlines the relevance of our present study and makes further studies of the western Ethiopian lowlands highly pertinent for the understanding of African plant geography.

4

This project: background; research questions; methods and terminology; what has been done since Friis et al. (2010)?

After the conclusion of several projects relating to the western Ethiopian woodlands, particularly the gathering of data in connection with the fire projects mentioned in Section 3.5 and 3.9, a temporary politically stable situation in parts of the western Ogaden, in areas where the species-rich *Acacia-Commiphora* bushland had previously been poorly studied, suddenly opened new opportunities to work there. Field work in Ethiopia in general, and in the Ogaden and in the Somalia region in particular, requires both a stable security situation in the study area and the approval of our work by the authorities, local authorities as well as the regional police. The favourable opportunities for work in the western Ogaden in 2014, 2015 and 2017 were utilized for studies of the *Acacia-Commiphora* bushland in that part of the country (see Friis et al. 2016, 2017; Paton et al. 2018; Friis et al. 2019; Friis 2019), and a paper on the new vegetation type, the *Transitional Semi-evergreen Bushland* (TSEB; Breugel 2016b). However, after 2017 the opportunity to work in the Ogaden ended, and it was timely to return to general studies of the western woodlands.

Apart from the intention to present a significantly better set of data on the woody species of the western woodlands than previously, we have wanted to ask a few scientific questions in relation to the vegetation of that area, questions that were touched upon in the previous studies described in Section 3:

1. Can we confirm the extent of the *Combretum-Terminalia* woodlands? – Was it defined correctly as a Potential Natural Vegetation in Friis et al. (2010), including its delimitations in the north and in the south?
2. How do the *Combretum-Terminalia* woodlands relate to the Sudanian woodlands between the western Ethiopian border and the Atlantic Ocean? Do the widely distributed species follow the same pattern across the continent? And how do the western Ethiopian woodlands relate to the Zambezian woodlands?
3. How uniform is the *Combretum-Terminalia* woodland in Ethiopia? Does the vegetation and flora in the western lowlands and on the western Ethiopian escarpment represent one or several ecological or floristic units?
4. If the *Combretum-Terminalia* woodland is not uniform, is there a gradual transition from lowland to highland and/or from south to north? – Or, if the vegetation and flora represent two or more notably distinct units, then what species or what environmental parameters can be used to define these units?
5. Can we point to important drivers with regard to variation (continuous or discontinuous) along the increasing latitudes and/or altitude in the western woodlands?

We will provide data and discussions in the following Sections 5-10 to address these questions and sum up our conclusions in Section 11. The first step towards finding solutions to these questions was to produce an updated version of the vegetation map in Friis et al. (2010) by adding the new information that had become available since 2010. As described in Section 3, several papers on vegetation had been published after 2010, requiring modifications of the extension of various vegetation types, in particular the mapping of additional vegetation types. The VECEA map (Lillesø et al. 2011) introduced true *Desert* (DES) as distinct from the *Semi-desert scrubland* (SDS); further studies of the fire ecology of Ethiopia modified the extent of the *Transitional rain forest* (TRF) (Breugel et al. 2016a); other studies resulted in the establishment of the additional vegetation types *Transitional semi-evergreen bushland* (TSEB) (Breugel et al. 2016b) and the *Intermediate evergreen Afromontane Forest* (IAF) (Abiyot Berhanu et al. 2018). Because of these modifications a new map has been compiled (Fig. 4-1) and used as the base for our analyses and discussions in this work, in particular the distribution maps in Section 6.

We have recorded the woody species in 151 relevés located at different altitudes from the border with Kenya to the border with Eritrea, including relevés in the accessible parts of the deep river valleys that cut into the highlands from the Nile Basin (Fig. 4-1). An overview of the distribution of the relevés on altitudinal zones and the size of the areas they characterize is presented in Table 7-1. During these studies we found 169 woody species inside the relevés. We also recorded longitude, latitude and elevation of the relevés and estimated several environmental parameters, such as slope, degree of burning, biomass available for burning, and soil. For some of these, the altitude, slope, aspect, biomass, and degree of burning, we were able to test the field observations against data from environmental databases (see Appendix 2 and 3).

One question left open in Friis et al. (2010) and Breugel et al. (2016a) was to discuss the northern and southern limits of the CTW-area. When selecting the profiles and relevés to study (see description in Section 5) we endeavoured to cover as wide a range of sites as possible inside our CTW-area, political situation and security permitting. In order to specify this research question further and make sure that we cover areas with suitable environmental parameters, we have modelled the climate range represented by these relevés with DIVA-GIS (Hijmans et al. 2005a), using its bioclimatic variables and the software Bioclim (Hijmans et al. 2005b). Given that our relevés represent the climate of the CTW-area the models would show if it is likely that there are significant areas with similar climate that are not covered by our relevés. Two climatic models were produced with DIVA-GIS, one with four, the widest number of potential climatic parameters (Fig. 4-2, above) and one with the narrowest number of potential climatic parameters (all 19 in Bioclim) (Fig. 4-2, below). Both models show a slight overspill of the potential CTW into Eritrea in the north and hardly reaching Lake Turkana and Lake Chew Bahir in the south. We take this as indication of a reasonably good fit between the covering of our relevés and the extent of the modelled climate they represent. These findings will also be mentioned in Section 11.

Of necessity the distribution of the relevés is uneven, which makes it difficult to state in detail how many of our 169 species occur in a given predefined area, for example in one degree squares or smaller unites. In Fig. 4-5, an attempt has been made to show how the number of species varies when recoded on one degree squares (1/1), on quarter degree squares (1/4), on sixteenth degree squares (1/16) and on sixty-fourth degree squares (1/64). The map with species-richness on one degree squares shows more than 86 species in a number of areas: along the borders in the eastern Gamo-Gofa (GG), in and at the Gibe Gorge in the north-eastern Kefa (KF) floristic region, along the border between the Ilubabor (IL) and Welega (WG) floristic region north of the town of Gambela, between Kurmuk, Gizen and the Abay River (Blue Nile), generally around Lake Tana and in the Tacazze Valley where it is crossed by the road between Gondar and Enda Selassie. This pattern is repeated on the map with quarter degree squares, but the highest richness values have moved away from eastern GG, the Gibe Valley and the surroundings of Lake Tana, while a new square with high richness turned up where the road from Bure to Nekemt crosses the Abay Gorge. With our map of richness on sixteenth degree squares, a square at the Gibe Gorge, several squares at Kurmuk and Gizen, squares in the Abay and Tacazze Gorges and near Lake Tana reappear, and a somewhat similar result is seen with the distribution on sixty-fourth degree squares.

These squares represent areas that are potentially species-rich in woody plants of the western woodlands, but more detailed studies of the areas between the roads are needed to further specify real species-richness. In Section 2.9 and 11.6 we argue that the areas, at Kurmuk and Gizen in the WG floristic region, may be areas with high species-richness and should be considered as potential location of a new protected area (See also Section 11.6).

For the 169 species of woody plants observed in our relevés, we have recorded the holdings of specimens from the whole of Ethiopia in the herbaria of C, ETH, FT and K. These specimens in herbaria were supplemented with scanned images mainly from WAG and BR, obtained via GBIF (GBIF.org 2020). The acronyms of the herbaria are those of *Index Herbariorum* (http://sweetgum.nybg.org/science/ih/), see also the list of acronyms at the end of this work. We mapped the distribution of all the species based on our field observations, the georeferenced herbarium material and verified records from international databases (Fig. 4-4, map to the right). The data shows that the species in our relevés are not restricted to the CTW-vegetation, but may also occur in a number of other vegetation types. We have tried to classify the phytogeography of each of these species in Section 6, and in Section 7 analysed their distribution with phytogeographical methods as discussed in Section 4.2. Finally, we have in Section 10 analysed the data from the 151 relevés with cluster and principal component analyses in order to explore the continuity of the vegetation.

4.1. Materials and methods: general considerations on the study of vegetations in the tropics, our collection of data in the field and herbarium studies; analyses and software

The most comprehensive method in the study of tropical vegetation is the study of very large research relevés or plots, varying in size from ca. 10 to over 100 ha. Such plot studies have chiefly been located in the rain forests of Latin America and tropical Asia, but also in rain forests of Nigeria, Cameroon and D.R. Congo (Chase 2014; a list provided in https://forestgeo.si.edu/sites-all). It has been impossible for us to use the same methodology of traditional plot studies in Ethiopia because of the topography, the amount of human influence on the vegetation and the unstable administrative situation that makes it problematic to maintain permanent relevés. Even semi-permanent relevés are difficult to maintain; Guo et al. (2018) had established such relevés in dry Afromontane forests in north-eastern Tigray, a location which now, in 2021, is deeply affected by armed conflict. Studies of non-permanent smaller relevés scattered over large areas were used in an interesting, previously mentioned study by Kershaw (1968), demonstrating the high local variability of the Sudanian woodlands.

We have used inventories of species observed in relevés of unspecified size and have aimed at a non-hierarchical characterization of the vegetation. Traditionally there has been disagreement in vegetation science between on one side the school of Zürich-Montpellier, and on the other side the school of Grenoble-Toulouse. The former has used strictly defined relevés and hierarchically classified vegetation types and applied a strict nomenclature for the vegetation, epitomized by the publications by Braun-Blanquet (1921, 1964). The latter, using fewer rigid methods, is represented in the tropics by the publications by Dobremez, but also by followers in tropical Africa and elsewhere outside the tropics. The work of Dobremez is of particular relevance here because of his classification and mapping of the vegetation of Nepal (Dobremez 1976; Lillesø et al. 2005), like Ethiopia a topographically highly diverse country with climates varying from tropical to subarctic (or in Nepal even arctic) and complex and not well known vegetation.

Dobremez (1976: 127) has stated how his method was different from that of the Zürich-Montpellier school. His emphasis was on descriptions where the potential vegetation in each zone was scored independently within environmentally defined limits that made ecological sense in the field, rather than on attempts with methodically placed relevés, defining a rigid, hierarchical description of the vegetation. In these aspects, our work is similar to that of Dobremez and differs from that of Braun-Blanquet.

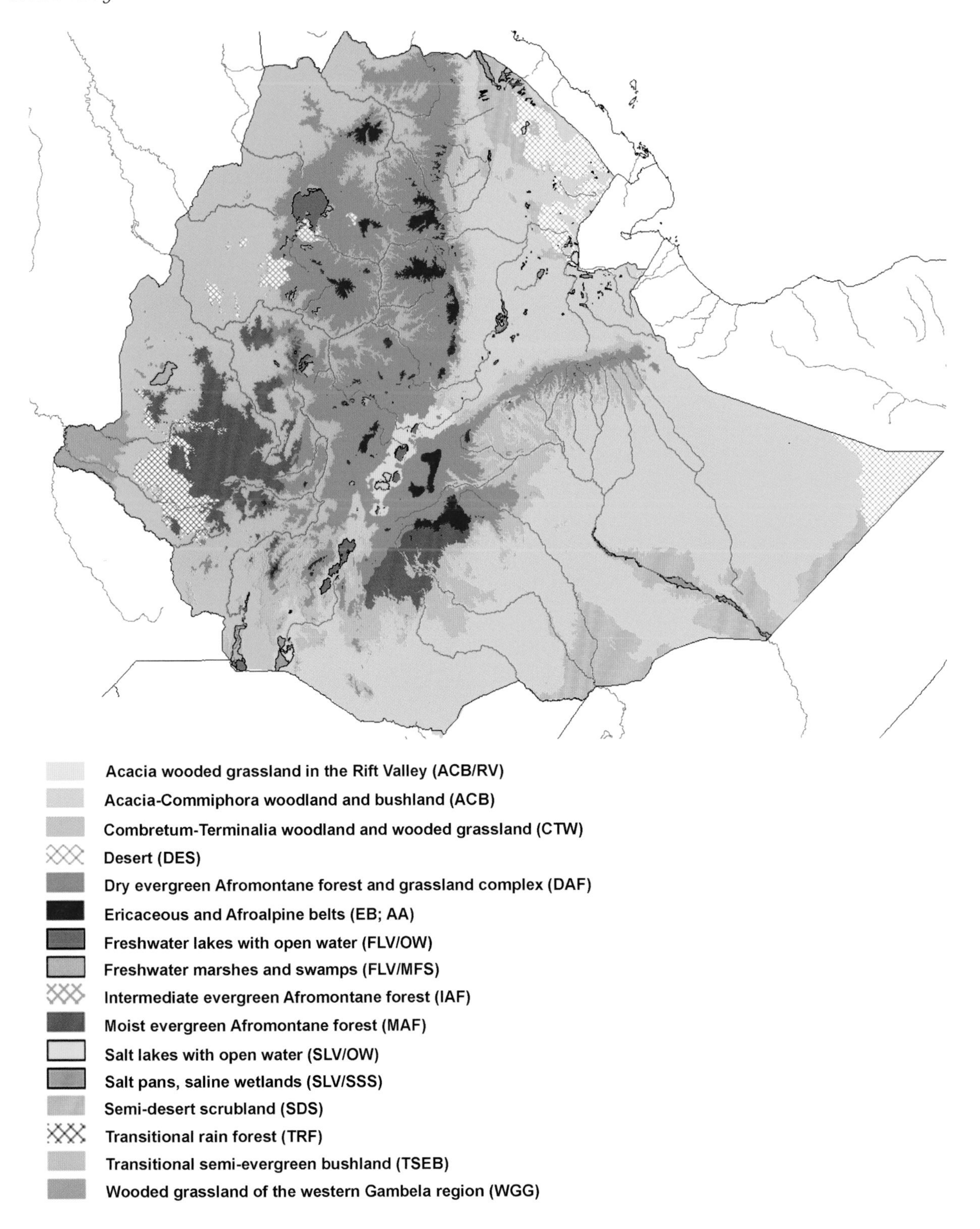

Fig. 4-1. A new map of the potential natural vegetation of Ethiopia, redrafted for this work. Based on the map in Friis et al. (2010) and recompiled after improvements by Lillesø et al. (2011), Breugel et al. (2016a, 2016b), and Abiyot Berhanu et al. (2018).

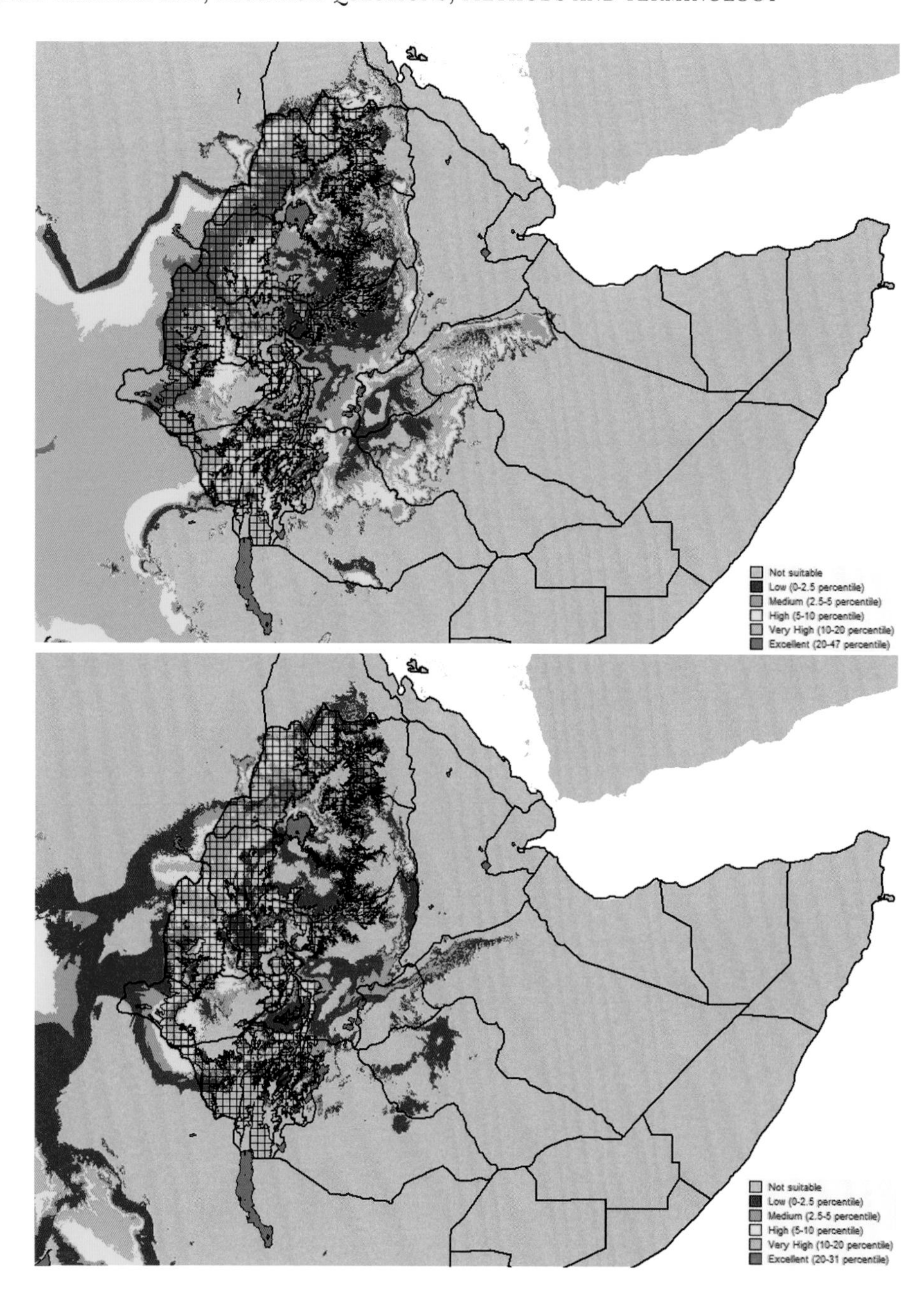

Fig. 4-2. Modelled distributions of climates like those observed in our 151 relevés. The coordinates of the relevés were recorded in the field. The modelling was done with DIVA-GIS, using its bioclimatic variables and the software Bioclim. Image above: For each relevé, the model uses the following environmental parameters for the site: Bio1 (Annual mean temperature), Bio12 (Annual precipitation), Bio13 (Precipitation in wettest month) and Bio14 (Precipitation in driest month). Image below: For the same relevés, the model uses all the 19 climatic parameters in Bioclim. The figure shows the floristic regions used in the *Flora of Ethiopia and Eritrea* and *Flora of Somalia* (black lines) and lakes (blue). The area of the modelled *Combretum-Terminalia woodland and wooded grassland* (CTW) is cross-hatched.

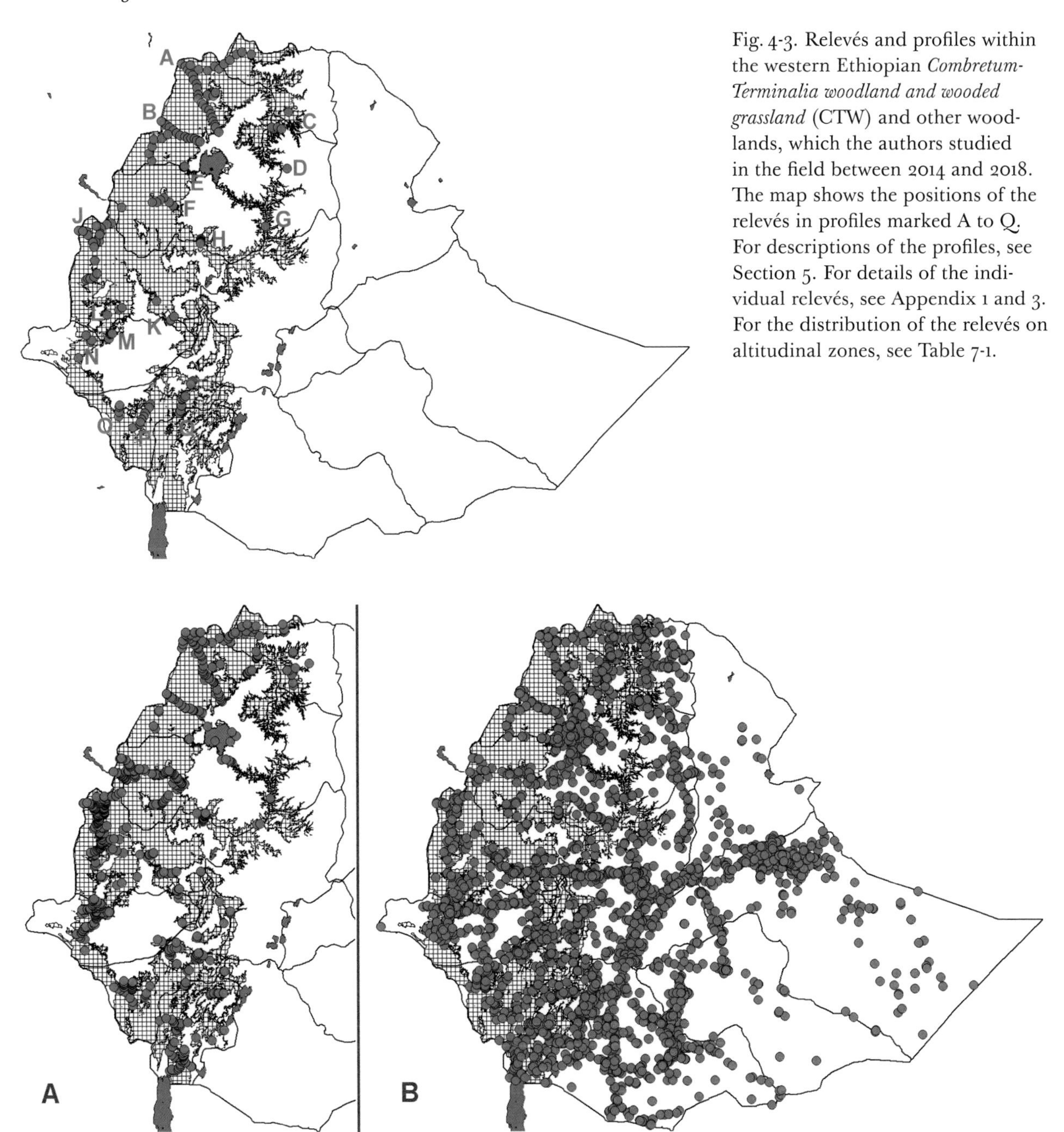

Fig. 4-3. Relevés and profiles within the western Ethiopian *Combretum-Terminalia woodland and wooded grassland* (CTW) and other woodlands, which the authors studied in the field between 2014 and 2018. The map shows the positions of the relevés in profiles marked A to Q. For descriptions of the profiles, see Section 5. For details of the individual relevés, see Appendix 1 and 3. For the distribution of the relevés on altitudinal zones, see Table 7-1.

Fig. 4-4. Additional localities and records relevant for this work. A (left map). Localities in the Ethiopian *Combretum-Terminalia* woodland inside and outside the relevés and studied by us in the years 1970-2018. The localities include both places where we have collected specimens and made observations with records. B (right map). All records from the whole of Ethiopia of species of woody plants observed in the relevés. This map shows the sites of all specimen-records used for mapping in the Section 6. The *Combretum-Terminalia* woodland as defined in this work is marked with cross hatching.

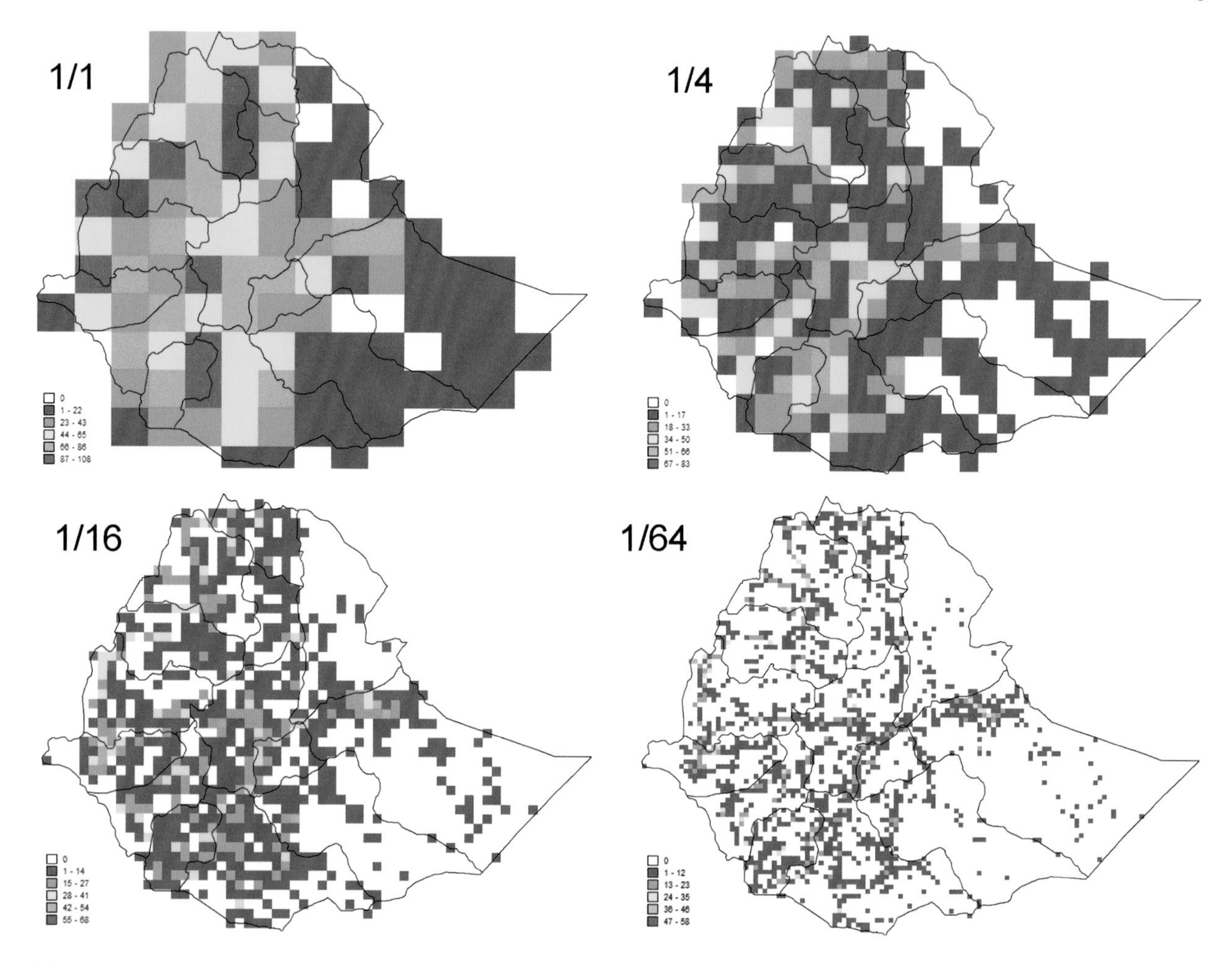

Fig. 4-5. Species-richness of the 169 species of woody plants found in our relevés in the western Ethiopian woodlands, based on the data in Fig. 4-4B. 1/1, using one degree squares. 1/4, using quarter degree squares. 1/16, using sixteenth degree squares. 1/64, using sixty-fourth degree squares. High species-richness in the western woodland is observed in the Gambela lowlands, in the areas around Kurmuk and Gizen, around Lake Tana and where the road between Gondar and Enda-Selassie crosses the Tacazze River.

Methods like those of Dobremez have also been used in pioneer vegetation mapping in many parts of tropical Africa, for example the vegetation map of Uganda by Langdale-Brown et al. (1964) and that of Zambia by Trapnell (2001), both incorporated in the VECEA map of eastern Africa described in Section 3.7, and comparable methods have been used elsewhere in the tropics (see Blasco 1988). Due to detailed knowledge of the vegetation, it has been possible to map the complex vegetation in South Africa, Lesotho and Swaziland (Mucina et al. 2006), classifying biomes, ecoregions and bioregions.

We have used the concept of potential natural vegetation both in Friis et al. (2010), in Breugel et al. (2016b) and in this work. Chiarucci et al. (2010) have claimed that the concept of potential natural vegetation is plagued with problems and that the concept should be abandoned because it is a static concept characterized by reference to existing mature vegetation, and because it seems to ignore that vegetation is dynamic. We are aware that vegetation is dynamic, but it must be described in space and time. We have therefore used the concept of potential natural vegetation in a wider sense than Chiarucci et al. (2010), allowing within our types of potential natural vegetation alternatively stable states or states that vary within definable limits. Within the limits of potential natural vegetation, we allow vegetation types that are recognizable within the range of variation generally accepted for the tropical African vegetation types that have been used in the pioneer studies discussed above. We have also taken note of the scale of the vegetation considered, so that the definition of potential natural vegetation is allowed a variation within limits, taking into consideration the heterogeneity of the landscape.

We do agree with Chiarucci et al. (2010) that models for long-term vegetation dynamics cannot be fully deterministic because of the existence of environmental heterogeneity and random processes, and we accept that modelling vegetation dynamics must include complex simulations with many factors involved. As mentioned, Dobremez expressed the view that his altitudinal zones and vegetation types should be conceived as 'containers', outer frames of different ecosystems, natural, semi-natural or man-made, and that the natural vegetation should be conceived as one of the possible contents in that container (Dobremez 1976: 127; discussed by Lillesø et al. 2005). If our vegetation types in the western woodlands of Ethiopia (*Combretum-Terminalia* woodland, *Acacia seyal-Balanites aegyptiaca* woodland, *Borassus aethiopum* stands, etc.) are not the ideal types of potential natural vegetation, they will at least agree with Dobremez's views referred to here, and the frequency with which we have observed variants of these types far from each other supports our view that they do indeed represent potential vegetation types.

This is not to discard the methods of Braun-Blanquet from the study of Ethiopian vegetation on a smaller scale. Zerihun Woldu & Feoli (2001) studied the bushland vegetation around Adwa in Tigray with the method of Braun-Blanquet, resulting in the recognition of three major types of bushland (respectively characterised by *Acacia etbaica*, *Euclea racemosa* subsp. *schimperi* and *Senna obtusifolia*), but we think it would be impractical to apply the same methods on an Ethiopian scale.

In this work we seek to combine a broad review of what we have been able to trace of existing knowledge in the literature of the western woodlands of Ethiopia with the results from our own field observations, herbarium studies and analyses. The sources of the data in Section 6 are described in the following, first the field data, afterwards the specimen data.

The collection of field data (during the years 2014-2018) and the identification of the species were thus planned: We selected the profiles (sequences of relevés) along motorable roads from our previous experience with field work in Ethiopia since 1970, and with consideration to logistic possibilities and political stability. The locations of the individual relevés were selected at a distance of more or less 10 km, but occasionally, if it was necessary to work along very long profiles and very little accommodation, we used a distance between the relevés of ca. 20 km. Two data sets were obtained: floristic data and environmental data. Although it was difficult because of restrictions on traffic and access, we also attempted to ensure that the relevés were spread out altitudinally; the altitudinal distribution can be seen in Table 7-1, where the areas of the altitudinal zones with relevés have also been calculated.

Floristic data (simple presence/absence) were recorded only for woody plants and supplemented with general notes on grasses or herbs in the undergrowth. Due to the highly varied topography, and sometimes intense cultivation in the areas we wanted to study, it was not possible to use quadrates as relevés for the recording of species. Instead, we followed an adaptation of the method used in Breugel et al. (2016b); we attempted to set a time limit of 45 minutes, during which we collected floristic data. Occasionally we allowed more time in particularly difficult terrain, if for example we were moving steeply uphill or downhill, which prevented us from making observations some of the time while we were moving, or if for some reason we had to move at the same slow pace on level ground. We tried to localize areas with vegetation that appeared as natural and homogenous as possible, and walked around in this vegetation, looking for and recording as many species of woody plants as possible during the allotted time. Most species were readily and easily identifiable in the field. Samples of taxonomically difficult taxa were collected, pressed and dried, and taken to the National Herbarium of Ethiopia (ETH), or sent as loan to the Herbarium at the Royal Botanic Gardens, Kew (K), for the best possible identification. Despite this, some difficulties with identification of specimens remained. Six instances required a judgement call. They are as follows:

Identification of immature and sterile plants of two species of *Acacia* caused difficulties: *Acacia hecatophylla* (leaflets 1.25-2.5 (-3.5) mm wide, greyish beneath; in Ethiopia recorded from ca. 1400-1700 m a.s.l., but also cited from much lower altitudes as a component of the *Balanites-Acacia* vegetation on black clays by Harrison & Jackson (1958)) and *Acacia polyacantha* subsp. *campylacantha* (leaflets 0.4-0.9 (-1.25) mm wide, not conspicuously pale beneath; in Ethiopia recorded from 500-1600 m a.s.l.). The former species has previously only been recorded from north of the Abay River, where we recorded it frequently, but we think that some sterile material from the Omo River system may also be this species. *Acacia polyacantha* subsp. *campylacantha* is known from the border with Eritrea to the border with Kenya. Our identifications of *Acacia hecatophylla* from south of the Abay River may require confirmation.

Identification of specimens of two species of *Lannea* with very young leaves and no flowers or fruits caused difficulties: *Lannea barteri* (indumentum distinctly pilose; in Ethiopia recorded at 500-1600 m a.s.l.) and *Lannea schimperi* (indumentum distinctly floccose or sometimes nearly absent; in Ethiopia recorded at (800-) 1050-1750 m a.s.l.). *Lannea schimperi* is recorded from the Eritrean to the Kenyan border, while *L. barteri* is only known from the western woodlands south of Lake Tana. We have particularly identified records from low altitudes in the Abay River system as *Lannea barteri*. The leaves of these plants were not mature when we studied them, and we observed no flowers or fruits.

Identification of young specimens without fully developed leaves and inflorescences of two species of *Protea* caused difficulties: *Protea gaguedi* (leaves usually less than 3 cm wide, bracts pubescent, usually less than 15 mm wide; in Ethiopia recorded from 1400-3000 m a.s.l.) and *Protea madiensis* (leaves usually more than 4 cm wide, bracts glabrous, except for ciliate margin, usually more than 20 mm wide; in Ethiopia recorded from 1200-1700 m a.s.l.). We attempted to identify our records to one of these two species according to the above criteria.

Identification of sterile specimens of species in the genus *Rhus* caused difficulty: We decided that there was only one species of *Rhus* in our relevés, but because the plants were mostly sterile, it was difficult to identify that species with certainty to any of three rather similar species, *Rhus longipes, Rhus natalensis* and *Rhus vulgaris*. *Rhus longipes* has stems with dark grey to brown or reddish bark, reticulate tertiary veins prominent, panicles with flowers evenly spaced, and is in Ethiopia recorded from 1200-1600 m a.s.l. *Rhus natalensis* has stems with grey bark, reticulate tertiary veins obscure, panicles with flowers evenly spaced, and is in Ethiopia recorded from (700-) 1200-2200 m a.s.l. *Rhus vulgaris* has stems with dark grey to brown or reddish bark, reticulate tertiary veins closely, but not conspicuously reticulate, panicles with flowers, at least the male ones, cluster into separate globose groups, and is in Ethiopia recorded from 1500-2800 m a.s.l. According to these criteria, all records have been identified as *Rhus vulgaris.*

Identification of sterile and sometimes partly leafless specimens of species to three potential species of the genus *Sterculia* caused difficulty. *Sterculia africana* (leaves sparsely puberulous to pubescent, stipule-scars (2-) 3.5-5 mm wide; in Ethiopia recorded from 0-1775 m a.s.l.), *Sterculia setigera* (leaves greyish pubescent to tomentose, stipule-scars 1-2 mm wide; in Ethiopia recorded from 700-1900 m a.s.l.), and *Sterculia cinerea* (leaves whitish lanate beneath, stipule-scars 1-2 mm wide; in Ethiopia this species is present in the western woodlands of the Gambela Region, but is rare, it is recorded from 0-650 m a.s.l.). Our records documented with specimens were identified as *Sterculia africana*. Pichi Sermolli (1957) and Harrison and Jackson (1958) have identified some specimens of *Sterculia* from along the Ethiopia-Sudan border as *Sterculia cinerea*, but from the above criteria we have decided to identify all our doubtful records as *Sterculia africana*. This identification needs further study.

Identification of sterile or partly leafless specimens of two possible species of *Terminalia* caused difficulty: *Terminalia laxiflora* (bark on second year's growth becoming corky, mature leaves hairy to glabrous, fruits 6.5-10.5 cm long; in Ethiopia recorded from 450 to 1500 m a.s.l.), *Terminalia schimperiana* (bark on second year's growth remaining fibrous, mature leaves tomentose or pubescent, sometimes glabrous, fruits (2.5-) 4-7 cm long; in Ethiopia recorded from (1000-) 1300-2200 m). When in doubt, because the characters did not suffice, we identified the records from below 1100 m a.s.l. as *Terminalia laxiflora*.

For a few relevés, the floristic information was obtained from earlier studies published in the literature (Jensen & Friis 2001; Abeje Eshete Wassie 2011; Mindaye Teshome et al. 2013, 2017). For most of these, it seems certain that a much longer time than our ca. 45 minutes has been used for data collection in each relevé, resulting in a longer species list that is almost certainly more complete than the ones we recorded. Based on our general floristic knowledge of western Ethiopia, we have had reasons to question a few of the identifications in these relevés. Our alternative identifications are indicated clearly in Appendix 1.

Environmental data: The environmental data obtained for each relevé consists of direct observations and environmental parameters directly available or calculated from data available on the internet databases. At each relevé, latitude, longitude, and altitude of sites were recorded with a Garmin GPSMAP 72, and subjective notes on slope, soil and grazing were made. Soil types in descriptive terms could not be based on the soil classes described in Section 2.5, but were later converted to a sequence of 11 numbered classes, based on the weathering of the rocks from which the soils were derived, the soil colour and the grain size. The classes are shown in Fig. 2-19.

The altitude, recorded in the field in m a.s.l., was verified with data from a digital elevation model, the 90 m SRTM DEM (CGIAR-CSI 2008; https://cgiarcsi.community/data/srtm-90m-digital-elevation-database-v4-1/). Slope and Aspect were estimated in the field, but also computed on data from the 90 m SRTM DEM. The Topographical or Terrain Wetness Index (TWI) was computed from the 90 m SRTM DEM. Fire frequency (FF or Firefreq, number of fires in the pixel) was calculated from data recorded between 2003 and 2013 by the MODIS programme (Breugel et al. 2016a). Average fire frequency for 2003-2013 of all cells in a radius of 2 cells around the focal pixel, Median fire frequency for the period 2003-2013 of all cells in a radius of 2 cells around the focal pixel, and Maximum fire frequency for the period 2003-2013 of all cells in a radius of 2 cells around the focal pixel, were all calculated using data from the same source as Firefreq. Parameters relating to climate were derived from the BIOCLIM data set (http://www.diva-gis.org/climate) from ca. 1950 to 2000. All derived environmental parameters are specified in Appendix 3.

Specimen data from herbaria: For the study of the distribution of the individual species in Ethiopia, which is presented in Section 6, we used the data from the field studies, supplemented with specimens of the relevant species studied at the herbaria of C, ETH, FT, and K. This was supplemented further with data from imaged or photographed specimens held at WAG, P and BR (Gbif.org 2020). Data from records in GBIF and other databases, where we have not been able to verify the identity from images have not been used; Stropp et al. (2016) found that 70% of the available species-occurrence data contained erroneous or incomplete information. Our data, assembled in an Excel spreadsheet, was thus gathered from 1965 observations of species in the relevés, 8279 herbarium collections that were or could be georeferenced (duplicates in different herbaria only counted once). Specimens without coordinates indicated by the collectors were, whenever possible, georeferenced by Ib Friis, using his personal experience of Ethiopian geography and supplemented with data obtained from Google Earth Pro (7.3.3.7699 (64-bit); downloaded from https://www.google.com/earth/versions/).

For our analyses we produced two data sets with Excel, one based on data from our relevés, another on data from the relevés, the herbarium material and verified records from databases mentioned above. The data set from the relevés consists of two matrices, one with presence and absence of the 169 species recorded from the 151 relevés and another with environmental data from field observations and public databases, all listed in Appendix 3. These two matrices have been used for the analyses in Section 10, and the methods used for the analyses are described in more detail there. A subset of the data set from the relevés, consisting of a matrix with the presence and absence of the 169 × 151 presence-absence records and supplemented with rounded off altitudes was produced for the data in Table 7-1 and the altitudinal cluster in Fig. 7-5. The other data set (from relevés, georeferenced herbarium material and georeferenced and verified records from databases) were used for the distribution maps and other information for Section 6, as well as for the maps in Fig. 4-3, 4-4 and 4-5. Derived from this second data set were spreadsheets with records for each species of the phytogeographical elements and the distribution on predefined areas in Ethiopia (see Section 4 and 7 and Appendix 4), as well as information about altitudinal ranges of the species (compilation of information in Section 6), distribution on continental scale and records for each species of a number of ecologically important parameters (height of plants, leaves simple or compound, leaf-size classes, leaf persistence, fire resistance and adaptations to deter browsers; see Section 8 and Appendix 5).

For the African distributions studied in Section 9, we were kindly sent georeferenced distributional data for the species of Combretaceae by Jan J. Wieringa, Naturalis Biodiversity Center, Leiden, the Netherlands, which included further supporting material from Carel C.H. Jonkind, Botanic Garden Meise, Belgium. We have checked the data for African countries other than Ethiopia against the records available in the RAINBIO database (Dauby et al. 2016; Sosef et al. 2017). According to Sosef et al. (2017: 3) the contribution of Ethiopian records to RAINBIO has come from publicly available datasets, mainly GBIF. For the Ethiopian records we have only included specimens that we have seen, either as specimens or as digital images.

The distributional data for each species was converted to shapefiles with the software DIVA-GIS 7.5 (http://www.diva-gis.org/). For the mapping presented in Section 6, the following external shapefiles were used, apart from the digital elevation model mentioned above: The alignment of country borders and rivers are from https://www.naturalearthdata.com/downloads/50m-cultural-vectors/50m-admin-0-countries-2/ and https://www.naturalearthdata.com/downloads/10m-physical-vectors/10m-rivers-lake-centerlines/.

Information in detail about the methods used in the cluster-analyses and principal component analyses used in Section 10 is given at the beginning of that Section and in Appendix 6. Photographs and digital maps were finalized with Adobe Photoshop CS6.

4.2. Phytogeographical classification, concepts and terms

In this Section we present and define our use of the basic ideas about the area- and taxon-based plant geography relevant to this work. We review a few of the concepts from the beginning of phytogeography in the first decades of the 19th century, to the debates about them in recent years, and the earlier work on the Ethiopia forests (Friis 1992) that also used these concepts. Because of the current computerised studies of African plant geography described in Section 3.12, we feel that such a broad review is pertinent.

Having in Section 6 produced maps of the 169 taxa of woody plants observed in the relevés, we classified the taxa based on comparisons of their ecology and the patterns they form on the distribution maps. When meeting difficulties in classifying the distributions of the taxa on the data in Section 6, we have taken note of how they were distributed in neighbouring countries, or sometimes even in Africa as a whole, and then argued under each species in support of our conclusion.

We have been concerned about the sufficiency of the information on the distribution maps, as many species show a scattered distribution and are notably recorded along the main roads. In order to explore the potential distributions of the species in Ethiopia, but also outside the country, we have in Section 9 modelled the entire African distribution of twelve typical woodland species that have mostly been recorded in many relevés and can be used to characterize the CTW-area. Using the entire African distribution, we hope to have produced as adequate models as possible. Most of these twelve species occur between Ethiopia and the Atlantic Ocean in the area referred to by White (1983) as the Sudanian Regional Centre of Endemism. The model predictions show that suitable habitats are expected to exist in the gaps between point records. In our phytogeographical classification of the species, we consider suitable habitats predicted by the models to be part of the actual distribution areas of these species. Species with most of their records and predicted suitable habitat in the CTW-area are assigned to the local phytochorion CTW (*e.g.*, *Combretum collinum* and *Terminalia schimperiana*; see discussion of the concept local phytochorion later in this Section), while species with most of their records and predicted suitable habitat outside the CTW-area are assigned to a different local phytochorion (e.g., *Terminalia brownii*).

In a work on the forests and forest trees of North-East Tropical Africa, Friis (1992) selected some basic phytogeographic and ecological concepts and terms, which we will also use in this work. Their theoretical background will therefore be briefly revisited in the following. The definition of phytogeographical regions, the principle phytochorion, follows the early ideas about phytogeography of De Candolle (1820: 410-412) and Schouw (1822). They considered phytogeographical regions to represent the common area of distribution of taxa that are either restricted to the regions (endemic taxa) or have more or less identical patterns of distribution within the region.

De Candolle and Schouw recommended that phytogeographical regions (phytochoria) should be based on the observation of congruent distribution areas for as many species as possible. Schouw pointed out that in practice it could be difficult to distinguish boundaries between two closely adjacent phytochoria because many taxa, although with generally congruent distributions, might not quite reach or perhaps just transgress the borders of the phytochorion and thus blur the pattern. He therefore suggested that the extent of the phytochoria should be delimited primarily based on directly observed homogeneity of their vegetation, and the analyses of their flora carried out on the more centrally placed parts of the areas. He also suggested guidelines for how to describe phytochoria at various hierarchical levels, not just in the meaning of phytogeographical regions as defined by White (1965) and Wickens (1976). A good description of a phytogeographical region (phytochorion) *sensu* Schouw should include an account of the endemic taxa, a description of the characteristic vegetation, the predominant life-forms, etc.

In his study from 1992, Friis defined his forest types on the Horn of Africa as local phytogeographical regions, phytochoria, because they represented shared distributions of several species and were represented in areas of recognizable and relatively well defined vegetation. Friis (1992) did not make any attempt to classify his local forest phytochoria on the hierarchical scale of Wickens (1976: 38), but according to that system they would be categorized at the lowest ranks, such as sectors or districts.

Following the example of Friis (1992) in the present study, we will define phytochoria based on our sample from the relevés. Here, 89 out of 169 taxa are restricted to the phytochorion occupied by the vegetation type *Combretum-Terminalia woodland and wooded grassland* (CTW), at least within Ethiopia. For this work we will therefore refer to the area of the vegetation type of CTW as a local phytochorion of unspecific rank. In the following we will use the designation 'CTW' as a common marker in the discussion of the vegetation type *Combretum-Terminalia* woodland, of the area in western Ethiopia occupied by this vegetation type, of the local phytochorion and of the assembly of species which is mainly distributed within this vegetation or area. We will similarly use the designation ACB as a common marker for the vegetation type *Acacia-Commiphora bushland* (ACB) for the area in eastern Ethiopia occupied by this vegetation type, and for the assembly of species which is mainly distributed within this vegetation or area. We will specify if we mean the CTW-vegetation, the CTW-area (phytochorion) or the CTW-species.

The traditional term for a group of species belonging to a common distribution type and thus a common phytochorion is 'element', the other basic concept in regional phytogeography. Although first used in the 19th century, this term has remained controversial since the beginning of the vicariance biogeography around 1970 (Humphries 2001), as will be discussed further in this Section.

The term 'element' has normally been considered to be coined by Christ (1867), who used it in the study of the Alpine flora of Central Europe to designate a group of species with a reasonably uniform geographical distribution, e.g., restricted to a single phytogeographical region or phytochorion. However, Fattorini (2015, 2016, see further below) has pointed out that Christ may not be the first to use the term 'element'. In the same year, Areschoug (1867), in a study of the Scandinavian flora, used the same term to indicate a group of plants with distributions extending in a particular direction from Scandinavia. Wickens (1976: 38) defined 'element' as a group of taxa belonging to a particular phytogeographical region or domain (a phytochorion at the hierarchical level below region), but he also cited a more specific definition proposed by Braun-Blanquet and reworded by Cain (1944): 'the floristic and phytosociological expression of a territory of limited extent; it includes the taxonomic units and the phytogeographic groups of taxa characteristic of a given region.' In our present study, the 'territory of limited extent' is taken to include a phytochorion of low rank, for example the area of the vegetation type *Combretum-Terminalia woodland and wooded grassland* (CTW), and, in order to indicate the limited size of the phytochoria used to define the element, we use the term 'local element.' We also use the term 'local sub-element' for species restricted to the CTW-area, but with a specific distribution in relation to that, for example either widespread in the CTW-area (CTW[all]), northern (CTW[n]), central (CTW[c], southern (CTW[s], or any combination of these. Local sub-elements marked with 'ext' are transgressing species that have an extended distribution into the *Transitional semi-evergreen bushland* (TSEB) in the Rift Valley or the south-eastern slope of the Ethiopian Highlands towards the eastern lowlands. Additional elements, the distribution of which are not identical with a vegetation type or a phytochorion, have also been used and will be discussed further in this Section. Here we will use only a few (MI: marginal intruders; TRG: transgressors, and other species that transgress vegetational or phytogeographical boundaries).

Some authors have expanded the concept of 'element' much beyond this to describe shared attributes other than purely geographical distribution, accepting also as elements groups of taxa that to some extent transgress the boundaries of a phytogeographical region in agreement with certain theories about origin, migration routes, etc. Areschoug (1867) used the term to theorize about the immigration of the Scandinavian flora after the latest glacial period, and Eig (1931a, 1931b) used the term in a similar context to theorize about the origin of the Palestinian flora. According to this widening of the concept of 'element,' several qualifying adjectives were added, providing such terms as genetic (group of species having related species in other areas), historic element (group of species having an assumed common history, having entered their present area at the same time), migratory element (group of species having followed a common migratory route), etc. (White 1970).

In this work we will restrict the use of the term 'element' to the strictly descriptive: taxa that share a distribution within a phytochorion, here the CTW-area, or taxa that clearly transgress the boundaries of our CTW-area from a wider distribution in other phytochoria. Already mentioned above, these are: Marginal intruding species (MI), which have their main distribution within one phytochorion or vegetation type and marginally intrude across the border into another phytochorion or vegetation type. Apart from the broadly defined MI, we have already mentioned the intruders of the '+ext' type, which strictly are marginal intruders into other phytochoria to the east of the CTW-area. Transgressing species (TRG) are widely distributed in two or more phytochoria and can be distinguished by being polyregional, that is being equally widely distributed in two or several phytochoria. Finally, for this work we have identified a small group of species, mainly belonging to the family Capparaceae, all sharing a wider distribution in the *Acacia-Commiphora* bushland (and hence the ACB phytochorion) but do intrude into the driest part of the *Combretum-Terminalia* woodlands in the north or in the south or in both. This element could be classified as Marginal Intruders, but unlike the largest number of such species, which are Marginal Intruders into the *Combretum-Terminalia* woodlands from the Ethiopian Highlands without a specific area into which they intrude, these are restricted to limited areas in the north or the south of the area of the CTW or both; this element is therefore referred to as ACB[-CTW], indicating a main distribution in the ACB-area, and a specific transgression into an ecologically defined area of the CTW.

Similar principles of regional phytogeography were developed and employed in White's and Wickens' works on Africa. In the discussion of his vegetation map and phytochoria, White (1976) addressed the old question of coincidence between areas defined on similarity in the appearance of vegetation (physiognomy) and the floristically defined phytochoria. He restated that there is no *a priori* reason why the pattern lines of a vegetation map based on appearance of the vegetation should chiefly coincide with those of a chorological map based on overlapping distribution patterns of an element of the flora. But all practical experience had, he argued, so far suggested that if chorological maps were based on chorological data alone, rather than on patterns derived from the appearance of vegetation, the resulting phytochoria would not be significantly different. Phytogeographical regions could therefore, as shown empirically, be based on vegetation maps but had to be checked against distribution maps of a relevant number of individual species.

Some controversy in the branch of plant geography studying areas of distribution (chorology) has existed between the school basing their classification of distribution patterns on inspection, versus those employing computerized classification based on presence and absence within strictly predefined boundaries such as degree squares, and this difference of opinion is as old as the computer. Recognizing patterns of distribution by visual inspection of many distributional maps of individual taxa, in combination with data on vegetation, has been the subject of criticism by a number of authors. Jardine (1972) strongly advocated the use of computers in phytogeographical analyses and summed up the most frequently advanced critique: "the human eye may discover groups even when none is present; there is a tendency to select particularly striking distributions and arrange the remaining around them, and phytogeographical classification is often biased because of preconceived ideas about factors determining the distribution of plants".

Clayton & Hepper (1974) argued against this criticism, but supported the use of computers to test the phytogeographical conclusions: it is not surprising that the findings of chorology and ecology (in the sense of the study of the distribution of vegetation types) coincide, for the distribution of a floristic pool must obviously conform to the ultimate limitation of the range of its members imposed by major environmental factors. The selection of species with the same distribution due to common ecological tolerance or adaptation is also the mechanism which has created vegetation, and it is therefore not surprising that phytogeographical regions often agree with ecologically defined categories.

These basic concepts in the terminology used in biogeography have recently been reviewed by Passalaqua (2015) and Fattorini (2015, 2016). Passalaqua (2015) reopened the long-standing debate about the phytogeographic terminology by discussing the well-known term 'element' (and the recent term 'chorotype'), and concluded that "systematic biogeography should regard 'element' as referring to a group of taxa defined according to the biogeographical area they occupy," which is the first and oldest usage of the term. Fattorini (2015, 2016) maintained classical concepts and terms, and stated that descriptive biogeography is mostly identifying regions ('phytochoria') characterized by particular species assemblages, but also grouping species together by the common areas in which they are found (in this work called 'elements'). According to Fattorini, 'chorology' is the identification of areas with recurrent patterns of distribution, which may produce hierarchical classifications of phytochoria in regionalisation systems (e.g., realms, regions, dominions, provinces and districts), and recognition of species assemblages based on recurrent, similar species ranges often referred to as 'chorological categories.' The former agrees with 'phytochoria', the latter with 'elements.'

According to Fattorini, the confusion in descriptive biogeography primarily exists regarding the second category, the 'element', which he distinguishes from 'chorotype.' Again, the basic use of the term 'element' is for a group of organisms that share a common general or local distribution, but Fattorini (2016) listed the confusing variety of definitions of the term 'element' based on various centre-of-origin-dispersal models. To allow for speciation by vicariance and rid biogeography of the narrow focus on dispersalism that has become associated with the term 'element', Fattorini proposed to replace 'element' with the 'chorotype' as a term which should stay free of historical assumptions. Fattorini (2015, 2016) also pointed out that he reckoned with two different uses of the term 'chorotype.' One according to which species with roughly similar overall ranges are classified to the same 'global chorotype', another according to which species with roughly similar distributions within a certain region (and not based on their overall ranges) are referred to a specific 'regional chorotype.'

Because in this work we mostly focus on Ethiopia, Eritrea and the Horn of Africa, we generally use concepts and terms in agreement with the 'regional' (or even a 'local') version of Fattorini's concept of chorotype. However, not included in Fattorini's review is our involvement with phytogeographical terminology for the Horn of Africa, to which we will now return. In the study of the forests of the Horn of Africa Friis (1992: 54-60) used the term 'local phytochoria' for areas of similar vegetation and congruent species distribution. However, in Friis (1992) the terminology for groups of species with congruent distribution was confused. Friis used the term 'element' in the traditional way in the introduction (Friis 1992: 54-60), but the controversial ideas of dispersalism, already much disapproved of then by the supporters of vicariance biogeography, caused him to shy away from using this term in the systematic enumeration of the species (Friis 1992: 89-269). In that part of the book, Friis used the neutral term 'distribution type.' Choosing our terminology for this work, we find that both the continuation of the practice by Friis (1992) and the introduction of the term 'chorotype' would add to the confusion. Therefore, for Friis' 'distribution type' and Fattorini's 'regional chorotypes' we will use the terms 'local element' or 'local sub-element.' The latter is part of a local element, always comprising fewer taxa that the local element that either has a more restricted shared distribution inside the local phytochorion than the local element, e.g., restricted to the southern part of the CTW-area, CTW[s], or is part of a local element that shares a common transgression into another phytochorion, e.g., ACB-species that transgress into the northern and southern part of the CTW-area.

See more about this in Section 7, where our phytogeographical data are analysed. More data on the elements in relation to altitudinal and geographical distributions can be seen in Fig. 7-3, 7-6 and 7-9. Phytochoria are normally treated as non-overlapping (while species may transgress their boundaries), but for the statistics of environmental adaptations in Section 8 we needed to count the species (or in some cases the subspecies) with specific characteristics within separately defined areas, simply recording taxa with particular ecological adaptations in these areas, irrespectively of whether they also occurred in other areas or not. The areas used for these records are also those shown in Fig. 7-2. They are WWn, northern part of the western woodlands. WWc, central part of the western woodlands. WWs, southern part of the western woodlands. EWW, all areas east of the western woodlands, except the areas with *Acacia-Commiphora* bushland (ACB). ASO, areas east of the western woodlands with *Acacia-Commiphora* bushland (ACB), mainly in Afar and Ethiopian Somalia, i.e., the AF-, SD-, BA- and HA-floristic regions). They must not be confused with phytogeographical categories.

4.3. Comparison with the *Combretum-Terminalia* woodland (CTW) as described in Friis et al. (2010)

When preparing the *Atlas of the potential Vegetation of Ethiopia*, we compiled a list of species recorded from the vegetation type defined as *Combretum-Terminalia* woodland (Friis at al. 2010: Appendix 3, pp. 177-237), based on the literature and observations in the field. In Appendix 2, this list has been compared with our findings in this work in order to estimate the level of agreement between the two works.

In the 2010-list, 112 species were considered "trees" or "trees or shrubs" and were associated with the CTW-vegetation (although some only tentatively), either as exclusive CTW-species or as occurring in CTW but with a different vegetation type as their characteristic vegetation. Of the 169 species seen in our relevés for this work, 95 were in 2010 categorized as CTW-species, (including the relatively narrowly distributed species *Dombeya buettneri, Meyna tetraphylla, Ozoroa pulcherrima,* which was listed as a "shrub", and *Grewia flavescens*, which was listed as a "shrub or liana").

The cumulative total of taxa between the 112 CTW trees and/or shrubs of the 2010 publication and the 169 taxa under study here is 195, of which 158 are associated with the CTW (as occurring in CTW as its main vegetation type or extending into it) in either one of the lists. Five species out of the 169 taxa under study here were not treated in the 2010 publication, while 26 species in the list in the 2010 publication were not found in our relevés. Comparing the 164 taxa in common between both lists, there is agreement concerning their phytogeographical affiliation for 60-69% of the taxa (depending on the level of detail). The discrepancies, the 31-40%, consist of taxa that have been recategorized here following the extensive field work in large parts of the CTW-area between 2014 and 2018. Thirty eight species (23%) were not associated with CTW at all in 2010, but are here categorized either as CTW local elements (sub-element with or without extension), as ACB[-CTW] or as ACB[-CTW+ext]. In the 2010 publication thirteen species (8%) were associated with CTW (as the characteristic vegetation type or in combination with other vegetation types) and are here in this work considered to be Marginal Intruders (MI) or Transgressors (TRG). *Acacia persiciflora* was listed in the 2010-publication without associated vegetation type; it is here considered a Marginal Intruder (MI). Fifteen species have been recategorized from being typical of the ACB and reaching into CTW to being typical of CTW and extending into other vegetation types.

The 26 taxa associated with CTW in the 2010 publication but not found in any of our relevés are listed at the bottom of the table in Appendix 2. Nine of them are indicated as strict CTW-elements in the 2010 list, of which three deserve special mentioning: *Dombeya longibracteolata* is a rare species in the southernmost part of the CTW-area in the KF and GG floristic regions, transgressing to the Rift Valley in the SD floristic region. *Euphorbia nigrispinoides* has been found only in the northern part of the Rift Valley in a vegetation very similar to *Combretum-Terminalia* woodland, but in a border region between *Acacia wooded grassland in the Rift Valley* (ACB/RV) and *Transitional semi-evergreen bushland* (TSEB). *Euphorbia venefica,* previously known as *E. venenifica,* was only known from CTW-area west of the WG floristic region, outside Ethiopia, but was found in 2018 by us along the route of profile P, near but outside relevé no. 95 (Weber et al. 2020). The absence of three of the other species listed in Friis et al. (2010), but not seen in our relevés, may represent problems with identification of sterile material.

In conclusion, we find that there is a reasonably good agreement between the species recorded from the CTW-area in the 2010-publication, particularly in the more widely distributed species, those found in many relevés and considered typical CTW-species. The new field observations do also refine information about the woody flora of the *Combretum-Terminalia* woodland and other western woodlands, and document the overlap with the flora of the Ethiopian Highlands with the many highland species that occur in a few relevés at the highest altitudes of the CTW-area.

5
Descriptions of the profiles

The profiles studied during our field work in 2014-2018 are marked in Fig. 4-3, together with results from our studies in the Gambela lowlands in 1996-1999 and the data sets cited from research in natural frankincense populations mentioned in Section 3.10 (see information about all the relevés in Appendix 1). All places where Ib Friis, Sebsebe Demissew and Odile Weber have made observations and collected specimens between 1970 and 2018 in connection with the collections in the series 'Friis et al.' are marked on Fig. 4-4 (map to the left), and in Fig. 4-4 (map to the right) all the collecting sites that are used for the maps in Section 6 are mapped together.

In the following we give an overview of our descriptive field observations, organized in agreement with the profiles from the studies in 2014-2018 and arranged in segments marked with the letters A-Q as in Fig. 4-3. In this overview, the information from the individual relevés in Appendix 1 has been summarized so that the study areas are described in context; we hope that this will make it easier to understand our data and help any reconstruction of our observations that might be necessary in the future.

Profile R is a review of data from the literature, as we were not able to work in the Omo Valley in 2014-2018, but had visited the areas to the east of the Omo River for plant-collecting on several previous occasions.

Floristically, the *Combretum-Terminalia woodland and wooded grassland* (CTW) is a moderately diverse vegetation type, with species, subspecies and varieties that do not occur elsewhere in Ethiopia, ca. 40%. A number of woodland and wooded grassland types are associated, notably types with *Acacia senegal*, *A. seyal* and *Acacia-Balanites* wooded grassland. The total number of woody taxa in various profiles that contain more than a few relevés is surprisingly similar. The profile with most taxa, more than 60, is the long and much branched Profile J in central Ethiopia with high rainfall and highest fire-frequencies, while the equally long Profile A between Humera and Gondar in the north, dry and with low fire-frequencies, had about 50 woody species.

As seen on the new vegetation map (Fig. 4-1), the woodlands usually form large, coherent areas, including long strips of woodland along the deep river valleys, but there are areas where the topographical situation is more complex. One of these is the fragmented area with woodlands on the shores of Lake Tana (in the GD and GJ floristic regions). The water level of Lake Tana fluctuates, but always at altitudes around or slightly below the limit here defined as the upper limit for the *Combretum-Terminalia* woodland and hence the lower limit for the Afromontane vegetation. Most of the area below the 1800 m contour around Lake Tana is in fact made up by temporary floodplains, but unflooded slopes and hillsides are often, according to Pichi Sermolli (1938), Friis et al. (2010) and our recent field observations, covered by wooded grassland with *Stereospermum kunthianum* (Bignoniaceae) and other species elsewhere associated with *Combretum-Terminalia* woodland.

Another part of Ethiopia with a complex topography that leads to isolated patches of *Combretum-Terminalia* woodland is the complex of river valleys along the Omo and Sagan Rivers in the extreme south of Ethiopia, as can also be seen in Fig. 5-92. For various reasons, we were unable to record profiles at Lake Tana or at the Omo and Sagan Rivers.

A distinct variant of *Combretum-Terminalia* woodland and wooded grassland with very little representation of the genera *Combretum* and *Terminalia* exists in the WG floristic region near the upper limit of the vegetation type, where it forms a transition to *Moist evergreen Afromontane forest* (MAF); it is dominated by the tree *Syzygium guineense* subsp. *macrocarpum*

(Myrtaceae), which forms a 5-8 metres high tree stratum, although the growth of the individual plants is often shrubby. This situation is mentioned under Section 2.6., and it is further described under Profile J and shown in Fig. 5-52.

A: Profile of the escarpment and the north-western lowlands between Humera and Gondar

The long profile A from Humera on the Ethiopian-Eritrean border to the watershed above the town of Gondar includes our relevés no. 1-23 (see Appendix 1), studied on 20.-24.10.2014. From this main profile, the altitudinal range of which is 600-2450 m a.s.l., two smaller profiles branch out, one towards the highland to the east of the main profile, starting from the little town of Dansha (relevés 23-25), altitudinal range 900-1300 m a.s.l., and another, longer branch from just south of Humera towards Shire (previously Enda Selassie; relevés 26-35), altitudinal rage 650-1700 m a.s.l. From earlier visits we also had observations from the Tacazze Valley on the Enda Selassie-Gondar road (Fig. 5-10, Fig. 5-11), but for political reasons we could not record data for relevés in that part of the Tacazze Valley in 2014-2018. During earlier visits to the Tacazze Valley south of Enda Selassie, we observed flowering populations of *Acacia amythethophylla* (Fig. 5-11, Fig. 5-12), a species we later observed in the Abay Valley but did not photograph there.

The course of the main profile follows rather closely the route taken by Pichi Sermolli in 1937 and gives us an opportunity to compare observations with a difference of ca. 80 years (Fig. 3-1 and Fig. 3-2). Along the first ca. 35 km from Humera towards Gondar the terrain was flat, mostly ca. 600 m a.s.l., and the soil was a vertisol, which in 2014-2018 was intensively cultivated, mainly with sesame (*Sesamum indicum; Pedaliaceae*), a non-indigenous annual crop grown for seeds and the oil derived from them. Even in areas with what seemed close to natural vegetation, the ground-cover of grasses and herbs was rarely taller than 10 cm, and there was no sign of burning. As already observed by Pichi Sermolli, the dominant tree in these woodlands was *Balanites aegyptiaca*, which was mostly left standing in the fields, probably because the fruit of this tree is edible, while lower trees and shrubs were cleared away (Fig. 5-1). In places where the potential natural vegetation was preserved, there was *Acacia mellifera, A. seyal, A. senegal* and *A. oerfota* in the lower strata, as also mentioned by Pichi Sermolli and Harrison and Jackson (Fig. 5-2). In areas liable to flooding during the rains, the *Acacia seyal* was replaced with low specimens of *A. hecatophylla*. Occasionally within this expanse of vertisol there were patches with low rocks, which were porous, probably basaltic, but resembling grey limestone. In these areas, which could not be cultivated, there was scrub of *Dalbergia melanoxylon, Dichrostachys cinerea, Acacia seyal, Acacia mellifera, Boscia coriacea*, which was not seen in our relevés, and *B. senegalensis*, with some *Balanites aegyptiaca* where the soil was deeper. These species also occurred on flat ground, where the vertisol had been covered by layers of gravel or windblown sand. Where there were more rocks, there was a dense or open woodland of *Anogeissus leiocarpa, Boswellia papyrifera, Combretum collinum, C. hartmannianum* (Fig. 5-3), *C. molle, Dalbergia melanoxylon, Dichrostachys cinerea, Grewia flavescens, Lannea fruticosa, Ochna leucophloeos, Sclerocarya birrea, Sterculia africana, Strychnos innocua, Terminalia brownii, T. laxiflora, Ximenia americana, and Ziziphus spina-christi.*

Approximately 40 km south of Humera towards Gondar the terrain began to rise, soon reaching 800 m a.s.l., and the extensive plains with vertisol gave way to rocky ground with brown, grey, and reddish soils in between the rocks. Again, the ground-cover of grasses and herbs was rarely taller than 10 cm, and there was no sign of burning. On the rocky ground, the vegetation was dominated by *Anogeissus leiocarpa, Combretum hartmannianum* and *Acacia seyal*, very often with *Ziziphus abyssinica* and *Z. spina-christi* in the lower strata.

Above 800 m a.s.l., 50-70 km south of Humera, we observed on local patches of vertisol the first *Piliostigma thonningii* and *Ficus sycomorus,* while in rocky areas *Pterocarpus lucens* was dominant and, growing directly on the rocks, *Ficus glumosa* and *F. ingens.* At the same altitudes, but about 80 km south of Humera and on brown clayey soil with many stones, we noted the additional species *Flueggea virosa* and *Lonchocarpus laxiflora*.

Many small temporary streams had a line of trees along them, mainly of *Tamarindus indica*, *Diospyros mespiliformis* and *Ficus sur*. In bare, overgrazed patches, there were frequently large shrubs of *Calotropis procera*.

Although the altitude was steadily becoming higher, there were sometimes flat areas surrounded by a brim of rock, so that vertisol could develop, but again the ground-cover of grasses and herbs was rarely taller than 10 cm, and there was no sign of burning. A particularly large area of this type, at the time of our visit very muddy and probably completely flooded in the rainy season, was found ca. 110 km south of Humera at altitudes of ca. 900 m a.s.l. It contained the following woody species, some of which were elsewhere associated with rocky ground, *Acacia hecatophylla, Anogeissus leiocarpa, Combretum hartmannianum, Dichrostachys cinerea, Ficus sycomorus, Flueggea virosa, Gardenia ternifolia, Lannea schimperi, Pterocarpus lucens, Terminalia laxiflora,* and *T. macroptera*.

Although it was difficult to identify the smaller rivers mentioned by Pichi Sermolli (1940), we found it easy to recognize the largest, the Angereb River, which he, on the other hand, did not mention. The modern road crossed this river at the little town of Asheri, south of which there were prominent and rather steep, low hills with a ground-cover of grasses and herbs rarely taller than 10 cm and with no sign of burning. The woody species were *Anogeissus leiocarpa, Boswellia papyrifera, B. pirottae, Calotropis procera, Combretum hartmannianum, Dalbergia melanoxylon, Ficus glumosa, Flueggea virosa, Grewia flavescens, Lannea fruticosa, Lonchocarpus laxiflorus, Maytenus senegalensis, Pterocarpus lucens, Sterculia africana,* and *Ziziphus spina-christi*. This was the first time we observed the two species of *Boswellia* growing together, the relatively widespread *B. papyrifera* and the rarer *B. pirottae*, which is endemic to northern Ethiopia. Surprisingly, only 10 km further to the south this rocky landscape was followed by a flat area with extensive areas of vertisol, where *Acacia hecatophylla* and *Acacia seyal* were common. Near the riverine forest of a tributary to the Angereb River, but bordering intensely cultivated and irrigated farmland, there was a stand of *Kigelia africana* (Fig. 5-15).

A notable difference from the observations made by Pichi Sermolli was that we never, along the Humera-Gondar profile, observed large stands of the lowland bamboo, *Oxytenanthera abyssinica*. Patches of this species were first observed on rocky slopes about 140 km south of Humera together with *Anogeissus leiocarpa, Boswellia papyrifera, Combretum collinum, Lonchocarpus laxiflorus,* and *Sterculia africana*. Not far from there, we observed slender trees of *Acacia hockii* on rocky ground, and again rock-bordered, flat patches of ground with vertisol, *Acacia hecatophylla, Anogeissus leiocarpa,* and *Terminalia laxiflora*.

Above ca. 1400 m a.s.l., ca. 175 km south of Humera, the vegetation began to contain elements from Afromontane bushland and woodland, such as the dense woodland or forest around the church of Beit Michael near Pichi Sermolli's Tucùr-Dinghià, today known as Tekeldengy. The vegetation around the church was protected from grazing and burning and contained a mixture of predominantly lowland and highland trees and shrubs, *Acacia polyacantha* subsp. *campylacantha* (low), *Anogeissus leiocarpa* (low), *Bridelia scleroneura* (low), *Carissa spinarum* (low), *Combretum molle* (low), *Cordia africana* (high), *Croton macrostachyus, Diospyros mespiliformis, Ficus sycomorus, Flacourtia indica* (all high), *Maytenus senegalensis* (low), *Oncoba spinosa* (high), *Schrebera alata* (high), *Terminalia schimperiana* (low), and *Vitex doniana* (low).

Further south, near Gondar and the watershed at 2450 m a.s.l., there was a transition between *Combretum-Terminalia* woodland and degraded Afromontane semi-evergreen bushland with trees and very low, probably grazed ground-cover. The bushland and grassland were alternating with small patch of rather intensively cultivated farmland on dark brown and stony soil, which is very similar to the common highland soil. There were no signs of burning of this vegetation.

The total list of woody plants along the ca. 170 km long main profile from Humera to Gondar, with exception of the mainly or completely Afromontane relevés no. 21-22 at 2050-2450 m a.s.l., contained the following species: *Acacia polyacantha* subsp. *campylacantha* (Fig. 5-4), *Acacia hecatophylla, Acacia hockii, Acacia mellifera, Acacia oerfota, Acacia senegal, Acacia seyal* (Fig. 5-8), *Albizia malacophylla, Anogeissus leiocarpa* (Fig. 5-5), *Balanites aegyptiaca* (Fig. 5-1), *Boscia senegalensis, Boswellia papyrifera* (Fig. 5-6, Fig. 5-7, Fig. 5-9), *Boswellia pirottae, Bridelia scleroneura, Calotropis procera* (Fig. 5-14), *Carissa spinarum, Combretum collinum* (Fig. 5-9), *Combretum hartmannianum* (Fig. 5-3), *Combretum molle, Combretum rochetianum, Cordia africana, Croton macrostachyus, Dalbergia melanoxylon, Dichrostachys cinerea, Diospyros mespiliformis, Ficus glumosa, Ficus sycomorus* (Fig. 5-5), *Flacourtia indica, Flueggea virosa, Gardenia ternifolia, Grewia flavescens, Kigelia africana* (Fig. 5-15), *Lannea fruticosa* (Fig. 5-9, Fig. 5-10), *Lonchocarpus laxiflorus, Maerua oblongifolia, Maytenus senegalensis, Ochna leucophloeos, Oncoba spinosa, Oxytenanthera abyssinica, Piliostigma thonningii, Pterocarpus lucens, Schrebera alata, Sclerocarya birrea, Sterculia africana, Strychnos innocua, Terminalia brownii, Terminalia laxiflora, Terminalia schimperiana, Vitex doniana, Ximenia americana, Ziziphus abyssinica,* and *Ziziphus spina-christi.*

A.1. Branch from the Humera-Gondar road towards the north-east

To the east of the Humera-Gondar road are several chains of mountains from around the small towns of Birkutan and Adi Remot [Adi Ramez] and southwards, reaching an altitude of above 2000 m a.s.l. On a new road from the small town of Dansha [Dansheha] at ca. 900 m a.s.l. we studied a branch of the main profile up to an altitude of 1300 m a.s.l. but unfortunately we had to return before reaching the limit of the highlands. The woodlands from 900 m to 1300 m a.s.l. were all typical of the rocky landscapes with red or brown soils and many stones, with the grasses on the ground-cover mostly ca. 10 cm tall, occasionally up to 30 cm, but with no sign of burning. The vegetation contained the following woody species: *Acacia polyacantha* subsp. *campylacantha, Anogeissus leiocarpa, Combretum hartmannianum, Entada abyssinica, Erythrina abyssinica, Ficus glumosa, Lannea fruticosa, Lannea schimperi, Lonchocarpus laxiflorus, Maytenus senegalensis, Piliostigma thonningii, Pterocarpus lucens, Sterculia africana, Stereospermum kunthianum, Terminalia schimperiana, Vangueria madagascariensis,* and *Ziziphus spina-christi.*

A.2. Branch from Humera to Shire

Almost parallel with the border between Eritrea and Ethiopia there is a road branching off from the Humera-Gondar road ca. 5 km south of Humera at ca. 600 m a.s.l. and ending at the town of Shire [Enda Selassie] in the Tigray highlands at about 2000 m a.s.l. This road allowed us to make a profile in a west-east direction and to follow the southern boundary of the Kafta-Shiraro National Park. The profile began, as did the one towards Gondar, on vertisol with *Balanites aegyptiaca* woodland and an undergrowth of *Acacia* scrub (Fig. 5-1, where the scrub had been cleared; for the scrub, see Fig. 5-2). Further from Humera and at altitudes between 600 and 1100 m a.s.l., where there were flat rocky outcrops, large specimens of Baobab (*Adansonia digitata*) occurred, growing directly in cracks in the rock. Throughout this branch of the profile, although the layer of woody plants might be well developed (Fig. 5-9), there was either no ground-cover or a scarce ground-cover of mainly annual grasses 10-20 cm tall. The woody species were *Adansonia digitata, Albizia amara, Anogeissus leiocarpa, Boswellia papyrifera* (Fig. 5-7), *Capparis decidua* (Fig. 5-13), *Combretum collinum, C. hartmannianum, C. molle, Dalbergia melanoxylon, Dichrostachys cinerea, Ficus glumosa, Grewia bicolor, G. flavescens, G. tenax, G. villosa, Lannea fruticosa* (Fig. 5-9), *Ozoroa insignis, Pterocaropus lucens, Sclerocarya birrea, Sterculia africana, Terminalia brownii, T. schimperiana,* and *Ziziphus spina-christi.* As along the main profile from Humera to Gondar, there were areas on this branch with vertisol and *Acacia* bushland alternating with areas with rocky outcrops and woodlands dominated by a variety of species. Here there were also almost pure populations of *Calotropis procera* in areas that had obviously been very heavily grazed.

At about 85 km from Humera and 1100 m a.s.l. we observed areas with almost pure stands of *Boswellia papyrifera,* usually with marks on the trunk from harvesting of resin used for frankincense, but it was not possible to say if the trees formed a naturally pure population or whether species other than *Boswellia papyrifera* had been cut (Fig. 5-7).

Near Shire [Enda Selassie] there were rather intensively cultivated areas with few patches of remaining natural vegetation consisting of a mixture of Afromontane species and species typical of *Combretum-Terminalia* woodlands: *Albizia amara, Cordia africana, Croton macrostachyus, and Euphorbia abyssinica.* The remaining natural vegetation was only preserved on the rocky slopes, where there was also invasion of *Otostegia fruticosa* and *Dodonaea angustifolia.*

B: Profile of the escarpment and the north-western lowlands between Metema and Aykel

Profile B includes our relevés no. 36-53 (see Appendix 1), studied 24.-27.2.2015, and covering a line between the border with Sudan and the watershed at Aykel, with a branch from Genda Wuha (also known as Maganan, previously Shehedi or Shehibi) towards Tewodros Ketema (last town on the route of our survey on this profile). The road used for the branch from Gender Wuha past Tewodros Ketema continued ultimately to Shahura west of Lake Tana, where we studied another profile, recorded as Profile E. During our recording of relevés in this profile at the height of the dry season in February 2015 we noticed many fire-scars, and images of the woodland in this area are therefore included with the text on fires (Fig. 2-20, Fig. 2-21). Included in Profile B is also relevé 148, studied at Metema by Abeje Eshete Wassie (2011) and Haile Adamu Wale et al. (2012). The altitudinal range of the profile is 650-2100 m a.s.l. Also included, as relevé 148, is a dataset from Abeje Eshete Wassie (2011), recorded near Metema at 750 m a.s.l.

Near Metema, at ca. 700 m a.s.l., there were relatively flat areas with vertisols and dark brown, stony soils, both with *Balanites aegyptiaca* woodland as described from Profile A. The ground-cover consisted of short perennial grasses (*Hyparrhenia*, < 1 m tall) or the natural vegetation had been replaced by *Sesamum* fields. On the low hills and slopes there were short annual grasses (< 20 cm tall). There were burnt patches in the natural vegetation on the vertisol, but burning was not intense before we were nearer to the mountains beyond Genda Wuha. Due to the more various soils and the occasional rocky outcrops the area has a richer woody flora, containing: *Acacia polyacantha* subsp. *campylacantha, Acacia hecatophylla, A. hockii, A. seyal, Albizia malacophylla, Anogeissus leiocarpa, Balanites aegyptiaca, Boswellia papyrifera, B. pirottae, Combretum adenogonium, C. collinum, C. collinum, C. hartmannianum* (Fig. 5-17, Fig. 5-18), *C. molle, Cordia africana, Dalbergia melanoxylon, Dichrostachys cinerea, Diospyros mespiliformis, Entada africana, Ficus glumosa, F. sycomorus, Flueggea virosa, Gardenia ternifolia, Grewia bicolor, Lannea fruticosa, L. schimperi, Lonchocarpus laxiflorus* (Fig. 5-20), *Maytenus senegalensis, Ochna leucophloeos, Piliostigma thonningii, Pterocarpus lucens, Sterculia africana, Stereospermum kunthianum, Strychnos innocua, Terminalia laxiflora, Ximenia americana, Ziziphus abyssinica,* and *Ziziphus spina-christi* (Fig. 5-19).

Already a few kilometres east of Genda Wuha (Shehedi), the landscape became more undulating and the dominant vegetation was woodland on hilly ground with brown, stony soil, which verged into vertisol at the base of the hill-slopes. This pattern continued to ca. 70 km east of Genda Wuha (Shehedi), when the landscape at altitudes of ca. 1200 m a.s.l. became more mountainous and a steeper ascent towards the highland began. About 30 km to the east of Genda Wuha (Shehedi) the natural ground-cover was taller and with more perennial grass up to ca. 8 cm tall. Here and there were fire-scars in the natural ground-cover. Here *Pterocarpus lucens* became very common on the rocky ground and *Pseudocedrela kotschyi, Dombeya quinqueseta* and *Syzygium guineense* subsp. *macrocarpum* were seen for the first time in the relevés from northern Ethiopia: *Acacia hecatophylla, A. seyal, A. sieberiana, Anogeissus leiocarpa, Balanites aegyptiaca, Dichrostachys cinerea, Diospyros mespiliformis, Dombeya quinqueseta, Ficus ingens, Ficus sycomorus, Flueggea virosa, Gardenia ternifolia, Lannea fruticosa, Maytenus arbutifolia, Oncoba spinosa, Piliostigma thonningii, Pseudocedrela kotschyi,*

Pterocarpus lucens, Sterculia africana, Stereospermum kunthianum, Strychnos innocua, Syzygium guineense subsp. *macrocarpum, Terminalia laxiflora, T. schimperiana, Vernonia auriculifera, Vitex doniana, Ziziphus abyssinica,* and *Z. spina-christi.*

Fig. 5-1. Open fallow vegetation dominated by *Balanites aegyptiaca* at ca. 650 m a.s.l. (GD floristic region) on vertisol. Site south of Humera towards Gondar. The vegetation was previously open deciduous woodland with clumps of *Acacia* bushland, as could be seen from a few uncultivated patches, but the area is now used for cultivation of sesame (*Sesamum indicum* L.). Early dry season. Image 8359 by Ib Friis. Oct. 2014.

Fig. 5-2. *Acacia senegal* dominated scrub with lower stratum of *Acacia mellifera* at ca. 650 m a.s.l. (GD floristic region). Site with vertisol just south of Humera on the road to Gondar. Early dry season. Image 8356 by Ib Friis. Oct. 2014.

Fig. 5-3. Well-developed tree of *Combretum hartmannianum* south of Humera towards Gondar at ca. 700 m a.s.l. (GD floristic region). The smaller trees and shrubs in the background are *Ziziphus spina-christi*. Early dry season. Image 8361 by Ib Friis. Oct. 2014.

Fig. 5-4. Fruiting tree of *Acacia polyacantha* subsp. *campylacantha* south of Humera towards Gondar at ca. 700 m a.s.l. (GD floristic region). The vegetation is open wooded grassland dominated by *Acacia polyacantha* subsp. *campylacantha*. Early dry season. Image 8367 by Ib Friis. Oct. 2014.

Fig. 5-5. Mixed open woodland south of Humera towards Gondar at ca. 800 m a.s.l. (GD floristic region). The tree in the middle (middle ground) is *Ficus sycomorus*, to the right and left trees of *Anogeissus leiocarpa*. Early dry season. Image 8394 by Ib Friis. Oct. 2014.

Fig. 5-6. Open and slightly mixed *Boswellia papyrifera* woodland south of Humera at ca. 800 m a.s.l. (GD floristic region). The site is on the Humera-Gondar road. The substrate is reddish, rocky soil with very sparse ground-cover. Early dry season. Image 8428 by Ib Friis. Oct. 2014.

Fig. 5-7. Open and slightly mixed *Boswellia papyrifera* woodland east of Humera at ca. 800 m a.s.l. (GD floristic region). The site is on the Humera-Shire (Enda Selassie) road. The substrate is pale reddish soil, the ground-cover is sparse and grazed by goats. Some of the trees scarred from extraction of resin (frankincense). Early dry season. Image 8436 by Ib Friis. Oct. 2014.

Fig. 5-8. *Acacia seyal* scrub south of Humera towards Gondar at ca. 800 m a.s.l. (GD floristic region). The trees show reddish inner bark. The substrate is grey soil. Early dry season. Image 8444 by Ib Friis. Oct. 2017.

Fig. 5-9. Mixed open woodland at the Tacazze River to the east of Humera at ca. 750 m a.s.l. (GD floristic region). The site is at the bridge on the new road between Humera and Shire (Enda Selassie). *Combretum collinum*, *Boswellia papyrifera* and *Lannea fruticosa* common. The substrate is pale, grey soil. Early dry season. Image 8498 by Ib Friis. Oct. 2014.

Fig. 5-10. Mixed open woodland and wooded grassland in the Tacazze Valley at ca. 1450 m a.s.l. (TU floristic region). The site is on the Shire (Enda Selassie)-Gondar road. The substrate is pale and stony red soil. In the foreground shrubs or small trees of *Lannea fruticosa*. Early dry season. Image 8825 by Ib Friis. Oct. 2009.

Fig. 5-11. Mixed open woodland and wooded grassland south of Enda Selassie at ca. 1350 m a.s.l. (TU floristic region). The site is on the upper south-facing slopes of the Tacazze Valley on the Shire (Enda Selassie)-Gondar road. The small tree in the centre foreground is *Acacia amythethophylla*. Dry season. Image 1475A by Ib Friis. Oct. 2005.

Fig. 5-12. Flowering *Acacia amythethophylla* south of Enda Selassie at ca. 1350 m a.s.l. (TU floristic region). The site is on the upper south-facing slopes of the Tacazze Valley on the Shire (Enda Selassie)-Gondar road. Image 8832 by Ib Friis. Oct. 2009.

Fig. 5-13. The almost leafless *Capparis decidua* in flower at ca. 800 m a.s.l. (TU floristic region). Early dry season. Image 5497 by Ib Friis. Oct. 2006.

Fig. 5-14. Flowering *Calotropis procera* at ca. 800 m a.s.l. (TU floristic region). Image 5667 by Ib Friis. Dry season. Nov. 2006.

Fig. 5-15. An old fruiting tree of *Kigelia africana* on alluvial soil near the Angereb River at ca. 850 m a.s.l. (GD floristic region). The site is near the bridge where the road Humera-Gondar crosses the Angereb River. The fruits of *Kigelia africana* are 30-60 cm long, partly lignified berries hanging in elongated peduncles. Early dry season. Image 8784 by Ib Friis. Nov. 2014.

4

5

6

7

8

9

10

11

12

13

14

15

The last ascent to Aykel was steep, and at ca. 2000 m a.s.l. and above the woodland was Afromontane with these woody plants: *Acacia abyssinica, Bersama abyssinica, Brucea antidysenterica, Croton macrostachyus, Euphorbia ampliphylla, Ficus ingens, Ficus sycomorus, Maytenus arbutifolia,* and *Vernonia auriculifera.*

As a supplement to this profile, we have added a photograph of the woodlands on the northern shore of Lake Tana, which we had visited on a previous trip, but where we were not able to record a relevé (Fig. 5-16). The area is intensively cultivated, there is an old monastery, the Mandaba Monastery, which has both protected the woody plants and cultivated fields in the natural woodland for centuries; the woody vegetation is now dominated by species of *Acacia, Rhus* and *Maytenus.*

B.1. Branch to the south of the Metema-Aykel road beyond Tewodros Ketema

This branch of our profile B.1 between Metema and Aykel, with relevés 45-50, followed part of the road passable by vehicles through the frontier district of Qwara, forming a big curve through the area, starting from Genda Wuha (also known as Shehedi or Maganan) and ending up at Shahura in the district of Alefa to the west of Lake Tana. For practical reasons we could not work further than ca. 15 km south of Tewodros Ketema, ca. 120 km from Genda Wuha (Shehedi), with a relevé near the border with the Beni Shangul-Gumuz Regional State. The other end of the road west of Shahura was covered by our Profile E. The altitudinal range of the entire profile was 600-900 m a.s.l.

Most of the area consists of a mosaic of large, flat areas with vertisol or brown, stony soil and rocky areas of very variable sizes, from small outcrops to large, flat rocks. In the rocky areas the ground-cover consisted of perennial grasses (< 50 cm tall), and on the level ground with vertisol there were perennial grasses (ca. 80-150 cm tall). In the areas with tall grass, there were scars of burning, some very extensive. The complexity of this mosaic makes it difficult to give separate species lists for the vertisols and the rocks: *Acacia hecatophylla, Acacia nilotica, A. seyal, A. sieberiana, Adansonia digitata, Anogeissus leiocarpa, Balanites aegyptiaca, Boswellia papyrifera, Combretum collinum, C. hartmannianum, C. molle, C. rochetianum, Dalbergia melanoxylon, Entada africana, Ficus glumosa, F. ingens, F. sycomorus, Grewia flavescens, Lannea fruticosa, L. schimperi, Lonchocarpus laxiflorus, Piliostigma thonningii, Pterocarpus lucens, Sterculia africana, Stereospermum kunthianum, Strychnos innocua, Terminalia laxiflora, T. schimperiana,* and *Ziziphus spina-christi.*

Fig. 5-16. Mixed open, partly cultivated woodland and wooded grassland on the north shore of Lake Tana at ca. 1780 m a.s.l. (GD floristic region). The site, which is not included in the analyses in this work, is on the slopes towards the lake to the west of Gorgora. On the peninsula in the lake the monastery of Mandaba (14th century). Dry season. Image 4680 by Ib Friis. Dec. 2011.

Fig. 5-17. Well-developed tree of *Combretum hartmannianum* south-west of Shehedi at ca. 750 m a.s.l. (GD floristic region). The tree, which grows in heavily grazed mixed open woodland and wooded grassland, is partly losing its leaves. Height of dry season. Image 9303 by Ib Friis. Feb. 2015.

Fig. 5-18. Leaves (image above) and fruits (image below) of *Combretum hartmannianum* south-west of Shehedi at ca. 750 m a.s.l. (GD floristic region). Height of dry season. Image 9305 and 9306 by Ib Friis. Feb. 2015.

Fig. 5-19. Shrub of *Ziziphus spina-christi* south-west of Shehedi ca. 750 m a.s.l. (GD floristic region). The shrub forms clumps in mixed open woodland and wooded grassland. Height of dry season. Image 9312 by Ib Friis. Feb. 2015.

Fig. 5-20. Flowering (image above) and almost leafless (image below) specimens of *Lonchocarpus laxiflorus* south-west of Shehedi at ca. 750 m a.s.l. (GD floristic region). The tree flowers mainly in the dry season. The image above shows detail of canopy. The vegetation is mixed open woodland or wooded grassland on grey soil. The short undergrowth was heavily grazed by cows. Note the thickened, fire resistant bark on the trunk of *Lonchocarpus.* Both images taken at the height of dry season. Image 9325 & 9327 by Ib Friis. Feb. 2015.

16

17

18

19

20

C: Profile of the Tacazze Valley north-west of Sekota

Profile C in the upper part of the Tacazze Valley includes our relevés no. 144-146 (see Appendix 1), studied 20.11.2018 between Sekota and the Tacazze River. Included here is also relevé no. 147 from near Abergelle north of Sekota and just north of the Tacazze Valley, studied by Abeje Eshete Wassie in 2011. The altitudinal range is 1150-1600 m a.s.l. Beginning from the bridge on the Tacazze River leading from Sekota to the Semien Mountains, the vegetation was very open and poor in species, in many places verging towards subdesert scrubland with trees. The soil was nearly white and very hard, with many stones. The ground-cover was sparse, always less than ca. 20 cm tall, and there were no signs of burning (Fig. 5-23).

From the shores of the river and upwards, the vegetation was dominated by *Acacia asak* on moderately steep slopes. Other species, few of them typical of the *Combretum-Terminalia* woodlands, were: *Acacia mellifera, Adansonia digitata, Albizia amara, Balanites aegyptiaca, Boscia angustifolia, B. senegalensis, Calotropis procera, Capparis decidua, Ficus cordata* subsp. *salicifolia, Grewia tenax, Salvadora persica, Stereospermum kunthianum, Terminalia brownii* and *Ziziphus spina-christi.*

Above the relatively steep-sided inner gorge, there was a wide outer gorge, which was partly cultivated with drought-resistant crops, mainly of *Sorghum*. In the fields there were trees remaining from the woodland, mainly *Albizia amara* and *Stereospermum kunthianum* (Fig. 5-22). At the highest level of the gorge the soil was almost completely bare, and only very few trees were seen, for example isolated trees of *Combretum molle* (Fig. 5-21).

At Abergele at 1600 m a.s.l. and ca. 45 km north of the above locality, the woody flora was richer, but still with relatively few species typical of the *Combretum-Terminalia* woodlands: *Acacia abyssinica, A. etbaica, A. mellifera, A. oerfota, Boswellia papyrifera, Capparis decidua, Combretum hartmannianum, Commiphora africana, Dichrostachys cinerea, Grewia erythraea, G. villosa, Lannea fruticosa, L. triphylla, Maerua angolensis, Salvadora persica, Senna singueana, Stereospermum kunthianum,* and *Terminalia brownii.*

Given the relatively few species typical of the *Combretum-Terminalia* woodlands in these relevés, it might seem more natural to place them in the vegetation type 'Semi-Desert Bushland.' However, noting the presence of such species typical of the *Combretum-Terminalia* woodland as *Albizia amara, Balanites aegyptiaca, Stereospermum kunthianum,* and *Terminalia brownii,*

Fig. 5-21. *Combretum molle* at the upper edge of the Tacazze Valley near Sekota at ca. 1600 m a.s.l. (WU floristic region). The vegetation is completely degraded woodland. The substrate is a thin layer of reddish soil over basaltic rocks. Image 4797 by Ib Friis. Early dry season. Nov. 2018.

Fig. 5-22. Open, partly cultivated woodland below the upper edge of the Tacazze Valley at ca. 1350 m a.s.l. (WU floristic region). The site is on the track between Sekota and a bridge on the Tacazze River. The dominant tree is *Albizia amara*. Early dry season. Image 4814 by Ib Friis. Nov. 2018.

Fig. 5-23. Lowest part of the Tacazze Valley near the bridge on the road from Sekota to the Semien Mountains at ca. 1150 m a.s.l. (WU floristic region). Scattered clumps and shrubs of *Acacia mellifera* and others woody species; here with a large tree of *Adansonia digitata*. Early dry season. Image 4806 by Ib Friis. Nov. 2018.

Fig. 5-24. Degraded woodland at the uppermost reaches of the Tacazze Valley at ca. 1800 m a.s.l. (WU floristic region). This site with woodland and wooded grassland is south of Lalibela on the road towards Gashena. In foreground *Acacia etbaica, A. tortilis, A. seyal*, etc.; in the background the same species, but with *Combretum molle, C. collinum*, and *Stereospermum kunthianum*. The substrate is a thin layer of grey, stony soil over rocks. Early dry season. Image 4793 by Ib Friis. Nov. 2018.

Fig. 5-25. Woodland on slope of the western escarpment at Alefa, ca. 1400 m a.s.l. (GD floristic region). The trees in foreground are *Boswellia pirottae*, the white-flowered subshrubs below the trees are *Barleria grandis* (Acanthaceae), an endemic in the woodlands of the large, western river valleys. The substrate is a thin layer of grey, stony soil over basaltic rocks. Early dry season. Image 4765 by Ib Friis. Nov. 2018.

21

22

23

24

25

we find it more logical to classify the vegetation as a very dry and open form of the *Combretum-Terminalia* woodland (CTW). The very dry relevés in the upper reaches of the Tacazze and Abay River Valleys, including no. 143 in profile D, form the third cluster in our analyses in Section 10.

D: Upper reach of Tacazze Valley near Lalibela

Profile D from the uppermost reaches of the Tacazze Valley included only one relevé, no. 143 (see Appendix 1), which was studied 18.11.2018, where the road from the main road (between Woldiya and Debre Tabor) towards Lalibela crosses the valley. Altitude 1850 m a.s.l.

The vegetation in the immediate surrounding of the river was disturbed by cutting of trees and cattle being taken for watering at the river. The sparse ground was therefore heavily grazed, and there were no signs of burning. Among the few trees were: *Acacia abyssinica, A. etbaica, A. seyal, A. tortilis, Senna petersiana,* and *Ziziphus spina-christi*, very few of them typical of *Combretum-Terminalia* woodland (Fig. 5-24).

The narrow valley was surrounded by very open scrub with evergreen shrubs like *Euclea racemosa* subsp. *schimperi* and *Dodonaea angustifolia.* At higher altitudes and further away from the river, *Acacia etbaica* was also very common. In many places between Lalibela and Sekota this semi-evergreen bushland has a very scattered tree-component of *Combretum molle* and (less common) *Combretum collinum*, *Stereospermum kunthianum*, *Lannea fruticosa, Cordia africana* and *Croton macrostachyus,* indicating a mixture between species normally found in *Afromontane evergreen bushland* and species normally occurring in *Combretum-Terminalia* woodland. As for Profile C, it might seem more natural to place this relevé in the vegetation type 'Semi-Desert Bushland.' However, noting the presence in the Tacazze Valley near Lalibela of such species typical of the *Combretum-Terminalia* woodland as *Combretum molle*, *C. collinum*, *Stereospermum kunthianum*, and *Lannea fruticosa*, we find it more logical to classify the vegetation as a very dry and open form of the *Combretum-Terminalia* woodland.

E: Profile of the western escarpment at Shahura

Profile E includes our relevés no. 137-142 (see Appendix 1), studied 15.11.2018 between Shahura in the Alefa Highlands west of Lake Tana and the lowlands towards Tewodros Ketema and to the south of Genda Wuha (Shehedi), which was studied as part of our Profile B. Altitudinal range 1100-1600 m a.s.l. There is about 70 km between the relevé at 750 m a.s.l. southeast of Tewodros Ketema in the sequence of relevés no. 45-50, and the relevé no. 137 at 1100 m a.s.l. at the lowest parts of the western escarpment in this profile.

The vegetation at the lowest parts of the western escarpment at Shahura was a relatively dense woodland on hilly ground. The ground-cover consisted mainly of annual grasses and herbs, up to ca. 20 cm tall, and there was no sign of burning. The woody flora included mainly species typical of *Combretum-Terminalia* woodlands: *Acacia polyacantha* subsp. *campylacantha, A. seyal, Anogeissus leiocarpa, Combretum collinum, Ficus sycomorus, Flueggea virosa, Lannea schimperi, Lonchocarpus laxiflorus, Pterocarpus lucens, Sterculia africana, Terminalia schimperiana,* and *Ziziphus spina-christi.* New in comparison with the profiles further north was *Allophylus rubifolius,* which is also represented in profiles at higher altitudes and further south.

In the relevés at 1300-1500 m a.s.l. the ground-cover consisted in places of more than 2 m tall grasses (*Loudetia*, *Hyparrhenia*) and herbs, and the vegetation was highly prone to fierce grass fires (Fig. 5-25). The woody vegetation at the relevés up to 1500 m a.s.l. was relatively rich in woody species typical of the *Combretum-Terminalia* woodlands: *Acacia polyacantha* subsp. *campylacantha, A. seyal, Albizia malacophylla, Allophylus rubifolius, Anogeissus leiocarpa, Boswellia papyrifera, B. pirottae, Combretum collinum, C. hartmannianum, Cordia africana, Erythrina abyssinica, Ficus sycomorus, Flueggea virosa, Hymenodictyon floribundum, Lannea fruticosa, Lonchocarpus laxiflorus, Piliostigma thonningii, Pterocarpus lucens, Rhus vulgaris, Rumex abyssinicus, Sarcocephalus latifolius, Sterculia africana, Syzygium guineense* subsp. *macrocarpum, Terminalia schimperiana, Trema orientalis, Vitex doniana, Ziziphus*

Fig. 5-26. Clump of the palm *Borassus aethiopum* west of Mambuk (sometimes spelt Mambruk) at ca. 950 m a.s.l. (GJ floristic region). The site is on the road between Chagni and Guba (Mankush). *Borassus aethiopum* is most frequent in moist places in mixed woodland. To the left various species of *Acacia*, the dominant group of plants in the area. Early dry season. Image 3437 by Ib Friis. Jun. 2017.

Fig. 5-27. Clump of the palm *Hyphaene thebaica* in mixed woodland west of Mambuk (sometimes spelt Mambruk) at ca. 950 m a.s.l. (GJ floristic region). The site is on the road between Chagni and Guba (Mankush). The tall trees to the right are *Anogeissus leiocarpa*. Early dry season. Image 1673 by Ib Friis. Oct. 2010.

Fig. 5-28. Mixed open woodland west of Mambuk (sometimes spelt Mambruk) at ca. 950 m a.s.l. (GJ floristic region). The site is on the road between Chagni and Guba (Mankush). In the foreground a small tree of *Balanites aegyptiaca*; most of the small trees with reddish bark are *Acacia seyal*. The forking palm is *Hyphaene thebaica*. The substrate is pale reddish soil with low, already withered ground-cover. Early dry season. Image 1603 by Ib Friis. Oct. 2010.

Fig. 5-29. Dense mixed woodland on red, stony soil west of Mambuk (sometimes spelt Mambruk) at ca. 950 m a.s.l. (GJ floristic region). The site is on the road between Chagni and Guba (Mankush). Early dry season. Image 3429 by Ib Friis. Jun. 2017.

Fig. 5-30. Hill with mixed woodland in the lowland near Guba at ca. 700 m a.s.l. (GJ floristic region). The site is on the road between Chagni and Guba (Mankush). In the foreground to the left a small clump of bamboo, *Oxytenanthera abyssinica*. Although still early after the end of the rains, the ground-cover is already withering and the leaves of deciduous trees on the hill are turning yellow or brown. Early dry season. Image 1685 by Ib Friis. Oct. 2010.

Fig. 5-31. Flowers of *Boswellia papyrifera* near Guba at ca. 700 m a.s.l. (GJ floristic region). The site is on the road between Chagni and Guba (Mankush). *Boswellia papyrifera* flowers when leafless during the dry season. Early dry season. Image 1635 by Ib Friis. Oct. 2010.

Fig. 5-32. *Adansonia digitata* in the lowland near Guba at ca. 700 m a.s.l. (GJ floristic region). The site is on the road between Chagni and Guba (Mankush). The vegetation is mixed woodland; most of the ground-cover is dry, but green in the shade under the tree. Early dry season. Image 1690 by Ib Friis. Oct. 2010.

Fig. 5-33. *Terminalia macroptera* near Mambuk (sometimes spelt Mambruk) at ca. 950 m a.s.l. (GJ floristic region). Image above: Detached fruiting branchlet. Image below: Intact tree near human habitation. The site is on the road between Chagni and Guba (Mankush). The vegetation is partly farmed mixed woodland on reddish soil. Early dry season. Image 3403 and 3405 by Ib Friis. Jun. 2017.

Fig. 5-34. Fruiting palm of *Borassus aethiopum* near Debre Zeyit (Wenbera) at ca. 950 m a.s.l. (GJ floristic region). The site is on the road between Chagni and Debre Zeyit (Wenbera). The vegetation is in mixed open woodland and wooded grassland. Early dry season. Image 1418 by Ib Friis. Oct. 2010.

Fig. 5-35. Fruiting trees and shrubs of *Sarcocephalus latifolius* near Debre Zeyit (Wenbera) at ca. 1400 m a.s.l. (GJ floristic region). Image above: Branches with fruiting inflorescences (syncarps). Image below: Fruiting shrub. The site is on the road between Chagni and Debre Zeyit (Wenbera). The vegetation is mixed open woodland and wooded grassland. Early dry season. Image 1472 and 1475 by Ib Friis. Oct. 2010.

Fig. 5-36. Fruiting *Terminalia laxiflora* near Debre Zeyit (Wenbera) at ca. 1400 m a.s.l. (GJ floristic region). The site is on the road between Chagni and Debre Zeyit (Wenbera). The vegetation was mixed open woodland and wooded grassland. Early dry season. Image 1481 by Ib Friis. Oct. 2010.

Fig. 5-37. Much degraded, open woodland between Mekane Selam and Mertule Maryam at ca. 1550 m a.s.l. (near border between WU and GJ floristic region). The site is located on the west-facing escarpment of the Abay Valley in the big bend of the river. *Albizia amara* is the only common tree in this vegetation. Rainy season. Image 3248 by Ib Friis. Jun. 2017.

26

27

28

29

30

31

32

33

34

35

37

abyssinica, Z. mucronata, and Z. spina-christi. Woodfordia uniflora occurred on the steep rocky slopes and was not seen in other profiles. *Sarcocephalus latifolius* was a new species in relation to the profiles studied further north but was seen to be common in the basin of the Abay River.

Throughout the profile, the vegetation looked rather degraded, at least at some levels; this was probably due to heavy grazing, which was confirmed by the frequent presence of *Calotropis procera. Rumex abyssinicus*, which is rather common in degraded areas in the Ethiopian Highlands. Here it occurred as an invasive species in several relevés, like in places in the Abay valley in profile G. *Lantana camara* was observed, but only as a casual, invasive weed and not recorded.

At 1600 and 1800 m a.s.l., relevés were studied a few hundred meters below a village. These relevés were the highest with representative species of the *Combretum-Terminalia* woodlands in the profile, but the combretaceous species were mixed with Afromontane species, and not prone to grass fires: *Anogeissus leiocarpa, Cordia africana, Dichrostachys cinerea, Diospyros abyssinica, Embelia schimperi, Ficus vasta, Flueggea virosa, Sarcocephalus latifolius, Syzygium guineense* subsp. *macrocarpum,* and *Terminalia schimperiana.*

F: Profile of the western escarpment at Manduria and continuation towards the lowlands along the Abay River

Profile F includes our relevés no. 54-59 (see Appendix 1), studied 22.6.2017 on the road between Chagni in the Ethiopian Highlands and Mankush (Goba) near the border with Sudan, passing the area of Manduria, with a sizeable village with the same name. When we had planned to work along this road, it was partly inaccessible due to both the intensive work and the strict security on the building site of the GERD Dam on the Abay River. The altitudinal range of the entire profile was 930-1400 m a.s.l.

As it was impossible for political reasons to reach the lowest point near the Sudan border, the observations in this profile were started from the top, west of the village of Manduria at 1400 m a.s.l., where the vegetation was open woodland with *Sarcocephalus latifolius* (Fig. 5-35) as the completely dominant woody species, growing with a number of typical woodland species on a moderately steep rocky slope with brown soil rich in stones. The ground-cover consisted of low perennial grasses and other annual and perennial herbs, in most areas not very tall or dense due to shading from the dense bush of *Sarcocephalus*. There was no sign of burning. The woody species were: *Albizia amara, Anogeissus leiocarpa, Cordia africana, Ficus sur, Flueggea virosa, Lannea schimperi, Lonchocarpus laxiflorus, Oxytenanthera abyssinica, Piliostigma thonningii, Pterocarpus lucens, Sarcocephalus latifolius,* and *Terminalia schimperiana.*

The next level downwards was on the river-bank of the Gilgil Beles River, at 1050 m a.s.l., where the absolutely dominant species was *Acacia polyacantha* subsp. *campylacantha;* apart from that species, only a few specimens of *Cordia africana* and *Terminalia laxiflora* were observed. Here the soil was dark brown to almost black and, due to the proximity of the river, it seemed as if it would remain damp for most of the year. The ground-cover consisted of short annual and perennial grasses (< 50 cm tall), with no sign of burning.

The subsequent four relevés were all from west of the village of Mambuk, from 7 km to 37 km to the west on relatively flat terrain, the altitudes of which varied from 1150 to 930 m a.s.l. The soils were vertisol, dark brown soil, but there were also areas with little soil at rocky outcrops (Fig. 5-26 to Fig. 5-29). The ground-cover varied remarkably from relevé to relevé, partly due to the terrain, partly due to human interference. In areas with extensive cultivation there were a few patches with natural vegetation where the ground-cover consisted of tall perennial grasses ca. 80 cm tall, and there were scars from burning. In areas with rocky outcrops the ground-cover was sparse, with hardly any grass, but the climber *Tylosema fassoglensis* was common, and there was no sign of burning. The woody vegetation consisted of the relatively typical *Combretum-Terminalia* species, with the highest diversity on the rocky outcrops. The presence of the palms

Borassus aethiopum (Fig. 5-26, Fig. 5-34) and *Hyphaene thebaica* (Fig. 5-27, Fig. 5-28) was characteristic of the better preserved natural vegetation, but no palm species occurred in our relevés:

Acacia hecatophylla, Acacia seyal (Fig. 5-28), *Albizia amara, Annona senegalensis, Anogeissus leiocarpa* (Fig. 5-27), *Balanites aegyptiaca* (Fig. 5-28), *Boswellia pirottae, Combretum collinum, Cordia africana, Crossopteryx febrifuga, Entada africana, Ficus sur, F. sycomorus, Flueggea virosa, Gardenia ternifolia, Grewia mollis, Lannea schimperi, Lonchocarpus laxiflorus, Maytenus senegalensis, Ochna leucophloeos, Oxytenanthera abyssinica, Piliostigma thonningii, Polyscias farinosa, Pseudocedrela kotschyi, Pterocarpus lucens, Sarcocephalus latifolius* (Fig. 5-35), *Securidaca longipedunculata, Sterculia africana, Stereospermum kunthianum, Terminalia laxiflora* (Fig. 5-36), *T. macroptera* (Fig. 5-33), *T. schimperiana*, and *Ziziphus abyssinica*. Not previously observed in relation to the more northern profiles are *Annona senegalensis, Crossopteryx febrifuga,* and *Polyscias farinosa*, and, outside the relevés, the palms *Borassus aethiopum* and *Hyphaene thebaica*.

From visits to the area around Guba (Mankush) in 2010, where the vegetation appeared more open and drier then near Mambuk (sometimes spelt Mambruk; Fig. *5-30*), we had observed *Adansonia digitata* (Fig. 5-32) and *Oxytenanthera abyssinica* (Fig. 5-30) in approximately the same area, as well as large stands of *Boswellia papyrifera* (Fig. 5-31).

In 2017 it was not possible to repeat a visit made in 2010 to the road between Chagni and Debre Zeyit (Wenbera), where we had observed stands of *Sarcocephalus latifolius* (Fig. 5-35), *Borassus aethiopum* (Fig. 5-34) and *Terminalia laxiflora* (Fig. 5-36).

G: Abay Gorge between Mekane Selam and Mertule Maryam

Profile G includes only one relevé, no. 90 (see Appendix 1), studied 9.6.2017 on the eastern slope of the Abay Gorge between Mekane Selam to the east of the river and Mertule Maryam to the west of the river. Altitude 1550 m a.s.l. The road across this eastern part of the Abay Gorge was mostly following the highlands as close to the river as possible, but descended from the eastern side into the valley of a tributary to the Abay with an open, uniform woodland vegetation completely dominated by *Albizia amara*, before finally reaching the bridge on the Abay River via a sequence of hairpins on very steep slopes, where road work had destroyed the woody vegetation. Therefore, we were only able to record the vegetation of one relevé, typical of the valley of the tributary at ca. 1550 m a.s.l.

The vegetation was open woodland on rocky slopes, dominated by *Albizia amara* (Fig. 5-37). There was hardly any soil or ground-cover, except for large specimens of *Rumex* (*R. nervosus, R. abyssinicus*). There was no sign of burning. The woody species observed in this vegetation were *Acacia mellifera, A. senegal, Albizia amara, and Boscia salicifolia.*

H: Profile of the Abay Gorge between Bure and Nekemt

Profile H includes our relevés no. 126-136 (see Appendix 1), studied 10.-12.11.2018 on the northern slope of the Abay Gorge. Altitudinal range 850-1850 m a.s.l. This gorge has a wide upper part with deciduous woodland, villages, and farming at the altitudinal range of 1650-1850 m a.s.l. (relevés no. 126-128), and a narrow inner gorge with deciduous woodland at an altitudinal range of 850-1650 m a.s.l. The relevés in the inner gorge fall in three groups; from the upper edge to ca. half way down (relevés no. 129-131), from ca. half way down to the steep slopes above the bottom (relevés 132-134), and finally the rocky slopes near the river (relevés no. 135-136). Because of our approach to the gorge from the highlands, the profile was studied in the sequence from the highest to the lowest parts.

H.1. The wide upper gorge

The vegetation in the uppermost part of the gorge consisted of mixed, open or sometimes dense woodland on gently sloping terrain with many rocks and larger areas of rocky outcrops (Fig. 5-38). There were several villages with extensive farmland at this level of the gorge. The soil was red and stony, and the ground-cover consists of short perennial grasses up to 20 cm tall, probably grazed, but with no sign of burning. The flora of woody species was rich, mainly in species of typical *Combretum-Terminalia* woodland, but also with some species that are also widespread in Afromontane vegetation: *Acacia amythethophylla, A. gerrardii* (Fig. 5-38), *A. persiciflora, A. seyal, Albizia malacophylla, Anogeissus leiocarpa* (Fig. 5-38), *Capparis tomentosa, Combretum collinum, C. molle, Cordia africana, Croton macrostachyus, Cussonia ostinii, Dichrostachys cinerea, Dombeya quinqueseta, Erythrina abyssinica, Ficus glumosa, F. sycomorus, F. thonningii, F. vasta, Flueggea virosa, Gardenia ternifolia, Grewia ferruginea, G. trichocarpa, Heteromorpha arborescens, Hymenodictyon floribundum, Lannea fruticosa, L. schimperi, Lonchocarpus laxiflorus, Maytenus senegalensis, Ormocarpum pubescens, Ozoroa insignis, Pavetta oliveriana, Piliostigma thonningii, Polyscias farinosa, Premna schimperi, Rhus vulgaris, Schrebera alata, Senna petersiana, Sterculia africana, Strychnos innocua,* and *Vangueria madagascariensis.*

H.2. The upper part of the inner gorge from the upper edge to ca. half way down

The vegetation of this part of the gorge was studied on steeply sloping ground with grey, stony soil, consisting of fine gravel derived from crystalline rocks. The non-native and invasive subshrub *Chamaecrista rotundifolia* (Pers.) Greene (native of Central America, new to Ethiopia) formed almost monospecific stands in the ground-cover, while other places had patches of up to 2 m tall perennial grasses. However, there was no sign of burning. The layer of woody plants was a mixed, open woodland with the trees mostly typical of the *Combretum-Terminalia* woodland. A new CTW-element in relation to the profiles further north was *Acacia venosa* (which we had previously collected as far north as in woodland near Lake Tana, but not observed in Profile E). However, in some relevés all species of the genera *Combretum* and *Terminalia* were absent, although *Anogeissus leiocarpa* was present. The woody species in this part of the profile were: *Acacia persiciflora, A. seyal, A. venosa, Anogeissus leiocarpa, Boswellia papyrifera, B. pirottae, Combretum collinum, Dichrostachys cinerea, Entada africana, Faurea rochetiana, Ficus glumosa, F. sycomorus, Flueggea virosa, Hymenodictyon floribundum, Lannea fruticosa, Maerua angolensis, Ochna leucophloeos, Ozoroa pulcherrima, Rhus vulgaris, Senna petersiana, Stereospermum kunthianum, Strychnos innocua, Terminalia laxiflora, Vangueria madagascariensis,* and *Ziziphus mucronata.*

H.3. The lower part of the inner gorge from ca. halfway down to near the river

This part of the gorge consists of steep rocky slopes with grey, stony soil, supporting an open woodland (Fig. 5-39). The ground-cover consisted of ca. 2 m tall perennial grasses (*Hyparrhenia* and *Loudetia*). Burning of this ground-cover is liable to produce fierce grass fires, but no fire-scar was observed in this part of the gorge. New woody plants in relation to the specie in the profiles further north were *Maerua angolensis* and *Ormocarpum pubescens.* The woody species on these steep slopes were: *Acacia seyal, Albizia amara, Anogeissus leiocarpa, Boswellia papyrifera, B. pirottae* (both species of *Boswellia* seen from this site in Fig. 5-42), *Combretum collinum, C. molle, Dalbergia melanoxylon, Dichrostachys cinerea, Erythrina abyssinica, Ficus glumosa, Flueggea virosa, Grewia trichocarpa, Lannea fruticosa, Lonchocarpus laxiflorus, Maerua angolensis, Ormocarpum pubescens, Senna petersiana, Sterculia africana* (Fig. 5-40, Fig 5-41), *Stereospermum kunthianum, Terminalia schimperiana,* and *Ziziphus spina-christi. Ziziphus abyssinica* was also seen, but not in the relevés (Fig. 5-43).

H.4. The lowest level of the inner gorge

This part of the gorge consists of steep rocky slopes with very thin soil over the rocks and an open woodland. The ground-cover consisted of ca. 2 m tall grasses, *Hyparrhenia* and *Loudetia.* Our observations were made early in the fire season, but the grass was already burnt in one spot, leaving a very large fire-scar. During earlier visits, which we had made at other times of the year, nearly all grass in the lowermost part of the Abay gorge had been burnt. Some woody species, which seemed distinct from those listed here, were leafless and sterile and out of reach on the steep slopes. The woody species that could be identified were: *Acacia amythethophylla, Anogeissus leiocarpa, Boswellia papyrifera, Calotropis procera, Cassia arereh, Combretum collinum, C. molle, Dalbergia melanoxylon, Dichrostachys cinerea, Ficus glumosa, Ficus cordata* subsp. *salicifolia, Flueggea virosa, Grewia trichocarpa, Hymenodictyon floribundum, Lannea barteri* (Fig. 5-41), *L. fruticosa, L. schimperi, Lonchocarpus laxiflorus, Pterocarpus lucens, Sterculia africana* (Fig. 5-41), *Stereospermum kunthianum, Strychnos innocua, Tamarindus indica, Terminalia laxiflora, T. schimperiana,* and *Ziziphus spina-christi.*

J: Profiles radiating from Assosa towards the Abay River, Gizen, Kurmuk, Bambasi and Tongo

This profile J includes three branches radiating in different directions from the town of Assosa.

A north-eastern branch, J.1, including our relevés no. 60-68 (see Appendix 1), studied 23.-25.6.2017 between the GERD Dam and the new lower bridge on the Abay River (partly inaccessible due to construction work and measures for tight security) via Sherkole, Mengi and Homesha towards Assosa (Fig. 5-45 to Fig. 5-46, Fig. 5-49, Fig. 5-50, Fig. 5-53 to Fig. 5-56, Fig. 5-61), but also including the road to the border crossing into the Sudan at Gizen (Fig. 2-26), which was visited in 2005. However, no relevés from near Gizen were studied in 2017, but included here are also three relevés (no. 80-82) with information gathered during field work in 2015 by Mindaye Teshome and Abeje Eshete and published two years later (Mindaye Teshome et al. 2017): relevé no. 80, at Ashefabego west of the road from Mengi to the Sudan border at Gizen, relevé no. 81 at Arenja, on the new road from Mengi to Guba (Mankush), just across the Abay River (altitude 920 m), and relevé no. 82 at Baneshegol, on the road north of Mengi towards Gizen (altitude 700 m). Altitudinal range 675-1600 m a.s.l.

A south-western branch, J.2, including our relevés no. 69-73 (see Appendix 1), studied 26.6.2017 between

Fig. 5-38. Fruiting trees in the outer gorge of the Abay Gorge south of Bure at ca. 1700 m a.s.l. (GJ floristic region). The tree to the right is *Anogeissus leiocarpa.* The tree to the left is *Acacia gerrardii.* The vegetation of rather dense and mixed woodland occurs on the south-facing side of the gorge. Early dry season. Image 4674 by Ib Friis. Nov. 2018.

Fig. 5-39. The inner Abay Gorge on the road between Bure and Nekemt at ca. 900 m a.s.l. Image above: the inner gorge on the north-facing side just above the riverbed, with mixed open woodland and wooded grassland. The tree to the left in the upper image is *Lannea schimperi.* Image below: the riverbed with mixed open woodland and wooded grassland, seen from the north-facing side of the gorge. The river forms the boundary between the floristic regions GJ (to the left) and WG (to the right). Image 1367 & 1382 by Ib Friis. Early dry season. Oct. 2010.

Fig. 5-40. *Sterculia africana* in mixed open woodland near the Abay Gorge at ca. 950 m a.s.l.(WG floristic region). The site is just above the inner gorge on the side facing north on the road between Bure and Nekemt. Image 1394 by Ib Friis. Oct. 2010.

Fig. 5-41. Open *Lannea barteri-Sterculia africana* dominated woodland at the Abay River at ca. 850 m a.s.l. (GJ floristic region). Image above: woodland on crests of the low hills above the river. Image below: woodland near the river. The site is south of Bure on the road between Bure and Nekemt. The vegetation on the south-facing side of the gorge consists of a mixture of evergreen and deciduous species on rocks with very thin, grey soil and slight ground-cover that does burn in the dry season. Image 4707 & 4723 by Ib Friis. Nov. 2018.

Fig. 5-42. *Boswellia papyrifera* (left tree) and *B. pirottae* (right tree) in the Abay Gorge south of Bure at ca. 950 m a.s.l. (GJ floristic region). The two species grow together on the south-facing side of the gorge in mixed open woodland in the lower part of the inner gorge and reach about the same height. The site is near the bottom of the Abay Gorge south of Bure on the road between Bure and Nekemt. Early dry season. Image 4727 by Ib Friis. Nov. 2018.

Fig. 5-43. *Ziziphus abyssinicus* on the road between Bure and Nekemt at ca. 1100 m a.s.l. (WG floristic region). The vegetation is mixed open woodland and wooded grassland just above the inner Abay Gorge on the north-facing side of the gorge. Early dry season. Image 1391 by Ib Friis. Oct. 2010.

38

39

40

41

42

43

Tongo, Bambasi, Begi on one side and Assosa on the other (Fig. 5-52, Fig. 5-60). Deviating is the woodland dominated by *Syzygium guineense* subsp. *macrocarpum* near Begi, area no. 73. Included is also relevé no. 79, which was studied 28.6.2017 south of Assosa at the bamboo thicket called Anbessa Chaka (Fig. 5-51). Altitudinal range 1375-1600 m a.s.l.

A north-western branch, J.3, including our relevés no. 74-78 (see Appendix 1), studied 27.6.2017 towards the Sudan between Assosa and Kurmuk (Fig. 5-44, Fig. 5-47, Fig. 5-48, Fig. 5-57, Fig. 5-58, Fig. 5-59). Included here are also three relevés (no. 83-84) with information gathered during field work in 2015 by Mindaye Teshome and Abeje Eshete Wassie and published two years later (Mindaye Teshome et al. 2017): relevé no. 83 at Kurmuk (altitude 675 m), and relevé no. 84 east of Gulashe, between Assosa and Kurmuk (altitude 820 m). Altitudinal range 675-1450 m a.s.l.

J.1. Branches between Assosa, Gizen, and the Abay River, via Baneshegol, Arenja, Sherkole, Mengi and Homesha)

This branch, J.1, is somewhat heterogenous regarding landscape, ranging from flat terrain with few rocks to gently or rather steeply sloping terrain with scattered, large rocks. The altitudinal range of the profile is wide, 675-1600 m a.s.l. The soils were pale reddish and sandy on much of the flat ground and in upper part of slope, but fine-grained, almost vertisol at base of slopes. In some places with extensive farmland on level terrain there was dark red soil without rocks, gravel, or sand. The ground-cover of the uncultivated patch consisting of tall perennial grasses (mostly 100-180 cm tall), which may burn fiercely. Fire-scars were seen in several relevés. New woody species in relation to profiles further to the north are *Cassia arereh* (Fig. 5-61), *Dalbergia boehmii* and *Fadogia cienkowskii*. The woody vegetation was mixed open woodland with a rich flora of species typical of *Combretum-Terminalia* woodland [note that the identifications of Mindaye Teshome et al. (2017) have sometimes been suggested altered]: *Acacia polyacantha* subsp. *campylacantha, A. hecatophylla, A. senegal, A. seyal, A. venosa, Albizia amara, A. malacophylla, Allophylus rubifolius* (by Mindaye Teshome et al. (2017) named as *A. abyssinicus*, which is a forest species not seen by us elsewhere in the western woodlands), *Annona senegalensis* (Fig. 5-54), *Anogeissus leiocarpa* (Fig. 5-45), *Balanites aegyptiaca, Borassus aethiopum, Boswellia papyrifera* (Fig. 5-47), *B. pirottae, Bridelia micrantha, Cassia arereh* (Fig. 5-61), *Combretum adenogonium, C. collinum, C. molle, Commiphora pedunculata, Cordia africana, Crossopteryx febrifuga, Dalbergia boehmii, D. melanoxylon, Dichrostachys cinerea, Dombeya quinqueseta* (by Mindaye Teshome et al. (2017) named as the Afromontane *D. torrida*, but almost certainly *D. quinqueseta*), *Entada africana, Fadogia cienkowskii, Ficus glumosa, F. ingens, F. sycomorus, F. thonningii, Flacourtia indica, Flueggea virosa, Gardenia ternifolia, Grewia mollis, G. velutina, Hyphaene thebaica, Kigelia africana, Lannea fruticosa* (Fig. 5-48), *L. schimperi, L. schweinfurthii* (by Mindaye Teshome et al. (2017) named as *L. welwitschii*, but almost certainly *L. schweinfurthii*), *Lonchocarpus laxiflorus* (Fig. 5-51), *Maytenus senegalensis, Ochna leucophloeos, Oxytenanthera abyssinica* (Fig. 5-51), *Ozoroa insignis* (Fig. 5-55), *O. pulcherrima, Piliostigma thonningii, Pseudocedrela kotschyi, Psorospermum febrifugum, Pterocarpus lucens, Rhus vulgaris, Sarcocephalus latifolius, Sclerocarya birrea, Securidaca longipedunculata* (Fig. 5-53), *Senna singueana, Steganotaenia araliacea, Sterculia africana* (by Mindaye Teshome et al. (2017) named as *S. setigera*; this may be possible, but we have not observed that species in the area), *Stereospermum kunthianum* (Fig. 5-56), *Strychnos innocua, Syzygium guineense* subsp. *macrocarpum* (Fig. 5-52), *Terminalia laxiflora, T. macroptera, T. schimperiana, Vernonia amygdalina, Vitex doniana, Ximenia americana, Ziziphus abyssinica, Ziziphus mauritiana,* and *Ziziphus spina-christi.*

J.2. Branch between Assosa and Tongo, via Anbessa Chaka and Bambasi.

This branch, J.2, is relatively homogenous regarding landscape, with mainly flat terrain with few steep hillsides to gently undulating terrain with or without scattered, large rocks. The soils were dark to pale reddish, and cultivation was frequent. The altitudinal range is 1350-1550 m a.s.l. In one relevé near Bambasi (no. 73) the level terrain seemed highly liable to flooding during the rains, and the soil was dark brown to black and cracking, like vertisol. In Anbessa Chaka (no. 79) the soil was brown and shallow over large rocks and generally rocky ground. In all relevés but one (no. 70), the ground-cover was moderately dense and consisted of low to high perennial grasses (ca. 50 cm tall on damp ground, up to 2 m tall in areas with rocks). In relevé no. 70, the canopy of the woodland was so dense that it had shaded out almost all ground-cover. Extensive signs of burning on the hillsides and rocky terrain, not many fire-scars in the cultivated areas. New in relation to profiles further to the north are *Dombeya buettneri*, *Fadogia cienkowskii*, and *Zanthoxylum gilletii* (Fig. 5-60). The latter was only seen south of Begi near the extensive swamps that form the source of the Dabus River. The vegetation was mainly a mixed open woodland with a rich woody flora of species typical of *Combretum-Terminalia* woodland: *Acacia polyacantha* subsp. *campylacantha, A. seyal, Albizia malacophylla, Annona senegalensis, Bridelia scleroneura, Carissa spinarum, Combretum collinum, C. molle, Cordia africana, Croton macrostachyus, Dichrostachys cinerea, Dombeya buettneri, D. quinqueseta, Entada abyssinica, Fadogia cienkowskii, Faurea rochetiana, Ficus sycomorus, Flacourtia indica, Flueggea virosa, Gardenia ternifolia, Grewia mollis, Lannea schimperi, Lonchocarpus laxiflorus* (Fig. 5-51), *Maytenus senegalensis, Oxytenanthera abyssinica* (Fig. 5-51), *Ozoroa insignis* (Fig. 5-55), *Piliostigma thonningii, Polyscias farinosa, Psorospermum febrifugum, Pterocarpus lucens, Rhus vulgaris, Securidaca longipedunculata* (Fig. 5-53), *Steganotaenia araliacea, Stereospermum kunthianum* (Fig. 5-56), *Strychnos innocua, Syzygium guineense* subsp. *macrocarpum* (Fig. 5-52), *Terminalia laxiflora, T. macroptera, T. schimperiana, Vitex doniana,* and *Zanthoxylum gilletii* (Fig. 5-60).

J.3. Branch between Assosa and Kurmuk

The landscape of this branch, J.3, varies considerably from the higher ground near Assosa towards the lower ground at Kurmuk on the Sudan border. At the border, in the immediate surroundings of Kurmuk, the landscape is flat to slightly sloping, but along the route towards Assosa the landscape soon changes to hilly terrain. From ca. 22 km from Kurmuk the landscape again becomes flat (relevé no. 77) and soon after slightly sloping to hilly again. The soils were mostly deep red and gravely or stony and finally almost black (in no. 78). In almost all relevés the ground-cover consisted of perennial grass 1-2 m tall and liable to fierce burning. The altitudinal range of the entire J.3-profile is 650-1450 m a.s.l. The vegetation was almost consistently an open, mixed woodland with a rich woody flora of species typical of *Combretum-Terminalia* woodland: *Acacia polyacantha* subsp. *campylacantha, A. hecatophylla, A. hockii, A. senegal, A. seyal, A. venosa, Albizia malacophylla, Allophylus rubifolius* (in the publication of Mindaye Teshome et al. (2011) and Abeje Eshete Wassie (2011) named as *A. abyssinicus*, which is a forest species not seen by us elsewhere in the western woodlands), *Annona senegalensis, Anogeissus leiocarpa, Balanites aegyptiaca, Boswellia papyrifera* (Fig. 5-47), *B. pirottae, Combretum adenogonium, C. collinum, C. hartmannianum, Commiphora pedunculata, Cordia africana, Cussonia arborea* (Fig. 5-59), *Dalbergia boehmii, D. melanoxylon, Dichrostachys cinerea, Dombeya quinqueseta, Entada abyssinica, E. africana, Faurea rochetiana, Ficus sycomorus, F. thonningii, Gardenia ternifolia, Grewia mollis, Hyphaene thebaica, Lannea fruticosa* (Fig. 5-48), *L. schimperi, Lonchocarpus laxiflorus* (Fig. 5-51), *Maytenus senegalensis, Ochna leucophloeos, Oxytenanthera abyssinica* (Fig. 5-51), *Ozoroa insignis, O. pulcherrima, Piliostigma thonningii, Protea gaguedi* (Fig. 5-58), *Pseudocedrela kotschyi, Pterocarpus lucens, Sclerocarya birrea, Securidaca longipedunculata* (Fig. 5-53), *Sterculia africana, Stereospermum kunthianum, Strychnos innocua, Terminalia laxiflora, T. macroptera, T. schimperiana, Ximenia americana, Ziziphus abyssinica, Z. mucronata,* and *Z. spina-christi*. A small tree of *Maerua angolensis* was observed and photographed near Kurmuk (Fig. 5-57), but was not recorded in the relevés; it is illustrated here because it occurred in relevés in Profile H, the Abay Gorge south of Bure.

Fig. 5-44. Mixed open woodland and wooded grassland between Assosa and Kurmuk at ca. 1000 m a.s.l. (WG floristic region). Note the lush green colour of the ground-cover, which consists of sprouting perennial grasses. Rainy season. Image 3464 by Ib Friis. Jun. 2017.

Fig. 5-45. Mixed open woodland between Mengi and Abay River at ca. 700 m a.s.l. (WG floristic region). The site is on the road between Mengi and the GERD Dam on the Abay River. *Anogeissus leiocarpa* is here the dominant tree. Rainy season. Image 3482 by Ib Friis. Jun. 2017.

Fig. 5-46. Mixed open woodland between Mengi and Gizen at ca. 750 m a.s.l. (WG floristic region). The site is on the road between Mengi and Gizen, near Mengi. Image 3729 by Ib Friis. Early dry season. Nov. 2005.

Fig. 5-47. Flowering and leafless tree of *Boswellia papyrifera* near Kurmuk at ca. 700 m a.s.l. (WG floristic region). The site is near Kurmuk on the road between Assosa and Kurmuk. In the foreground upgrowth of *Lannea fruticosa*. Early dry season. Image 3783 by Ib Friis. Nov. 2005.

Fig. 5-48. Mixed open woodland between Assosa and Kurmuk at ca. 750 m a.s.l. (WG floristic region). To the left small trees of *Lannea fruticosa* with partly withering leaves. Early dry season. Image 5801 by Ib Friis. Nov. 2006.

Fig. 5-49. Mixed, open woodland between Mengi and Gizen at ca. 1000 m a.s.l. (WG floristic region). The site is near Mengi. Early dry season. Image 6191 by Ib Friis. Nov. 2012.

Fig. 5-50. Hill with mixed, open woodland between Mengi and Gizen at ca. 1000 m a.s.l. (WG floristic region). The site is near Mengi. Substrate reddish soil; ground-cover very tall. Early dry season. Image 6203 by Ib Friis. Nov. 2012.

Fig. 5-51. The bamboo thicket called Anbessa Chaka (meaning the 'Lion Forest' or 'Lion thicket') south of Assosa at ca. 1450 m a.s.l. on the road between Bambasi and Assosa. (WG floristic region). Left image: view of the edge of the *Oxytenanthera abyssinica* thicket (the bamboo is to the left in the image) and *Lonchocarpus laxiflorus* to the right in the image. Early dry season. Nov. 2005. Right image a view inside the bamboo thicket with field assistants showing size. Early dry season. Image 3958 & 3965 by Ib Friis. Nov. 2005.

Fig. 5-52. Low woodland dominated by *Syzygium guineense* subsp. *macrocarpum* near Bambasi at ca. 1450 m a.s.l. (WG floristic region). The site is between Mendi and Bambasi. The trees in *Syzygium guineense* subsp. *macrocarpum* woodland never reach a hight greater than here. Early dry season. Image 6302 by Ib Friis. Nov. 2012.

Fig. 5-53. Flowering *Securidaca longipedunculata* in mixed open woodland at ca. 700 m a.s.l. (WG floristic region). The site is between Mengi and the GERD Dam on the Abay River. Rainy season. Image 3479 by Ib Friis. Jun. 2017.

Fig. 5-54. Fruiting *Annona senegalensis* in mixed dense woodland between Assosa and Mengi at ca. 1100 m a.s.l. (WG floristic region). The site is just south of Mengi. Rainy season. Image 3510 by Ib Friis. Jun. 2017.

Fig. 5-55. Flowering, narrow leaved form of *Ozoroa insignis* in mixed dense woodland at ca. 1100 m a.s.l. (WG floristic region). The site is between Assosa and Mengi, south of Mengi. Rainy season. Image 3511 by Ib Friis. Jun. 2017.

Fig. 5-56. Flowering tree of *Stereospermum kunthianum* in mixed open woodland at ca. 1100 m a.s.l. (WG floristic region). The site is between Mengi and Gizen, nearest to Mengi. Early dry season. Image 3719 by Ib Friis. Nov. 2005.

Fig. 5-57. Small flowering tree of *Maerua angolensis* near Kurmuk at ca. 900 m a.s.l. (WG floristic region). The site is on the road between Assosa and Kurmuk. Early dry season. Image 3768 by Ib Friis. Nov. 2005.

Fig. 5-58. Flowering tree of *Protea gaguedi* near Kurmuk at ca. 900 m a.s.l. (WG floristic region). The site is on the road between Assosa and Kurmuk. Early dry season. Image 3799 by Ib Friis. Nov. 2005.

Fig. 5-59. Tree of *Cussonia arborea* near Kurmuk at ca. 950 m a.s.l. (WG floristic region). The site is on the road between Assosa and Kurmuk. Early dry season. Image 3806 by Ib Friis. Nov. 2005.

Fig. 5-60. Flowering trees of *Zanthoxylum gilletii* in open woodland south of Begi at ca. 1550 m a.s.l. (WG floristic region). Image above: fruiting branches in open woodland south of Begi towards Gidami. Early dry season. Nov. 2005. Image below: flowering tree in partly cleared open woodland between Tonga and Bambasi. Rainy season. Image 2005 & 3518 by Ib Friis. Jun. 2017.

44

45

46

47

48

49

50

51

52

53

54

55

56

57

58

59

60

K: Profile of the Didessa Valley between Bedele and Nekemt and Gimbi and Nekemt

Profile K includes our relevés no. 119-121 (see Appendix 1), studied on 25.10.2018 between Bedele and Nekemt (altitudinal range 1250-1400 m a.s.l.), and relevé 125, studied on 31.10.2018 near the Didessa River (altitude 1300 m a.s.l.). Due to civil unrest in 2018 it was not possible to study the profile between Gimbi and Nekemt in detail, and much of the woodland along that part of the road had, since the resettlement programmes in the 1980s, been deeply influenced by the settlements of highland people and their plantations in and around Arjo near the Didessa River (not to be confused with the much older settlement named Arjo on the Bedele-Nekemt road). The slopes of the Didessa Valley between Bedele and Nekemt are gently undulating, as is most of the valley bottom, except for the level ground just around the river (Fig. 5-62).

Large areas of the Didessa Valley have been converted to farmland, partly providing raw materials for the Arjo DIdessa sugar factory, partly converted to mango plantations. The soil is brown and stony. Between the cultivated areas the ground-cover mostly consists of perennial grasses up to 2 m tall, which may burn fiercely. The vegetation was almost consistently an open, mixed woodland with a rich woody flora of species typical of *Combretum-Terminalia* woodland: *Albizia malacophylla, Bridelia scleroneura, Combretum collinum, Cordia africana, Cussonia arborea, Dodonaea angustifolia, Entada abyssinica, Ficus sur, F. sycomorus, Grewia mollis, Lonchocarpus laxiflorus, Oxytenanthera abyssinica, Piliostigma thonningii* (Fig. 5-64), *Rhus vulgaris, Stereospermum kunthianum, Terminalia schimperiana,* and *Vitex doniana*. Outside the relevés *Syzygium guineense* subsp. *macrocarpum* had been photographed on a visit in 2000 (Fig. 5-63).

L: Profile between Dembidolo and Gimbi with mixture of woodland and forest and one relevé in the Didessa Valley near Gimbi

Profile L includes our relevés no. 122-124 (see Appendix 1), studied 28.10.2018 along the road between Dembidolo and Gimbi and no. 125, studied 31.10.2018 in the Didessa Valley just east of Gimbi. Altitudinal range 1400-1600 m a.s.l. and 1300 m a.s.l. in the Didessa Valley. The profile crosses an undulating landscape with degraded patches of forest and open, but now partly cultivated deciduous woodland (Fig. 5-65), particularly around the villages of Chanka and Alem Teferi. The soil was typically brown and stony. Along the road, between relevés no. 122 and 123 there were many areas at lower altitudes, in which the species occurring there were more typical of *Transitional Rain Forest* (TRF) than of the western Ethiopian woodlands. Examples were specimens of *Albizia grandibracteata, Cordia africana* and *Trichilia dregeana,* the latter sometimes dominant. The presence of *Albizia grandibracteata* and *Trichilia dregeana* clearly indicate that the area contains species from past *Transitional Rain Forest* (TRF). Nearly all areas with forest trees had under-planting of *Coffea arabica*, and even the massive epiphytic fern *Platycerium elephantotis* was growing on some of them. This fern is also a species associated with the *Transitional Rain Forest* (TRF). In contrast the open, cultivated land here and there had trees typical of the *Combretum-Terminalia* woodland.

During field trips in 2006 and 2012 to areas around Gidami (between ca. 1700 and 1800 m) we observed a similar mixture of *Combretum-Terminalia* woodland and *Transitional Rain Forest* (TRF), with some Afromontane species. In 2018 we observed from the ridges between Dembidolo and Alem Teferi lower-lying, extensive areas of farmland in which many trees had been left, mainly *Anogeissus leiocarpa*. The woody species observed in our relevés 122-124 included species typical of *Combretum-Terminalia* woodland, but also Afromontane species and species typical of the *Transitional Rain Forest* (TRF). In order to show the mixed relation to woodlands and forests in western

Ethiopia, we will already here in the list mention the phytogeographical categories to which they are referred in Section 6, listing only the species occurring in open woodland:

Acacia abyssinica (Afromontane MI), *A. seyal* (CTW-species), *Albizia grandibracteata* (Afromontane MI, also in TRF), *A. malacophylla* (CTW-species), *Bersama abyssinica* (Afromontane MI), *Bridelia micrantha* CTW-species), *Clausena anisata* (Afromontane MI), *Combretum collinum* (Fig. 5-66) (CTW-species), *C. molle* (CTW-species), *Cordia africana* (transgressor TRG), *Croton macrostachyus* (transgressor TRG), *Entada abyssinica* (Fig. 5-67) (CTW-species), *Ficus lutea* (Afromontane MI), *F. sur* (transgressor TRG), *F. sycomorus* (CTW-species), *F. vasta* (Afromontane MI), *Flacourtia indica* (Afromontane MI), *Flueggea virosa* (CTW-species), *Gardenia ternifolia* (CTW-species), *Maytenus senegalensis* (transgressor TRG), *Piliostigma thonningii* (CTW-species), *Rhus vulgaris* (transgressor TRG), *Stereospermum kunthianum* (CTW-species), *Syzygium guineense* subsp. *macrocarpum* (CTW-[sub] species), *Terminalia schimperiana* (CTW-species), and *Trema orientalis* (Afromontane MI). The relevé no. 125 does not really belong with this profile, but agrees better with the pure deciduous woodlands in Profile K, the Didessa Valley south of Nekemt. It is included here because we studied it rather quickly on our way to Nekemt after having been detained in Gimbi for a couple of days by civil unrest.

M: Profile of the western escarpment below Bure and towards the Gambela lowlands

Profile M included our relevés no. 114-118 (see Appendix 1), studied 24.10.2018 along the road from Metu to Gambela (Fig. 5-68, Fig. 5-69). Bure, not to be confused with Bure in the GJ floristic region, is the last village or small town before one reaches the Bure Escarpment, on which the main road steeply and with many hairpins leads from the highlands around Bure towards the Gambela lowlands. Our profile did not reach the lowermost part of the steep escarpment due to restrictions caused by security regulations. Altitudinal range 800-1300 m a.s.l. On the entire part of the escarpment that we studied, the landscape consisted of steeply sloping ground and the soil was light brown to brown and stony. The ground-cover was dense and consisted of 2-2.5 m tall, perennial grasses (*Hyparrhenia*, *Loudetia*), which is prone to fierce burning. During previous visits covering much more of the escarpment we had seen the landscape completely covered with ashes after fires. Neither before, nor in 2018 we noticed any clearly marked subdivision of the profile, and most of the species were typical species of the *Combretum-Terminalia* woodlands.

A new species in relation to the profiles further to the north was *Ficus ovata*, otherwise a lowland forest species; its presence on the Bure Escarpment could be due to its habitat at the base of a small slope produced by the roadbuilding, which would provide shade and moisture. The vegetation was a typical mixed, open woodland. The woody species include: *Acacia polyacantha* subsp. *campylacantha, A. seyal, Albizia malacophylla, Annona senegalensis, Balanites aegyptiaca, Bridelia scleroneura, Calotropis procera, Combretum collinum, C. molle, Cordia africana, Cussonia arborea, Dombeya quinqueseta, Entada africana, Ficus ovata, F. sycomorus, Flueggea virosa, Grewia mollis, Harrisonia abyssinica, Lannea fruticosa, Lonchocarpus laxiflorus, Piliostigma thonningii, Pseudocedrela kotschyi, Pterocarpus lucens* (Fig. 5-69), *Rhus vulgaris* and *Terminalia laxiflora.*

N: Relevés of woodlands in the Gambela lowlands

Profile N included our relevés no. 85-89 (see Appendix 1) and is part of the Gambela lowlands that we included in the western woodlands of Ethiopia. Further away from the escarpment, the vegetation is open *Acacia* wooded grassland which we have referred to the *Wooded grassland of the western Gambela region* (WGG)). The lowlands were inaccessible during the years 2014-2018, when the major part of these studies was made. The earlier studies were carried out together with another research group (Ib Friis, Michael Jensen, Menassie Gashaw) during the years 1992-1993 and published by Jensen & Friis (2001).

This allowed us to supplement the data as follows. Relevé no. 85, studied inside extensive fenced area with woodland around Gambela Airport. Altitude 600 m a.s.l. The mixed woodland had in the early 1990s been fenced off for more than 10 years due to the safety of the airport. Ground-cover of perennial herbs up to 1 m tall, but hardly any grasses. No cultivation and no signs of grazing. According to local informants, the vegetation had not burnt for a significant length of time, probably many years.

Relevé no. 86, studied at the foot of the Bure Escarpment, ca. 5 km towards Gambela from where the road from Bure to Gambela crosses the Baro River, with mixed, relatively open woodland on terrain gently sloping away from the escarpment, soil with many stones or loose gravel over a brown to reddish soil; altitude 650 m a.s.l. The ground-cover consisted of patchy areas of tall grasses (ca. 2 m tall), which burnt regularly once or twice a year. The images (Fig. 5-70, Fig. 5-71) show views of the hills near the escarpment before and after grass fires. After burning, the perennial grasses sprout again, and several trees provide new leaves.

Relevé no 87, studied between 10 and 12 km on the road towards Akobo, south of the bridge on the Baro River next to the town of Gambela, with mixed woodland with fairly dense canopy on slightly undulating terrain, low ridges with coarser sand and lower ground with finer brown to red soil; altitude 550 m a.s.l. Ground-cover of tall grasses (ca. 2 m tall). These areas were burning regularly about once a year, except for the areas close to Gambela Airport, where grass fires were put out whenever possible to protect the installations of the airport. This meant that the woodlands were rather dense near the airport (Fig. 5-72).

Relevé no 88, studied between 80 and 85 km on the road to the south across the bridge on the Baro River next to the town of Gambela; between Akobo and Pugnido, with mixed, relatively open woodland slightly undulating terrain, low ridges with coarser and lower ground with finer yellowish-brown soil; altitude 450 m a.s.l. Ground-cover of tall grasses (ca. 2 m tall). Burning at least once or twice a year.

Relevé no. 89, studied 22 km west of Gambela towards Itang, with very open mixed woodland or wooded grassland (canopy cover ca. 25%) on flat terrain, with dark brown, almost black soil (black cotton soil).; altitude 500 m a.s.l. (Fig. 5-73 to Fig. 5-76). Ground-cover of tall annual and perennial grasses (> 2 m tall, often up to 3 m). Burning regularly.

A difference from previous relevés was the frequent presence of *Harrisonia abyssinica,* which especially occurred near termite mounds. Other elements, which were not observed in other relevés, were *Erythrococca trichogyne* and *Erythroxylum fischeri,* two species that are more associated with lowland forests than woodlands. They also occurred in evergreen vegetation on termite mounds, but only rarely.

The floristic composition of the woody vegetation of these five sites was so similar that a combined list is given here: *Acacia senegal, Allophylus rubifolius, Annona senegalensis, Anogeissus leiocarpa, Balanites aegyptiaca* (Fig. 5-73), *Bridelia scleroneura, Cadaba farinosa, Combretum adenogonium, C. collinum, C. molle, Crossopteryx febrifuga, Dichrostachys cinerea, Diospyros mespiliformis, Entada africana, Erythrococca trichogyne, Erythroxylum fischeri, Ficus sycomorus, Flueggea virosa, Gardenia ternifolia, Grewia mollis, Grewia tenax, Grewia velutina, Harrisonia abyssinica, Lannea barteri, L. fruticosa, Lonchocarpus laxiflorus, Maerua oblongifolia, M. triphylla, Maytenus senegalensis, Meyna tetraphylla, Ochna leucophloeos, Pterocarpus lucens, Sterculia africana, Stereospermum kunthianum, Strychnos innocua, Tamarindus indica, Terminalia laxiflora, Vangueria madagascariensis, Ximenia americana, Ziziphus abyssinica, Z. mauritiana,* and *Z. pubescens.*

O: Profile from Gurefada towards the Akobo Valley near Dima

Profile O included our relevés no. 102-105 (see Appendix 1), studied on 16.10.2018, covering the slope from Gurefada and the southern edge of the Bebeka Forest down to the Akobo Valley, where the village Dima was located just above the river. Altitudinal range 625-1300 m a.s.l. The profile started from the lowest altitude at Dima village and the Akobo Valley along the border with South Sudan.

Fig. 5-61. Fruiting trees of *Cassia arereh* at ca. 650 m a.s.l. (WG floristic region). Both images from trees between Mengi and the GERD Dam on the Abay River. Left image: a fruiting tree at the edge of woodland. Rainy season. Jun. 2017. Right image: fruiting branches in dense woodland. Early dry season. Image 3485 & 6247 by Ib Friis. Nov. 2012.

Fig. 5-62. Mixed open woodland in the Didessa Valley at ca. 1400 m a.s.l. (IL floristic region). The site is on the north-facing side of the Didessa Valley between Bedele and Nekemt. Image LUXA0948 by Odile Weber. Early dry season. Oct. 2018.

Fig. 5-63. Flowering *Syzygium guineense* subsp. *macrocarpum* in the Didessa Valley at ca. 1400 m a.s.l. (WG floristic region). The site is in mixed open woodland near the bottom of the valley between Bedele and Nekemt. Dry season. Image 8602 by Ib Friis. Feb. 2000.

Fig. 5-64. Fruiting *Piliostigma thonningii* in mixed woodland in the Didessa Valley at ca. 1400 m a.s.l. (WG floristic region). The site is near the bottom of the valley between Bedele and Nekemt. *Piliostigma thonningii* may produce flowers and fruits both as a shrub and as a small tree. Dry season. Image 8759 by Ib Friis. Jan. 2000.

Fig. 5-65. Mixed and very open woodland between Dembidolo and Gimbi at ca. 1450 m a.s.l. (WG floristic region). The site is near Alem Teferi on the road between Dembidolo and Gimbi; the substrate is reddish soil. The area has probably been cultivated or heavily grazed. Image LUXA1057 by Odile Weber. Early dry season. Oct. 2018.

Fig. 5-66. Fruiting *Combretum collinum* in open woodland between Dembidolo and Gimbi at ca. 1550 m a.s.l. (WG floristic region). The site is near Chanka on the road between Dembidolo and Gimbi. Early dry season. Image Nov_2013_009 by Sebsebe Demissew. Nov. 2013.

Fig. 5-67. Fruiting *Entada abyssinica* in open woodland between Dembidolo and Gimbi at ca. 1550 m a.s.l. (WG floristic region). The site is near Chanka on the road between Dembidolo and Gimbi. Early dry season. Image Nov_2013_411 by Sebsebe Demissew. Nov. 2013.

Fig. 5-68. Mixed open woodland on the Bure Escarpment at ca. 900 m a.s.l. (IL floristic region). The site is near Bure on the road between Metu and Gambela. The undergrowth of grasses is very tall and provides fierce grass fires. Early dry season. Image LUXA0824 by Odile Weber. Oct. 2018.

Fig. 5-69. Mixed open woodland on the Bure Escarpment at ca. 900 m a.s.l. (IL floristic region). The site is near Bure on the road between Metu and Gambela. The undergrowth of grasses is very tall and provides fierce grass fires. To the left a large tree of *Pterocarpus lucens*. Early dry season. Image LUXA0854 by Odile Weber. Oct. 2018.

Fig. 5-70. Baro River and surrounding woodlands at ca. 550 m a.s.l. (IL floristic region). Image above: open woodland on hill. Image below: Dense woodland near river. Two sites were studied in this vegetation along the road between the Bure Escarpment and Gambela. Late rainy season. Image BSC5804 & BSC5805 by Mekuria Argaw. Sep. 2017.

Fig. 5-71. Baro River and surrounding woodlands at ca. 550 m a.s.l. (IL floristic region). Site between the Bure Escarpment and Gambela. Resprouting tufts of perennial grasses and trees appeared after grass fire. The vegetation is open deciduous woodland. Height of dry season. Image BSC06346 by Mekuria Argaw. Feb. 2018.

Fig. 5-72. Woodland around the Gambela Airport, south of Gambela and Baro River at ca. 600 m a.s.l. (IL floristic region). The woodland is unusually dense due to active protection against grass fires near the airport. Rainy season. Image from a video by Dr. Obang Metho, reproduced here with his permission. 2019.

Fig. 5-73. Wetland or wet wooded grasslands near Itang on the Baro River west of Gambela at ca. 450 m a.s.l. (IL floristic region). The vegetation is damp, often with puddles of water. In the foreground *Balanites aegyptiaca*. Dry season. Image BSC06265 by Mekuria Argaw. Feb. 2018. Mekuria Argaw phot.

Fig. 5-74. Wetland or wet wooded grasslands near Itang on the Baro River west of Gambela at ca. 450 m a.s.l. (IL floristic region). The vegetation is flooded (image below) and unflooded (image above) deciduous woodland or wooded grassland. Cows graze newly sprouted, short grass while dry grass still surrounds the base of trees and shrubs. Dry season. BSC06248 & BSC06253 by Mekuria Argaw. Feb. 2018.

61

62

63

64

65

66

67

68

69

70

71

72

73

74

The landscape gently rises from the Akobo River near Dima at 615 m a.s.l. to the border of the Bebeka Forest at ca. 1300 m a.s.l. The landscape is a gently sloping or slightly undulating terrain with brown soil, sometimes stony or very stony. The ground-cover was variable, mostly consisting of up to 2 m tall perennial grassed (*Hyparrhenia*, *Loudetia*), sometimes of perennial herbs (e.g., *Sida*); the former liable to burning, the latter not.

The vegetation along this profile was in the lower part a mixed, open woodland (Fig. 5-77, Fig. 5-78, Fig. 5-85, Fig. 5-86). Near Dima, the vegetation had been much influenced by cultivation. There was no clearly marked subdivision of the profile, and most of the species are typical species of the *Combretum-Terminalia* woodlands. The woody species included: *Acacia polyacantha* subsp. *campylacantha, A. seyal* (Fig. 5-77), *Annona senegalensis, Balanites aegyptiaca, Boscia salicifolia* (Fig. 5-85), *Bridelia scleroneura* (Fig. 5-81), *Combretum collinum, Cordia africana, Dombeya quinqueseta, Entada africana, Erythrina abyssinica, Ficus sur, F. sycomorus, F. vasta, Grewia mollis, Harrisonia abyssinica* (Fig. 5-86), *Lannea fruticosa, L. schimperi, Lonchocarpus laxiflorus, Maytenus senegalensis* (Fig. 5-84), *Ozoroa insignis, Piliostigma thonningii, Pseudocedrela kotschyi* (Fig. 5-83), *Sclerocarya birrea, Sterculia africana* (Fig. 5-78), *Stereospermum kunthianum, Terminalia brownii* (Fig. 5-80), *T. schimperiana, Trema orientalis,* and *Vitex doniana*. Outside the relevés the following species were photographed in 2000 and 2009: *Acacia hockii* (Fig. 5-82) and *Tamarindus indica* (Fig. 5-79).

P: Profile from forest south of Shewa Gimira through woodland-forest mosaic to Maji

Profile P includes our relevés no. 106-113 (see Appendix 1), studied 19.-20.10.2018 along the road from Mezan Teferi to Maji via Shewa Gimira, Bechuma, Djemu and Tum. Because of the mosaic of open woodland and patches with forest around Djemu, we have divided the description of the profile into two parts, the first with the relevés 106-107, covering the areas between the Afromontane forests from Bechuma to the tiny patches of *Transitional Rain Forest* (TRF) around Djemu, and the second with relevés no. 108-113, covering the area from Djemu to the base of the mountain on which the small town of Maji is located. Altitudinal range for first part 1350-1780 m a.s.l. a.s.l., for second part 1060-1550 m a.s.l.

P.1. From Shewa Gimira, to Bechuma and Djemu

The first relevé of this profile, P.1, was no. 106 at 1780 m a.s.l. It was untypical for the *Combretum-Terminalia* woodlands. The vegetation could best be described as scarcely wooded grassland on steep slopes. Of the few trees, the dominant was *Cordia africana*. The soil of the slope was brown and stony, and the ground-cover consisted of moderately tall grass, up to 1 m tall in places but usually much lower due to grazing. Burning might be possible, but there was no sign of it, and local informants did not confirm that any burning of the slope took place. The woody species included species typical of *Combretum-Terminalia* woodland and were: *Acacia polyacantha* subsp. *campylacantha, Albizia schimperiana, Combretum molle, Cordia africana, Entada abyssinica, Ficus thonningii, F. vasta, Rhus vulgaris* (Fig. 5-87), *Terminalia schimperiana,* and *Vitex doniana*.

Our study of the next relevé, no. 107, suggested that it was at the border between the western woodlands and the *Transitional Rain Forest* (TRF), which was represented in a small river valley (tributary to Akobo). On both sides of the river towards Djemu, there was much degraded lowland forest, mostly restricted to the slopes of the valley. Pioneering forest species like *Trema orientalis* (Fig. 5-88) and *Albizia grandibracteata* were the most common. Better preserved patches with tall forest, presumably also *Transitional Rain Forest* (TRF), was seen around 1300 m a.s.l.

The woodland-component of the vegetation was mixed, open woodland on steep sloping ground. The ground-cover in the woodland was short perennial grass, heavily grazed and 10-20 cm tall; unlikely to burn. The typical woodland species were: *Combre-*

tum molle, Cordia africana, Ficus thonningii, Terminalia schimperiana, and *Vitex doniana.*

P.2. Profile from Djemu to the village of Tum and the mountain of Maji

In this part of the profile, P.2, to the south of Djemu, the vegetation pattern was more typical of the *Combretum-Terminalia* woodland than between Shewa Gimira and Djemu. Just south of Djemu, at 1350 m a.s.l., the vegetation was a mixed, open and species-rich woodland with signs of scattered present and past cultivation and dark brown soil with stones. Probably due to influence of agriculture, the ground-cover was very low, with perennial grasses 10-20 cm tall and with no sign of burning.

Further south, ca. 15 km south of Djemu, the vegetation changed to mixed, open woodland on gently undulating ground, the same type of soil and the ground-cover moderately tall grass, up to 1 m tall and a clear potential for burning, but no sign of fires.

Again, further south, ca. 32 km south of Djemu, the woodland was partly replaced with farmland and a few species associated with montane evergreen bushland were observed on or near termite mounds, for example *Euclea racemose* subsp. *schimperi* and *Acacia dolichocephala*. The area was regularly subject to burning, as could be seen from fire-scars and blackening of the bark of trees. Several on-going grass fires were also observed.

At the foot of the Maji Mountains and around the village of Tum below the mountain, there was mixed, open woodland on flat terrain, surrounded by farmland and surrounding a *Scleria* swamp on deep brown soil. The ground-cover consisted of tall, perennial grasses up to 1.5 m tall, but no sign of burning. Surprisingly, no woodland species of *Combretum* was observed. Very soon after the ascent to Maji had started, the vegetation changed to Moist Afromontane Forest and derived evergreen bushland.

The typical woodland species along this part of profile south of Djemu were: *Acacia polyacantha* subsp. *campylacantha, A. dolichocephala* (Fig. 5-91), *A. seyal,*

Fig. 5-75. Mixed wooded grassland west of Gambela at ca. 450 m a.s.l. (IL floristic region). Very tall ground-cover, providing fierce grass fires. Early dry season. Reproduced from Friis et al. (2010: Fig. 17C), ca. 1997.

Fig. 5-76. Mixed wooded grassland near Gambela at ca. 450 m a.s.l. (IL floristic region). Mixed woodland and wooded grassland with more than three-meter-tall grass stratum in the undergrowth, providing fierce grass fires. Field assistant showing size. Early season. Reproduced from Friis et al. (2010: Fig. 17D), ca. 1997.

Fig. 5-77. Mixed open woodland near Dima at ca. 650 m a.s.l. (KF floristic region). The site is south of Bebeka towards Dima, along the road from Bebeka to Dima. *Acacia seyal* is dominant here, and the bark of the trees is grey after burning. Dry season. Image 8985 by Ib Friis. Jan. 2000.

Fig. 5-78. Tree of *Sterculia africana* in mixed open woodland near Dima at ca. 650 m a.s.l. (KF floristic region). The site is on the Bebeka-Dima road, just north of Dima. Late rainy season. Image LUXA0459 by Odile Weber. Oct. 2018.

Fig. 5-79. Fruiting tree of *Tamarindus indica* in mixed open woodland near Dima at ca. 650 m a.s.l. (KF floristic region). The site is on the Bebeka-Dima road north of Dima. Image 6701 by Ib Friis. Dry season. Jan. 2009.

Fig. 5-80. Fruiting *Terminalia brownii* in transitional semi-evergreen bushland east of Ginir at ca. 1400 m a.s.l. (BA floristic region). Image above: close-up of fruiting branch. Image below: Full grown tree. Early dry season. Image 8256 & 8258 by Ib Friis. Nov. 2014.

Fig. 5-81. Fruiting branch of *Bridelia scleroneura* in mixed open woodland near Dima at ca. 650 m a.s.l. (KF floristic region). The site is on the Bebeka-Dima road north of Dima. Dry season. Image 8956 by Ib Friis. Jan. 2000.

Fig. 5-82. Fruiting branch of *Acacia hockii* in mixed open woodland near Dima at ca. 650 m a.s.l. (KF floristic region). The site is on the Bebeka-Dima road north of Dima. Dry season. Image 8967 by Ib Friis. Jan. 2000.

Fig. 5-83. Fruiting branch of *Pseudocedrela kotschyi* in mixed open woodland near Dima at ca. 650 m a.s.l. (KF floristic region). The site is on the Bebeka-Dima road north of Dima. Dry season. Image 8978 by Ib Friis. Jan. 2000.

75

76

78

79

80

81

82

83

Annona senegalensis, Bridelia scleroneura, Combretum collinum, C. molle, Cordia africana, Cussonia arborea, Dichrostachys cinerea, Dombeya quinqueseta (Fig. 5-89), *Entada abyssinica, E. africana, Euclea racemosa* subsp. *schimperi, Faurea rochetiana* (Fig. 5-90), *Ficus platyphylla, F. sur, F. sycomorus, Gardenia ternifolia, Gnidia lamprantha, Grewia mollis, Heteromorpha arborescens, Lannea schimperi, Maytenus senegalensis, Ozoroa insignis, Pavetta crassipes, Piliostigma thonningii, Protea gaguedi, P. madiensis, Psorospermum febrifugum, Rhus vulgaris* (Fig. 5-87), *Stereospermum kunthianum, Terminalia brownii, T. schimperiana, Vitex doniana,* and *Ziziphus abyssinica.*

Q: Profile between the Gojeb Valley south of Jimma and the Omo Valley

Profile Q included our relevés no. 91-101 (see Appendix 1), and was studied 9.-13.10.2018. The profile reached in two sequences from the highlands south of Jimma across the Gojeb Valley and the Gojeb River to the highlands of the Chebera-Churchura National Park and again from the park to the Omo Valley. Altitudinal range 750-1630 m a.s.l.

Q.1. Gojeb Valley

At these sites in the profile Q.1, the limits in the Gojeb Valley between the Afromontane forest and the woodland are at ca. 1750 m a.s.l., but between 1750 and 1500 m a.s.l. (our relevés no. 91-94), the vegetation has been disturbed by plantations of *Cupressus*, and therefore the highest area with mixed, open woodland is at ca. 1500 m a.s.l. The landscape there was represented by flat or slightly sloping ground, with deep brown soil between stones or brown soil without stones. One relevé south of the river was on a steep slope with deep, dark brown soil, few stones and a 10-20 cm tall ground-cover of perennial grasses unlikely to burn. The ground-cover consisted mainly of grasses up to 1 m tall (short *Hyparrhenia* and *Loudetia*). In the following three relevés there were no sign of recent burning, one relevé next to Gojeb River seemed to have regular burning, but most of the ground-cover was represented by subshrubs unlikely to burn, like *Desmodium* sp., *Acalypha* sp. and *Hoslunda* sp. The woody species in this part of the profile were: *Acacia hecatophylla, Acacia seyal, Annona senegalensis, Bridelia scleroneura, Combretum molle, Cordia africana, Cussonia arborea, Dichrostachys cinerea, Entada abyssinica, Entada africana, Ficus sur, Ficus sycomorus, Flueggea virosa, Gardenia ternifolia, Grewia mollis, Heteromorpha arborescens, Lannea schimperi, Lonchocarpus laxiflorus, Piliostigma thonningii, Rhus vulgaris, Steganotaenia araliacea, Stereospermum kunthianum, Terminalia schimperiana, Vernonia amygdalina,* and *Vitex doniana.*

Q.2. From the Omo Valley to the forests in the Chebera-Churchura National Park

This part of the profile, referred to as Q.2, was studied from the Omo River at and upwards, beginning with mixed, very open woodland on hilly terrain just above the river with and deep, red soil with few stones (Fig. 5-93).

In the first relevés near the Omo River the ground-cover was perennial grasses up to 1-1.5 m tall (*Hyparrhenia*, *Loudetia*), but also many subshrubby legumes, *Desmodium*, etc., and the vegetation seemed prone to burning. In places where the canopy was almost closed there was a shaded and weakly developed undergrowth. Very interesting new records made outside the relevés were populations of *Euphorbia venefica*, hitherto only known from the area around Mt. Fazugli near the Abay in Sudan (Weber et al. 2020).

At about 15-18 km from the Omo River, at 950 m a.s.l., the vegetation was a dense mixed woodland next to past and present farmland on undulating ground and with dense *Vernonia* scrub in the ground-cover. This vegetation formed a transition to areas with steeply sloping mixed woodland at the forest edge at the Chebera-Churchura National Park (forest rather like *Transitional Rain Forest* (TRF) with *Polyscias ferruginea, Vernonia amygdalina, Phoenix reclinata, Maesa lanceolata* and *Albizia grandibracteata* (Fig. 5-92).

After a small patch of forest, the vegetation again changed to a mosaic of mixed, open woodland and forest on sloping ground, marking the lower boundary of the next forest zone (in which occurred *Pouteria altissima, Albizia grandibracteata, Malacantha alnifolia, Ficus vallis-choudae, Ficus lutea*, etc.).

Inside the forest boundary, and generally in many places in the Chebera-Churchura National Park there were small areas with mixed open woodland on flat terrain, again with deep, dark brown soil with few stones and ground-cover of short grass and little burning, but almost certainly possibility for fires.

The woodlands in this mosaic contained these species typical of the *Combretum-Terminalia* woodland: *Annona senegalensis, Bridelia scleroneura, Combretum collinum* (Fig. 5-96), *C. molle* (Fig. 5-95), *Cordia africana, Crossopteryx febrifuga, Dombeya quinqueseta, Entada abyssinica, E. africana, Ficus platyphylla, F. sur, F. sycomorus, Gardenia ternifolia, Gnidia lamprantha, Lannea fruticosa, L. schimperi, Lonchocarpus laxiflorus, Maesa lanceolata, Oxytenanthera abyssinica, Pavetta crassipes, Piliostigma thonningii* (Fig. 5-94), *Protea madiensis, Pseudocedrela kotschyi, Rhus vulgaris, Stereospermum kunthianum, Terminalia laxiflora* (Fig. 5-97), *T. schimperiana*, and *Vitex doniana* (Fig. 5-98). The road from the Omo Valley met the older road-system between Bonga and Sodu near the small town of Ameya, which is in the relatively moist highlands and surrounded by moist Afromontane forest.

R: South of Maji and towards the lower Omo Valley

Information about the vegetation between Maji and the lower Omo Valley is limited, and we have never been able to visit the Omo Valley to the west of the river. A profile in the Omo National Park was planned for this work, but it was not possible to study this profile due to logistic and political difficulties.

Of our own observations, we have only been able to make use of old and scattered notes from the area between Lake Chew Bahir and the Omo River at Omorate. The available sources of information are the study of the vegetation along the lower Omo River itself from a publication by Carr (1998) and studies of the vegetation in the Omo National park by Jacobs and Schloeder (Schloeder 1999; Jacobs & Schloeder 2002).

Fig. 5-84. Flowering *Maytenus senegalensis* in mixed open woodland near Dima at ca. 650 m a.s.l. (KF floristic region). The site is on the Bebeka-Dima road north of Dima. Dry season. Image 8997 by Ib Friis. Jan. 2000.

Fig. 5-85. Tree of *Boscia salicifolia* in mixed open woodland near Dima at ca. 650 m a.s.l. (KF floristic region). The site is on the Bebeka-Dima road, just north of Dima. Late rainy season. Image LUXA0452 by Odile Weber. Oct. 2018.

Fig. 5-86. Shrub of *Harrisonia abyssinica* in mixed open woodland near Dima at ca. 650 m a.s.l. (KF floristic region). The site is on the Bebeka-Dima road, just north of Dima. Late rainy season. Image LUXA0470 by Odile Weber. Oct. 2018.

Fig. 5-87. Shrub or small tree of *Rhus vulgaris* in the southern Rift Valley at ca. 1450 m a.s.l. (GG floristic region). The vegetation is *Transitional Semi-Evergreen Bushland* (TSEB). Early dry season. Image 3440 by Ib Friis. Nov. 2011.

Fig. 5-88. Small tree of *Trema orientalis* in the southern Rift Valley at ca. 1300 m a.s.l. (GG floristic region). The vegetation is *Transitional Semi-Evergreen Bushland* (TSEB) with patches of forest. Early dry season. Image 3471 by Ib Friis. Nov. 2011.

Fig. 5-89. Tree of *Dombeya quinqueseta* between Djemu and Maji at ca. 1300 m a.s.l. (KF floristic region). The vegetation is open woodland and wooded grassland. Early dry season. Image LUXA0608 by Odile Weber. Oct. 2018.

Fig. 5-90. Tree of *Faurea rochetiana* in woodland between Djemu and Maji at ca. 1500 m a.s.l. (KF floristic region). The vegetation is open woodland and wooded grassland. *Faurea rochetiana* was the dominant tree in this woodland. Early dry season. Image LUXA0769 by Odile Weber. Oct. 2018.

Fig. 5-91. Tree of *Acacia dolichocephala* between Djemu and Maji at ca. 1500 m a.s.l. (KF floristic region). The vegetation is open woodland and wooded grassland. Early dry season. Image LUXA0776 by Odile Weber. Oct. 2018.

Fig. 5-92. Mosaic of forest and woodland between Ameya and the Omo Valley at ca. 1000 m a.s.l. (KF floristic region). The site is in the southern part of the Chebera-Churchura National Park, past the highest point and towards the Omo River. The vegetation is a mosaic of *Transitional Rain Forest* (with some *Moist Afromontane Forest*) and mixed open woodland. Early dry season. Image 4590 by Ib Friis. Oct. 2018.

84

85

86

87

88

89

90

91

92

Carr's study focussed on the vegetation along the river and has only limited information about the plains away from it, but she stated: "Broadly defined to include the levee backslopes and adjacent mudflats (or ancient floodplains), the riverine zone in the lower Omo basin supports a relatively luxuriant vegetation compared with the dry grasslands in the surrounding plains environments." Jacobs & Schloeder's studies covered a zone of grasslands and wooded grasslands on either side of the park headquarters on the Mui River ca. 70 km in a north-south and 40 km in an east-west direction. Solomon Tilahun et al. (1996) characterized the main features of the Omo National Park thus: "The major land features of the park are the Omo River on the east, the Maji Mountains, the Sharum and Sai Plains in the north and west and the Lilibai Plain and the Birga Hills to the south. ... The Park is crossed by several rivers, all of which drain into the Omo. The most important is Mui which runs across the middle of the park. Much of the park is around 800 m a.s.l. but the southern part by the river drops to 450 m a.s.l. The highest peak in the Maji Mountains is 1541 m a.s.l. ... The heaviest rain falls in March and April, less in September and October. The park encompasses an extensive open grassland interspersed with various stands of woodland species, bush and riverine vegetation. The plains are covered in black cotton soils and the hills by lighter coloured soils."

Carr (1998) stated about the areas away from the river:

> Much of the upland plains [above the river basin in the strict sense] is severely overgrazed due to the overcrowding of livestock that has resulted from a radical reduction of the local pastoralists' land access caused by governmental decisions taken over the past fifty years. Large expanses of open deciduous thicket and semidesert scrub, characterized by low forage value species, now predominate, and vast areas near the largescale human settlements that have formed in recent years are virtually devoid of plant life. Stemming from the complex geomorphic history of the area, ecologically anomalous localities with intrazonal vegetation are scattered throughout the lower basin. Most notable among these localities are basaltic and rhyolitic volcanic highlands, such as Mt. Nakwa and Mt. Nkalabong, which support a variety of distinct vegetation types; salt springs with dense stands of palm grassland with *Hyphaene thebaica* most conspicuous; and a large area of uplifted and eroded sands, silts and clays, the Omo Beds (geographic focus of the Omo Expedition's paleontological studies), where succulent and thorn thicket and dry scrub are characteristic.

These studies indicate that the vegetation in large areas around the Omo River away from the riverine vegetation is dry grassland and bushland with very little woodland, similar to what we have observed from the eastern side of the river (Fig. 5-99). We have recorded that from the east side of the Omo and up to ten to forty km to the east of the river, depending on the topography, the grasslands and bushlands are replaced by *Acacia-Commiphora bushland* (ACB), as defined in Friis et al. (2010).

On the banks in the Omo delta there was, according to Carr (1998), riparian bushland consisting of *Cordia sinensis*, which was strongly dominant, as well as *Acacia mellifera*, *Ziziphus mauritiana*, and *Ficus sycomorus*, with significant numbers of *Ximenia caffra* [not *X. americana*] and *Grewia fallax*. The best developed riverine forest included these woody species, a number of which are not known from the western woodlands of Ethiopia further to the north or west, but are instead distributed in *Acacia-Commiphora* woodland or bushland to the east: *Acacia tortilis, Allophylus macrobotrys* [not *A. rubifolius*], *Celtis toka* [as *C. integrifolia*], *Cordia sinensis, Crateva adansonii, Ficus sycomorus, Grewia fallax* [considered synonym of *Grewia arborea* (Forssk.) Lam.], *Maytenus senegalensis*, *Glenniea africana* [as *Melanodiscus oblongus*, not otherwise recorded from Ethiopia], *Flueggea virosa* [as *Securinega virosa*], *Tamarindus indica, Tapura fischeri*, *Trichilia emetica* [as *T. roka*], *Uvaria* sp., *Ximenia caffra*, and *Ziziphus pubescens*. On silt berms, the following woody species were observed: *Ziziphus mauritiana*, *Ximenia caffra*, *Harrisonia abyssinica*, *Cordia sinensis*, *Maytenus senegalensis* and seedlings of *Ficus*. In transitional vegetation to mudflats further away from the river occurred: *Acacia mellifera*, *Acacia reficiens*, *Acacia sieberiana*, *Acacia tortilis*, *Allophylus macrobotrys, Cadaba rotundifolia*, *Capparis sp., Cordia sinensis, Flueggea virosa, Grewia fallax, Maytenus senegalensis, Salvadora persica, Terminalia brevipes, Ximenia caffra,* and *Ziziphus* spp.

Apart from the above-mentioned woody plants, Carr (1998) includes a list of species collected in the lower Omo River basin, some of these away from the river, but it is unfortunately not stated how at what distance, or whether they were collected from east or west of the river. There is a much higher number of species belong to the family Burseraceae, particularly the genus *Commiphora*, than that normally found in typical *Combretum-Terminalia* woodland. In the following list the species marked with an asterisk are mentioned in Friis et al. (2010) as typical of the *Acacia-Commiphora bushland* (ACB): *Acacia brevispica, A. drepanolobium*, A. horrida*, A. oerfota* [as *A. nubica*], *A. paolii*, A. reficiens*, A. senegal, A. seyal, A. sieberiana, Adenium obesum, Balanites aegyptiaca*, B. rotundifolia** [as *B. orbicularis*], *Boscia angustifolia, B. coriacea, Boswellia neglecta** [as *B. hildebrandtii*], *Cadaba farinosa, C. gillettii, C. glandulosa, Calotropis procera, Capparis fasicularis, C. tomentosa*, Celtis toka* [as *C. integrifolia*], *Combretum aculeatum*, Commiphora africana, C. boiviniana*, C. habessinica* [as *C. madagascariensis*], *Cordia crenata, Crateva adansonii, Delonix elata, Dichrostachys cinerea, Diospyros scabra, Dobera glabra, Euphorbia grandicornis, Euphorbia tirucalli, Grevia tenax, G. bicolor, G. villosa*, Haplocoelum foliolosum, Harrisonia abyssinica, Lannea rivae* [as *L. floccosa*], *Lepisanthes senegalensis, Maerua crassifolia*, M. oblongifolia, M. subcordata, Ormocarpum trichocarpum*, Ozoroa insignis* [as *Heeria reticulata*], *Premna resinosa*, Rhus natalensis, Salvadora persica, Sesamothamnus busseanus, Sterculia sp.* (= *Carr* 340), *Suaeda monoica, Tamardindus indica, Tapura fischeri, Tarenna graveolens, Terminalia brevipes, Uvaria leptocladon, Ximenia americana, Zanthoxylum chalybeum* [as *Fagara chalybea*], *Ziziphus mauritiana,* and *Z. mucronata.*

Although not all mentioned by Friis et al. (2010) in their species lists for the *Acacia-Commiphora bushland* (ACB), several other species in Carr's list from the Omo Valley are more associated with *Acacia-Commiphora bushland* (ACB) than with *Combretum-Terminalia woodland and wooded grassland* (CTW), for example *Delonix elata, Lannea rivae,* and *Sesamothamnus busseanus*, while *Balanites aegyptiaca* occurs in both vegetation types.

Schloeder (1999) recorded the woody plants listed below from the area they studied in the Omo National Park. There is a much higher number of species belonging to the family Capparaceae (*Boscia, Cadaba, Capparis,* and *Maerua*) than in typical *Combretum-Terminalia* woodland. Again, the species marked with an asterisk are mentioned in Friis et al. (2010) as typical of the *Acacia-Commiphora bushland* (ACB):

Acacia brevispica, A. ehrenbergiana, A. etbaica, A. horrida*, A. mellifera, A. nilotica, A. paolii*, A. reficiens*, A. senegal, A. seyal, Adenium obesum, Albizia anthelmintica, Balanites aegyptiaca, B. rotundifolia*, Boscia angustifolia, Cadaba farinosa, C. gillettii, C. glandulosa, C. mirabilis, C. rotundifolia, Capparis tomentosa, Combretum aculeatum, C. adenogonium, C. hereroense*, C. molle, Commiphora africana, C. bruceae, Cordia sinensis* [as *Cordia gharaf*], *Dichrostachys cinerea, Diospyros scabra, Dobera glabra, Euphorbia breviarticulara, Ficus vasta, Grewia tenax, G. velutina, G. villosa, Harrisonia abyssinica, Lannea fruticosa, Lannea rivae, L. schweinfurthii, Maerua aethiopica, M. angolensis, M. crassifolia, M. oblongifolia, M. pseudopetalosa, M. subcordata*, Maytenus senegalensis, Ormocarpum trichocarpum, Ozoroa insignis, Pappea capensis, Premna resinosa*, Rhus natalensis, Salvadora persica, Sclerocarya birrea, Sterculia africana, Stereospermum kunthianum, Tapura fischeri, Terminalia brownii, T. spinosa*.*

The above lists show a floristic composition of the lower Omo Valley, at least on the plains at least to 40 km on either side of the river (but away from the riverine forest), that contain a notable element of the *Acacia-Commiphora bushland* (ACB), but also some species of the *Combretum-Terminalia woodland and wooded grassland* (CTW), such as *Lannea fruticosa, Sclerocarya birrea, Stereospermum kunthianum,* and *Terminalia brownii.* In Friis et al. (2010: Plate B4) we have drawn the border between the *Combretum-Terminalia woodland and wooded grassland* (CTW) and the *Acacia-Commiphora bushland* (ACB) to the east of the lower Omo River at Omorate and towards Lake Chew Bahir. It appears from the species-lists above that it would be more floristically correct to draw the CTW-ACB dividing-line somewhere between the mountain of Maji and the lower Omo Valley.

There is not yet sufficient data to specify the more precise position of that line, but from our experience with floristic boundaries elsewhere in Ethiopia it is possible that the line would follow a contour, possibly touching the eastern slopes of Mt. Nkalabong towards the lower Omo Valley. See further comparison with earlier proposed boundaries in Section 11.1 and a preliminary map in Fig. 11-1.

Fig. 5-93. Hill with open woodland and wooded grassland south of Ameya towards the Omo Valley at ca. 950 m a.s.l. (KF floristic region). The site is below the highest point in the Chebera-Churchura National Park towards the Omo River. Early dry season. Image LUXA0258 by Odile Weber. Oct. 2018.

Fig. 5-94. Shrub of *Piliostigma thonningii* in mixed open woodland in the Gojeb Valley at ca. 1100 m a.s.l. (KF floristic region). The site is on the north-facing slopes of the Gojeb Valley south of Jimma towards Ameya. The vegetation is open woodland and wooded grassland. Early dry season. Image LUXA0149 by Odile Weber. Oct. 2018.

Fig. 5-95. Fruiting tree of *Combretum molle* in mixed open woodland south of Ameya towards the Omo Valley at ca. 1000 m a.s.l. (KF floristic region). The site is below the highest point in the Chebera-Churchura National Park towards the Omo River. The vegetation is open woodland and wooded grassland. Early dry season. Image LUXA0365 by Odile Weber. Oct. 2018.

Fig. 5-96. Fruiting tree of *Combretum collinum* in mixed open woodland south of Ameya towards the Omo Valley at ca. 1000 m a.s.l. (KF floristic region). The site is below the Chebera-Churchura National Park towards the Omo River. The vegetation is open woodland and wooded grassland. Early dry season. Image LUXA0290 by Odile Weber. Oct. 2018.

Fig. 5-97. Fruiting tree of *Terminalia laxiflora* in mixed open woodland south of Ameya towards the Omo Valley at ca. 1000 m a.s.l. (KF floristic region). The site is below the Chebera-Churchura National Park towards the Omo River. The vegetation is open woodland and wooded grassland. *Terminalia laxiflora* develops a corky bark on branches after second year's growth. Early dry season. Image LUXA0309 by Odile Weber. Oct. 2018.

Fig. 5-98. Tree of *Vitex doniana* in mixed open woodland south of Ameya towards the Omo Valley at ca. 1000 m a.s.l. (KF floristic region). The site is below the Chebera-Churchura National Park towards the Omo River. The vegetation is open woodland and wooded grassland. Early dry season. Image LUXA0320 by Odile Weber. Oct. 2018.

Fig. 5-99. Low bushland and grassland at the lower reaches of the Omo River at ca. 500 m a.s.l. (GG floristic region). The alluvial plains around the river are mainly covered by grasslands with scattered scrub, but here, near the small town of Omorate, there are more extensive bushlands. The woody vegetation is dominated by *Commiphora habessinica* (*C. kua*). Image 3598 by Ib Friis. Rainy season. Nov. 2011.

94

95

96

97

98

99

6

The woody plants in the sample; their distribution, ecological range and floristic element

Introduction

This Section presents maps and information from (1) the records made for each species during our field work (ca. 1965 observations in 151 relevés), (2) our specimens collected during field work in Ethiopia since 1970, information from herbarium studies at ETH, FT and K and the study of scanned specimens at BR and WAG (ca. 8280 georeferenced collections), and (3) information from the literature for the same species, which has been compiled from the most recent floristic works on the Horn of Africa (Eritrea, Somalia, Kenya, Uganda, Sudan and South Sudan) and adjacent countries, particularly those of East Africa (Uganda, Kenya, Tanzania).

In this Section, where we repeatedly refer to the same works for species after species, we have for brevity used a shorthand defined in Table 6-1 for standard floristic works of the region. For the *Flora of Tropical East Africa*, which has appeared in many fascicles with one family in each, the exact reference to each fascicle can be seen in Table 6-2. We have followed the classification in these works, particularly the *Flora of Ethiopia and Eritrea* and the *Flora of Tropical East Africa,* as closely as possible, despite recent taxonomic changes. The phytogeography and ecology of the species is the major point here, not the taxonomy, and following the most used floristic works will make the presentation easier to use for people who are familiar with the standard works. In case newer taxonomic views have resulted in name changes in relation to the standard works, we have added information about this. In *The Plants of Sudan and South Sudan*, the family classification and the sequence of families follows APG III (2009), and we have noted where this has affected family delimitation and names. Regarding the genus *Acacia*, we have followed the traditional view, but have added the new names in the genera *Vachellia* and *Senegalia*.

Under the name, synonymy, and references to the literature for each species, we have several subheadings: *Taxonomy, Description, Habitat, Distribution in Ethiopia, Altitudinal range in Ethiopia, General distribution,* and *Chorological classification.* The sources, type and use of this information are the following:

Taxonomy: This part presents the accepted name and reviews subjects of taxonomic importance. We have listed infraspecific taxa and made assumptions about the importance of infraspecific variation. Wherever possible, all African infraspecific taxa have been mentioned, and it has been assessed whether they are relevant for this study, or if it has been possible to use these taxa in the mapping and if they are relevant for our ecological considerations.

Description: A short synthesis of characters that appear relevant for the ecology of the species, particularly the habit and height of the plant, whether it has special features such as thorns, thick bark, large underground tubers or other feature that can protect the plant against grass fires, etc., leaf size, and whether the plant is deciduous or evergreen. Unfortunately, information about the latter is very deficient, both in the literature and on labels on herbarium specimens, and in most cases, it has been necessary to rely on our field experience and assumptions based on this. It is important to note that the distinction between 'evergreen' and 'deciduous' is not so sharp in the tropical regions as it is in the temperate regions. Some genera may shed all their leaves on one side of the tree while the leaves remain on the other side, a phenomenon we have often seen on specimens in the genus *Ficus*, which are here considered to be deciduous. The leaf sizes are classified according to Raunkiær's leaf-size classes (Raunkiær 1934): Leptophyll: less than

25 mmsq; Nanophyll: 25-225 mmsq; Microphyll: 225-2,025 mmsq; Mesophyll: 2,025-18,225 mmsq; Macrophyll: 18,225-164,025 mmsq; Megaphyll: greater than 164,025 mmsq). This data has mainly been gathered from the FEE and FTEA by multiplying median length of leaves and leaflets with their median width. The data sets are used in an ecological analysis of the species observed in our relevés, see Section 8.

Habitat: Information about the habitats of the species has been compiled from the floristic works of Ethiopia, Eritrea, Somalia, Sudan and South Sudan (FEE; FS; FTNA; PSS) and compared with our own observations from the field or from our herbarium studies. The information has been used is our assessment of the 'Distribution type' (phytogeographical element or sub-element, see definitions in Section 4.2) and in a phytogeographical analysis of the species observed in our relevés, see Section 7.

Distribution in Ethiopia: A review of the information from the *Flora of Ethiopia and Eritrea* (FEE), compared with the information from our own observations from the field and the herbarium studies. Our information about the distribution in western Ethiopia is almost always more detailed than the information cited from the FEE, particularly regarding records from the previously relatively poorly known WG floristic region, and the entry is a supplement to the distribution maps. Observations that require future confirmation are pointed out, and particularly doubtful records cited.

Altitudinal range in Ethiopia: A review of the information from the *Flora of Ethiopia and Eritrea* (FEE), compared with our own observations based on field observations and herbarium studies (marked as 'our records'). The information has been used is our assessment of the phytogeographical element or sub-element and used for analyses in Section 7 and 8.

General distribution: A critical review and summary of the information from the literature cited at the accepted name. The information from this data set has been used for an analysis of the distributions and phytogeographical element or sub-element in Section 7.2

Chorological classification: Under this heading, which only refers to the distribution in Ethiopia, are two sets of classifications and terms are used here, one that records actual presence and absence of species in specific areas, irrespective of the general distribution of the species, and another that follows the ideas about phytogeographical elements based phytochoria approximately congruent with vegetation types and named after them.

The phytogeographical elements or sub-elements apply to the distribution in Ethiopia. A review of the phytogeographical concepts behind this classification is presented in Section 4.2 and 7.1.2. See the definition of the categories in Table 6-3. Where necessary, we have argued for our choice of phytogeographical category under the individual species, using field experience and the mapped distribution of the species as the criteria, in the same way as the tree species have been categorized into phytogeographical elements for the evergreen forests of Malawi (White 1970; Dowsett-Lemaire 2001) and phytogeographical elements for the forests of the Horn of Africa (Friis 1992). The classification is used in a phytogeographical analysis of the species observed in our relevés is described further in Section 4.

The definitions of categories based on actual presence and absence of species in specific areas is given in Table 6-3. This classification is used in an ecological analysis of the species observed in our relevés, see further on that in Section 8.

Table 6-1. Abbreviations for floristic works. Used in the headings and synonym blocks in Section 6. Unless referred to in the running text, these works are not included in the List of References.

BS 51	I. Friis & K. Vollesen. (1998). Flora of the Sudan-Uganda border area east of the Nile, 1. Biologiske Skrifter 51(1): 1-389 & I. Friis & K. Vollesen. (2005). Flora of the Sudan-Uganda border area east of the Nile, 2. Biologiske Skrifter 51(2). The two volumes have continuous pagination.
FEE	Various editors, 1989-2009. *Flora of Ethiopia and Eritrea* (Vol. 3 as *Flora of Ethiopia*). Published in eight volumes with separate pagination. National Herbarium of Ethiopia, Addis Ababa University, and Institute of Systematic Botany, Uppsala University.
FS	M. Thulin (ed.), 1993-2006. *Flora of Somalia*. Published in four volumes with separate pagination. Royal Botanic Gardens, Kew.
FTEA	Various editors and authors (see below), 1952-2012. *Flora of Tropical East Africa*. Published in one or more parts for each family, 263 parts in all. Separate pagination. Various publishers. The families relevant here are listed in Table 6-2.
FTNA	I. Friis, 1992. Forests and forest trees of Northeast Tropical Africa, their habitats and distribution patterns in Ethiopia, Djibouti, and Somalia. *Kew Bulletin Additional Series* 15: i-iv, 1-396.
PSS	I. Darbyshire, M. Kordofani, I. Farag, R. Candiga & H. Pickering, 2015. *The Plants of Sudan and South Sudan. An annotated Checklist*. Royal Botanic Gardens, Kew. 400 pp.

Table 6-2. Fascicles of family accounts in the Flora of Tropical East Africa. Used in the headings and synonym blocks in Section 6 as 'FTEA, [abbreviated family name].' Unless referred to in the running text, these works are not included in the List of References.

Anacardiaceae	Kokwaro, J.O. (1986). Flora of Tropical East Africa, Anacardiaceae. A.A. Balkema, Rotterdam.
Annonaceae	Verdcourt, B. (1971). Flora of Tropical East Africa, Annonaceae. Crown Agent, London.
Apiaceae (Umbelliferae)	Townsend, C.C. (1989). Flora of Tropical East Africa, Umbelliferae. A.A. Balkema, Rotterdam.
Apocynaceae, sensu lat.	Omino, E.A. (2002). Flora of Tropical East Africa, Apocynaceae, 1. A.A. Balkema, Rotterdam. Goyder, D., Harris, T., Masinde, S., Meve, U. & Venter, J. (2012). Flora of Tropical East Africa, Apocynaceae, 2. Royal Botanic Gardens, Kew.
Araliaceae	Tennant, J.R. (1968). Flora of Tropical East Africa, Araliaceae. Crown Agent, London.
Arecaceae	Dransfield, J. (1986). Flora of Tropical East Africa, Palmae. A.A. Balkema, Rotterdam.
Asteraceae	Beentje, H.J. (ed.) (2000). Flora of Tropical East Africa, Compositae, 1. A.A. Balkema, Rotterdam.
Balanitaceae	Sands, M.J.S. (2003). Flora of Tropical East Africa, Balanitaceae. A.A. Balkema, Rotterdam.
Bignoniaceae	Bidgood, S., Verdcourt, B. & Vollesen, K. (2006). Flora of Tropical East Africa, Bignoniaceae, & Verdcourt, B. Cobaeaceae. Royal Botanic Gardens, Kew.
Bombacaceae	Beentje, H.J. (1989). Flora of Tropical East Africa, Bombacaceae. A.A. Balkema, Rotterdam.
Boraginaceae	Verdcourt, B. (1991). Flora of Tropical East Africa, Boraginaceae. A.A. Balkema, Rotterdam.
Burseraceae	Gillett, J.B. (1991). Flora of Tropical East Africa, Burseraceae. A.A. Balkema, Rotterdam.

Capparaceae	Elffers, J., Graham, R.A. & De Wolf, G.P. (1964). Flora of Tropical East Africa, Capparidaceae. Crown Agent, London.
Celastraceae	Robson, N.K.B., Hallé, N., Mathew, B. & Blakelock, R. (1994). Flora of Tropical East Africa, Celastraceae. A.A. Balkema, Rotterdam.
Combretaceae	Wickens, G.E. (1973). Flora of Tropical East Africa, Combretaceae. Crown Agent, London.
Ebenaceae	White, F. & Verdcourt, B. (1996). Flora of Tropical East Africa, Ebenaceae. A.A. Balkema, Rotterdam.
Erythroxylaceae	Verdcourt, B. (1984). Flora of Tropical East Africa, Erythroxylaceae. A.A. Balkema, Rotterdam.
Euphorbiaceae	Carter, S. & Radcliffe-Smith, A. (1988). Flora of Tropical East Africa, Euphorbiaceae (Part 2). A.A. Balkema, Rotterdam.
Euphorbiaceae	Radcliffe-Smith, A. (1987). Flora of Tropical East Africa, Euphorbiaceae (Part 1). A.A. Balkema, Rotterdam.
Flacourtiaceae	Sleumer, H. (1993). Flora of Tropical East Africa, Flacourtiaceae. Crown Agent, London.
Guttiferae	Bamps, P., Robson, N. & Verdcourt, B. (1978). Flora of Tropical East Africa, Guttiferae. Crown Agent, London.
Hypericaceae	Milne-Redhead, E. (1953). Flora of Tropical East Africa, Hypericaceae. Crown Agent, London.
Icacinaceae	Lucas, G.L. (1968). Flora of Tropical East Africa, Icacinaceae. Crown Agent, London.
Leguminosae subfam. Caesalpinioideae	Brenan, J.P.M. (1967). Flora of Tropical East Africa, Leguminosae subfamily Caesalpinioideae. Crown Agent, London.
Leguminosae subfam. Mimosoideae	Brenan, J.P.M. (1959). Flora of Tropical East Africa, Leguminosae subfamily Mimosoideae. Crown Agent, London.
Leguminosae subfam. Papilionoideae	Gillett, J.B., Polhill, R.M. & Verdcourt, B. (1971). Flora of Tropical East Africa, Leguminosae subfamily Papilionoideae (1) & (2). Crown Agent, London.
Loganiaceae	Bruce, E.A. & Lewis, J. (1960). Flora of Tropical East Africa, Loganiaceae. Crown Agent, London.
Lythraceae	Verdcourt, B. (1994). Flora of Tropical East Africa, Lythraceae. A.A. Balkema, Rotterdam.
Meliaceae	Styles, B.T. & White, F. (1991). Flora of Tropical East Africa, Meliaceae. A.A. Balkema, Rotterdam.
Melianthaceae	Verdcourt, B. (1958). Flora of Tropical East Africa, Melianthaceae. Crown Agent, London.
Moraceae	Berg, C.C. (1989). Flora of Tropical East Africa, Moraceae. A.A. Balkema, Rotterdam.
Myrsinaceae	Halliday, P. (1984). Flora of Tropical East Africa, Myrsinaceae. A.A. Balkema, Rotterdam.
Myrtaceae	Verdcourt, B. (2001). Flora of Tropical East Africa, Myrtaceae. A.A. Balkema, Rotterdam.
Ochnaceae	Verdcourt, B. (2005). Flora of Tropical East Africa, Ochnaceae. Royal Botanic Gardens, Kew.
Olacaceae	Lucas, G.L. (1968). Flora of Tropical East Africa, Olacaceae. Crown Agent, London.
Oleaceae	Turrill, W.B. (1952). Flora of Tropical East Africa, Oleaceae. Crown Agent, London.
Poaceae (Gramineae)	Clayton, W.D. (1970). Flora of Tropical East Africa, Gramineae, 1. Crown Agent, London.
Polygalaceae	Paiva, J. (2004). Flora of Tropical East Africa, Polygalaceae. Royal Botanic Gardens, Kew.
Polygonaceae	Graham, R.A. (1958). Flora of Tropical East Africa, Polygonaceae. Crown Agent, London.
Proteaceae	Brummitt, R.K. & Marner, S.K. (1993). Flora of Tropical East Africa, Proteaceae. A.A. Balkema, Rotterdam.
Rhamnaceae	Johnston, M.C. (1972). Flora of Tropical East Africa, Rhamnaceae. Crown Agent, London.
Rubiaceae	Bridson, D. & Verdcourt, B. (1988). Flora of Tropical East Africa, Rubiaceae, 2. A.A. Balkema, Rotterdam. Verdcourt, B. & Bridson, D. (1991). Flora of Tropical East Africa, Rubiaceae, 3. A.A. Balkema, Rotterdam.

Rutaceae	Kokwaro, J.O. (1982). Flora of Tropical East Africa, Rutaceae. A.A. Balkema, Rotterdam.
Salvadoraceae	Verdcourt, B. (1968). Flora of Tropical East Africa, Salvadoraceae. Crown Agent, London.
Sapindaceae	Davies, F.G. & Verdcourt, B. (1998). Flora of Tropical East Africa, Sapindaceae. A.A. Balkema, Rotterdam.
Simaroubaceae	Stannard, B. (2000). Flora of Tropical East Africa, Simaroubaceae. A.A. Balkema, Rotterdam.
Sterculiaceae	Cheek, M. & Dorr, L. (2007). Flora of Tropical East Africa, Sterculiaceae. Royal Botanic Gardens, Kew.
Thymelaeaceae	Peterson, B. (1978). Flora of Tropical East Africa, Thymelaeaceae. Crown Agent, London.
Tiliaceae	Whitehouse, C., Cheek, M., Andrews, S. & Verdcourt, B. (2001). Flora of Tropical East Africa, Tiliaceae & Muntingiaceae. A.A. Balkema, Rotterdam.
Ulmaceae	Polhill, R.M. (1966). Flora of Tropical East Africa, Ulmaceae. Crown Agent, London.
Verbenaceae	Verdcourt, B. (1992). Flora of Tropical East Africa, Verbenaceae. A.A. Balkema, Rotterdam.

Table 6-3. The chorological categories based on element and sub-element. Used for phytogeographical analyses (non-overlapping; only one category per taxon possible).

CTW[all]	Distributed in the *Combretum-Terminalia* woodlands, from north to south or almost so; no eastward extension.
CTW[all+ext]	Distributed in the *Combretum-Terminalia* woodlands, from north to south or almost so, with extension to the Rift Valley and east of the Rift.
CTW[n]	Distributed in the northern part of the *Combretum-Terminalia* woodlands, mainly north of the Abay River; no eastward extension.
CTW[n+c]	Distributed in the northern and central part of the *Combretum-Terminalia* woodlands, reaching the Gambela Region in the south; no eastward extension.
CTW[s]	Distributed in the southern part of the *Combretum-Terminalia* woodlands, mainly south of the Abay River; no eastward extension.
CTW[s+ext]	Distributed in the southern part of the *Combretum-Terminalia* woodlands, mainly south of the Abay River and with extension to the Rift Valley and east of the Rift.
CTW[c]	Distributed in the central part of the *Combretum-Terminalia* woodlands, mainly south of the Abay River; no eastward extension.
CTW[c+s+ext]	Distributed in the central and southern part of the *Combretum-Terminalia* woodlands, mainly south of the Abay River; with extension to the Rift Valley and east of the Rift.
ACB[-CTW]	Distributed in the *Acacia-Commiphora bushland* (ACB), marginally intruding in the *Combretum-Terminalia woodland and wooded grassland* (CTW) in the north or in the south, or both.
MI	Marginally intruding species in the western woodlands, mainly with a main distribution in the highlands; usually with one or few records in our relevés.
TRG	Transgressing species, occurs similarly in both the *Combretum-Terminalia woodland and wooded grassland* (CTW) and in other vegetation types, particularly *Dry evergreen Afromontane forest* (DAF).

Table 6-4. Geographical areas used to record distribution based on direct observation. The observations have been made on the distribution maps in Section 6 and are used for the ecological analyses in this study. It is possible to record species in one or more areas for this.

WWn	Occurs in the northern part of the western woodlands
WWc	Occurs in the central part of the western woodlands
WWs	Occurs in the northern part of the western woodland
EWW	Occurs in areas to the east of the western woodland (mainly in the *Transitional semi-evergreen bushland* (TSEB)
ASO	Occurs in areas dominated by *Acacia-Commiphora bushland* (ACB), mainly in the Afar and Somalia regions

6.1. Annonaceae

FTEA, Annonaceae: 1-132 (1971); *BS* 51: 63-66 (1998); FEE 2(1): 3-12 (2000); PSS: 71-73 (2015).
Other characteristic western woodland species in this family occur in Ethiopia but were not seen in our relevés. They are: **Hexalobus monopetalus** (A. Rich.) Engl. & Diels (1901), which occurs in tropical Africa from Senegal to Ethiopia and south to Angola and South Africa, and **Xylopia longipetala** De Wild. & T. Dur. (1899) (previously known as **Xylopia parviflora** (A. Rich.) Benth. (1862), *nom. Illeg.*), also in tropical Africa from Sierra Leone to Ethiopia and south to South Africa. Both are known from the woodlands of the Gambela Region (FEE) and their distribution type in Ethiopia would be CTW[c].

6.*1.1.* **Annona senegalensis** *Pers. (1806).*

FTEA, Annonaceae: 113 (1971); *BS* 51: 63 (1998); FEE 2(1): 11 (2000); PSS: 71 (2015).
Taxonomy: No infraspecific taxa.
Description: Shrub or small tree (0.8-) 1.5-10 m tall. No thorns. Leaves simple, apparently deciduous. Mesophyll-Macrophyll. The rough and corrugated bark on old trunks may be an adaptation for fire resistance.
Habitat: 'Wooded grassland on greyish clay, mostly in combretaceous woodland' (FEE). 'Wooded grassland' (PSS). In relatively humid and species-rich woodlands on greyish or brownish soils (our observation).
Distribution in Ethiopia: GJ WG IL KF GG BA (FEE). The isolated FEE record from BA is based *Mesfin Tadesse et al.* 5221 (ETH) from near the Harenna Forest (Fig. 6-1).
Altitudinal range in Ethiopia: 500-1500 m a.s.l. (FEE). 500-1600 m a.s.l. (our records).
General distribution: From Senegal to Ethiopia, south to Angola and northern South Africa; also in Madagascar, Comoro, and Cape Verde Is.
Chorological classification: Distribution according to direct observation. [WWn], WWc, WWs, EWW. *Local phytogeographical element and sub-element.* CTW[c+s+ext]. Widely distributed in the central and southern part of the *Combretum-Terminalia* woodlands, extending to the Rift Valley and to the east of the Rift.

6.2. Capparaceae

FTEA, Capparidaceae: 1-88 (1964), as **CAPPARIDACEAE**; FS 1: 37-60 (1993), as Capparaceae; *BS* 51: 76-81 (1998); FEE 2(1): 11-120 (2000), as CAPPARIDACEAE (*Capparaceae*); PSS: 262-264 (2015).
Several species of **Capparis** other than *C. decidua* and *C. tomentosa* are known from the western woodlands, including *C. erythrocarpos* Isert (1789), in tropical Africa from Guinea-Bissau to Ethiopia and south to Angola, and *C. sepiaria* L. (1762), in tropical Africa from Senegal to Ethiopia and south to South Africa, but these species were not observed in our relevés.

6.2.1. **Boscia angustifolia** *A. Rich. (1831).*

FTEA, Capparidaceae: 55 (1964); FS 1: 46 (1993); FEE 2(1): 114 (2000); PSS: 262 (2015).
Taxonomy: In East Africa two varieties, *Boscia angustifolia* var. *angustifolia* and *B. angustifolia* var. *corymbosa* (Gilg) De Wolf, only distinguished on slight differences in the leaf-indumentum. Recognition of infraspecific taxa not relevant here.
Description: Shrub or small tree to 8 m tall. No thorns.

Leaves simple, presumably evergreen. Microphyll-mesophyll. The rough bark on old trunks may be an adaptation for fire resistance.
Habitat: 'Deciduous open woodland and bushland, and grassland with scattered trees, scrub' (FEE). 'Deciduous bushland, also on rocky and stony soil' (FS). 'Wooded grassland' (PSS). Dry woodland, open woodland, or bushland (our observation).
Distribution in Ethiopia: AF TU GD SD GG (FEE). Restricted to the driest areas in the north and the south (Fig. 6-2).
Altitudinal range in Ethiopia: (50-) 1000-1900 m a.s.l. (FEE, including range in Eritrea). 550-1350 (-1600) m a.s.l. (our records).
General distribution: From Senegal to Ethiopia, Eritrea and Somalia, south to Mozambique, north to Egypt; also in Arabia.
Chorological classification: Distribution according to direct observation. **WWn, WWs, ASO**. *Local phytogeographical element and sub-element.* ACB[-CTW]. The species occurs in the southern *Acacia-Commiphora bushlands* (ACB) but is absent from the south-eastern lowlands. However, due to its distribution in the Sahel across Africa it has been classified as ACB[-CTW].

6.2.2. **Boscia salicifolia** *Oliv. (1868).*

FTEA, Capparidaceae: 52 (1964); *BS* 51: 76 (1998); FEE 2(1): 116 (2000); PSS: 262 (2015).
Taxonomy: No infraspecific taxa.
Description: Large shrub or tree to 10 (-14) m tall. No thorns. Leaves simple, presumably evergreen. Microphyll-mesophyll. Bark on old trunks thickened, but probably not adaptation for fire resistance.
Habitat: 'Dense savanna or bushland, *Dodonaea-Euclea-Olea* scrub, grassland with scattered trees, on rocky slopes' (FEE). 'Wooded grassland' (PSS). Mainly in dry bushland, common in the dry grasslands in the Rift Valley (our observation).
Distribution in Ethiopia: TU WU SU AR KF SD GG HA (FEE, including range in Eritrea). The FEE records from TU and WU are not confirmed, but records from Eritrea makes it likely that it also occurs there (Fig. 6-3).
Altitudinal range in Ethiopia: 800-1900 m a.s.l. (FEE). 500-1500 (-1800) m a.s.l. (our records).
General distribution: From Senegal to Ethiopia and Eritrea, south to Zimbabwe and Mozambique.
Chorological classification: Distribution according to direct observation. **[WWn], WWc, WWs, EWW, ASO.** *Local phytogeographical element and sub-element.* **CTW[c+s+ext]**. The distribution in dry *Combretum-Terminalia woodland and wooded grassland* (CTW) is too wide and the distribution in the *Acacia-Commiphora bushland* is too restricted to classify it as ACB[-CTW]. It extends to the Rift Valley and to the east of the Rift.

6.2.3. **Boscia senegalensis** *Lam. (1819).*

FEE 2(1): 116 (2000); PSS: 262 (2015).
Taxonomy: No infraspecific taxa.
Description: Shrub or small tree to 5 m tall. No thorns. Leaves simple, presumably evergreen. Microphyll-mesophyll. Apparently, no adaptation for fire resistance.
Habitat: 'Deciduous woodland, semi-desert scrub, riverine formations on rocky outcrops or on sandy soil' (FEE). 'Semi-desert scrub' (PSS). Mainly in dry bushland and woodland, and in semi-desert scrub in the AF floristic region (our observation).
Distribution in Ethiopia: SD BA [but with records from EW] (FEE). Also in the floristic regions TU, GD and AF (Fig. 6-4).
Altitudinal range in Ethiopia: 700-1300 m a.s.l. (FEE, including range in Eritrea). 550-1650 m a.s.l. (our records).
General distribution: From Senegal to Ethiopia and Eritrea; also in North Africa.
Chorological classification: Distribution according to direct observation. **WWn, ASO.** *Local phytogeographical element and sub-element.* **ACB[-CTW].** The species is largely absent from the south-eastern lowlands and absent from Somalia. Due to its distribution in the Sahel across Africa, it has been classified as ACB[-CTW].

6.2.4. **Cadaba farinosa** *Forssk. (1775).*

FTEA, Capparidaceae: 75 (1964); FS 1: 50 (1993); FEE 2(1): 88 (2000); *BS* 51: 76 (1998); PSS: 262 (2015).

Fig. 6-1. *Annona senegalensis*. Distribution in Ethiopia. Observed records are marked with red dots on vegetation map reproduced from Fig. 4-1.

Fig. 6-2. *Boscia angustifolia*. Distribution in Ethiopia. Observed records are marked with red dots on vegetation map reproduced from Fig. 4-1.

Fig. 6-3. *Boscia salicifolia*. Distribution in Ethiopia. Observed records are marked with red dots on vegetation map reproduced from Fig. 4-1.

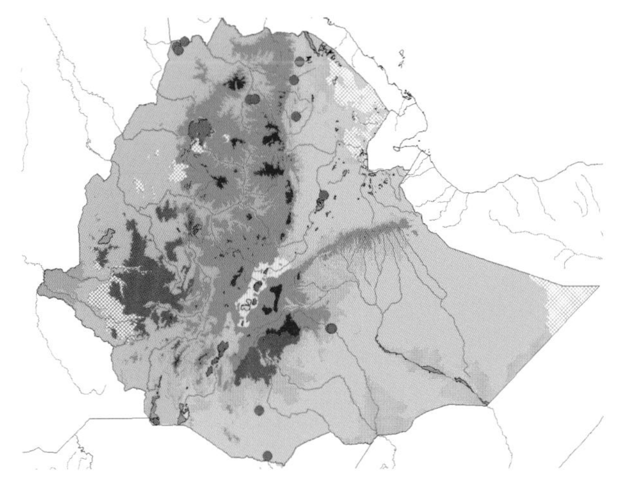

Fig. 6-4. *Boscia senegalensis*. Distribution in Ethiopia. Observed records are marked with red dots on vegetation map reproduced from Fig. 4-1.

Cadaba farinosa subsp. **farinosa**
FTEA, Capparidaceae: 75 (1964); FS 1: 50 (1993); FEE 2(1): 88 (2000); *BS* 51: 76 (1998); PSS: 262 (2015).
Taxonomy: Two subspecies have been recognized, *Cadaba farinosa* subsp. *farinosa* and *C. farinosa* subsp. *adenotricha* (Gilg. & Gilg.-Ben.) R.A. Graham (1963). According to PSS, *C. farinosa* subsp. *farinosa* is recorded from drier habitats than the other subspecies, and only *C. farinosa* subsp. *farinosa* is known from Ethiopia. Recognition of infraspecific taxa not relevant here.
Description: Shrub 1-5 m tall. No thorns. Leaves simple, presumably evergreen. Microphyll. No obvious adaptations for fire resistance.
Habitat: 'Grasslands with scattered trees, deciduous bushland, semi-desert scrub, riverine formations' (FEE). '*Acacia-Commiphora* bushland, semidesert open bushland, riverine formations, dry riverbeds, on a range of soils' (FS for *Cadaba farinosa* subsp. *farinosa*). 'Grassland, bushland, semi-desert' (PSS for *Cadaba farinosa* subsp. *farinosa*). Very common in many types of dry habitats, in riparian formations in the big river valleys only in the dry transition zone (our observation).
Distribution in Ethiopia: AF TU GD GJ WU SU AR WG IL KF GG SD BA [FEE]. Also records from HA (Fig. 6-5).
Altitudinal range in Ethiopia: Sea level to 2000 m a.s.l. (FEE, including range in Eritrea). 600-1800 m a.s.l. (our records).
General distribution (species as a whole): Widespread across tropical Africa from Senegal to Ethiopia, Eritrea, and Somalia, south to Angola and Tanzania; also in Egypt, Arabia, Iran and Pakistan.
Chorological classification: Distribution according to direct observation. **WWn, WWc, WWs, ASO.** *Local phytogeographical element and sub-element.* **ACB[-CTW]**. The species occurs in the drier parts of the western lowlands in the north, in the Afar Region and the upper reaches of the Abay Gorge; it is extremely common in the drier parts of the Rift Valley, and has scattered occurrences in the south-eastern lowlands. The records from the Gambela lowlands do not agree with the normal pattern for **ACB[-CTW]**; however, due to the major part of its area of distribution, it has been classified as **ACB[-CTW]**.

6.2.5. Capparis decidua *(Forssk.) Edgew. (1862).*

FS 1: 52 (1993); FEE 2(1): 94 (2000); PSS: 263 (2015).
Taxonomy: No infraspecific taxonomy.
Description: Large shrub to 4 m tall. Stipular thorns straight, up to 4 mm long, branchlets become spine-tipped when old. Seemingly leafless but does have minute leaves (Nanophyll) which are soon falling (deciduous). No obvious adaptations for fire resistance.
Habitat: 'Grassland with scattered trees, *Acacia* scrub, semidesert scrub, on sandy or gravelly plains or on rocky ground' (FEE). 'Wooded grassland, semi-desert' (PSS). Only in very dry habitats, and mainly on sandy soils (our observation).
Distribution in Ethiopia: AF TU GD HA (FEE). Also according to our records (Fig. 6-6).
Altitudinal range in Ethiopia: Sea level to 1600 m a.s.l. (FEE, including range in Eritrea). 500-1550 m a.s.l. (our records).
General distribution: From Mauritania and Senegal to Ethiopia and Eritrea [not yet found in Somalia, where it according to FS is expected to occur], also in Socotra, North Africa, Arabia, Egypt (Sinai), Iran, Pakistan, and India.
Chorological classification: Distribution according to direct observation. **WWn, ASO.** *Local phytogeographical element and sub-element.* **ACB[-CTW]**. The species occurs in the driest parts of the northern woodlands and is recorded from Afar (AF), but it is absent from the south-western lowlands, where ACB species often occur. It occurs in the *Transitional Semi-Evergreen Bushland* (TSEB) in eastern Ethiopia and in the Ogaden but is so far unknown from Somalia. With its distribution in the Sahel across Africa and despite its limited distribution on the Horn of Africa, it has been classified as **ACB[-CTW]**.

6.2.6. Capparis tomentosa *Lam. (1785).*

FTEA, Capparidaceae: 62 (1964); FS 1: 52 (1993); *BS* 51: 78 (1998); FEE 2(1): 95 (2000); PSS: 263 (2015).

Taxonomy: Although FTEA states that there is considerable variation within this species as to the degree of pubescence and the size of the leaves, the variation is continuous. Recognition of infraspecific taxa not relevant here.
Description: Shrub, often scrambling, to 4 m tall. Hooked stipular thorns up to 6 mm. Leaves simple, some falling early, but mature leaves presumably evergreen. Microphyll-mesophyll. No obvious adaptations for fire resistance.
Habitat: 'Wooded grassland, *Acacia-Commiphora* shrub, forest margins, often in riverine forest and at temporary ponds' (FEE). 'Deciduous bushland, riverine woodland' (FS). 'Bushland, wooded grassland' (PSS). In many types of scrub-vegetation, often together with other evergreen shrubs (our observation).
Distribution in Ethiopia: AF GD WU SU AR WG IL KF GG SD HA (FEE). Also in the TU floristic region (Fig. 6-7).
Altitudinal range in Ethiopia: 500-2200 m a.s.l. (FEE, including range in Eritrea). 500-2000 m a.s.l. (our records).
General distribution: From Senegal to Ethiopia, Eritrea, and Somalia, south to Namibia and South Africa; also in Arabia and the Mascarenes.
Chorological classification: Distribution according to direct observation. **WWn, WWc, WWs, ASO.** *Local phytogeographical element and sub-element.* **ACB[-CTW].** The species occurs in the northern and southern parts of the western lowlands but is rather rare in the central part of the *Combretum-Terminalia woodland and wooded grassland* (CTW). It is very common in parts of the Rift Valley and to the east of the Rift. With its distribution in the Sahel across Africa and despite its limited distribution on the Horn of Africa, it has been classified as **ACB[-CTW]**.

6.2.7. Maerua angolensis *DC. (1824).*

FTEA, Capparidaceae: 28 (1964); FS 1: 40 (1993); *BS* 51: 79 (1998); FEE 2(1): 108 (2000); PSS: 264 (2015).
Taxonomy: Another subspecies, *Maerua angolensis* subsp. *socotrana* (Schweinf. ex Balf.f.) Kers, is endemic on Socotra. The Ethiopian material, *Maerua angolensis* subsp. *angolensis*, is homogenous. Recognition of infraspecific taxa not relevant here.
Description: Shrub or slender tree to 7 m tall. No thorns. Leaves simple, probably evergreen. Microphyll-mesophyll. No obvious adaptations for fire resistance.
Habitat: 'Montane *Acacia* woodland with tall grass, shrubland and bushland, on rocky slopes, in soil derived from limestone, granite or volcanic rocks' (FEE for *Maerua angolensis* subsp. *angolensis*), 'open *Acacia-Commiphora* bushland, thorn bush, in shallow soil on limestone and red sand' *Maerua angolensis* subsp. *socotrana*). 'Deciduous bushland' (FS for species as a whole). 'Bushland, wooded grassland' (PSS). In a wide range of vegetation types together with other bushes or small trees (our observation).
Distribution in Ethiopia: TU GD WU SU KF SD BA HA (FEE for species as a whole). Also in the GJ (middle Abay Gorge) and WG floristic regions (Fig. 6-8).
Altitudinal range in Ethiopia: (500-) 1200-1900 m a.s.l. (FEE, including range in Eritrea). 600-2000 m a.s.l. (our records).
General distribution (species as a whole): Widespread across tropical Africa from Senegal to Ethiopia, Eritrea, and Somalia, south to Angola and South Africa; also on Socotra.
Chorological classification: Distribution according to direct observation. **WWn, WWc, WWs, ASO.** *Local phytogeographical element and sub-element.* **ACB[-CTW].** The species occurs in the northern and southern parts of the western woodlands and in parts of the south-eastern lowlands. Apart from two records near Kurmuk in the WG floristic region, it is lacking from the central part of the *Combretum-Terminalia woodland* (CTW). With its distribution in the Sahel across Africa and despite its limited distribution on the Horn of Africa, it has been classified as **ACB[-CTW]**.

6.2.8. Maerua oblongifolia *(Forssk.) A. Rich. (1847).*

FTEA, Capparidaceae: 37 (1964); FS 1: 43 (1993); *BS* 51: 79 (1998); FEE 2(1): 112 (2000); PSS: 264 (2015).
Taxonomy: FTEA states that *Maerua oblongifolia* is

Fig. 6-5. *Cadaba farinosa*. Distribution in Ethiopia. Observed records are marked with red dots on vegetation map reproduced from Fig. 4-1.

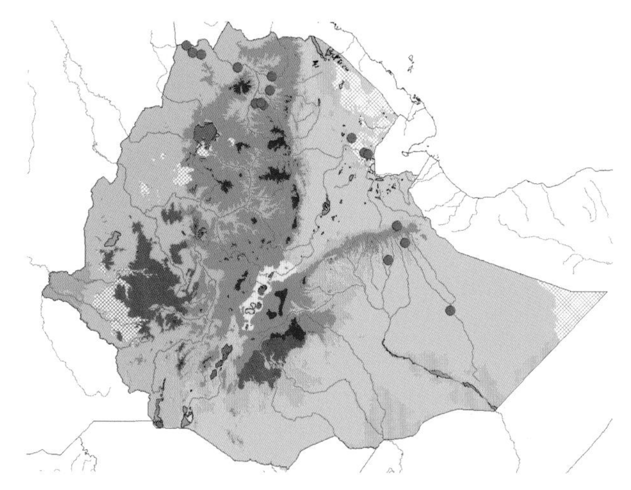

Fig. 6-6. *Capparis decidua*. Distribution in Ethiopia. Observed records are marked with red dots on vegetation map reproduced from Fig. 4-1.

Fig. 6-7. *Capparis tomentosa*. Distribution in Ethiopia. Observed records are marked with red dots on vegetation map reproduced from Fig. 4-1.

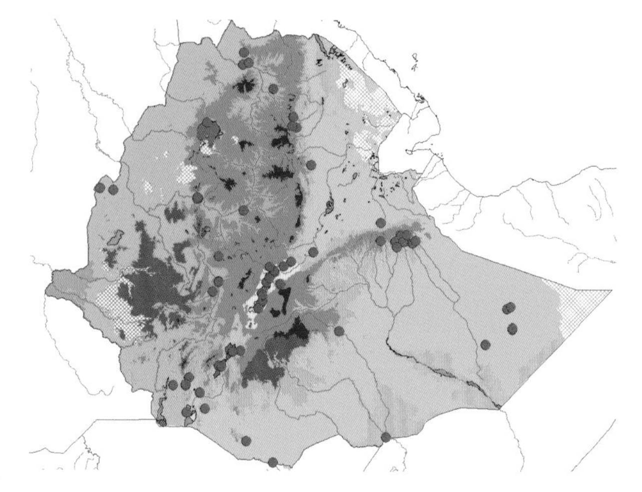

Fig. 6-8. *Maerua angolensis*. Distribution in Ethiopia. Observed records are marked with red dots on vegetation map reproduced from Fig. 4-1.

an exceedingly variable species, but the variation is continuous and no infraspecific taxa are recognized. Recognition of infraspecific taxa not relevant here.
Description: Scrambling shrub to 3 m tall or a dwarf shrub sprouting from woody root crown. No thorns. Leaves simple, presumably evergreen. Microphyll. The ability to sprout from woody root crown is presumably an adaptation for fire resistance.
Habitat: 'Deciduous bushland with scattered trees, grassland and scrub, usually dry, stony sandy places, sometimes in wadis' (FEE). 'Deciduous to semi-evergreen bushland' (FS). 'Bushland, semi-desert' (PSS). In a wide range of vegetation type together with other bushes or small trees (our observation).
Distribution in Ethiopia: GD SU IL KF GG SD HA (FEE). Also in the TU (Tacazze Valley) and BA floristic regions (Fig. 6-9).
Altitudinal range in Ethiopia: Sea level-1800 m a.s.l. (FEE, including range in Eritrea). 600-1700 m a.s.l. (our records).
General distribution: From Senegal to Ethiopia, Eritrea, and Somalia, south to Uganda and Kenya; also in Egypt and Arabia.
Chorological classification: Distribution according to direct observation. **WWn, WWc, WWs, ASO.** *Local phytogeographical element and sub-element.* ACB[-CTW]. The species occurs in the north and the south of the western woodlands, but, like *Cadaba farinosa,* it is recorded from the lowlands of the Gambela Region. It is common in the Rift Valley, and east of the Rift, with a few records in the south of the *Acacia-Commiphora bushland* (ACB). With its distribution in the Sahel across Africa and despite its limited distribution on the Horn of Africa, it has been classified as ACB[-CTW].

6.2.9. **Maerua triphylla** *A. Rich. (1847).*

FTEA, Capparidaceae: 43 (1964); FS 1: 40 (1993); *BS* 51: 80 (1998); FEE 2(1): 105 (2000); PSS: 264 (2015).
Taxonomy: *Maerua triphylla* var. *calophylla* (Gilg) DeWolf may occur in slightly moister habitats than *M. triphylla* var. *johannis* (Volkens & Gilg) DeWolf (1962) and *M. triphylla* var. *pubescens* (Klotsch) DeWolf (1962), but we have not been able to distinguish these taxa in our Ethiopian material. Recognition of infraspecific taxa might be relevant, but has not been possible here.
Description: Small tree to 9 m tall. No thorns. Leaves simple or trifoliolate, probably evergreen. Leaves or leaflets mesophyll. No obvious adaptation for fire resistance.
Habitat: 'Limestone slopes, roadsides, beside rivers and in dry river beds, *Combretum-Terminalia* woodland near river on gravelly, shallow soil, open *Acacia-Commiphora* or tree *Euphorbia* woodland, montane evergreen shrub, on rocks or on heavy, loamy or alluvial soil' (FEE for species as a whole, compiled information). 'Semi-evergreen bushland' (FS *Maerua triphylla* var. *calophylla* and *Maerua triphylla* var. *pubescens*), 'rocky hillside' (FS for *Maerua triphylla* var. *johannis*). 'Bushland, riverine forest margins, often on rocky ground' (PSS for *Maerua triphylla* var. *calophylla*); 'bushland, grassland' (PSS, for *Maerua triphylla* var. *triphylla*). In a wide range of vegetation type together with other bushes or small trees, ranging from sometimes dry riverine habitat to dry bushland (our observation).
Distribution in Ethiopia: GJ SU AR WG(?) IL GG SD BA HA (FEE). The FEE records of this species in the GJ or WG floristic regions are not confirmed, but records from the Abay Valley may be the source of the record from GJ. Additionally, we have recorded it from the KF floristic region (Fig. 6-10).
Altitudinal range in Ethiopia: 500-1500 (-2100) m a.s.l. (FEE). 600-1700 m a.s.l. (our records).
General distribution (species as a whole): From Sudan and South Sudan to Ethiopia and Somalia, south to D.R. Congo and Tanzania; also in Yemen.
Chorological classification: Distribution according to direct observation. **[WWn], WWc, WWs, ASO.** *Local phytogeographical element and sub-element.* ACB[-CTW]. The species is absent from northern Ethiopia north of the Abay Gorge, it is common in the lower Omo Valley and at Lake Chew-Bahir. It occurs in the Rift Valley and east of the Rift. The species has been observed to be riverine, but also occur in deciduous bushland; these two extremes may represent infraspecific taxa. It is possible that further studies may show that infraspecific taxa belong to different local elements or sub-elements, but our present information does not

Fig. 6-9. *Maerua oblongifolia*. Distribution in Ethiopia. Observed records are marked with red dots on vegetation map reproduced from Fig. 4-1.

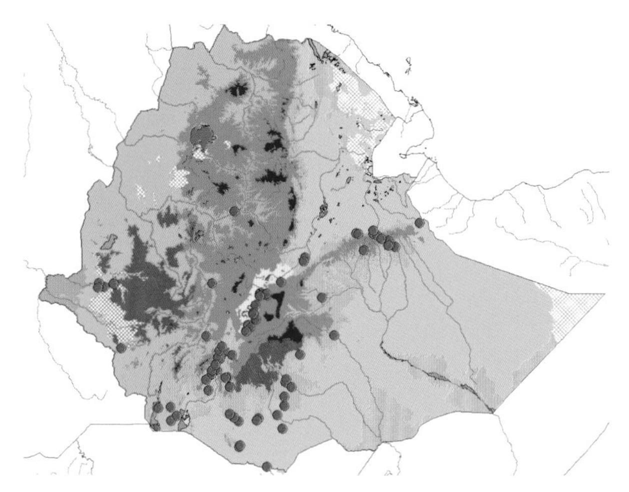

Fig. 6-10. *Maerua triphylla*. Distribution in Ethiopia. Observed records are marked with red dots on vegetation map reproduced from Fig. 4-1.

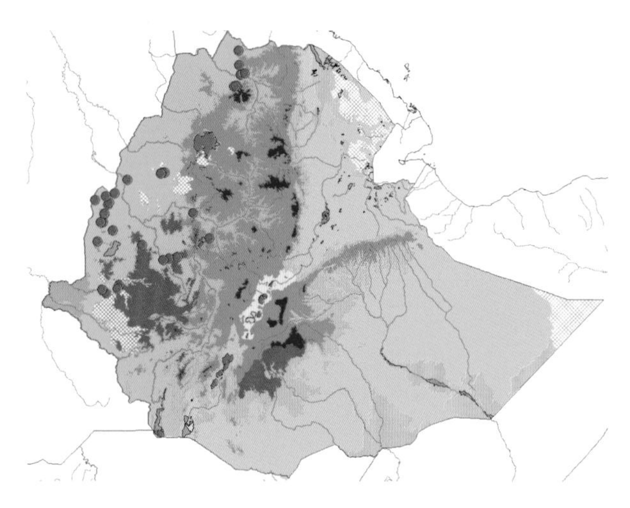

Fig. 6-11. *Securidaca longipedunculata*. Distribution in Ethiopia. Observed records are marked with red dots on vegetation map reproduced from Fig. 4-1.

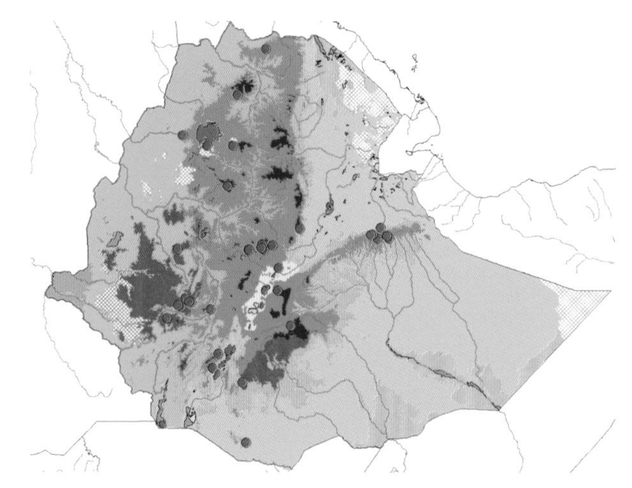

Fig. 6-12. *Rumex abyssinicus*. Distribution in Ethiopia. Observed records are marked with red dots on vegetation map reproduced from Fig. 4-1.

allow further study, and it has been decided here to assign the species as a whole to **ACB[-CTW]**.

6.3. Polygalaceae

BS 51: 83-85 (1998); FEE 2(1): 177-188 (2000); FTEA, Polygalaceae: 1-61 (2004); PSS: 198-200 (2015).

6.3.1. **Securidaca longipedunculata** *Fresen. (1837).*

BS 51: 85 (1998); FEE 2(1): 177 (2000); FTEA, Polygalaceae: 6 (2004); PSS: 199 (2015).
Taxonomy: *Securidaca longipedunculata* var. *longipedunculata* is widespread; another variety, *S. longipedunculata* var. *parvifolia* Oliv. (1868; in FEE as *S. longipedunculata* var. *parvifolia* Johnson, without indication of year of publication) from South-Central Africa. The Ethiopian material is homogenous. Recognition of infraspecific taxa not relevant here.
Description: Shrub or small tree to 6 m tall. No thorns. Leaves simple, probably deciduous. Microphyll. Although growing in fire-swept wooded grasslands, it seems to have no obvious adaptations for fire resistance.
Habitat: 'Woodland or wooded grassland with *Terminalia, Combretum, Protea, Gardenia*, etc., subject to burning' (FEE). 'Woodland, wooded grassland' (PSS, for *Securidaca longipedunculata* var. *longipedunculata*). In typical *Combretum-Terminalia* woodland of the northern and central woodlands with tall groundcover and frequent burning (our observation).
Distribution in Ethiopia: TU GD GJ WG SU IL GG (FEE). The FEE record from the GG floristic region has not been confirmed (Fig. 6-11).
Altitudinal range in Ethiopia: 500-1750 m a.s.l. (FEE). 500-1900 m a.s.l. (our records).
General distribution (of species as a whole): Widespread across tropical Africa from Senegal to Ethiopia, south to Namibia and South Africa.
Chorological classification: Distribution according to direct observation. **WWn, WWc, WWs.** *Local phytogeographical element and sub-element.* **CTW[n+c]**. The species occurs in north-western Ethiopia, avoiding the driest areas. In west-central Ethiopia, it occurs in the central western woodlands from the Sudan border to the Abay and the Didessa Valleys. It is represented in the woodlands of the Gambela lowlands, but does not occur in south-west Ethiopia; it has no eastwards extension.

6.4. Polygonaceae

FTEA, Polygonaceae: 1-40 (1958); *BS* 51: 93-94 (1998); FEE 2(1): 336-347 (2000); PSS: 271-273 (2015).

6.4.1. **Rumex abyssinicus** *Jacq. (1776).*

FTEA, Polygonaceae: 7 (1958); *BS* 51: 94 (1998); FEE 2(1): 339 (2000); PSS: 273 (2015).
Taxonomy: No infraspecific taxa.
Description: Woody based perennial herb to 4 m tall. No thorns. Leaves simple, leaves probably deciduous. Mesophyll-Macrophyll. The massive woody rhizome may be an adaptation for fire resistance.
Habitat: 'A common and tolerated weed in fields and plantations, by paths and in secondary scrub' (FEE). 'Forest margins, weed of cultivation and fallow' (PSS). Common in degraded vegetation, typically in mixed bushland (our observation).
Distribution in Ethiopia: TU GD GJ SU AR WG KF IL GG SD BA HA (FEE). No additional records (Fig. 6-12).
Altitudinal range in Ethiopia: 1200-3300 m a.s.l. (FEE, including range in Eritrea). 1300-3100 m a.s.l. (our records).
General distribution: From Nigeria to Ethiopia and Eritrea, south to Angola and Mozambique; also in Madagascar.
Chorological classification: Distribution according to direct observation. **WWn, WWc, EWW.** *Local phytogeographical element.* **MI**. Marginally intruding species with a main distribution in the highlands and in the *Transitional semi-evergreen bushland* (TSEB); it ascends almost to the lower limit of the *Ericaceous belt* (EB).

6.5. Lythraceae

FTEA, Lythraceae: 1-8 (1994); *BS* 51: (1998); FEE 2(1): 394-408 (2000), as **LYTHRACEAE (incl. *Punicaceae*)**; PSS: 237-238 (2015).

6.5.1. Woodfordia uniflora *(A. Rich.) Koehne (1881).*

FTEA, Lythraceae: 4 (1994); FEE 2(1): 396 (2000); PSS: 238 (2015).
Taxonomy: No infraspecific taxa.
Description: Shrub or small tree to 3 (-5) m tall. No thorns. Leaves simple, probably deciduous. Microphyll-mesophyll. No obvious adaptations for fire resistance.
Habitat: 'Rocky slopes in shade, often riverine, on basalt or limestone' (FEE). 'Rocky streamsides, rocky slopes, woodland' (PSS). In mixed bushland on rocks and rocky slopes, sometimes on rocks along rivers and on moist, old stone walls (our observation).
Distribution in Ethiopia: TU GD GJ WU SU AR KF SD BA HA (FEE, including range in Eritrea). No additional records (Fig. 6-13).
Altitudinal range in Ethiopia: 1250-2500 m a.s.l. (FEE). (850-) 1200-2100 m a.s.l. (our records).
General distribution: From Nigeria to Ethiopia and Eritrea, south to Uganda and Kenya.
Chorological classification: Distribution according to direct observation. **WWn, WWc, EWW.** *Local phytogeographical element.* **MI.** Marginally intruding species in the western woodlands with a main distribution in the highlands and in *Transitional semi-evergreen bushland* (TSEB). The record from a rocky streambed near Metema (*Friis et al.* 6969) at 850 m a.s.l., is a surprising outlier.

6.6. Thymelaeaceae

FTEA, Thymelaeaceae: 1-36 (1978); *BS* 51: 110-111 (1998); FEE 2(1): 429-435 (2000); PSS: 260-261 (2015).
Gnidia kraussiana Meisn. (1843; *BS* 51: 110 (1998); PSS: 260 (2015)) is another woody species that is likely to be found in western woodlands with grass fires, as it is found in such habitats in Sudan and South Sudan, but it has not been recorded from Ethiopia in FEE, nor was it seen in our relevés.

6.61. Gnidia lamprantha *Gilg (1894).*

FTEA, Thymelaeaceae: 31 (1978); *BS* 51: 111 (1998); FEE 2(1): 435 (2000); PSS: 261 (2015).
Taxonomy: No infraspecific taxa.
Description: Much-branched shrub or small tree to 3 (-5) m tall. No thorns. Leaves simple, probably evergreen. Microphyll. No obvious adaptations for fire resistance.
Habitat: 'Grassland, sometimes fire-swept, bushland, dry grassy forests and hillsides, red or black soil' (FEE). 'Montane wooded grassland and bushland' (PSS). In wooded grassland with tall ground-cover liable to burning, along forest margins and in clearings (our observation).
Distribution in Ethiopia: WG KF GG SD (FEE). No additional records (Fig. 6-14).
Altitudinal range in Ethiopia: 1200-2285 m a.s.l. (FEE). 1200-1900 m a.s.l. (our records).
General distribution: Sudan, Ethiopia, Uganda, Kenya, and Tanzania.
Chorological classification: Distribution according to direct observation. **WWc, WWs, EWW.** *Local phytogeographical element.* **MI.** Marginally intruding species in the western woodlands with a main distribution in the highlands, in *Transitional rain forest* (TRF) (documented by *Friis et al.* 3942 from Bebeka south of Mezan Teferi. Our sight record from near the southern forest boundary in the Chebera-Churchura National Park is from a similar habitat), and *Transitional semi-evergreen bushland* (TSEB). A record from inside *Acacia-Commiphora bushland* (ACB) is from a hill with dense woodland.

6.7. Proteaceae

FTEA, Proteaceae: 1-31 (1993); *BS* 51: 113-114 (1998); FEE 2(1): 436-438 (2000); PSS: 154 (2015).

6.7.1. **Faurea rochetiana** *(A. Rich.) Chiov. ex Pic. Serm. (1950).*

FTEA, Proteaceae: 3 (1993); *BS* 51: 113 (1998); PSS: 154 (2015).
Faurea speciosa Welw. (1869).
FEE 2(1): 436 (2000).
Taxonomy: No infraspecific taxa. The basionym of *F. rochetiana* is *Leucospermum rochetianum* A. Rich. (1851). Due to an oversight, the name *F. speciosa* was erroneously taken up in the *Flora of Ethiopia and Eritrea*.
Description: Medium sized tree to 10 m tall. No thorns. Leaves simple, evergreen. Mesophyll. Bark rough with vertical ridges, often blackened; the thick and fibrous bark may be an adaptation for fire resistance.
Habitat: 'Upland wooded grassland, upland evergreen bushland' (FEE). 'Woodland and wooded grassland' (PSS). Woodland and wooded grassland with tall ground-cover liable to burning (our observation).
Distribution in Ethiopia: GD GJ SU KF SD BA (FEE). No additional records (Fig. 6-15).
Altitudinal range in Ethiopia: 1200-2100 m a.s.l. (FEE, including range in Eritrea). 850-1800 m a.s.l. (our records).
General distribution: From Togo to Nigeria, also in South Sudan, Ethiopia, and Eritrea, south to Angola and South Africa.
Chorological classification: Distribution according to direct observation. **[WWn], WWc, WWs, EWW**. *Local phytogeographical element and sub-element.* **CTW[c+s+ext]**. Distributed in the southern part of the *Combretum-Terminalia* woodlands, mainly south of the Abay River, extending to the Rift Valley and to the east of the Rift. There is one northern record in Profile E.

6.7.2. **Protea gaguedi** *J.F. Gmel. (1796).*

FTNA: 99 (1992); FTEA, Proteaceae: 17 (1993); *BS* 51: 113 (1998); FEE 2(1): 438 (2000); PSS: 154 (2015).
Taxonomy: No infraspecific taxa.
Description: Small tree with thick, branched stems, up to 9 (-10) m tall. No thorns. Leaves simple, evergreen. Mesophyll. The thick bark, often blackened, may be an adaptation for fire resistance.
Habitat: 'Dry upland forest with *Juniperus* and *Hagenia*, mostly along edges and in clearings, upland evergreen bushland and wooded grassland' (FEE). 'Dry single-dominant Afromontane forest ..., mainly at forest edges and in clearings; chiefly outside closed forest, in fireswept grassland and woodland' (FTNA). 'Wooded grassland' (PSS). *Dry evergreen Afromontane forest* (DAF), particularly at edges and in clearings, evergreen bushland, and wooded grassland (our observation).
Distribution in Ethiopia: TU GD GJ SU AR WG IL KF GG SD BA HA (FEE). No additional records (Fig. 6-16).
Altitudinal range in Ethiopia: 1400-3000 m a.s.l. (FEE, including range in Eritrea). 1600-3000 m a.s.l. (FTNA, including the range in the entire Horn of Africa). 1400-2700 m a.s.l. (our records).
General distribution: From Ethiopia and Eritrea south to Angola and Mozambique.
Chorological classification: Distribution according to direct observation. **WWn, WWc, WWs, EWW**. *Local phytogeographical element.* **MI**. Marginally intruding species in the western woodlands, with a main distribution in the highlands and *Transitional semi-evergreen bushland* (TSEB). One record from inside *Acacia-Commiphora bushland* (ACB) is from the Gerire Highlands (*Benardelli & Reghini* 51). The record from east of Kurmuk is a sight record from our relevé 78 at 1450 m a.s.l.; it may have been *P. madiensis,* but *P. gaguedi* has elsewhere been documented at similar altitudes.

6.7.3. **Protea madiensis** *Oliv. (1875).*

FTEA, Proteaceae: 10 (1993); *BS* 51: 114 (1998); FEE 2(1): 438 (2000); PSS: 154 (2015).
Taxonomy: *Protea madiensis* subsp. *madiensis* occurs in eastern and south-eastern Africa. The West African subspecies, *Protea madiensis* subsp. *occidentalis* (Beard) Chisumpa & Brummitt, occurs from the Guinean Republic to Cameroon. Recognition of infraspecific taxa not relevant here.
Description: Shrub or small tree to 4 m tall. No thorns. Leaves simple, evergreen. Mesophyll-Macrophyll. The thick bark, often blackened, may be an adaptation for fire resistance.

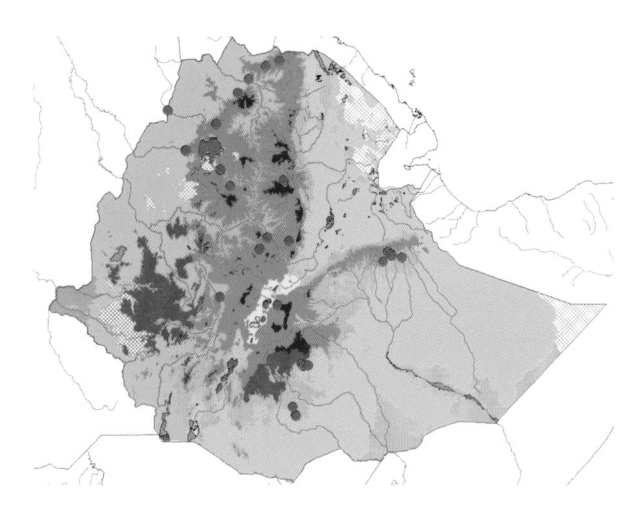

Fig. 6-13. *Woodfordia uniflora*. Distribution in Ethiopia. Observed records are marked with red dots on vegetation map reproduced from Fig. 4-1.

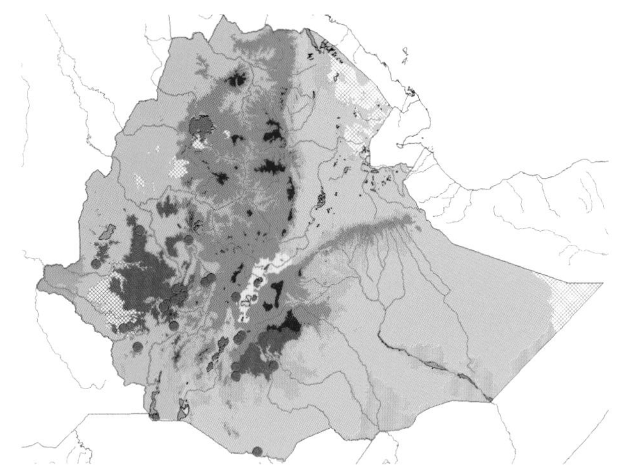

Fig. 6-14. *Gnidia lamprantha*. Distribution in Ethiopia. Observed records are marked with red dots on vegetation map reproduced from Fig. 4-1.

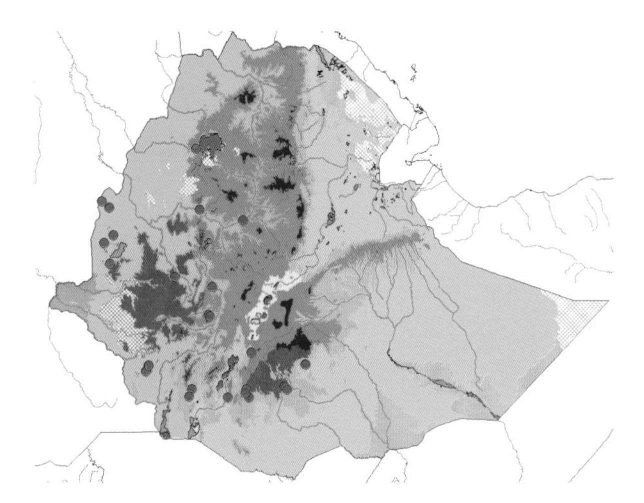

Fig. 6-15. *Faurea rochetiana*. Distribution in Ethiopia. Observed records are marked with red dots on vegetation map reproduced from Fig. 4-1.

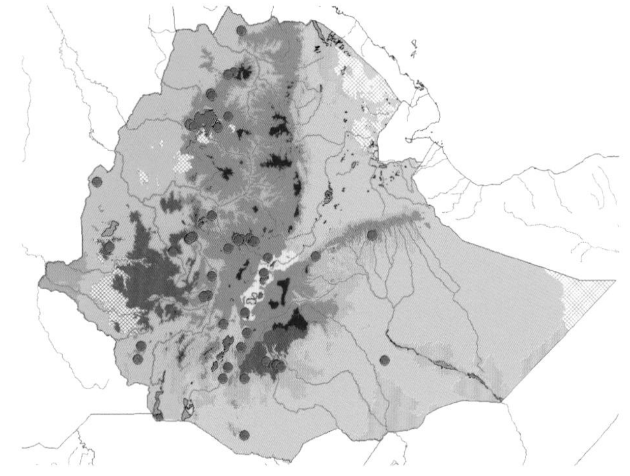

Fig. 6-16. *Protea gaguedi*. Distribution in Ethiopia. Observed records are marked with red dots on vegetation map reproduced from Fig. 4-1.

Habitat: 'Wooded grassland' (FEE). 'Upland grassland' (PSS). Wooded grassland with tall groundcover, liable to burning (our observation).
Distribution in Ethiopia: GD/Sudan WG KF GG SD BA (FEE). No additional records (Fig. 6-17).
Altitudinal range in Ethiopia: 1200-1700 m a.s.l. (FEE). 1100-1800 m a.s.l. (our records).
General distribution: From Guinean Republic to Cameroon (*Protea madiensis* subsp. *occidentalis*) and from Nigeria to Ethiopia, south to Angola and Mozambique (*Protea madiensis* subsp. *madiensis*).
Chorological classification: Distribution according to direct observation. **WWc, WWs, EWW**. *Local phytogeographical element and sub-element.* **CTW[c+s+ext]**. Widely distributed in the central and southern part of the *Combretum-Terminalia* woodlands, mainly south of the Abay River; one record to the east of the Rift in *Transitional semi-evergreen bushland* (TSEB), *Friis et al.* 3469 (C, ETH, K) at Harenna Forest.

6.8. Flacourtiaceae

FS 1: 203-207 (1993); *BS* 51: 115-117 (1998); FEE 2(1): 443-450 (2000); PSS: 229 (2015), as "**SALICACEAE.**" The definition of the Flacourtiaceae follows here the definition in the FTEA, Flacourtiac. (1993) and other floristic works before the PSS.

6.*8.1.* **Flacourtia indica** *(Burm. f.) Merr. (1917).*

FTNA: 104 (1992); FTEA, Flacourtiaceae: 57 (1993); FS 1: 206 (1993); *BS* 51: 116 (1998); FEE 2(1): 446 (2000); PSS: 229 (2015).
Taxonomy: No infraspecific taxa.
Description: Shrub or small tree to about 15 m tall. Thorns present. Leaves simple, probably evergreen. Mesophyll. Apparently, no adaptation for fire resistance.
Habitat: 'Understorey in upland *Olea* and *Podocarpus* forest, bush-clumps in upland grassland (forest remnants), forest edges, riverine forest and scrub, *Acacia* bushland' (FEE). 'Undifferentiated Afromontane forest and single-dominant Afromontane forest, often along forest edges and in clearings and persisting in gully forest; also in secondary montane evergreen bushland, in humid wooded grassland and in riparian woodland' (FTNA). 'Semi-evergreen open forest' (FS). 'Dry forest and forest margins, riverine forest, montane bushland' (PSS). Typically a forest or bushland plant, in woodland nearly always in association with clumps of evergreen trees or shrubs (our observation).
Distribution in Ethiopia: GD GJ SU AR WG IL KF GG SD BA HA (FEE). No additional records (Fig. 6-18).
Altitudinal range in Ethiopia: (1100-) 1500-2350 m a.s.l. (FEE). 1200-2400 m a.s.l. (FTNA, including the range in the entire Horn of Africa). 1100-2350 m a.s.l. (our records).
General distribution: From Senegal to Ethiopia and Somalia, south to Botswana and South Africa; also in Madagascar, the Mascarenes, the Seychelles, India, and Indonesia.
Chorological classification: Distribution according to direct observation. **WWn, WWc, WWs, EWW**. *Local phytogeographical element.* **MI**. Marginally intruding species in the western woodlands with main distribution in the highlands and in *Transitional semi-evergreen bushland* (TSEB). Although there are several records in areas of the *Combretum-Terminalia woodland and wooded grassland* (CTW), most are at the upper limit of that vegetation. One, however, *Friis et al.* 15934 was collected from relevé 137 at the lowermost part of the escarpment at 1100 m a.s.l. west of Chagni.

6.*8.2.* **Oncoba spinosa** *Forssk. (1775).*

FTEA, Flacourtiaceae: 16 (1975); FTNA: 105 (1992); FS 1: 205 (1993); *BS* 51: 117 (1998); FEE 2(1): 443 (2000); PSS: 229 (2015).
Taxonomy: No infraspecific taxa.
Description: Small tree to 10 m tall. Thorns present. Leaves simple, apparently evergreen. Mesophyll. Apparently, no adaptation for fire resistance.
Habitat: 'Riverine forest, *Combretum-Terminalia* woodland' (FEE). 'Riverine forest; also in moist woodland' (FTNA). 'Riverine forest and bushland' (FS). 'Woodland, riverine forest and margins, rocky hillslopes' (PSS). Typically a riparian forest or evergreen bush-

land plant, in woodland nearly always along streams or in association with clumps of evergreen trees or shrubs (our observation).
Distribution in Ethiopia: TU SU AR WG IL KF SD BA HA (FEE). Also recorded from the GD floristic region (Fig. 6-19).
Altitudinal range in Ethiopia: 400-1800 m a.s.l. (FEE, including range in Eritrea). 30-1500 m a.s.l. (FTNA, including the range in the entire Horn of Africa). 1100-2350 m a.s.l. (our records).
General distribution: From Senegal to Ethiopia, Eritrea, and Somalia, south to Botswana and South Africa; also in Arabia.
Chorological classification: Distribution according to direct observation. **WWn, WWc, WWs, EWW**. *Local phytogeographical element.* **MI**. Marginally intruding species in the western woodlands, with a main distribution in the highlands, rarely in *Riparian vegetation* (RV) and *Transitional rain forest* (TRF). The few records in our relevés are all from the upper margins of the *Combretum-Terminalia* woodlands.

6.9. Ochnaceae

FEE 2(2): 66-69 (1995); *BS* 51: 124-126 (1998); FTEA, Ochnaceae: 1-60 (2005); PSS: 223-224 (2015).

6.*9.1.* **Ochna leucophloeos** *Hochst. ex A. Rich. (1847).*

FEE 2(2): 67 (1995); FTEA, Ochnaceae: 21 (2005); PSS: 224 (2015).
Taxonomy: Two subspecies are recognized, *Ochna leucophloeos* subsp. *leucophloeos* in north-eastern tropical Africa and *O. leucophloeos* subsp. *ugandensis* Verdc. in northern Uganda only. Recognition of infraspecific taxa not relevant here.
Description: Often a shrub to ca. 3.5 m tall, but occasionally a tree to 8 m tall. No thorns. Leaves simple, probably evergreen. Mesophyll. The scaly and powdery bark may be an adaptation for fire resistance.
Habitat: '*Combretum-Terminalia* and *Pterocarpus-Terminalia* woodland on rocky slopes' (FEE). 'Woodland on rocky slopes' (PSS, for *Ochna leucophloeos* subsp. *leucophloeos*). A typical woodland plant associated with rocky outcrops (our observation).
Distribution in Ethiopia: TU GD GJ WG IL (FEE). No additional records (Fig. 6-20).
Altitudinal range in Ethiopia: 400-1500 m a.s.l. (FEE, including range in Eritrea). 600-1400 m a.s.l. (our records).
General distribution (species as a whole): Sudan, Ethiopia, Eritrea, south to northern Uganda.
Chorological classification: Distribution according to direct observation. **WWn, WWc**. *Local phytogeographical element and sub-element.* **CTW[n+c]**. Wide distribution in the *Combretum-Terminalia* woodlands, from the north to the Akobo River in the Gambela lowlands.

6.10. Myrtaceae

FEE 2(2): 71-106 (1995); *BS* 51: 126-127 (1998); FTEA, Myrtaceae: 1-89 (2001); PSS: 240 (2015).

6.*10.1.* **Syzygium guineense** *DC. subsp.* **macrocarpum** *(Engl.) F. White (1962)*

FEE 2(2): 78 (1995); FTEA, Myrtaceae: 79 (2001); PSS: 240 (2015).
Later used name: **Syzygium pratense** Byng, Global Fl. 4: 132 (2018).
Taxonomy: This taxon has in most modern floras, including the ones cited above, been accepted as a subspecies of *Syzygium guineense.* Ecologically, it seems clearly distinct from other subspecies, e.g. *S. guineense* subsp. *guineense* and *S. guineense* subsp. *afromontanum* F. White, but herbarium material of these three taxa is not always possible to identify. Here, the taxon is recognized as a subspecies, but lately the taxon has been considered a distinct species, named **Syzygium pratense** Byng.
Description: Large shrub or small tree to 8 (-12) m tall. No thorns. Leaves simple, evergreen. Mesophyll. No apparent adaptation for fire resistance.
Habitat: 'Restricted to woodlands' (FEE for *Syzygium guineense* subsp. *macrocarpum*). 'Fire-prone woodland and wooded grassland' (PSS for *Syzygium guineense* subsp. *macrocarpum*). Forming extensive mixed or al-

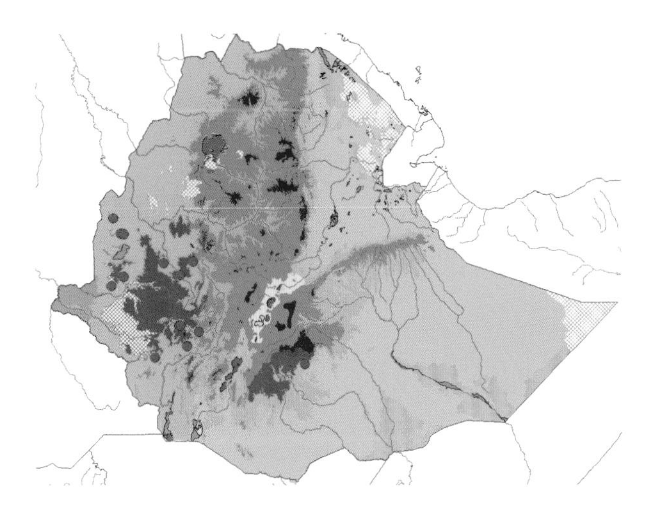

Fig. 6-17. *Protea madiensis*. Distribution in Ethiopia. Observed records are marked with red dots on vegetation map reproduced from Fig. 4-1.

Fig. 6-18. *Flacourtia indica*. Distribution in Ethiopia. Observed records are marked with red dots on vegetation map reproduced from Fig. 4-1.

Fig. 6-19. *Oncoba spinosa*. Distribution in Ethiopia. Observed records are marked with red dots on vegetation map reproduced from Fig. 4-1.

Fig. 6-20. *Ochna leucophloeos*. Distribution in Ethiopia. Observed records are marked with red dots on vegetation map reproduced from Fig. 4-1.

most monospecific woodlands in the upper zones of the woodlands in the central parts of western Ethiopia (our observation).
Distribution in Ethiopia: SU WG KF SD BA (FEE). Also recorded from the GD and GJ floristic regions (Fig. 6-21).
Altitudinal range in Ethiopia (only *Syzygium guineense* subsp. *macrocarpum*): 1400-2500 m a.s.l. (FEE). 1200-2100 (-2500) m a.s.l. (our records).
General distribution (of *Syzygium guineense* subsp. *macrocarpum*): From Senegal through South Sudan to Ethiopia, south to Angola, Zimbabwe, and possibly Mozambique. The distribution across Africa from Senegal to Ethiopia is not well known.
Chorological classification: Distribution according to direct observation. **WWc, WWs.** *Local phytogeographical element and sub-element.* **CTW[c]**. Apart from a few records from the western escarpment in the GD and GJ flora regions, the general distribution of this taxon is in the central part of the western woodlands. A few records from the Rift Valley indicate a slight eastward extension from the *Combretum-Terminalia* woodlands but there is no extension to the east of the Rift, and it has therefore not been reflected in the distribution types.

6.11. Combretaceae

FTEA, Combretaceae: 1-100 (1973); FS 1: 247-254 (1993); FEE 2(2): 115-132 (1995); *BS* 51: 130-135 (1998); PSS: 234-237 (2015).
The tree *Combretum nigricans* Guill. & Perr. (1833), distributed from Senegal to Ethiopia (FEE 2(2): 118 (1995) has been recorded (*Kuls* 298, 302) from *Combretum-Terminalia* woodland the GJ floristic region at an altitude of ca. 1900 m; it has not been observed in our relevés.
The scrambling and climbing shrub *Combretum aculeatum* Vent (1808; FEE 2(2): 120 (1995)), distributed from Senegal to Eritrea, Ethiopia and Somalia and south to Tanzania, occurs both in association with the *Combretum-Terminalia* woodland in the west and in the eastern *Acacia-Commiphora* bushlands. Apart from that species that occurs equally in the western and eastern lowlands and *Terminalia brownii*, that is a marginally intruding species in the CTW-area, the Ethiopian specie of *Combretum* and *Terminalia* can be classified as associated either with the *Combretum-Terminalia* woodland in the west (and may also occur in some highland vegetation; these species are included here) or with the eastern *Acacia-Commiphora* bushlands; examples of the latter are *Combretum contractum* Engl. & Diels (1907), *C. hereroense* Schinz (1888), *Terminalia orbicularis* Engl. & Diels (1900), *T. brevipes* Pampan. (1915), *T. spinosa* Engl. (1895), *T. prunoides* Laws. (1871), *T. polycarpa* Engl. & Diels (1900), and *T. basilei* Chiov. (1929).

6.*11.1.* Anogeissus leiocarpa *(DC.) Guill. & Perr. (1832).*

FEE 2(2): 130 (1995); PSS: 234 (2025).
Later used name: **Terminalia leiocarpa** (DC.) Baill. (1876).
Taxonomy: No infraspecific taxa.
Description: Tree to 15 m tall. No thorns. Leaves simple, apparently evergreen. Mesophyll. Bark flaking in rectangular patches, probably an adaptation for fire resistance.
Habitat: '*Combretum-Terminalia* and *Anogeissus-Pterocarpus* woodland, wooded grassland and bushland, *Acacia-Lannea* bushland, often a dominant species, riverbanks' (FEE). 'Woodland, wooded grassland and bushland, sometimes dominant' (PSS). Often dominant species in the western woodlands (our observation).
Distribution in Ethiopia: TU GD GJ SU WG IL (FEE). No additional records (Fig. 6-22).
Altitudinal range in Ethiopia: 450-1900 m a.s.l. (FEE, including range in Eritrea). 500-1750 m a.s.l. (our records).
General distribution: From Mauritania and Senegal to Ethiopia and Eritrea, south to D.R. Congo.
Chorological classification: Distribution according to direct observation. **WWn, WWc**. *Local phytogeographical element and sub-element.* **CTW[n+c]**. Widely distributed in the northern and central part of the *Combretum-Terminalia* woodlands, reaching into the valleys of the Tacazze and Abay Rivers and common in the Gambela

Region in the south. Harrison & Jackson (1958) indicated the presence of *Anogeissus leiocarpa-Combretum hartmannianum* woodland south of the Akobo River on the Sudan side of the border, but we have not recorded any observation of this species south of the Gambela lowlands.

6.*11.2*. Combretum adenogonium *Steud. ex A. Rich. (1848).*

FEE 2(2): 117 (1995); *BS* 51: 130 (1998); PSS: 234 (2015).
Combretum fragrans F. Hoffm. (1889).
FTEA, Combretaceae: 29 (1973).
Taxonomy: No infraspecific taxa.
Description: Tree to 12 m tall. No thorns. Leaves simple, apparently evergreen. Mesophyll-Macrophyll. Bark smooth or flaking in rectangular patches, probably an adaptation for fire resistance.
Habitat: '*Combretum-Terminalia* woodland and wooded grassland on alluvial clay soil or on rocky slopes' (FEE). 'Woodland and wooded grassland' (PSS). *Combretum-Terminalia* woodland (our observation).
Distribution in Ethiopia: TU GD GJ SU WG IL GG SD (FEE). No additional records (Fig. 6-23).
Altitudinal range in Ethiopia: 500-2000 (-2300) m a.s.l. (FEE, including range in Eritrea). 500-1800 m a.s.l. (our records).
General distribution: From Senegal to Ethiopia and Eritrea, south to Botswana, Zimbabwe, and Mozambique.
Chorological classification: Distribution according to direct observation. **WWn, WWc, WWs, EWW.** *Local phytogeographical element and sub-element.* **CTW[all+ext].** Widely distributed in the western woodlands, from north to south or almost so and with extension to the Rift Valley and east of the Rift.

6.*11.3*. Combretum collinum *Fresen. (1837).*

FTEA, Combretaceae: 24 (1973); FEE 2(2): 116 (1995); PSS: 234 (2015).
Taxonomy: The material dealt with here includes several subspecies, which it was impossible to distinguish during our field work: *Combretum collinum* subsp. *collinum* (FTEA, Combretaceae: 26 (1973); FEE 2(2): 116 (1995); *BS* 51: 131 (1998); PSS: 234 (2015)), *Combretum collinum* subsp. *binderianum* (Kotschy) Okafor (FTEA, Combretaceae: 24 (1973); FEE 2(2): 116 (1995); PSS: 234 (2015)), *Combretum collinum* subsp. *elgonense* (Exell) Okafor (FTEA, Combretaceae: 25 (1973); FEE 2(2): 116 (1995); *BS* 51: 131 (1998); PSS: 235 (2015)), and *Combretum collinum* subsp. *hypopilinum* (Diels) Okafor (FTEA, Combretaceae: 25 (1973); FEE 2(2): 117 (1995); PSS: 235 (2015)). Characteristically, the subspecies all occur in wooded grasslands and woodlands with overlapping altitudinal ranges. Recognition of infraspecific taxa not possible here.
Description: Tree to (10-) 15 m tall. No thorns. Leaves simple, probably evergreen. Mesophyll-Macrophyll. Bark thick, rough and fissured, probably an adaptation for fire resistance.
Habitat: '*Combretum, Combretum-Terminalia, Combretum-Terminalia-Stereospermum* and *Anogeissus-Combretum* woodland and wooded grassland, undergrowth with tall grass cover and subject to regular burning' (FEE, compilation for all infraspecific taxa). 'Woodland and wooded grassland' (PSS, compilation for all infraspecific taxa). In a wide range of woodlands and semi-evergreen bushland (our observation).
Distribution in Ethiopia: TU GD GJ SU WG IL KF GG SD BA HA (FEE). No additional records (Fig. 6-24).
Altitudinal range in Ethiopia: (450-)1100-2200 m a.s.l. (FEE, including range in Eritrea). 600-1800 m a.s.l. (our records).
General distribution (species as a whole): From Guinean Republic to Ethiopia and Eritrea, south to South Africa.
Chorological classification: Distribution according to direct observation. **WWn, WWc, WWs, EWW**. *Local phytogeographical element and sub-element.* **CTW[all+ext].** Widely distributed in the western woodlands, from north to south or almost so (not recorded from the lower Omo Basin), with extension to the Rift Valley and east of the Rift.

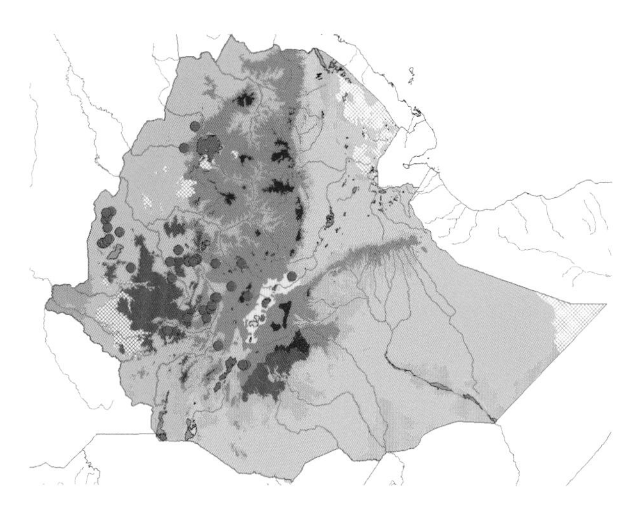

Fig. 6-21. *Syzygium guineense* subsp. *macrocarpum*. Distribution in Ethiopia. Observed records are marked with red dots on vegetation map reproduced from Fig. 4-1.

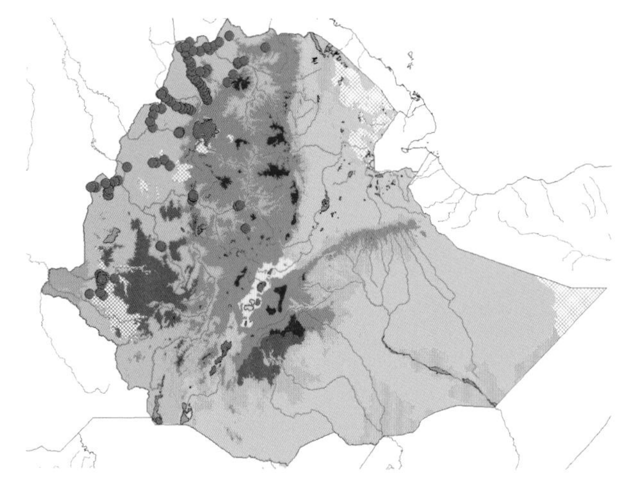

Fig. 6-22. *Anogeissus leiocarpa*. Distribution in Ethiopia. Observed records are marked with red dots on vegetation map reproduced from Fig. 4-1.

Fig. 6-23. *Combretum adenogonium*. Distribution in Ethiopia. Observed records are marked with red dots on vegetation map reproduced from Fig. 4-1.

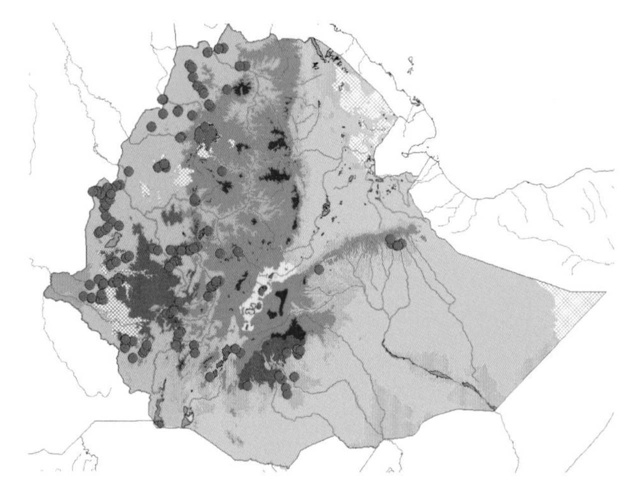

Fig. 6-24. *Combretum collinum*. Distribution in Ethiopia. Observed records are marked with red dots on vegetation map reproduced from Fig. 4-1.

6.*11.4*. Combretum hartmannianum *Schweinf. (1867).*

FEE 2(2): 117 (1995); PSS: 235 (2015).
Taxonomy: No infraspecific taxa.
Description: Tree to 25 m tall. No thorns. Leaves simple, probably deciduous. Mesophyll. Bark smooth, apparently no adaptation for fire resistance.
Habitat: '*Combretum-Terminalia* woodland and wooded grassland on alluvial soil, often reported to be dominant' (FEE). 'Woodland and wooded grassland' (PSS). Woodland and wooded grassland, mainly on vertisol and with tall ground-cover liable to burn (our observation).
Distribution in Ethiopia: TU GD (FEE). Also recorded from the WG floristic region (Fig. 6-25).
Altitudinal range in Ethiopia: 500-1200 m a.s.l. (FEE, including range in Eritrea). 650-1200 m a.s.l. (our records).
General distribution: Sudan (doubtfully South Sudan?), Ethiopia and Eritrea.
Chorological classification: Distribution according to direct observation. **WWn, WWc**. *Local phytogeographical element and sub-element.* CTW[n]. Widely distributed in the northern part of the western woodlands, mainly north of the Abay River; no eastward extension. Harrison & Jackson (1958) indicated the presence of *Anogeissus leiocarpa-Combretum hartmannianum* woodland south of the Akobo River on the Sudan side of the border, but we have recorded no observation of *Combretum hartmannianum* south of Kurmuk.

6.*11.5*. Combretum molle *R. Br. ex G. Don (1827).*

FTEA, Combretaceae: 33 (1973); FS 1: 249 (1993); FEE 2(2): 118 (1995); *BS* 51: 131 (1998); PSS: 235 (2015).
Taxonomy: Stated by FTEA to be an extremely variable species regarding indumentum leaf-shape and size of fruits. However, many intermediate specimens were observed. Recognition of infraspecific taxa not relevant here.
Description: Tree to 15 (-20) m tall. No thorns. Leaves simple, probably evergreen. Mesophyll. Bark smooth to thick, rough and fissured, probably an adaptation for fire resistance.
Habitat: 'A wide variety of *Combretum* and *Combretum-Terminalia* woodland and wooded grassland, often on rocky slopes, usually with ground-cover of tall grass and subject to regular burning, often a dominant species, penetrating into riverine forest, dry *Juniperus* forest, ground-water forest and even lowland rain forest' (FEE). 'Deciduous or semi-evergreen bushland or woodland, usually on rocky slopes and outcrops' (FS). 'Woodland and wooded grassland where often dominant, dry forest' (PSS). We can confirm this very wide habitat range, which also includes clumps of trees in semi-evergreen bushland. The highest records are from river valleys in the highlands (our observation).
Distribution in Ethiopia: TU GD GJ WU SU AR WG IL KF GG SD BA HA (FEE). No additional records (Fig. 6-26).
Altitudinal range in Ethiopia: 500-2200 (-2500) m a.s.l. (FEE, including range in Eritrea). 600-2100 m a.s.l. (our records).
General distribution: From Senegal to Ethiopia, Eritrea, and Somalia, south to Angola and South Africa; also in Madagascar and Arabia.
Chorological classification: Distribution according to direct observation. **WWn, WWc, WWs, EWW**. *Local phytogeographical element and sub-element.* **CTW[all+ext]**. Widely distributed in the western woodlands, from north to south or almost so, also common in the Rift Valley and with extension to the east of the Rift in *Transitional semi-evergreen bushland* (TSEB). The two record (*Benardelli & Reghini* 48 (FT) and *Friis et al.* 15537 (C, ETH, K) from the BA floristic region, apparently surrounded by *Acacia-Commiphora bushland* (ACB), are from the Gerire Hills, outlying mountains with *Transitional semi-evergreen bushland* (TSEB).

6.*11.6*. Combretum rochetianum *A. Rich. ex A. Juss. (1851).*

FEE 2(2): 117 (1995); PSS: 236 (2015).
Taxonomy: No infraspecific taxa.
Description: Tree to 5 m tall. No thorns. Leaves simple,

probably evergreen. Mesophyll. Bark not known, but no other obvious adaptation for fire resistance.
Habitat: '*Combretum-Terminalia* woodland and bushland on alluvial plains, often dominant' (FEE). 'Woodland and bushland' (PSS). We have no additional observations.
Distribution in Ethiopia: TU GD GJ (FEE). Additional records from the SU (*WJJO de Wilde et al.* 10120) and GG (*Friis et al.* 11372) floristic regions. These two records require confirmation (Fig. 6-27).
Altitudinal range in Ethiopia: 1000-1900 m a.s.l. (FEE, including range in Eritrea). (650-) 1100-1900 m a.s.l. (our records).
General distribution: Sudan, South Sudan, Ethiopia, and Eritrea.
Chorological classification: Distribution according to direct observation. **WWn, [WWc, WWs]**. *Local phytogeographical element and sub-element.* **CTW[n]**. Widely distributed in the western woodlands, from north to south or almost so; no eastward extension. Both distributional classifications assume that the two records from SU and GG are correctly identified, which to some extent is in doubt.

6.*11.7.* Terminalia brownii *Fresen. (1837).*

FTEA, Combretaceae: 90 (1993); FS 1: 253 (1993); FEE 2(2): 127 (1995); *BS* 51: 133 (1998); PSS: 236 (2015).
Taxonomy: No infraspecific taxa.
Description: Tree to 10 (-15) m tall. No thorns. Leaves simple, probably evergreen. Mesophyll. Bark smooth to thick, fissured, probably an adaptation for fire resistance.
Habitat: '*Acacia, Acacia-Commiphora, Acacia-Combretum, Combretum-Terminalia, Terminalia* and *Anogeissus* woodland, wooded grassland and bushland on a wide variety of soils, but usually in rocky places, relic tree in farmland, river banks, dry riverine forest, often a dominant species' (FEE). 'Woodland or bushland' (FS). 'Woodland, wooded grassland, bushland and dry riverine forest' (PSS). In mixed woodland and as dominant species in a range of drier woodland habitats (our observation). *Terminalia brownii* differs from most other CTW-trees in also being a cultivated tree with the Konso people in the GG floristic region. Although said by Vollesen (1995a: 128) in general terms that *T. brownii* is left as a relic tree in cultivated areas, we have seen it planted and carefully attended to in intensively cultivated fields in Konso farmland, where it is trimmed into a characteristic shape; Engels & Goettsch (1991: 179) agree, suggesting that the Konso people actively cultivate the tree in their terraced fields, and state that the leaves and young branches are cut and used as fodder.
Distribution in Ethiopia: TU GD WU SU AR GG SD BA HA (FEE). No additional records (Fig. 6-28).
Altitudinal range in Ethiopia: 300-2000 m a.s.l. (FEE, including range in Eritrea). 700-1850 m a.s.l. (our records).
General distribution: From Nigeria to Ethiopia, Eritrea, and Somalia, south to Tanzania; also in Arabia.
Chorological classification: Distribution according to direct observation. **WWn, WWs, ASO**. *Local phytogeographical element and sub-element.* **ACB[-CTW]**. The species has a limited distribution in the *Acacia-Commiphora bushland* (ACB), but it is common in the Rift Valley in *Acacia wooded grassland of the Rift Valley* (ACB/RV) and in the *Transitional semi-evergreen bushland* (TSEB) to the east of the Rift. The distribution agrees best with the local element and sub-element **ACB[-CTW]**, particularly by the fact that it is only marginally intruding in the *Combretum-Terminalia woodland and wooded grassland* (CTW) in the north and in the south.

6.*11.8.* Terminalia laxiflora *Engl. & Diels (1900).*

FTEA, Combretaceae: 98 (1973); FEE 2(2): 128 (1995); *BS* 51: 134 (1998); PSS: 236 (2015).
Taxonomy: No infraspecific taxa.
Description: Tree to 15 m tall. No thorns. Leaves simple, probably evergreen. Mesophyll-Macrophyll. Bark smooth to thick, rough and fissured, probably an adaptation for fire resistance.
Habitat: '*Combretum-Terminalia, Anogeissus-Pterocarpus* woodland and wooded grassland on grey, sandy or alluvial soils or on rocky slopes' (FEE). 'Woodland and wooded grassland' (PSS). Mostly in the lower

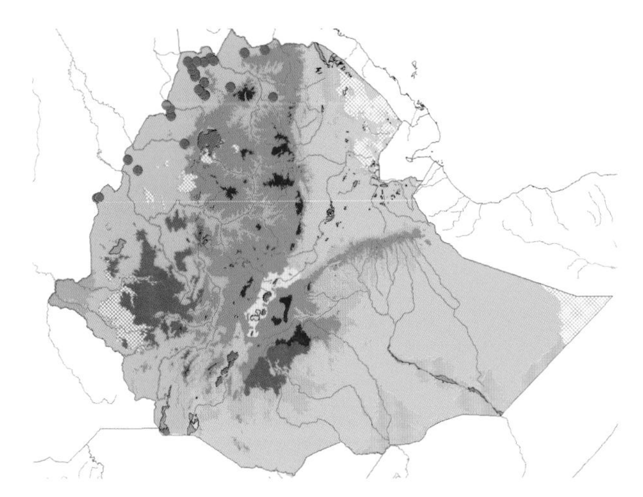

Fig. 6-25. *Combretum hartmannianum*. Distribution in Ethiopia. Observed records are marked with red dots on vegetation map reproduced from Fig. 4-1.

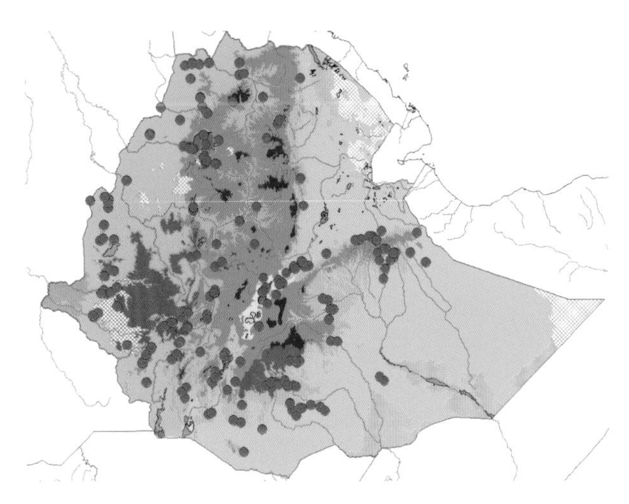

Fig. 6-26. *Combretum molle*. Distribution in Ethiopia. Observed records are marked with red dots on vegetation map reproduced from Fig. 4-1.

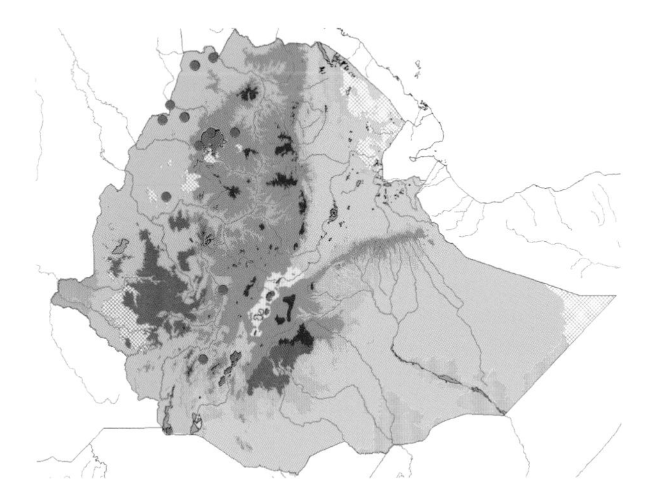

Fig. 6-27. *Combretum rochetianum*. Distribution in Ethiopia. Observed records are marked with red dots on vegetation map reproduced from Fig. 4-1.

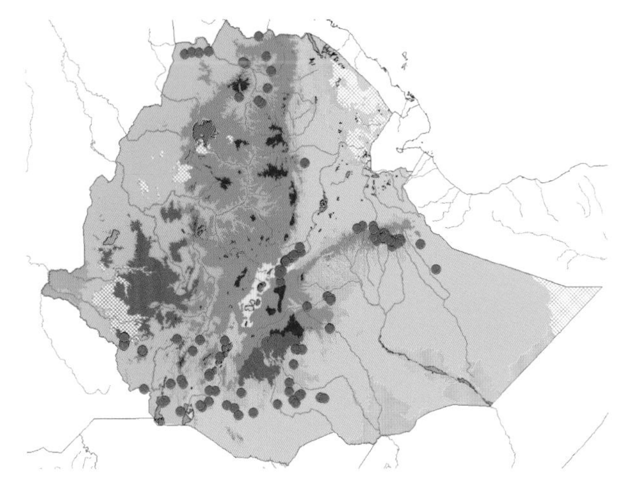

Fig. 6-28. *Terminalia brownii*. Distribution in Ethiopia. Observed records are marked with red dots on vegetation map reproduced from Fig. 4-1.

zones of the *Combretum-Terminalia* woodland (our observation).
Distribution in Ethiopia: GD GJ IL (FEE). Also recorded from the TU, WG and KF floristic regions (Fig. 6-29).
Altitudinal range in Ethiopia: 450-1500 m a.s.l. (FEE). 500-1550 m a.s.l. (our records).
General distribution: From Senegal to Ethiopia and Eritrea, south to D.R. Congo, Uganda, Kenya and Tanzania.
Chorological classification: Distribution according to direct observation. **WWn, WWc, WWs.** *Local phytogeographical element and sub-element.* **CTW[all].** Widely distributed in the western woodlands, from north to south or almost so; no eastward extension.

6.*11.9.* Terminalia macroptera *Guill. & Perr. (1832).*

FTEA, Combretaceae: 87 (1973); FEE 2(2): 127 (1995); PSS: 236 (2015).
Taxonomy: No infraspecific taxa.
Description: Tree to 12 m tall. No thorns. Leaves simple, probably evergreen. Macrophyll. Bark thick, rough and fissured, probably an adaptation for fire resistance.
Habitat: '*Combretum-Terminalia* wooded grassland with ground-cover of tall grasses on black cotton soil and rocky slopes' (FEE). 'Woodland and wooded grassland' (PSS). Mostly in *Combretum-Terminalia* woodland on vertisol and dark brown soil or rocks, common in the central part of the woodlands (our observation).
Distribution in Ethiopia: WG (FEE). Also recorded from the GD, GJ, WG and IL floristic regions (Fig. 6-30).
Altitudinal range in Ethiopia: 1300-1400 m a.s.l. (FEE). 600-1700 m a.s.l. (our records).
General distribution: From Senegal to Ethiopia, south to D.R. Congo and Uganda.
Chorological classification: Distribution according to direct observation. **WWn, WWc.** *Local phytogeographical element and sub-element.* **CTW[n+c].** Widely distributed in the northern and central part of the western woodlands, reaching the Gambela lowlands in the south.

6.*11.10.* Terminalia schimperiana *Hochst. (1844).*

FEE 2(2): 128 (1995); *BS* 51: 134 (1998); PSS: 236 (2015).
Terminalia glaucescens Planch. ex Benth.
FTEA, Combretaceae: 89 (1973).
Taxonomy: No infraspecific taxa.
Description: Tree to 10 m tall. No thorns. Leaves simple, probably evergreen. Mesophyll-Macrophyll. Bark smooth to thick, rough and fissured, probably an adaptation for fire resistance.
Habitat: '*Combretum-Terminalia-Stereospermum-Piliostigma* woodland and wooded grassland on rocky slopes, black clay or sandy soil, often as a dominant, often persisting in cultivated areas' (FEE). 'Woodland and wooded grassland' (PSS). In the *Combretum-Terminalia* woodland in a wide range of habitats (our observation).
Distribution in Ethiopia: TU GD GJ SU WG IL KF GG SD BA (FEE). No additional records (Fig. 6-31).
Altitudinal range in Ethiopia: (1000-) 1300-2200 m a.s.l. (FEE). (700-) 900-2000 m a.s.l. (our records).
General distribution: From Guinea-Bissau to Ethiopia and Eritrea, south to Uganda and Tanzania.
Chorological classification: Distribution according to direct observation. **WWn, WWc, WWs, EWW.** *Local phytogeographical element and sub-element.* **CTW[all+ext].** Widely distributed in the western woodlands, from north to south or almost so, extending to the Rift Valley and to east of the Rift.

6.12. Guttiferae

FTEA, Hypericaceae: 1-23 (1953); FTEA, Guttiferae: 1-35 (1978); FEE 2(2): 135-143 (1995), as "Guttiferae (Clusiaceae)"; *BS* 51: 136-138 (1998), as "Clusiaceae"; PSS: 232, as "Hypericaceae".

Fig. 6-29. *Terminalia laxiflora*. Distribution in Ethiopia. Observed records are marked with red dots on vegetation map reproduced from Fig. 4-1.

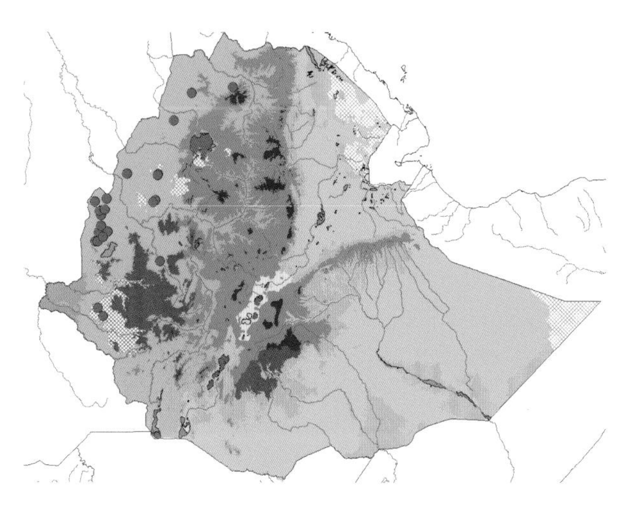

Fig. 6-30. *Terminalia macroptera*. Distribution in Ethiopia. Observed records are marked with red dots on vegetation map reproduced from Fig. 4-1.

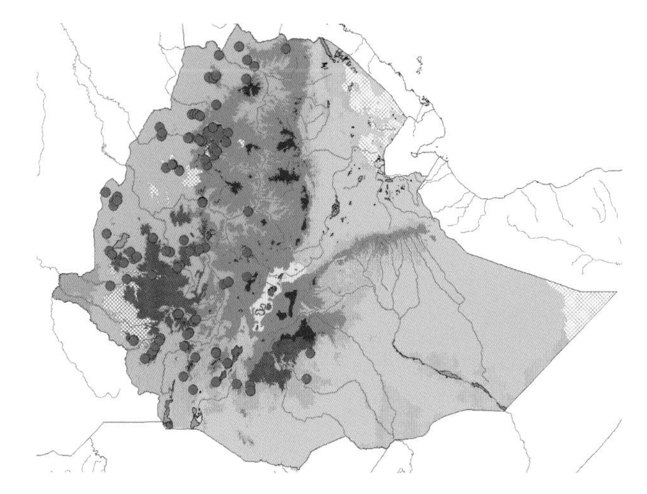

Fig. 6-31. *Terminalia schimperiana*. Distribution in Ethiopia. Observed records are marked with red dots on vegetation map reproduced from Fig. 4-1.

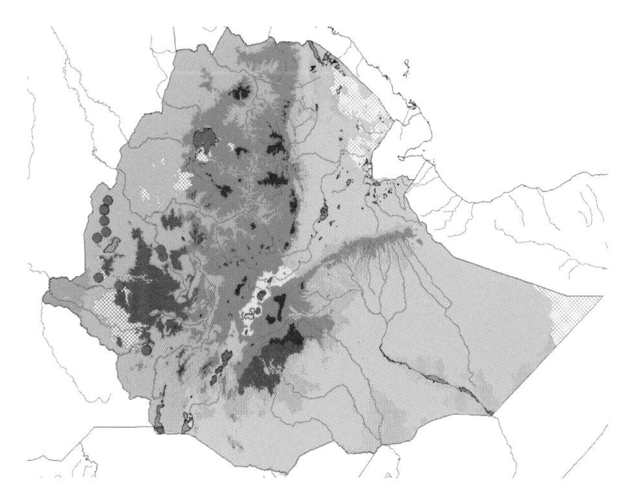

Fig. 6-32. *Psorospermum febrifugum*. Distribution in Ethiopia. Observed records are marked with red dots on vegetation map reproduced from Fig. 4-1.

6.*12.1.* Psorospermum febrifugum *Spach (1836).*

FTEA, Hypericaceae: 16 (1953); FTEA, Guttiferae: 32 (1978); FEE 2(2): 140 (1995); *BS* 51: 138 (2015); PSS: 232 (2015).
Taxonomy: No infraspecific taxa.
Description: Shrub, small tree or tree to 6 (-10) m tall. No thorns. Leaves simple, probably evergreen. Mesophyll. Bark corky, thick, rough and fissured, flaking, probably an adaptation for fire resistance.
Habitat: 'In *Protea-Combretum* wooded grassland on black clay' (FEE). 'Woodland' (PSS). In *Combretum-Terminalia* woodland with tall ground-cover liable to burning, also on rocky outcrops (our observation).
Distribution in Ethiopia: WG IL (FEE). Also recorded from the KF floristic region (Fig. 6-32).
Altitudinal range in Ethiopia: 1500-1800 m a.s.l. (FEE). 700-1800 m a.s.l. (our records).
General distribution: From Sierra Leone to Ethiopia, south to Angola and South Africa.
Chorological classification: Distribution according to direct observation. WWc, [WWs]. *Local phytogeographical element and sub-element.* CTW[c]. Widely distributed in the central part of the western woodlands, the southernmost record is a sight record in our relevé 108 near Djemu; no eastward extension.

6.13. Tiliaceae

FEE 2(2): 145-164 (1995); *BS* 51: 138-143 (1998); FS 2: 5-21 (1999); FTEA, Tiliaceae & Muntingiaceae: 1-120 (2001); PSS: 256-258 (2015), as "Malvaceae: Grewioideae (Sparrmanniaceae)".

6.*13.1.* Grewia bicolor *Juss. (1804).*

FEE 2(2): 146 (1995); *BS* 51: 139 (1998); FS 2: 6 (1999); FTEA, Tiliaceae: 42 (2001); PSS: 256 (2015).
Taxonomy: No infraspecific taxa.
Description: Shrub or small tree to 6 m tall. No thorns. Leaves simple, probably evergreen. Mesophyll. Bark apparently smooth, probably no adaptation for fire resistance.
Habitat: '*Acacia* woodland, wooded grassland, along riverbeds and streams, on sandy soil and rocky areas' (FEE). '*Acacia* woodland, wooded grassland on rocky limestone hills or on sandy soil' (FS). 'Woodland, bushland' (PSS). In a wide range of habitats (our observation)
Distribution in Ethiopia: TU GD GJ WU SU WG IL KF GG SD BA HA (FEE). The FEE records from the WG and IL floristic region have not been confirmed (Fig. 6-33).
Altitudinal range in Ethiopia: 500-1800 m a.s.l. (FEE, including range in Eritrea). 700-1600 m a.s.l. (our records).
General distribution: From Senegal to Ethiopia, Eritrea and Somalia, south to Angola, Botswana and Zimbabwe; also in Arabia and India.
Chorological classification: Distribution according to direct observation. WWn, [WWc], WWs, ASO. *Local phytogeographical element and sub-element.* ACB[-CTW]. The species is not widely distribution in the *Acacia-Commiphora bushland* (ACB) but is common in the Rift Valley and to the east of the Rift. However, the species is only marginally intruding in the *Combretum-Terminalia woodland and wooded grassland* (CTW) in the north and in the south, and has therefore been classified as ACB[-CTW].

6.*13.2.* Grewia erythraea *Schweinf. (1868).*

FEE 2(2): 152 (1995); FE 2: 11 (1999); FTEA, Tiliaceae: 18 (2001); PSS: 256 (2015).
Taxonomy: No infraspecific taxa.
Description: Shrub to 2 (-3) m tall. No thorns. Leaves simple, probably evergreen. Microphyll. Bark smooth, probably no adaptation for fire resistance.
Habitat: '*Acacia-Commiphora* woodland, among lava and limestone rocks and in stony soil' (FEE). '*Acacia-Commiphora* woodland, on lava and limestone rocks, and in stony soil' (FS). '*Acacia-Commiphora* bushland' (PSS). In a wide range of dry habitats, mainly in *Acacia-Commiphora* bushland (our observation).
Distribution in Ethiopia: TU WU SU GG SD HA (FEE). Also recorded from the AF floristic region, the FEE record from SU has not been confirmed (Fig. 6-34).

Altitudinal range in Ethiopia: 450-1650 m a.s.l. (FEE, including range in Eritrea). 700-1850 m a.s.l. (our records).
General distribution: Sudan, Ethiopia, Eritrea and Somalia, south to Kenya; also in Egypt, Arabia and Afghanistan.
Chorological classification: Distribution according to direct observation. **WWn, WWs, ASO**. *Local phytogeographical element and sub-element.* **ACB[-CTW]**. The species only intrudes into the *Combretum-Terminalia woodland and wooded grassland* (CTW) in upper reaches of the Tacazze Valley. It occurs in the Rift Valley and to the east of the Rift both in the *Transitional semi-evergreen bushland* (TSEB) and the *Acacia-Commiphora bushland* (ACB). Despite its absence from south-western Ethiopia, the most suitable local element and sub-element has been assessed to be ACB[-CTW].

6.*13.3*. Grewia ferruginea *Hochst. ex A. Rich. (1847).*

FEE 2(2): 150 (1995); FTEA, Tiliaceae: 22 (2001); PSS: 256 (2015).
Taxonomy: No infraspecific taxa.
Description: Shrub or small tree to 6 m tall. No thorns. Leaves simple, probably evergreen. Mesophyll. Bark smooth, probably no adaptation for fire resistance.
Habitat: 'In gallery forest near lakes, along rivers and in open *Acacia-Combretum* woodland, on dark brown soil' (FEE). 'Riverine forest, woodland' (PSS). In a wide range of habitats, including Afromontane evergreen bushland on bare rocks and at forest margins (our observation).
Distribution in Ethiopia: TU GD GJ WU SU AR WG IL KF SD BA HA (FEE). No additional records (Fig. 6-35).
Altitudinal range in Ethiopia: 1350-2700 m a.s.l. (FEE, including range in Eritrea). 1500-2500 (-3000) m a.s.l. (our records).
General distribution: Sudan, Ethiopia, Eritrea, south to northern Kenya.
Chorological classification: Distribution according to direct observation. **WWn, WWc, EWW**. *Local phytogeographical element and sub-element.* **CTW[n+c+ext]**. Widely distributed, but scattered, in the western woodlands, from north to south or almost so, but always near the upper limits of the woodlands and extending into the highlands, thus the local element could also with some justification be classified as **MI**. It seems to be absent from the south-western woodlands but extends to the Rift Valley and to the east of the Rift.

6.*13.4*. Grewia flavescens *Juss. (1804).*

FEE 2(2): 149 (1995); FTEA, Tiliaceae: 33 (2001); PSS: 256 (2015).
Taxonomy: No infraspecific taxa.
Description: Shrub or small tree, sometimes scandent, up to 7.5 m tall. No thorns. Leaves simple, probably evergreen. Mesophyll. Bark smooth, probably no adaptation for fire resistance.
Habitat: 'Open *Acacia-Terminalia* woodland, on sandy, gravelly or loamy clay soils' (FEE). 'Woodland' (PSS). In a wide range of habitats, mainly with dry vegetation (our observation).
Distribution in Ethiopia: TU GD GJ WU SU IL GG SD BA HA (FEE). No additional records (Fig. 6-36).
Altitudinal range in Ethiopia: 900-1900 (-2300) m a.s.l. (FEE, including range in Eritrea). 650-1850 m a.s.l. (our records).
General distribution: From Mauritania and Senegal to Ethiopia and Eritrea, south to Namibia and South Africa; also in Arabia and India.
Chorological classification: Distribution according to direct observation. **WWn, WWc, [WWs], EWW**. *Local phytogeographical element and sub-element.* **CTW[all+ext]**. Widely distributed in the western woodlands, from north to south or almost so, but poorly represented in the south, extending to the Rift Valley and to the east of the Rift. One record from near the border with Kenya (*C Puff et al.* 870430-1/7) in the ACB-area of the SD floristic region may be from an outlying hill with *Transitional semi-evergreen bushland* (TSEB).

Fig. 6-33. *Grewia bicolor*. Distribution in Ethiopia. Observed records are marked with red dots on vegetation map reproduced from Fig. 4-1.

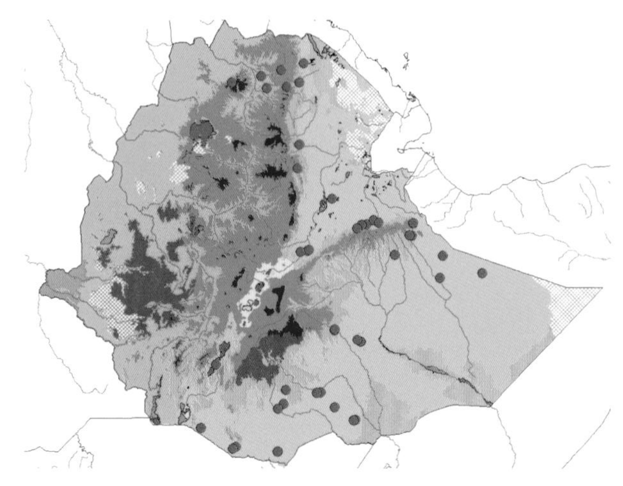

Fig. 6-34. *Grewia erythraea*. Distribution in Ethiopia. Observed records are marked with red dots on vegetation map reproduced from Fig. 4-1.

Fig. 6-35. *Grewia ferruginea*. Distribution in Ethiopia. Observed records are marked with red dots on vegetation map reproduced from Fig. 4-1.

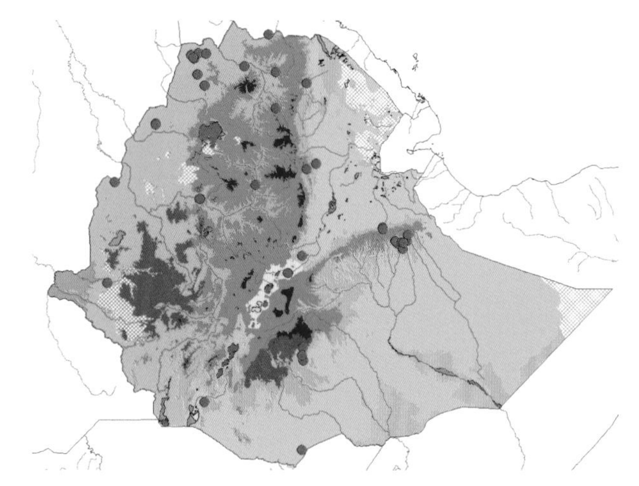

Fig. 6-36. *Grewia flavescens*. Distribution in Ethiopia. Observed records are marked with red dots on vegetation map reproduced from Fig. 4-1.

6.*13.5*. Grewia mollis *Juss. (1804).*

FEE 2(2): 146 (1995); *BS* 51: 139 (1998); FS 2: 9 (1999); FTEA, Tiliaceae: 44 (2001); PSS: 256 (2015).
Taxonomy: No infraspecific taxa.
Description: Shrub or small tree to 7 m tall. No thorns. Leaves simple, probably evergreen. Mesophyll. Bark smooth, probably no adaptation for fire resistance.
Habitat: 'Common in *Acacia-Commiphora* woodland' (FEE). 'In ravine with *Pistacia* and *Buxus*' [one record only] (FS). 'Woodland and wooded grassland' (PSS). In a wide range of habitats, also common in typical *Combretum-Terminalia* woodland with tall ground-cover and liable to burning (our observation).
Distribution in Ethiopia: TU GD GJ WU SU AR WG IL KF GG SD BA HA (FEE). Also recorded from the AF floristic region (Fig. 6-37).
Altitudinal range in Ethiopia: 650-1900 (-2390) m a.s.l. (FEE). 450-1950 m a.s.l. (our records).
General distribution: From Senegal to Ethiopia, Eritrea and Somalia, south to Zambia; also in Arabia.
Chorological classification: Distribution according to direct observation. **WWn, WWc, WWs, EWW**. *Local phytogeographical element and sub-element.* **CTW[all+ext]**. Widely distributed in the western woodlands, from north to south or almost so, but rare in the north and the south; extending to the Rift Valley and to the east of the Rift.

6.*13.6*. Grewia tenax *(Forssk.) Fiori (1912).*

FEE 2(2): 152 (1995); *BS* 51: 140 (1998); FS 2: 11 (1999); FTEA, Tiliaceae: 14 (2001); PSS: 257 (2015).
Taxonomy: No infraspecific taxa.
Description: Shrub to 5 m tall. No thorns. Leaves simple, probably deciduous. Microphyll-mesophyll. Bark smooth, probably no adaptation for fire resistance.
Habitat: '*Acacia-Terminalia-Combretum* woodland, between lava rocks and in sandy soil' (FEE). '*Acacia-Commiphora* scrub, *Acacia* woodland, seasonally flooded bushland and limestone slopes, on alluvial sandy plains' (FS). 'Dry wooded grassland and bushland' (PSS). Most frequent in a range of dry habitats or habitats with sharply contrasting seasonality of rainfall (our observation).
Distribution in Ethiopia: TU WU SU AR KF GG SD BA HA (FEE). Also recorded from the IL floristic region. This may seem surprising because of the relatively moist habitat there in comparison with the rest of the country, but the records are confirmed by independently collected and identified specimens (*Friis et al.* 7209, *Tesfaye Awash et al.* 309, *Sebsebe Demissew et al.* 1452), and according to PSS it also occurs nearby on the other side of the South Sudan border (Fig. 6-38).
Altitudinal range in Ethiopia: Sea level-1450 (-1800) m a.s.l. (FEE, including range in Eritrea). 450-1450 m a.s.l. (our records).
General distribution: From Mauritania and Senegal to Ethiopia, Eritrea and Somalia, south to Namibia and South Africa; also in Arabia and Sri Lanka.
Chorological classification: Distribution according to direct observation. **WWn, [WWc], WWs, ASO**. *Local phytogeographical element and sub-element.* **ACB[-CTW]**. This species is intruding into the western woodlands in the north and in the south and in the dry river valleys of Tacazze and the Abay River, but the pattern is not typical due to records from the Gambela lowlands. Due to its wide distribution in the ACB-area in the Somali region of Ethiopia and in Somalia it is classified as **ACB[-CTW]**.

6.*13.7*. Grewia trichocarpa *Hochst. ex A. Rich. (1847).*

FEE 2(2): 148 (1995); *BS* 51: 140 (1998); FS 2: 9 (1999); FTEA, Tiliaceae: 45 (2001); PSS: 257 (2015).
Taxonomy: No infraspecific taxa.
Description: Shrub or small tree to 6 m tall. No thorns. Leaves simple, probably evergreen. Mesophyll. Bark smooth, probably no adaptation for fire resistance.
Habitat: '*Acacia-Commiphora* woodland and riverine forest ... on sandy and rocky soil' (FEE). 'Probably *Acacia-Commiphora* woodland at intermediate altitudes' (FS). 'Riverine forest, wooded grassland' (PSS). The range of habitats indicated in the citation is considerable, from vegetation dominated by *Acacia-Commiphora* bushland to *Riparian vegetation* (RV). We

have no additional observations but see discussion below under 'Distribution type.'
Distribution in Ethiopia: TU GD GJ WU SU AR WG IL KF GG SD BA HA (FEE). No additional records (Fig. 6-39).
Altitudinal range in Ethiopia: 650-1900 (-2390) m a.s.l. (FEE, including range in Eritrea). 850-2000 (-2350) m a.s.l. (our records).
General distribution: Sudan, South Sudan, Ethiopia, Eritrea and Somalia, south to Tanzania; also in Arabia.
Chorological classification: Distribution according to direct observation. **WWn, WWc, WWs, EWW**. *Local phytogeographical element and sub-element.* **CTW[all+ext]**. Distributed mainly in the central-southern part of the western woodlands, with a few records in the Abay Valley and also at lower altitudes of the Ethiopian Highlands in the north, and has, for a local CWT-element absent an unusual distribution in the north; it is absent from large areas of *Combretum-Terminalia* woodland, apart from dry river valleys; extending to the Rift Valley and to the east of the Rift. One record in the *Acacia-Commiphora bushland* (ACB) in the SD floristic region (*Mesfin Tadesse* 1260 (ETH), 15 km north of Moyale) seems to require confirmation but may be from a hill with outlying *Transitional semi-evergreen bushland* (TSEB). The phytogeographical classification of this species is slightly doubtful.

6.*13.8*. Grewia velutina *(Forssk.) Vahl (1790)*

FEE 2(2): 148 (1995); FS 2: 9 (1999); FTEA, Tiliaceae: 44 (2001); PSS: 257 (2015).
Taxonomy: No infraspecific taxa.
Description: Large shrub or tree to 8 m tall. No thorns. Leaves simple, probably evergreen. Microphyll-mesophyll. Bark smooth, probably no adaptation for fire resistance.
Habitat: 'Open *Acacia* woodland on gravelly granite soil' (FEE). 'Open *Acacia-Buxus* bushland on granite or limestone' (FS). 'Rocky hillsides, open woodland' (PSS). We have no further observations, see discussion below under 'Distribution type.'
Distribution in Ethiopia: TU GJ WU SU AR IL KF GG SD BA HA (FEE). Also recorded from the AF floristic region; the FEE records from the TU and GJ floristic regions have not been confirmed (Fig. 6-40).
Altitudinal range in Ethiopia: 550-2450 m a.s.l. (FEE). 600-1750 m a.s.l. (our records).
General distribution: Sudan, Ethiopia, Eritrea and Somalia, south to Uganda, Kenya; also in Arabia.
Chorological classification: Distribution according to direct observation. **WWc, WWs, EWW**. *Local phytogeographical element and sub-element.* **CTW[c+s+ext]**. Widely distributed in the southern part of the western woodlands, mainly south of the Abay River, but with one record from near Kurmuk, otherwise with records from the Gambela lowlands and the lower Omo Valley, extending to the Rift Valley and to the east of the Rift. Also with some records from the *Acacia-Commiphora bushland* (ACB) in the SD floristic region, probably from hills with outlying *Transitional semi-evergreen bushland* (TSEB).

6.*13.9*. Grewia villosa *Willd. (1804)*.

FEE 2(2): 152 (1995); *BS* 51: 141 (1998); FS 2: 14 (1999); FTEA, Tiliaceae: 30 (2001); PSS: 257 (2015).
Taxonomy: No infraspecific taxa.
Description: Shrub to 4 m tall. No thorns. Leaves simple, probably deciduous. Mesophyll-Macrophyll. Bark smooth, probably no adaptation for fire resistance.
Habitat: '*Acacia-Terminalia-Combretum* woodland on rocky and sandy soils and limestone areas' (FEE). '*Acacia-Terminalia-Combretum* bushland and wooded grassland, often among boulders, on rocky ground or in riverine areas' (FS). 'Dry rocky hillsides, woodland' (PSS). Generally in dry habitats (our observation).
Distribution in Ethiopia: TU WU SU IL KF GG SD BA HA (FEE). The FEE record from the IL floristic region has not been confirmed (Fig. 6-41).
Altitudinal range in Ethiopia: 400-1800 m a.s.l. (FEE, including range in Eritrea). 450-2000 m a.s.l. (our records).
General distribution: From Mauritania and Senegal to Ethiopia, Eritrea and Somalia, south to Namibia and South Africa; also in Arabia and India.
Chorological classification: Distribution according to direct

Fig. 6-37. *Grewia mollis*. Distribution in Ethiopia. Observed records are marked with red dots on vegetation map reproduced from Fig. 4-1.

Fig. 6-38. *Grewia tenax*. Distribution in Ethiopia. Observed records are marked with red dots on vegetation map reproduced from Fig. 4-1.

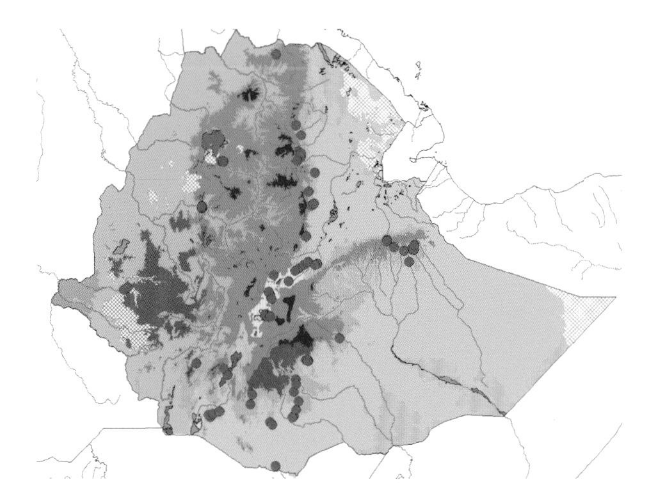

Fig. 6-39. *Grewia trichocarpa*. Distribution in Ethiopia. Observed records are marked with red dots on vegetation map reproduced from Fig. 4-1.

Fig. 6-40. *Grewia velutina*. Distribution in Ethiopia. Observed records are marked with red dots on vegetation map reproduced from Fig. 4-1.

observation. **WWn, WWs, ASO.** *Local phytogeographical element and sub-element.* **ACB[-CTW].** Widely distributed in *Acacia-Commiphora bushland* (ACB), marginally intruding in the *Combretum-Terminalia woodland and wooded grassland* (CTW) in the north and in the south, extending to the Rift Valley and to the east of the Rift.

6.14. Sterculiaceae

FEE 2(2): 165-185 (1995); *BS* 51: 143-147 (1998); FS 2: 21-37 (1999); FTEA, Sterculiaceae: 1-134 (2007); PSS: 260 (2015), as for **Sterculia**, "Malvaceae: Sterculioideae (Sterculiaceae)" and, as for **Dombeya**, "Malvaceae: Dombeyoideae (Pentapetaceae)".
The definition of the Sterculiaceae here follows that of the FTEA, Sterculiaceae (2007) and earlier floristic works.

6.*14.1.* Dombeya buettneri *K. Schum. (1893).*

FEE 2(2): 168 (1995); *BS* 51: 144 (1998); FTEA, Sterculiaceae: 63 (2007); PSS: 258 (2015).
Taxonomy: No infraspecific taxa.
Description: Shrub or small tree to 6 (-9) m tall, often with several unbranched stems from base. No thorns. Leaves simple, probably evergreen. Mesophyll-Macrophyll. Bark smooth, very tough and fibrous, exuding sticky sap when cut; no obvious adaptation for fire resistance (but new stems may shoot from the rootstock after fires).
Habitat: '*Combretum-Terminalia* wooded grassland with tall undergrowth of *Hyparrhenia*, on rocky hillsides and outcrops' (FEE). 'Wooded grassland, secondary forest, rocky hillslopes' (PSS). *Combretum-Terminalia* woodland with tall ground-cover, likely to burn (our observation).
Distribution in Ethiopia: IL KF (FEE). Also recorded from the WG floristic region (Fig. 6-41).
Altitudinal range in Ethiopia: 1300-1800 m a.s.l. (FEE). 1200-1800 m a.s.l. (our records).
General distribution: From Guinean Republic to Ethiopia, south to Zimbabwe and Zambia.
Chorological classification: Distribution according to direct observation. **WWc, WWs.** *Local phytogeographical element and sub-element.* **CTW[c+s].** Widely distributed in the central and southern part of the western woodlands, mainly south of the Abay River; no eastward extension.

6.*14.2.* Dombeya quinqueseta *(Delile) Exell (1935).*

FEE 2(2): 170 (1995); *BS* 51: 144 (1998); FTEA, Sterculiaceae: 68 (2007); PSS: 258 (2015).
Taxonomy: No infraspecific taxa.
Description: Shrub or tree to 8 m tall. No thorns. Leaves simple, probably evergreen. Mesophyll-Macrophyll. Bark smooth, probably no adaptation for fire resistance.
Habitat: '*Combretum-Terminalia-Stereospermum* and *Pterocarpus* woodland, wooded grassland and bushland, mostly on rocky slopes' (FEE). 'Wooded grassland' (PSS). *Combretum-Terminalia* woodland with tall ground-cover, likely to burn (our observation).
Distribution in Ethiopia: TU GD GJ SU WG IL KF GG (FEE). Also recorded from SD floristic region (Fig. 6-43).
Altitudinal range in Ethiopia: 850-1900 (-2200) m a.s.l. (FEE, including range in Eritrea). 700-1800 m a.s.l. (our records).
General distribution: From Senegal to Ethiopia and Eritrea, south to Uganda and Kenya.
Chorological classification: Distribution according to direct observation. **WWn, WWc, WWs, EWW.** *Local phytogeographical element and sub-element.* **CTW[all+ext].** Widely distributed in the western woodlands, from north to south or almost so, extending to the east of the Rift.

6.*14.3.* Sterculia africana *(Lour.) Fiori (1912).*

FEE 2(2): 184 (1995); FS 2: 35 (1999); FTEA, Sterculiaceae: 18 (2007); PSS: 260 (2015).
Taxonomy: No infraspecific taxa. To distinguish in the field between *Sterculia africana*, *Sterculia cinerea* A. Rich. and *Sterculia setigera* Delile was not always possible (see Section 4.1, particularly difficulties in the absence of fruits or sometimes also leaves; misidentifications may therefore have occurred.

Fig. 6-41. *Grewia villosa*. Distribution in Ethiopia. Observed records are marked with red dots on vegetation map reproduced from Fig. 4-1.

Fig. 6-42. *Dombeya buettneri*. Distribution in Ethiopia. Observed records are marked with red dots on vegetation map reproduced from Fig. 4-1.

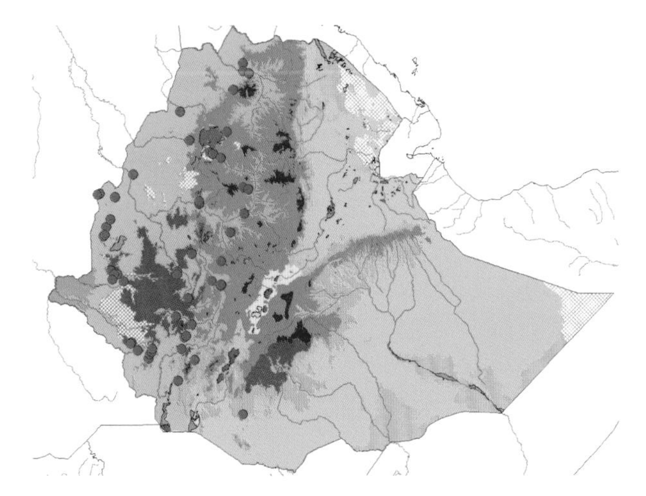

Fig. 6-43. *Dombeya quinqueseta*. Distribution in Ethiopia. Observed records are marked with red dots on vegetation map reproduced from Fig. 4-1.

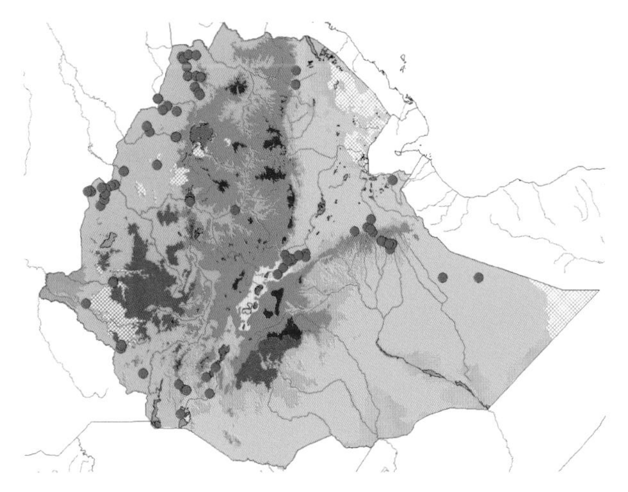

Fig. 6-44. *Sterculia africana*. Distribution in Ethiopia. Observed records are marked with red dots on vegetation map reproduced from Fig. 4-1.

Description: Tree to 10 (-15) m tall. No thorns. Leaves simple, probably deciduous. Mesophyll-Macrophyll. Bark peeling in greyish to pale purplish flakes, probably an adaptation for fire resistance.
Habitat: '*Acacia*, *Acacia-Commiphora* and *Acacia-Terminalia* woodland and bushland on gravelly to stony soil, often on rocky slopes and lava-flows' (FEE). 'Deciduous woodland or bushland, often on rocky slopes' (FS). 'Dry bushland and woodland, often on rocky hillslopes' (PSS). Dry and open *Combretum-Terminalia* woodland, often on rocky slopes (our observation).
Distribution in Ethiopia: AF WU SU IL GG BA HA (FEE). Also recorded from the GD, GJ, WG and KF floristic regions; the FEE records from the WU and BA floristic regions have not been confirmed. These differences may be due to the complications distinguishing between *S. africana* and *S. setigera* mentioned in Section 4 (Fig. 6-44).
Altitudinal range in Ethiopia: Sea level to 1775 m a.s.l. (FEE, including range in Eritrea). 550-1550 m a.s.l. (our records).
General distribution: Sudan, Ethiopia, Eritrea and Somalia, [apparently lacking in Uganda and Kenya], south to Namibia, Botswana and Mozambique; also in Egypt.
Chorological classification: Distribution according to direct observation. **WWn, WWc, WWs, EWW**. *Local phytogeographical element and sub-element.* **CTW[all+ext]**. Widely distributed in the western woodlands, from north to south or almost so, with extension to the Rift Valley and to the east of the Rift; also with an extension to the *Acacia-Commiphora bushland* (ACB). According to FS, it occurs in Somalia but only in the northern part.

6.15. Bombacaceae

FTEA, Bombacaceae: 1-9 (1989); FEE 2(2): 186-189 (1995); *BS* 51: 147 (1998); FS 2: 38-40 (1999); PSS: 249 (2015), as "**Malvaceae: Bombacoideae (*Bombacaceae*)**".
The definition of the Bombacaceae here follows that of the FTEA, Bombacac. (1989) and other floristic works before the PSS.

6.15.1. **Adansonia digitata** *L. (1753).*

FTEA, Bombacaceae: 4 (1989); FEE 2(2): 186 (1995); FS 2: 39 (1999); PSS: 249 (2015).
Taxonomy: No infraspecific taxa.
Description: Tree with massive conical trunk to 20 m tall, up to 5 m or more in diameter. No thorns. Leaves compound (digitately 5-6 foliolate, deciduous. Macrophyll (leaflets mesophyll). Bark smooth, with moisture-storing tissue below the bark, probably an adaptation to drought and fire resistance.
Habitat: '*Acacia-Balanites-Adansonia* woodland and wooded grassland on sandy soil over granite, rocky outcrops, riverbanks' (FEE). 'Deciduous bushland and woodland' (FS). 'Dry woodland, wooded grassland, semi-desert' (PSS). In a range of dry habitats, from *Acacia-Commiphora* bushland to *Combretum-Terminalia* woodland, often growing on dry rocks (our observation).
Distribution in Ethiopia: TU GD SD (sight record) (FEE). Also recorded from the WU and WG floristic region; the sight record from SD has not been confirmed (Fig. 6-45).
Altitudinal range in Ethiopia: 700-1700 m a.s.l. (FEE, including range in Eritrea). 600-1550 m a.s.l. (our records).
General distribution: From Mauritania and Senegal to Ethiopia, Eritrea and Somalia, south to Namibia and South Africa; also in Madagascar.
Chorological classification: Distribution according to direct observation. **WWn, [WWc]**. *Local phytogeographical element and sub-element.* **CTW[n]**. Widely distributed in the northern part of the western woodlands, mainly north of the Abay River, but with a few records from near Gizen (*Sebsebe Demissew* 5965 (ETH); *Friis et al.* 9220 (C, ETH, K)).

6.16. Erythroxylaceae

FTEA, Erythroxylaceae: 5 (1984); FEE 2(2): 264 (1995); *BS* 51: 154 (1998); PSS: 215 (2015).

6.*16.1.* Erythroxylum fischeri *Engl. (1895).*

FTEA, Erythroxylaceae: 5 (1984); FTNA: 123 (1992); FEE 2(2): 264 (1995); *BS* 51: 154 (1998); PSS: 215 (2015).
Taxonomy: No infraspecific taxa.
Description: Small tree to 9 (-18) m tall. No thorns. Leaves simple, evergreen. Mesophyll. Apparently, no adaptations for fire resistance.
Habitat: 'Evergreen forest, stream-banks' (FEE). 'Dry peripheral semi-evergreen Guineo-Congolian rain forest; also in lowland forest-woodland mosaic' (FTNA). 'Forest, dense woodland' (PSS). In lowland forest, but also along edges where the forest meets with *Combretum-Terminalia* woodland, and then often with other evergreen shrubs on termite mounds (our observation).
Distribution in Ethiopia: IL (FEE). No additional records (Fig. 6-46).
Altitudinal range in Ethiopia: 500-600 m a.s.l. (FEE). Ca. 550 m a.s.l. (FTNA, including the range in the entire Horn of Africa). 500-600 m a.s.l. (our records).
General distribution: South Sudan, Ethiopia, south to D.R. Congo and Tanzania.
Chorological classification: Distribution according to direct observation. **WWc.** *Local phytogeographical element.* **MI.** Distribution restricted to *Combretum-Terminalia woodland and wooded grassland* (CTW), *Transitional rain forest* (TRF), and *Riparian vegetation* (RV) in the Gambela lowlands. Because this species is mostly considered a species of lowland forest, we have decided to classify it as a marginal intruder in *Combretum-Terminalia woodland and wooded grassland* (CTW) from areas of moist lowlands. This is the only species observed in our relevés, which we consider a MI-local element intruding from other lowland vegetation types.

6.17. Euphorbiaceae

FTEA, Euphorbiac. (Part 1): 1-408 (1987); FTEA, Euphorbiac. (Part 2): 409-599 (1988); FS 1: 267-339 (1993); FEE 2(2): 265-380 (1995); *BS* 51: 154-175 (1998). The definition of the Euphorbiaceae follows here the delimitations in FTEA, Euphorbiac. 1-2 (1987, 1988), *BS* 51 (1998) and FEE 2(2) (1995).

A significant taxonomic redefinition has taken place of the taxa included in the Euphorbiaceae since these works. PSS: 215-227 (2015) has this distribution on families of the species here included in the Euphorbiaceae: P. 215, as "*Peraceae*" (*Clutia abyssinica*); p. 215-223, as "Euphorbiaceae" (*Croton macrostachyus, Erythrococca trichogyne, Euphorbia abyssinica, Euphorbia ampliphylla*); pp. 224-227, as "Phyllantaceae" (*Bridelia micrantha, Bridelia scleroneura, Flueggea virosa*).

6.*17.1.* Bridelia micrantha *Baill. (1862).*

FTEA, Euphorbiac. 1: 127 (1987); FTNA: 127 (1992); FEE 2(2): 269 (1995); PSS: 224 (2015).
Taxonomy: No infraspecific taxa.
Description: Small tree or tree, 3-15 (-27) m tall. Thorns present. Leaves simple, evergreen. Mesophyll. Bark smooth or somewhat rough and only slightly fissured, probably no adaptation for fire resistance.
Habitat: 'In forests around rivers and lakes, less often away from water in higher rainfall areas' (FEE). 'Afromontane rain forest and undifferentiated Afromontane forest ..., especially in moist places, persisting in gully forests; riverine forest; also in montane secondary evergreen bushland on moist ground' (FTNA). 'Open woodland, forest margins and associated bushland' (PSS). In Moist Afromontane Forest (MAF), along margins and bushland, and in moist *Combretum-Terminalia* woodland (our observation).
Distribution in Ethiopia: GD GJ SU AR WG IL KF GG SD BA (FEE). No additional records (Fig. 6-47).
Altitudinal range in Ethiopia: (1050-) 1250-2200 m a.s.l. (FEE). 1200-2300 m a.s.l. (FTNA, including the range in the entire Horn of Africa). 850-2000 (-2100) m a.s.l. (our records).
General distribution: From Senegal to Ethiopia, south to Angola and South Africa; also in Réunion.
Chorological classification: Distribution according to direct observation. **WWn, WWc, WWs, EWW.** *Local phytogeographical element and sub-element.* **CTW[all+ext].** Widely distributed in the central and southern part of the western woodlands, mainly south of the Abay River, but with extension into the *Intermediate evergreen Afromontane forest* (IAF) around the Lake Tana

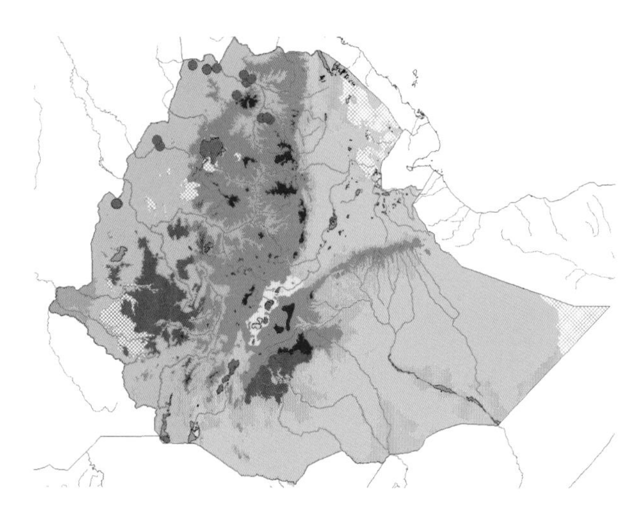

Fig. 6-45. *Adansonia digitata*. Distribution in Ethiopia. Observed records are marked with red dots on vegetation map reproduced from Fig. 4-1.

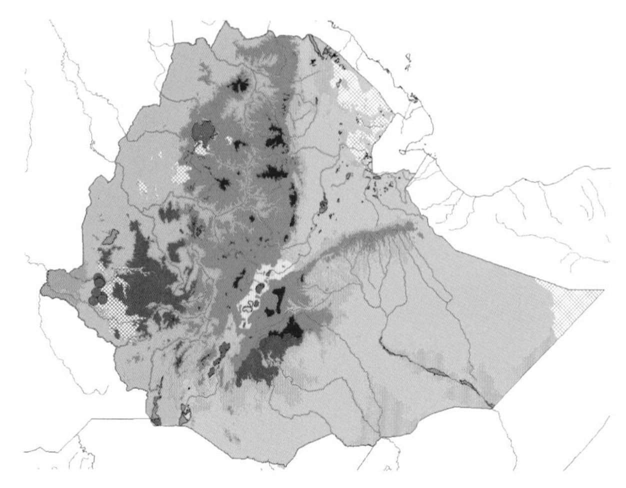

Fig. 6-46. *Erythroxylum fischeri*. Distribution in Ethiopia. Observed records are marked with red dots on vegetation map reproduced from Fig. 4-1.

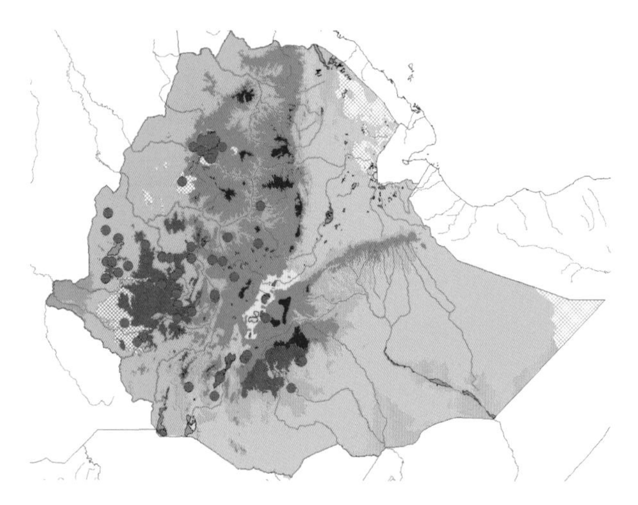

Fig. 6-47. *Bridelia micrantha*. Distribution in Ethiopia. Observed records are marked with red dots on vegetation map reproduced from Fig. 4-1.

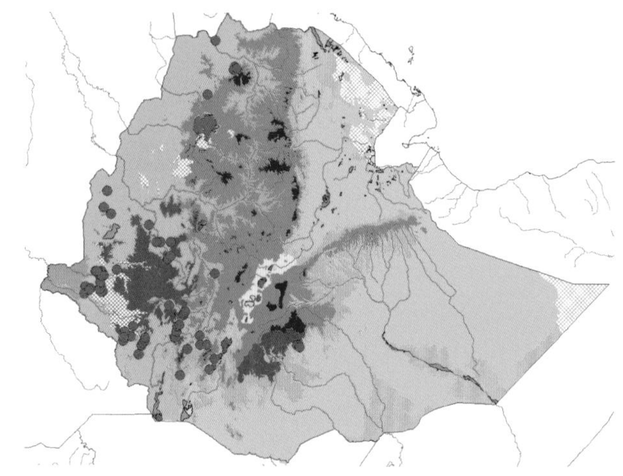

Fig. 6-48. *Bridelia scleroneura*. Distribution in Ethiopia. Observed records are marked with red dots on vegetation map reproduced from Fig. 4-1.

and further south into *Dry evergreen Afromontane forest* (DAF); the species extends to the Rift Valley and to the east of the Rift.

6.17.2. Bridelia scleroneura *Müll.Arg. (1864).*

FTEA, Euphorbiac. 1: 122 (1987); FEE 2(2): 267 (1995); *BS* 51: 160 (1998); PSS: 225 (2015).
Taxonomy: Two subspecies, the widespread *Bridelia scleroneura* subsp. *scleroneura* and *B. scleroneura* subsp. *angolensis* (Müll. Arg.) Radcl.-Sm., restricted to southwestern Africa, are currently accepted. Recognition of infraspecific taxa not relevant here.
Description: Shrub or small tree to 5 (-10) m tall. Thorns formed by branchlets sometimes present. Leaves simple, evergreen. Mesophyll. Bark thickened and fissured, possibly an adaptation for fire resistance.
Habitat: 'Open woodland or wooded grassland, dry riverine forest' (FEE). 'Open woodland' (PSS, for *Bridelia scleroneura* subsp. *scleroneura*). *Combretum-Terminalia* woodland with well-developed ground-cover, liable to burn (our observation).
Distribution in Ethiopia: GD WG IL KF GG BA (FEE). Also recorded from the TU and SU floristic regions (Fig. 6-48).
Altitudinal range in Ethiopia: 440-1800 m a.s.l. (FEE). 550-1700 m a.s.l. (our records).
General distribution (species as a whole): From Ghana to Ethiopia and Eritrea, south to Angola and D.R. Congo; also in Arabia.
Chorological classification: Distribution according to direct observation. **WWn, WWc, WWs, EWW**. *Local phytogeographical element and sub-element.* **CTW[all+ext]**. Widely distributed in the western woodlands, from north to south or almost so, with a notable gap in and around the GJ floristic region; it extends to the Rift Valley and to the east of the Rift.

6.17.3. Clutia abyssinica *Jaub. & Spach (1855).*

FTEA, Euphorbiac. 1: 333 (1987); FS 1: 279 (1993); FEE 2(2): 286 (1995); *BS* 51: 161 (1998); PSS: 215 (2015).
Taxonomy: Three varieties are currently accepted in this species. *Clutia abyssinica* var. *abyssinica* in Sudan and Ethiopia south to Angola and South Africa, *Clutia abyssinica* var. *usambarica* Pax & Hoffm. (1911), in Kenya and Tanzania, and *Clutia abyssinica* var. *pedicellaris* (Pax) Pax (1911) from D.R. Congo to Somalia and Kenya and south to Malawi. The Ethiopian material belongs to *Clutia abyssinica* var. *abyssinica*, but collections from the Mega Plateau in southernmost Ethiopia have more hairy stems and more elliptical leaves than typical *Clutia abyssinica* var. *abyssinica*; this deviating form is known from only one locality. Hence the recognition of infraspecific taxa not relevant here.
Description: Shrub to 2 (-4) m tall. No thorns. Leaves simple, evergreen. Mesophyll. Bark smooth, no adaptation for fire resistance.
Habitat: 'Evergreen bushland and margins of *Juniperus* and *Podocarpus* forest, mostly in disturbed sites' (FEE). 'Evergreen bushland and margins of *Juniperus* forest, mostly in disturbed sites' (FS for *Clutia abyssinica* var. *abyssinica*). 'Wooded grassland, forest margins and secondary bushland' (PSS for species as a whole). Mainly at forest margins and secondary bushland, occasionally with evergreen species in *Combretum-Terminalia* woodland (our observation).
Distribution in Ethiopia: TU GD GJ WU SU AR KF SD BA HA (FEE). No additional records (Fig. 6-49).
Altitudinal range in Ethiopia: 1450-2950 m a.s.l. (FEE). 1300-2800 m a.s.l. (our records).
General distribution (species as a whole): D.R. Congo to Ethiopia, Eritrea and Somalia, south to South Africa.
Chorological classification: Distribution according to direct observation. **WWn, WWc, WWs, EWW**. *Local phytogeographical element.* **MI**. Marginally intruding species, with a main distribution in the highlands and in *Transitional semi-evergreen bushland* (TSEB); only a few records in our relevés.

6.17.4. Croton macrostachyus *Delile (1848).*

FTEA, Euphorbiaceae: 149 (1987); FTNA: 128 (1992); FEE 2(2): 326 (1995); *BS* 51: 161 (1998); PSS: 217 (2015).
Taxonomy: No infraspecific taxa.
Description: Large shrub, small or large tree to 15 (-25) m tall. No thorns. Leaves simple, apparently deciduous, but only for a short time. Mesophyll-Macrophyll.

Bark thickened and longitudinally fissured, this, in combination with the massive size of old trunks, possibly an adaptation for fire resistance.
Habitat: 'Forest margins and secondary woodlands, extending into disturbed areas and along edges of roads, mostly in soils of volcanic origin' (FEE). 'Afromontane rain forest, undifferentiated Afromontane forest ... and single-dominant Afromontane forest, usually most frequent in secondary formations; also in secondary montane evergreen bushland and in moist woodland' (FTNA). 'Forest margins' (PSS). In a wide range of habitats, from Moist and Dry evergreen Afromontane forest (MAF and DAF) to moist woodlands (our observation).
Distribution in Ethiopia: TU GD GJ WU SU AR WG IL KF SD BA HA (FEE). No additional records (Fig. 6-50).
Altitudinal range in Ethiopia: (500-) 1050-2350 m a.s.l. (FEE, including range in Eritrea). 1300-2500 m a.s.l. (FTNA, including the range in the entire Horn of Africa). 500-2500 m a.s.l. (our records).
General distribution: From Guinean Republic to Ethiopia and Eritrea, south to Angola and Mozambique; also in Madagascar and Arabia.
Chorological classification: Distribution according to direct observation. **WWn, WWc, [WWs], EWW.** *Local phytogeographical element.* **TRG.** Transgressing species, occurs in both the *Combretum-Terminalia woodland and wooded grassland* (CTW) and in a range of other vegetation types, particularly *Dry evergreen Afromontane forest* (DAF), but also common in the wetter forest types and in *Transitional semi-evergreen bushland* (TSEB).

6.*17.5.* Erythrococca trichogyne *(Müll. Arg.) Prain (1911)*

FTEA, Euphorbiaceae: 274 (1987); FEE 2(2): 298 (1995); *BS* 51: 163 (1998); PSS: 218 (2015).
Taxonomy: No infraspecific taxa.
Description: Shrub or small tree to 6 m tall. No thorns. Leaves simple, evergreen. Mesophyll. Bark smooth, no obvious adaptation for fire resistance.
Habitat: 'Moist evergreen forest, sometimes persisting in more open degraded forest' (FEE). 'Forest and associated bushland' (PSS). Forest and clumps of bushland with other evergreen species in woodland (our observation).
Distribution in Ethiopia: GJ SU WG IL KF BA (FEE). Also recorded from the AR floristic region (Fig. 6-51).
Altitudinal range in Ethiopia: 1400-2200 m a.s.l. (FEE). 600-2200 m a.s.l. (our records).
General distribution: Ethiopia, South Sudan, south to Angola, Zimbabwe and Mozambique.
Chorological classification: Distribution according to direct observation. **[WWn], WWc, WWs, EWW.** *Local phytogeographical element.* **MI.** Marginally intruding species in the western woodlands, with a main distribution in the highlands and in *Transitional semi-evergreen bushland* (TSEB), rarely in *Transitional rain forest* (TRF); only a few records in our relevés.

6.*17.6.* Euphorbia abyssinica *J.F. Gmel. (1791).*

FS 1: 325 (1993); FEE 2(2): 334 (1995); PSS: 218 (2015).
Taxonomy: No infraspecific taxa.
Description: Succulent, later woody tree to 9 m tall with erect branches forming an obconical crown. Thorns present, on spine-shields along the 4-8 prominent ribs on the stout stems. Leaves well developed in seedlings, but later scale-like and falling soon (Leptophyll). Bark thickened, fissured, probably an adaptation to drought and fire resistance.
Habitat: 'Locally abundant on steep rocky hillsides, sometimes forming pure stands, often around churches; used for live fencing at higher altitudes' (FEE). 'On stony soils of the hills of the north-facing escarpment, often dominant' (FS). 'Rocky hillslopes' (PSS). Rocky slopes on the border between highland vegetation and *Combretum-Terminalia* woodland (our observation).
Distribution in Ethiopia: TU GD GJ SU SD HA (FEE). Also recorded from KF floristic region (Fig. 6-52).
Altitudinal range in Ethiopia: 1900-2400 m a.s.l. (FEE, including range in Eritrea). 1300-2250 m a.s.l. (our records).
General distribution: Sudan, Ethiopia, Eritrea and Somalia.
Chorological classification: Distribution according to direct

Fig. 6-49. *Clutia abyssinica*. Distribution in Ethiopia. Observed records are marked with red dots on vegetation map reproduced from Fig. 4-1.

Fig. 6-50. *Croton macrostachyus*. Distribution in Ethiopia. Observed records are marked with red dots on vegetation map reproduced from Fig. 4-1.

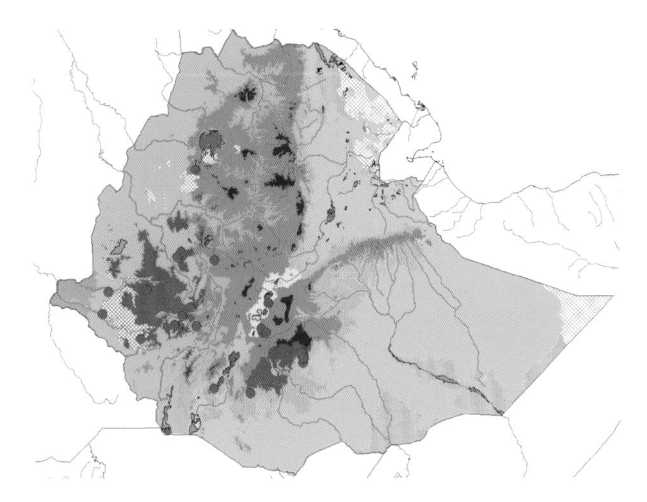

Fig. 6-51. *Erythrococca trichogyne*. Distribution in Ethiopia. Observed records are marked with red dots on vegetation map reproduced from Fig. 4-1.

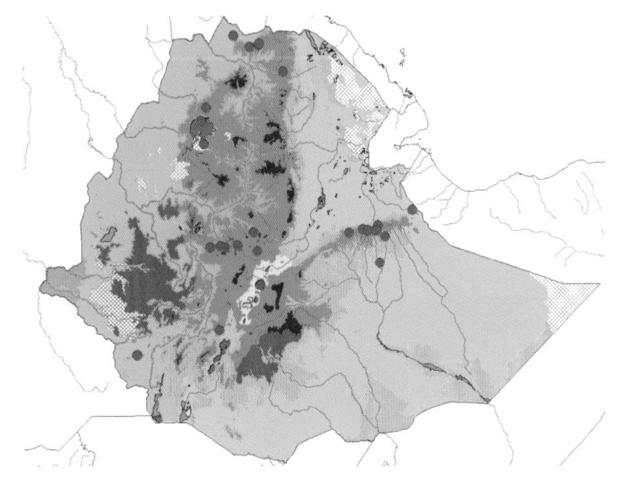

Fig. 6-52. *Euphorbia abyssinica*. Distribution in Ethiopia. Observed records are marked with red dots on vegetation map reproduced from Fig. 4-1.

observation. WWn, [WWc], WWs, EWW. *Local phytogeographical element*. MI. Marginally intruding species in the western woodlands, with a main distribution in the highlands or in *Transitional semi-evergreen bushland* (TSEB); only a few records in our relevés.

6.17.7. Euphorbia ampliphylla *Pax (1895).*

FTEA, Euphorbiaceae: 483 (1988), as *E. obovalifolia* A. Rich. (1851); FTNA: 130 (1992); FEE 2(2): 334 (1995); *BS* 51 163 (1998); PSS: 218 (2015).
Taxonomy: The type of the traditionally used name, *E. obovalifolia* A. Rich., belongs unfortunately to another species, possibly *E. abyssinica*. Recognition of infraspecific taxa not relevant here.
Description: Succulent tree to 30 m tall with candelabrum-shaped branching. Thorns present, but weak, on spine-shields along the 3-4 prominent ribs or wings on the stout stems. Leaves on the young parts of the stems, simple, deciduous. Mesophyll. Bark on old plants dark, thickened and fissured, probably an adaptation for fire resistance.
Habitat: 'Moister montane forest, often left after clearance also extensively used in higher rainfall areas for live fencing' (FEE). 'Afromontane rain forest and undifferentiated Afromontane forest ..., chiefly in disturbed areas, in clearings and along edges; also in secondary montane evergreen bushland. Frequent cultivation of this species may have extended its range' (FTNA). 'Montane forest on rocky hillslopes' (PSS). Rocky slopes on the border between highland vegetation and *Combretum-Terminalia* woodland (our observation).
Distribution in Ethiopia: TU GD GJ WU SU IL KF SD HA (FEE). The FEE records from TU, GJ, and WU floristic region have not been confirmed (Fig. 6-53).
Altitudinal range in Ethiopia: (1299-) 1700-2700 m a.s.l. (FEE). 1600-2700 m a.s.l. (FTNA, including range in the entire Horn of Africa). 900-2500 m a.s.l. (our records).
General distribution: South Sudan, Ethiopia and Eritrea, south to Zambia and Malawi.
Chorological classification: Distribution according to direct observation. WWn, WWc, [WWs], EWW. *Local phytogeographical element*. MI. Marginally intruding species in the western woodlands, with a main distribution in the highlands or in *Transitional semi-evergreen bushland* (TSEB), rarely in *Transitional rain forest* (TRF); only a few records in our relevés.

6.17.8. Flueggea virosa *(Willd.) Royle (1836).*

FTEA, Euphorbiaceae: 1: 68 (1987); FS 1: 271 (1993); FEE 2(2): 272 (1995); *BS* 51: 167 (1998); PSS: 225 (2015), all as *Flueggea virosa* (Willd.) Voigt (1845).
Taxonomy: In the distribution area from West Africa to China, this species is represented by three subspecies, *Flueggea virosa* subsp. *virosa* in most of Africa and Asia, *F. virosa* subsp. *himalaica* D.G. Long in Nepal to Myanmar and *F. virosa* subsp. *melanthesoides* (F. Muell.) G.L. Webster on New Guinea and in Australia. Recognition of infraspecific taxa not relevant here.
Description: Shrub or small tree to 4.5 (-6) m tall. No thorns. Leaves simple, deciduous. Microphyll. Bark smooth or only slightly fissured, no obvious adaptation for fire resistance.
Habitat: 'Mostly in open *Acacia-Commiphora* woodland or riverine forest on alluvial flats, black cotton soil and well-drained rocky slopes, often locally abundant' (FEE). 'Mostly in open woodland or in clearings within riverine forest on alluvial flats, often subject to seasonal flooding' (FS). 'Woodland, grassland, disturbed ground' (PSS). In a wide range of habitats from *Acacia-Commiphora* to *Combretum-Terminalia* woodlands, often very common on the edge of *Riparian vegetation* (RV) and alluvial flats (our observation).
Distribution in Ethiopia: TU GD GJ WU SU WG IL KF GG SD BA HA (FEE). The FEE record from the WU floristic region has not been confirmed (Fig. 6-54).
Altitudinal range in Ethiopia: 400-2050 m a.s.l. (FEE, including range in Eritrea). 400-1800 m a.s.l. (our records).
General distribution: From Senegal to Ethiopia, Eritrea and Somalia, south to Namibia and South Africa; also in Madagascar and Arabia and in Asia as far east as China (in Himalaya and Myanmar, New Guinea and Australia replaced by other subspecies).
Chorological classification: Distribution according to direct

observation. **WWn, WWc, WWs, EWW.** *Local phytogeographical element and sub-element.* **CTW[all+ext].** Widely distributed in the western woodlands, from north to south or almost so (lacking in the extreme south-west, reappears in the lower Omo Valley), with extension to the Rift Valley and to the east of the Rift.

6.18. Leguminosae subfam. Caesalpinioideae

FTEA, Leguminosae-Caesalpinioideae: 1-231 (1967); FEE 3: 49-70 (1989); *BS* 51: 179-184 (1998), as "**Fabaceae subfam. Caesalpinioideae**"; FS 1: 342-361 (1993), as "**Fabaceae subfam. Caesalpinioideae**"; PSS: 189-192 (2015), as "**Leguminosae: Caesalpinioideae (Fabaceae: Caesalpinioideae)**".
Here the classification of genera and species in this subfamily follows FEE 3 (1989) and PSS: 189-192 (2015), but later names in current use are cited.

6.*18.1.* Cassia arereh *Delile (1826).*

FEE 3: 56 (1989); PSS: 189 (2015).
Taxonomy: No infraspecific taxa.
Description: Shrub or tree to 10 m tall. No thorns. Leaves pinnate, probably deciduous. Macrophyll (leaflets microphyll-mesophyll). Bark not known, but seen in the field to be blackened, probably an adaptation for fire resistance.
Habitat: 'Deciduous woodland and wooded grassland, often in rocky places or riparian' (FEE). 'Wooded grassland' (PSS). *Combretum-Terminalia* woodland with tall ground-cover, liable to burning (our observation).
Distribution in Ethiopia: TU GD GJ SU KF (FEE). Also recorded from the WG floristic region (Fig. 6-55).
Altitudinal range in Ethiopia: 1200-2000 m a.s.l. (FEE, including range in Eritrea). 650-1750 m a.s.l. (our records).
General distribution: From Nigeria to Ethiopia and Eritrea.
Chorological classification: Distribution according to direct observation. **WWn, WWc.** *Local phytogeographical element and sub-element.* **CTW[n+c].** Widely distributed in the northern and central part of the western woodlands, reaching the Gambela Region and the Gibe Gorge in the south with one record from each; no eastwards extension.

6.*18.2.* Piliostigma thonningii *(Schumach.) Milne-Redh. (1947).*

FTEA, Leguminosae-Caesalpinioideae: 206 (1967); FEE 3: 68 (1989); *BS* 51: 181 (1998); PSS: 191 (2015).
Later used name: **Bauhinia thonningii** Schumach. (1827).
Taxonomy: No infraspecific taxa.
Description: Shrub or small tree to 10 m tall. No thorns. Leaves simple, but conspicuously bilobed, deciduous. Mesophyll-Macrophyll. Bark thickened, rough and fissured, clearly an adaptation for fire resistance.
Habitat: 'Deciduous woodland and wooded grassland, often in river valleys' (FEE). 'Wooded grassland' (PSS). In a wide range of habitats within *Combretum-Terminalia* woodland (our observation).
Distribution in Ethiopia: TU GD GJ WU SU IL KF SD BA (FEE). Also recorded from the WG and IL floristic regions; the FEE record from the WU floristic region has not been confirmed (Fig. 6-56).
Altitudinal range in Ethiopia: 500-2000 m a.s.l. (FEE, including range in Eritrea). 500-1900 m a.s.l. (our records).
General distribution: From Senegal to Ethiopia and Eritrea, south to Namibia and South Africa.
Chorological classification: Distribution according to direct observation. **WWn, WWc, WWs, EWW.** *Local phytogeographical element and sub-element.* **CTW[all+ext].** Widely distributed in the western woodlands, from north to south or almost so and with extension to the Rift Valley and to the east of the Rift.

6.*18.3.* Senna petersiana *(Bolle) Lock (1988).*

FEE 3: 59 (1989); *BS* 51: 183 (1998); PSS: 192 (2015).
Cassia petersiana Bolle (1861).
FTEA, Leguminosae-Caesalpinioideae: 72 (1967); FTNA: 140 (1992).

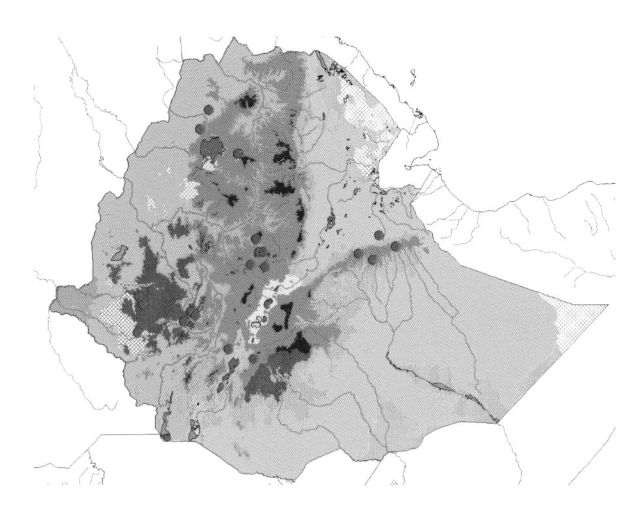

Fig. 6-53. *Euphorbia ampliphylla*. Distribution in Ethiopia. Observed records are marked with red dots on vegetation map reproduced from Fig. 4-1.

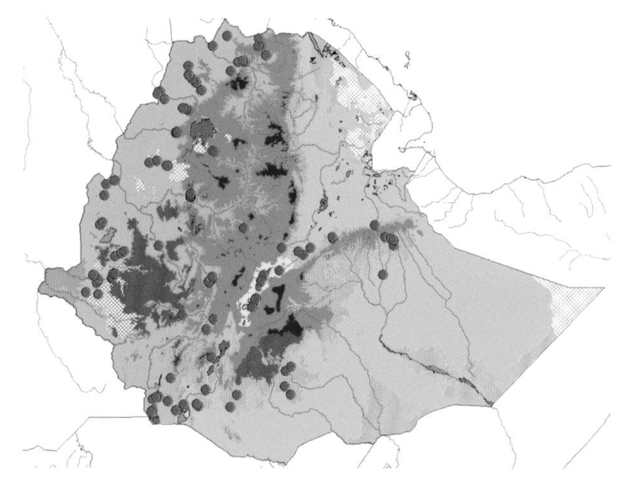

Fig. 6-54. *Flueggea virosa*. Distribution in Ethiopia. Observed records are marked with red dots on vegetation map reproduced from Fig. 4-1.

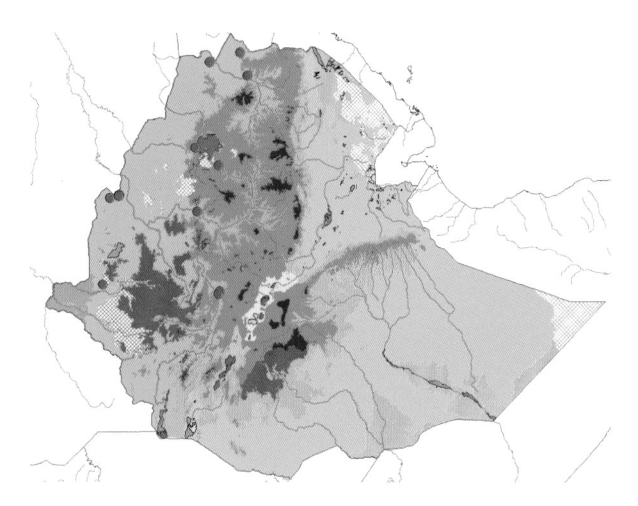

Fig. 6-55. *Cassia arereh*. Distribution in Ethiopia. Observed records are marked with red dots on vegetation map reproduced from Fig. 4-1.

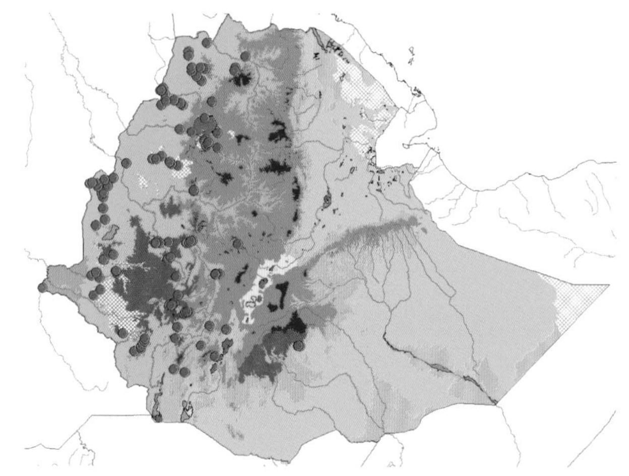

Fig. 6-56. *Piliostigma thonningii*. Distribution in Ethiopia. Observed records are marked with red dots on vegetation map reproduced from Fig. 4-1.

Taxonomy: No infraspecific taxa.
Description: Shrub or small tree to 8 (-12) m tall. No thorns. Leaves pinnately compound; probably evergreen. Mesophyll-Macrophyll (leaflets microphyll-mesophyll). Bark not known, probably no adaptation for fire resistance.
Habitat: 'Forest edges, evergreen bushland and thicket, deciduous woodland, sometimes riparian' (FEE). 'Afromontane rain forest and undifferentiated Afromontane forest ..., mostly in clearings and along edges; riverine forest; frequently also found in primary and secondary montane evergreen or semi-deciduous bushland' (FTNA). 'Bushland' (PSS). Forest edges, mainly of *Dry evergreen Afromontane forest* (DAF), and in clumps of evergreen shrubs in *Combretum-Terminalia* woodland (our observation).
Distribution in Ethiopia: WU GD WG SU IL KF SD (FEE). Also recorded from the GJ and HA floristic regions (Fig. 6-57).
Altitudinal range in Ethiopia: 1200-2000 m a.s.l. (FEE & FTNA, including the range in the entire Horn of Africa). 1500-2000 m a.s.l. (our records).
General distribution: From Cameroon to Ethiopia, south to South Africa.
Chorological classification: Distribution according to direct observation. **WWn, WWc, WWs, EWW**. *Local phytogeographical element and sub-element.* **CTW[all+ext]**. Widely distributed in the western woodlands, from north (where restricted to higher altitudes) to the south or almost so, with extension to the Rift Valley and to the east of the Rift.

6.*18.4.* Senna singueana *(Delile) Lock (1988).*

FEE 3: 59 (1989); *BS* 51: 183 (1998); PSS: 192 (2015)
Cassia singueana Delile (1826).
FTEA, Leguminosae-Caesalpinioideae: 73 (1967).
Taxonomy: No infraspecific taxa.
Description: Shrub or small tree to 15 m tall. No thorns. Leaves compound (pinnate), probably deciduous. Macrophyll (leaflets microphyll). Bark not known, probably no adaptation for fire resistance.
Habitat: 'Deciduous woodland, bushland and wooded grassland of various types, often riparian' (FEE). 'Woodland' (PSS). In evergreen and semi-evergreen bushland and a range of habitats in *Combretum-Terminalia* woodland (our observation).
Distribution in Ethiopia: TU GD GJ WU SU SD (FEE). Also recorded from the WG and GG floristic regions (Fig. 6-58).
Altitudinal range in Ethiopia: 1500-2000 (-2400) m a.s.l. (FEE, including range in Eritrea). 1200-2300 m a.s.l. (our records).
General distribution: From Mali and Ivory Coast to Ethiopia and Eritrea, south to South Africa; also in the Comoro Islands.
Chorological classification: Distribution according to direct observation. **WWn, WWc, WWs, EWW**. *Local phytogeographical element and sub-element.* **CTW[all+ext]**. Widely distributed in the western woodlands, from north to south or almost so; with extension to the Rift Valley and to the east of the Rift.

6.*18.5.* Tamarindus indica *L. (1753).*

FTEA, Leguminosae-Caesalpinioideae: 153 (1967); FEE 3: 66 (1989); FTNA: 141 (1992); FS 1: 358 (1993); *BS* 51: 183 (1998); PSS: 192 (2015).
Taxonomy: No infraspecific taxa.
Description: Tall tree to about 20 (-24) m tall. No thorns. Leaves compound (pinnate); evergreen. Mesophyll (leaflets microphyll). Bark smooth, no obvious adaptation for fire resistance.
Habitat: 'Grassland, woodland and *Combretum* bushland, frequently riparian' (FEE). 'Grassland, woodland and *Acacia-Commiphora* bushland, frequently riparian' (FS). 'Riverine forest and ground water forest; often transgressing into woodland and wooded grassland' (FTNA). 'Wooded grassland' (PSS). In a range of open habitats, mainly in damp places in *Acacia-Commiphora bushland* (ACB) and *Combretum-Terminalia woodland* (CTW) and in *Riparian vegetation* (our observation).
Distribution in Ethiopia: TU GD GJ WU SU IL KF GG SD (FEE). Also recorded from the BA and HA floristic regions (Fig. 6-59).
Altitudinal range in Ethiopia: Sea level to 1500 m a.s.l. (FEE, including range in Eritrea). Sea level to 1300 m

a.s.l. (FTNA, including the range in the entire Horn of Africa). 550-1600 m a.s.l. (our records).
General distribution: From Senegal to Ethiopia, Eritrea and Somalia, south to South Africa; also in tropical Asia, including Arabia. This species, used for its edible fruits, is so widely planted and naturalized that its natural range is difficult to establish.
Chorological classification: Distribution according to direct observation. **WWn, WWc, WWs, EWW.** *Local phytogeographical element and sub-element.* **CTW[all+ext].** Widely distributed in the western woodlands, from north to south or almost so (often associated with *Riparian vegetation* (RV)); with extension to the Rift Valley and to the east of the Rift.

6.19. Leguminosae subfam. Mimosoideae

FTEA, Leguminosae-Mimosoideae: 1-173 (1959); FEE 3: 71-96 (1989); FS 1: 361-387 (1993), as "**Fabaceae subfam. Mimosoideae**"; *BS* 51: 184-198 (1998), as "**Fabaceae subfam. Mimosoideae**"; PSS: 192-198 (2015), as "**Leguminosae: Mimosoideae (Fabaceae: Mimosoideae)**".
The authors are aware of the new distinction of species that previously had names in *Acacia* but after Kyalangalilwa et al. (2013) all the indigenous species in Ethiopia are now placed in *Senegalia* and *Vachellia*. Here, where the emphasis is as much or more on ecological characteristics than on taxonomic terminology, we have found it acceptable to follow the older concept, on which so many terms of African vegetation is based. Therefore, the classification of genera and species in *Acacia* follows FEE 3 (1989) and PSS: 192-198 (2015), but names in current use in *Senegalia* and *Vachellia* are also cited.

6.*19.1.* Acacia abyssinica *Hochst. ex Benth. (1846).*

FTEA, Leguminosae-Mimosoideae: 112 (1959); FEE 3: 89 (1989); FTNA: 142 (1992); *BS* 51: 184 (1998); PSS: 192 (2015).
Later used name: **Vachellia abyssinica** (Hochst. ex Benth.) Kyal. & Boatwr. (2013).
Taxonomy: Two subspecies are recognized, subsp. *abyssinica*, restricted to Eritrea and Ethiopia, and *Acacia abyssinica* subsp. *calophylla* Brenan (1957), distributed in Ethiopia and further south in Africa; all Ethiopian material is subsp. *abyssinica*. Recognition of infraspecific taxa not relevant here.
Description: Tree to 15 (-20) m tall, with flat-topped crown. Thorns present (but small). Leaves compound (bipinnate), apparently evergreen or only shortly deciduous. Mesophyll (leaflets Leptophyll). Bark dark, thickened and fissured, probably an adaptation for fire resistance.
Habitat: 'Woodland, wooded grassland, forest margins and along reams and rivers' (FEE). 'Afromontane rain forest and undifferentiated Afromontane forest, especially along edges and in clearings; also in wooded grassland and woodland (sometimes very dense, forming a pioneer stage in the regrowth of Afromontane forest), and in secondary montane evergreen bushland' (FTNA). 'Wooded grassland' (PSS for *Acacia abyssinica* subsp. *calophylla*). In a wide range of habitats, in the highlands forming nearly monospecific, forest-like dense woodland or scattered trees in wooded grassland, also in the uppermost zone of *Combretum-Terminalia* woodland (our observation).
Distribution in Ethiopia: TU GD GJ WU SU WG AR IL KF SD BA (FEE). Also recorded from the HA floristic region (Fig. 6-60).
Altitudinal range in Ethiopia: 1500-2800 m a.s.l. (FEE, including range in Eritrea). 1500-2900 m a.s.l. (FTNA, including the range in the entire Horn of Africa). 1300-2600 m a.s.l. (our records).
General distribution (species as a whole): D.R. Congo to Ethiopia and Eritrea, south to Zimbabwe and Mozambique. The distribution of the subspecies overlaps.
Chorological classification: Distribution according to direct observation. **WWn, WWc, EWW.** *Local phytogeographical element.* **MI.** Marginally intruding species in the western woodlands, with a main distribution in the highlands and in *Transitional semi-evergreen bushland* (TSEB); only a few records in our relevés.

Fig. 6-57. *Senna petersiana*. Distribution in Ethiopia. Observed records are marked with red dots on vegetation map reproduced from Fig. 4-1.

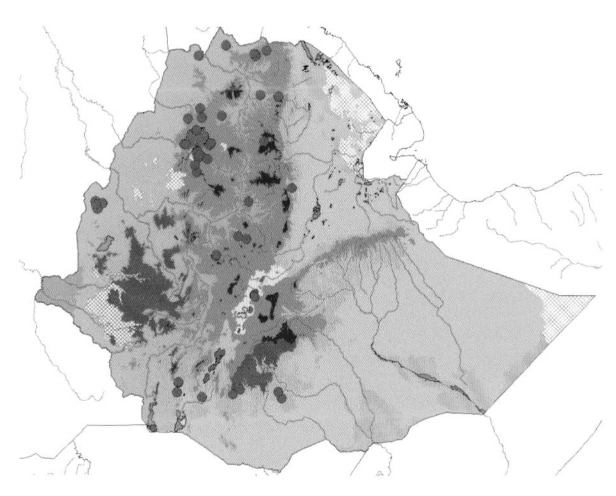

Fig. 6-58. *Senna singueana*. Distribution in Ethiopia. Observed records are marked with red dots on vegetation map reproduced from Fig. 4-1.

Fig. 6-59. *Tamarindus indica*. Distribution in Ethiopia. Observed records are marked with red dots on vegetation map reproduced from Fig. 4-1.

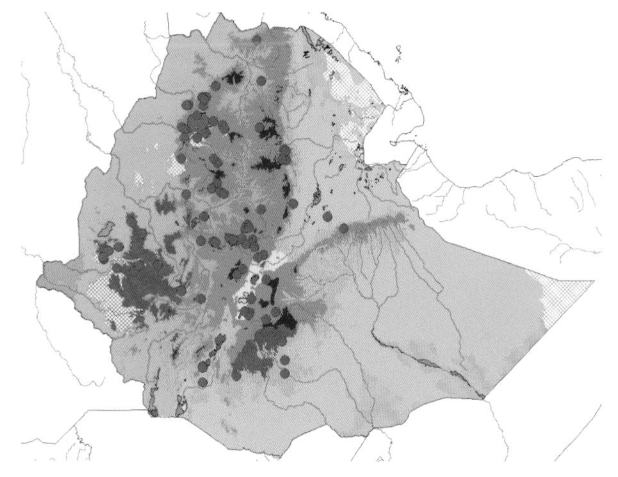

Fig. 6-60. *Acacia abyssinica*. Distribution in Ethiopia. Observed records are marked with red dots on vegetation map reproduced from Fig. 4-1.

6.*19.2*. Acacia amythethophylla *Steud. ex A. Rich. (1847).*

FEE 3: 86 (1989); *BS* 51: 185; PSS: 192 (2015).
Acacia macrothyrsa Harms (1900).
FTEA, Leguminosae-Mimosoideae: 101 (1959).
Later used name: **Vachellia amythethophylla** (Steud. ex A. Rich.) Kyal. & Boatwr. (2013).
Taxonomy: No infraspecific taxa.
Description: Tree to 12 m tall. Stipular spines to 0.8 (-1.6) cm. Leaves compound (bipinnate), probably deciduous. Macrophyll (leaflets Nanophyll). Bark grey, thickened and fissured, probably an adaptation for fire resistance.
Habitat: 'Woodland' (FEE). 'Woodland' (PSS). Open, rather dry woodland in the upper zone of *Combretum-Terminalia* woodland (our observation).
Distribution in Ethiopia: TU GD (FEE). Also recorded from the GJ floristic region (Fig. 6-61).
Altitudinal range in Ethiopia: 1300-1450 m a.s.l. (FEE, including range in Eritrea). 950-1850 m a.s.l. (our records).
General distribution: From Mali and Ghana to Ethiopia and Eritrea, south to Angola and Mozambique.
Chorological classification: Distribution according to direct observation. **WWn**. *Local phytogeographical element and sub-element.* **CTW[n]**. Widely distributed in the northern part of the western woodlands, mainly north of the Abay River; no eastward extension.

6.*19.3*. Acacia asak *(Forssk.) Willd. (1806).*

FEE 3: 80 (1989); PSS: 192 (2015).
Later used name: **Senegalia asak** (Forssk.) Kyal. & Boatwr. (2013).
Taxonomy: No infraspecific taxa.
Description: Shrub or tree to 10 m tall. Small thorns present, single or in threes, the central one curved downwards. Leaves bipinnate, probably deciduous. Mesophyll (leaflets Nanophyll). Bark dark, thickened and fissured, probably an adaptation for fire resistance.
Habitat: 'By water courses, on rocky ground and in deciduous bushland' (FEE). 'Bushland' (PSS). Dry bushland and open, dry woodland (our observation).
Distribution in Ethiopia: TU WU HA (FEE). Also recorded from the AF, GD and SU floristic regions; the records from KF (*M. Kaji s.n.* (ETH)), Omo National Park, Mui River) and GG (*Teshome Soromessa et al.* 199 (ETH)), south-west of Lake Chamo) need confirmation (Fig. 6-62).
Altitudinal range in Ethiopia: Ca. 400-1900 m a.s.l. (FEE, including range in Eritrea). 500-1900 m a.s.l. (our records).
General distribution: Sudan, Ethiopia and Eritrea; also in Egypt and Arabia.
Chorological classification: Distribution according to direct observation. **WWn, WWs, ASO**. *Local phytogeographical element and sub-element.* **ACB[-CTW]**. In the driest parts of the *Acacia-Commiphora bushland* (ACB) along the western and southern edge of the AF floristic region, marginally intruding in the *Combretum-Terminalia woodland and wooded grassland* (CTW) in the Tacazze Valley the north and the Omo Valley in the south (*Kaji* s.n. (ETH)).

6.*19.4*. Acacia polyacantha *Willd. (1806).*

FTEA, Leguminosae-Mimosoideae: 87 (1959); FEE 3: 81 (1989); *BS* 51: 189 (1998); PSS: 194 (2015).
Acacia polyacantha subsp. **campylacantha** (Hochst. ex A. Rich.) Brenan (1956).
FTEA, Leguminosae-Mimosoideae: 87 (1959); FEE 3: 81 (1989); *BS* 51: 189 (1998); PSS: 194 (2015).
Later used name: **Senegalia polyacantha** (Willd.) Seigler & Ebinger (2009) and **Senegalia polyacantha** subsp. **campylacantha** (Hochst. ex A. Rich.) Kyal. & Boatwr. (2013).
Taxonomy: Two subspecies are generally recognized; in Ethiopia, and Africa in general, *Acacia polyacantha* subsp. *campylacantha* (Hochst. ex A. Rich.) Brenan (1956); *Acacia polyacantha* subsp. *polyacantha* is restricted to India. Recognition of infraspecific taxa not relevant here.
Description: Tree to 20 m tall. Thorns up to 12 mm long present in pairs below the leaves or leaf-scars. Leaves bipinnate, probably deciduous. Mesophyll (leaflets Nanophyll). Bark pale yellow to grey, pealing to show

paler inner layers, ultimately fissured, probably an adaptation for fire resistance.

Habitat: 'Wooded grassland, deciduous woodland and bushland, riverine and ground water forest' (FEE). 'Woodland and wooded grassland, often riverine' (PSS for *Acacia polyacantha* subsp. *campylacantha*). In a wide range of habitats in *Combretum-Terminalia* woodland, sometimes dominant in *Riparian vegetation* (RV) (our observation).

Distribution in Ethiopia: TU GD GJ SU IL KF GG SD (FEE). Also recorded from the WU and WG floristic regions (Fig. 6-63).

Altitudinal range in Ethiopia: 500-1600 m a.s.l. (FEE, including range in Eritrea). 550-1850 m a.s.l. (our records).

General distribution (subsp. *campylacantha*): From Senegal to Ethiopia and Eritrea, south to Angola and South Africa. (subsp. *polyacantha*): India.

Chorological classification: Distribution according to direct observation. **WWn, WWc, WWs, EWW**. *Local phytogeographical element and sub-element.* **CTW[all+ext]**. Widely distributed in the western woodlands, from north to south, with extension to the Rift Valley and to the east of the Rift.

6.*19.5*. Acacia dolichocephala *Harms (1897).*

FTEA, Leguminosae-Mimosoideae: 79 (1959); FEE 3: 85 (1989): *BS* 51: 186 (1998); PSS: 193 (2015).

Later used name: **Vachellia dolichocephala** (Harms) Kyal. & Boatwr. (2013).

Taxonomy: No infraspecific taxa.

Description: Tree to 7.5 m tall. Thorns up to 5 (-15) mm long (stipular spines). Leaves bipinnate, probably deciduous. Mesophyll (leaflets Nanophyll). Bark of trunk not known in detail, blackened in the field, probably an adaptation for fire resistance.

Habitat: 'Woodland and grassland, often at lake margins and riversides' (FEE). 'Riverine woodland, rocky outcrops' (PSS). Woodland with tall ground-cover, liable to burning (our observation).

Distribution in Ethiopia: GJ SU AR GG SD (FEE). Also recorded from GD, WU, IL, KF, SD and BA floristic regions (Fig. 6-64).

Altitudinal range in Ethiopia: 1100-2130 m a.s.l. (FEE). 500-1800 m a.s.l. (our records).

General distribution: South Sudan, Ethiopia, Uganda, Kenya and Tanzania.

Chorological classification: Distribution according to direct observation. **WWn, WWc, WWs, EWW**. *Local phytogeographical element and sub-element.* **CTW[all+ext]**. Widely distributed in the western woodlands (lacking in the extreme north, more common in the south), with extension to the Rift Valley and to the east of the Rift.

6.*19.6*. Acacia etbaica *Schweinf. (1867).*

FTEA, Leguminosae-Mimosoideae: 114 (1959); FEE 3: 88 (1989); FS 1: 380 (1993); *BS* 51: 186 (1998); PSS: 193 (2015).

Acacia etbaica subsp. **etbaica**

FEE 3: 88 (1989); FS 1: 381 (1993); PSS: 193 (2015).

Acacia etbaica subsp. **uncinata** Brenan (1957).

FTEA, Leguminosae-Mimosoideae: 114 (1959); FEE 3: 88 (1989); FS 1: 381 (1993).

Acacia etbaica subsp. **platycarpa** Brenan (1957)

FTEA, Leguminosae-Mimosoideae: 114 (1959); FEE 3: 88 (1989).

Later used name for the species: **Vachellia etbaica** (Schweinf.) Kyal. & Boatwr. (2013).

Taxonomy: Four subspecies are currently recognized in this species, *Acacia etbaica* subsp. *etbaica* (in Sudan, possibly also South Sudan, Eritrea and Somalia; also in Egypt and Arabia), *Acacia etbaica* subsp. *australis* Brenan (in Kenya and Tanzania), *Acacia etbaica* subsp. *platycarpa* Brenan (from southern Ethiopia to Tanzania), and *Acacia etbaica* subsp. *uncinata* Brenan (Eritrea, central and southern Ethiopia, Somalia, Uganda, Kenya; also in Arabia). The Ethiopian material dealt with here has been identified as subsp. *uncinata*. We have not been able to make a clear distinction between the northern subsp. *etbaica* (in FEE only recorded from Eritrea) and the southern subsp. *uncinata* in our mapping, and we are not able to make a distinction between the infraspecific taxa here.

Description: Tree to 12 m tall. Thorns up to 6 cm long (stipular spines short and hooked, or straight and up

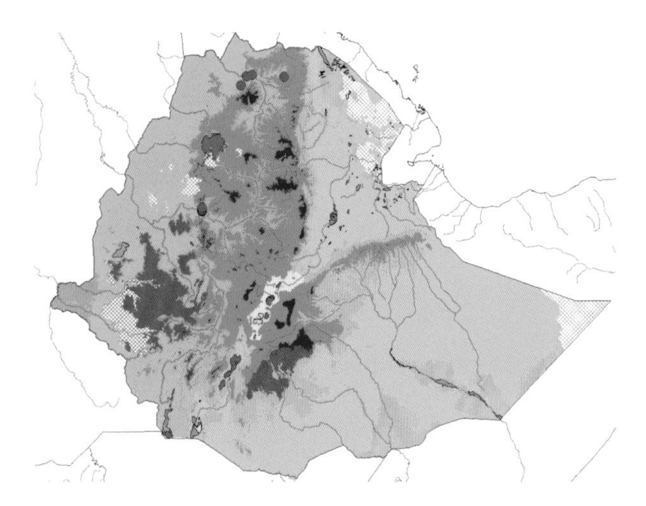

Fig. 6-61. *Acacia amythethophylla*. Distribution in Ethiopia. Observed records are marked with red dots on vegetation map reproduced from Fig. 4-1.

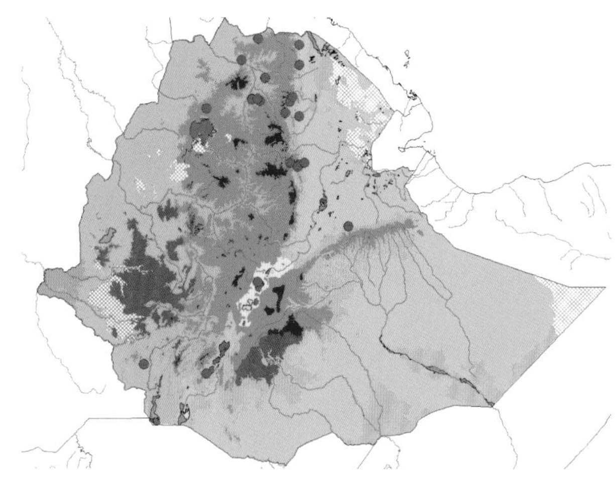

Fig. 6-62. *Acacia asak*. Distribution in Ethiopia. Observed records are marked with red dots on vegetation map reproduced from Fig. 4-1.

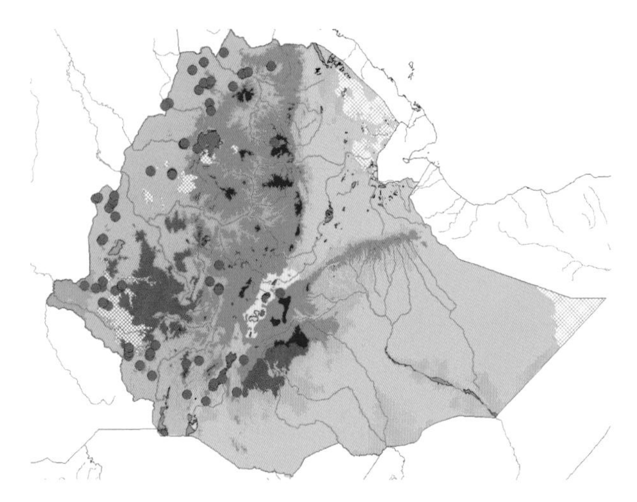

Fig. 6-63. *Acacia polyacantha* subsp. *campylacantha*. Distribution in Ethiopia. Observed records are marked with red dots on vegetation map reproduced from Fig. 4-1.

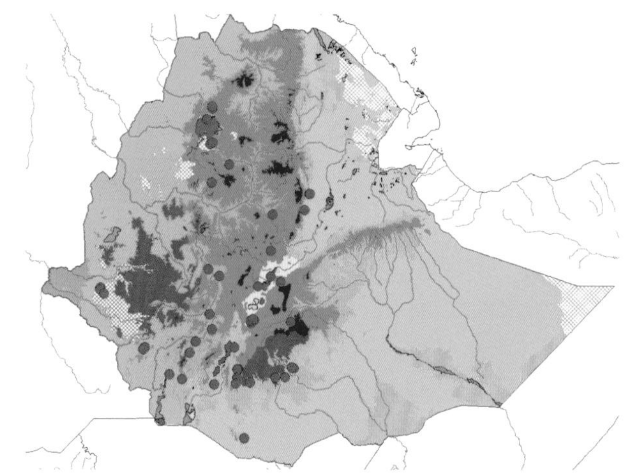

Fig. 6-64. *Acacia dolichocephala*. Distribution in Ethiopia. Observed records are marked with red dots on vegetation map reproduced from Fig. 4-1.

to 7 mm long, sometimes 6 cm long spines intermixed or sometimes all spines long and straight). Leaves bipinnate, probably deciduous. Microphyll to mesophyll (leaflets Nanophyll). Bark of trunk rough and almost black in the field, probably an adaptation for fire resistance.
Habitat: 'Wooded grassland' (FEE for *Acacia etbaica* subsp. *uncinata*), 'bushland' (FEE for *Acacia etbaica* subsp. *platycarpa*). '*Juniperus* forest' (FS for *Acacia etbaica* subsp. *etbaica*), 'Woodland, bushland' (FS for *Acacia etbaica* subsp. *uncinata*). 'Bushland' (PSS for *Acacia etbaica* subsp. *etbaica*). Mainly in dry bushland or at forest margins in the lower zone of the highlands and in *Acacia-Commiphora* bushland (our observation, not including *Benedetto* 382, see below).
Distribution in Ethiopia: WU AR SD HA (FEE *Acacia etbaica* subsp. *uncinata* and subsp. *platycarpa*). Species as a whole also recorded from the TU, AF, GD, GJ, SU, KF, GG, and BA floristic regions. This study adds several records in northern and southern Ethiopia, which may require confirmation; however, particularly the record from WG (*Benedetto* 382 (FT), Gidami) is outside the area one should expect, and does require confirmation (Fig. 6-65).
Altitudinal range in Ethiopia: 600-2000 m a.s.l. (FEE for species as a whole; including range in Eritrea). 500-2250 m a.s.l. (our records).
General distribution (species as a whole): Sudan (possibly also South Sudan), Ethiopia, Eritrea and Somalia, Uganda, Kenya, Tanzania.
Chorological classification: Distribution according to direct observation. **WWn, WWc, WWs, ASO**. *Local phytogeographical element and sub-element.* **ACB[-CTW]**. In the *Acacia-Commiphora bushland* (ACB) and the dry part of the highlands, marginally intruding in the *Combretum-Terminalia woodland and wooded grassland* (CTW) in the north and in the south (see note about the record, *Benedetto* 382, from Gidami in the WG floristic region).

6.*19.7.* Acacia gerrardii *Benth. (1875).*

FTEA, Leguminosae-Mimosoideae: 119 (1959); FEE 3: 88 (1989); *BS* 51: 186 (1998); PSS: 193 (2015).

Acacia gerrardii subsp. **gerrardii** var. **gerrardii**
FTEA, Leguminosae-Mimosoideae: 119 (1959); FEE 3: 88 (1989); *BS* 51: 186 (1998); PSS: 193 (2015).
Later used name: **Vachellia gerrardii** (Benth.) P.J.H. Hurter (2008).
Taxonomy: Two subspecies and four varieties of this species are currently accepted, but in Ethiopia (and in large parts of the African distribution area) only subsp. *gerrardii* var. *gerrardii*. Recognition of infraspecific taxa not relevant here.
Description: Tree to 8 (-15) m tall. Thorns up to ca. 1 cm long, rarely to 6 cm long, straight or slightly recurved, not hooked (all thorns are stipular spines). Leaves bipinnate, probably deciduous. Mesophyll (leaflets Nanophyll). Bark of trunk rough, dark grey to almost black, probably an adaptation for fire resistance.
Habitat: 'Woodland and wooded grassland' (FEE for the species as a whole). 'Woodland and wooded grassland' (PSS for *Acacia gerrardii* subsp. *gerrardii* var. *gerrardii*). A rather rare species in woodland and bushland, mainly known from the deep river valleys in the highlands (our observation).
Distribution in Ethiopia: SU (FEE). Also recorded from the GJ and HA floristic region (Fig. 6-66).
Altitudinal range in Ethiopia: Ca. 1500-2000 m a.s.l. (FEE). 1500-2100 m a.s.l. (our records).
General distribution (species as a whole): Burkina Faso, Benin and Nigeria to Ethiopia, south to South Africa; also in Egypt (Sinai), Arabia, Syria and Iraq (var. *gerrardii* distributed in the tropical African part of this area).
Chorological classification: Distribution according to direct observation. **[WWn], WWc, EWW**. *Local phytogeographical element.* **MI**. Marginally intruding species in the western woodlands, but only in the large river valleys, with a main distribution in the highlands and in *Transitional semi-evergreen bushland* (TSEB); only a few records in our relevés.

6.*19.8.* Acacia hecatophylla *Steud. ex A. Rich. (1847).*

FTEA, Leguminosae-Mimosoideae: 87 (1959); FEE 3: 84 (1989); *BS* 51: 187 (1998); PSS: 193 (2015).

Later used name: **Senegalia hecatophylla** (Steud. ex A. Rich.) Kyal. & Boatwr. (2013).
Taxonomy: No infraspecific taxa.
Description: Tree to 5 (-8) m tall. Thorns (not stipular spines) absent or present, up to 5 under each leaf or leaf-scar, spreading or hooked, 3-6 mm long. Leaves bipinnate, probably deciduous. Mesophyll (leaflets Nanophyll). Bark of trunk not known, blackened in the field, probably an adaptation for fire resistance.
Habitat: 'Woodland and wooded grassland' (FEE). 'Woodland' (PSS). Rather common in the woodlands, particularly on vertisol, in the northern part of the *Combretum-Terminalia* woodlands (our observation).
Distribution in Ethiopia: TU GD GJ (FEE). Also recorded from WG and KF floristic regions. The record from KF (*Friis et al.* sight record, Gojeb Valley) requires confirmation (Fig. 6-67).
Altitudinal range in Ethiopia: Ca. 1450-1700 m a.s.l. (FEE, including range in Eritrea). 600-1700 m a.s.l. (our records).
General distribution: D.R. Congo, Sudan, South Sudan to Ethiopia and Eritrea, south to Uganda.
Chorological classification: Distribution according to direct observation. **WWn, WWc, [WWs]**. *Local phytogeographical element and sub-element.* **CTW[n+c]**. Widely distributed in the northern part of the western woodlands, mainly north of the Abay River (see note above about the record from the Gojeb Valley in KF); no eastwards extension.

6.*19.9*. Acacia hockii *De Wild. (1913).*

FTEA, Leguminosae-Mimosoideae: 104 (1959); FEE 3: 86 (1989); *BS* 51: 187 (1998); PSS: 193 (2015).
Later used name: **Vachellia hockii** (De Wild.) Seigler & Ebinger (2010).
Taxonomy: No infraspecific taxa. The species may previously have been confused with *Acacia seyal*, from which *A. hockii* differs most distinctively by its peeling bark on the trunk, while the bark in *A. seyal* is powdery (Fig. 6-68).
Description: Tree to 6 (-12) m tall. Thorns (stipular spines) straight, short or up to 2 (-4) cm long. Leaves bipinnate, probably deciduous. Mesophyll (leaflets Nanophyll). Bark of young and older branches peeling, bark of trunk not known, probably the peeling is an adaptation for fire resistance.
Habitat: 'Deciduous woodland and wooded grassland' (FEE). 'Woodland' (PSS). In a wide range of habitats in open woodland and bushland (our observation).
Distribution in Ethiopia: GG (FEE). Also recorded from the GD, GJ, SU, WG, IL, KF, SD, BA and HA floristic regions (Fig. 6-68).
Altitudinal range in Ethiopia: Ca. 700 m a.s.l. (FEE). 750-2050 m a.s.l. (our records).
General distribution: From Guinean Republic to Ethiopia and Eritrea, south to Zimbabwe; also in Arabia.
Chorological classification: Distribution according to direct observation. **WWn, WWc, WWs, EWW**. *Local phytogeographical element and sub-element.* **CTW[all+ext]**. Widely distributed in the western woodlands, but only from the GD floristic province and southwards, with extension to the Rift Valley and to the east of the Rift.

6.*19.10*. Acacia mellifera *(Vahl) Benth. (1842).*

FTEA, Leguminosae-Mimosoideae: 84 (1959); FEE 3: 81 (1989); FS 1: 376 (1993); *BS* 51: 188 (1998); PSS: 194 (2015).
Later used name: **Senegalia mellifera** (Vahl) L.A. Silva & J. Freitas (2014). The combination proposed by Seigler & Ebinger (2010), *Phytologia* 92(1): 94. (2010) was invalid, as the basionym was not cited (Melbourne ICBN: Art. 41.5; Shenzhen ICNPAF: Art. 41.5). Correct citation *Senegalia mellifera* (Vahl) L.A.Silva & J.Freitas, *Phytotaxa* 184(5): 296. (2014).
Taxonomy: Two subspecies currently recognized, *Acacia mellifera* subsp. *mellifera* through most of the distribution area, *Acacia mellifera* subsp. *detinens* (Burch.) Brenan (1956) in southern Africa. All Ethiopian material seems homogenous. Recognition of infraspecific taxa not relevant here.
Description: Shrub or small tree to 6 (-9) m tall. Thorns (not stipular spines) paired below each leaf or leaf-scar, hooked, up to 5 (-6) mm long. Leaves bipinnate, probably deciduous. Microphyll (leaflets Nanophyll). Bark of trunk thick, with longitudinal fissure, probably an adaptation for fire resistance.

Fig. 6-65. *Acacia etbaica*. Distribution in Ethiopia. Observed records are marked with red dots on vegetation map reproduced from Fig. 4-1.

Fig. 6-66. *Acacia gerrardii*. Distribution in Ethiopia. Observed records are marked with red dots on vegetation map reproduced from Fig. 4-1.

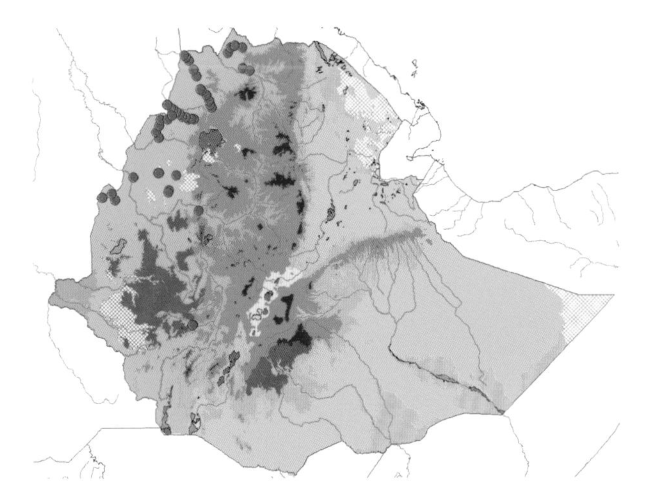

Fig. 6-67. *Acacia hecatophylla*. Distribution in Ethiopia. Observed records are marked with red dots on vegetation map reproduced from Fig. 4-1.

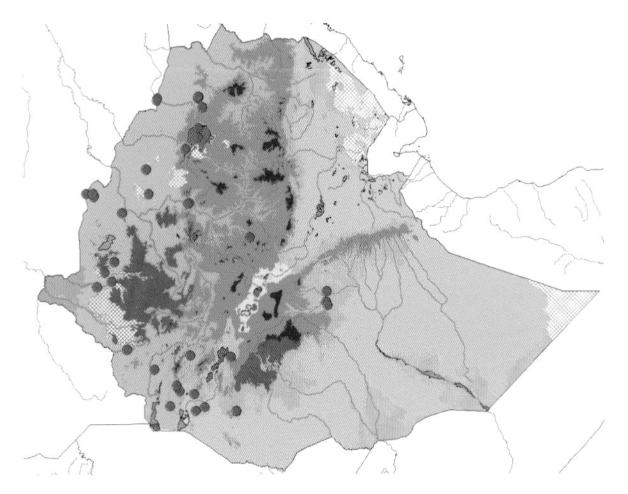

Fig. 6-68. *Acacia hockii*. Distribution in Ethiopia. Observed records are marked with red dots on vegetation map reproduced from Fig. 4-1.

Habitat: 'Deciduous bushland, dry scrub, often forming thickets and an indicator of overgrazing' (FEE). 'Deciduous bushland, dry scrub' (FS). 'Bushland' (PSS for *Acacia mellifera* subsp. *mellifera*). Mainly in deciduous bushland and dry scrub (our observation).
Distribution in Ethiopia: WU SU KF GG SD HA (FEE). Also recorded from the TU, AF, GD and BA floristic regions (Fig. 6-69).
Altitudinal range in Ethiopia: 400-2500 m a.s.l. (FEE, including range in Eritrea). 650-1750 m a.s.l. (our records).
General distribution: (*Acacia mellifera* subsp. *mellifera*): Sudan, Ethiopia, Eritrea and Somalia, south to Namibia, Botswana and South Africa. (*Acacia mellifera* subsp. *detinens*) Tanzania and Angola to South Africa.
Chorological classification: Distribution according to direct observation. **WWn, WWs, ASO**. *Local phytogeographical element and sub-element.* ACB[-CTW]. Widely distributed in the *Acacia-Commiphora bushland* (ACB), marginally intruding in the *Combretum-Terminalia woodland and wooded grassland* (CTW) in the north and the south; common in the southern Rift Valley and to the east of the Rift.

6.*19.11*. Acacia nilotica *(L.) Willd. ex Delile (1813).*

FTEA, Leguminosae-Mimosoideae: 109 (1959); FEE 3: 86 (1989); FS 1: 379 (1993); *BS* 51: 188 (1998); PSS: 194 (2015).
Acacia nilotica subsp. **nilotica**
FEE 3: 87 (1989); PSS: 194 (2015).
Acacia nilotica subsp. **subalata** (Vatke) Brenan (1957).
FTEA, Leguminosae-Mimosoideae: 110 (1959); FEE 3: 87 (1989); *BS* 51: 1888 (1998); PSS: 194 (2015).
Acacia nilotica subsp. **kraussiana** (Benth.) Brenan (1957).
FTEA, Leguminosae-Mimosoideae: 110 (1959); FEE 3: 87 (1989).
Acacia nilotica subsp. **leiocarpa** Brenan (1957)
FTEA, Leguminosae-Mimosoideae: 111 (1959); FEE 3: 87 (1989); FS 1: 379 (1993).
Later used name for species: **Vachellia nilotica** (L.) P.J.H. Hurter & Mabb. (2008).

Taxonomy: At least nine subspecies are recognized within this species, mainly distinguished on characters from the pods. It has not been possible to observe a suitable number of plants with ripe pods for this stud. Recognition of infraspecific taxa not possible here.
Description: Tree to 24 m tall. Thorns (stipular spines) paired at each leaf or leaf-scar, straight, spreading or sometimes somewhat reflexed, up to 8 cm long. Leaves bipinnate, probably deciduous. Mesophyll (leaflets Nanophyll). Bark of trunk with longitudinal fissure, probably an adaptation for fire resistance.
Habitat: 'Woodland and scrub' (FEE for all subspecies of *Acacia nilotica* in Ethiopia). 'Woodland and scrub' (FS for *Acacia nilotica* subsp. *leiocarpa*), 'woodland and semi-evergreen bushland' (FS for *Acacia nilotica* subsp. *subalata*). 'Woodland' (PSS for most taxa); 'bushland and wooded grassland' (PSS for *Acacia nilotica* subsp. *subalata*). Frequent in various types of bushland and woodland, also riparian woodland, but not common in the *Combretum-Terminalia* woodland (our observation).
Distribution in Ethiopia: GG (FEE for *Acacia nilotica* subsp. *subalata*), KF GG SD (FEE for *Acacia nilotica* subsp. *kraussiana*), SU AR HA (FEE for *Acacia nilotica* subsp. *leiocarpa*). Species as a whole also recorded from the TU, AF, GD, GJ, and WG floristic regions (Fig. 6-70).
Altitudinal range in Ethiopia: 900-1000 m a.s.l. (FEE for *Acacia nilotica* subsp. *subalata*), ca. 700-1700 m a.s.l. (FEE for *Acacia nilotica* subsp. *kraussiana*), 1100-1700 m a.s.l. (FEE for *Acacia nilotica* subsp. *leiocarpa*). 750-1950 m a.s.l. (our records).
General distribution (species as a whole): From Guinean Republic to Ethiopia and Eritrea, south to South Africa; also in Arabia and India.
Chorological classification: Distribution according to direct observation. **WWn, WWc, WWs, ASO**. *Local phytogeographical element and sub-element.* ACB[-CTW]. Widely distributed in the *Acacia-Commiphora bushland* (ACB), in the Rift Valley and in *Transitional semi-evergreen bushland* (TSEB), marginally intruding in the *Combretum-Terminalia woodland and wooded grassland* (CTW) in the north or in the south. A few records in the central part the western woodlands, at least some of which are

from *Riparian vegetation* (RV), might confuse the classification in phytogeographical elements, but considering the wide distribution in the *Acacia-Commiphora bushland* (ACB), the species has here been classified as ACB[-CTW].

6.*19.12.* Acacia oerfota *(Forssk.) Schweinf. (1896).*

FEE 3: 90 (1989); FS 1: 382 (1993); *BS* 51: 181 (1998); PSS: 194 (2015).
Acacia nubica Benth. (1842).
FTEA, Leguminosae-Mimosoideae: 129 (1959).
Later used name: **Vachellia oerfota** (Forssk.) Kyal. & Boatwr. (2013)
Taxonomy: No infraspecific taxa.
Description: Shrub, mostly branched from the base, rarely small tree to 5 m tall. Thorns (stipular spines) paired at each leaf or leaf-scar, straight, up to 2.5 cm long. Leaves bipinnate, probably deciduous. Microphyll to Mesophyll (leaflets Nanophyll). Bark of trunk smooth, probably no adaptation for fire resistance.
Habitat: 'Deciduous bushland and semi-desert scrub' (FEE). 'Deciduous bushland and semi-desert scrub' (FS). 'Bushland' (PSS). Dry bushland and semi-desert scrub (our observation).
Distribution in Ethiopia: TU WU AF SU SD BA HA (FEE). Also recorded from the GG floristic region (Fig. 6-71).
Altitudinal range in Ethiopia: 100-1600 m a.s.l. (FEE, including range in Eritrea). 250-1600 m a.s.l. (our records).
General distribution: From Chad to Sudan, Ethiopia, Eritrea and Somalia, south to Tanzania; also in Egypt and Arabia.
Chorological classification: Distribution according to direct observation. **WWn, WWs, ASO**. *Local phytogeographical element and sub-element.* **ACB[-CTW]**. Widely distributed in the *Acacia-Commiphora bushland* (ACB), marginally intruding in the *Combretum-Terminalia woodland and wooded grassland* (CTW) in the north and in the south, also in a few localities in the Rift Valley and to the east of the Rift.

6.*19.13.* Acacia persiciflora *Pax (1907).*

FTEA, Leguminosae-Mimosoideae: 86 (1959); FEE 3: 81 (1989); *BS* 51: 189 (1998); PSS: 194 (2015).
Later used name: **Senegalia persiciflora** (Pax) Kyal. & Boatwr. (2013)
Taxonomy: No infraspecific taxa.
Description: Tree to 9 (-15) m tall. Thorns few (not stipular spines) paired at each leaf or leaf-scar, curved, up to 3 mm long. Leaves bipinnate, probably deciduous. Mesophyll (leaflets Nanophyll). Bark of trunk not known, sometimes blackened in the field, probably an adaptation for fire resistance.
Habitat: 'Woodland and wooded grassland' (FEE). 'Bushland and wooded grassland' (PSS). Woodlands and bushland in the lower zone of the highland vegetation (our observation).
Distribution in Ethiopia: GD GJ SU KF SD HA (FEE). Also recorded from the WU floristic region (Fig. 6-72).
Altitudinal range in Ethiopia: Ca. 1700-2100 m a.s.l. (FEE). 1400-2000 m a.s.l. (our records).
General distribution: D.R. Congo, Sudan, Ethiopia, Uganda, Kenya.
Chorological classification: Distribution according to direct observation. **WWn, WWc, EWW**. *Local phytogeographical element.* **MI**. Marginally intruding species in the western woodlands, with a main distribution in the highlands and in *Transitional semi-evergreen bushland* (TSEB); only one or few records in our relevés.

6.*19.14.* Acacia senegal *(L.) Willd. (1806).*

FTEA, Leguminosae-Mimosoideae: 93 (1959); FEE 3: 78 (1989); FS 1: 370 (1993); *BS* 51: 190 (1998); PSS: 195 (2015).
Acacia senegal var. **senegal**
FTEA, Leguminosae-Mimosoideae: 93 (1959); PSS: 195 (2015).
Later used name: **Senegalia senegal** (L.) Britton (1930).
Taxonomy: Three varieties in Ethiopia, the widespread *Acacia senegal* var. *senegal*, *Acacia senegal* var. *kerensis* Schweinf. (1896), recorded mostly from Eritrea and a few parts of eastern Ethiopia and *Acacia senegal* var.

Fig. 6-69. *Acacia mellifera*. Distribution in Ethiopia. Observed records are marked with red dots on vegetation map reproduced from Fig. 4-1.

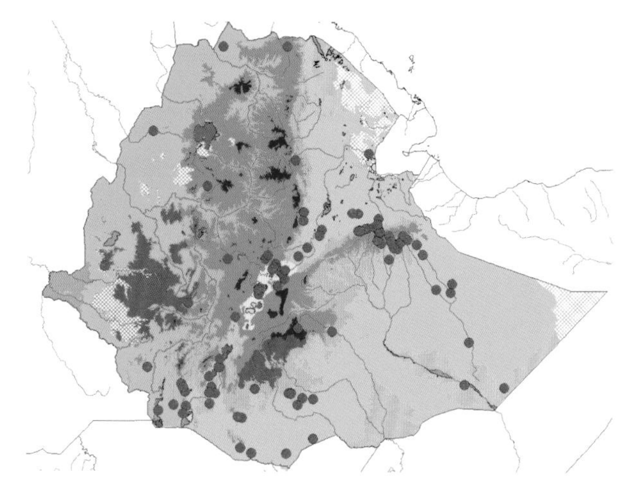

Fig. 6-70. *Acacia nilotica*. Distribution in Ethiopia. Observed records are marked with red dots on vegetation map reproduced from Fig. 4-1.

Fig. 6-71. *Acacia oerfota*. Distribution in Ethiopia. Observed records are marked with red dots on vegetation map reproduced from Fig. 4-1.

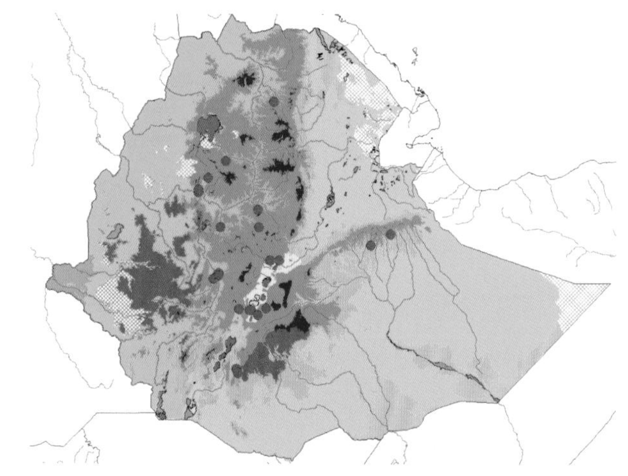

Fig. 6-72. *Acacia persiciflora*. Distribution in Ethiopia. Observed records are marked with red dots on vegetation map reproduced from Fig. 4-1.

leiorachis Brenan (1953; syn.: *Acacia circummarginata* Chiov. (1915)), which is restricted to southern Ethiopia and Somalia (considered a distinct species, *A. circummarginata* Chiov. (1915), in FS). Recognition of infraspecific taxa not possible here.
Description: Shrub or small tree to 12 m tall. Thorns (not stipular spines) usually in threes at each leaf or leaf-scar, up to 7 mm long, the central one curved downwards, the two lateral ones curved upwards, or the lateral ones may be missing. Leaves bipinnate, probably deciduous. Mesophyll (leaflets Nanophyll). Bark of trunk scaly and rough, probably an adaptation for fire resistance.
Habitat: 'Wooded grassland, deciduous bushland, dry scrub' (FEE for *Acacia senegal* var. *senegal*), 'dry bushland' (FEE for *Acacia senegal* var. *leiorachis*). '*Acacia-Commiphora* bushland, dry scrub' (FS for species as a whole, excluding material referred to *A. circummarginata*). 'Wooded grassland, bushland' (PSS for *Acacia senegal* var. *senegal*). In a wide range of dry woodlands and bushlands (our observation).
Distribution in Ethiopia: AF WU SU AR BA GG SD (FEE for *Acacia senegal* var. *senegal*), SU HA (FEE for *Acacia senegal* var. *leiorachis*). Species as a whole also recorded from the GJ, WG, IL, and KF floristic regions (Fig. 6-73).
Altitudinal range in Ethiopia: Ca. 600-1700 m a.s.l. (FEE for *Acacia senegal* var. *senegal*, including range in Eritrea), ca. 1000-1700 m a.s.l. (FEE for *Acacia senegal* var. *leiorachis*). 380-1750 m a.s.l. (our records).
General distribution (species as a whole): From Senegal to Ethiopia, Eritrea and Somalia, south to Tanzania; also in Egypt and Arabia.
Chorological classification: Distribution according to direct observation. **WWn, WWc, WWs, EWW.** *Local phytogeographical element and sub-element.* **CTW[all+ext]**. Distributed in parts of the *Combretum-Terminalia* woodlands, in the north, the central and the southern part, with extension to the Rift Valley and to the east of the Rift. It is difficult to decide if this species should be classified as **CTW[all+ext]** or as **ACB[-CTW]**, as it is distributed in the *Acacia-Commiphora bushland* (ACB). However, because the species is widespread in the western woodland and is known also to be widespread on the Sudan side of the border (Harrison & Jackson 1958), is has been classified as **CTW[all+ext]**.

6.*19.15*. Acacia seyal *Delile (1813)*.

FTEA, Leguminosae-Mimosoideae: 103 (1959); FEE 3: 85 (1989); FS 1: 379 (1993); *BS* 51: 190 (1998); PSS: 195 (2015).
Later used name: **Vachellia seyal** (Delile) P.J.H. Hurter (2008).
Acacia seyal var. **seyal**
FTEA, Leguminosae-Mimosoideae: 103 (1959); FEE 3: 86 (1989); *BS* 51: 190 (1998); PSS: 195 (2015).
Acacia seyal var. **fistula** (Schweinf.) Oliv. (1871).
FTEA, Leguminosae-Mimosoideae: 103 (1959); FEE 3: 86 (1989); FS 1: 379 (1993); *BS* 51: 190 (1998); PSS: 195 (2015).
Taxonomy: Two varieties are currently recognized, *Acacia seyal* var. *seyal* and *Acacia seyal* var. *fistula* (Schweinf.) Oliv., basically recognized on their spines forming ant galls. Both varieties occur in Ethiopia, and we have not always been able distinguish between them in the field or in the herbaria. Recognition of infraspecific taxa not possible here. Material from Somalia has all been named *Acacia seyal* var. *fistula*.
Description: Tree to 9 (-12) m tall. Thorns (stipular spines) two at each leaf or leaf-scar, up to 8 mm long, sometimes forming hollow "ant galls" (*Acacia seyal* var. *fistula*). Leaves bipinnate, probably deciduous. Mesophyll (leaflets Nanophyll). Bark of branchlets and branches becomes reddish and flakes, exposing a yellowish or reddish powdery underbark, bark on trunk similar, yellowish or reddish, probably an adaptation for fire resistance.
Habitat: 'Woodland, wooded grassland' (FEE for both varieties of *Acacia seyal*). 'Woodland, wooded grassland' (FSS for *Acacia seyal* var. *fistula*). 'Wooded grassland' (PSS for *Acacia seyal* var. *seyal*), 'wooded grassland, bushland' (PSS for *Acacia seyal* var. *fistula*). A very common species in a wide range of woodlands and bushlands, in the *Combretum-Terminalia* sometimes forming pure or almost pure stands (our observation).
Distribution in Ethiopia: TU WU AF GD GJ SU AR KF IL SD HA (FEE for *Acacia seyal* var. *seyal*), SU IL

SD HA (FEE for *Acacia seyal* var. *fistula*). Species as a whole also recorded from the WG and BA floristic regions (Fig. 6-74).
Altitudinal range in Ethiopia: 1200-2400 m a.s.l. (FEE for *Acacia seyal* var. *seyal*, including range in Eritrea), 1200-2100 m a.s.l. (FEE for *Acacia seyal* var. *fistula*, including range in Eritrea). 550-2100 m a.s.l. (our records).
General distribution: (*Acacia seyal* var. *seyal*): from Senegal to Ethiopia and Eritrea, south to Tanzania; also in Egypt and Arabia. (*Acacia seyal* var. *fistula*): Ethiopia and Somalia, south to Zambia and Mozambique.
Chorological classification: Distribution according to direct observation. **WWn, WWc, WWs, EWW.** *Local phytogeographical element and sub-element.* **CTW[all+ext].** Widely distributed in the western woodlands, from north to south, with extension to the Rift Valley and to the east of the Rift. Also scattered in *Acacia-Commiphora bushland* (ACB) and in woodlands in the highlands. Like with *Acacia senegal*, it is difficult place this species in only one category, as it may fit with **CTW[all+ext]** or be regarded as a transgressing species, **TRG**. However, from our data and map, it looks as the main distribution is in the western woodlands, and the species has therefore been classified as CTW[all+ext].

6.19.16. Acacia sieberiana *DC. (1825).*

FTEA, Leguminosae-Mimosoideae: 127 (1959); FEE 3: 89 (1989); *BS* 51: 191 (1998); PSS: 195 (2015).
Later used name: **Vachellia sieberiana** (DC.) Kyal. & Boatwr. (2013).
Taxonomy: Three infraspecific taxa are currently accepted: *Acacia sieberiana* var. *sieberiana*, from Senegal to Ethiopia, south to Mozambique, *Acacia sieberiana* var. *villosa* A. Chev. (1927), from Senegal to Sudan, and *Acacia sieberiana* var. *woodii* (Burt-Davy) Keay & Brenan (1951), from Senegal to Ethiopia, south to South Africa. In Ethiopia *Acacia sieberiana* var. *sieberiana* and *Acacia sieberiana* var. *woodii* have strongly overlapping distribution; the infraspecific variation seems according to FEE mainly to be in indumentum and growth form of the crown, both characters which were not easy to record from herbarium material or the field studies. Recognition of infraspecific taxa not possible here.
Description: Tree to 18 m tall. Thorns (stipular spines) two at each leaf or leaf-scar, up to 9 (-12.5) cm long. Leaves bipinnate, probably deciduous. Mesophyll (leaflets Nanophyll). Bark of branchlets and branches flake, exposing an olive or yellowish underbark; bark on trunk pale grey, rough, probably an adaptation for fire resistance.
Habitat: 'Deciduous woodland and riverine forest' (FEE for *Acacia sieberiana* var. *sieberiana*), 'woodland and wooded grassland' (FEE for *Acacia sieberiana* var. *woodii*). 'Woodland' (PSS for *Acacia sieberiana* var. *sieberiana*), 'woodland and bushland' (PSS for *Acacia sieberiana* var. *woodii*). In various types of woodland and bushland (our observation).
Distribution in Ethiopia: WU SU WG IL KF SD (FEE for *Acacia sieberiana* var. *sieberiana*), TU GD WU SU?AR KF SD (FEE for *Acacia sieberiana* var. *woodii*). Also recorded from the GJ floristic region (Fig. 6-75).
Altitudinal range in Ethiopia: 500-2200 m a.s.l. (FEE for *Acacia sieberiana* var. *sieberiana*, including range in Eritrea), ca. 1700-2100 m a.s.l. (FEE for *Acacia sieberiana* var. *woodii*). 500-1950 m a.s.l. (our records).
General distribution (species as a whole): From Senegal to Ethiopia and Eritrea, south to South Africa.
Chorological classification: Distribution according to direct observation. **WWn, WWc, WWs, EWW.** *Local phytogeographical element and sub-element.* **CTW[all+ext].** Widely distributed in the western woodlands, from the north nearly to the south, with some extension to highlands, eastwards to the Rift Valley and to the east of the Rift. A single record (*Cufodontis* 744 (FT), Moyale) in the *Acacia-Commiphora bushland* (ACB) probably represents a hill with *Transitional semi-evergreen bushland* (TSEB).

6.19.17. Acacia tortilis *(Forssk.) Hayne (1827).*

FTEA, Leguminosae-Mimosoideae: 117 (1952); FEE 3: 87 (1989); FS 1: 380 (1993); *BS* 51: 192 (1998), PSS: 195 (2015).
Acacia tortilis subsp. **spirocarpa** (Hochst. ex A. Rich.) Brenan (1957).

FTEA, Leguminosae-Mimosoideae: 117 (1952); FEE 3: 87 (1989); FS 1: 380 (1993); *BS* 51: 192 (1998), PSS: 195 (2015).
Later used name: **Vachellia tortilis** (Forssk.) Galasso & Banfi (2008).
Taxonomy: Five infraspecific taxa are currently recognized: *Acacia tortilis* subsp. *tortilis*, in Egypt, Sudan, possibly South Sudan, Uganda, Somalia, Yemen and possibly Eritrea, *Acacia tortilis* subsp. *raddiana* (Savi) Brenan (1957), from Egypt and Sudan, Somalia, Kenya, south to Namibia, Botswana and Mozambique, *Acacia tortilis* subsp. *spirocarpa* (Hochst. ex A. Rich.) Brenan (1957), from Sudan, Ethiopia, Somalia, south to Angola, Zimbabwe and Mozambique, *Acacia tortilis* subsp. *heteracantha* (Burch.) Brenan (1957), in southern Africa from Angola and Mozambique to South Africa, and *Acacia tortilis* subsp. *campoptila* (Schweinf.) Boulos (1995), restricted to Yemen (the subspecies also have names in *Vachellia*). In Ethiopia only *Acacia tortilis* subsp. *spirocarpa*. Recognition of infraspecific taxa not relevant here.
Description: Tree to 21 m tall. Thorns (stipular spines), usually two at each leaf or leaf-scar, usually hooked and up to 5 mm long, mixed with straight ones up to 10 cm long. Leaves bipinnate, probably deciduous. Mesophyll (leaflets Nanophyll). Bark on trunk grey, rough, thickened and fissured, probably an adaptation for fire resistance.
Habitat: 'Woodland and wooded grassland, dry scrub' (FEE for *Acacia tortilis* subsp. *spirocarpa*). 'Woodland, wooded grassland, dry scrub' (FS for *Acacia tortilis* subsp. *spirocarpa*). 'Wooded grassland, bushland' (PSS for *Acacia tortilis* subsp. *spirocarpa*). In dry habitats in various types of woodland and bushland, also in the driest parts of the *Acacia-Commiphora* bushland (our observation).
Distribution in Ethiopia: TU AF WU SU AR BA HA (FEE for *Acacia tortilis* subsp. *spirocarpa*). Also recorded from the GG and SD floristic regions (Fig. 6-76).
Altitudinal range in Ethiopia: Ca. 600-1900 m a.s.l. (FEE for *Acacia tortilis* subsp. *spirocarpa*, including range in Eritrea). 650-1850 m a.s.l. (our records).
General distribution (species as a whole): Sudan, Ethiopia, Eritrea and Somalia, south to South Africa; also in North Africa, the Middle East and Arabia.
Chorological classification: Distribution according to direct observation. **WWn, WWs, ASO**. *Local phytogeographical element and sub-element.* **ACB[-CTW]**. Widely distributed in the *Acacia-Commiphora bushland* (ACB), marginally intruding in the *Combretum-Terminalia woodland and wooded grassland* (CTW) in the north and in the south, and in dry parts of the upper Tacazze Valley.

6.*19.18.* Acacia venosa *Hochst. ex Benth. (1846).*

FEE 3: 81 (1989); PSS: 195 (2015).
Later used name: **Senegalia venosa** (Hochst. ex Benth.) Kyal. & Boatwr. (2013).
Taxonomy: No infraspecific taxa.
Description: Tree to 6 m tall. Thorns (not stipular spines), usually two at each leaf or leaf-scar, but sometimes absent, hooked and up to 5 mm long. Leaves compound (bipinnate), probably deciduous. Mesophyll (leaflets Nanophyll). Bark on trunk not known, blackened in the field, probably with adaptation for fire resistance.
Habitat: 'Woodland, evergreen bushland' (FEE). 'Woodland, bushland' (PSS). In woodland and bushland in the transition between the *Combretum-Terminalia* woodland and the lowermost zone of the highland vegetation, particularly common in the woodlands around Lake Tana (our observation).
Distribution in Ethiopia: TU GD (FEE). Also recorded from the WU, GJ and WG floristic regions (Fig. 6-77).
Altitudinal range in Ethiopia: 1890-2400 m a.s.l. (FEE, including range in Eritrea). 1050-1900 m a.s.l. (our records).
General distribution: Sudan (?), Ethiopia, Eritrea.
Chorological classification: Distribution according to direct observation. **WWn, WWc**. *Local phytogeographical element and sub-element.* **CTW[n+c]**. Widely distributed in the northern and central parts of the western woodlands, mainly north of the Abay River and east of Kurmuk, with one record (*Mercier* V 152 (FT), Alamata-Kobbo) in Transitional Semi-Evergreen bushland in the WU floristic region.

Fig. 6-73. *Acacia senegal*. Distribution in Ethiopia. Observed records are marked with red dots on vegetation map reproduced from Fig. 4-1.

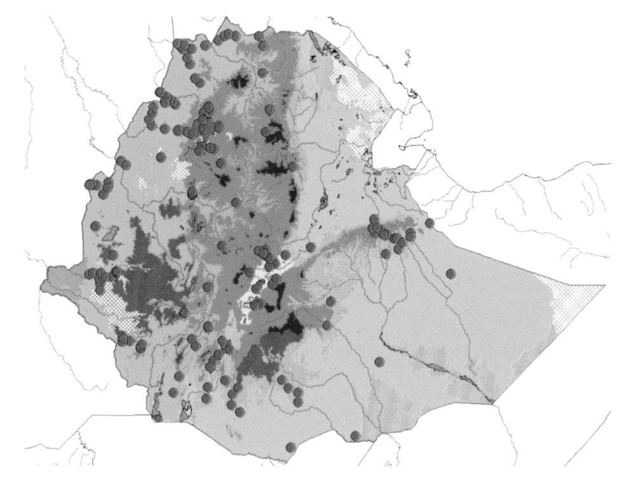

Fig. 6-74. *Acacia seyal*. Distribution in Ethiopia. Observed records are marked with red dots on vegetation map reproduced from Fig. 4-1.

Fig. 6-75. *Acacia sieberiana*. Distribution in Ethiopia. Observed records are marked with red dots on vegetation map reproduced from Fig. 4-1.

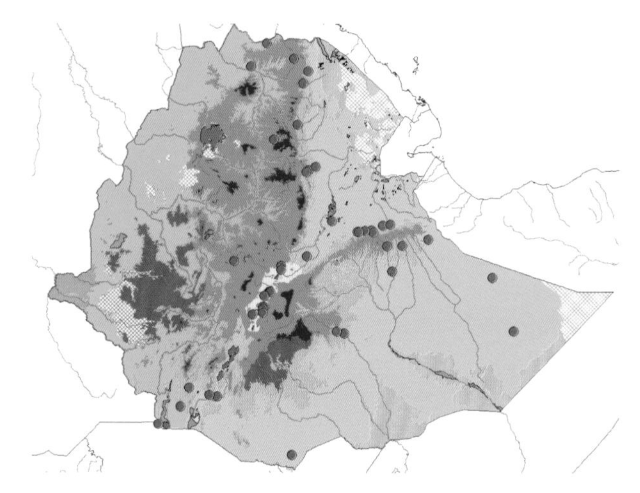

Fig. 6-76. *Acacia tortilis*. Distribution in Ethiopia. Observed records are marked with red dots on vegetation map reproduced from Fig. 4-1.

6.*19.19*. Albizia amara *(Roxb.) Boiv. (1834).*

FTEA, Leguminosae-Mimosoideae: 152 (1959); FEE 3: 94 (1989); FS 1: 385 (1993); *BS* 51: 192 (1998); PSS: 195 (2015).
Albizia amara subsp. **sericocephala** (Benth.) Brenan (1955).
FTEA, Leguminosae-Mimosoideae: 151 (1959); FEE 3: 94 (1989); *BS* 51: 192 (1998); PSS: 195 (2015).
Taxonomy: Two subspecies are currently recognized. In Ethiopia only *Albizia amara* subsp. *sericocephala* (Benth.) Brenan (1955); *Albizia amara* subsp. *amara* occurs in India and Sri Lanka. Recognition of infraspecific taxa not relevant here.
Description: Tree to 12 (-15) m tall. No thorns. Leaves compound (bipinnate), apparently deciduous. Mesophyll (leaflets Nanophyll). Bark on trunk dark grey, rough, thickened and fissured, probably with adaptation for fire resistance.
Habitat: 'Wooded grassland, scrub' (FEE for *Albizia amara* subsp. *sericocephala*). '*Acacia-Commiphora* bushland, semi-evergreen bushland and woodland' (FS for species as a whole). 'Wooded grassland' (PSS for *Albizia amara* subsp. *sericocephala*). Dry or very dry woodland, wooded grassland or bushland, sometimes forming pure or almost pure stands (our observation).
Distribution in Ethiopia: TU GD SU SD (FEE for *Albizia amara* subsp. *sericocephala*). Also recorded from the GJ, WU, WG, and GG floristic regions (Fig. 6-78).
Altitudinal range in Ethiopia: 1400-1700 m a.s.l. (FEE for *Albizia amara* subsp. *sericocephala*, including range in Eritrea). 775-2000 m a.s.l. (our records).
General distribution (species as a whole): Sudan, South Sudan, Ethiopia, Eritrea and Somalia, south to Botswana and South Africa; also in India and Sri Lanka.
Chorological classification: Distribution according to direct observation. WWn, WWc, WWs, EWW. *Local phytogeographical element and sub-element.* CTW[all+ext]. The records from southern Ethiopia (*Friis et al.* 14203 (C, ETH, K), south of Key Afer; *Teshome Soromessa et al.* 245 (ETH), at Key Afer) may from the map look as they were found in *Transitional Semi-Evergreen Bushland* (TSEB), but do in fact represent presence in *Combretum-Terminalia woodland and wooded grassland* (CTW) in the south, and it has therefor been classified as CTW[all+ext].

6.*19.20*. Albizia grandibracteata *Taub. (1895).*

FTEA, Leguminosae-Mimosoideae: 161 (1959); FEE 3: 96 (1989); FTNA: 144 (1992); *BS* 51: 193 (1998); PSS: 196 (2015).
Taxonomy: No infraspecific taxa.
Description: Tree to 25 (-30) m tall, with flat-topped crown. No thorns. Leaves compound (bipinnate), deciduous. Macrophyll (leaflets microphyll). Bark on trunk usually smooth, but ultimately becoming rough, probably an adaptation for fire resistance.
Habitat: 'Rain forest, riverine forest' (FEE). 'Transitional rain forest and Afromontane rain forest; also in secondary montane evergreen bushland, sometimes left as a single tree in derived grassland or farmland' (FTNA). 'Forest, often riverine' (PSS). Mainly in forests, but sometimes left as a single tree in woodland (our observation).
Distribution in Ethiopia: WG SU IL KF SD (FEE). Also recorded from the GG floristic region. The isolated record in the WG floristic region, *Sebsebe Demissew et al.* 5961 (ETH), 19 km from Assosa towards Komosha, needs confirmation (Fig. 6-79).
Altitudinal range in Ethiopia: 1200-1700 m a.s.l. (FEE). 1050-2200 m a.s.l. (FTNA, including the range in the entire Horn of Africa). 500-1850 m a.s.l. (our records).
General distribution: D.R. Congo to Ethiopia, south to Tanzania.
Chorological classification: Distribution according to direct observation. WWc, WWs, EWW. *Local phytogeographical element.* MI. Marginally intruding species in the western woodlands, with main distribution in the *Moist evergreen Afromontane forest* (MAF) and the *Transitional rain forest* (TRF); only a few records in our relevés.

6.*19.21*. Albizia malacophylla *(A. Rich.) Walp. (1852).*

FTEA, Leguminosae-Mimosoideae: 145 (1959); FEE 3: 93 (1989); PSS: 196 (2015).

Albizia malacophylla var. **malacophylla**
FEE 3: 93 (1989).
Albizia malacophylla var. **ugandensis** Baker f. (1930).
FTEA, Leguminosae-Mimosoideae: 145 (1959); FEE 3: 93 (1989); PSS: 196 (2015).
Taxonomy: Two varieties, *Albizia malacophylla* var. *malacophylla* is restricted to northern Ethiopia; *Albizia malacophylla* var. *ugandensis* occurs in the remaining large distribution area outside northern Ethiopia; we have found no simple and workable characters for identification during our field work. Recognition of infraspecific taxa not possible here.
Description: Tree to 6 (-12) m tall. No thorns. Macrophyll (leaflets Nanophyll). Leaves compound (bipinnate), apparently deciduous. Bark on trunk usually smooth, but ultimately becoming rough, probably an adaptation for fire resistance.
Habitat: 'Wooded grassland' (FEE for *Albizia malacophylla* var. *malacophylla*), 'wooded grassland, riverine forest' (FEE for *Albizia malacophylla* var. *ugandensis*). 'Wooded grassland' (PSS for *Albizia malacophylla* var. *ugandensis*). A fairly common species in the *Combretum-Terminalia* woodland and wooded grassland (our observation).
Distribution in Ethiopia: TU GD (FEE for *Albizia malacophylla* var. *malacophylla*), GD GJ WG IL (FEE for *Albizia malacophylla* var. *ugandensis*). Species as a whole also recorded from the SU floristic province (Fig. 6-80).
Altitudinal range in Ethiopia: Ca. 1400-1900 m a.s.l. (FEE for *Albizia malacophylla* var. *malacophylla*, including range in Eritrea), 550-2200 m a.s.l. (FEE for *Albizia malacophylla* var. *ugandensis*). 500-2100 m a.s.l. (our records).
General distribution (species as a whole): From Senegal to Ethiopia and Eritrea, south to Uganda.
Chorological classification: Distribution according to direct observation. **WWn, WWc.** *Local phytogeographical element and sub-element.* **CTW[n+c]**. Widely distributed in the northern and central part of the western woodlands, reaching the Gambela Region in the south, no eastern extension.

6.*19.22.* Albizia schimperiana *Oliv. (1871).*

FTEA, Leguminosae-Mimosoideae: 154 (1959); FEE 3: 96 (1989); FTNA: 147 (1992); *BS* 51: 194 (1998); PSS: 196 (2015).
Taxonomy: Two infraspecific taxa recognized, *Albizia schimperiana* var. *schimperiana*, and *Albizia schimperiana* var. *tephrocalyx* Brenan (1955). The infraspecific taxa are not recognized in FEE, in spite of the slightly different habitats indicated for the two varieties in PSS. Recognition of infraspecific taxa not relevant here.
Description: Tree to 30 m tall, with flat-topped crown. No thorns. Macrophyll (leaflets Nanophyll). Leaves compound (bipinnate), apparently evergreen. Bark on trunk usually smooth, but ultimately becoming rough, probably an adaptation for fire resistance.
Habitat: 'Upland forest, evergreen bushland' (FEE, species as a whole). 'Afromontane rain forest, undifferentiated Afromontane forest ... and riverine forest; also in humid woodland and secondary montane evergreen bushland; often left as single tree in farmland, used for hanging beehives' (FTNA for *Albizia schimperiana* var. *schimperiana*). 'Forest, including disturbed areas' (PSS for *Albizia schimperiana* var. *schimperiana*), 'riverine forest and forest margins' (PSS for *Albizia schimperiana* var. *tephrocalyx*). Mainly distributed in forests, chiefly in *Moist evergreen Afromontane forest* (MAF), and in bushland along forest margins in the highlands (our observation).
Distribution in Ethiopia:?TU GD WU GJ WG SU AR IL KF GG SD (FEE, without infraspecific distinction). Also recorded from the HA floristic region (Fig. 6-81).
Altitudinal range in Ethiopia: 1600-2600 m a.s.l. (FEE, without infraspecific distinction). 1550-2800 m a.s.l. (FTNA, including the range in the entire Horn of Africa). 1750-2800 m a.s.l. (our records).
General distribution (species as a whole): D.R. Congo to Ethiopia (? possibly also in Eritrea), south to Zimbabwe and Mozambique.
Chorological classification: Distribution according to direct observation. **WWn, WWc, WWs, EWW.** *Local phytogeographical element.* **MI**. Marginally intruding species in the western woodlands, with main distribution in the

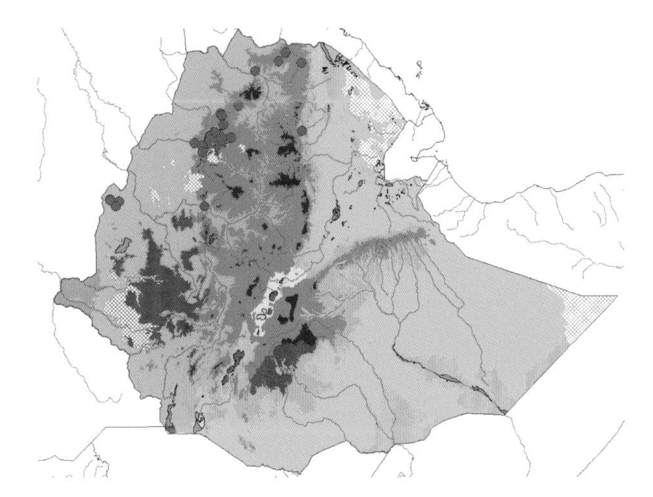

Fig. 6-77. *Acacia venosa*. Distribution in Ethiopia. Observed records are marked with red dots on vegetation map reproduced from Fig. 4-1.

Fig. 6-78. *Albizia amara*. Distribution in Ethiopia. Observed records are marked with red dots on vegetation map reproduced from Fig. 4-1.

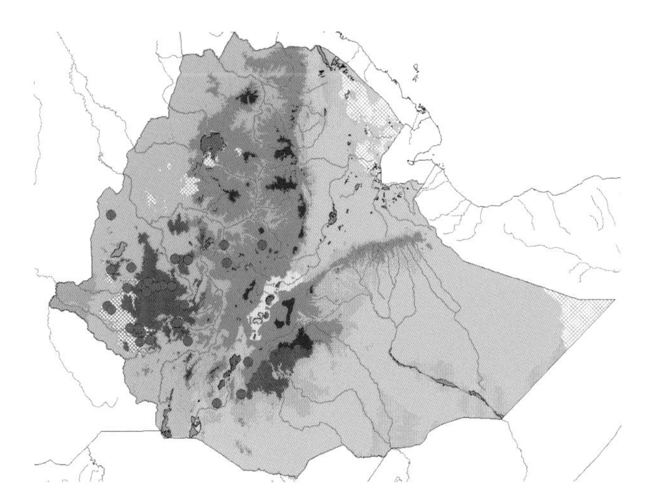

Fig. 6-79. *Albizia grandibracteata*. Distribution in Ethiopia. Observed records are marked with red dots on vegetation map reproduced from Fig. 4-1.

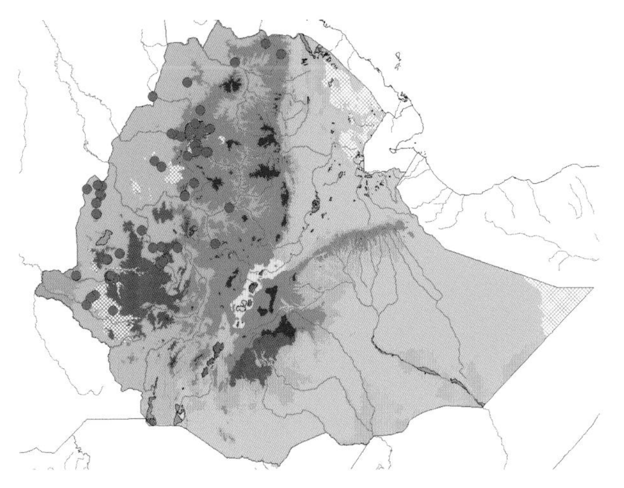

Fig. 6-80. *Albizia malacophylla*. Distribution in Ethiopia. Observed records are marked with red dots on vegetation map reproduced from Fig. 4-1.

highlands and in *Transitional semi-evergreen bushland* (TSEB); only a few records in our relevés.

6.19.23. Dichrostachys cinerea *(L.) Wight & Arn. (1834).*

FTEA, Leguminosae-Mimosoideae: 36 (1959); FEE 3: 74 (1989); FS 1: 366 (1993); *BS* 51: 195 (1998); PSS: 196 (2015).
Taxonomy: No infraspecific taxa recognized.
Description: Shrub or small tree to 8 (-12) m tall. Thorns present, formed by small, lateral twigs ending in spines. Leaves compound (bipinnate), deciduous. Macrophyll (leaflets microphyll). Bark on trunk rough, probably an adaptation for fire resistance.
Habitat: 'Woodland and bushland' (FEE). '*Acacia-Commiphora* bushland, rocky hillsides, also on alluvial soil' (FS). 'Woodland and bushland, often dominating in overgrazed areas' (PSS). A somewhat weedy species in a wide range of habitats in woodland and bushland (our observation).
Distribution in Ethiopia: TU WU GJ WG SU AR KF GG SD BA HA (FEE). Also recorded from the GD, IF, and AR floristic regions (Fig. 6-82).
Altitudinal range in Ethiopia: 450-2000 m a.s.l. (FEE, including range in Eritrea). 450-1950 m a.s.l. (our records).
General distribution: From Senegal to Ethiopia, Eritrea and Somalia, south to Namibia and South Africa; also in the Cape Verde Islands and in tropical Asia, including Arabia, and in Australia.
Chorological classification: Distribution according to direct observation. **WWn, WWc, WWs, EWW**. *Local phytogeographical element.* **TRG**. Transgressing species, occurs in both the *Combretum-Terminalia woodland and wooded grassland* (CTW) and in other vegetation types, particularly in the lower zones of the *Dry evergreen Afromontane forest* (DAF), in the Rift Valley, the *Transitional semi-evergreen bushland* (TSEB), and the *Acacia-Commiphora bushland* (ACB).

6.19.24. Entada abyssinica *Steud. ex A. Rich. (1847).*

FTEA, Leguminosae-Mimosoideae: 13 (1959); FEE 3: 72 (1989); FS 1: 363 (1993); *BS* 51: 96 (1998); PSS: 197 (2015).
Taxonomy: No infraspecific taxa.
Description: Tree to 10 (-15) m tall. No thorns. Leaves compound (bipinnate), deciduous. Macrophyll (leaflets microphyll). Bark on trunk thick, smooth or rough, probably an adaptation for fire resistance.
Habitat: 'Woodland, wooded grassland and scrub' (FEE). 'Woodland, evergreen bushland' (FS). 'Woodland and wooded grassland' (PSS). A widespread species in the *Combretum-Terminalia* woodland, mostly at higher altitudes than *Entada africana* (our observation).
Distribution in Ethiopia: TU GJ WG SU IL KF GG SD HA (FEE, including range in Eritrea). Also recorded from the GD, WG and BA floristic regions (Fig. 6-83).
Altitudinal range in Ethiopia: 1300-2050 m a.s.l. (FEE). 950-2000 m a.s.l. (our records).
General distribution: From Guinean Republic to Ethiopia, Eritrea and Somalia, south to Angola and Mozambique.
Chorological classification: Distribution according to direct observation. **WWn, WWc, WWs, EWW**. *Local phytogeographical element and sub-element.* **CTW[all+ext]**. Widely distributed in the western woodlands, from north to south, extending to the Rift Valley and to the east of the Rift.

6.19.25. Entada africana *Guill. & Perr. (1832).*

FTEA, Leguminosae-Mimosoideae: 12 (1959); FEE 3: 72 (1989); *BS* 51: 196 (1998); PSS: 197 (2015).
Taxonomy: No infraspecific taxa.
Description: Tree to 10 m tall. No thorns. Leaves compound (bipinnate), deciduous. Macrophyll (leaflets microphyll). Bark on trunk thick and rough, probably an adaptation for fire resistance.
Habitat: 'Woodland' (FEE). 'Woodland' (PSS). A widespread species in the *Combretum-Terminalia* woodland, mostly at lower altitudes than *Entada abyssinica* (our observation).

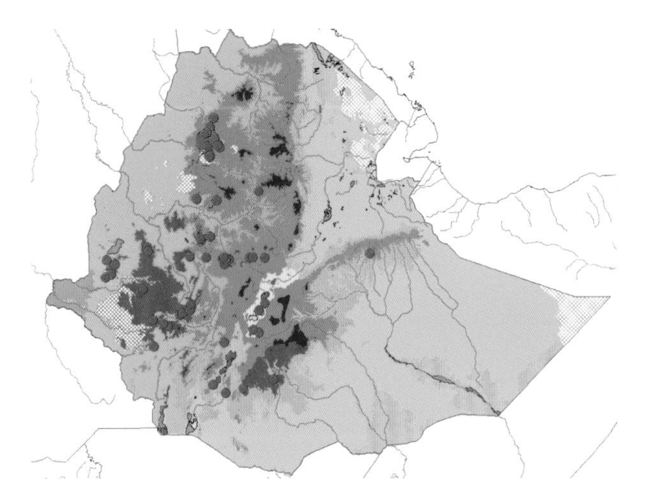

Fig. 6-81. *Albizia schimperiana*. Distribution in Ethiopia. Observed records are marked with red dots on vegetation map reproduced from Fig. 4-1.

Fig. 6-82. *Dichrostachys cinerea*. Distribution in Ethiopia. Observed records are marked with red dots on vegetation map reproduced from Fig. 4-1.

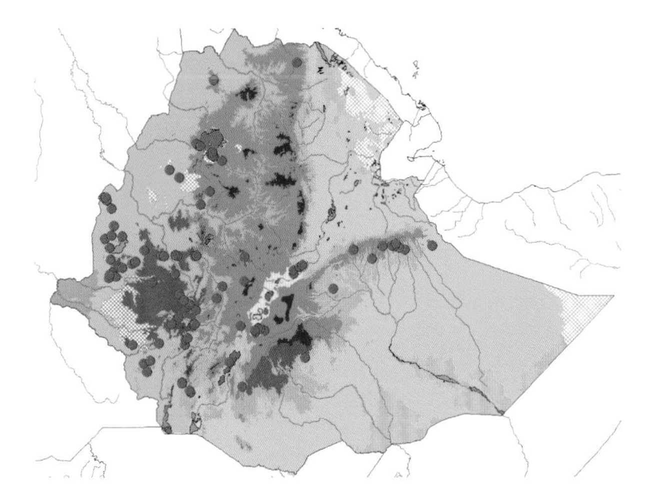

Fig. 6-83. *Entada abyssinica*. Distribution in Ethiopia. Observed records are marked with red dots on vegetation map reproduced from Fig. 4-1.

Fig. 6-84. *Entada africana*. Distribution in Ethiopia. Observed records are marked with red dots on vegetation map reproduced from Fig. 4-1.

Distribution in Ethiopia: GD GJ IL (FEE). Also recorded from WG and KF floristic region (Fig. 6-84).
Altitudinal range in Ethiopia: 450-1100 m a.s.l. (FEE). 450-1750 m a.s.l. (our records).
General distribution: From Senegal to Ethiopia, south to D.R. Congo and Uganda.
Chorological classification: Distribution according to direct observation. **WWn, WWc, WWs.** *Local phytogeographical element and sub-element.* **CTW[all].** Widely distributed in the western woodlands, but not to the extreme north or the extreme south; no eastern extension.

6.20. Leguminosae subfam. Papilionoideae

FTEA, Leguminosae-Papilionoideae, 1: 1-501 (1971); FTEA, Leguminosae-Papilionoideae, 2: 503-1109 (1971); *BS* 51: 198-237 (1998), as "**Fabaceae subfam. Faboideae**"; PSS: 161-189 (2015), as "**Leguminosae: Papilionoideae (Fabaceae: Faboideae)**".

6.20.1. **Calpurnia aurea** *(Ait.) Benth. (1837).*

FTEA, Leguminosae-Papilionoideae, 1: 47 (1971); FEE 3: 104 (1989); FTNA: 150 (1992); *BS* 51: 202 (1998); PSS: 163 (2015).
Calpurnia aurea subsp. **aurea**
FTEA, Leguminosae-Papilionoideae, 1: 47 (1971); FEE 3: 104 (1989); *BS* 51: 202 (1998); PSS: 163 (2015).
Taxonomy: The species is divided into three subspecies. Tropical African plants belong to *Calpurnia aurea* subsp. *aurea*; *Calpurnia aurea* subsp. *indica* Brummitt (1967) is restricted to southern India and *Calpurnia aurea* subsp. *sylvatica* (Burch.) Brummitt (1967) is restricted to the Cape region of South Africa. *However,* Beaumont et al. (1999) has merged *Calpurnia aurea* subsp. *aurea* and *Calpurnia aurea* subsp. *sylvatica*). Recognition of infraspecific taxa not relevant here.
Description: Shrub or small tree to 10 m tall. No thorns. Leaves compound (pinnate), apparently evergreen. Mesophyll (leaflets microphyll). Bark on trunk smooth, no adaptation for fire resistance.
Habitat: 'Forest margins, bushland or grassland, favoured by overgrazing' (FEE). 'Afromontane rain forest, undifferentiated Afromontane forest ... and single-dominant Afromontane forest ..., especially in forest clearings and along forest margins; sometimes in riverine forest; also in secondary montane evergreen bushland' (FTNA). 'Bushland, forest margins' (PSS). A species mainly distributed at forest margins and in bushland in the lower zone of the highland vegetation, both in *Dry and Moist Evergreen Afromontane forest* (DAF and MAF) (our observation).
Distribution in Ethiopia: TU GD GJ WU WG SU AR KF GG SD BA HA (FEE). No additional records (Fig. 6-85).
Altitudinal range in Ethiopia: 1650-2550 m a.s.l. (FEE, including range in Eritrea). 1400-2500 m a.s.l. (FTNA, including the range in the entire Horn of Africa). 1550-2550 m a.s.l. (our records).
General distribution (species as a whole) Central African Republic to Ethiopia and Eritrea, south to Angola and South Africa; also in India.
Chorological classification: Distribution according to direct observation. **WWn, WWc, WWc, EWW.** *Local phytogeographical element.* **MI.** Marginally intruding species in the western woodlands, with a main distribution in the highlands and in *Transitional semi-evergreen bushland* (TSEB); only a few records in our relevés.

6.20.2. **Dalbergia boehmii** *Taub. (1895).*

FTEA, Leguminosae-Papilionoideae, 1: 106 (1971); FEE 1: 216 (2009); PSS: 169 (2015).
Taxonomy: Two infraspecific taxa currently accepted. The widespread *Dalbergia boehmii* subsp. *boehmii* and *Dalbergia boehmii* subsp. *stuhlmannii* (Taub.) Polhill (1969) is known only from Tanzania. Recognition of infraspecific taxa not relevant here.
Description: Tree to 10 (-21) m tall. No thorns. Leaves compound (pinnate), apparently deciduous. Mesophyll (leaflets microphyll). Bark on trunk ultimately rough and flaking in small pieces, probably an adaptation for fire resistance.
Habitat: 'Riverine vegetation dominated by *Anogeissus leiocarpa, Diospyros mespiliformis* and *Tamarindus indica*, and in broadleaved woodland dominated by *Combretum collinum, Terminalia laxiflora, Albizia mala-*

cophylla, Pterocaropus lucens and *Ozoroa insignis'* (FEE). 'Bushland, woodland, riverbanks' (PSS). In the central part of the *Combretum-Terminalia* woodland with tall undergrowth liable to burning (our observation).
Distribution in Ethiopia: WG (FEE). Also recorded from the GJ floristic region (Fig. 6-86).
Altitudinal range in Ethiopia: 755-1250 m a.s.l. (FEE). 650-1400 m a.s.l. (our records).
General distribution: Scattered distribution from Senegal to Sudan and Ethiopia, south to Angola, Zimbabwe and Mozambique.
Chorological classification: Distribution according to direct observation. **WWc.** *Local phytogeographical element and sub-element.* **CTW[c]**: Widely distributed in the central part of the western woodlands, mainly south of the Abay River; no extension to the Rift Valley or to the east of the Rift.

6.20.3. Dalbergia melanoxylon *Guill. & Perr. (1832).*

FTEA, Leguminosae-Papilionoideae, 1: 100 (1971); FEE 3: 105 (1989); *BS* 51: 210 (1998); PSS: 169 (2015).
Taxonomy: No infraspecific taxa.
Description: Tree to 12 (-30) m tall. No thorns. Leaves compound (pinnate), apparently deciduous. Mesophyll (leaflets Microphyll). Bark on trunk ultimately rough and flaking in small pieces, probably an adaptation for fire resistance.
Habitat: 'Deciduous woodland, wooded grassland or bushland, often in rocky places or valleys of impeded drainage' (FEE). 'Woodland and wooded grassland' (PSS). A common species in the northern parts of the *Combretum-Terminalia* woodland and in the northern river valleys (our observation).
Distribution in Ethiopia: TU GD (FEE). Also recorded from the GJ and WG floristic regions (Fig. 6-87).
Altitudinal range in Ethiopia: 600-1900 m a.s.l. (FEE, including range in Eritrea). 650-1250 m a.s.l. (our records).
General distribution: From Senegal to Ethiopia and Eritrea, south to Angola and South Africa.
Chorological classification: Distribution according to direct observation. **WWn, WWc.** *Local phytogeographical element and sub-element.* **CTW[n+c]**. Widely distributed in the northern part of the western woodlands, mainly north of the Abay River, southernmost records at Kurmuk and in the Abay Gorge; no eastwards extension.

6.20.4. Erythrina abyssinica *DC. (1825).*

FTEA, Leguminosae-Papilionoideae, 2: 555 (1971); FEE 3: 161 (1989); *BS* 51: 215 (1998); PSS: 171 (2025).
Taxonomy: Two infraspecific taxa have been recognized for either the northern end of the distribution area (*Erythrina abyssinica* subsp. *abyssinica*) or the southern end (*Erythrina abyssinica* subsp. *suberifera* (Welw. ex Baker) Verdc. (1970)). However, this distinction is not generally upheld, no infraspecific taxa are recognized in FEE and BS. Recognition of infraspecific taxa not relevant here.
Description: Tree to 15 m tall. Branches with many scattered thorns. Leaves compound (trifoliolate), deciduous (tree flowering when leafless). Mesophyll (leaflets Mesophyll). Bark on trunk ultimately thick and corky, probably an adaptation for fire resistance.
Habitat: 'Grassland, woodland, forest edges, rocky places, widely used to form live fences around homesteads' (FEE). 'Wooded grassland' (PSS). In a range of habitats in woodlands (our observation).
Distribution in Ethiopia: TU GD GJ WU SU AR WG IL KF GG SD (FEE). Also recorded from the AF floristic region (Fig. 6-88).
Altitudinal range in Ethiopia: 1300-2400 m a.s.l. (FEE, including range in Eritrea). 650-2600 m a.s.l. (our records).
General distribution: D.R. Congo to Ethiopia and Eritrea, south to Angola and Mozambique.
Chorological classification: Distribution according to direct observation. **WWn, WWc, WWs, EWW.** *Local phytogeographical element and sub-element.* **CTW[all+ext]**. Widely distributed in the western woodlands, from north to south, extending to the Rift Valley and to the east of the Rift; also extension into the lower zones of the *Dry evergreen Afromontane forest* (DAF), where it occurs along forest margins.

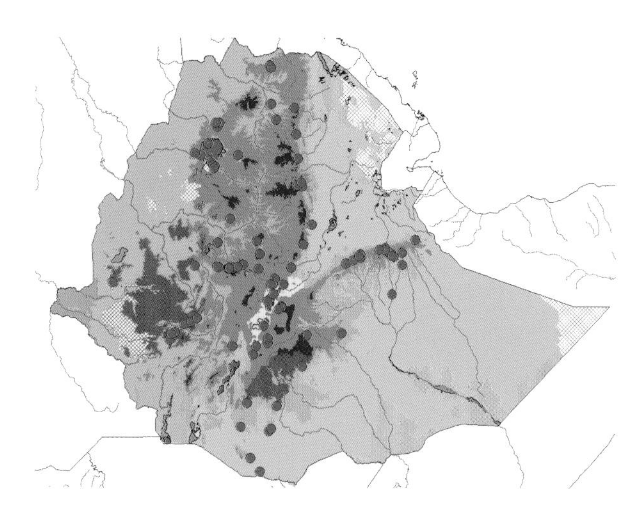

Fig. 6-85. *Calpurnia aurea*. Distribution in Ethiopia. Observed records are marked with red dots on vegetation map reproduced from Fig. 4-1.

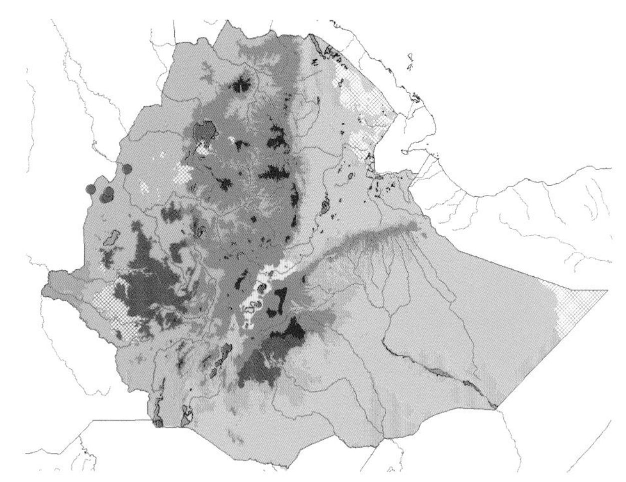

Fig. 6-86. *Dalbergia boehmii*. Distribution in Ethiopia. Observed records are marked with red dots on vegetation map reproduced from Fig. 4-1.

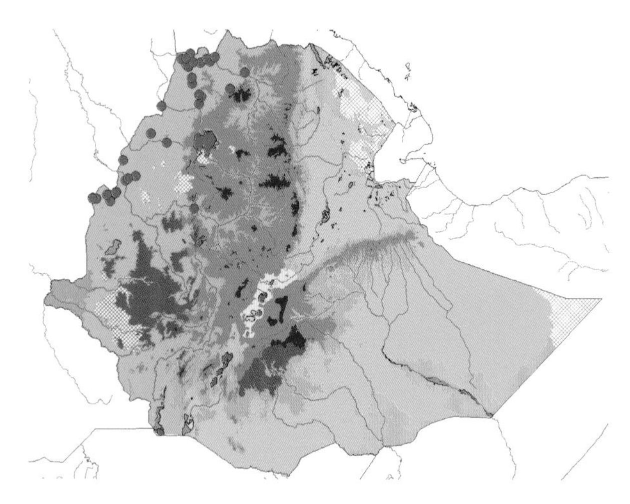

Fig. 6-87. *Dalbergia melanoxylon*. Distribution in Ethiopia. Observed records are marked with red dots on vegetation map reproduced from Fig. 4-1.

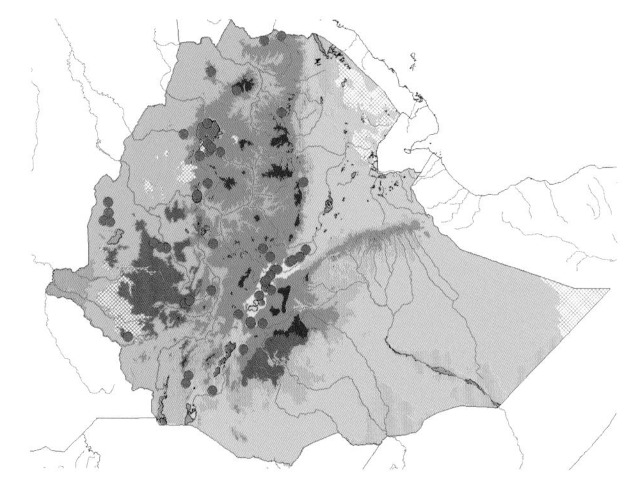

Fig. 6-88. *Erythrina abyssinica*. Distribution in Ethiopia. Observed records are marked with red dots on vegetation map reproduced from Fig. 4-1.

6.20.4. Lonchocarpus laxiflorus *Guill. & Perr. (1832).*

FTEA, Leguminosae-Papilionoideae, 1: 67 (1971); FEE 3: 104 (1989); 222; *BS* 51: (1998).
Philenoptera laxiflora (Guill. & Perr.) Roberty (1954). PSS: 180 (2015)
Taxonomy: No infraspecific taxa.
Description: Tree to 12 m tall. No thorns. Leaves compound (pinnate), deciduous (tree flowering when leafless). Macrophyll (leaflets Mesophyll). Bark on trunk ultimately thick and flaking, probably an adaptation for fire resistance.
Habitat: 'Wooded grassland' (FEE). 'Woodland, often in rocky areas' (PSS). A very common and widespread species in *Combretum-Terminalia* woodland (our observation).
Distribution in Ethiopia: TU GD WG SU IL KF BA (FEE). Also recorded from the GJ and GG floristic regions (Fig. 6-89).
Altitudinal range in Ethiopia: 450-2150 m a.s.l. (FEE, including range in Eritrea). 550-2100 m a.s.l. (our records).
General distribution: From Senegal to Ethiopia and Eritrea, south to Uganda.
Chorological classification: Distribution according to direct observation. WWn, WWc, WWs. *Local phytogeographical element and sub-element.* CTW[all]. Widely distributed in the western woodlands, from north to south, in the extreme south with some slight extension into the *Transitional semi-evergreen bushland* (TSEB); no other eastward extension.

6.20.5. Ormocarpum pubescens *(Hochst.) Cufod. ex Gillett (1966).*

FEE 3: 143 (1989); PSS: 180 (2015).
Taxonomy: No infraspecific taxa.
Description: Small tree to 3 m tall (*Friis et al.* 15930 was ca. 3 m tall; other specimens may probably be slightly higher). No thorns. Leaves compound (pinnate), deciduous. Mesophyll (leaflets Nanophyll). Bark on trunk not known in detail, thickened and blackened in the field, probably an adaptation for fire resistance.
Habitat: 'Woodland or bushland' (FEE). 'Woodland, bushland' (PSS). Apparently a rather rare species in *Combretum-Terminalia* woodland, and with scattered distribution (our observation).
Distribution in Ethiopia: TU GD (FEE). Also recorded from the GJ floristic region (Fig. 6-90).
Altitudinal range in Ethiopia: 1000-2100 m a.s.l. (FEE, including range in Eritrea). 1200-1950 m a.s.l. (our records).
General distribution: From Senegal to Cameroon, Sudan, Ethiopia and Eritrea.
Chorological classification: Distribution according to direct observation. WWn. *Local phytogeographical element and sub-element.* CTW[n]. Widely distributed in the northern part of the western woodlands, north of or at the Abay River; no eastward extension.

6.20.6. Pterocarpus lucens *Lepr. ex Guill. & Perr. (1832).*

FTEA, Leguminosae-Papilionoideae, 1: 82 (1971); FEE 3: 105 (1989); *BS* 51: 225 (1998); PSS: 181 (2015).
Taxonomy: Two infraspecific taxa are currently recognized, the widespread *Pterocarpus lucens* subsp. *lucens*, which is widespread from Senegal to Ethiopia, and the completely disjunct *Pterocarpus lucens* subsp. *antunesii* (Taub.) Rojo (1972), restricted to Namibia and South Africa. No infraspecific taxa accepted in FEE and *BS* 51. Recognition of infraspecific taxa not relevant here.
Description: Tree to 18 m tall. No thorns. Leaves compound (pinnate), deciduous. Macrophyll (leaflets Mesophyll). Bark on trunk ultimately rough, probably an adaptation for fire resistance.
Habitat: 'Wooded grassland, particularly on rocky hills' (FEE). 'Wooded grassland' (PSS for *Pterocarpus lucens* subsp. *lucens*). A very common and widespread species in *Combretum-Terminalia* woodland, often on rocks (our observation).
Distribution in Ethiopia: TU GD GJ WG IL (FEE). No additional records (Fig. 6-91).
Altitudinal range in Ethiopia: 550-1520 m a.s.l. (FEE, including range in Eritrea). 600-1400 m a.s.l. (our records).

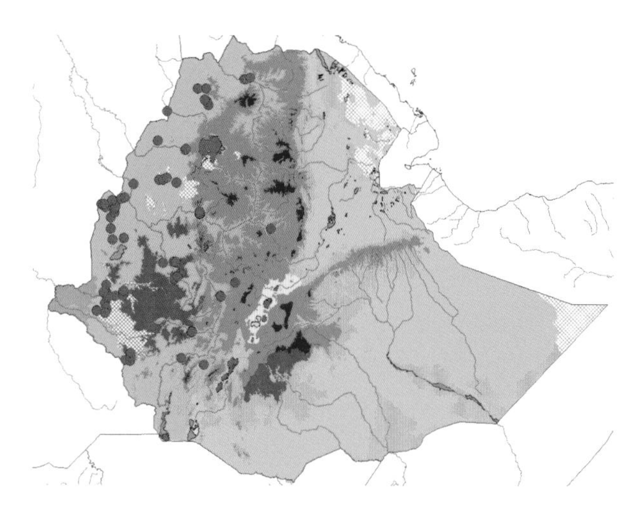

Fig. 6-89. *Lonchocarpus laxiflorus*. Distribution in Ethiopia. Observed records are marked with red dots on vegetation map reproduced from Fig. 4-1.

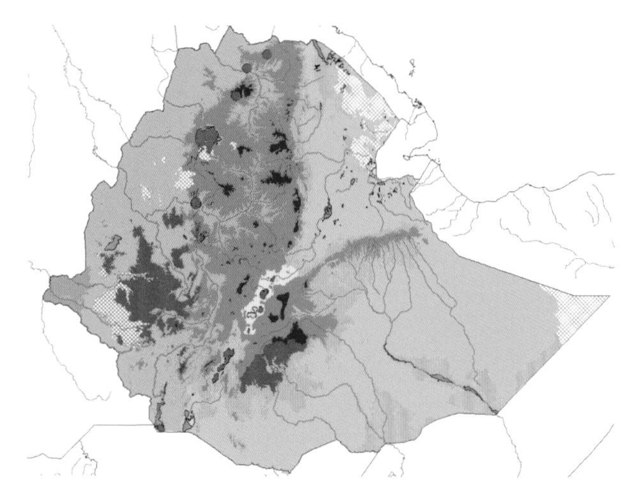

Fig. 6-90. *Ormocarpum pubescens*. Distribution in Ethiopia. Observed records are marked with red dots on vegetation map reproduced from Fig. 4-1.

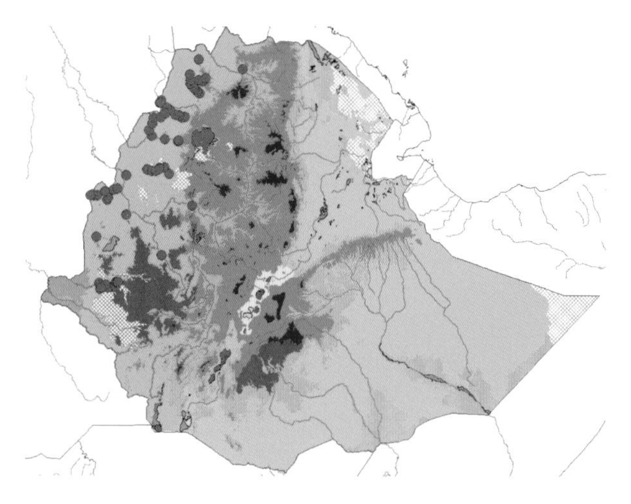

Fig. 6-91. *Pterocarpus lucens*. Distribution in Ethiopia. Observed records are marked with red dots on vegetation map reproduced from Fig. 4-1.

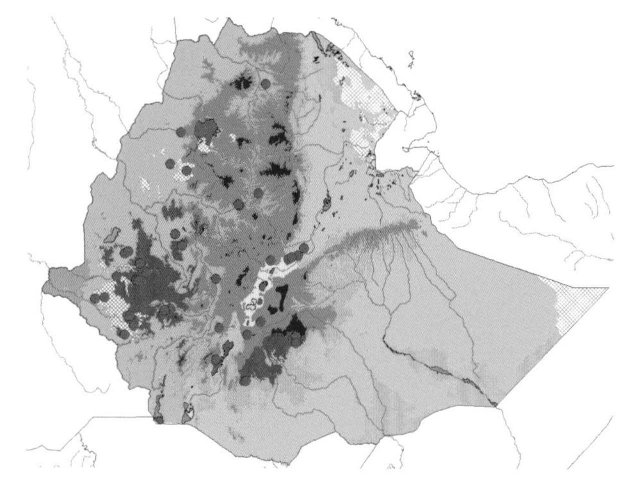

Fig. 6-92. *Trema orientalis*. Distribution in Ethiopia. Observed records are marked with red dots on vegetation map reproduced from Fig. 4-1.

General distribution: From Senegal to Ethiopia and Eritrea, south to Uganda.
Chorological classification: Distribution according to direct observation. **WWn, WWc**. *Local phytogeographical element and sub-element.* **CTW[n+c]**. Widely distributed in the northern and central part of the western woodlands, reaching the Gambela Region in the south; no eastward extension.

6.21. Ulmaceae

FTEA, Ulmaceae: 1-15 (1966); FEE 3: 266-269 (1989); *BS* 51: 238-241 (1998); FS 2: 88-91 (1999), incl. ***Celtidaceae***; PSS: 202 (2015), where the genus **Trema** is referred to the **Cannabaceae**.

6.*21.1.* **Trema orientalis** *(L.) Blume (1852).*

FTEA, Ulmaceae: 10 (1966); FEE 3: 268 (1989); FTNA: 159 (1992); *BS* 51: 240 (1998); FS 2: 91 (1999); PSS: 202 (2015).
Taxonomy: No infraspecific taxa.
Description: Large shrub or small tree to 12 (-15) m tall. No thorns. Leaves simple, apparently evergreen. Mesophyll. Bark of trunk smooth, probably no adaptation for fire resistance.
Habitat: 'Pioneer of forest edges and clearing, riverine forest, rarely dry evergreen bushland' (FEE). 'Afromontane rain forest, undifferentiated Afromontane forest ... and dry single-dominant Afromontane forest ..., especially along margins and in clearings; riverine forest; also in secondary montane evergreen bushland' (FTNA). 'Forest clearings, riverine woodland' (FS). 'Pioneer of forest margins and clearings, persisting in secondary bushland' (PSS). Forest pioneer along margins and clearings of *Dry* and *Moist Evergreen Afromontane Forest* (DAF and MAF), also where forest meets *Combretum-Terminalia* woodland (our observation).
Distribution in Ethiopia: GD GJ WG SU AR IL KF BA (FEE). Also recorded from the TU, AF, WU, GG, and SD floristic regions (Fig. 6-92).
Altitudinal range in Ethiopia: (500-) 900-1800 m a.s.l. (FEE, including range in Eritrea). 500-1800 m a.s.l. (FTNA, including the range in the entire Horn of Africa). 600-1800 m a.s.l. (our records).
General distribution: From Senegal to Ethiopia, Eritrea and Somalia, south to South Africa, also in Madagascar, the Mascarenes, Arabia and in tropical Asia.
Chorological classification: Distribution according to direct observation. **WWn, WWc, EWW**. *Local phytogeographical element.* **MI**. Marginally intruding species in the western woodlands, with main distribution in and at forests and bushland in the highlands, *Riparian vegetation* (RV) or in *Transitional rain forest* (TRF). The records from the Awash Valley in the Rift (SU and AF floristic regions) are from *Riparian vegetation* (RV).

6.22. Moraceae

FTEA, Moraceae: 1-96 (1989); FEE 3: 271-301 (1989); *BS* 51: 241-250 (1998); FS 2: 91 (1999); PSS: 202-206 (2015).

6.*22.1.* **Ficus glumosa** *Delile (1826).*

FTEA, Moraceae: 65 (1989); FEE 3: 293 (1989); *BS* 51: 245 (1998); FS 2: 101 (1999); PSS: 204 (2015).
Taxonomy: No infraspecific taxa.
Description: Shrub or tree to 10 (-15) m tall. No thorns. Leaves simple, deciduous. Mesophyll. Bark of trunk smooth, probably no adaptation for fire resistance.
Habitat: 'On rocks, boulders and rocky slopes in deciduous woodland or bushland' (FEE). 'On rocks, boulders and rocky slopes in deciduous woodland or bushland' (FS). 'Rock outcrops and screes, open wooded grassland' (PSS). A rather common species on rocky outcrops in woodland or bushland, sometimes at edges of *Dry evergreen Afromontane forest* (DAF) (our observation).
Distribution in Ethiopia: TU GD GJ IL SD HA (FEE). Also recorded from the AF, WG, SU, and GG floristic regions (Fig. 6-93).
Altitudinal range in Ethiopia: 500-1650 m a.s.l. (FEE, including range in Eritrea). 550-1850 m a.s.l. (our records).
General distribution: From Senegal to Ethiopia, Eritrea

and Somalia, south to Namibia and South Africa, also in Arabia.
Chorological classification: Distribution according to direct observation. **WWn, WWc, WWs, EWW**. *Local phytogeographical element and sub-element.* **CTW[all+ext]**. Widely distributed in the western woodlands, from north to south, with extension to the lower zones of the *Dry evergreen Afromontane forest* (DAF), the Rift Valley and to the east of the Rift.

6.22.2. Ficus ingens *(Miq.) Miq. (1867).*

FTEA, Moraceae: 60 (1989); FEE 3: 291 (1989); *BS* 51: 245 (1998); FS 2: 100 (1999); PSS: 204 (2015).
Taxonomy: No infraspecific taxa.
Description: Shrub or tree to 15 (-18) m tall. No thorns. Leaves simple, deciduous. Mesophyll. Bark of trunk smooth, probably no adaptation for fire resistance.
Habitat: 'Rocks, rocky slopes, deciduous woodland, evergreen bushland, edge of dry montane forest' (FEE). 'Rocks, rocky slopes in bushland, edges of dry montane forest' (FS). 'Wooded grassland, rock outcrops, riverine forest margins' (PSS). A rather common species on rocky outcrops in woodland or bushland, also on rocks along rivers in woodlands (our observation).
Distribution in Ethiopia: TU GD GJ SU IL SD HA (FEE). Also recorded from the AF and WG floristic region (Fig. 6-94).
Altitudinal range in Ethiopia: 600-2050 m a.s.l. (FEE, including range in Eritrea). 775-2000 m a.s.l. (our records).
General distribution: From Senegal to Ethiopia, Eritrea and Somalia, south to South Africa, also in Arabia.
Chorological classification: Distribution according to direct observation. **WWn, WWc, WWs, EWW**. *Local phytogeographical element and sub-element.* **CTW[all+ext]**. Widely distributed in the western woodlands, from north almost to the south (the species grows on rocky outcrops and is missing in areas with few rocks, particularly in the Gambela lowlands and the Omo Valley), and with extension to the lower zones of the *Dry evergreen Afromontane forest* (DAF), to the Rift Valley and to the east of the Rift.

6.22.3. Ficus lutea *Vahl (1805).*

FTEA, Moraceae: 69 (1989); FEE 3: 294 (1989); FTNA: 162 (1992); *BS* 51: 245 (1998); PSS: 204 (2015).
Taxonomy: No infraspecific taxa.
Description: Large tree to 20 m tall. No thorns. Leaves simple, deciduous. Macrophyll. Bark of trunk smooth, probably no adaptation for fire resistance (old trees are probably fire resistant due to their size).
Habitat: 'Riverine forest, upland rain forest or evergreen bushland' (FEE). 'Transitional rain forest, Afromontane rain forest and riverine forest' (FTNA). 'Forest, persisting in secondary bush' (PSS). Mainly in forests, but occasionally in woodland and bushland (our observation).
Distribution in Ethiopia: IL KF GG?SD (FEE). Also recorded from the GD (Getachew et al. 964; requires confirmation), WG and HA floristic regions; the FEE record from SD is confirmed (Fig. 6-95).
Altitudinal range in Ethiopia: 900-1500 m a.s.l. (FEE). 800-1300 m a.s.l. (FTNA, including the range in the entire Horn of Africa). 775-2000 m a.s.l. (our records).
General distribution: From Senegal to Ethiopia, south to Angola and South Africa; also in the Cape Verde Islands, Madagascar and the Seychelles.
Chorological classification: Distribution according to direct observation. **WWn, WWc, EWW**. *Local phytogeographical element.* **MI**. Marginally intruding species in the western woodlands, with main distribution in the highlands, in *Transitional semi-evergreen bushland* (TSEB), *Transitional rain forest* (TRF) or in *Riparian vegetation* (RV).

6.22.4. Ficus ovata *Vahl (1805).*

FTEA, Moraceae: 81 (1989); FEE 3: 292 (1989); FTNA: 163 (1992); *BS* 51: 246 (1998); PSS: 204 (2015).
Taxonomy: No infraspecific taxa.
Description: Tree to 25 m tall. No thorns. Leaves simple, only briefly deciduous. Macrophyll. Bark of trunk smooth, probably no adaptation for fire resistance (old trees are probably fire resistant due to their size).
Habitat: 'Riverine forest, upland rain forest, evergreen bushland, sometimes persisting in cultivation follow-

Fig. 6-93. *Ficus glumosa*. Distribution in Ethiopia. Observed records are marked with red dots on vegetation map reproduced from Fig. 4-1.

Fig. 6-94. *Ficus ingens*. Distribution in Ethiopia. Observed records are marked with red dots on vegetation map reproduced from Fig. 4-1.

Fig. 6-95. *Ficus lutea*. Distribution in Ethiopia. Observed records are marked with red dots on vegetation map reproduced from Fig. 4-1.

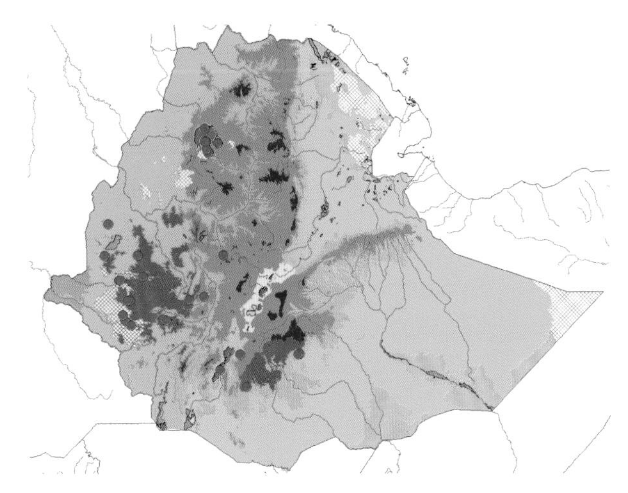

Fig. 6-96. *Ficus ovata*. Distribution in Ethiopia. Observed records are marked with red dots on vegetation map reproduced from Fig. 4-1.

ing forest clearing' (FEE). 'Afromontane rain forest, undifferentiated Afromontane forest and riverine forest; also in riparian woodland, occasionally as a single tree in secondary evergreen bushland or derived montane grassland' (FTNA). 'Forest, often riverine, woodland and wooded grassland' (PSS). Mainly in *Dry* and *Moist evergreen Afromontane forests* (DAF and MAF), but occasionally in woodland and bushland (our observation).
Distribution in Ethiopia: GD GJ WG KF GG SD (FEE). Also recorded from the BA floristic region (Fig. 6-96).
Altitudinal range in Ethiopia: 1300-2000 m a.s.l. (FEE, including range in Eritrea). 1850-2100 m a.s.l. (FTNA, including the range in the entire Horn of Africa). 900-1900 m a.s.l. (our records).
General distribution: From Senegal to Ethiopia and Eritrea, south to Angola and Mozambique.
Chorological classification: Distribution according to direct observation. **WWn, WWc, [WWs], EWW**. *Local phytogeographical element.* **MI**. Marginally intruding species in the western woodlands, with main distribution in the highlands, in *Transitional semi-evergreen bushland* (TSEB), *Transitional rain forest* (TRF) and *Riparian vegetation* (RV).

6.22.5. Ficus platyphylla *Delile (1826).*

FTEA, Moraceae: 64 (1989); FEE 3: 296 (1989); *BS* 51: 246 (1998); FS 2: 102 (1999); PSS: 204 (2015).
Taxonomy: No infraspecific taxa.
Description: Tree to 20 m tall. No thorns. Leaves simple, only briefly deciduous. Macrophyll. Bark of trunk smooth, probably no adaptation for fire resistance (old trees are probably fire resistant due to their size).
Habitat: 'Deciduous woodland' (FEE). 'Bushland, rocky places' (FS). 'Riverine forest and scrub, rocky areas in woodland or grassland' (PSS). A typical woodland species that grows to a fairly large tree (our observation).
Distribution in Ethiopia: IL GG (FEE). Also recorded from the WG and KF floristic regions (Fig. 6-97).
Altitudinal range in Ethiopia: 600 m a.s.l. (FEE). 450-1400 m a.s.l. (our records).
General distribution: From Senegal to Ethiopia and Somalia, south to Uganda.
Chorological classification: Distribution according to direct observation. **WWc, WWs**. *Local phytogeographical element and sub-element.* **CTW[c+s]**. Widely distributed in the southern part of the western woodlands, mainly south of the Abay River; the records from Ethiopia are from woodland, not from *Riparian vegetation* (RV), as is the case in the South Sudan according to PSS; no eastern extension.

6.22.6. Ficus cordata *Thunb. subsp.* salicifolia *(Vahl) C.C. Berg (1988).*

FTEA, Moraceae: 60 (1989); *BS* 51: 244 (1998); FS 2: 99 (1999); PSS: 203 (2015).
Ficus salicifolia Vahl (1790).
FEE 3: 290 (1989).
Taxonomy: Currently two subspecies are recognized; in Ethiopia only *Ficus cordata* subsp. *salicifolia*. *Ficus cordata* subsp. *cordata* is restricted to southern Africa from Angola to South Africa, and possibly also Cameroon. Recognition of infraspecific taxa not relevant here.
Description: Shrub or small tree to 8 (-15) m tall. No thorns. Leaves simple, only shortly deciduous. Mesophyll. Bark of trunk smooth, probably no adaptation for fire resistance (partly protected by growing on rocky outcrops with very little inflammable biomass).
Habitat: 'On rocks, rocky slopes and lava flows, on granite, lava or limestone, often on rocks in riverine scrub or along wadis' (FEE). 'On rocks and rocky slopes, often growing directly from crevices of bare rocks' (FS). 'Rocky hillslopes, riverine forest and scrub, woodland' (PSS). A rather rare species, at least at lower altitudes, more common near the lower zone of the highland vegetation, always growing on rocky outcrops in woodland or bushland (our observation).
Distribution in Ethiopia: TU GD GJ SU HA (FEE). Also recorded from the AF and GG floristic regions; the FEE record from GD is not confirmed (Fig. 6-98).
Altitudinal range in Ethiopia: 900-2300 m a.s.l. (FEE, including range in Eritrea). 1000-2250 m a.s.l. (our records).

General distribution (species as a whole): Cameroon to Ethiopia, Eritrea and Somalia, south to South Africa; also in Egypt and Arabia.
Chorological classification: Distribution according to direct observation. **WWc, WWs, EWW.** *Local phytogeographical element.* **MI.** Marginally intruding species in the western woodlands, with main distribution in the highlands; marginally intruding into the *Acacia-Commiphora bushland* (ACB).

6.22.7. Ficus sur *Forssk. (1775).*

FTEA, Moraceae: 56 (1989); FEE 3: 287 (1989); FTNA: 164 (1992); *BS* 51: 247 (1998); PSS: 204 (2015).
Taxonomy: No infraspecific taxa.
Description: Tree to 30 m tall. No thorns. Leaves simple, only shortly deciduous. Macrophyll. Bark of trunk smooth, apparently no adaptation for fire resistance (old trees are probably fire resistant due to their size).
Habitat: 'Upland rain forest and riverine forest, mountain grassland or secondary scrub, mainly at streams' (FEE). 'Transitional rain forest, Afromontane rain forest, undifferentiated Afromontane forest ... and riverine forest; sometimes as an isolated tree in secondary montane evergreen bushland and in moist, derived montane grassland' (FTNA). 'Forest, including riverine fringes, often in clearings, wooded grassland' (PSS). A common and widespread species in a wide range of relatively moist habitats, including in the southern part of the *Combretum-Terminalia* woodland (our observation).
Distribution in Ethiopia: TU GD GJ WU WG SU AR IL KF SD HA (FEE). Also recorded from the GG and BA floristic regions (Fig. 6-99).
Altitudinal range in Ethiopia: 1400-2500 m a.s.l. (FEE, including range in Eritrea). 1250-2800 m a.s.l. (FTNA, including the range in the entire Horn of Africa). 875-2450 m a.s.l. (our records).
General distribution: From Senegal to Ethiopia and Eritrea, south to Angola and South Africa; also in the Cape Verde Islands and Arabia.
Chorological classification: Distribution according to direct observation. **WWn, WWc, WWs, EWW.** *Local phytogeographical element.* **TRG.** Transgressing species, occurs similarly in both the *Combretum-Terminalia woodland and wooded grassland* (CTW) and in other vegetation types, particularly in the *Dry evergreen Afromontane forest* (DAF), but also in *Transitional rain forest* (TRF), *Transitional semi-evergreen bushland* (TSEB) and one record in *Acacia-Commiphora bushland* (ACB). In the driest habitats the species is associated with *Riparian vegetation* (RV).

6.22.8. Ficus sycomorus *L. (1753).*

FTEA, Moraceae: 54 (1989); FEE 3: 285 (1989); FTNA: 165 (1992); *BS* 51: 247 (1998); FS 2: 99 (1999); PSS: 205 (2015).
Ficus sycomorus subsp. *gnaphalocarpa* (Miq.) C.C. Berg (1980).
FEE 3: 285 (1989); FS 2: 99 (1999).
Taxonomy: *Ficus sycomorus* subsp. *gnaphalocarpa*, which was recognized in FEE, is now considered a synonym of the species autonym, *Ficus sycomorus* subsp. *sycomorus*, because the assumed diagnostic characters of both can be seen on the same trees. Recognition of infraspecific taxa not relevant here.
Description: Tree to 20 (-30) m tall, mostly with rather short trunk and spreading branches. No thorns. Leaves simple, only shortly deciduous. Mesophyll-Macrophyll. Bark of trunk smooth, presumably no adaptation for fire resistance (old trees are probably fire resistant due to their size).
Habitat: 'River and lake margins, woodland, wooded grassland, evergreen bushland, forest edges and clearings, "coffee forest"' (FEE for both *Ficus sycomorus* subsp. *sycomorus* and *Ficus sycomorus* subsp. *gnaphalocarpa*). 'Afromontane rain forest and undifferentiated Afromontane rain forest, especially along edges and in clearings; riverine forest; also in riparian woodland and secondary evergreen bushland; left as single tree in farmland; occasionally single tree on rocky outcrop' (FTNA for species as a whole). 'Riverine vegetation, bushland, wooded grassland, especially at wells' (FS for species as a whole). 'Riverine and upland forest margins, lakesides, rocky slopes, wooded grassland with localised moisture' (PSS for species as a whole).

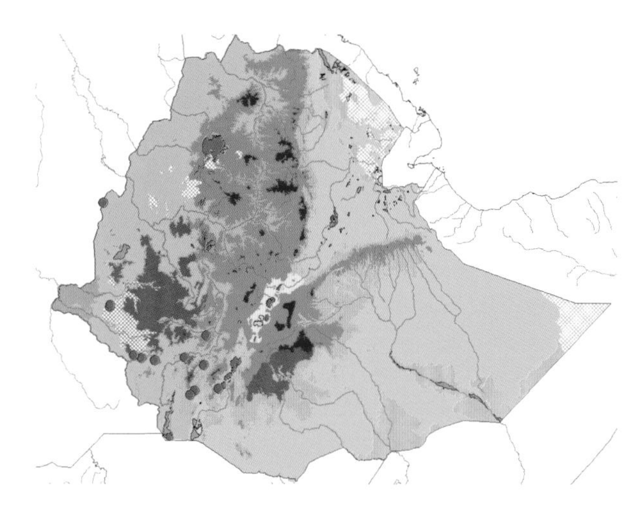

Fig. 6-97. *Ficus platyphylla*. Distribution in Ethiopia. Observed records are marked with red dots on vegetation map reproduced from Fig. 4-1.

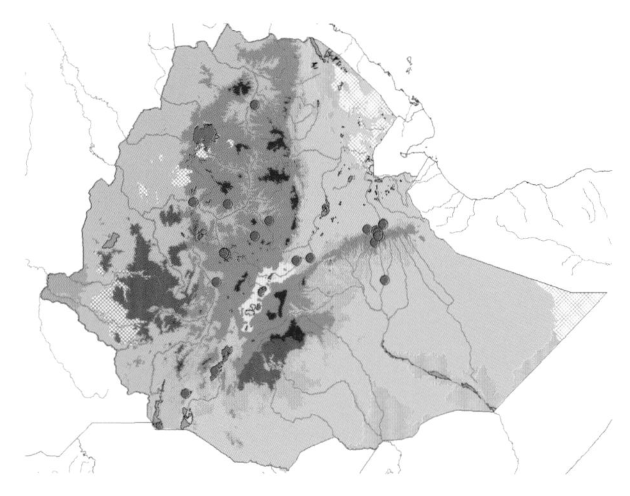

Fig. 6-98. *Ficus cordata* subsp. *salicifolia*. Distribution in Ethiopia. Observed records are marked with red dots on vegetation map reproduced from Fig. 4-1.

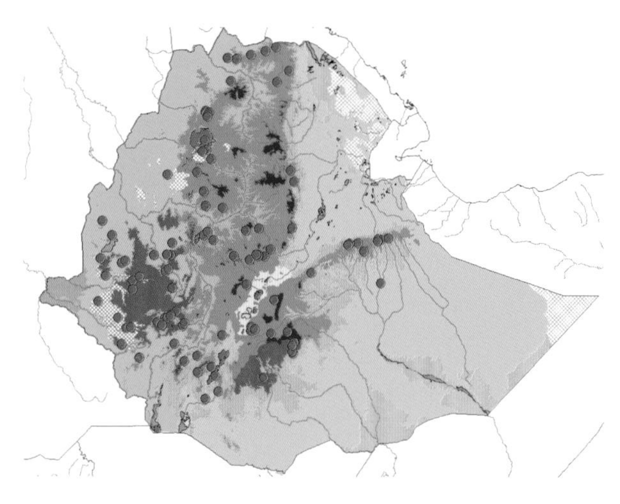

Fig. 6-99. *Ficus sur*. Distribution in Ethiopia. Observed records are marked with red dots on vegetation map reproduced from Fig. 4-1.

Fig. 6-100. *Ficus sycomorus*. Distribution in Ethiopia. Observed records are marked with red dots on vegetation map reproduced from Fig. 4-1.

A common and widespread species in a wide range of relatively moist habitats, including along edges of *Dry evergreen Afromontane forest* (DAF) and all parts of the *Combretum-Terminalia* woodland; the old trees have cultural importance because of their wide crowns and the shade they provide, and the species may be spread and encouraged by humans (our observation).
Distribution in Ethiopia: TU GD GJ WU SU IL KF GG SD HA (FEE for *Ficus sycomorus* subsp. *sycomorus*), TU GD GJ SU IL KF SD HA (FEE for *Ficus sycomorus* subsp. *gnaphalocarpa*). Also recorded from the AF floristic region; although no record has been made, it is unlikely that the species does not occur in BA (Fig. 6-100).
Altitudinal range in Ethiopia: 500-2000 m a.s.l. (FEE for *Ficus sycomorus* subsp. *sycomorus*, including range in Eritrea), 500-1900 m a.s.l. (FEE for *Ficus sycomorus* subsp. *gnaphalocarpa*). 50-2000 m a.s.l. (FTNA for species as a whole and including the range in the entire Horn of Africa). 550-2350 m a.s.l. (our records).
General distribution (species as a whole): From Senegal to Ethiopia, Eritrea and Somalia, south to Namibia and South Africa; also in the Cape Verde Islands, Madagascar, the Comoro Islands, Egypt and Arabia.
Chorological classification: Distribution according to direct observation. **WWn, WWc, WWs, EWW.** *Local phytogeographical element and sub-element.* **CTW[all+ext].** Widely distributed in the western woodlands, from north to south, extending to the Rift Valley and the lower zones of the highlands (in *Dry evergreen Afromontane forest* (DAF) and *Transitional semi-evergreen bushland* (TSEB)).

6.22.9. Ficus thonningii *Blume (1836).*

FTEA, Moraceae: 73 (1989); FEE 3: 298 (1989); FTNA: 166 (1992); *BS* 51: 248 (1998); FS 2: 103 (1999); PSS: 205 (2015).
Ficus hochstetteri (Miq.) A. Rich. (1851).
FEE 3: 299 (1989).
Ficus ruspolii Warb. (1905).
FEE 3: 299 (1989).
Taxonomy: No infraspecific taxa. *Ficus hochstetteri* A. Rich. and *F. ruspolii* Warb., which were both recognized in FEE with some questioning, are now in the FTEA, Moraceae, considered synonyms of *Ficus thonningii*, and these views are accepted here.
Description: Tree to 15 (-30) m tall (starting typically as an epiphytic "strangler" fig and sometimes remaining so). No thorns. Leaves simple, only shortly deciduous. Mesophyll. Bark of trunk smooth, presumably no adaptation for fire resistance (old trees are probably fire resistant due to their size).
Habitat: 'Rain forest, "coffee forest", dry evergreen forest, lake shores, evergreen bushland, or persisting as single tree in farmland following forest clearing or in villages' (FEE). 'Afromontane rain forest and undifferentiated Afromontane forest ..., sometimes persisting in gully forests; riverine forest; also in secondary montane evergreen bushland, and often as a single tree in derived grassland or farmland' (FTNA). 'Not known, specimen cited from the mountains of northern Somalia' (FS). 'Forest, including riverine strips, rock outcrops, wooded grassland' (PSS). A common and widespread species in a wide range of relatively moist habitats, including the central and southern parts of the *Combretum-Terminalia* woodland; the species may be spread by humans (our observation).
Distribution in Ethiopia: TU GD WU WG SU AR IL KF GG SD BA HA (FEE). No additional records (Fig. 6-101).
Altitudinal range in Ethiopia: 1300-2200 m a.s.l. (FEE, including range in Eritrea). 1500-2500 m a.s.l. (FTNA, including the range in the entire Horn of Africa). 675-2350 m a.s.l. (our records).
General distribution: From Senegal to Ethiopia, Eritrea and Somalia, south to Angola and South Africa; also in the Cape Verde Islands.
Chorological classification: Distribution according to direct observation. **WWn, WWc, WWs, EWW.** *Local phytogeographical element.* **MI.** Marginally intruding species in the western woodlands, with main distribution in the *Dry evergreen Afromontane forest* (DAF) in the highlands, in *Transitional semi-evergreen bushland* (TSEB), *Riparian vegetation* (RV) and *Transitional rain forest* (TRF); only few records in our relevés.

6.*22.10*. Ficus vasta *Forssk. (1775)*

FTEA, Moraceae: 64 (1989); FEE 3: 296 (1989); *BS* 51: 249 (1998); FS 2: 102 (1999); PSS: 205 (2015).
Taxonomy: Currently no infraspecific taxa are recognized in this species. To the south it is replaced by the related *Ficus wakefieldii* Hutch. (1915), distributed from D.R. Congo to Kenya and south to Zambia.
Description: Tree to 25 m tall. No thorns. Leaves simple, only shortly deciduous. Macrophyll. Bark of trunk smooth, probably no adaptation for fire resistance (old trees probably fire resistant due to their size).
Habitat: 'On rocks, in forest clearings and riverine forest, at forest margins and lake shores, occasionally in open woodland, in secondary bushland and farmland, often left or planted as a single big tree in village marketplaces' (FEE). 'On rocks, in woodland, bushland often remaining as a single big tree on or at the base of rocks, sometimes in vegetation along seasonal streams' (FS). 'Forest and forest margins, often riverine, rocky outcrops, secondary bushland' (PSS). A widespread species in a wide range of relatively moist habitats, including large parts of the *Combretum-Terminalia* woodland; the old trees have cultural importance because of their wide crowns and the shade they provide, and the species may be spread and encouraged by humans (our observation).
Distribution in Ethiopia: TU GD GJ WU SU AR IL KF GG SD HA (FEE). Also recorded from the WG floristic region (Fig. 6-102).
Altitudinal range in Ethiopia: 1000-2400 m a.s.l. (FEE, including range in Eritrea). 500-2300 m a.s.l. (our records).
General distribution: Sudan, Ethiopia, Eritrea and Somalia, Uganda and Kenya; also in Arabia.
Chorological classification: Distribution according to direct observation. **WWn, WWc, WWs, EWW**. *Local phytogeographical element.* **MI**. Marginally intruding species in the western woodlands, with main distribution in the highlands, in *Transitional semi-evergreen bushland* (TSEB), *Riparian vegetation* (RV) and *Transitional rain forest* (TRF); only few records in our relevés.

6.23. Celastraceae

FEE 3: 331-347 (1989), as **Celastraceae** (incl. ***Hippcrateaceae***); FTEA, Celastraceae: 1-78 (1994); *BS* 51: 255-260 (1998); FS 2: 109-120 (1999), as **Celastraceae** (incl. ***Hippcrateaceae***); PSS: 212-213 (2015).
The generic name **Gymnosporia** has been re-erected for Old-World species that were once transferred from a genus with that name to the genus **Maytenus**, but the latter is now mainly considered to be restricted to the New World (see for example Jordaan & Wyk 2006). The names used in the FEE have been maintained here.

6.*23.1*. Maytenus arbutifolia *(A. Rich.) Wilczek (1960)*.

FEE 3: 335 (1989); FTNA: 173 (1992); FTEA, Celastraceae: 12 (1994); FS 2: 110 (1999).
Later used name: **Gymnosporia arbutifolia** (A. Rich.) Loes. (1893).
Taxonomy: In the FEE two infraspecific taxa are recognized, *Maytenus arbutifolia* var. *arbutifolia* and *Maytenus arbutifolia* var. *sidamoensis* Sebsebe (1985), the latter restricted to southern Ethiopia and later raised to rank of subspecies (*Gymnosporia arbutifolia* subsp. *sidamoensis* (Sebsebe) Jordaan (2006)). The main distinction between the two infraspecific taxa is according to FEE the indumentum of the leaves. There seems to be no or limited ecological relevance in the distinction of the two taxa and it was not possible to distinguish them in the field observations and material used for this study.
Description: Shrub or small tree, in forest to 10, rarely 12 m tall, outside forests considerably less. Thorns usually present. Leaves simple, evergreen. Microphyll. Bark smooth, probably no adaptation for fire resistance.
Habitat: 'Forest, forest margins, degraded forest, riverbanks, grassland and scattered thickets and along roads' (FEE for *Maytenus arbutifolia* var. *arbutifolia*), 'forest and forest margins' (FEE for *Maytenus arbutifolia* var. *sidamoensis*). 'Open *Juniperus* forest, evergreen scrub' (FS). 'Undifferentiated Afromontane forest ... and single-dominant Afromontane forest ..., especially

along forest margins; very common in secondary montane evergreen bushland' (FTNA). [No information from the Sudan and South Sudan, from where the species is not recorded]. A species mainly of forest margins and clearings associated with the border zone between the highland vegetation and the *Combretum-Terminalia* woodland and the semi-evergreen bushland (our observation).
Distribution in Ethiopia: TU GD GJ WU WG SU AR IL KF BA HA (FEE for *Maytenus arbutifolia* var. *arbutifolia*), SU SD (FEE for *Maytenus arbutifolia* var. *sidamoensis*). Species as a whole also recorded from the GG floristic region (Fig. 6-103).
Altitudinal range in Ethiopia: 1200-3000 m a.s.l. (FEE for *Maytenus arbutifolia* var. *arbutifolia*, including range in Eritrea), 1699-2150 m a.s.l. (FEE for *Maytenus arbutifolia* var. *sidamoensis*). 1200-3000 m a.s.l. (FTNA for species as a whole and including the range in the entire Horn of Africa). 1200-2650 m a.s.l. (our records).
General distribution (species as a whole): D.R. Congo to Ethiopia, Eritrea and Somalia, south to Tanzania; also in Arabia.
Chorological classification: Distribution according to direct observation. **WWn, WWc, WWs, EWW**. *Local phytogeographical element.* **MI**. Marginally intruding species in the western woodlands, with main distribution in the highlands and in *Transitional semi-evergreen bushland* (TSEB); only few records in our relevés.

6.23.2. Maytenus senegalensis *(Lam.) Exell (1952).*

FEE 3: 336 (1989); FTEA, Celastraceae: 17 (1994); *BS* 51: 257 (1998); FS 2: 112 (1999).
Name in previous and later use: **Gymnosporia senegalensis** (Lam.) Loes. (1893).
PSS: 212 (2015).
Taxonomy: No infraspecific taxa.
Description: Shrub or tree to 8 (-13) m tall. Unarmed or with branchlets modified as thorns up to 5 cm long. Leaves simple, apparently evergreen. Mesophyll. Bark smooth, probably no adaptation for fire resistance.
Habitat: 'Deciduous woodland, open dry scrub, grassland, dry mountain slopes, riverbanks, edges of lakes' (FEE). 'Deciduous woodland and bushland, also on gypsum' (FS). 'Woodland and bushland, riverbanks' (PSS). The species has a wide distribution in the *Combretum-Terminalia* woodland, but is also widely distributed at the edges of *Dry evergreen Afromontane forest* (DAF) and bushland in the highland vegetation (our observation).
Distribution in Ethiopia: TU GD GJ WU WG SU AR IL KF GG SD BA HA (FEE). Also recorded from the AF floristic region (Fig. 6-104).
Altitudinal range in Ethiopia: Ca. 380-2440 m a.s.l. (FEE, including range in Eritrea). 450-2500 m a.s.l. (our records).
General distribution: From Senegal to Ethiopia, Eritrea and Somalia, south to Namibia and South Africa; also in Madagascar, Spain, Morocco, Algeria, Egypt, Arabia and in Asia from Afghanistan to India and Bangladesh.
Chorological classification: Distribution according to direct observation. **WWn, WWc, WWs, EWW**. *Local phytogeographical element.* **TRG**. Transgressing species, occurs similarly in both the *Combretum-Terminalia woodland and wooded grassland* (CTW) and in other vegetation types, particularly *Dry evergreen Afromontane forest* (DAF), extending into the Rift Valley and into *Transitional semi-evergreen bushland* (TSEB); also a few records from *Acacia-Commiphora bushland* (ACB) and *Semi-desert scrubland* (SDS), the latter probably along wadis.

6.24. Icacinaceae

FTEA, Icacinaceae: 1-18 (1968); (1968); FEE 3: 348-352 (1989); *BS* 51: 260-261 (1998); PSS: 288-289 (2015).

6.24.1. Apodytes dimidiata *E. Mey. ex Arn. (1840).*

FTEA, Icacinaceae: 4 (1968); FEE 3: 348 (1989); FTNA: 177 (1992); *BS* 51: 260 (1998); PSS: 288 (2015).
Apodytes dimidiata var. acutifolia (Hochst. ex A. Rich.) Boutique (1960).

Fig. 6-101. *Ficus thonningii*. Distribution in Ethiopia. Observed records are marked with red dots on vegetation map reproduced from Fig. 4-1.

Fig. 6-102. *Ficus vasta*. Distribution in Ethiopia. Observed records are marked with red dots on vegetation map reproduced from Fig. 4-1.

Fig. 6-103. *Maytenus arbutifolia*. Distribution in Ethiopia. Observed records are marked with red dots on vegetation map reproduced from Fig. 4-1.

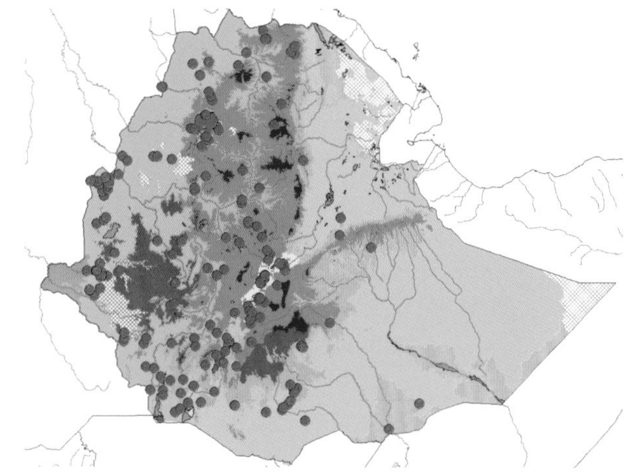

Fig. 6-104. *Maytenus senegalensis*. Distribution in Ethiopia. Observed records are marked with red dots on vegetation map reproduced from Fig. 4-1.

FTEA, Icacinaceae: 4 (1968); FEE 3: 348 (1989).
Apodytes dimidiata subsp. **acutifolia** (Hochst. ex A. Rich.) Cufod. (1958)
PSS: 288 (2015).
Taxonomy: In the FEE two infraspecific taxa are recognized, there as variety, while in PSS as subspecies. The nominal infraspecific taxon is not supposed to occur in Ethiopia and South Sudan (apart from one intermediate specimen mentioned in FEE). The distinction between var. or subsp. *dimidiata*, which is widespread outside Ethiopia, and var. or subsp. *acutifolia* is in the indumentum of the ovary does not seem to be ecologically relevant. Recognition of infraspecific taxa not relevant here.
Description: Large shrub or tree to 25 m tall. No thorns. Leaves simple, apparently evergreen. Mesophyll. Bark smooth, probably no adaptation for fire resistance.
Habitat: '*Podocarpus-Olea-Syzygium* upland rain forest, often remaining as a solitary tree in grassland after forest clearing, secondary forest, riverine forest, secondary *Acacia-Combretum* wooded grassland' (FEE for *Apodytes dimidiata* subsp. *acutifolia*). 'Afromontane rain forest and undifferentiated Afromontane forest ..., often in clearings and at forest edges; also in secondary montane evergreen bushland or left as isolated tree in farmland' (FTNA for *Apodytes dimidiata* subsp. *acutifolia*). 'Montane forest' (PSS for *Apodytes dimidiata* subsp. *acutifolia*). Although mainly a forest species in the highlands (*Dry* and Moist *evergreen Afromontane forest* (DAF and MAF), there are a few records from *Combretum-Terminalia* woodlands (our observation).
Distribution in Ethiopia: TU GD GJ WG SU AR IL KF GG SD BA HA (FEE for *Apodytes dimidiata* subsp. *acutifolia*). The FEE record from TU floristic region is not confirmed here (Fig. 6-105).
Altitudinal range in Ethiopia: 1350-2600 m a.s.l. (FEE for *Apodytes dimidiata* subsp. *acutifolia*, including range in Eritrea). 1200-3500 m a.s.l. (FTNA for *Apodytes dimidiata* subsp. *acutifolia* and including the range in the entire Horn of Africa). 1500-3200 m a.s.l. (our records).
General distribution (species as a whole): Nigeria to Ethiopia and Eritrea, south to South Africa; also in Madagascar and tropical Asia from India and Sri Lanka to China.
Chorological classification: Distribution according to direct observation. **WWn, WWc, WWs, EWW**. *Local phytogeographical element.* **MI**. Marginally intruding species in the western woodlands, with main distribution in the highlands and in *Transitional semi-evergreen bushland* (TSEB), only few records in our relevés.

6.25. Salvadoraceae

FTEA, Salvadoraceae: 1-10 (1968); FEE 3: 353-355 (1989); FS 2: 122-126 (1999); PSS: 261-262 (2015).

6.25.1. Salvadora persica *L. (1753)*.

FTEA, Salvadoraceae: 7 (1968); FEE 3: 354 (1989); FS 2: 126 (1999); PSS: 261 (2015).
Salvadora persica var. **persica**
FTEA, Salvadoraceae: 7 (1968); FEE 3: 354 (1989); FS 2: 126 (1999); PSS: 261 (2015).
Taxonomy: Only one infraspecific taxon, *Salvadora persica* var. *persica* is known from Ethiopia. *Salvadora persica* var. *crassifolia* Verdc. (1964) is known from Somalia and Socotra, *Salvadora persica* var. *pubescens* Brenan (1949) occurs in East Africa south to Angola and *Salvadora persica* var. *cyclophylla* (Chiov.) Cufod. (1958) occurs along the coast of Kenya and Somalia; *Salvadora persica* var. *tuticornica* T.A. Rao & Chakraborti (1996) occurs in India. Recognition of infraspecific taxa not relevant here.
Description: Shrub or small tree to 6 m tall. No thorns. Leaves simple, evergreen, sub-succulent. Mesophyll. Bark smooth, probably no adaptation for fire resistance.
Habitat: 'Fringing *Acacia-Commiphora* woodland, dense thorn scrub and thickets, especially associated with Capparaceae, *Hyphaene* associations near springs and desert flood plains mostly on saline, sandy or loamy soils' [in Eritrea also coastal] (FEE). 'Desert plains, rocky hillsides, bushland, often along wadis and near springs, and mostly on saline, sandy or loamy soils' (FS *Salvadora persica* var. *persica*). 'Dense thornbush, fringing woodland, open dry sandy plains, costal

scrub, sometimes dominant' (PSS). Mainly a species of very dry habitats, typically Semi-Desert Scrub or *Acacia-Commiphora* bushland, it occurs in a few localities within the area included in the *Combretum-Terminalia* woodland here (our observation).
Distribution in Ethiopia: AF SU KF GG SD HA (FEE). Also recorded from the TU and BA floristic regions (Fig. 6-106).
Altitudinal range in Ethiopia: Sea level to 1300 m a.s.l. (FEE, including range in Eritrea). 250-1600 m a.s.l. (our records).
General distribution (species as a whole): From Mauritania to Ethiopia, Eritrea and Somalia, south to Namibia and Zimbabwe; also in Arabia and Asia to India and Sri Lanka.
Chorological classification: Distribution according to direct observation. **WWn, WWs, [EWW], ASO.** *Local phytogeographical element and sub-element.* **ACB[-CTW].** Widely distributed in the *Acacia-Commiphora bushland* (ACB), marginally intruding in the *Combretum-Terminalia woodland and wooded grassland* (CTW) in the north and in the south, and extending to the Rift Valley and into *Transitional semi-evergreen bushland* (TSEB), particularly in the south and to the east of the Rift.

6.26. Olacaceae

FTEA, Olacaceae: 1-16 (1968); FEE 3: 356-360 (1989); *BS* 51: 261-262 (1998); FS 2: 126-128 (1999); PSS: 268 (2015).

6.*26.1.* Ximenia americana *L. (1753).*

FTEA, Olacaceae: 3 (1968); FEE 3: 356 (1989); *BS* 51: 261 (1998); FS 2: 127 (1999); PSS: 268 (2015).
Ximenia americana var. **americana**
PSS: 268 (2015).
Taxonomy: The species includes three varieties: the widespread *Ximenia americana* var. *americana; Ximenia americana* var. *argentinensis* DeFilipps (1969) in South America and *Ximenia americana* var. *microphylla* Welw. ex Oliver (1868), which occurs in southcentral Africa. Hence all Ethiopian material belongs to *Ximenia americana* var. *americana*. Recognition of infraspecific taxa not relevant here.
Description: Shrub or small tree to 7 m tall. Thorns present, formed by side-shoots up to 3 cm long. Leaves simple, probably deciduous. Mesophyll. Bark smooth, probably no adaptation for fire resistance.
Habitat: '*Acacia* woodland, *Acacia-Balanites* woodland, *Combretum-Terminalia* wooded grassland' (FEE). 'Wooded grassland, deciduous woodland, riverine and coastal formations' (FS). 'Dry woodland, bushland and wooded grassland' (PSS). A widespread species that occurs in several habitats, including *Combretum-Terminalia* woodland and semi-evergreen bushland (our observation).
Distribution in Ethiopia: TU GD GJ WU WG SU AR IL KF GG SD BA HA (FEE). Also recorded from the AF floristic region (Fig. 6-107).
Altitudinal range in Ethiopia: 500-2100 (-2450) m a.s.l. (FEE, including range in Eritrea). 500-1900 m a.s.l. (our records).
General distribution (species as a whole): From Senegal to Ethiopia, Eritrea and Somalia, south to South Africa; also in tropical Asia and America.
Chorological classification: Distribution according to direct observation. **WWn, WWc, WWs, EWW.** *Local phytogeographical element and sub-element.* **CTW[all+ext].** Widely distributed in the western woodlands, from north to south, extending to the Rift Valley and to the east of the Rift.

6.27. Rhamnaceae

FTEA, Rhamnaceae: 1-41 (1972); FEE 3: 385-398 (1989); *BS* 51: 266-269 (1998); FS 2: 150-157 (1999); PSS: 201 (2015).

6.*27.1.* Ziziphus abyssinica *Hochst. ex A. Rich. (1847).*

FTEA, Rhamnaceae: 27 (1972); FEE 3: 396 (1989); *BS* 51: 268 (1998); PSS: 201 (2015).
Taxonomy: No infraspecific taxa.
Description: Shrub or small tree to 6 (-13) m tall. Thorns (stipular spines) present. Leaves simple, deciduous.

Fig. 6-105. *Apodytes dimidiata*. Distribution in Ethiopia. Observed records are marked with red dots on vegetation map reproduced from Fig. 4-1.

Fig. 6-106. *Salvadora persica*. Distribution in Ethiopia. Observed records are marked with red dots on vegetation map reproduced from Fig. 4-1.

Fig. 6-107. *Ximenia americana*. Distribution in Ethiopia. Observed records are marked with red dots on vegetation map reproduced from Fig. 4-1.

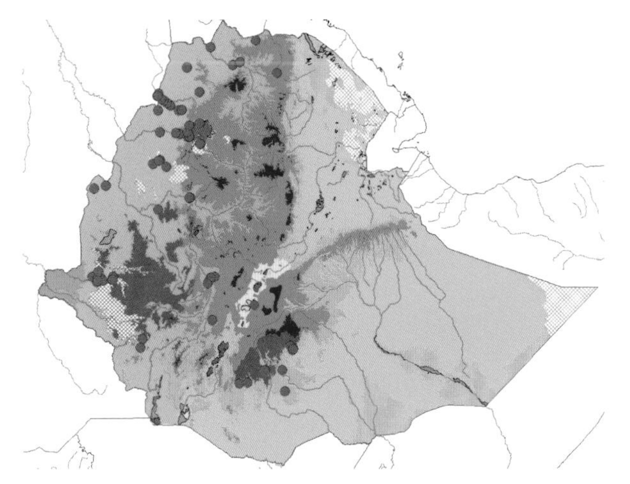

Fig. 6-108. *Ziziphus abyssinica*. Distribution in Ethiopia. Observed records are marked with red dots on vegetation map reproduced from Fig. 4-1.

Mesophyll. Bark smooth to reticulately fissured with thickening, probably an adaptation for fire resistance.
Habitat: '*Combretum-Terminalia* and *Acacia* woodland, wooded grassland and bushland' (FEE). 'Woodland and wooded grassland, margins of riverine forest, rock outcrops' (PSS). In a range of woodlands and bushlands, including bushlands in the lower zone of the highland vegetation (our observation).
Distribution in Ethiopia: TU GD GJ WG SU IL KF GG SD BA (FEE). The FEE record from the GG floristic region is not confirmed (Fig. 6-108).
Altitudinal range in Ethiopia: 450-2000 m a.s.l. (FEE, including range in Eritrea). 600-1800 m a.s.l. (our records).
General distribution: From Senegal to Ethiopia, Eritrea and Somalia, south to Angola and Mozambique.
Chorological classification: Distribution according to direct observation. **WWn, WWc, WWs, EWW.** *Local phytogeographical element and sub-element.* **CTW[all+ext].** Widely distributed in the western woodlands, from north almost to south (not recorded south of the Maji Mountain), extending to the Rift Valley and to the east of the Rift.

6.27.2. **Ziziphus mauritiana** *Lam. (1789).*

FTEA, Rhamnaceae: 29 (1972); FEE 3: 396 (1989); *BS* 51:269 (1998); FS 2: 156 (1999); PSS: 201 (2015).
Taxonomy: No infraspecific taxa.
Description: Shrub or tree to 15 m tall. Thorns (stipular spines) present, rarely absent. Leaves simple, deciduous. Mesophyll. Bark smooth to reticulately fissured when thickened, probably an adaptation for fire resistance.
Habitat: 'Riverine thickets and riverbanks, *Acacia* woodland on alluvial soil, *Acacia* wooded grassland' (FEE). 'Bushland and woodland, often along rivers and wadis' (FS). '*Acacia* woodland, termite mounds in wooded grassland, riverbanks' (PSS). In a range of woodlands and bushlands, including *Combretum-Terminalia* woodland (our observation).
Distribution in Ethiopia: IL GG SD BA HA (FEE). Also recorded from the GJ and WG floristic regions (Fig. 6-109).
Altitudinal range in Ethiopia: 400-1600 m a.s.l. (FEE). 250-1600 m a.s.l. (our records).
General distribution: From Senegal to Ethiopia and Somalia, south to Tanzania; widely cultivated elsewhere and possibly indigenous in parts of Asia.
Chorological classification: Distribution according to direct observation. **WWc, WWs, EWW.** *Local phytogeographical element and sub-element.* **CTW[c+s+ext].** Widely distributed in the central and southern part of the western woodlands, mainly south of the Abay River; extension to the Rift Valley and to the east of the Rift; a few records from *Acacia-Commiphora bushland* (ACB) and *Semi-desert scrubland* (SDS), probably along wadis.

6.27.3. **Ziziphus mucronata** *Willd. (1809).*

FTEA, Rhamnaceae: 25 (1972); FEE 3: 393 (1989); FS 2: 155 (1999); PSS: 201 (2015).
Taxonomy: The species contains two infraspecific taxa, *Ziziphus mucronata* subsp. *mucronata*, and *Ziziphus mucronata* subsp. *rhodesica* Drummond (1965), which occurs from Zambia and southwards in southern Africa. The Ethiopian material seems homogenous. Recognition of infraspecific taxa not relevant here.
Description: Shrub or tree to 15 (-20) m tall. Thorns (stipular spines) present, rarely absent. Leaves simple, deciduous. Microphyll. Bark smooth to rough corrugated, probably an adaptation for fire resistance.
Habitat: '*Acacia-Terminalia, Acacia-Balanites* and *Anogeissus-Boswellia* woodland and bushland, *Acacia* woodland on alluvial soil' (FEE). 'Woodland and bushland, often along rivers and wadis' (FS). 'Bushland and woodland, dry riverine forest' (PSS). In a range of woodlands and bushlands, including *Combretum-Terminalia* woodland (our observation).
Distribution in Ethiopia: GD GJ WU SU AR IL KF GG SD BA HA (FEE). Also recorded from the WG (sight record by Mindaye T. & Abeje E.; may need confirmation) and AF floristic regions; the FEE record from GD has not been confirmed here (Fig. 6-110).
Altitudinal range in Ethiopia: 100-1950 (-2100) m a.s.l. (FEE, including range in Eritrea). 550-1900 m a.s.l. (our records).
General distribution (species as a whole): From Senegal

to Ethiopia, Eritrea and Somalia, south to South Africa; also in Madagascar and Arabia.
Chorological classification: Distribution according to direct observation. [WWn], WWc, WWs, EWW. *Local phytogeographical element and sub-element.* CTW[c+s+ext]: Widely distributed in the central and southern part of the western woodlands, mainly at or south of the Abay River; with extension to the Rift Valley and to the east of the Rift; also records from *Acacia-Commiphora bushland* (ACB).

6.27.4. Ziziphus pubescens *Oliv. (1887).*

FTEA, Rhamnaceae: 24 (1972); FEE 3: 393 (1989); FTNA: 180 (1992); *BS* 51: 269 (1998); PSS: 201 (2015).
Taxonomy: No infraspecific taxa.
Description: Shrub, occasionally scrambling, but growing to a tree to 15 m tall. No thorns or stipular spines. Leaves simple, deciduous. Microphyll. Bark rough and reticulate fissured when thickened, probably an adaptation for fire resistance.
Habitat: 'Lowland dry semi-evergreen forest, groundwater forest, riverine forest' (FEE). 'Dry peripheral semi-evergreen Guineo-Congolian forest, ground water forest and riverine forest' (FTNA). 'Dry forest including riverine fringes, woodland' (PSS). Occurring within the area of the *Combretum-Terminalia* woodland, but within that mostly associated with lowland dry forest and riverine forest (our observation).
Distribution in Ethiopia: IL GG (FEE). No additional records (Fig. 6-111).
Altitudinal range in Ethiopia: 400-650 m a.s.l. (FEE). 400-600 m a.s.l. (FTNA, including the range in the entire Horn of Africa). 450-700 m a.s.l. (our records).
General distribution: South Sudan, Ethiopia, south to Angola and Zimbabwe.
Chorological classification: Distribution according to direct observation. WWc, WWs. *Local phytogeographical element and sub-element.* CTW[c+s]. Widely distributed in the southern part of the western woodlands, entirely south of the Abay River, but not a typical species of the western woodlands, occurring also in the lowermost zone of the *Transitional rain forest* (TRF) and very prominent in *Riparian vegetation* (RV); no eastward extension.

6.27.5. Ziziphus spina-christi *(L.) Desf. (1798).*

FTEA, Rhamnaceae: 24 (1972); FEE 3: 395 (1989); FS 2: 155 (1999); PSS: 201 (2015).
Taxonomy: The widespread taxon is *Ziziphus spina-christi* var. *spina-christi*. *Ziziphus spina-christi* var. *microphylla* A. Rich. (1847) was described from northern Ethiopia, but no infraspecific taxa maintained in FEE and PSS. Recognition of infraspecific taxa not relevant here.
Description: Shrub or tree to 15 m tall. Thorns (short stipular spines) present, rarely absent. Leaves simple, deciduous. Microphyll. Bark smooth, apparently no adaptation for fire resistance.
Habitat: 'Wooded grassland on limestone slopes, *Acacia* bushland on alluvial soils, in or along dry riverbeds, edges of cultivations and gardens' (FEE). 'Woodland or bushland, often along wadis' (FS). 'Woodland and bushland, rocky hillslopes' (PSS). In the drier parts of the *Combretum-Terminalia* woodland and semi-evergreen bushland (our observation).
Distribution in Ethiopia: TU GD WU AF SU GG BA HA (FEE). Also recorded from the GJ, WG and IL floristic regions (Fig. 6-112).
Altitudinal range in Ethiopia: Sea level to 1900 (-2400) m a.s.l. (FEE, including range in Eritrea). 600-1850 m a.s.l. (our records).
General distribution: From Mauritania to Egypt, the Sudan, Ethiopia, Eritrea and Somalia, south to Tanzania; also in Arabia and Asia to Pakistan. The species is widespread in Ethiopia; in FEE it is suggested that at least part of the distribution is due to dispersal by humans, plant grow from fruits eaten and stones discarded by both local inhabitants and travellers. Some populations along wadis and riverbeds in Ethiopia are supposed to represent indigenous populations.
Chorological classification: Distribution according to direct observation. WWn, WWc, WWs, EWW. *Local phytogeographical element and sub-element.* CTW[all+ext]. Widely distributed in the western woodlands, from north to south or almost so (not represented by records from the Gambela region), with extension to the Rift Valley and to the east of the Rift.

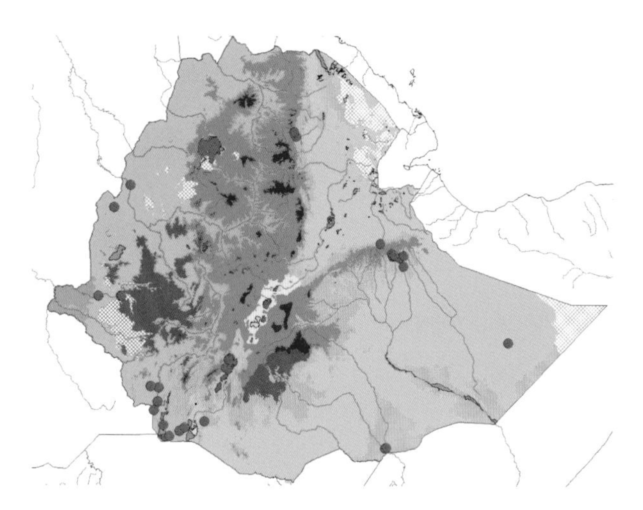

Fig. 6-109. *Ziziphus mauritiana*. Distribution in Ethiopia. Observed records are marked with red dots on vegetation map reproduced from Fig. 4-1.

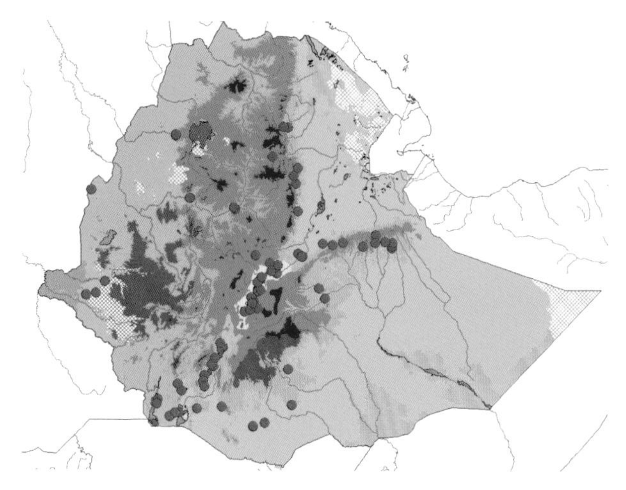

Fig. 6-110. *Ziziphus mucronata*. Distribution in Ethiopia. Observed records are marked with red dots on vegetation map reproduced from Fig. 4-1.

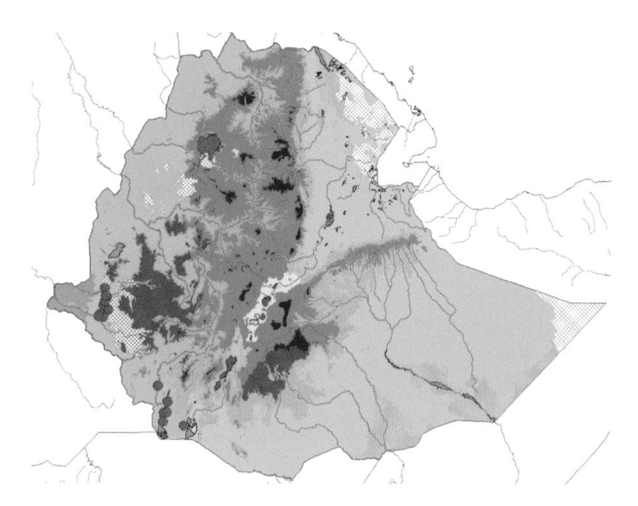

Fig. 6-111. *Ziziphus pubescens*. Distribution in Ethiopia. Observed records are marked with red dots on vegetation map reproduced from Fig. 4-1.

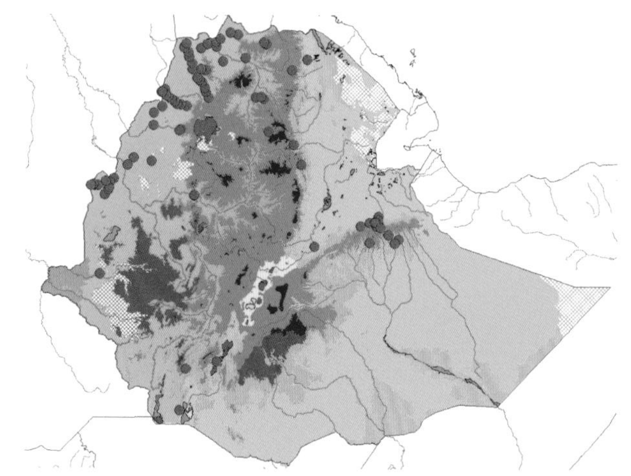

Fig. 6-112. *Ziziphus spina-christi*. Distribution in Ethiopia. Observed records are marked with red dots on vegetation map reproduced from Fig. 4-1.

6.28. Rutaceae

FTEA, Rutaceae: 1-52 (1982); FEE 3: 419-432 (1989); *BS* 51: 275-278 (1998); FS 2: 171-182 (1999); PSS: 246-247 (2015).
In FS and PSS the genus **Harrisonia** has been transferred to Rutaceae.

6.*28.1.* **Clausena anisata** *(Willd.) Benth. (1849).*

FTEA, Rutaceae: 49 (1982); FEE 3: 429 (1989); FTNA: 180 (1992); *BS* 51: 275 (1998); PSS: 246 (2015).
Taxonomy: Several varieties have been recognized in this species, apart from the widespread *Clausena anisata* var. *anisata,* for example from central and southern Africa, *C. anisata* var. *mollis* Engl. (1895), *C. anisata* var. *multijuga* Welw. ex Hiern (1896) and *C. anisata* var. *paucijuga* (Kurz) Milono (1996), but the recognition of these taxa is not general. No infraspecific taxa are recognized in FEE and *BS* 51. Recognition of infraspecific taxa not relevant here.
Description: Shrub or small tree to 6 (-10) m tall. No thorns. Leaves compound (pinnate), deciduous. Mesophyll (leaflets microphyll). Bark smooth, probably no adaptation for fire resistance.
Habitat: 'Montane forest margins, sometimes forming dense thickets or understory trees in tall moist forest, also common in secondary bushland' (FEE). 'Afromontane rain forest, undifferentiated Afromontane forest ... and dry single-dominant Afromontane forest ..., particularly along edges and in clearings; also in secondary montane evergreen bushland' (FTNA). 'Montane forest margins' (PSS for *Clausena anisata* var. *anisata*). A species mainly associated with margins of *Dry and Moist evergreen Afromontane forest* (DAF and MAF),Afromontane forest margins and montane evergreen bushland, but a few records are from *Combretum-Terminalia* woodland (our observation).
Distribution in Ethiopia: TU GD GJ SU WG IL KF SD (FEE). The FEE record from TU has not been confirmed (Fig. 6-113).
Altitudinal range in Ethiopia: 1500-2300 m a.s.l. (FEE & FTNA, including the range in the entire Horn of Africa). 1400-2300 m a.s.l. (our records).
General distribution (species as a whole): From Sierra Leone and Guinean Republic to Ethiopia, south to South Africa.
Chorological classification: Distribution according to direct observation. **WWn, WWc, WWs, EWW**. *Local phytogeographical element.* **MI**. Marginally intruding species in the western woodlands, with main distribution in the highlands and in *Transitional semi-evergreen bushland* (TSEB); only a few records in our relevés.

6.*28.2.* **Zanthoxylum gilletii** *(De Wild.) P.G. Waterman (1975).*

FTEA, Rutaceae: 38 (1982); *BS* 51: 278 (1998); FEE 1: 228 (2009); PSS: 247 (2015).
Taxonomy: No infraspecific taxa.
Description: Tree to 10(-35) m tall. Branches and stems densely beset with straight woody thorns up to 1-3 cm long. Leaves compound (pinnate), deciduous. Megaphyll (leaflets Macrophyll). Bark grey and smooth, probably no adaptation for fire resistance.
Habitat: 'Riverine forest' (FEE). 'Lowland and mid-altitude forest' (PSS). In Tropical East Africa this species occurs in 'Rain-forest' (FTEA, Rutaceae). Recorded from riverine forest, farmland and *Combretum-Terminalia* woodland (our observation).
Distribution in Ethiopia: WG (FEE). No additional records (Fig. 6-114).
Altitudinal range in Ethiopia: Ca. 1350 m a.s.l. (FEE). 1350-1400 m a.s.l. (our records).
General distribution: From Sierra Leone to Ethiopia, south to Zimbabwe.
Chorological classification: Distribution according to direct observation. **WWc**. *Local phytogeographical element and sub-element.* **CTW[c]**. Distributed in the central part of the *Combretum-Terminalia* woodlands. The species occurs in *Riparian vegetation* (RV) and as single trees in farmland that was formerly woodland around the upper Dabus River. From its distribution outside Ethiopia, one should expect the species to occur in *Transitional rain forest* (TRF), but no record has yet been made from that vegetation type; no eastward extension.

6.29. Balanitaceae

FEE 3: 433-436 (1989); *BS* 51: 102-103 (1998); FS 2: 168-171 (1999); FTEA, Balanitaceae: 1-15 (2003); PSS: 159 (2015), as part of "Zygophyllaceae".

6.*29.1.* Balanites aegyptiaca *(L.) Delile (1813).*

FEE 3: 433 (1989); *BS* 51: 102 (1998); FS 2: 169 (1999); FTEA, Balanitaceae: 6 (2003); PSS: 159 (2015).
Balanites aegyptiaca var. **aegyptiaca**
FEE 3: 433 (1989); FS 2: 169 (1999); FTEA, Balanitaceae: 6 (2003); PSS: 159 (2015).
Balanites aegyptiaca var. **pallida** Sands (1983).
FEE 3: 434 (1989); FS 2: 169 (1999); FTEA, Balanitaceae: 6 (2003).
Taxonomy: No infraspecific taxa in *BS* 51; in FEE two varieties, *Balanites aegyptiaca* var. *aegyptiaca* and *Balanites aegyptiaca* var. *pallida*. The latter mainly in Somalia, and in FEE with only two collections from the Awash Valley and Afar. The characters used to separate the two varieties relate to the density of the indumentum and qualitative characters. Recognition of infraspecific taxa not relevant here.
Description: Tree to 12 (-15) m tall. Usually spiny, with 2-8 cm long, stout thorns, attached 0.5-1 cm above the leaves or leaf-scars. Leaves compound (2-foliolate), deciduous. Mesophyll (leaflets Mesophyll). Bark on older trunks corky, deeply fissured, apparently an adaptation for fire resistance.
Habitat: 'Dry savanna or *Acacia* woodland, rarely in *Terminalia-Combretum* wooded grassland, on a variety of soils' (FEE for *Balanites aegyptiaca* var. *aegyptiaca*), 'open scrub' (FEE for *Balanites aegyptiaca* var. *pallida*). 'Woodland, wooded grassland' (FS for *Balanites aegyptiaca* var. *aegyptiaca*), 'Open *Acacia-Commiphora* bushland' (FS for *Balanites aegyptiaca* var. *pallida*). 'Dry wooded grassland and bushland, wadis' (PSS for *Balanites aegyptiaca* var. *aegyptiaca*). Occurs in the whole of the *Combretum-Terminalia* woodland and semi-evergreen bushland, but also in a range of other dry habitats (our observation).
Distribution in Ethiopia: TU WU SU AR IL GG SD HA (FEE for *Balanites aegyptiaca* var. *aegyptiaca*), AF SU (FEE for *Balanites aegyptiaca* var. *pallida*). Also recorded from the GD, GJ, WG, KF and BA floristic regions (Fig. 6-115).
Altitudinal range in Ethiopia: 700-1800 m a.s.l. (FEE for *Balanites aegyptiaca* var. *aegyptiaca*, including range in Eritrea), 1250 m a.s.l. (FEE for *Balanites aegyptiaca* var. *pallida*). 450-2000 m a.s.l. (our records).
General distribution (species as a whole): From Mauritania and Senegal to Ethiopia, Eritrea and Somalia, south to Zimbabwe, also in Egypt, Israel and Arabia.
Chorological classification: Distribution according to direct observation. **WWn, WWc, WWs, EWW, [ASO].**
Local phytogeographical element and sub-element. **CTW[all+ext].** Widely distributed in the western woodlands, from north to south, with extension to the Rift Valley and to the east of the Rift; also a few records from *Acacia-Commiphora bushland* (ACB), where the species apparently occurs in or near patches of *Transitional semi-evergreen bushland* (TSEB).

6.30. Simaroubaceae

FEE 3: 437-441 (1989); *BS* 51: 278-279 (1998); FS 2: 171-182 (1999); FTEA, Simaroubaceae: 1-15 (2000); PSS: 247 (2015).
In FS, the genus **Harrisonia** has been moved to Rutaceae, and in PSS to Sapindaceae.

6.*30.1.* Brucea antidysenterica *J.F. Mill. (1779).*

FEE 3: 440 (1989); FTNA: 196 (1992); *BS* 51: 278 (1998); FTEA, Simaroubaceae: 7 (2000); PSS: 247 (2015).
Taxonomy: No infraspecific taxa.
Description: Shrub or small tree up to 5 (-10) m tall. No thorns. Leaves compound (pinnate), presumably evergreen. Macrophyll (leaflets Mesophyll). Bark smooth, apparently no adaptation for fire resistance.
Habitat: 'Montane forest, evergreen forest, forest margins, secondary growth in deforested area, montane grassland' (FEE). 'Afromontane rain forest and undifferentiated Afromontane forest, often at margins, also in secondary montane evergreen bushland and montane grassland' (FTNA). 'Upland forest, evergreen

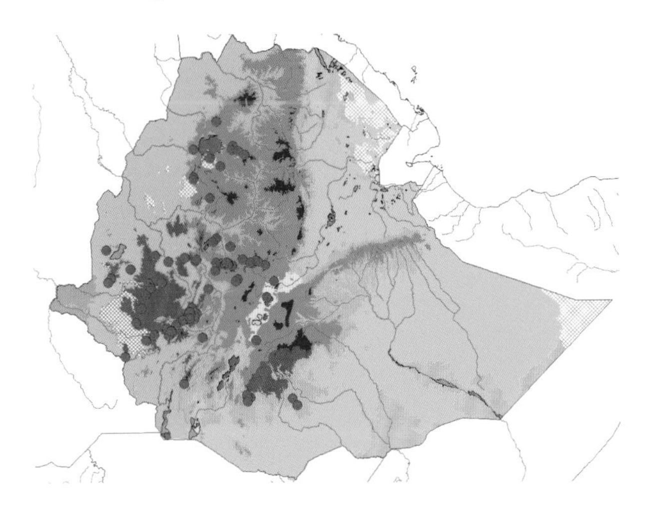

Fig. 6-113. *Clausena anisata*. Distribution in Ethiopia. Observed records are marked with red dots on vegetation map reproduced from Fig. 4-1.

Fig. 6-114. *Zanthoxylum gilletii*. Distribution in Ethiopia. Observed records are marked with red dots on vegetation map reproduced from Fig. 4-1.

Fig. 6-115. *Balanites aegyptiaca*. Distribution in Ethiopia. Observed records are marked with red dots on vegetation map reproduced from Fig. 4-1.

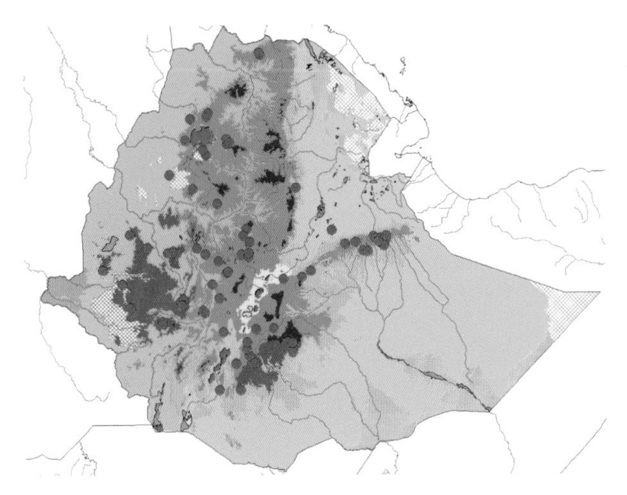

Fig. 6-116. *Brucea antidysenterica*. Distribution in Ethiopia. Observed records are marked with red dots on vegetation map reproduced from Fig. 4-1.

bushland' (PSS). Occurring in *Dry* and *Moist evergreen Afromontane forest* (DAF and MAF), particularly along forest margins, and in various types of bushland, but also in localities in *Combretum-Terminalia* woodland (our observation).
Distribution in Ethiopia: TU GD GJ? WG SU AR IL KF SD BA HA (FEE). The FEE record is confirmed for the GJ floristic region (Fig. 6-116).
Altitudinal range in Ethiopia: 1650-2800 m a.s.l. (FEE, including range in Eritrea). (1000-) 1650-2900 m a.s.l. (FTNA, including the range in the entire Horn of Africa). 1000-2750 m a.s.l. (our records).
General distribution: From Guinean Republic to Ethiopia and Eritrea, south to Angola and Zambia; also in Arabia.
Chorological classification: Distribution according to direct observation. **WWn, WWc, [WWs], EWW.** *Local phytogeographical element.* **MI.** Marginally intruding species in the western woodlands, with main distribution in the highlands and in *Transitional semi-evergreen bushland* (TSEB); only one or few records in our relevés.

6.30.2. **Harrisonia abyssinica** *Oliv. (1868).*

FEE 3: 437 (1989); *BS* 51: 279 (1998); FS 2: 173 (1999); FTEA, Simaroubaceae: 1 (2000); PSS: 247 (2015).
Taxonomy: No infraspecific taxa.
Description: Shrub or small tree to 13 m tall. Bark often with scattered thorns or wart-like outgrowths, mostly paired at the base of the leaves or leaf-scars. Leaves compound (pinnate), presumably evergreen. Macrophyll (leaflets Mesophyll). Bark smooth, apparently no adaptation for fire resistance.
Habitat: 'Dry evergreen forest, forest edges, also wooded grassland, thickets and riverine vegetation' (FEE). 'Riverine forest and woodland, sometimes forming thickets' (FS). 'Dry evergreen forest, thickets and riverine vegetation' (PSS). In *Combretum-Terminalia* woodland, mainly on termite mounds, but also in semi-evergreen bushland (our observation).
Distribution in Ethiopia: SU/KF IL GG SD BA (FEE). The FEE record SU/KF is to be corrected to KF; from near the border between the SU and the KF floristic regions, we have seen *Friis et al.* 14288 (C, ETH, K) from the KF side, another record from the Gibe River is *KL Brehme in HF Mooney* 9080 (K, ETH), which is from "S side of bridge", again on the KF side. We have not seen records from the SU side (Fig. 6-117).
Altitudinal range in Ethiopia: Sea level to 1700 m a.s.l. (FEE). 600-1750 m a.s.l. (our records).
General distribution: From Guinean Republic to Cameroon, from Sudan and South Sudan to Ethiopia and Somalia, south to Mozambique.
Chorological classification: Distribution according to direct observation. **WWc, WWs, EWW.** *Local phytogeographical element and sub-element.* **CTW[c+s+ext].** Widely distributed in the southern part of the western woodlands, mainly south of the Abay River, common in the lower Omo Valley and extending to the southern part of the Rift Valley and to the east of the Rift.

6.31. Burseraceae

FEE 3: 442-478 (1989); FTEA, Burseraceae: 1-95 (1991); *BS* 51: 280-281 (1998); FS 2: 183-228 (1999); PSS: 242 (2025).

6.31.1. **Boswellia papyrifera** *(Delile) Hochst. (1843).*

FEE 3: 443 (1989); FTEA, Burseraceae: 5 (1991); *BS* 51: 280 (1998); PSS: 242 (2015).
Taxonomy: No infraspecific taxa.
Description: Tree to 10 (-12) m tall. No thorns. Leaves compound (imparipinnate), deciduous (flowering when leafless). Macrophyll (leaflets Mesophyll). Bark smooth, peeling in large, thin, papery flakes or scrolls, but this is apparently no adaptation for fire resistance (we have observed in the field that the papery flakes or scrolls burn easily).
Habitat: 'Dry *Acacia-Commiphora* woodland and wooded grassland, *Pterocarpus* woodland, often dominant on steep, rocky slopes, also on lava flows' (FEE). 'Dry woodland and wooded grassland' (PSS). In the dry northern and north-western parts of the *Combretum-Terminalia* woodlands (our observation).
Distribution in Ethiopia: TU GD GJ SU (FEE). Also recorded from the WG floristic region. The record

from the KF-side of the Gibe Gorge (*Aman Dekebo* D4 (ETH), Abelti) is probably based on a misidentification of a specimen of *B. pirottae*. (Fig. 6-118).
Altitudinal range in Ethiopia: 950-1800 m a.s.l. (FEE, including range in Eritrea). 750-1600 m a.s.l. (our records).
General distribution: From Nigeria to Ethiopia and Eritrea, south to Uganda.
Chorological classification: Distribution according to direct observation. **WWn, WWc**. *Local phytogeographical element and sub-element.* CTW[n+c]. Widely distributed in the northern part of the western woodlands, mainly north of the Abay River. The distribution is unusual because here a northern species extends to the Rift Valley; it does not extend to the east of the Rift.

6.*31.2*. Boswellia pirottae *Chiov. (1911).*

FEE 3: 443 (1989).
Taxonomy: No infraspecific taxa.
Description: Tree to 10 m tall. No thorns. Leaves compound (imparipinnate), deciduous (flowering when leafless). Macrophyll (leaflets Mesophyll). Bark rugose to reticulately fissured when thickened, not peeling, probably an adaptation for fire resistance.
Habitat: '*Commiphora-Boswellia, Combretum* and *Acacia-Lannea* woodland on steep, rocky slopes' (FEE). [No information from the Sudan, as the species does not occur in that country according to PSS]. Mainly in the northern part of the *Combretum-Terminalia* woodland in Ethiopia (our observation).
Distribution in Ethiopia: GD GJ WU SU KF (FEE). Also recorded from the WG floristic region. The FEE record from the KF side of the Gibe Gorge is based on MG Gilbert et al. 8349 (ETH, K) from near Abelti, which seems correct (Fig. 6-119).
Altitudinal range in Ethiopia: 1200-1800 m a.s.l. (FEE). 750-1750 m a.s.l. (our records).
General distribution: Restricted to Ethiopia.
Chorological classification: Distribution according to direct observation. **WWn, WWc**. *Local phytogeographical element and sub-element.* CTW[n+c]. Widely distributed in the northern and central part of the western woodlands; no eastward extension.

6.*31.3*. Commiphora africana *(A. Rich.) Engl. (1883).*

FEE 3: 454 (1989); FTEA, Burseraceae: 44 (1991); *BS* 51: 280 (1998); FS 2: 202 (1999); PSS: 242 (2015).
Commiphora africana var. **africana**
FEE 3: 454 (1989); FTEA, Burseraceae: 46 (1991); *BS* 51: 280 (1998); PSS: 242 (2015).
Commiphora africana var. **rubriflora** (Engl.) Wild (1959).
FEE 3: 457 (1989); FTEA, Burseraceae: 47 (1991); *BS* 51: 280 (1998).
Taxonomy: Two infraspecific taxa recorded in FEE, *Commiphora africana* var. *africana* and *Commiphora africana* var. *rubriflora*, which also occurs in East Africa south to Zimbabwe, but it is specifically stated that the ecology of the two varieties is identical. Recognition of infraspecific taxa not relevant here.
Description: Shrub or small tree to 10 m tall. Trunk and branches mostly beset with stiff thorns (spiny branchlets). Leaves compound (3-foliolate), deciduous. Mesophyll (leaflets Mesophyll). Bark peeling and regenerating quickly, probably an adaptation for fire resistance.
Habitat: '*Acacia*, *Acacia-Commiphora*, *Commiphora-Boswellia*, *Combretum-Terminalia* woodland, wooded grassland and bushland, often on rocky slopes in areas with basement rocks but also on level sandy to loamy soil and recent lava flows' (FEE for both varieties). '*Acacia-Commiphora* bushland or woodland, also on coastal dunes' (FS). 'Bushland and wooded grassland' (PSS). In *Acacia-Commiphora* bushland, only a few records from *Combretum-Terminalia* woodland (our observation).
Distribution in Ethiopia: TU GD AF WU SU KF GG SD BA HA (FEE for *Commiphora africana* var. *africana*), GG SD (FEE for *Commiphora africana* var. *rubriflora*). No additional records (Fig. 6-120).
Altitudinal range in Ethiopia: (150-) 500-1900 (-2100) m a.s.l. (FEE for *Commiphora africana* var. *africana*, including range in Eritrea), 700-1900 m a.s.l. (FEE for *Commiphora africana* var. *rubriflora*). 550-1650 m a.s.l. (our records).
General distribution (species as a whole) From Mauri-

Fig. 6-117. *Harrisonia abyssinica*. Distribution in Ethiopia. Observed records are marked with red dots on vegetation map reproduced from Fig. 4-1.

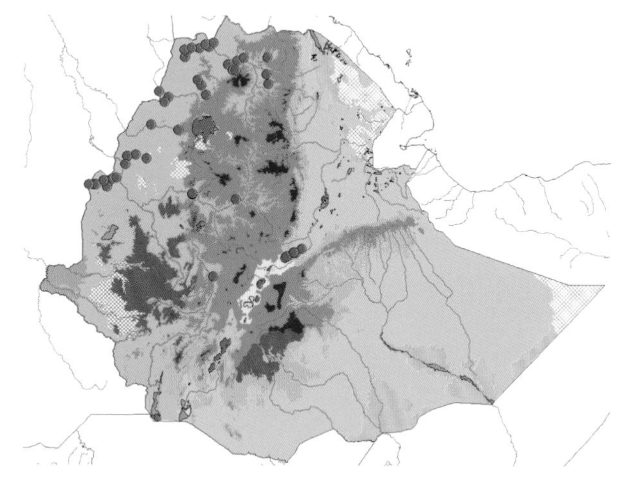

Fig. 6-118. *Boswellia papyrifera*. Distribution in Ethiopia. Observed records are marked with red dots on vegetation map reproduced from Fig. 4-1.

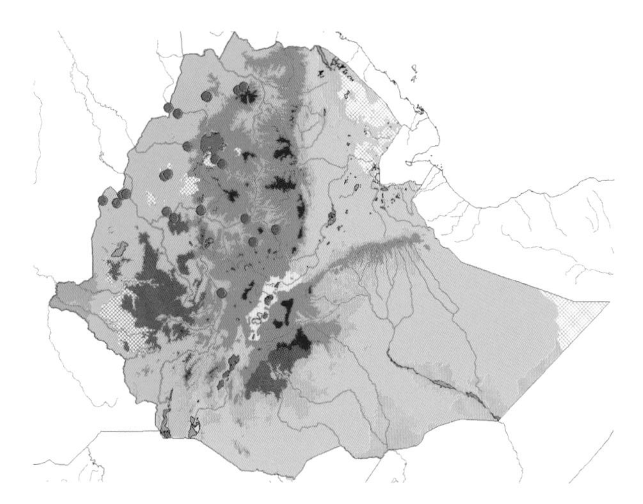

Fig. 6-119. *Boswellia pirottae*. Distribution in Ethiopia. Observed records are marked with red dots on vegetation map reproduced from Fig. 4-1.

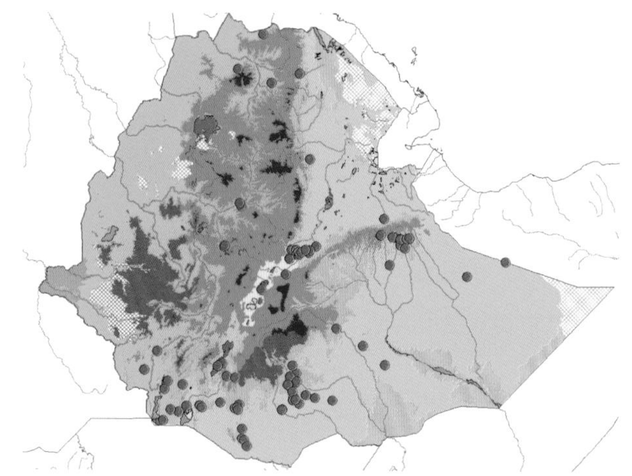

Fig. 6-120. *Commiphora africana*. Distribution in Ethiopia. Observed records are marked with red dots on vegetation map reproduced from Fig. 4-1.

tania and Senegal to Ethiopia, Eritrea and Somalia, south to Angola and South Africa.
Chorological classification: Distribution according to direct observation. **WWn, WWs, ASO**. *Local phytogeographical element and sub-element.* **ACB[-CTW]**. Widely distributed in the *Acacia-Commiphora bushland* (ACB), marginally intruding in the *Combretum-Terminalia woodland and wooded grassland* (CTW) in the north and in the south.

6.*31.4.* Commiphora pedunculata *(Kotschy & Peyr.) Engl. (1883).*

FEE 3: 457 (1989); FTEA, Burseraceae: 73 (1991); PSS: 242 (2025)
Taxonomy: No infraspecific taxa.
Description: Shrub or small tree to 5 m tall. Trunk and branches sometimes with stiff thorns (spiny branchlets). Leaves compound (3-foliolate), deciduous. Mesophyll (leaflets Mesophyll). Bark peeling and regenerating quickly, probably an adaptation for fire resistance.
Habitat: '*Combretum-Terminalia* woodland and wooded grassland' (FEE). 'Woodland and wooded grassland' (PSS). In *Combretum-Terminalia* woodland (our observation).
Distribution in Ethiopia:?GD?GJ?WG (FEE, no confirmed collection from Ethiopia had been made then). Here confirmed records from the GJ and WG floristic regions. Records from the GD floristic provinces, with its relatively dry habitats, are less likely (Fig. 6-121).
Altitudinal range in Ethiopia: Up to ca. 1000 m a.s.l. (FEE). 650-950 m a.s.l. (our records).
General distribution: From Senegal and Mali to Ethiopia, Tanzania, Malawi and Zambia.
Chorological classification: Distribution according to direct observation. **WWc**. *Local phytogeographical element and sub-element.* **CTW[c]**. Distributed in the central part of the western woodlands south of the Abay River; no eastern extension.

6.32. Meliaceae

FEE 3: 479-489 (1989); FTEA, Meliaceae: 1-68 (1991); *BS* 51: 281-284 (1998); PSS: 247-249 (2015).

6.*32.1.* Pseudocedrela kotschyi *(Schweinf.) Harms (1895).*

FEE 3: 479 (1989); FTEA, Meliaceae: 56 (1991); *BS* 51: 283 (1998); PSS: 248 (2015).
Taxonomy: No infraspecific taxa.
Description: Tree to 12 (-18) m tall. No thorns. Leaves compound (pinnate), deciduous. Macrophyll (leaflets Mesophyll). Bark thick, rough, deeply and regularly fissured, probably adaptation for fire resistance.
Habitat: 'Combretaceous wooded grassland' (FEE). 'Woodland and wooded grassland' (PSS). In Ethiopia restricted to *Combretum-Terminalia* woodland (our observation).
Distribution in Ethiopia: GD IL (FEE). Also recorded from the GJ, WG, KF and GG floristic regions (Fig. 6-122).
Altitudinal range in Ethiopia: Ca. 500 m a.s.l. (FEE). 500-1200 m a.s.l. (our records).
General distribution: From Senegal to Ethiopia, south to Uganda.
Chorological classification: Distribution according to direct observation. **WWn, WWc, WWs**. *Local phytogeographical element and sub-element.* **CTW[all]**. Widely distributed in the western woodlands, from north to south, apart from the extreme north and the extreme south; no eastern extension.

6.33. Sapindaceae

FEE 3: 490-510 (1989); FTEA, Sapindaceae: 1-109 (1998); *BS* 51: 285-288 (1998); FS 2: 239-253 (1999); PSS: 244-246 (2015).

6.*33.1.* Allophylus rubifolius *(Hochst. ex A. Rich.) Engl. (1892).*

FEE 3: 502 (1989); FTEA, Sapindaceae: 88 (1998); *BS* 51: 285 (1998); FS 2: 243 (1999); PSS: 245 (2015).

Taxonomy: Friis et al. (1996) established several infraspecific taxa of this species outside the present study area, where the only infraspecific taxon is *Allophylus rubifolius* var. *rubifolius*. Recognition of infraspecific taxa not relevant here.
Description: Shrub or medium sized tree to 7 (-12) m tall. No thorns (spiny branchlets). Leaves compound (trifoliolate), presumably deciduous. Macrophyll (leaflets Mesophyll). Bark smooth or very slightly rough, probably no adaptation for fire resistance.
Habitat: '*Acacia-Commiphora, Acacia-Combretum* and *Combretum-Terminalia* woodland and bushland, alluvial *Acacia* woodland, riverine forest, termite mounds' (FEE). 'Edges of *Juniperus* forest, evergreen bushland, often near rocks' (FS for *Allophylus rubifolius* var. *rubifolius*). 'Woodland, bushland, grassland and riverine forest' (PSS for *Allophylus rubifolius* var. *rubifolius*). Occurs in a range of woodlands and bushland, along forest margins, mainly *Dry evergreen Afromontane forest* (DAF), and in riverine forest (our observation).
Distribution in Ethiopia: TU GD GJ SU IL GG SD BA HA (FEE). Also recorded from the AF, WG and KF floristic regions (Fig. 6-123).
Altitudinal range in Ethiopia: 400-1800 (-2000) m a.s.l. (FEE, including range in Eritrea). 500-1850 (-2000) m a.s.l. (our records).
General distribution: (of species as a whole): Sudan and South Sudan to Ethiopia, Eritrea and Somalia, south to South Africa; also in Arabia and on Socotra.
Chorological classification: Distribution according to direct observation. **WWn, WWc, WWs, EWW**. *Local phytogeographical element and sub-element.* **CTW[all+ext]**. Widely distributed in the western woodlands, from north to south, extending to the Rift Valley and to the east of the Rift. One record, *Mooney* 9003 (ETH), from Woliso (SU floristic region) extends to the edge of the *Dry evergreen Afromontane forest* (DAF).

6.33.2. Dodonaea angustifolia *L. f. (1781).*

FTNA: 190 (1992); FEE 3: 491 (1989).
Dodonaea viscosa (L.) Jacq. (1760).
FTEA, Sapindaceae: 8 (1998); PSS: 245 (2015).
Dodonaea viscosa var. **angustifolia** (L. f.) Benth. (1863).
FS 2: 240 (1999).
Taxonomy: The species concept in the works cited above change between considering *Dodonaea viscosa* a widely conceived species including both costal plants (*Dodonaea viscosa* in the strict sense) and inland forms (*Dodonaea angustifolia*) or making a distinction on the species level between the two. According to FEE, there is only the latter form in Ethiopia. Recognition of infraspecific taxa not relevant here.
Description: Small tree to 8 m tall. No thorns. Leaves simple, evergreen. Microphyll. Bark smooth, probably no adaptation for fire resistance.
Habitat: 'Edges of upland forest, upland bushland and grassland, secondary forest and scrub, invading recently cleared areas of forest and invading overgrazed *Acacia-Commiphora* bushland, also in cultivated areas' (FEE). 'Edges of dry montane forest, upland evergreen bushland and scrub, invading *Acacia-Commiphora* bushland' (FS). 'Undifferentiated Afromontane forest ... and single-dominant Afromontane forest ..., mainly along edges and in clearings; also in upland and montane evergreen bushland and secondary growth in abandoned cultivations in forested areas; also sometimes transgressing into deciduous bushland' (FTNA). 'Secondary bushland, grassland and forest margins' (PSS for *Dodonaea viscosa*). In semi-evergreen bushland and in bushland in the lower zone of the highland vegetation, but also in a few localities in *Combretum-Terminalia* woodland, mainly on termite mounds (our observation).
Distribution in Ethiopia: TU GD GJ WU SU AR WG KF GG SD BA HA (FEE). Also recorded from the AF floristic region (Fig. 6-124).
Altitudinal range in Ethiopia: (500-) 1000-2600 (-2900) m a.s.l. (FEE, including range in Eritrea). 800-2650 m a.s.l. (FTNA, including the range in the entire Horn of Africa). 1450-2650 m a.s.l. (our records).
General distribution: Pantropical and subtropical.
Chorological classification: Distribution according to direct observation. **WWn, WWc, WWs, EWW**. *Local phytogeographical element.* **MI**. Marginally intruding species

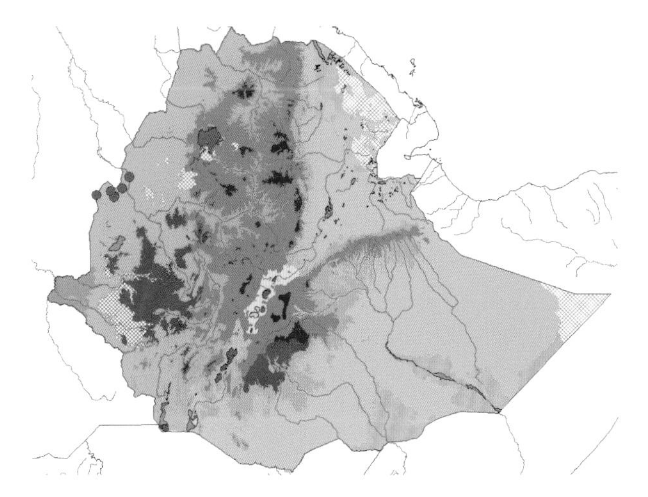

Fig. 6-121. *Commiphora pedunculata*. Distribution in Ethiopia. Observed records are marked with red dots on vegetation map reproduced from Fig. 4-1.

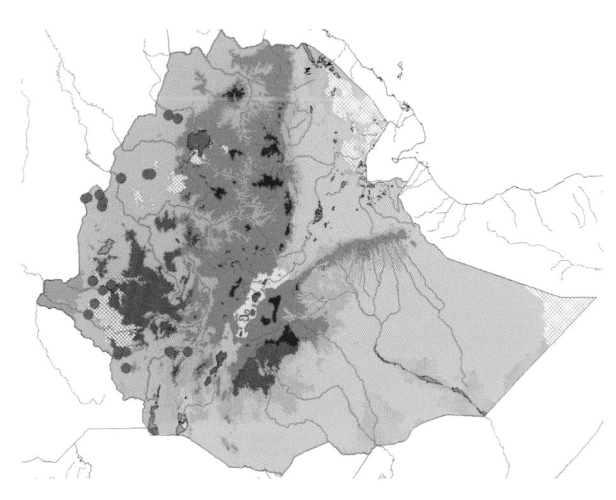

Fig. 6-122. *Pseudocedrela kotschyi*. Distribution in Ethiopia. Observed records are marked with red dots on vegetation map reproduced from Fig. 4-1.

Fig. 6-123. *Allophylus rubifolius*. Distribution in Ethiopia. Observed records are marked with red dots on vegetation map reproduced from Fig. 4-1.

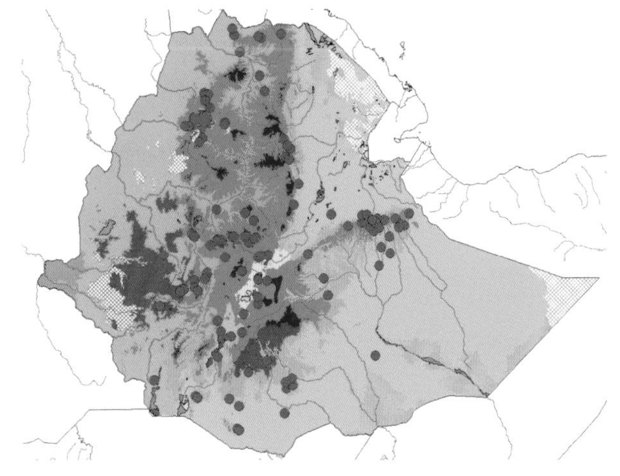

Fig. 6-124. *Dodonaea angustifolia*. Distribution in Ethiopia. Observed records are marked with red dots on vegetation map reproduced from Fig. 4-1.

in the western woodlands, with main distribution in the highlands, the records from inside the *Acacia-Commiphora bushland* (ACB) are from small areas on hills with *Transitional semi-evergreen bushland* (TSEB); only a few records in our relevés.

6.34. Melianthaceae

FTEA, Melianthaceae: 1-7 (1958); FEE 3: 511-512 (1989); *BS* 51: 288-289 (1998); PSS: 234 (2015).
The genus **Bersama** has recently been transferred to the family Francoaceae but is maintained in the Melianthaceae in PSS.

6.*34.1.* **Bersama abyssinica** *Fresen. (1837).*

FTEA, Melianthaceae: 2 (1958); FTNA: 202 (1992); FEE 3: 511 (1989); *BS* 51: 288 (1998); PSS: 234 (2015).
Taxonomy: Many infraspecific taxa have been proposed in this species, but according to FEE only *Bersama abyssinica* subsp. *abyssinica* is known from Ethiopia. Recognition of infraspecific taxa not relevant here.
Description: Shrub or small tree to 15 m tall. No thorns. Leaves compound (pinnate), evergreen. Macrophyll (leaflets Mesophyll). Bark on trunk smooth to moderately rough, probably no adaptation for fire resistance.
Habitat: 'Essentially in *Juniperus-Podocarpus* forest and degraded remnants of it, often in thickets and copses in grassland and open woodland, and even cultivations, often on slopes and hills, also in gallery forest and montane scrub' (FEE). 'Afromontane rain forest, undifferentiated Afromontane forest ... and single-dominant Afromontane forest ..., often at forest margins; riverine forest; also in secondary montane evergreen bushland, occasionally a single in derived montane grassland' (FTNA). 'Montane forest and forest margins, secondary bushland' (PSS). Mainly a species of Afromontane forests and forest margins and montane evergreen bushland, but a few records are recorded from *Combretum-Terminalia* woodland (our observation).
Distribution in Ethiopia: TU GD GJ WU SU AR WG IL KF SD BA HA (FEE). No additional records (Fig. 6-125).
Altitudinal range in Ethiopia: 1700-2715 m a.s.l. (FEE, including range in Eritrea). 1250-2800 m a.s.l. (FTNA). 1050-2600 m a.s.l. (our records).
General distribution (species as a whole): From Cameroon to Ethiopia and Eritrea, south to Zimbabwe; also in Arabia.
Chorological classification: Distribution according to direct observation. **WWn, WWc, [WWs], EWW.** *Local phytogeographical element.* **MI**. Marginally intruding species in the western woodlands, with main distribution in the highlands and in *Transitional semi-evergreen bushland* (TSEB), the records from the *Acacia-Commiphora bushland* (ACB) are from small areas on hills with *Transitional semi-evergreen bushland* (TSEB); only a few records of this taxon in our relevés.

6.35. Anacardiaceae

FTEA, Anacardiaceae: 1-59 (1986); FEE 3: 513-532 (1989); *BS* 51: 289-293 (1998); FS 2: 254-267 (1999); PSS: 242-244 (2015).

All potential western woodland species in Ethiopia are accounted for here.

6.*35.1.* **Lannea barteri** *(Oliv.) Engl. (1897).*

FTEA, Anacardiaceae: 21 (1986); FEE 3: 518 (1989); *BS* 51: 289 (1998); PSS: 242 (2015).
Taxonomy: No infraspecific taxa. Closely related to *Lannea schimperi,* from which it can be distinguished by the stiff, erect hairs on the underside of the leaves, as opposed to the pinkish or rusty coloured indumentum of crisped hairs on the underside of the leaves of *L. schimperi*. See Section 4 about the consequences of these difficulties for this study.
Description: A spreading tree to 15 (-18) m tall with a straight and distinct bole. No thorns. Leaves compound (5-9 (-13)-foliolate), deciduous. Macrophyll (leaflets Mesophyll). Bark on trunk smooth or slightly fissures on old trunks, probably no adaptation for fire resistance.
Habitat: 'Lowland woodland on clay soil and along river valleys' (FEE). 'Woodland, wooded grassland,

Fig. 6-125. *Bersama abyssinica*. Distribution in Ethiopia. Observed records are marked with red dots on vegetation map reproduced from Fig. 4-1.

Fig. 6-126. *Lannea barteri*. Distribution in Ethiopia. Observed records are marked with red dots on vegetation map reproduced from Fig. 4-1.

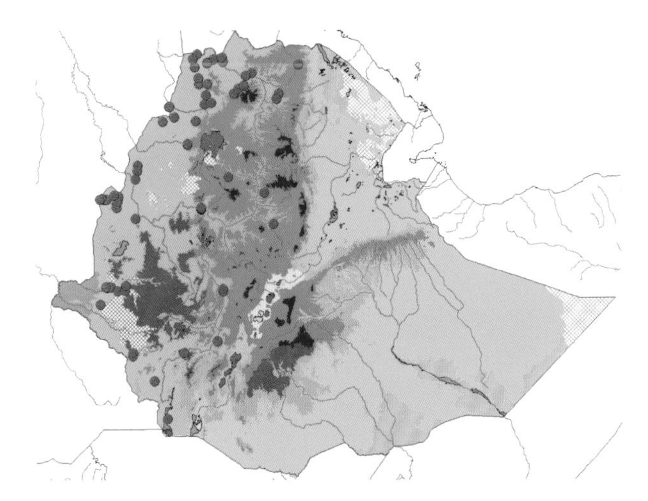

Fig. 6-127. *Lannea fruticosa*. Distribution in Ethiopia. Observed records are marked with red dots on vegetation map reproduced from Fig. 4-1.

Fig. 6-128. *Lannea schimperi*. Distribution in Ethiopia. Observed records are marked with red dots on vegetation map reproduced from Fig. 4-1.

riverine forest margins' (PSS). 'Wooded grassland, forest edges, near rivers' (FTEA, Anacardiaceae). Lowest zone of *Combretum-Terminalia* woodland (our observation).
Distribution in Ethiopia: GD SU IL KF GG (FEE). Also recorded from the GJ and WG floristic regions (Fig. 6-126).
Altitudinal range in Ethiopia: 500-1600 m a.s.l. (FEE). 600-1650 m a.s.l. (our records).
General distribution: From Guinean Republic to Ethiopia, south to D.R. Congo and Uganda.
Chorological classification: Distribution according to direct observation. **WWn, WWc, [WWs]**. *Local phytogeographical element and sub-element.* **CTW[n+c]**. Widely distributed in the western woodlands, from north to south or almost so, but with rather few records, probably due to difficulty with identification (see Section 4); extending to the western side of the Rift Valley but no extension to the east of the Rift.

6.35.2. Lannea fruticosa *(A. Rich.) Engl. (1897).*

FTEA, Anacardiaceae: 23 (1986); FEE 3: 518 (1989); *BS* 51: 289 (1998); PSS: 243 (2015).
Taxonomy: No infraspecific taxa.
Description: Tree to 10 (-12) m tall. No thorns. Leaves compound (11-15 (-17)-foliolate), deciduous. Macrophyll (leaflets Mesophyll). Bark on trunk rough and deeply fissured when thickened, probably an adaptation for fire resistance.
Habitat: 'Broadleaved deciduous woodland, extending up along river valleys, locally common' (FEE). 'Woodland' (PSS). Common and widespread in *Combretum-Terminalia* woodland (our observation).
Distribution in Ethiopia: TU GD WU SU KF GG SD (FEE). Also recorded from the GJ and WG floristic regions (Fig. 6-127).
Altitudinal range in Ethiopia: 600-2100 m a.s.l. (FEE, including range in Eritrea). 500-1600 m a.s.l. (our records).
General distribution: From Nigeria to Ethiopia and Eritrea, south to D.R. Congo and Uganda; also in Arabia.
Chorological classification: Distribution according to direct observation. **WWn, WWc, WWs**. *Local phytogeographical element and sub-element.* **CTW[all]**. Widely distributed in the western woodlands, from north to south, no eastward extension.

6.35.3. Lannea schimperi *(A. Rich.) Engl. (1897).*

FTEA, Anacardiaceae: 19 (1986); FEE 3: 516 (1989); BS51: 290 (1998); PSS: 243 (2015).
Taxonomy: No infraspecific taxa.
Description: Small or medium tree to 9 (-15) m tall. No thorns. Leaves compound (5-11 (-13)-foliolate), deciduous. Macrophyll (leaflets Mesophyll). Bark on trunk smooth to rough and reticulately peeling, probably an adaptation for fire resistance.
Habitat: 'Deciduous woodland on rocky slopes and outcrops, on volcanic, limestone and basement complex, or in lowland woodland with *Pterocarpus* etc.' (FEE). 'Woodland' (PSS). *Combretum-Terminalia* woodland and semi-evergreen bushland (our observation).
Distribution in Ethiopia: TU GD GJ SU WG KF SD BA HA (FEE). Also recorded from the IL and GG floristic regions (Fig. 6-128).
Altitudinal range in Ethiopia: (800-) 1050-1750 (-2300) m a.s.l. (FEE, including range in Eritrea). 600-1900 m a.s.l. (our records).
General distribution: From Nigeria to Ethiopia and Eritrea, south to Zambia and Mozambique.
Chorological classification: Distribution according to direct observation. **WWn, WWc, WWs, EWW**. *Local phytogeographical element and sub-element.* **CTW[all+ext]**. Widely distributed in the western woodlands, from north to south, with extension to the Rift Valley and to the east of the Rift.

6.35.4. Lannea triphylla *(A. Rich.) Engl. (1897).*

FTEA, Anacardiaceae: 13 (1986); FEE 3: 514 (1989); FS 2: 256 (1999).
Taxonomy: No infraspecific taxa.
Description: Shrub or small tree to 5 m tall. No thorns. Leaves compound (trifoliolate(-5-foliolate), decidu-

ous. Mesophyll (leaflets Mesophyll). Bark on trunk rough, probably an adaptation for fire resistance.
Habitat: '*Acacia-Commiphora* bushland on well drained sandy soil and gentle rocky slopes' (FEE). '*Acacia-Commiphora* bushland, often on rocky slopes' (FS). [Not recorded in PSS]. *Acacia-Commiphora* bushland, a few records also in *Combretum-Terminalia* woodland (our observation).
Distribution in Ethiopia: TU GG SD BA HA (FEE). Also recorded from the KF floristic region (Fig. 6-129).
Altitudinal range in Ethiopia: 300-1350 (-1500) m a.s.l. (FEE, including range in Eritrea). 450-1600 (-1900) m a.s.l. (our records).
General distribution: Ethiopia, Eritrea and Somalia, south to Uganda and Kenya; also in Arabia.
Chorological classification: Distribution according to direct observation. **WWn, WWs, ASO**. *Local phytogeographical element and sub-element.* **ACB[-CTW]**. Widely distributed in the *Acacia-Commiphora bushland* (ACB), marginally intruding in the *Combretum-Terminalia woodland and wooded grassland* (CTW) in the north and the south.

6.35.5. Lannea schweinfurthii *(Engl.) Engl. (1897).*

FTEA, Anacardiaceae: 25 (1986); FEE 3: 518 (1989); *BS* 51: 290 (1998); PSS: 243 (2015).
Taxonomy: Some records from the publication of Mindaye Teshome et al. (2017) were named *Lannea welwitschii* (Hiern) Engl. (1898). However, *L. welwitschii* is a forest species restricted to southwestern Ethiopia, and we have therefore accepted the records as misidentifications of the similar *L. schweinfurthii*. This species is divided into the following infraspecific taxa, *L. schweinfurthii* var. *schweinfurthii*, *L. schweinfurthii* var. *stuhlmannii* (Engl.) Kokwaro (1980), *L. schweinfurthii* var. *acutifolia* (Engl.) Kokwaro (1980) and *L. schweinfurthii* var. *tomentosa* (Dunkley) Kokwaro (1980). According to FEE and FS all the Ethiopian and Somali material is intermediate between var. *schweinfurthii* and var. *stuhlmannii*. Recognition of infraspecific taxa not relevant here.
Description: Shrub or tree to 15 (-22) m tall. No thorns. Leaves compound (3-9 (-13)-foliolate), deciduous. Mesophyll-Macrophyll (leaflets Mesophyll). Bark on trunk rough, flaking off in long flakes, probably an adaptation for fire resistance.
Habitat: 'Along water courses in open *Acacia* woodland, open deciduous woodland and wooded grassland on stony soil or on termite mounds' (FEE for *Lannea schweinfurthii* var. *schweinfurthii*). 'Deciduous or semi-evergreen bushland or woodland, often on rocky outcrops' (FS). 'Woodland and wooded grassland' (PSS for *Lannea schweinfurthii* var. *schweinfurthii*). Central and southern parts of *Combretum-Terminalia* woodland and semi-evergreen bushland (our observation).
Distribution in Ethiopia: IL GG SD BA HA (FEE). Also recorded from the WG and KF floristic regions (Fig. 6-130).
Altitudinal range in Ethiopia: 600-1500 m a.s.l. (FEE for *Lannea schweinfurthii* var. *schweinfurthii*). 500-1600 m a.s.l. (our records).
General distribution (species as a whole): Sudan (?), South Sudan, Ethiopia and Somalia, south to Namibia and South Africa.
Chorological classification: Distribution according to direct observation. **WWc, WWs, EWW**. *Local phytogeographical element and sub-element.* **CTW[c+s+ext]**. Widely distributed in the central and southern part of the western woodlands, mainly south of the Abay River, extending to the Rift Valley and to the east of the Rift.

6.35.6. Ozoroa insignis *Delile (1843).*

FTEA, Anacardiaceae: 7 (1986); FEE 3: 521 (1989); FS 2: 260 (1999); PSS: 243 (2015).
Ozoroa insignis subsp. **insignis**
FTEA, Anacardiaceae: 7 (1986); FEE 3: 521 (1989); PSS: 243 (2015).
Taxonomy: Three subspecies are recognized in this species, the widespread *Ozoroa insignis* subsp. *insignis, O. insignis* subsp. *latifolia* (Engl.) R. & A. Fern. (1965) in western tropical Africa and *O. insignis* subsp. *reticulata* (Baker f.) Gillett (1980), which occur from Kenya to South Africa. The Ethiopian material, particularly

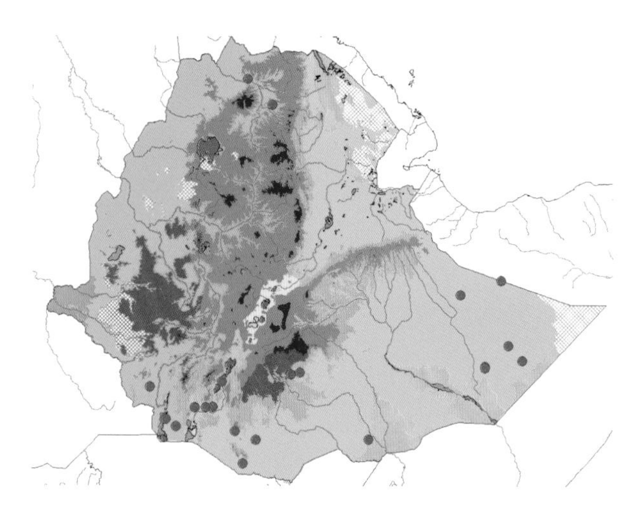

Fig. 6-129. *Lannea triphylla*. Distribution in Ethiopia. Observed records are marked with red dots on vegetation map reproduced from Fig. 4-1.

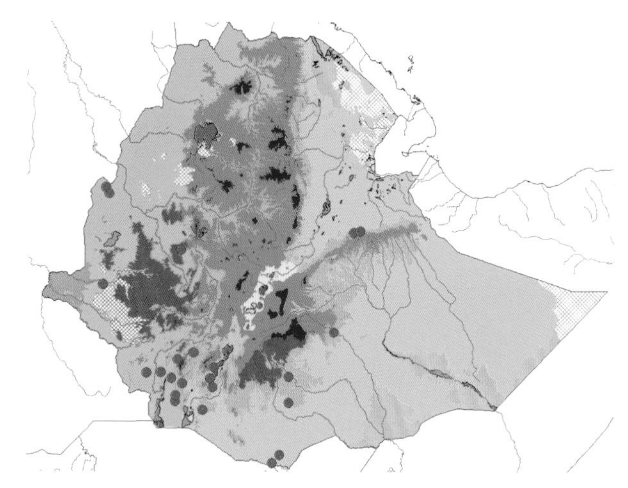

Fig. 6-130. *Lannea schweinfurthii*. Distribution in Ethiopia. Observed records are marked with red dots on vegetation map reproduced from Fig. 4-1.

Fig. 6-131. *Ozoroa insignis*. Distribution in Ethiopia. Observed records are marked with red dots on vegetation map reproduced from Fig. 4-1.

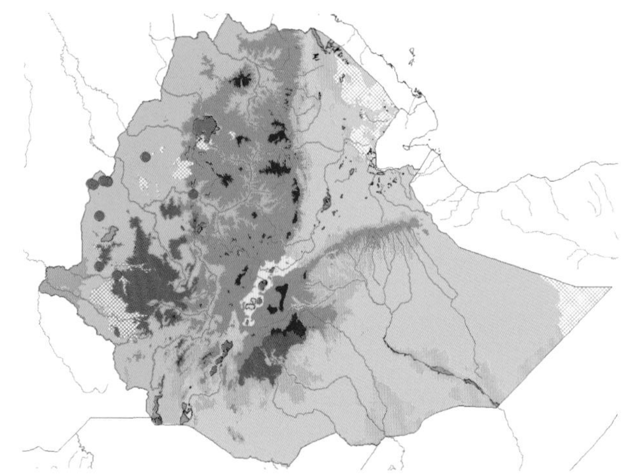

Fig. 6-132. *Ozoroa pulcherrima*. Distribution in Ethiopia. Observed records are marked with red dots on vegetation map reproduced from Fig. 4-1.

from the southwestern part of the country, have larger and more reticulate leaves than material from further north or east, but with no simple discontinuity. Recognition of infraspecific taxa not relevant here.
Description: Shrub or tree to 15 m tall. No thorns. Leaves simple, probably deciduous. Mesophyll. Bark on trunk corky and much fissured, probably an adaptation for fire resistance.
Habitat: 'Broadleaved deciduous woodland, often with *Terminalia*, on well drained rocky slopes, sometimes in *Acacia* bushland or woodland overlying limestone' (FEE for *Ozoroa insignis* subsp. *insignis*). 'Deciduous to semi-evergreen bushland or woodland' (FS). 'Woodland, often on rocky hillslopes' (PSS for *Ozoroa insignis* subsp. *insignis*). In *Combretum-Terminalia* woodland and semi-evergreen bushland (our observation).
Distribution in Ethiopia: TU GD SU GG SD BA HA (FEE). No additional records (Fig. 6-131).
Altitudinal range in Ethiopia: 1350-2000 m a.s.l. (FEE, including range in Eritrea). 650-1900 m a.s.l. (our records).
General distribution (species as a whole): From Senegal to Ethiopia, Eritrea and Somalia, south to South Africa.
Chorological classification: Distribution according to direct observation. **WWn, WWc, WWs, EWW**. *Local phytogeographical element and sub-element.* **CTW[all+ext]**. Widely distributed in the western woodlands, from north to south, extending to the Rift Valley and to the east of the Rift.

6.35.7. Ozoroa pulcherrima *(Schweinf.) R. & A. Fernandes (1965).*

FTEA, Anacardiaceae: 8 (1986); FEE 3: 521 (1989); PSS: 243 (2015).
Taxonomy: No infraspecific taxa.
Description: Shrub to 4 (-6) m; stems shooting from woody rootstock. No thorns. Leaves simple, probably deciduous. Mesophyll. Bark thin and smooth, probably presenting no adaptation for fire resistance, but the woody rootstock, from which new shoots develop after fires, is an adaptation for fire resistance.
Habitat: 'Broadleaved deciduous woodland with *Terminalia*, *Combretum*, *Pterocarpus*, etc. on rocky or sandy soils' (FEE). 'Woodland and wooded grassland' (PSS). Central part of *Combretum-Terminalia* woodland (our observation).
Distribution in Ethiopia: GJ WG IL (FEE). No additional records (Fig. 6-132).
Altitudinal range in Ethiopia: 650-1400 m a.s.l. (FEE). 750-1500 m a.s.l. (our records).
General distribution: From Senegal to Ethiopia, south to Uganda.
Chorological classification: Distribution according to direct observation. **[WWn], WWc**. *Local phytogeographical element and sub-element.* **CTW[c]**. Distributed in the central part of the western woodlands, mainly south of the Abay River, no eastward extension.

6.35.8. Rhus vulgaris *Meikle (1951).*

FTEA, Anacardiaceae: 31 (1986); FEE 3: 529 (1989); FTNA: 206 (1992); *BS* 51: 292 (1998).
Rhus pyroides Burch. (1822)
FTEA, Anacardiaceae: 32 (1986); PSS: 244 (2015).
Later used name: **Searsia pyroides** (Burch.) Moffett (2007)
Taxonomy: With the wider taxonomic concept that includes *Rhus vulgaris* in *R. pyroides*, there are, apart from the widespread *R. pyroides* var. *pyroides*, three infraspecific taxa are recorded from Africa south of Angola and Tanzania, *R. pyroides* var. *dinteri* (Engl.) Moffett (1993), *R. pyroides* var. *glabrata* Sond. (1860) and *R. pyroides* var. *integrifolia* (Engl.) Moffett (1993). In Ethiopia only one infraspecific taxon. Recognition of infraspecific taxa not relevant here.
Description: Shrub or small tree to 10 m tall. No thorns. Leaves compound (trifoliolate), probably deciduous. Mesophyll (leaflets Mesophyll). Bark smooth, presumably no adaptation for fire resistance.
Habitat: 'Margins of *Podocarpus* and *Juniperus* forests and often as an important component of secondary bushland following clearance; evergreen bushland with *Euclea* ... or open deciduous woodland with *Combretum*, *Terminalia* and/or *Acacia*' (FEE without infraspecific taxonomic rank). 'Undifferentiated Afromontane forest ... and dry single-dominant Afromontane

forest ..., especially along forest margins; probably most common in secondary montane evergreen bushland and wooded upland grassland' (FTNA). 'Forest margins, montane bushland and woodland' (PSS for *Rhus pyroides* var. *pyroides*). Widely distributed in Afromontane forest, along forest margins and in Afromontane bushland, *Combretum-Terminalia* woodland and semi-evergreen bushland (our observation).
Distribution in Ethiopia: GD GJ WU SU AR WG KF SD BA HA (FEE). No additional records (Fig. 6-133).
Altitudinal range in Ethiopia: 1500-2800 m a.s.l. (FEE). 2000-3300 m a.s.l. (FTNA, including the range in the entire Horn of Africa). 1200-2550 m a.s.l. (our records).
General distribution (species as a whole): Cameroon to Ethiopia, south to South Africa.
Chorological classification: Distribution according to direct observation. **WWn, WWc, WWs, EWW**. *Local phytogeographical element.* **TRG**. Transgressing species, occurs in both the *Combretum-Terminalia woodland and wooded grassland* (CTW) and in other vegetation types, particularly *Dry evergreen Afromontane forest* (DAF).

6.35.9. Sclerocarya birrea *(A. Rich.) Hochst. (1844).*

FTEA, Anacardiaceae: 42 (1986); FEE 3: 519 (1989); *BS* 51: 292 (1998); PSS: 244 (2015).
Taxonomy: In tropical Africa south of Kenya also *Sclerocarya birrea* subsp. *caffra* (Sond.) Kokwaro (1980) and *S. birrea* subsp. *multifoliolata* (Engl.) Kokwaro (1980). In Ethiopia only *S. birrea* subsp. *birrea*. Recognition of infraspecific taxa not relevant here.
Description: A tree up to 18 m tall, with spreading crown. No thorns. Leaves compound (up to 37-foliolate), deciduous. Macrophyll (leaflets Mesophyll). Bark on trunk reticulate and flaking in large scales, probably an adaptation for fire resistance.
Habitat: 'Open deciduous woodland on rocky slopes' (FEE for *Sclerocarya birrea* subsp. *birrea*). 'Woodland and wooded grassland, often on rocky hillslopes' (PSS for *Sclerocarya birrea* subsp. *birrea*). 'Mixed deciduous woodland and wooded grassland; often on rocky hills' (FTEA, Anacardiac. for subsp. *birrea*). *Combretum-Terminalia* woodland and semi-evergreen bushland, for record from Semi-Desert Scrub, se below (our observation).
Distribution in Ethiopia: TU SU IL GG SD (FEE). Also recorded from the GD, WG and BA floristic regions. The isolated occurrence at the Webe Shebele River in the Ogaden is documented by a collection (*JJFE de Wilde* 5975 (WAG), Ogaden, ca. 7 km NE. of Gode experimental station), which seems correctly identified but could have originated from the experimental station (Fig. 6-134).
Altitudinal range in Ethiopia: 500-1700 m a.s.l. (FEE for *Sclerocarya birrea* subsp. *birrea*, including range in Eritrea). 500-1650 m a.s.l. (our records).
General distribution (species as a whole): From Senegal to Ethiopia and Eritrea, south to Angola and South Africa; also in Madagascar.
Chorological classification: Distribution according to direct observation. **WWn, WWc, WWs, EWW**. *Local phytogeographical element and sub-element.* **CTW[all+ext]**. Widely distributed in the western woodlands, from north to south (with considerable gaps), with extension to the Rift Valley and to the east of the Rift. The single record at the Webe Shebele River in the *Acacia-Commiphora* Bushland (ACB, see above), although correctly identified, requires further study (see above under '*Distribution in Ethiopia'*; the species has not been recorded from Somalia).

6.36. Araliaceae

FTEA, Araliaceae: 1-24 (1968); FEE 3: 537-542 (1989); *BS* 51: 295-297 (1998); PSS: 379 (2015).

6.36.1. Cussonia arborea *Hochst. ex A. Rich. (1847).*

FTEA, Araliaceae: 4 (1968); FEE 3: 538 (1989); *BS* 51: 295 (1998); PSS: 379 (2015).
Taxonomy: No infraspecific taxa.
Description: Tree to 13 m tall, with trunk up to 1 m or more in diameter and thick branches. No thorns. Leaves simple, deeply palmately lobed or digitately compound with 5-7 sessile leaflets, deciduous. Mac-

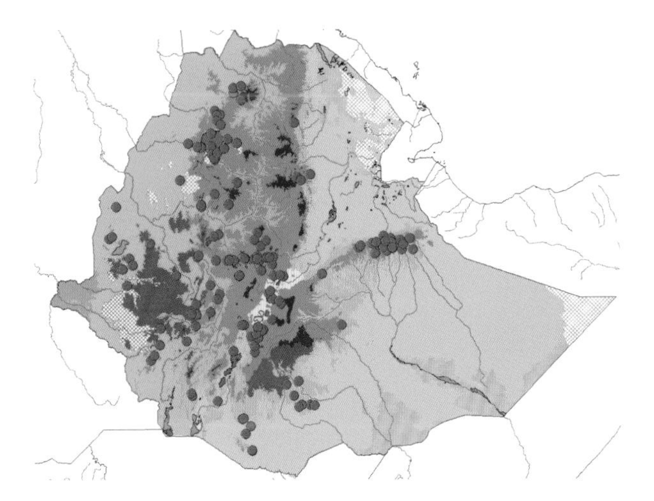

Fig. 6-133. *Rhus vulgaris*. Distribution in Ethiopia. Observed records are marked with red dots on vegetation map reproduced from Fig. 4-1.

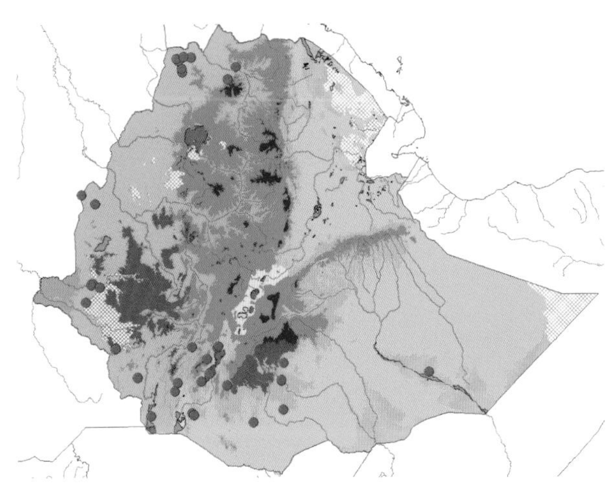

Fig. 6-134. *Sclerocarya birrea*. Distribution in Ethiopia. Observed records are marked with red dots on vegetation map reproduced from Fig. 4-1.

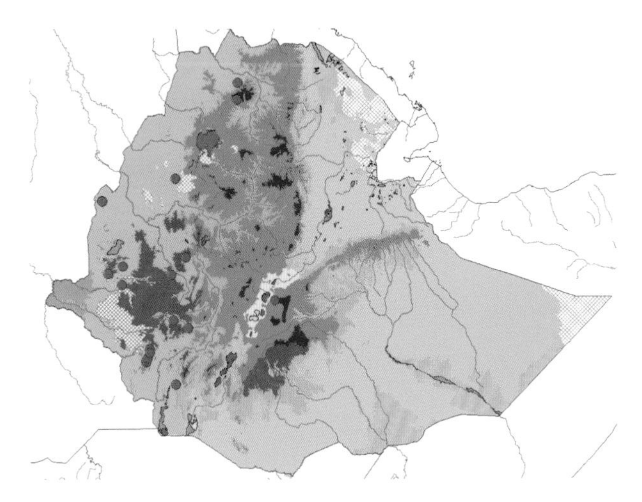

Fig. 6-135. *Cussonia arborea*. Distribution in Ethiopia. Observed records are marked with red dots on vegetation map reproduced from Fig. 4-1.

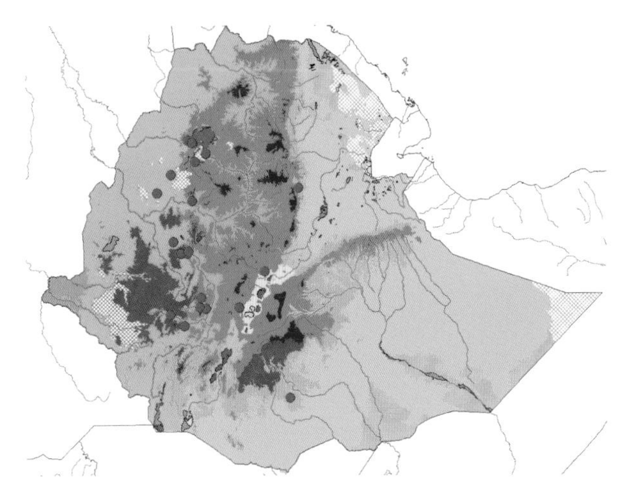

Fig. 6-136. *Cussonia ostinii*. Distribution in Ethiopia. Observed records are marked with red dots on vegetation map reproduced from Fig. 4-1.

rophyll (leaflets Mesophyll to Macrophyll). Bark on trunk thick and with large reticulate scales, an adaptation for fire resistance, particularly in combination with the thick and stunted trunk.
Habitat: 'Woodland, tree- or shrub-savanna, often in rocky places' (FEE). 'Wooded grassland, often in rocky areas' (PSS). *Combretum-Terminalia* woodland and one record from semi-evergreen bushland (our observation).
Distribution in Ethiopia: TU GD WG AR KF (FEE). Also recorded from the GJ, IL and GG floristic regions. It was not possible to establish the exact locality of two records from TU (*Schimper* (II) 1357 (K) and *Chiovenda* 450 (FT) (Fig. 6-135).
Altitudinal range in Ethiopia: Up to 2300 m a.s.l. (FEE, including range in Eritrea). 1150-1700 (-2200) m a.s.l. (our records).
General distribution: From Guinea-Bissau to Ethiopia and Eritrea, south to Zimbabwe and Mozambique.
Chorological classification: Distribution according to direct observation. **WWn, WWc, WWs, EWW**. *Local phytogeographical element and sub-element.* **CTW[all+ext]**. Widely distributed in the western woodlands, from north to south (with considerable gaps in the distribution), extending to the Rift Valley. Having large leaves and thick branches, the species is probably undercollected.

6.*36.2*. Cussonia ostinii *Chiov. (1911).*

FEE 4(1): 539 (2003).
Taxonomy: No infraspecific taxa.
Description: Tree to 7 m tall, with thick trunk and few branches. No thorns. Leaves palmately lobed, deciduous. Macrophyll. Bark on trunk thick and corky, with broad reticulate scales, clearly an adaptation for fire resistance.
Habitat: 'Upland savanna' (FEE). [Not in PSS]. Mostly found in the lower zone of open Afromontane vegetation, with a few records in *Combretum-Terminalia* woodland and semi-evergreen bushland (our observation).
Distribution in Ethiopia: GD GJ WG KF (FEE). Also recorded from the SU and SD floristic provinces. The record from SD (*Melesse M.* 103 (ETH), Negele Borana) requires confirmation (Fig. 6-136).
Altitudinal range in Ethiopia: 1500-2100 m a.s.l. (FEE). 1400-2000 m a.s.l. (our records).
General distribution: Restricted to Ethiopia and Eritrea.
Chorological classification: Distribution according to direct observation. **WWn, WWc, [WWs], EWW**. *Local phytogeographical element.* **MI**. Marginally intruding species in the *Combretum-Terminalia woodland and wooded grassland* (CTW), with main distribution in the wooded grasslands of the highlands and in *Transitional semi-evergreen bushland* (TSEB); nearly all records are from above 1500 m tall.

6.*36.3*. Polyscias farinosa *(Delile) Harms (1894).*

FEE 4(1): 537 (2003).
Taxonomy: No infraspecific taxa.
Description: Tree to 7 m tall. No thorns. Leaves pinnate, deciduous, flowering when leafless. Macrophyll (leaflets Mesophyll). Bark on trunk thin, but slightly corky, probably no adaptation for fire resistance.
Habitat: 'Open woodland, on slopes near streamsides' (FEE). [Not in PSS]. *Combretum-Terminalia* woodland and semi-evergreen bushland (our observation).
Distribution in Ethiopia: TU GD SU KF (FEE). Also recorded from the GJ, WG and GG floristic regions (Fig. 6-137).
Altitudinal range in Ethiopia: 1700-2000 m a.s.l. (FEE). (950-) 1700-2100 m a.s.l. (our records).
General distribution: Restricted to Ethiopia.
Chorological classification: Distribution according to direct observation. **WWn, WWc, WWs**. *Local phytogeographical element and sub-element.* **CTW[all]**. Widely distributed in the western woodlands, from north to south or almost so, but restricted to the upper zone of the *Combretum-Terminalia* woodland or *Transitional semi-evergreen bushland* (TSEB); no eastward extension.

6.37. Apiaceae

FTEA, Umbelliferae: 1-128 (1989); *BS* 51: 297-302 (1998); FS 2: 268-287 (1999), as **Apiaceae (Umbelliferae)**; FEE 4(1): 1-45 (2003); PSS: 379-381 (2015).

6.37.1. Heteromorpha arborescens *Cham. & Schltdl. (1826).*

FEE 4(1): 18 (2003); PSS: 381 (2015).
Heteromorpha trifoliata (H.L. Wendl.) Eckl. & Zeyh. (1837).
FTEA, Umbelliferae: 38 (1989); FTNA: 212 (1992); *BS* 51: 299 (1998).
Taxonomy: In southern tropical Africa a range of infraspecific taxa, including Heteromorpha arborescens var. arborescens. In Ethiopia only *H. arborescens* var. *abyssinica* (A. Rich.) H. Wollf (1910). Recognition of infraspecific taxa not relevant here.
Description: Shrub or small tree to 5 (-8) m tall. No thorns. Leaves simple or compound (trifoliolate, pentafoliolate), probably deciduous. Mesophyll (leaflets Microphyll to Mesophyll). Bark thin, flaking, probably no adaptation for fire resistance.
Habitat: 'In montane and riverine woodland margins and in secondary forest' (FEE for *Heteromorpha arborescens* var. *abyssinica*). 'Dry single-dominant Afromontane forest ... and undifferentiated Afromontane forest ..., especially along the margins and in clearings, persisting in gully forests; also in secondary montane evergreen bushland or left as a single shrub or small tree in derived montane grassland' (FTNA for species as a whole). 'Woodland and grassland on rocky slopes, forest margins' (PSS for *Heteromorpha arborescens* var. *abyssinica*). Mainly in the lower zone of open Afromontane vegetation, with a few records in *Combretum-Terminalia* woodland and semi-evergreen bushland (our observation).
Distribution in Ethiopia: GD GJ WU SU GG SD HA (FEE). Also recorded from the TU and KF floristic regions (Fig. 6-138).
Altitudinal range in Ethiopia: 1500-2500 m a.s.l. (FEE for *Heteromorpha arborescens* var. *abyssinica*, including range in Eritrea). 1600-2700 m a.s.l. (FTNA for species as a whole and including the range in the entire Horn of Africa). (1300-) 1500-2450 m a.s.l. (our records).
General distribution (species as a whole): Cameroon to Ethiopia and Eritrea, south to Namibia and South Africa; also in Arabia.
Chorological classification: Distribution according to direct observation. **WWn, WWc, WWs, EWW**. *Local phytogeographical element.* **MI**. Marginally intruding species in the *Combretum-Terminalia woodland and wooded grassland* (CTW), with main distribution in the wooded grasslands of the highlands and in *Transitional semi-evergreen bushland* (TSEB); nearly all records are from above 1500 m.

6.37.2. Steganotaenia araliacea *Hochst. (1844).*

FTEA, Umbelliferae: 115 (1989); *BS* 51: 301 (1998); FS 2: 283 (1999); FEE 4(1): 37 (2003); PSS: 381 (2015).
Taxonomy: No infraspecific taxa.
Description: Tree to 10 (-12) m tall. No thorns. Leaves compound (imparipinnate, with 3-4 pairs of leaflets), probably deciduous. Macrophyll (leaflets Mesophyll). Bark corky and thin outer bark flaking, probably an adaptation for fire resistance.
Habitat: 'Open woodland and on riverbanks' (FEE). 'Probably in deciduous bushland on hillslopes' (FS). 'Rocky woodland, gulleys, riverbanks' (PSS). Mostly found in *Combretum-Terminalia* woodland and semi-evergreen bushland, a few records in the lower zone of open Afromontane vegetation (our observation).
Distribution in Ethiopia: GD GJ SU AR WG SD HA (FEE). Also recorded from the TU, AF, KF, and GG floristic regions (Fig. 6-139).
Altitudinal range in Ethiopia: 1300-2100 m a.s.l. (FEE, including range in Eritrea). 750-2000 m a.s.l. (our records).
General distribution: From Guinean Republic to Ethiopia, Eritrea and Somalia, south to Namibia and South Africa.
Chorological classification: Distribution according to direct observation. **WWn, WWc, WWs, EWW**. *Local phytogeographical element and sub-element.* **CTW[all+ext]**. Widely distributed in the western woodlands, from north to south, extending to the Rift Valley and to the east of the Rift.

Fig. 6-137. *Polyscias farinosa*. Distribution in Ethiopia. Observed records are marked with red dots on vegetation map reproduced from Fig. 4-1.

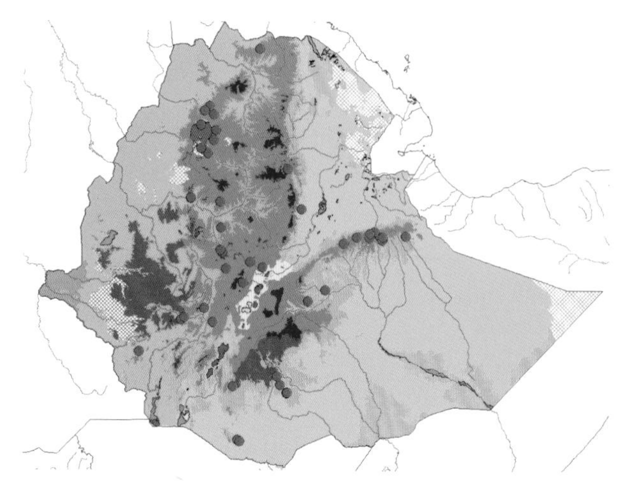

Fig. 6-138. *Heteromorpha arborescens*. Distribution in Ethiopia. Observed records are marked with red dots on vegetation map reproduced from Fig. 4-1.

Fig. 6-139. *Steganotaenia araliacea*. Distribution in Ethiopia. Observed records are marked with red dots on vegetation map reproduced from Fig. 4-1.

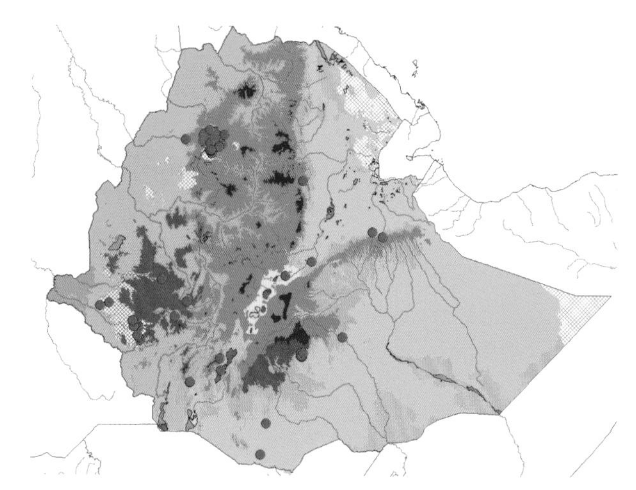

Fig. 6-140. *Diospyros abyssinica*. Distribution in Ethiopia. Observed records are marked with red dots on vegetation map reproduced from Fig. 4-1.

6.38. Ebenaceae

FTEA, Ebenaceae: 1-52 (1996); *BS* 51: 303-304 (1998); FEE 4(1): 49-53 (2003); FS 3: 6-11 (2006); PSS: 286-287 (2015).

6.38.1. Diospyros abyssinica *(Hiern) F. White (1956).*

FTNA: 215 (1992); FTEA, Ebenaceae: 22 (1996); FEE 4(1): 50 (2003); PSS: 286 (2015).

Taxonomy: Four infraspecific taxa are recognized within this species, the widespread *Diospyros abyssinica* subsp. *abyssinica* and in southern Africa south of Tanzania *D. abyssinica* subsp. *chapmaniorum* F. White (1980), *D. abyssinica* subsp. *attenuata* F. White (1988) and *D. abyssinica* subsp. *reticulata* F. White (1988). In Ethiopia only represented by *Diospyros abyssinica* subsp. *abyssinica*. Recognition of infraspecific taxa not relevant here.

Description: Shrub or tree to 15 (-36) m tall. No thorns. Leaves simple, evergreen. Microphyll. Bark on trunk rough, reticulate, scaling off in fibrous strips or oblong plates, probably an adaptation for fire resistance.

Habitat: 'Intermediate and montane humid forest, both broadleaved and dominated by *Podocarpus*' (FEE for *Diospyros abyssinica* subsp. *abyssinica*). 'Dry peripheral demi-evergreen Guineo-Congolian forest, transitional rain forest, Afromontane rain forest, undifferentiated Afromontane forest ...; common in the riparian forest around Lake Tana' (FTNA for *Diospyros abyssinica* subsp. *abyssinica*). 'Forest and montane bushland' (PSS for *Diospyros abyssinica* subsp. *abyssinica*). Mainly in Afromontane forest and montane bushland, a few records in *Combretum-Terminalia* woodland and semi-evergreen bushland (our observation).

Distribution in Ethiopia: TU GD GJ SU AR WG IL KF SD BA HA (FEE). The FEE records from the WG and AR floristic regions have not been confirmed but given how widespread the species is in the rest of the country, its absence from these two floristic provinces would be surprising. A single record from TU (*Chiovenda* 409 (FT), from locality in the surroundings of Axum) has not been possible to map (Fig. 6-140).

Altitudinal range in Ethiopia: 550-2500 m a.s.l. (FEE, including range in Eritrea). 550-2400 (FTNA). 500-1900 m a.s.l. (our records).

General distribution: From Guinean Republic to Ethiopia and Eritrea, south to Angola and Mozambique.

Chorological classification: Distribution according to direct observation. **WWn, WWc, WWs, EWW**. *Local phytogeographical element.* **MI**. Marginally intruding species in the western woodlands, with main distribution in the *Dry Afromontane Forest* complex (DAF) in the highlands and in *Transitional semi-evergreen bushland* (TSEB), the records from the *Acacia-Commiphora bushland* (ACB) are from small areas on hills of *Transitional semi-evergreen bushland* (TSEB); very rare in our relevés.

6.38.2. Diospyros mespiliformis *Hochst. ex A. DC. (1844).*

FTNA: 217 (1992); FTEA, Ebenaceae: 40 (1996); FEE 4(1): 50 (2003); PSS: 286 (2015).

Taxonomy: No infraspecific taxa.

Description: Shrub or small tree to 27 (-35) m tall. No thorns. Leaves simple, evergreen. Mesophyll. Bark on trunk very rough, longitudinally channelled, flaking in irregular scales, probably an adaptation for fire resistance.

Habitat: 'Riverine forest, open riparian woodland or scattered along temporary streams, sometimes at the base of rocky outcrops' (FEE). 'Riverine forest; transgressing into deciduous woodland, often associated with rocky outcrops' (FTNA). 'Riverine forest and woodland' (PSS). Mainly in riverine and Dry and Moist Afromontane forests (DAF and MAF) and montane bushland in the highlands, some records in *Combretum-Terminalia* woodland and semi-evergreen bushland may be away from rivers, but that has not always been specified on the record (our observation).

Distribution in Ethiopia: TU GD GJ SU WG IL KF SD BA (FEE). Also recorded from the AF and GG floristic regions (Fig. 6-141).

Altitudinal range in Ethiopia: 300-2000 m a.s.l. (FEE, including range in Eritrea & FTNA, including range in the whole Horn of Africa). 600-1800 m a.s.l. (our

records).
General distribution: From Senegal to Eritrea and Ethiopia, south to Namibia and South Africa; also in Arabia.
Chorological classification: Distribution according to direct observation. **WWn, WWc, WWs, EWW.** *Local phytogeographical element.* **MI.** Marginally intruding species in the western woodlands, with main distribution in the highlands and in *Transitional semi-evergreen bushland* (TSEB), the records from the *Acacia-Commiphora bushland* (ACB) are from small areas on hills with *Transitional semi-evergreen bushland* (TSEB); only a few records in our relevés.

6.38.3. Euclea racemosa *Murr. subsp.* schimperi *(A. DC.) F. White (1980).*

FTNA: 219 (1992); FTEA, Ebenaceae: 46 (1996); FEE 4(1): 51 (2003); FS 3: 10 (2006); PSS: 286 (2015).
Taxonomy: Seven infraspecific taxa are recognized in this species, all, apart from *Euclea racemosa* subsp. *schimperi* (A. DC.) F. White (1980) restricted to southern Africa from Mozambique and southwards (*E. racemosa* subsp. *bernardii* F. White (1980), *E. racemosa* subsp. *daphnoides* (Hiern) F. White (1980), *E. racemosa* subsp. *macrophylla* (E. Mey. ex A. DC.) F. White (1980), *E. racemosa* subsp. *racemosa*, *E. racemosa* subsp. *sinuata* F. White (1980) and *E. racemosa* subsp. *zuluensis* F. White (1980). Only one rather homogenous subspecies in Ethiopia, sometimes referred to by its basionym, *E. schimperi* A. DC. (1842). Recognition of infraspecific taxa not relevant here.
Description: Shrub or small tree to 10 (-15) m tall. No thorns. Leaves simple, evergreen. Microphyll. Bark on trunk smooth or minutely fissured, probably no adaptation for fire resistance.
Habitat: 'Open montane forest, usually *Juniperus-Podocarpus* forest, especially in clearings and along margins, in evergreen montane bushland as well as in *Buxus-Acokanthera* and *Acacia-Commiphora* bushland, also in riverine and in clumps of shrubs in montane grassland' (FEE for *Euclea racemosa* subsp. *schimperi*). 'Undifferentiated Afromontane forest ..., open, dry single-dominant Afromontane forest ... and transition between single-dominant Afromontane forest and East African evergreen and semi-evergreen bushland, especially along forest margins; frequent in primary and secondary Afromontane and East Africa evergreen and semi-evergreen bushland ...; also in riverine bushland and clumps in montane grassland' (FTNA). 'Evergreen bushland' (FS). 'Dry forest margins, bushland and grassland' (PSS for *Euclea racemosa* subsp. *schimperi*). Mainly in Dry Afromontane forest (DAF) and montane bushland, a few records in *Combretum-Terminalia* woodland and semi-evergreen bushland (our observation).
Distribution in Ethiopia: TU GD GJ WU SU KF SD BA HA (FEE). Also recorded from the AF floristic region (Fig. 6-142).
Altitudinal range in Ethiopia: (700-) 1000-2900 m a.s.l. (FEE, including range in Eritrea). 1000-2400 m a.s.l. (FTNA, including the range in the entire Horn of Africa). (700-) 1200-2350 m a.s.l. (our records).
General distribution (species as a whole): Sudan, South Sudan, Ethiopia, Eritrea and Somalia, south to South Africa, west to D.R. Congo; also in Egypt, Arabia and the Comoro Islands.
Chorological classification: Distribution according to direct observation. **WWn, WWs.** *Local phytogeographical element.* **MI.** Marginally intruding species in the western woodlands, with main distribution in the highlands and in *Transitional semi-evergreen bushland* (TSEB), the records from the *Acacia-Commiphora bushland* (ACB) are from small areas on hills of *Transitional semi-evergreen bushland* (TSEB); only a few records in our relevés.

6.39. Myrsinaceae

FTEA, Myrsinaceae: 1-23 (1984); *BS* 51: 309-310 (1998); FEE 4(1): 64-69 (2003); FS 3: 18-20 (2006); PSS: 287 (2015), as part of "**Primulaceae**".
The family Myrsinaceae is accepted as a distinct family in most floras cited here, but in PSS included in Primulaceae on pp. 287-288.

6.39.1. Embelia schimperi *Vatke (1876).*

FTEA, Myrsinaceae: 13 (1984); *BS* 51: 309 (1998); FEE 4(1): 69 (2003); PSS: 287 (2015).
Taxonomy: As pointed out by PSS, the genus *Embelia* in Africa needs taxonomic revision, but the Ethiopian material seems to be homogenous. Recognition of infraspecific taxa not relevant here.
Description: Scandent shrub or tree to 12 m tall. No thorns. Leaves simple, presumably evergreen. Mesophyll. Bark on trunk striate, fibrous, stripping off from older wood, probably no adaptation for fire resistance.
Habitat: 'Forest edges, secondary forest and scrub, along rivers' (FEE). 'Montane forest and bushland' (PSS). Mainly in Dry Afromontane forest (DAF) and montane bushland, a few records in *Combretum-Terminalia* woodland on rocky outcrops and in semi-evergreen bushland (our observation).
Distribution in Ethiopia: GD GJ SU AR WG IL KF GG SD BA (FEE). Also recorded from the HA floristic region (Fig. 6-143).
Altitudinal range in Ethiopia: 1700-2800 m a.s.l. (FEE). 1500-2700 m a.s.l. (our records).
General distribution: From Guinean Republic to Ethiopia, south to South Africa; also in Angola and Zimbabwe.
Chorological classification: Distribution according to direct observation. WWn, WWc, WWs, EWW. *Local phytogeographical element.* MI. Marginally intruding species in the western woodlands, with main distribution in the highlands and in *Transitional semi-evergreen bushland* (TSEB), the records from the *Acacia-Commiphora bushland* (ACB) are from small areas on hills of *Transitional semi-evergreen bushland* (TSEB); only a few records in our relevés.

6.39.2. Maesa lanceolata *Forssk. (1775).*

FTEA, Myrsinaceae: 3 (1984); FTNA: 228 (1992); *BS* 51: 309 (1998); FEE 4(1): 64 (2003); FS 3: 18 (2006); PSS: 288 (2015).
Taxonomy: No infraspecific taxa.
Description: Shrub or small tree to 20 (-24) m tall. No thorns. Leaves simple, evergreen. Macrophyll. Bark on trunk smooth or rough, probably no adaptation for fire resistance.
Habitat: 'Gallery forest, margin of evergreen forest, along riverbanks and streams, open woodland and valleys' (FEE). 'Afromontane rain forest, undifferentiated Afromontane forest ...and single-dominant Afromontane forest ..., particularly along edges and in clearings, persisting in gully forests; riverine forest; also in secondary montane evergreen bushland and in riparian woodland' (FTNA). '*Juniperus* forest and evergreen bushland, usually along streams' (FS). 'Montane forest, particularly along margins and in clearings, montane bushland and grassland' (PSS). Mainly in Dry and Moist Afromontane forest (DAF and MAF) and montane bushland, a few records in *Combretum-Terminalia* woodland and semi-evergreen bushland (our observation).
Distribution in Ethiopia: TU GD GJ WU SU AR WG IL KF BA HA (FEE). Also recorded from the GG and SD floristic regions (Fig. 6-144).
Altitudinal range in Ethiopia: 1350-3000 m a.s.l. (FEE). 950-3100 m a.s.l. (FTNA, including the range in the entire Horn of Africa). 1450-2950 m a.s.l. (our records).
General distribution: From Guinean Republic to Ethiopia, Eritrea and Somalia, south to South Africa; also in Madagascar and Arabia.
Chorological classification: Distribution according to direct observation. WWn, WWc, WWs, EWW. *Local phytogeographical element.* MI. Marginally intruding species in the western woodlands, with main distribution in the highlands, in *Transitional semi-evergreen bushland* (TSEB), *Riparian vegetation* (RV) and in *Transitional rain forest* (TRF), the records from inside the *Acacia-Commiphora bushland* (ACB) are from small areas on hills with *Transitional semi-evergreen bushland* (TSEB); only one or few records in our relevés.

6.40. Loganiaceae

FTEA, Loganiaceae: 1-47 (1960); *BS* 51: 311-312 (1998); FEE 4(1): 70-78 (2003); PSS: 203-304 (2015).

Fig. 6-141. *Diospyros mespiliformis*. Distribution in Ethiopia. Observed records are marked with red dots on vegetation map reproduced from Fig. 4-1.

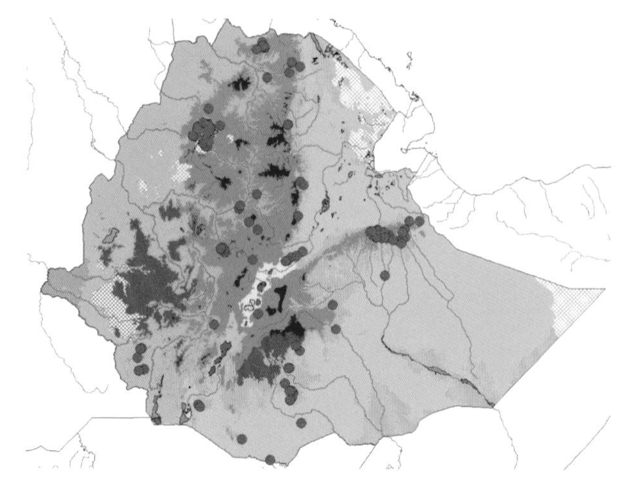

Fig. 6-142. *Euclea racemosa* subsp. *schimperi*. Distribution in Ethiopia. Observed records are marked with red dots on vegetation map reproduced from Fig. 4-1.

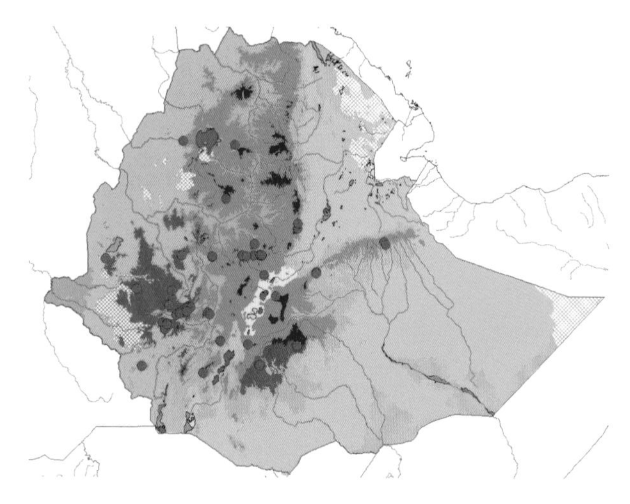

Fig. 6-143. *Embelia schimperi*. Distribution in Ethiopia. Observed records are marked with red dots on vegetation map reproduced from Fig. 4-1.

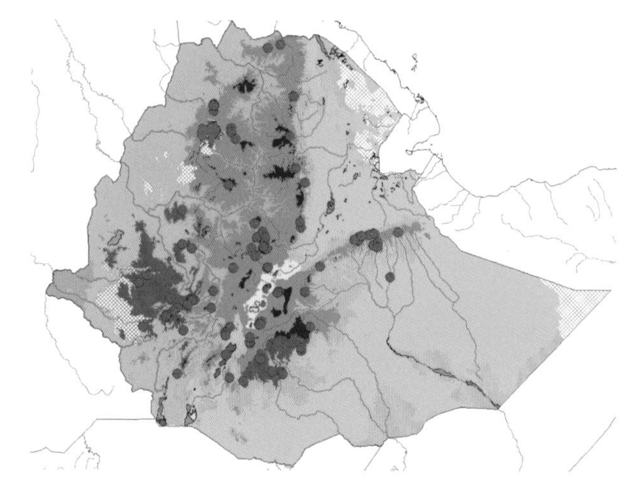

Fig. 6-144. *Maesa lanceolata*. Distribution in Ethiopia. Observed records are marked with red dots on vegetation map reproduced from Fig. 4-1.

6.*40.1.* Strychnos innocua *Delile (1826).*

FTEA, Loganiaceae: 25 (1960); *BS* 51: 312 (1998); FEE 4(1): 76 (2003); PSS: 304 (2015).
Taxonomy: From Tropical East Africa, FTEA, Loganiac. (1960) records *Strychnos innocua* subsp. *innocua* var. *innocua*, *S. innocua* subsp. *innocua* var. *pubescens* Solered. (1893), *S. innocua* subsp. *burtonii* (Baker) Bruce & Lewis (1956) var. *burtonii* and *S. innocua* subsp. *burtonii* (Baker) Bruce & Lewis (1956) var. *glabra* Bruce & Lewis (1956). None of these infraspecific taxa have been maintained for Ethiopian or Sudanese material. Recognition of infraspecific taxa not relevant here.
Description: Shrub or small tree to 12 m tall. No thorns. Leaves simple, probably evergreen. Mesophyll. Bark on trunk smooth, probably no adaptation for fire resistance.
Habitat: 'Deciduous woodland' (FEE). 'Woodland and wooded grassland' (PSS). In *Combretum-Terminalia* woodland (our observation).
Distribution in Ethiopia: TU GD GJ SU WG IL GG SD (FEE). Also recorded from the KF floristic region. The FEE record from the SU floristic region has not been confirmed (Fig. 6-145).
Altitudinal range in Ethiopia: 600-1400 (-1600) m a.s.l. (FEE, including range in Eritrea). 650-1400 m a.s.l. (our records).
General distribution: From Guinean Republic to Ethiopia and Eritrea, south to Angola, Zimbabwe and Mozambique.
Chorological classification: Distribution according to direct observation. **WWn, WWc, WWs.** *Local phytogeographical element and sub-element.* **CTW[all].** Widely distributed in the western woodlands, from north to south; a few records from *Transitional semi-evergreen bushland* (TSEB) on the western side of the Rift Valley, but no further eastwards extension.

6.41. Oleaceae

FTEA, Oleaceae: 1-31 (1952); *BS* 51: 313-316 (1998); FEE 4(1): 79-86 (2003); PSS: 325-326 (2015).

6.*41.1.* Olea europaea *L. subsp.* cuspidata *(Wall. & G. Don) Cif. (1942).*

FTNA: 236 (1992); *BS* 51: 314 (1998); FEE 4(1): 79 (2003); FS 3: 260 (2006); PSS: 325 (2015).
Olea chrysophylla Lam. (1791).
FTEA, Oleaceae: 9 (1952).
Taxonomy: Six infraspecific taxa are recognized within this species, the cultivated *Olea europaea* subsp. *europaea*, *O. europaea* subsp. *cuspidata* (Wall. & G. Don) Cif. (1912), which occurs in the mountains of Africa from Sudan and South Sudan, Ethiopia and Eritrea to Angola and South Africa, and also in Egypt and Asia from Arabia to China, *O. europaea* subsp. *cerasiformis* (Webb & Berthel.) G. Kunkel & Sunding (1972), restricted to the Mascarene Islands, *O. europaea* subsp. *laperrinei* (Batt. & Trab.) Cif. (1912), restricted to mountains in Sahara on the border between Algeria and Niger and possibly also in western Sudan, *O. europaea* subsp. *guanchica* P.Vargas, J. Hess, Muñoz Garm. & Kadereit (2001), restricted to the Mascarene Islands, and *O. europaea* subsp. *maroccana* (Greuter & Burdet) P. Vargas, J. Hess, Muñoz Garm. & Kadereit (2001), restricted to Morocco. The material from Ethiopia seems homogenous. Recognition of infraspecific taxa not relevant here, but the wild Ethiopian taxon has been referred to with its infraspecific name to distinguish it from the cultivated subspecies.
Description: Tree to 20 m tall, in open habitats with very broad crown. No thorns. Leaves simple, evergreen. Microphyll. Bark on trunk rough and thick, probably an adaptation for fire resistance.
Habitat: 'Evergreen *Juniperus-Podocarpus* forest' (FEE for *Olea europaea* subsp. *cuspidata*). 'Undifferentiated Afromontane forest ..., persisting in gully forests; also in secondary or high-altitude Afromontane evergreen bushland ... [TSEB], and as single trees in derived grassland and farmland' (FTNA for *Olea europaea* subsp. *cuspidata*). '*Juniperus* forest, evergreen bushland and woodland' (FS). '*Juniperus-Podocarpus* forest' (PSS for *Olea europaea* subsp. *cuspidata*). Mainly in Dry Afromontane forest (DAF), montane bushland and semi-evergreen bushland, a few records in *Combretum-Terminalia* woodland (our observation).

Distribution in Ethiopia: TU GD AF WU SU KF GG SD BA HA (FEE). Also recorded from the WG floristic region (Fig. 6-146)
Altitudinal range in Ethiopia: (1250-) 1700-2700 (-3000) m a.s.l. (FEE, including range in Eritrea). 1250-3100 m a.s.l. (FTNA, including the range in the entire Horn of Africa). 1400-2750 m a.s.l. (our records).
General distribution: See above under '*Taxonomy*.'
Chorological classification: Distribution according to direct observation. **WWn, WWc, WWs, EWW**. *Local phytogeographical element.* **MI**. Marginally intruding species in the western woodlands, with main distribution in the highlands and in *Transitional semi-evergreen bushland* (TSEB), the records from the *Acacia-Commiphora bushland* (ACB) are from small areas on hills (Gerire Hills) of *Transitional semi-evergreen bushland* (TSEB); only a few records in our relevés.

6.*41.2*. Schrebera alata *(Hochst.) Welw. (1869).*

FTEA, Oleaceae: 4 (1952); FTNA: 237 (1992); FEE 4(1): 82 (2003).
Taxonomy: No infraspecific taxa.
Description: Shrub or tree to 10 (-24) m tall. No thorns. Leaves compound (pinnate), evergreen. Mesophyll (leaflets mesophyll). Bark on trunk smooth, probably no adaptation for fire resistance.
Habitat: 'Dry and moist montane *Juniperus-Olea-Podocarpus* forest, particularly along margins and in clearings, secondary evergreen bushland, occasionally rocky savanna' (FEE). 'Afromontane rain forest, undifferentiated Afromontane forest ... and dry single-dominant Afromontane forest ..., particularly in clearings and along margins; sometimes as undergrowth in open *Juniperus* forest; also in secondary Afromontane evergreen bushland' (FTNA). [No record in PSS]. Mainly in Afromontane forest, montane bushland and semi-evergreen bushland, a few records in *Combretum-Terminalia* woodland (our observation).
Distribution in Ethiopia: TU GD SU AR IL KF SD BA HA (FEE). Also recorded from the GJ, WG, and GG floristic regions (Fig. 6-147).
Altitudinal range in Ethiopia: (1500-) 1700-2100 (-2500) m a.s.l. (FEE, including range in Eritrea). 1500-2500 m a.s.l. (FTNA, including the range in the entire Horn of Africa). 1400-2200 m a.s.l. (our records).
General distribution: Ethiopia and Eritrea, south to Angola and South Africa.
Chorological classification: Distribution according to direct observation. **WWn, WWc, WWs, EWW**. *Local phytogeographical element.* **MI**. Marginally intruding species in the western woodlands, with main distribution in the highlands and in *Transitional semi-evergreen bushland* (TSEB); only a few records in our relevés.

6.42. Apocynaceae

BS 51: 316-319 (1998); FTEA, Apocynac. 1: 1-116 (2002); FEE 4(1): 87-98 (2003); FS 3: 117-197 (2006), as **Apocynaceae (incl. *Asclepiadaceae*)**; PSS: 304-311 (2015), also including Asclepiadaceae.

6.*42.1*. Carissa spinarum *L. (1767).*

FTEA, Apocynac. 1: 12 (2002); FEE 4(1): 90 (2003); FS 3: 130 (2006); PSS: 305 (2015).
Carissa edulis (Forssk.) Vahl (1790).
BS 51: 317 (1998).
Taxonomy: No infraspecific taxa.
Description: Shrub to 3 m tall, sometimes scrambling to 20 m tall. With single or more often a pair of simple, rarely forked spines at the leaves or leaf-scars. Leaves simple, probably deciduous. Mesophyll. Bark smooth, probably no adaptation for fire resistance.
Habitat: 'In open *Acacia* woodland, often on termite mounds, and in riverine fringing vegetation' (FEE). 'Evergreen to semi-evergreen bushland, *Juniperus* forest' (FS). 'Wet forest' (PSS). Mainly in Afromontane forest, montane bushland and semi-evergreen bushland, a few records in *Combretum-Terminalia* woodland (our observation).
Distribution in Ethiopia: TU GD GJ AF WU SU AR WG KF GG SD BA HA (FEE). Also recorded from the IL floristic region (Fig. 6-148).
Altitudinal range in Ethiopia: (550-) 1000-2500 m a.s.l. (FEE, including range in Eritrea). 550-2400 m a.s.l. (our records).

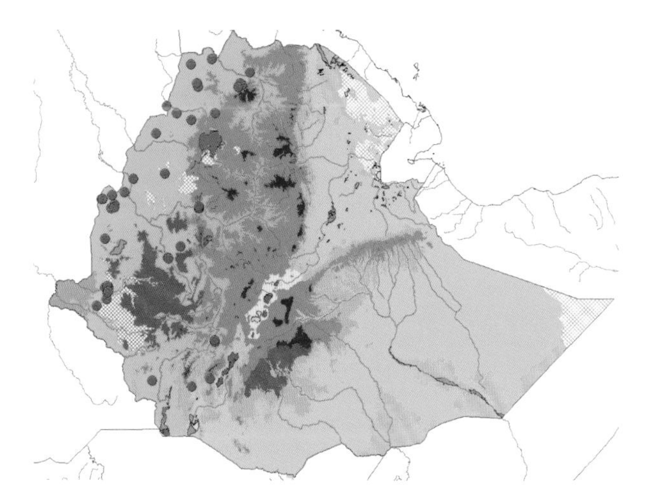

Fig. 6-145. *Strychnos innocua*. Distribution in Ethiopia. Observed records are marked with red dots on vegetation map reproduced from Fig. 4-1.

Fig. 6-146. *Olea europaea* subsp. *cuspidata*. Distribution in Ethiopia. Observed records are marked with red dots on vegetation map reproduced from Fig. 4-1.

Fig. 6-147. *Schrebera alata*. Distribution in Ethiopia. Observed records are marked with red dots on vegetation map reproduced from Fig. 4-1.

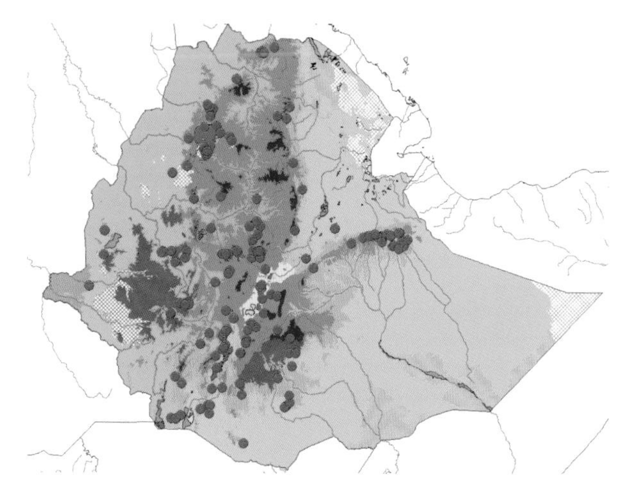

Fig. 6-148. *Carissa spinarum*. Distribution in Ethiopia. Observed records are marked with red dots on vegetation map reproduced from Fig. 4-1.

General distribution: From Mali to Ethiopia, Eritrea and Somalia, south to South Africa; also in Arabia and India, throughout tropical Asia to China and in New Guinea and Australia.
Chorological classification: Distribution according to direct observation. **WWn, WWc, WWs, EWW**. *Local phytogeographical element.* **MI**. Marginally intruding species in the western woodlands, particularly in clumps of evergreen and semi-evergreen shrubs on termite mounds, with main distribution in the highlands and in *Transitional semi-evergreen bushland* (TSEB); only a few records in our relevés.

6.43. Asclepiadaceae

BS 51: 319-326 (1998); FEE 4(1): 99-193 (2003); FS 3: 117-197 (2006), as part of **Apocynaceae**; FTEA, Apocynac. 2: 115-533 (2012), as part of **Apocynaceae**; PSS: 304-311 (2015), as part of **Apocynaceae**.

6.43.1. **Calotropis procera** *(Aiton) W.T. Aiton (1811).*

BS 51: 320 (1998); FEE 4(1): 119 (2003); FS 3: 144 (2006); FTEA, Apocynac. 2: 368 (2012); PSS: 305 (2015).
Taxonomy: No infraspecific taxa.
Description: Large shrub to 5 m tall, ultimate part of stems herbaceous. No thorns. Leaves simple, evergreen. Macrophyll. Bark on trunk smooth, probably no adaptation for fire resistance.
Habitat: 'Hot and dry country, common in dry riverbeds or disturbed areas' (FEE). 'Bushland, commonly in dry riverbeds and disturbed areas' (FS). 'Dry sandy areas' (PSS). In *Combretum-Terminalia* woodland, semi-evergreen bushland, *Acacia-Commiphora* bushland and Semi-Desert Scrub, mainly in overgrazed and degraded places (our observation).
Distribution in Ethiopia: TU GD AF WU SU AR IL GG SD BA HA (FEE). Also recorded from the GJ floristic region (Fig. 6-149).
Altitudinal range in Ethiopia: Sea level to 2250 m a.s.l. (FEE, including range in Eritrea). 700-2250 m a.s.l. (our records).
General distribution: From Mauritania and Senegal to Ethiopia, Eritrea and Somalia, south to Zimbabwe; also in North Africa and in Arabia, eastwards to India, introduced elsewhere.
Chorological classification: Distribution according to direct observation. **WWn, WWc, WWs, ASO**. *Local phytogeographical element and sub-element.* **ACB[-CTW]**. Some distribution in the *Acacia-Commiphora bushland* (ACB), marginally intruding in the *Combretum-Terminalia woodland and wooded grassland* (CTW) in the north or in the south and records from the Gambela lowlands. This species could possibly be classified as **TRG**, but we did not accept this because the distribution of *Calotropis procera* is in general restricted to only two vegetation types, *Combretum-Terminalia woodland and wooded grassland* (CTW) and *Acacia-Commiphora bushland* (ACB).

6.44. Rubiaceae

FTEA, Rubiac. 2: 415-748 (1988); FTEA, Rubiac. 3: 749-957 (1988); *BS* 51: 326-350 (1998); FEE 4(1): 194-282 (2003); FS 3: 61-110 (2006); PSS: 289-302 (2015).

6.44.1. **Crossopteryx febrifuga** *(G. Don) Benth. (1849).*

FTEA, Rubiac. 2: 457 (1988); *BS* 51: 330 (1998); FEE 4(1): 245 (2003); PSS: 290 (2015).
Taxonomy: No infraspecific taxa.
Description: Tree to 15 m tall. No thorns. Leaves simple, evergreen ('subpersistent' according to FTEA). Mesophyll. Bark on trunk scaly and reticulate, probably an adaptation for fire resistance.
Habitat: 'Deciduous woodland' (FEE). 'Woodland and wooded grassland' (PSS). Central part of the *Combretum-Terminalia* woodland (our observation).
Distribution in Ethiopia: WG IL (FEE). Also recorded from the GJ and KF floristic regions (Fig. 6-150).
Altitudinal range in Ethiopia: 500-1300 m a.s.l. (FEE). 500-1300 m a.s.l. (our records).
General distribution: From Senegal to Ethiopia, south to Namibia and South Africa.

Chorological classification: Distribution according to direct observation. [WWn], WWc. *Local phytogeographical element and sub-element.* CTW[c]. Widely distributed in the central part of the western woodlands, mainly south of the Abay River; no eastward extension.

6.44.2. Fadogia cienkowskii *Schweinf. (1868).*

FTEA, Rubiac. 3: 788 (1991); *BS* 51: 330 (1998); FEE 4(1): 275 (2003); PSS: 291 (2025).
Taxonomy: Two rather poorly defined infraspecific taxa (according to FEE) are recognized, the widespread *Fadogia cienkowskii* var. *cienkowskii* and *F. cienkowskii* var. *lanceolata* Robyns (1928), distributed from Cameroon to Tanzania, south to Angola. In Ethiopia only rather homogenous material of *F. cienkowskii* var. *cienkowskii*. Recognition of infraspecific taxa not relevant here.
Description: Subshrub to 1.5 m tall from woody rhizome. No thorns. Leaves simple, probably evergreen ('often persistent' according to FTEA). Mesophyll. Bark smooth, probably no adaptation for fire resistance, but the woody rootstock, from which new shoots may spout after fires, is an adaptation for fire resistance.
Habitat: 'Grassland, wooded grassland or woodland' (FEE for *Fadogia cienkowskii* var. *cienkowskii*). 'Grassland and wooded grassland' (PSS). Central part of the *Combretum-Terminalia* woodland (our observation).
Distribution in Ethiopia: WG (FEE). No additional records (Fig. 6-151).
Altitudinal range in Ethiopia: 1200-1600 m a.s.l. (FEE). 1200-1600 m a.s.l. (our records).
General distribution: From Mali to Ethiopia, south to Angola and South Africa.
Chorological classification: Distribution according to direct observation. WWc. *Local phytogeographical element and sub-element.* CTW[c]. Widely distributed in the central part of the western woodlands, mainly south of the Abay River; no eastward extension.

6.44.3. Gardenia ternifolia *Schumach. & Thonn. (1827).*

FTEA, Rubiac. 2: 509 (1988); *BS* 51: 332 (1998); FEE 4(1): 253 (2003): PSS: 292 (2025).
Gardenia ternifolia subsp. **ternifolia**
FTEA, Rubiac. 2: 509 (1988); *BS* 51: 332 (1998); FEE 4(1): 253 (2003): PSS: 292 (2025).
Gardenia ternifolia subsp. **jovis-tonantis** (Welw.) Verdc. (1979)
FTEA, Rubiac. 2: 509 (1988); FEE 4(1): 253 (2003): PSS: 292 (2025).
Taxonomy: Two subspecies, *Gardenia ternifolia* subsp. *ternifolia* and *G. ternifolia* subsp. *jovis-tonantis*, are recognized in this species, both recorded from Ethiopia. The general distribution of *G. ternifolia* subsp. *ternifolia* is from Senegal to Ethiopia, while the general distribution of the general distribution of *G. ternifolia* subsp. *jovis-tonantis* is from Nigeria and Cameroon to Ethiopia and from Ethiopia and Eritrea to Angola and Mozambique. A variety, *G. ternifolia* subsp. *jovis-tonantis* var. *goetzei* (Stapf & Hutch.) Verdc. (1979) occurs only south of Ethiopia and is rejected in FEE. The distinction between the two subspecies is based on slight differences in the indumentum, and although one seems to be restricted to the southwestern part of Ethiopia, there seems to be no relevance in the distinction of the two subspecies here.
Description: Shrub or small tree to 6 m tall. No thorns. Leaves simple, arranged ternately at the end of branches, probably evergreen. Mesophyll. Bark on trunk smooth, sometimes breaking down to form a yellowish or reddish powder, but without any protective layer, probably not an adaptation for fire resistance.
Habitat: 'Deciduous open woodland' (FEE for *Gardenia ternifolia* subsp. *ternifolia*), 'natural or disturbed woodland, grassland with scattered trees and scrub' (FEE for *Gardenia ternifolia* subsp. *jovis-tonantis*). 'Woodland and bushland' (PSS for *Gardenia ternifolia* subsp. *ternifolia*), 'wooded grassland and woodland' (PSS for *Gardenia ternifolia* subsp. *jovis-tonantis*). In *Combretum-Terminalia* woodland, also in secondary semi-evergreen vegetation in the lower Afromontane zone (our observation).

Fig. 6-149. *Calotropis procera*. Distribution in Ethiopia. Observed records are marked with red dots on vegetation map reproduced from Fig. 4-1.

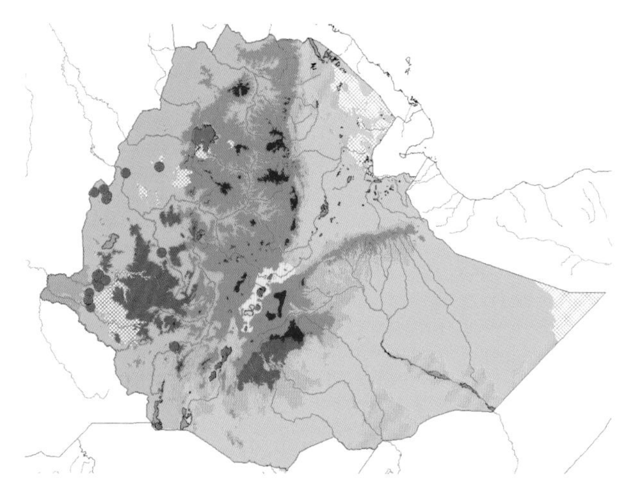

Fig. 6-150. *Crossopteryx febrifuga*. Distribution in Ethiopia. Observed records are marked with red dots on vegetation map reproduced from Fig. 4-1.

Fig. 6-151. *Fadogia cienkowskii*. Distribution in Ethiopia. Observed records are marked with red dots on vegetation map reproduced from Fig. 4-1.

Fig. 6-152. *Gardenia ternifolia*. Distribution in Ethiopia. Observed records are marked with red dots on vegetation map reproduced from Fig. 4-1.

Distribution in Ethiopia: IL KF WG (FEE for *Gardenia ternifolia* subsp. *ternifolia*), TU GD GJ WU SU AR WG IL KF SD BA HA (FEE for *Gardenia ternifolia* subsp. *jovis-tonantis*). Also recorded from the GG floristic region (Fig. 6-152).
Altitudinal range in Ethiopia: 900-1500 m a.s.l. (FEE for *Gardenia ternifolia* subsp. *ternifolia*), 900-2250 m a.s.l. (FEE for *Gardenia ternifolia* subsp. *jovis-tonantis*, including range in Eritrea). 500-2000 m a.s.l. (our records).
General distribution: See above under '*Taxonomy.*'
Chorological classification: Distribution according to direct observation. **WWn, WWc, WWs, EWW**. *Local phytogeographical element and sub-element.* **CTW[all+ext]**. Widely distributed in the western woodlands, from north to south, with extension to the Rift Valley and to east of the Rift.

6.44.4. Hymenodictyon floribundum *B.L. Rob. (1901).*

FTEA, Rubiac. 2: 452 (1988); *BS* 51: 333 (1998); FEE 4(1): 245 (2003); PSS: 293 (2015).
Taxonomy: No infraspecific taxa.
Description: Shrub or small tree to 9 m tall. No thorns. Leaves simple, deciduous, turning scarlet or crimson before falling. Mesophyll. Bark on main trunk rough and flaking reticulate, probably an adaptation for fire resistance.
Habitat: 'Rocky outcrops in woodland, occasionally along streams in forest' (FEE). 'Upland woodland, often on rocky outcrops' (PSS). Widespread, but closely associated with rocky outcrops in the *Combretum-Terminalia* woodland, also in secondary semi-evergreen vegetation in the lower Afromontane zone (our observation).
Distribution in Ethiopia: TU GD GJ SU WG IL GG SD (FEE). Also recorded from the KF floristic region (Fig. 6-153).
Altitudinal range in Ethiopia: 1100-2100 m a.s.l. (FEE, including range in Eritrea). 850-2000 m a.s.l. (our records).
General distribution: From Guinean Republic to Ethiopia and Eritrea, south to Angola and Zimbabwe.
Chorological classification: Distribution according to direct observation. **WWn, WWc, WWs, EWW**. *Local phytogeographical element and sub-element.* **CTW[all+ext]**. Widely distributed in the western woodlands, from north to south, mostly associated with rocky outcrops, with extension to the Rift Valley and to the east of the Rift.

6.44.5. Meyna tetraphylla *(Schweinf. ex Hiern) Robyns (1928).*

FTEA, Rubiac. 3: 859 (1991); *BS* 51: 335 (1998); FEE 4(1): 277 (2003); FS 3: 95 (2006); PSS: 294 (2015).
Taxonomy: Two infraspecific taxa are recognized in this species, the more northern *Meyna tetraphylla* subsp. *tetraphylla* and *M. tetraphylla* subsp. *comorensis* (Robyns) Verdc. (1982) in Kenya, Tanzania and on the Comoro Islands; the distinction of the latter is doubted in FEE. All Ethiopian material belongs to *M. tetraphylla* subsp. *tetraphylla*. Recognition of infraspecific taxa not relevant here.
Description: Shrub or tree to 9 m tall. Thorns present, up to 2.5 cm long, usually supraaxillary or distributed freely on the stems, formed by modified branches. Leaves simple, mostly on short-shoots, clustered with the flowers, probably evergreen. Mesophyll. Bark on trunk smooth or sometimes slightly longitudinally fissured, probably no adaptation for fire resistance.
Habitat: 'Woodland, often associated with thickets in gullies' (FEE). 'Deciduous bushland, often in seasonally flooded places or at base of granitic inselbergs' (FS). 'Woodland, riverine scrub' (PSS). Associated with rocky outcrops in the southern part of the *Combretum-Terminalia* woodland (our observation).
Distribution in Ethiopia: IL GG (FEE). No additional records (Fig. 6-154).
Altitudinal range in Ethiopia: 550-1500 m a.s.l. (FEE). 600-1240 m a.s.l. (our records).
General distribution (species as a whole): South Sudan, Ethiopia and Somalia, south to Tanzania; also in the Comoro Islands.
Chorological classification: Distribution according to direct observation. **WWc, WWs**. *Local phytogeographical element and sub-element.* **CTW[c+s]**. Widely distributed in

Fig. 6-153. *Hymenodictyon floribundum*. Distribution in Ethiopia. Observed records are marked with red dots on vegetation map reproduced from Fig. 4-1.

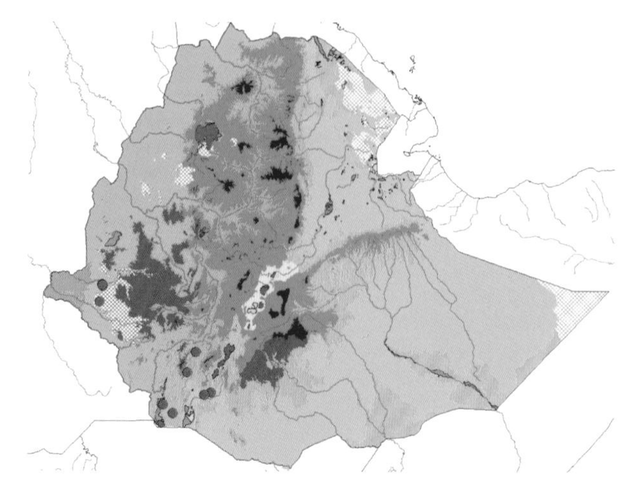

Fig. 6-154. *Meyna tetraphylla*. Distribution in Ethiopia. Observed records are marked with red dots on vegetation map reproduced from Fig. 4-1.

Fig. 6-155. *Pavetta crassipes*. Distribution in Ethiopia. Observed records are marked with red dots on vegetation map reproduced from Fig. 4-1.

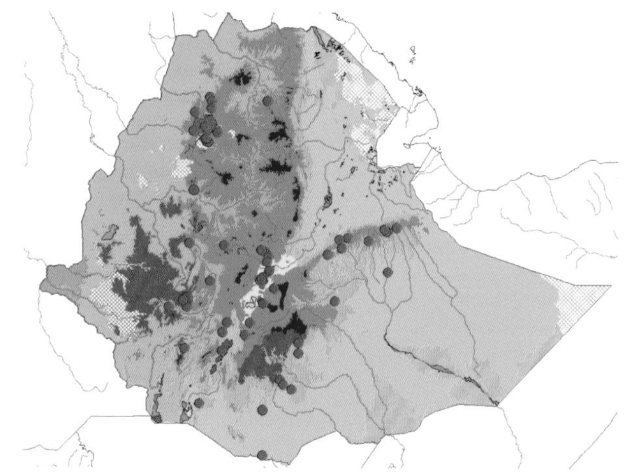

Fig. 6-156. *Pavetta oliveriana*. Distribution in Ethiopia. Observed records are marked with red dots on vegetation map reproduced from Fig. 4-1.

the southern part of the western woodlands, mainly south of the Abay River; no eastward extension.

6.44.6. Pavetta crassipes *K. Schum. (1895).*

FTEA, Rubiac. 2: 670 (1988); *BS* 51: 339 (1998); FEE 4(1): 265 (2003); PSS: 296 (2015).

Taxonomy: No infraspecific taxa.

Description: Shrub or small tree to 8 m tall. No thorns. Leaves clustered near the tip of the branches, simple, probably evergreen. Mesophyll. Bark on trunk with layers of corky, cracking bark, probably an adaptation for fire resistance.

Habitat: 'Deciduous woodland' (FEE). 'Woodland' (PSS). Widespread, but apparently not common in the *Combretum-Terminalia* woodland, also in secondary semi-evergreen vegetation (our observation).

Distribution in Ethiopia: KF GG SD (FEE). Also recorded from the GD and WG floristic regions (Fig. 6-155).

Altitudinal range in Ethiopia: 900-1500 m a.s.l. (FEE). 750-1500 m a.s.l. (our records).

General distribution: From Senegal to Ethiopia, south to Zambia and Mozambique.

Chorological classification: Distribution according to direct observation. [WWn], WWc, WWs, [EWW]. *Local phytogeographical element and sub-element.* CTW[c+s]. Widely distributed in the western woodlands, from north to south (or almost so), with marginal extension to the Rift Valley. The records of this species are few, particularly in the north; with its large, semi-succulent leaves it may be undercollected.

6.44.7. Pavetta oliveriana *Hiern (1877).*

FTEA, Rubiac. 2: 644 (1988); *BS* 51: 340 (1998); FEE 4(1): 263 (2003); PSS: 296 (2015).

Taxonomy: Two infraspecific taxa are currently accepted in this species, *Pavetta oliveriana* var. *oliveriana* and *P. oliveriana* var. *denudata* (Bremek.) Bridson (1978), which has been recorded from Ethiopia and Tanzania. However, FEE maintains that all the Ethiopian material that has been given the illegitimate name *Pavetta abyssinica* var. *cinerascens* A. Rich., should be placed in *Pavetta oliveriana* without formal infraspecific status. Recognition of infraspecific taxa not relevant here.

Description: Shrub or small tree to 4 (-7.5) m tall. No thorns. Leaves simple, probably evergreen. Mesophyll. Bark smooth, probably no adaptation for fire resistance.

Habitat: 'Riverine vegetation, open deciduous woodland, rocky outcrops with bushy clumps, montane scrub or forest remnants, *Olea-Podocarpus* forest' (FEE for *Pavetta oliveriana* var. *oliveriana*). 'Montane forest and grassland' (PSS for *Pavetta oliveriana* var. *oliveriana*). In bushland of the lower Afromontane zone and in secondary semi-evergreen vegetation, a few records in the *Combretum-Terminalia* woodland in the deep river valleys (our observation).

Distribution in Ethiopia: GD GJ WU SU AR WG IL KF GG SD BA HA (FEE for *Pavetta oliveriana* var. *oliveriana*). The FEE record from the IL floristic region has not been confirmed (Fig. 6-156).

Altitudinal range in Ethiopia: (1150-) 1450-2450 m a.s.l. (FEE for *Pavetta oliveriana* var. *oliveriana*, including range in Eritrea). 1500-2150 m a.s.l. (our records).

General distribution (species as a whole): South Sudan, Ethiopia and Eritrea, south to D.R. Congo and Tanzania.

Chorological classification: Distribution according to direct observation. WWn, WWc, WWs, EWW. *Local phytogeographical element.* MI. Marginally intruding species in the western woodlands, with main distribution in bushland in the highlands and in *Transitional semi-evergreen bushland* (TSEB); only few records in our relevés. The record in SD from inside *Acacia-Commiphora bushland* (ACB), *Friis et al.* 8753 (C, ETH,K) 20 km north-west of Moyale, is from a rocky outcrop with woodland.

6.44.8. Sarcocephalus latifolius *(Sm.) Bruce (1947).*

FTEA, Rubiac. 2: 439 (1988); *BS* 51: 347 (1998); FEE 4(1): 241 (2003); PSS: 300 (2015).

Later used name: **Nauclea latifolia** Sm. (1813).

Taxonomy: No infraspecific taxa.

Description: Small tree to 9 m tall. No thorns. Leaves simple, deciduous, becoming reddish before falling. Mesophyll (sometimes Macrophyll). Bark on trunk very fibrous and deeply fissured, probably an adaptation for fire resistance.
Habitat: 'Deciduous woodland, often in moist areas (near ponds, rivulets, etc.)' (FEE). 'Woodland and forest' (PSS). In *Combretum-Terminalia* woodland, with a few records from forest margins in the lower Afromontane zone (our observation).
Distribution in Ethiopia: GJ IL KF (FEE). Also recorded from the GD, WG and KF floristic regions (Fig. 6-157).
Altitudinal range in Ethiopia: 500-1500 m a.s.l. (FEE). 550-1600 m a.s.l. (our records).
General distribution: From Senegal to Ethiopia, south to Kenya and Tanzania.
Chorological classification: Distribution according to direct observation. WWn, WWc, [WWs]. *Local phytogeographical element and sub-element.* CTW[n+c]. Widely distributed in the northern and central part of the western woodlands, reaching the Gambela lowlands in the south; no eastward extension.

6.44.9. Vangueria madagascariensis *J.F. Gmel. (1791).*

FTEA, Rubiac. 3: 849 (1991); *BS* 51: 350 (1998); FEE 4(1): 279 (2003); PSS: 302 (2015).
Taxonomy: FEE proposed two infraspecific taxa, the widespread *Vangueria madagascariensis* var. *madagascariensis* and *V. madagascariensis* var. *abyssinica* (A. Rich.) Puff (2003), possibly restricted to Eritrea and the northernmost part of Ethiopia. The varieties are distinguished on differences in the indumentum, but FEE also claimed an ecological distinction, as *V. madagascariensis* var. *madagascariensis* is supposed to be widespread and associated with rocky outcrops in woodland or dry forest, while *Vangueria madagascariensis* var. *abyssinica* grows in *Riparian vegetation* (RV) in north-western Ethiopia. There is very little material from Ethiopia to support this; the type of var. *abyssinica* was collected from near the Tacazze River but may also have been from rocky outcrops. *Friis et al.* 6806, from near the Eritrean border and Sheraro, was collected in *Riparian vegetation* (RV) with *Tamarindus indica*, while *Friis et al.* 13554 from north of Guba and the Sudan border in Gojam was collected along a wadi on sandy soil. It seems that there may be a continuous ecological transition from riparian habitats to habitats at the base of rocks. Therefore, recognition of infraspecific taxa is not possible here.
Description: Shrub or tree to 15 m tall. No thorns. Leaves simple, deciduous. Macrophyll. Bark on trunk usually smooth and not powdery, probably no adaptation for fire resistance (according to FTEA a form in southern Tanzania has powdery bark).
Habitat: 'Associated with rocky outcrops in *Acacia-Commiphora* and other woodlands or in *Juniperus* forest, often at base of large boulders' (FEE for *Vangueria madagascariensis* var. *madagascariensis*), 'Riverine vegetation' (FEE for *Vangueria madagascariensis* var. *abyssinica*). 'Forest, woodland and bushland' (PSS for *Vangueria madagascariensis* without infraspecific taxa). In *Combretum-Terminalia* woodland, with a few records from forest margins in the lower Afromontane zone, often associated with rocky outcrops (our observation).
Distribution in Ethiopia: TU GD SU GG SD BA (FEE for *Vangueria madagascariensis* var. *abyssinica*), TU (FEE for *Vangueria madagascariensis* var. *abyssinica*). Also recorded from the GJ, WG and HA floristic provinces. The isolated records from SD are based on two collections from near Wachile (*Corradi* 2676 (FT), *Bally* 9265 (K) and one at Moyale (*Cufodontis* 731 (FT). One isolated record from HA is based on *Burger et al.* 1778 (WAG), near Gara Ades). These records require confirmation (Fig. 6-158).
Altitudinal range in Ethiopia: 1000-1600 m a.s.l. (FEE for *Vangueria madagascariensis* var. *madagascariensis*, including range in Eritrea), 450-1300 m a.s.l. (FEE for *Vangueria madagascariensis* var. *abyssinica*, including range in Eritrea). 600-1850 m a.s.l. (our records).
General distribution: From Ghana to Ethiopia and Eritrea, south to South Africa; also Madagascar, cultivated for its edible fruits in tropical Asia and the West Indies.
Chorological classification: Distribution according to direct

observation. **WWn, WWc, WWs, EWW**. *Local phytogeographical element and sub-element.* **CTW[all]**. Widely distributed in the western woodlands, from north to south or almost so. The extension to the Rift Valley and to the east of the Rift are doubtful (see above under '*Distribution in Ethiopia*'); therefore the classification **CTW[all+ext]** has not been applied here.

6.45. Asteraceae

BS 51: 351-386 (1998); FTEA, Compositae, 1: 1-315 (2000); FEE 4(2): 1-350 (2004); PSS: 358-378 (2015), as "Asteraceae (Compositae)".

6.45.1. **Vernonia amygdalina** *Delile (1826).*

FTNA: 256 (1992); *BS* 51: 381 (1998); FTEA, Compositae, 1: 178 (2000); FEE 4(2): 78 (2004); PSS: 376 (2015).
Later used name: **Gymnanthemum amygdalinum** (Delile) Sch. Bip. ex Walp. (1843).
Taxonomy: No infraspecific taxa.
Description: Shrub or small tree to 9 m tall. No thorns. Leaves simple, evergreen. Mesophyll. Bark on trunk faintly longitudinally fissured, probably no adaptation for fire resistance.
Habitat: '*Podocarpus* or *Aningeria* forest, usually in open spots near streams, or in fringe of glades, secondary forests, evergreen woodland or bushland, roadsides, wasteland, also grown in backyard gardens' (FEE). 'Afromontane rain forest, undifferentiated Afromontane forest ... and dry single-dominant Afromontane forest ...; also in secondary montane evergreen bushland, and sometimes forming clumps in upland wooded grassland' (FTNA). 'Forest margins, woodland, bushland and grassland' (PSS). In a wide range of habitats, from forest edges of *Dry* and *Moist evergreen Afromontane forest* (DAF and MAF) to moist or dry woodlands or *Acacia-Commiphora* bushland, and in gardens (our observation).
Distribution in Ethiopia: TU GD GJ?WU SU WG IL KF GG SD BA HA (FEE). A doubtful FEE record from WU has not been confirmed (Fig. 6-159).
Altitudinal range in Ethiopia: (650-) 1200-3000 m a.s.l. (FEE, including range in Eritrea). (600-) 1250-2800 m a.s.l. (FTNA, including the range in the entire Horn of Africa). 500-2350 m a.s.l. (our records).
General distribution: From Guinean Republic to Ethiopia and Eritrea, south to Botswana and South Africa; also in Arabia.
Chorological classification: Distribution according to direct observation. **WWn, WWc, WWs, EWW**. *Local phytogeographical element.* **TRG**. Transgressing species, occurs similarly in both the *Combretum-Terminalia woodland and wooded grassland* (CTW), where the records are scattered, and in other vegetation types, particularly *Dry evergreen Afromontane forest* (DAF), but also in *Moist evergreen Afromontane forest* (MAF), and in *Transitional semi-evergreen bushland* (TSEB) to the east of the Rift Valley.

6.45.2. **Vernonia auriculifera** *Hiern (1898).*

FTEA, Compositae, 1: 196 (2000); FEE 4(2): 83 (2004).
Later used name: **Gymnanthemum auriculiferum** (Hiern) Isawumi (2008).
Taxonomy: No infraspecific taxa.
Description: Shrub or small tree to 7.5 m tall. No thorns. Leaves simple, evergreen. Macrophyll. Bark on trunk smooth, probably no adaptation for fire resistance.
Habitat: 'Forest margins, moist woodland or grassland' (FEE). [No record in PSS]. In *Combretum-Terminalia* woodland, with a few records from forest margins of *Dry* and *Moist evergreen Afromontane forests* (DAF and MAF) in the lower Afromontane zone (our observation).
Distribution in Ethiopia: WG IL KF SD?BA (FEE). Also recorded from the GD and GG floristic regions; the doubtful FEE record from BA is confirmed (Fig. 6-160).
Altitudinal range in Ethiopia: 1200-2200 m a.s.l. (FEE). 1200-2450 m a.s.l. (our records).
General distribution: Nigeria, Cameroon, D.R. Congo, Ethiopia, Uganda, Kenya, Tanzania, Angola.
Chorological classification: Distribution according to direct observation. **WWn, WWc, WWs, EWW**. *Local phytogeographical element.* **MI**. Marginally intruding species in

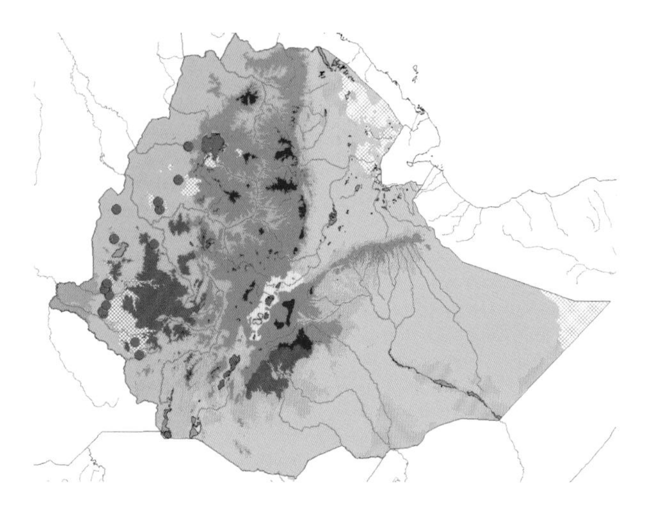

Fig. 6-157. *Sarcocephalus latifolius*. Distribution in Ethiopia. Observed records are marked with red dots on vegetation map reproduced from Fig. 4-1.

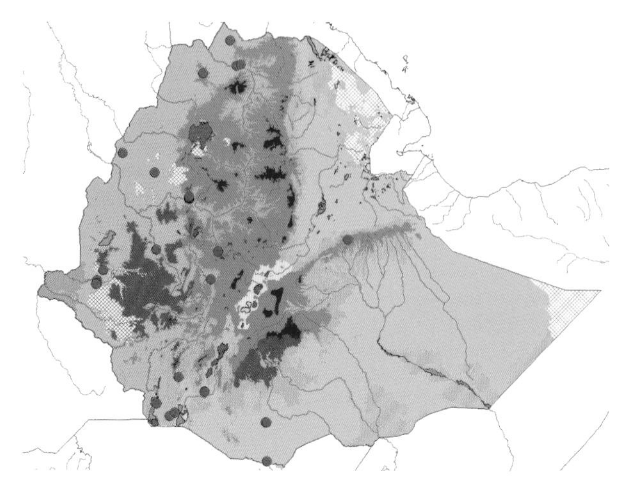

Fig. 6-158. *Vangueria madagascariensis*. Distribution in Ethiopia. Observed records are marked with red dots on vegetation map reproduced from Fig. 4-1.

Fig. 6-159. *Vernonia amygdalina*. Distribution in Ethiopia. Observed records are marked with red dots on vegetation map reproduced from Fig. 4-1.

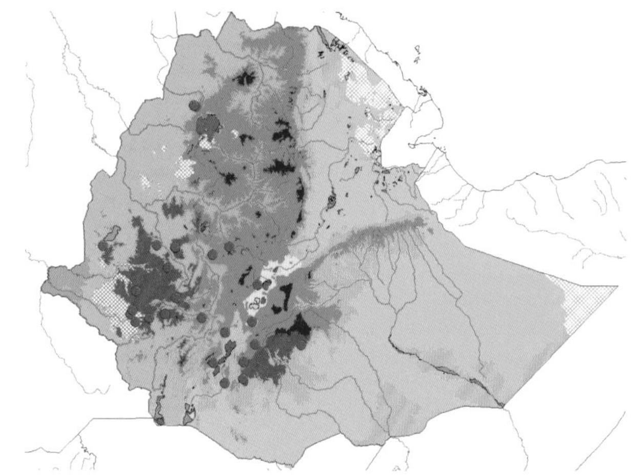

Fig. 6-160. *Vernonia auriculifera*. Distribution in Ethiopia. Observed records are marked with red dots on vegetation map reproduced from Fig. 4-1.

the western woodlands, with main distribution in the highlands and in *Transitional semi-evergreen bushland* (TSEB); only a few records in our relevés.

6.46. Boraginaceae

FTEA, Boraginaceae: 1-125 (1991), including ***Ehretiaceae***; *BS* 51: 409-412 (2005); FEE 5: 64-102 (2006), as **BORAGINACEAE**, including ***Ehretiaceae***; PSS: 312-315 (2015).

6.*46.1.* Cordia africana *Lam. (1792).*

FTEA, Boraginaceae: 31 (1991); FTNA: 258 (1992); *BS* 51: 409 (2005); FEE 5: 68 (2006); PSS: 312 (2015).
Taxonomy: No infraspecific taxa.
Description: Tree to 15 (-30) m tall. No thorns. Leaves simple, evergreen. Mesophyll. Bark on trunk rough and fibrous, peeling and longitudinally fissured, probably an adaptation for fire resistance.
Habitat: 'Open places in moist montane forest, forest edges, forest remnants around churches and other traditionally protected areas, as isolated trees in grassland and cultivated fields, in villages and public gardens' (FEE). 'Afromontane rain forest and undifferentiated Afromontane forest ..., usually along margins and in clearings. An early colonizer in forest regrowth. Often left when forest is cleared for cultivation, as the tree is an excellent shade tree for crops and, as the flowers yield plenty of nectar, beehives are often placed in the tree. Also in riverine forest, in secondary bushland and transgressing into humid types of woodland' (FTNA). 'Secondary and riverine forest and forest-wooded grassland transition' (PSS). A pioneer species in a wide range of habitats, from *Dry* and *Moist evergreen Afromontane forest* (DAF and MAF) to moist woodlands or *Acacia-Commiphora* bushland (our observation).
Distribution in Ethiopia: TU GD GJ WU SU AR WG IL KF GG SD BA HA (FEE). No additional records (Fig. 6-161).
Altitudinal range in Ethiopia: 700-2550 m a.s.l. (FEE, including range in Eritrea). 550-2600 m a.s.l. (FTNA). 500-2450 m a.s.l. (our records).
General distribution: From Guinean Republic to Ethiopia and Eritrea, south to Angola and South Africa; also in Arabia.
Chorological classification: Distribution according to direct observation. **WWn, WWc, WWs, EWW**. *Local phytogeographical element.* **TRG**. Transgressing species, occurs similarly in both the *Combretum-Terminalia woodland and wooded grassland* (CTW) and in other vegetation types, particularly in *Dry evergreen Afromontane forest* (DAF), *Moist evergreen Afromontane forest* (MAF) and *Transitional rain forest* (TRF).

6.47. Bignoniaceae

BS 51: 432-434 (1998); FTEA, Bignoniaceae: 1-52 (2006); FEE 5: 322-334 (2006); FS 3: 303-307 (2006); PSS: 354 (2015).

6.*47.1.* Kigelia africana *(Lam.) Benth. (1849).*

BS 51: 432 (2005); FEE 5: 325 (2006); FTEA, Bignoniaceae: 43 (2006); FS 3: 303 (2006); PSS: 354 (2015).
Taxonomy: The species includes two infraspecific taxa, *Kigelia africana* subsp. *africana* and *K. africana* subsp. *moosa* (Sprague) Bidgood & Verdc. (2006), the latter occurs from Sierra Leone to South Sudan and Kenya, southern Angola and Tanzania. The material from Ethiopia is all *K. africana* subsp. *africana*. Recognition of infraspecific taxa not relevant here.
Description: Shrub or tree to 18 (-24) m tall. No thorns. Leaves pinnate (3-8-jugate), probably evergreen. Macrophyll (leaflets Mesophyll). Bark on trunk smooth to rough or ridged, scaly and flaking, probably an adaptation for fire resistance.
Habitat: 'Woodland and grassland, frequently near water' (FEE). 'Woodland, usually near rivers' (FS). 'Wooded grassland, woodland, edges of riverine forest' (PSS). Widespread, but not common in *Combretum-Terminalia* woodland and semi-evergreen bushland, often associated with water or streams, one record from *Semi-Desert Scrubland* (SDS), presumably near a river (our observation).
Distribution in Ethiopia: GD GJ IL GG (FEE). Also recorded from the WG, KF and BA floristic regions.

The record from southern BA (*Ruspoli & Riva* 759-837 (FT)) is from the 'Web Ruspoli' River and requires confirmation, although *Kigelia africana* is known from the southern Somalia (FS) and the Northern Frontier District of Kenya (Fig. 6-162).
Altitudinal range in Ethiopia: 500-2000 m a.s.l. (FEE, including range in Eritrea). (250-) 500-1500 m a.s.l. (our records).
General distribution (species as a whole): From Senegal to Ethiopia, Eritrea and Somalia, south to Botswana and South Africa.
Chorological classification: Distribution according to direct observation. **WWn, WWc, WWs, EWW**. *Local phytogeographical element and sub-element.* **CTW[all+ext]**. Widely distributed in the western woodlands, from north (apparently lacking in TU, but occurs according to PSS in the adjacent parts of the Sudan) to south, extending to the Rift Valley and apparently one record along rivers in the *Semi-desert scrubland* (SDS) in the south-eastern part of the country (see above under '*Distribution in Ethiopia*').

6.*47.2*. **Stereospermum kunthianum** *Cham. (1832).*

BS 51: 434 (2005); FTEA, Bignoniaceae: 37 (2006); FEE 5: 323 (2006); PSS: 354 (2015).
Taxonomy: No infraspecific taxa.
Description: Tree to 20 m tall. No thorns. Leaves pinnate (5-7-foliolate), presumably deciduous. Macrophyll (leaflets Mesophyll). Bark on trunk flaking in plaques, probably an adaptation for fire resistance.
Habitat: 'Open woodland and savanna' (FEE). 'Wooded grassland and bushland' (PSS). In the *Combretum-Terminalia* woodland, semi-evergreen bushland and in the lower zone of open Afromontane vegetation (our observation).
Distribution in Ethiopia: TU GD GJ SU WG IL KF SD BA HA (FEE). Also recorded from the AF, WU and GG floristic regions (Fig. 6-163).
Altitudinal range in Ethiopia: 500-1950 m a.s.l. (FEE, including range in Eritrea). 550-2100 m a.s.l. (our records).
General distribution: From Senegal to Ethiopia and Eritrea, south to Zimbabwe and Mozambique; also in Arabia.
Chorological classification: Distribution according to direct observation. **WWn, WWc, WWs, EWW**. *Local phytogeographical element and sub-element.* **CTW[all+ext]**. Widely distributed in the western woodlands, from north to south, extending to the Rift Valley and to the east of the Rift.

6.48. Lamiaceae

FTEA, Verbenaceae: 1-156 (1992), including genera which in previous floras were referred to Verbenaceae; *BS* 51: 462-474 (1998), including genera which in previous floras were referred to Verbenaceae; *BS* 51: 457-462 (2005); FS 3: 308-355 (2006), including genera which in previous floras were referred to Verbenaceae (**Premna, Vitex**); FEE 5: 516-604 (2006); PSS: 331-340 (2015), including genera which in previous floras were referred to Verbenaceae (**Premna, Vitex**).

6.*48.1*. **Premna schimperi** *Engl. (1892).*

FTEA, Verbenaceae: 76 (1992); *BS* 51: 460 (2005); FEE 5: 519 (2006); FS 3: 310 (2006); PSS: 338 (2015).
Taxonomy: No infraspecific taxa.
Description: Shrub to 6 m tall. No thorns. Leaves simple, probably deciduous. Mesophyll. Bark on trunk not known, but probably no adaptation for fire resistance.
Habitat: 'Degraded and secondary forest, grassland and along paths in forests' (FEE). 'Evergreen bushland on granite' (FS). 'Degraded forest, secondary bushland and grassland' (PSS). In the lower zone of open Afromontane vegetation or along forest margins, mainly at *Dry evergreen Afromontane forest* (DAF), and in semi-evergreen bushland, also some records in *Combretum-Terminalia* woodland (our observation).
Distribution in Ethiopia: TU GD GJ SU AR WG KF SD HA (FEE). Also recorded from the WU, GG and BA floristic regions; the FEE record from TU is not confirmed (Fig. 6-164).
Altitudinal range in Ethiopia: 1350-2400 m a.s.l. (FEE). 1400-2200 m a.s.l. (our records).

Fig. 6-161. *Cordia africana*. Distribution in Ethiopia. Observed records are marked with red dots on vegetation map reproduced from Fig. 4-1.

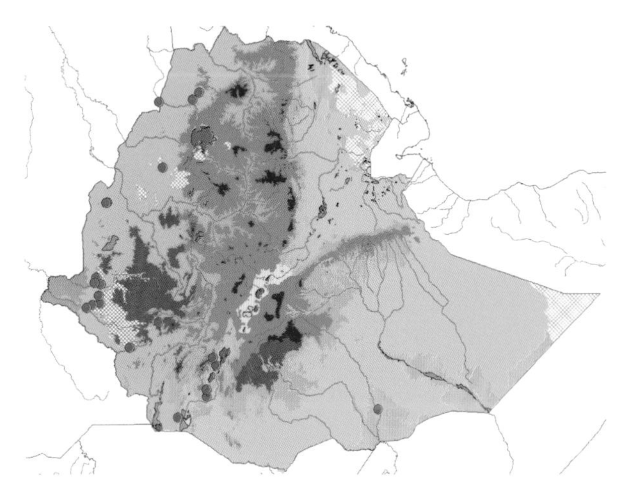

Fig. 6-162. *Kigelia africana*. Distribution in Ethiopia. Observed records are marked with red dots on vegetation map reproduced from Fig. 4-1.

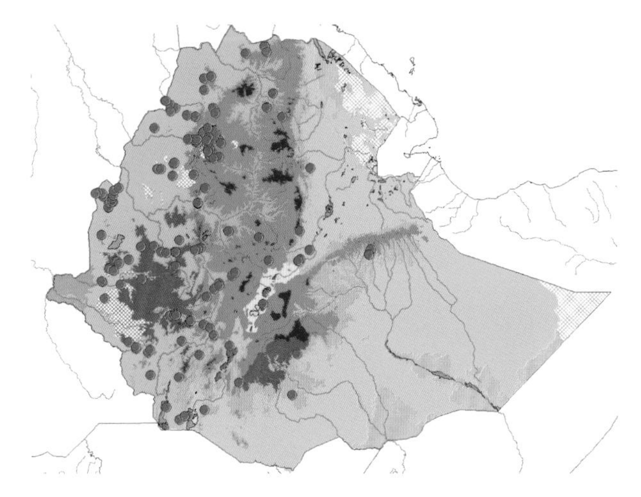

Fig. 6-163. *Stereospermum kunthianum*. Distribution in Ethiopia. Observed records are marked with red dots on vegetation map reproduced from Fig. 4-1.

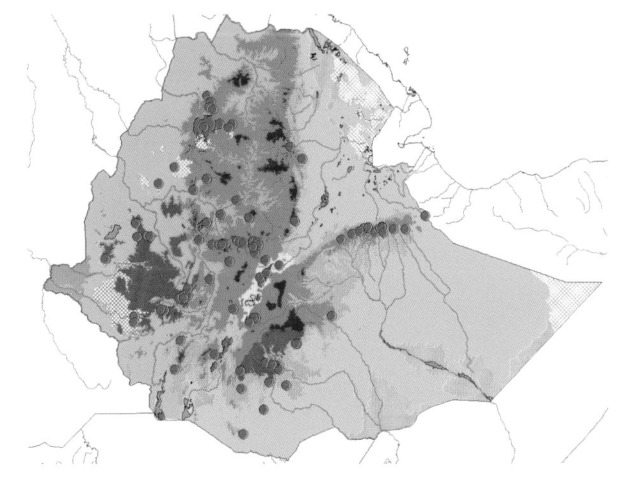

Fig. 6-164. *Premna schimperi*. Distribution in Ethiopia. Observed records are marked with red dots on vegetation map reproduced from Fig. 4-1.

General distribution: South Sudan, Ethiopia, Somalia, south to Tanzania.
Chorological classification: Distribution according to direct observation. **WWn, WWc, WWs, EWW**. *Local phytogeographical element.* **MI**. Marginally intruding species in the western woodlands, with main distribution in the highlands and in *Transitional semi-evergreen bushland* (TSEB); only few records in our relevés.

6.48.2. **Vitex doniana** *Sweet (1827).*

FTEA, Verbenaceae: 62 (1992); *BS* 51: 461 (2005); FEE 5: 521 (2006); PSS: 340 (2025).
Taxonomy: No infraspecific taxa.
Description: Tree to 12(-18) m tall. No thorns. Leaves digitately divided (5-7-foliolate), deciduous. Macrophyll (leaflets Mesophyll or Macrophyll). Bark on trunk thick and deeply fissured, probably an adaptation for fire resistance.
Habitat: '*Combretum-Terminalia* woodland, wooded grassland' (FEE). 'Wooded grassland, woodland, forest margins' (PSS). In *Combretum-Terminalia* woodland (our observation).
Distribution in Ethiopia: GJ WG KF SD (FEE). Also recorded from GD, IL and GG floristic regions (Fig. 6-165).
Altitudinal range in Ethiopia: 500-1960 m a.s.l. (FEE). 500-1950 m a.s.l. (our records).
General distribution: From Senegal to Ethiopia, south to Angola and Mozambique; also in the Comoro Islands.
Chorological classification: Distribution according to direct observation. **WWn, WWc, WWs**. *Local phytogeographical element and sub-element.* **CTW[all]**. Widely distributed in the western woodlands, from north (but apparently absent from the TU floristic region and according to PSS not recorded from the adjacent parts of the Sudan) to south; a slight extension into the *Transitional semi-evergreen bushland* (TSEB) to the west of the Rift Valley; no further eastward extension.

6.48.3. **Otostegia fruticosa** *(Forssk.) Schweinf. ex Penzig (1893).*

FEE 5: 540 (2006); PSS: 336 (2015).
Otostegia fruticosa subsp. **fruticosa**
FEE 5: 540 (2006); PSS: 336 (2015).
Taxonomy: Only one infraspecific taxon, *Otostegia fruticosa* subsp. *fruticosa,* has been recorded from Ethiopia, but a closely related taxon, *O. fruticosa* subsp. *schimperi* (Benth.) Sebald (1973), has been recorded from the eastern lowland of Eritrea. The distinction between the two infraspecific taxa is slight and quantitative and the ecology almost identical. Recognition of infraspecific taxa not relevant here.
Description: Shrub to 3 m tall. No thorns. Leaves simple, presumably evergreen. Mesophyll. Bark smooth, probably no adaptation for fire resistance.
Habitat: 'Rocky slopes, montane bushland, deciduous woodland' (FEE for *Otostegia fruticosa* subsp. *fruticosa*). 'Rocky slopes, bushland' (PSS for *Otostegia fruticosa* subsp. *fruticosa*). In the lower zone of open Afromontane vegetation or along margins of *Dry evergreen Afromontane forest* (DAF) and in semi-evergreen bushland, also a few records in *Combretum-Terminalia* woodland in the deep river valleys in the highlands (our observation).
Distribution in Ethiopia: TU GD SU (FEE). An isolated record from the BA floristic region would seem to require confirmation (*Sebsebe Demissew et al.* 4972 (ETH), 16 km east of Dodola) (Fig. 6-166).
Altitudinal range in Ethiopia: (500-) 1000-2500 (-3000) m a.s.l. (FEE for *Otostegia fruticosa* subsp. *fruticosa*, including range in Eritrea). 1500-2900 m a.s.l. (our records).
General distribution (species as a whole): Cameroon, Sudan, Ethiopia and Eritrea; also in Egypt and Arabia.
Chorological classification: Distribution according to direct observation. **WWn, EWW**. *Local phytogeographical element.* **MI**. Marginally intruding species in the western woodlands, with main distribution in the highlands; only a few records in our relevés.

6.49. Arecaceae

FTEA, Palmae: 1-56 (1986); FEE 6: 513-526 (1997); FS 4: 270-274 (1995); *BS* 51: 504-505 (1998); PSS: 94-95 (2015), as "Arecaceae (Palmae)".

6.*49.1.* Borassus aethiopum *Mart. (1838).*

FTEA, Palmae: 19 (1986); FEE 6: 518 (1997); *BS* 51: 504 (2005); PSS: 94 (2015).
Taxonomy: No infraspecific taxa.
Description: Tree to 30 m tall, with ventricose trunk. No thorns. Leaves simple, fan-shaped or slightly costapalmate, evergreen. Megaphyll. Bark on mature trunk smooth with annular leaf-scars, probably no adaptation for fire resistance.
Habitat: 'Forming stands at the edge of semi-deciduous lowland forest, in depressions in *Terminalia* woodland and along streams, also as isolated tree in fields' (FEE). 'Along watercourses and in drier wooded grassland, often forming dense colonies' (PSS). In the lower, central part of the *Combretum-Terminalia* woodland, often forming clumps near streams (our observation).
Distribution in Ethiopia: GJ IL ?SD (FEE). Also recorded from the WG floristic region; the doubtful FEE record from SD is not confirmed (Fig. 6-167).
Altitudinal range in Ethiopia: 400-950 m a.s.l. (FEE). 600-950 m a.s.l. (our records).
General distribution: From Senegal to Ethiopia, south to South Africa; also in Madagascar and the Comoro Islands.
Chorological classification: Distribution according to direct observation. **WWn, WWc**. *Local phytogeographical element and sub-element.* **CTW[n+c]**. Distributed in the central part of the western woodlands, mainly south of the Abay River; no eastward extension.

6.*49.2.* Hyphaene thebaica *(L.) Mart. (1838).*

FS 4: 273 (1995); FEE 6: 522 (1997); *BS* 51: 504 (2005); PSS: 94 (2015).
Taxonomy: No infraspecific taxa, but the delimitation of the species does not seem yet to be settled. In FEE the typical *Hyphaene thebaica* of the Nile Valley and adjacent pats of Ethiopia has been united with *H. dankaliensis* Becc. (1806) from the Afar region and the adjacent Eritrean coast and *H. nodularia* Becc. (1908) from the western lowlands of Eritrea, but *H. compressa* H. Wendl. (1878) from East Africa was kept as a distinct species, although with doubt about the classification of specimens from southern Ethiopia (particularly the Omo Valley and the Sagan Valley in Gamu Gofa). Because of increasing information about the variation of *H. thebaica* and despite the material from south-western Ethiopia does not show sufficient characteristics, we have decided here to refer this material to *H. thebaica*. The GG records are based on *Gereau et al.* 1408 (ETH), south of Omo River on nearly flat savannah, *Friis et al.* 8845 (C, ETH, K), Woyto Valley, and *Friis et al.* 9539 (C, ETH, K), 70 km south of the border of Mago National Park. Less clear is the identification of a specimen from near Mandera (*Friis et al.* 10020 (C, ETH, K)), which we for morphological reasons have also identified as *H. thebaica* although the material of *Hyphaene* from the southern part of Somalia, Kenya and further south is generally referred to *H. compressa*.
Description: Dichotomously branched tree to 20 m tall. Thorns present; no thorns on trunk, but petiole with recurved and pointed spines. Leaves simple, fan-shaped or slightly costapalmate, evergreen. Megaphyll. Bark on mature trunk smooth with annular leaf-scars that are frequently removed by fire, probably no adaptation for fire resistance.
Habitat: 'In the northwest and west of the Flora area (including Eritrea) mainly in river valleys and around oases, often in damp places in *Terminalia* woodland and on flood plains along rivers, in the Rift Valley and Afar often at hot springs' (FEE). 'Riverine fringes, oases, depressions in woodland' (PSS). In the lower zone of the *Combretum-Terminalia* woodland, rarely extending to semi-evergreen bushland, one record (see above under '*Taxonomy*') near river in *Semi-Desert Scrubland* (SDS) (our observation).
Distribution in Ethiopia: AF GD GJ IL (FEE). Also recorded from the WG, GG, SD and HA floristic regions; see also above under '*Taxonomy*' (Fig. 6-168).

Fig. 6-165. *Vitex doniana*. Distribution in Ethiopia. Observed records are marked with red dots on vegetation map reproduced from Fig. 4-1.

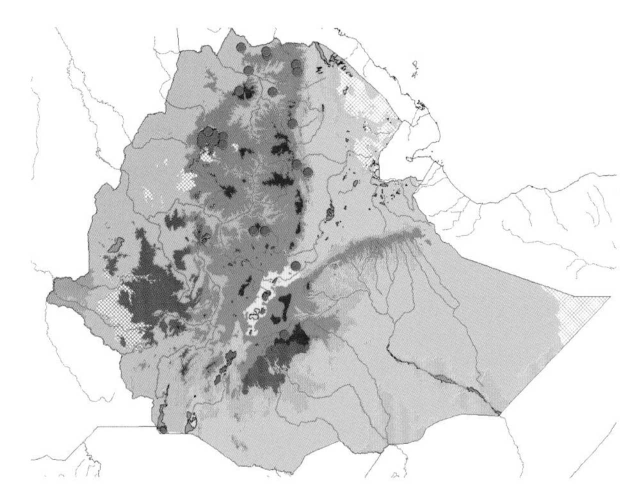

Fig. 6-166. *Otostegia fruticosa*. Distribution in Ethiopia. Observed records are marked with red dots on vegetation map reproduced from Fig. 4-1.

Fig. 6-167. *Borassus aethiopum*. Distribution in Ethiopia. Observed records are marked with red dots on vegetation map reproduced from Fig. 4-1.

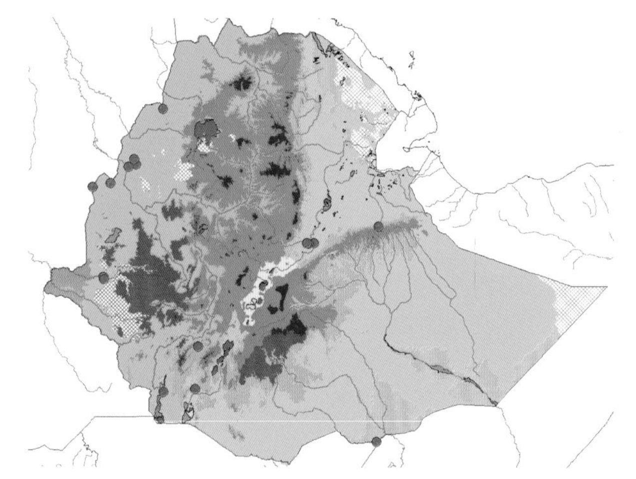

Fig. 6-168. *Hyphaene thebaica*. Distribution in Ethiopia. Observed records are marked with red dots on vegetation map reproduced from Fig. 4-1.

Altitudinal range in Ethiopia: From 100 m below sea level in Afar to 1000 m a.s.l. (FEE, including range in Eritrea). 500-900 m a.s.l. (our records).
General distribution: From Mauritania to Sudan, Ethiopia, Eritrea and Somalia; also Egypt and possibly also Arabia.
Chorological classification: Distribution according to direct observation. **WWn, WWc, WWs, EWW, [ASO].** *Local phytogeographical element and sub-element.* **CTW[all+ext]**. Not common, but widely distributed in the western woodlands, from north (no records from the TU floristic region, but the species is recorded from nearby in Eritrea and in the Sudan) to south, extending to the Rift Valley and to the east of the Rift.

6.50. Poaceae

FTEA, Gramineae, 1: 1-177 (1970); FEE 7: 1-368 (1995), as **POACEAE** (*Gramineae*); *BS* 51: 544-597 (1998); PSS: 117-151 (2015), as "Poaceae (Gramineae)".

6.50.1. Oxytenanthera abyssinica *(A. Rich.) Munro (1868)*

FTEA, Gramineae, 1: 11 (1970); FEE 7: 6 (1995); *BS* 51: 577 (2005); PSS: 138 (2015).
Taxonomy: No infraspecific taxa.
Description: Culms in dense clumps up to 10 m tall. No thorns. Leaves simple, evergreen. Mesophyll. Bark of the bamboo culms smooth and hard, probably no adaptation for fire resistance, but the massive underground rhizomes, from which new shoots may sprout after fires is an adaptation for fire resistance.
Habitat: 'Savanna woodland, favouring river valleys, often forming extensive stands' (FEE). 'Riverbanks in wooded grassland' (PSS). In *Combretum-Terminalia* woodland, often forming extensive stands on rocky slopes or in river valleys (our observation).
Distribution in Ethiopia: TU GD GJ WG (FEE). Also recorded from the SU floristic region (*Tewolde Berhan G.E. et al.* 1726 (ETH), at Fincha dam, below the power house) and the KF floristic region (*Friis et al.* (two sight records in relevés) between Ameya and Omo River) (Fig. 6-169).

Altitudinal range in Ethiopia: 1200-1800 m a.s.l. (FEE, including range in Eritrea). 750-1700 m a.s.l. (our records).
General distribution: From Senegal to Ethiopia and Eritrea, south to Angola and South Africa.
Chorological classification: Distribution according to direct observation. **WWn, WWc, WWs.** *Local phytogeographical element and sub-element.* **CTW[all]**. Widely distributed in the western woodlands, from north to south; no eastward extension.

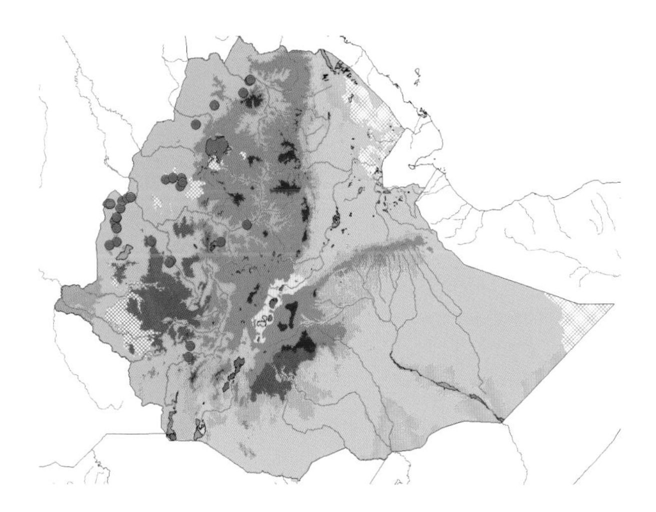

Fig. 6-169. *Oxytenanthera abyssinica*. Distribution in Ethiopia. Observed records are marked with red dots on vegetation map reproduced from Fig. 4-1.

7
Phytogeography and distribution (local, altitudinal, and general African)

The following studies are based on two sets of data (see Materials and Methods). One set is derived only from the relevés as presented in Appendix 1; in this Section it is used for the clustering of altitudinal zones in Fig. 7-5. The other set is derived from a combination of data from our relevés in combination with data from herbarium material and verified records from databases; this set is used for the remaining figures in this Section, Fig. 7-3 and 7-4 and Fig. 7-6 and 7-7. The totality of the relevés are mapped in Fig. 4-3, and the totality of the combined relevé, herbarium and data base records are mapped in Fig. 4-4. As mentioned in '4.1 Materials and Methods' we have attempted the best possible geographical and altitudinal distribution of the relevés. Most of our relevés are located between 500 and 1500 m a.s.l., at altitudes representing 73.7% of the entire CTW-area, while 25 relevés are situated between 1500 and 1999 m a.s.l., at altitudes representing 26.3% of the CTW-area, and only 3 are situated between 2000 and 2499 m a.s.l., representing a negligible part of the CTW-area (Table 7-1), which as stated in Section 3.12 is almost entirely restricted to areas below 1800 m a.s.l. The three altitudinal zones between 500 and 1999 m a.s.l. represent areas varying between ca. 60,000 and 85,000 sqkm. This represents our best approximation to an even geographical and altitudinal representation of the relevés. Our knowledge of the distribution in Ethiopia of the relevé-species outside the CTW-area is obviously defined by the available information in herbaria and databases (relevés, herbarium specimens, data from J.J. Wieringa and revised GBIF-data), and thus the intensity of collecting activity in the entire country. We will in Section 7.1 divide the sample of the 169 taxa into local elements and sub-elements as described in Section 4 and register the same taxa on geographical areas in Ethiopia and in Section 7.2 analyse the on a continental scale of the taxa that were recorded in the relevés.

Table 7-1. Relevés as distributed on altitudinal zones, and also indicating calculated areas of these altitudinal zones in sqkm and % of total area of CTW-phytochorion. The CTW-phytochorion is defined in Fig. 2-29, 4-1, 4-3 and 4-4. Calculation made with GIS, using altitudes from the SRTM DEM (CGIAR-CSI 2008).

Altitudes (m a.s.l.)	Number of relevés	Area of zone (sqkm) in the CTW	Area of zone in the CTW (%of total CTW)
0-499	1	12768.7	5.4
500-999	65	79326.7	33.3
1000-1499	60	83093.3	34.9
1500-1999	22	62652.8	26.3
2000-2499	3	48.6	0.0
Sum	151	237890.1	99.9

7.1. Defining the local floristic elements: geographical distribution, altitudinal ranges, limits and diversity

This Section deals with the number of taxa in the elements and sub-elements, with their altitudinal range, with their distribution on defined geographical areas in Ethiopia and with their distribution in Africa. The following Section 8 will illustrate the relation between the number of taxa in local floristic elements and sub-elements in relation to various ecological features that are assumed to represent adaptations to the environment in the CTW-area. We have used the methods with congruent distributions described in Section 4.2 to establish local phytochoria, local floristic elements and sub-elements.

In agreement with the concepts, terms and observations in Section 4.2 and 6, we will use the following terminology: we will consider the western woodlands one local phytochorion, the CTW-area; although it also includes small areas of other vegetation types (riverine forests, patches with *Transitional Rain Forest* (TRF) and isolated mountains with Afromontane vegetation). At this stage we will also consider the CTW-area as a relative homogenous vegetation as marked on the new vegetation map in Fig. 4-1. This is in spite of the variation observed on the very local scale, as demonstrated and discussed in Section 10. The local floristic element, the CTW-species, may be divided into sub-elements (see Fig. 7-1, data in Appendix 4), either widely distributed within the CTW-area or with restricted distribution inside the area in Ethiopia, as seen in Fig. 7-1 and 7-2. An overview of the local elements and sub-elements is given in the legend to Fig. 7-1. A number of the local CTW-species may transgress eastwards to the Rift Valley or further east along the margin between the Afromontane vegetation and the *Acacia-Commiphora bushland* (ACB), mainly into what is called *Transitional semi-evergreen bushland (TSEB)* on our new vegetation map in Fig. 4-1.

As mentioned in Section 4, a limited number of taxa have their main distribution in the whole or part of the *Acacia-Commiphora bushland* (ACB), but enter for a short distance into the driest northern or southern (or both) parts of the CTW-area. In the tables, diagrams and graphs, these taxa are referred to as ACB[-CTW], indicating their main distribution in *Acacia-Commiphora bushland* (ACB), but with distributional overlap with the northern or southern (or both) parts of the CTW-area. This overlap or close association between ACB and CTW elements in Tigray has been documented in the field in the north by Mehari Girmay et al. (2020) and in the south by Carr (1998), Schloeder (1999) and Jacobs & Schloeder (2002).

The taxa with marginal occurrence in the CTW-area and widespread at higher altitudes in the Ethiopian Highlands do in fact largely agree with the floristic element which White (1983) would consider Afromontane. These taxa are quite numerous but each taxon occurs only in one or very few relevés, and most often only the highest relevés in a profile, next to areas with Highland flora (see also Section 4.3). Such species were termed *Marginal Intruders* (MI) by White (1970), and the same term is used here. Taxa with their main distribution in several vegetation types or phytochoria, typically more than two, are termed *Transgressors* (TRG), a term also proposed by White (1970) and discussed here in Section 4.2. The phytogeographical classification of the taxa in the relevés is summarized in Fig. 7-1, where the relative size of the local elements and of the sub-elements can be seen.

Another way of looking at the data is by scoring taxa according to their distribution on arbitrarily defined geographical areas (see map in Fig. 7-2). A taxon can be scored several times, once for each area in which it occurs (data in Appendix 4). The number of taxa in each of these categories is shown in the top graph in Fig. 7-3. The number of taxa is highest in the area with highest rainfall (WWc), followed by the large, coherent but drier area in the north (WWn) and the more fragmented areas, with high rainfall in the south (WWs). Note that the number of taxa in the EWW-area includes only the taxa recorded both from the CTW-area and from the Ethiopian Highlands, not the entire flora of the Highlands, and it is therefore not surprising that the number of taxa recorded in EWW is lower than in any of the geographical areas WWn, WWc or WWs. Lowest is the number of taxa in ASO, which includes only the relatively few taxa that occur both in the lowland vegetation of *Acacia-Commiphora bushland* (ACB) and in the CTW-area, not the entire number of taxa in the *Acacia-Commiphora bushland* (ACB).

Counts of the taxa recorded from WWn, WWc, WWs, EWW and ASO on local elements and vice versa are shown in the two lower graphs of Fig. 7-3. The most widespread sub-element, CTW[all+ext], is well represented in all geographical areas; the same applies to the marginal intruders (MI). A number of sub-elements have a narrow distribution, such as the sub-element CTW[c] restricted to the not extensive central part of the CTW-area. Although the number of taxa is highest in WWc (because these taxa also occur elsewhere), the number of taxa restricted to the floristic sub-element CTW[c] is very low.

The lowermost graph of Fig. 7-3 documents the trivial observation that the sub-element CTW[all] has the same number of taxa as in each of the geographical areas WWn, WWc and WWs. It is also trivial that the sub-element CTW[all+ext] has the same number of taxa as the total number counted in the geographical areas WWn, WWc, WWs and EWW, etc. However, the local elements or sub-elements that transgress the boundary of CTW-area, that is ACB[-CTW], MI and TRG, occur with varying numbers in the different geographical areas. The column for ACB[-CTW] has a notably low number of taxa counted from WWc, but higher and almost identical numbers for the northern and southern areas, WWn and WWs, where the ACB[-CTW] sub-element is best represented. MI has a slightly lower number of taxa in WWs than in other geographical areas, perhaps because the mountains in south-western Ethiopia are lower than in the central and northern part of the country and hence there is a less prominent montane flora in the surrounding highlands of WWs.

In Fig. 7-4, we have compared the floristic contents of the areas WWn, WWc, WWs, EWW and ASO using a cluster analysis, presence-absence, Jaccard similarity coefficient, and flexible beta clustering. Data from outside the CTW-area are included to record presence in EWW and ASO. We have found a fairly well defined cluster consisting of the taxa occurring in the northern-central areas (WWn-WWc), joining the southern area (WWs), which is to be expected if the area of the western woodland really does represent as homogenous a vegetation as possible. The most meaningful finding in this clustering is that WWn-WWc cluster together, not WWc-WWs or WWn-WWs, and that this does not fully agree with the more detailed analyses shown in Fig. 10-6, where the relevé-data suggest that the some of the relevés in the WWc-area cluster with the 'blue' southern cluster and the relevés in the WWs-area join both the 'green' and the 'blue' cluster. The EWW and the ASO are not surprisingly quite distant. Detailed clustering of the data from the relevés is presented in Section 10. Here, it should be remembered that the same taxa may occur in several geographical areas, both in and outside the relevés, and support the linking of the areas in the cluster.

Studies of the altitudinal range of woodland vegetation and taxa are difficult to find in the literature, but there are a few observations of the limits and border-zones between lowland vegetation and montane forests. Hedberg (1951) studied the vegetation on the East African Mountains but only included montane forest and no woodlands; his lowermost limit of the montane forest belt was at 1700 m a.s.l., below which the forests gave place to a variety of vegetation types. Lind & Morrison (1974) gave no upper limit for the Combretaceous wooded grassland and woodland in East Africa, but suggested that the general uppermost limit of lowland vegetation, also including woodlands, should be set at ca. 2000 m a.s.l. In the mountains on the South Sudan-Uganda border, Friis & Vollesen (2005) suggested a general upper limit of the lowland vegetation at 1400-1800 m a.s.l., depending on exposure and rainfall. Friis et al. (2010) suggested that the highest record of vegetation that could be termed *Combretum-Terminalia Woodland* (CWT) in Ethiopia would be at 1800 m a.s.l. in the western part of the Tacazze Valley, while most of their other examples were from 1500 m or below.

Altitudinal limits between lowland and montane forests at a similar or slightly lower height have been reported. In the study of the forests on the Horn of Africa Friis (1992) found that there was a main discontinuity in the floristic composition between lowland forest and montane forest, and that this was located somewhere between 1220 and 1525 m a.s.l. For the whole of tropical Africa, White (1970) suggested a general limit between lowland and Afromontane forests somewhere between 4500 ft (1370 m) a.s.l. and 5000 ft (1525 m) a.s.l., while Hamilton (1975) rejected a sharp distinction between lowland and Afromontane forests in Uganda.

In order to see if we could detect discontinuity in the woodland flora resembling that found in the forest vegetation, we made a clustering analysis similar to the one by Friis (1992: Fig. 6), using altitudinal intervals of 500 m and calculated on similarity in species contents in the same way as in Friis (1992). The analysis is based on the relevé data on altitudinal zones used for Table 7-1. We found that the lowermost interval (0-499 m) emerged on a branch together with the highest interval, 2000-2499 m, while the intervals 1000-1499 m and 1500-1999 m came out extremely close together, pointing to the close floristic relation between these two intervals.

In spite of this, if we look closer at the species in the intervals 1000-1499 m a.s.l. and 1500-1999 m a.s.l. we see a slight difference with regard to floristic element. The species present in the interval 1000-1499 m a.s.l. and lower, but absent from the interval 1500-1999 m a.s.l. and higher, are *Combretum hartmannianum* (CTW[n]), *Crossopteryx febrifuga* (CTW[c]), *Dalbergia boehmii* (CTW[c]), *Ficus platyphylla* (CTW[c+s]), *Hyphaene thebaica* (CTW[all+ext]), *Pseudocedrela kotschyi* (CTW[all]) and *Zanthoxylum gilletii* (CTW[c]), while the species present in the interval 1500-1999 m a.s.l. but absent from the interval 1000-1499 m a.s.l. or lower are *Acacia gerrardii* (MI), *Albizia schimperiana* (MI), *Calpurnia aurea* (MI), *Embelia schimperi* (MI), *Polyscias farinosa* (CTW[all]), and *Schrebera alata* (MI). The species that come in from altitudes at ca. 1500 m a.s.l. are thus mostly marginal intruders with a main distribution above the CTW-area. The only species restricted to the interval 1000-1499 m a.s.l. is the rare *Zanthoxylum gilletii*. No species is restricted to the interval 1500-1999 m a.s.l.

The cluster consisting of the intervals between 1000 and 2000 m a.s.l. was found to be attached to the slightly more distant interval of 500-999 m a.s.l., representing the lowermost zone in the western woodlands with more than one relevé. In the interval of 0-499 m a.s.l. the number of taxa is much lower than for the other intervals, which may explain the unexpected position of this interval next to the transition zone to the highland vegetation at 2000-2499 m a.s.l.

Note that these results agree with the suggestion of Friis et al. (2010), where it was stated that the *Combretum-Terminalia woodland and wooded grassland* (CTW) reaches from the Gambela lowlands (at ca. 500 m a.s.l.) to their highest points at ca. 1800 m a.s.l. The existence of the highest cluster (2000-2499 m a.s.l.) is due to two relevés recorded above 2000 m a.s.l. in profile A between Humera and Gondar, where we continued to record data along the profile in relevés 21 and 22 right up into the mainly Afromontane vegetation above 2000 m a.s.l. In profile B between Metema and Aykel, the relevé 53 near Aykel, was also just above 2000 m a.s.l. and had Afromontane vegetation.

Unlike the results of Friis (1992) for the forest trees, we find that the woody species in the woodland flora of the intervals ranging from 500 to 2000 m a.s.l. do not fall into clearly separated groups. Based on this analysis we conclude that the floristic composition of the woodland vegetation between 500 and ca. 2000 m is relatively uniform, or at least that the altitudinal turn-over of the woodland species is gradual, with no sharp altitudinal discontinuity inside the altitudinal range of the CTW-area.

In Fig. 7-6 and Fig. 7-7 the columns show the number and proportion of taxa present in the altitudinal zones up to above 2500 m a.s.l. Here again the representation of taxa is quite uniform for the altitudinal intervals 500-1999 m a.s.l., and surprisingly uniform for the altitudinal intervals 1000-1499 m a.s.l., and 1500-1999 m a.s.l. The floristic sub-element ACB[-CTW] is most strongly represented below 500 m a.s.l., which is the altitude at which these species occur in the ACB-area, but above 500 m a.s.l. the number of taxa is relatively constant up to 2000 m a.s.l., both regarding numbers of taxa and the proportional representation of taxa. The marginal intruders (MI) are poorly represented below 1000 m a.s.l., but they increase in number and in proportional representation with altitude, being well represented above 2000 m a.s.l., where the Afromontane flora begins to dominate. This situation is comparable with the notable and very regular altitudinal increase in the Afromontane Forest element in Ethiopian forests studied by Friis (1992), where this increase is particularly prominent in his Fig. 7. The situation for TRG is comparable, but this element is poorly represented below 500 m a.s.l. In Fig. 7-7 we see that the proportional representation of the different geographical areas in Ethiopia is very similar at all altitudes.

In conclusion, and documented particularly by the clustering shown in Fig. 7-4, there is only a moderate spatial turnover of taxa or elements from CTW[n] to CTW[c], agreeing with our finding of a well-defined cluster consisting of the northern-central areas (WWn-WWc), and this is joined by the only slightly more distant southern area (WWs). This conclusion must be compared with the clustering of the relevés in Section 10, where the analyses in Fig. 10-6 reveal a northern and partly central (green) cluster and a mainly southern (blue) cluster. Many CTW-taxa transgress to the *Transitional semi-evergreen bushland* (TSEB) east of the Rift Valley, and the number of marginally intruding taxa into the highest zones of the western woodlands from the Ethiopian Highlands is significant. There is no sharply marked turnover in the CTW-elements between altitudes at 500 and 1999 m a.s.l., which just transgresses the upper limit of the western woodlands at 1800 m a.s.l. This may be slightly surprising, since the later study of the relevés in Section 10 reveals a difference in altitude between the relevés of the northern (green) cluster and those of the southern (blue) cluster, but this difference is variable from relevé to relevé. The methods used in this Section are coarser than the clustering methods used in Section 10, so although we have found no altitudinal limits within the CTW-area, the question is not yet settled. As should be expected, since they represent different local phytogeographical elements, there is notable turnover when comparing the CTW-element to the ACB[-CTW] sub-element, to the MI-element and to TRG-element.

Fig. 7-1. Local elements and sub-elements in the sample of 169 species observed in relevés. The phytogeographical classification of the species (in a few cases subspecies) that occur in the western woodlands. The inner ring indicates the elements, the outer ring the sub-elements. The sizes of the segments indicate the number of species or subspecies in each element or sub-element. The figure classifies the total sample of species into:

(i) Species (or subspecies) that are local elements in the CTW-area (with main Ethiopian distribution in that phytochorion), blue: The elements in the CTW-area are subdivided into sub-elements: widespread CTW-elements (CTW[all]), northern CTW-elements (CTW[n]), central CTW-elements (CTW[c]), southern CTW-elements (CTW[s]), and combinations of these. Elements marked with '+ext' have extension into the *Transitional semi-evergreen bushland* (TSEB) in the Rift Valley or the south-eastern slope of the Ethiopian Highlands towards the eastern lowlands.

(ii) Marginal intruders (MI), grey: MI stands for Marginally Intruding species or subspecies in the western woodlands, mainly with a main distribution in the highlands (one species, *Erythroxylum fischeri,* is marginally intruding from lowland forest; the rest could be called "highland intruders"); usually with one or few records in our relevés.

(iii) transgressors (TRG), orange yellow: TRG stands for TRansGressing species, occurs similarly widespread in both the *Combretum-Terminalia woodland and wooded grassland* (CTW) and in other vegetation types outside the western woodlands, particularly *Dry evergreen Afromontane forest* (DAF).

(iv) ACB[-CTW] are species or subspecies that are elements in ACB (a separate local phytochorion towards Somalia in *Acacia-Commiphora bushland* (ACB), with main Ethiopian distribution in that local phytochorion) – The species in the sub-element ACB[-CTW] have their main distribution in the ACB-area, but are intruding in the *Combretum-Terminalia woodland and wooded grassland* (CTW) in the dry parts with low ground biomass and low burning in the north or in the south, or both.

(v) Not marked with indication of local sub-elements in the outer ring on the figure are a few species with different (unusual) classification: Unmarked orange-brown segment in outer circle, included in the ACB-element in the inner circle: One taxon in ACB with extension into both the CTW-area and a significant extension into *Transitional semi-evergreen bushland* (TSEB). Unmarked dark blue segment in outer circle, included in CTW in the inner circle: Two taxa referred to the unusual sub-element CTW[n+c+ext]. Note that the phytogeographical classifications are based on the distribution of the species in the whole of Ethiopia.

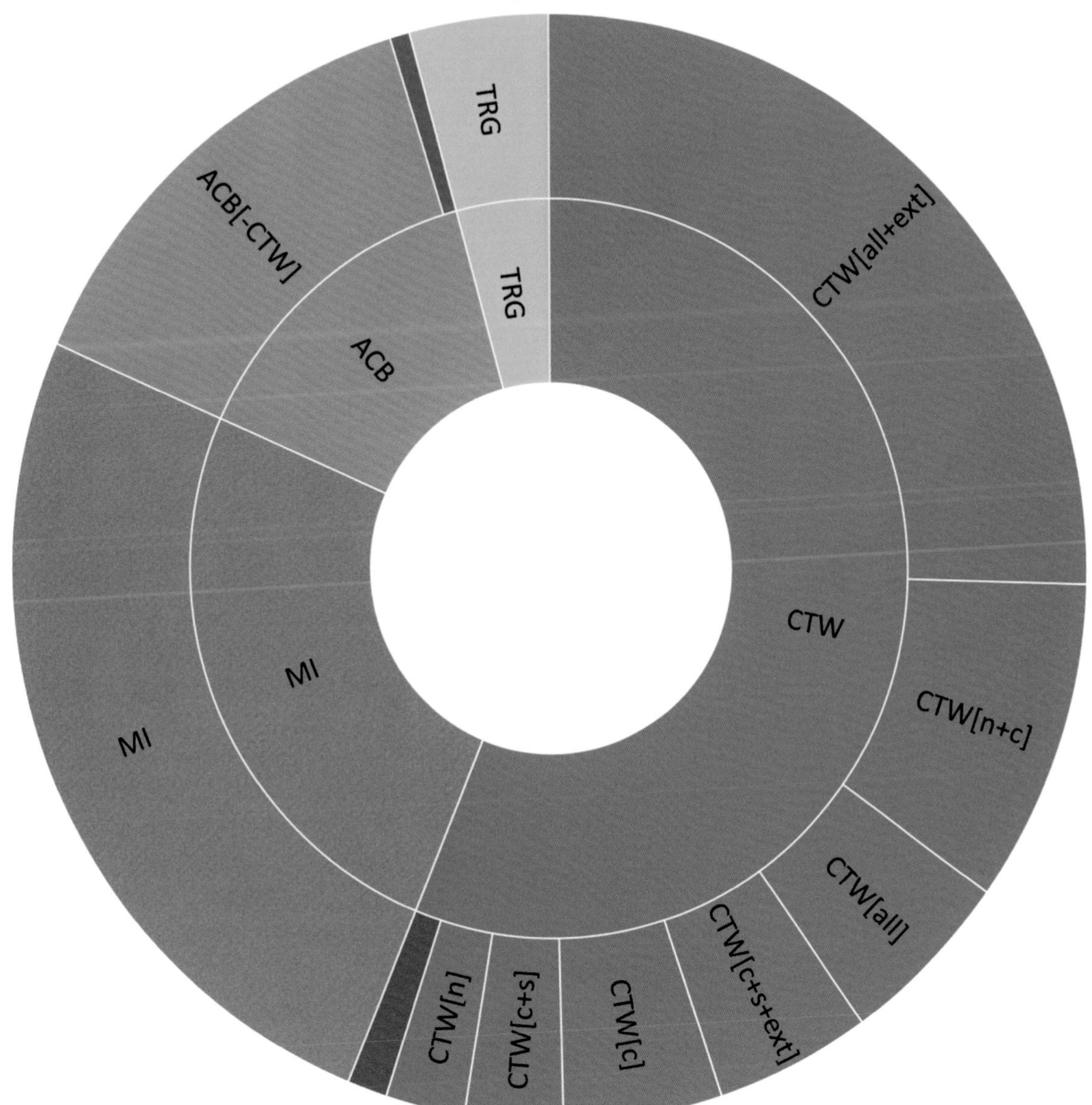
Taxa in the sample on local element
TRG
TRG
ACB[-CTW]
ACB
CTW[all+ext]
CTW
MI
MI
CTW[n+c]
CTW[all]
CTW[c+s+ext]
CTW[c]
CTW[c+s]
CTW[n]

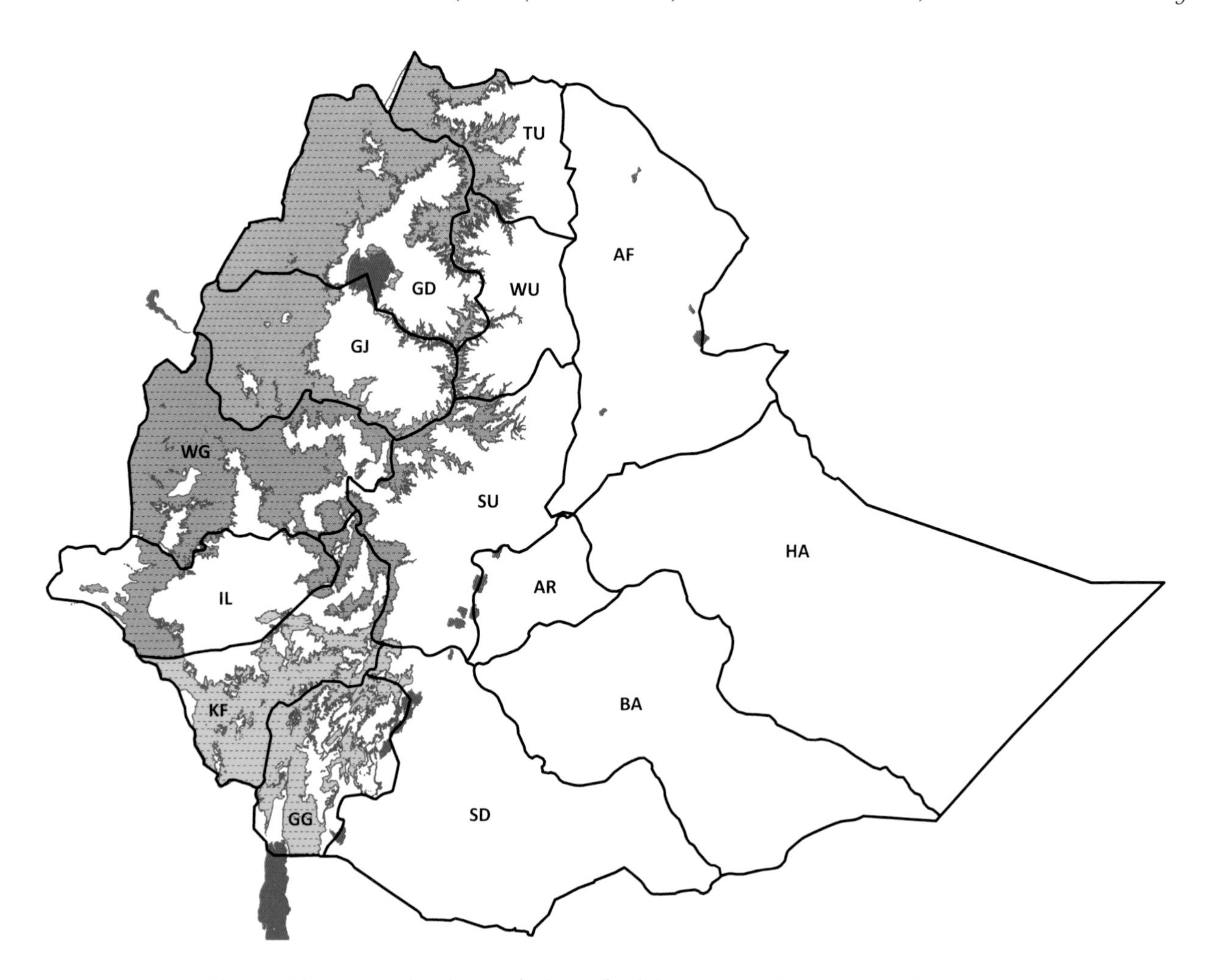

Fig. 7-2. The geographical areas (coloured) of the western Ethiopian woodlands (synonymous with the CTW-area and areas with associated vegetation types) used for recording presence and absence in WWn, WWc, and WWs and in other parts of Ethiopia. WWn (cyan) includes all of the CTW-area within the floristic provinces TU, WU, GD and GJ. WWc (green) includes all of the CTW-area within the floristic regions WG, SU, IL and the northern part of KF (to the north of the Gojeb Valley). WWs (yellow) includes all of the CTW-area within the floristic provinces of KF from and including the Gojeb Valley and GG. The categories EWW and ASO do not occupy such specific areas, but transgress the borders of the coloured area either widely to the east and mainly in the highlands (EWW) or mainly in the south-eastern Ethiopian lowlands towards Somalia, where the prominent vegetation is *Acacia-Commiphora* woodland or bushland (ASO).

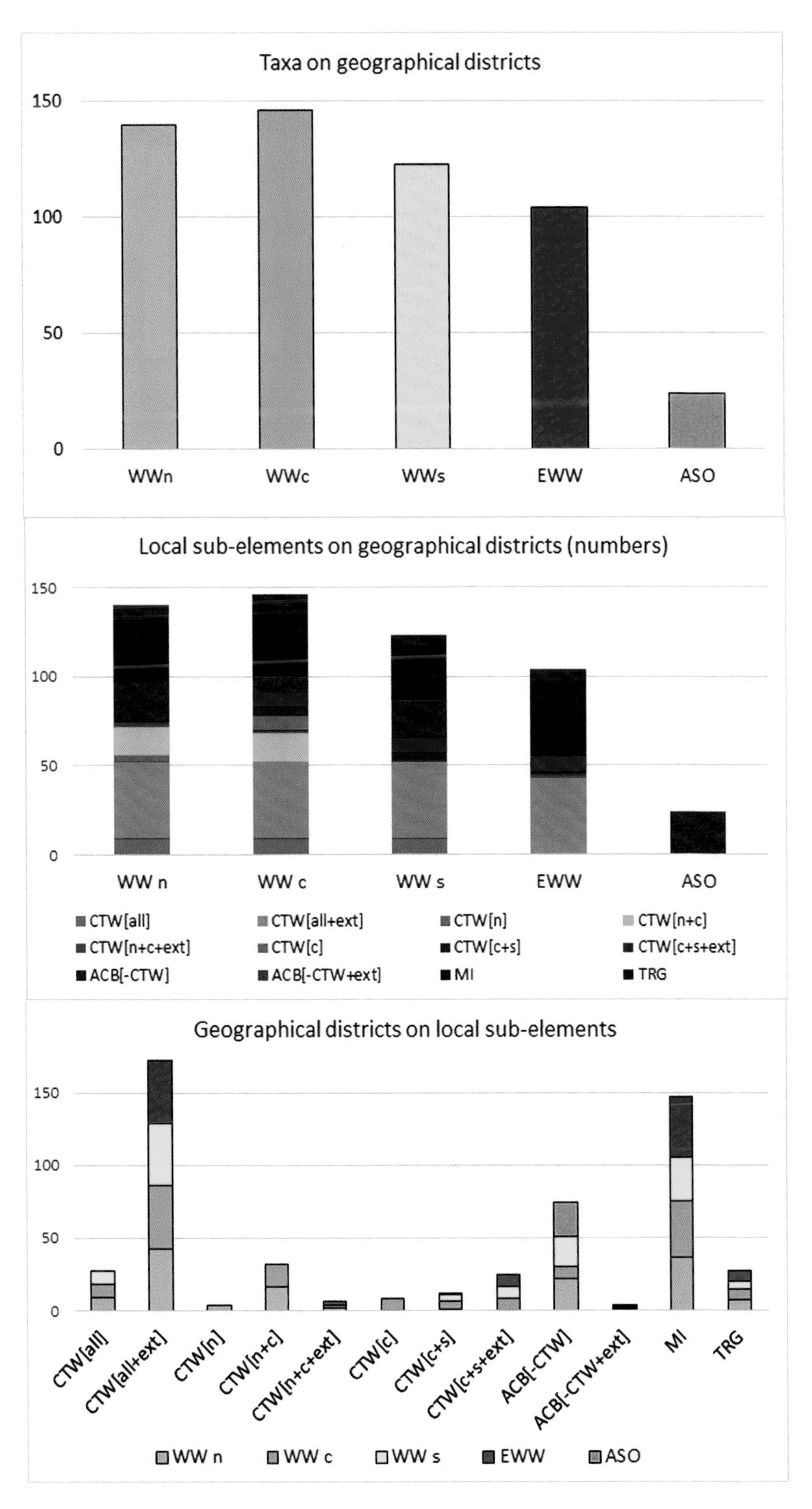

Fig. 7-3. The distribution of local elements on the geographical areas in Fig. 7-2. The top graph shows only the numbers of taxa recorded from the geographical areas. The middle graph shows the distribution of floristic elements and sub-elements on the geographical areas. The bottom graph shows the representation by number of taxa of the geographical areas on floristic elements and sub-elements. The legends for the floristic elements and sub-elements are as in Fig. 7-1. The designations for the geographical areas are: WWn, recorded from northern part of the western woodlands. WWc, recorded from central part of the western woodlands. WWs, recorded from southern part of the western woodlands. EWW, recorded from areas east of the western woodlands with Afromontane vegetation and *Transitional Semi-Evergreen Bushland* (TSEB) in or east of the Rift Valley. ASO, recorded from areas east of the western woodlands with *Acacia-Commiphora* bushland (ACB), mainly in Afar and Ethiopian Somalia, i.e., the AF-, SD-, BA- and HA-floristic regions).

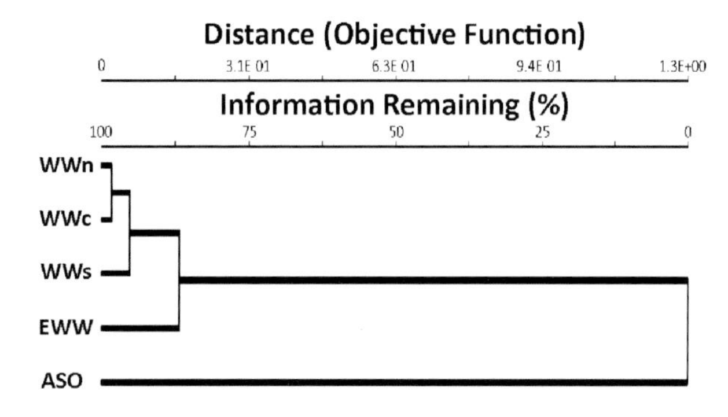

Fig. 7-4. Clustering of records from geographical areas in Ethiopia (no assignment to element). The designations for direct geographical observations in the areas are as in Fig. 7-2. The analysis is based on presence and absence in the geographical areas of Ethiopia of the 169 species recorded in this study. Jaccard's similarity quotient and Flexible beta clustering have been used.

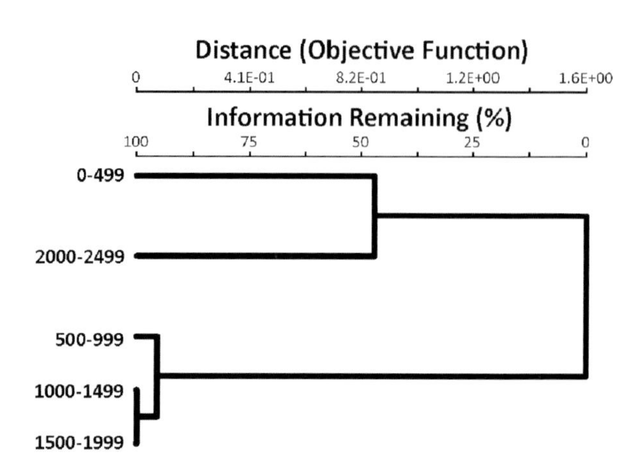

Fig. 7-5. Clustering of records from altitudinal intervals of 500 m each. The intervals are 0-499 m a.s.l., 500-999 m a.s.l., 1000-1499 m a.s.l., 1500-1999 m a.s.l. and 2000-2499 m a.s.l. The analysis is based on data of presence and absence from the 169 taxa recorded in the relevés in this study. Jaccard's similarity quotient and Flexible beta clustering have been used.

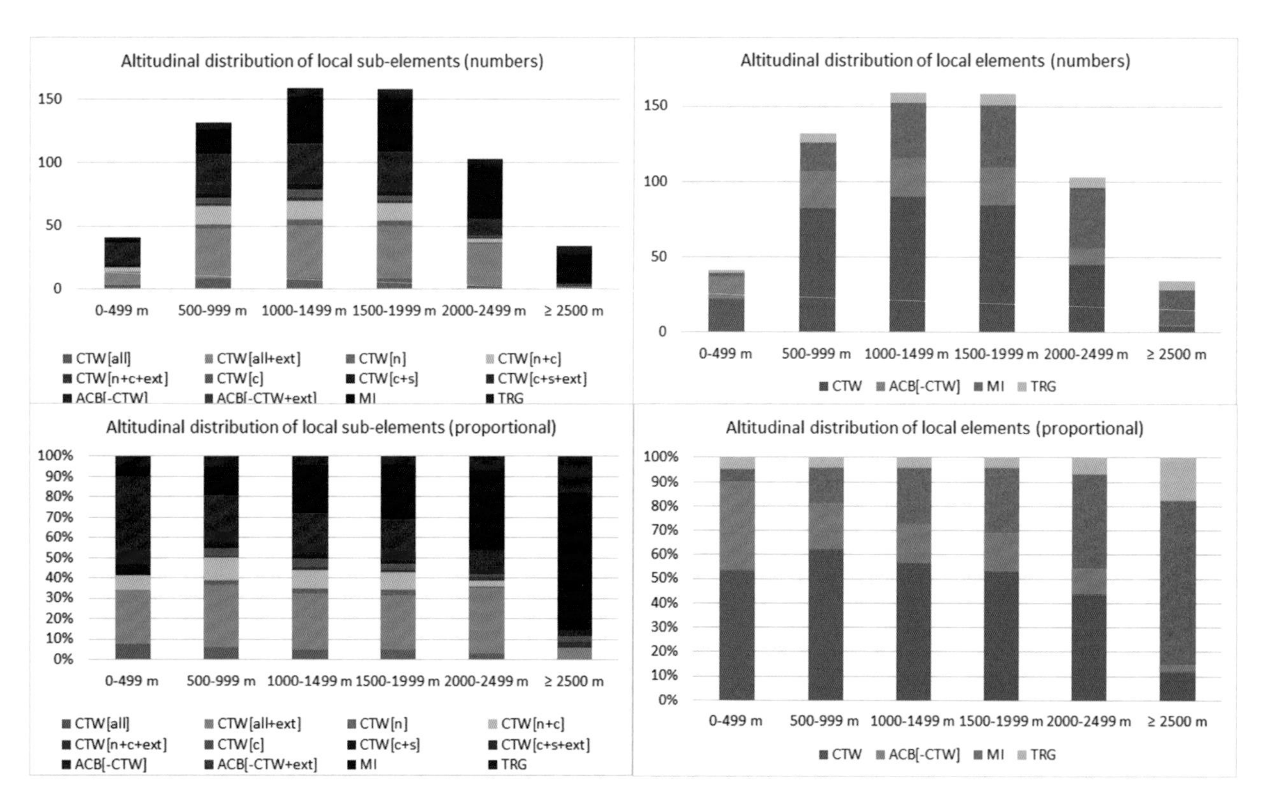

Fig. 7-6. Number of taxa (graphs above) and proportional representation (graphs below) in local elements and sub-elements, recorded on altitudinal zones. The columns are divided according to the local elements and sub-elements in Fig. 7-1. The graphs to the left show all sub-elements, the graphs to the right show only elements (CTW, ACB, MI and TRG). Note that the data are based on the altitudinal distribution of the species in the whole of Ethiopia.

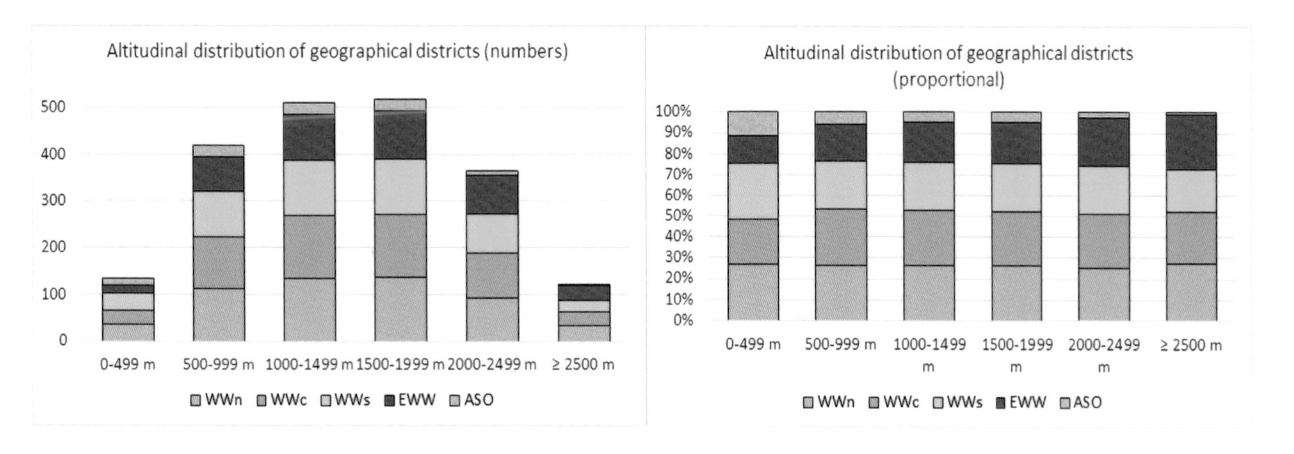

Fig. 7-7. Number of taxa (left graph) and proportional representation (right graph) in the geographical regions as defined in Fig. 7-2, recorded on altitudinal zones. Note that the taxa are counted each time they have been recorded as present in an altitudinal zone and a geographical region of Ethiopia.

7.2. Patterns of distribution on a continental African scale

The studies of Fayolle et al. (2018) have demonstrated a sharp floristic discontinuity in Africa between the N&W savannas and woodlands to the west of Ethiopia and the S&E savannas and woodlands in Ethiopia and southwards (see also Section 3.12). These results were obtained from samples that were mostly far from the western Ethiopian woodlands and do not illustrate the situation in the border zone between the areas to the N&W and the S&E. Here we will analyse our set of relevé-data with focus on the floristic discontinuity in Africa between the N&W savannas and woodlands to the west of Ethiopia and the S&E savannas and woodlands in Ethiopia and southwards. With our data we will try to explain the findings of Fayolle et al. (2018) with data from the border zone in which their floristic discontinuity is so apparent.

For an analysis of the taxa from our relevés, we have classified them according to two distributional patterns that overlap on the Horn of Africa. The first classification covers the pattern of distribution of the relevé-species in the direction west-east across northern tropical Africa between the Atlantic Ocean in the west and the Red Sea and the Indian Ocean in the east. The second classification covers the pattern of distribution of the relevé-species in the direction north-south between Sudan, Ethiopia and Eritrea in the north and South Africa in the south. In the west-east direction the categories are: first a group with taxa reaching all the way between the Atlantic Ocean and the Horn of Africa (WE), second a group reaching between Ghana, Nigeria or Cameroon and the Horn of Africa (HWE), and third a group restricted to the Horn of Africa in a wide sense, approximately east of the Nile (NEW). In the north-south direction we used four categories: first a group with taxa reaching from the Horn of Africa to Namibia and South Africa (ESA), secondly a group reaching from the Horn of Africa to the tropical countries in southern Africa (Angola, Zimbabwe, Malawi, Mozambique; EAZ), thirdly a group of taxa reaching from the Horn of Africa to the southern part of tropical East Africa, typically central or southern Tanzania (EEA), and finally a group that does not reach southwards beyond the northern Democratic Republic of Congo, northern Uganda and northern Kenya (NES). In direction (west-east and north-south) the taxa under study are only recorded in one category. The data for both the pattern across Africa from west to east and the pattern along Africa from north to south are documented in Table 7-2 and the results are shown in Fig. 7-8.

It appears that a majority of the relevé-taxa, 105 out of 169 (62%) have a wide distribution in the west-east direction across Africa, reaching from the Atlantic Ocean to the Horn of Africa, to which we can add 44 taxa reaching from somewhere in Central Africa to the Horn of Africa. Together these two groups make up a group of 149 taxa out of 169 (88%); these taxa span the discontinuity of Fayolle et al. (2018), reaching from West or Central Africa at least deeply into the river valleys in the Ethiopian Highlands.

It also appears from our data that a slightly smaller majority of the relevé-taxa, 66 out of 169 (39%), has a wide distribution in the north-south direction along Africa from South Africa to the Horn of Africa, and that 48 out of 169 (28%) reach the from the southern tropical African countries to the Horn of Africa. However, together these two groups make up 114 taxa (67%), a slightly higher but still comparable number of taxa when compared with the number of taxa reaching between the Horn of Africa and West Africa. A similar numbers of taxa, 27 (16%) and 28 (17%), reach between the Horn of Africa and eastern tropical Africa (normally Tanzania) or are restricted to the Horn of Africa in a wide sense.

If we look at a combination of the distributions in the two directions, we find that the highest number of taxa, 49 of 169 (29%), occur both from the Atlantic Ocean to the Horn of Africa and from the Horn of Africa to South Africa (WE-ESA). The second highest number, 35 taxa (21%), are slightly less widespread towards the south, occurring both from the Atlantic Ocean to the Horn of Africa and from the Horn to the tropical countries in southern Africa (WE-EAZ). These two groups together (84 taxa out of 169; nearly 50%), occurring both in the Sudanian and the Zambian regions, can therefore be termed Sudano-Zambesian species, and they occur in both in the N&W and in the S&E savannas and woodlands of Fayolle et al. (2018). A smaller number of taxa belongs to a group that is mostly restricted to the east of the Nile but extending into Tropical East Africa (NEW-EEA; they fall into the species of the S&E savannas and woodlands of Fayolle et al.). Two small groups occur either from the Atlantic Ocean to the Horn of Africa, true Sudanian species of the N&W savannas and woodlands of Fayolle et al., or are restricted to the Horn of Africa in the wide sense (NEW-NES). The remaining groups all have fewer than 10 taxa each and will not be discussed further.

We may conclude that the results of Fayolle et al. are correct if one contrasts the savanna floras from West Africa with the floras from savannas in East Africa south of the Horn of Africa. Our studies, however, indicate that the situation is more complex on a smaller scale, when one focusses on the western woodlands of Ethiopia. Those woodlands may be considered a transition zone between the N&W and the S&E savannas and woodlands, albeit with a more marked connection with the former than with the latter.

In Table 7-3 and Fig. 7-9 we have expanded the analysis on which Fig. 7-8 is based and included distributions of our relevé taxa on local elements. Among the taxa in the west-east distribution (top graph in Fig. 7-9), the highest number of WE-taxa belongs to the CTW-element, with 69 taxa out of 169 (41%), clearly indicating the close connection between the widespread Sudanian taxa and their distribution in the CTW-area in Ethiopia. All other floristic elements in the west-east distribution pattern have fewer than 20 taxa and are not further discussed here.

Among the taxa in the north-south distribution, the highest number of taxa, 32 of 169 (19%), is again found in the most widespread group, which consists of CTW-taxa that reach South Africa (ESA-CTW). However, with regard to number of taxa this group does not lead so notably over the other groups as did the leading group in the west-east pattern but it is followed rather closely by the number of CTW-taxa that reach the tropical countries in southern Africa, 29 taxa out of 169 (17%; EAZ-CTW). More than a third, 36%, of the CTW-species in Ethiopia reach far south in eastern Africa which underlines the connection between the western Ethiopian woodlands and the S&E savannas and woodlands of Fayolle et al. The remaining categories in the north-south pattern have low numbers of taxa and are not further discussed here, except that the combined number of marginal intruders (MI) that reach ESA and EAZ represent 31 taxa (18%). These taxa are typically widespread Afromontane taxa that intrude into the CTW-area in Ethiopia from higher altitudes. Again, this overlap between taxa widespread in eastern Africa and the western Ethiopian woodlands confirms that these woodlands may be considered a transition zone between the N&W and the S&E savannas and woodlands.

		North-south direction				Sum (proportion)
		ESA	EAZ	EEA	NES	
West-east direction	NEW	9	8	14	13	44 (26%)
	HWE	8	5	5	2	20 (12%)
	WE	49	35	8	13	105 (62%)
Sum (proportion)		66 (39%)	48 (28%)	27 (16%)	28 (17%)	169 (100%)

Table 7-2. Number of taxa in the western Ethiopian woodlands relevés on African distributional categories. Abbreviations in rows: NEW, species restricted to the Horn of Africa or near in the east-west direction. HWE, species distributed from approximately halfway to Atlantic Ocean in the west (typically with western boundary at Nigeria or Cameroon) to the Horn of Africa in the east. WE, species distributed from the Atlantic Ocean in the west to the Horn of Africa in the east. Abbreviations in columns: ESA, species distributed from the Horn of Africa in the north to South Africa and Namibia in the south. EAZ, species distributed from the Horn of Africa in the north to Angola, Zimbabwe and Mozambique in the south. EEA, species distributed from the Horn of Africa in the north to East Tropical Africa (normally Tanzania) in the south. NES, species restricted to the Horn of Africa or near in north-south direction. Note that the categories in the columns and in the rows are not overlapping with regard to their taxa.

	West-East			South-North				Sum (proportion of the 169 taxa in the relevés).
	WE	HWE	NEW	ESA	EAZ	EEA	NES	
CTW	69	8	17	32	29	13	20	94 (56%)
MI	15	9	19	20	11	7	5	43 (26%)
ACB[-CTW]	15	2	8	9	6	7	3	25 (15%)
TRG	6	1	0	5	2	0	0	7 (4%)

Table 7-3. Number of local phytogeographical elements on African distributional categories. Abbreviations for distribution across Africa as in Fig. 7-8 and Table 7-1, for local elements as in Section 7 and in Fig. 7-9. The sums in the right hand column are counted separately for West-East and South-North, and are the same for each.

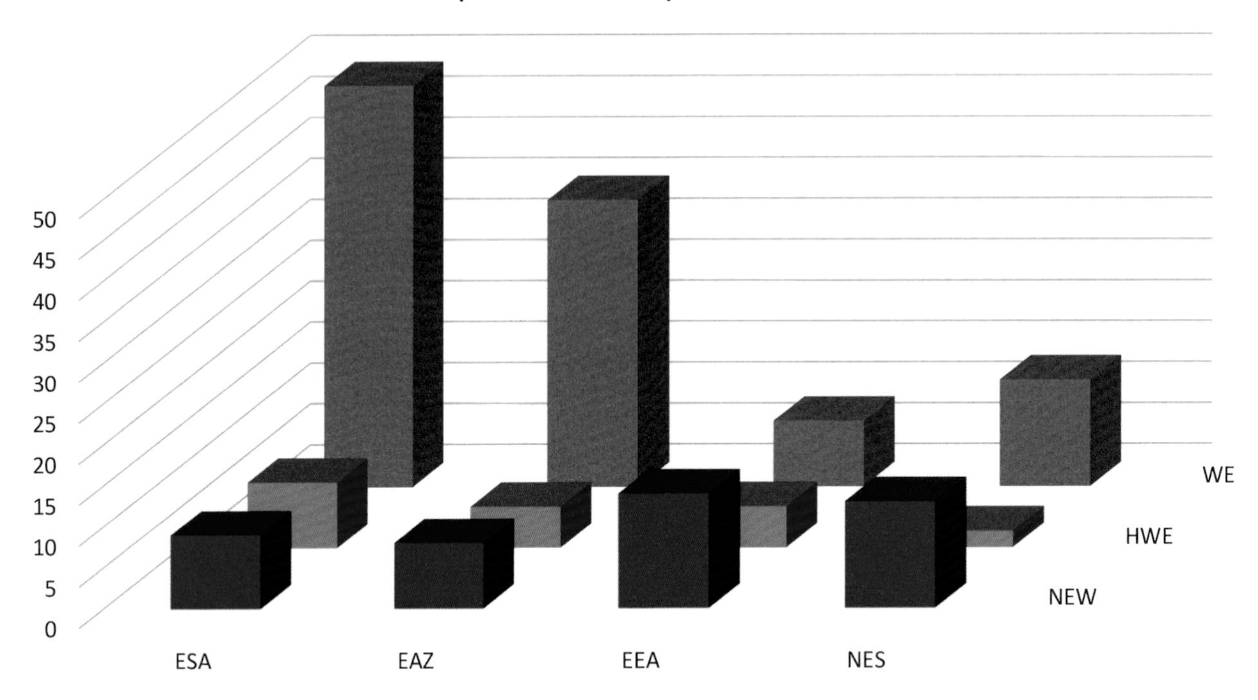

Fig. 7-8. The African distribution of the taxa in the sample of 169 taxa recorded in the relevés. (Rows from back to front) WE, species distributed from the Atlantic Ocean in the west to Ethiopia in the east. HWE, species distributed from approximately halfway to Atlantic Ocean in the west (typically with western boundary at Nigeria or Cameroon) to Ethiopia in the east. NEW, species restricted to the Horn of Africa or near in the east-west direction. (Rows from left to right). ESA, species distributed from the Horn of Africa in the north to South Africa and Namibia in the south. EAZ, species distributed from the Horn of Africa in the north to Angola, Zimbabwe and Mozambique in the south. EEA, species distributed from the Horn of Africa in the north to East Tropical Africa (normally Tanzania) in the south. NES, species restricted to the Horn of Africa or near in north-south direction. Note that the categories are not overlapping with regard to their taxa.

Opposite page:
Fig. 7-9. The African distribution of the local elements in the sample of 169 taxa recorded in the relevés. Above distribution of the local elements in west-east direction, below their distribution in north-south direction. The local floristic elements and the ranges of distribution are arranged differently above and below in order to allow all columns to be seen. The local elements are: CTW, species distributed in the western woodlands, some of which also occur in the *Transitional Semi-Evergreen Bushland* (TSEB) in the Rift Valley and to the east of the Rift. MI, marginal Intruders, mainly distributed in the Ethiopian Highlands. ACB[-CTW], *Acacia-Commiphora* Bushland species that enter the CTW-area in north and/or south. TRG, Transgressing species. The codes for the ranges of distribution in Africa are the same as in the previous figure. Note also the difference in vertical scale between the two parts of the illustration; the tallest column (WE-CTW) in the upper figure is twice the height of the tallest column (ESA-CTW) in the lower figure.

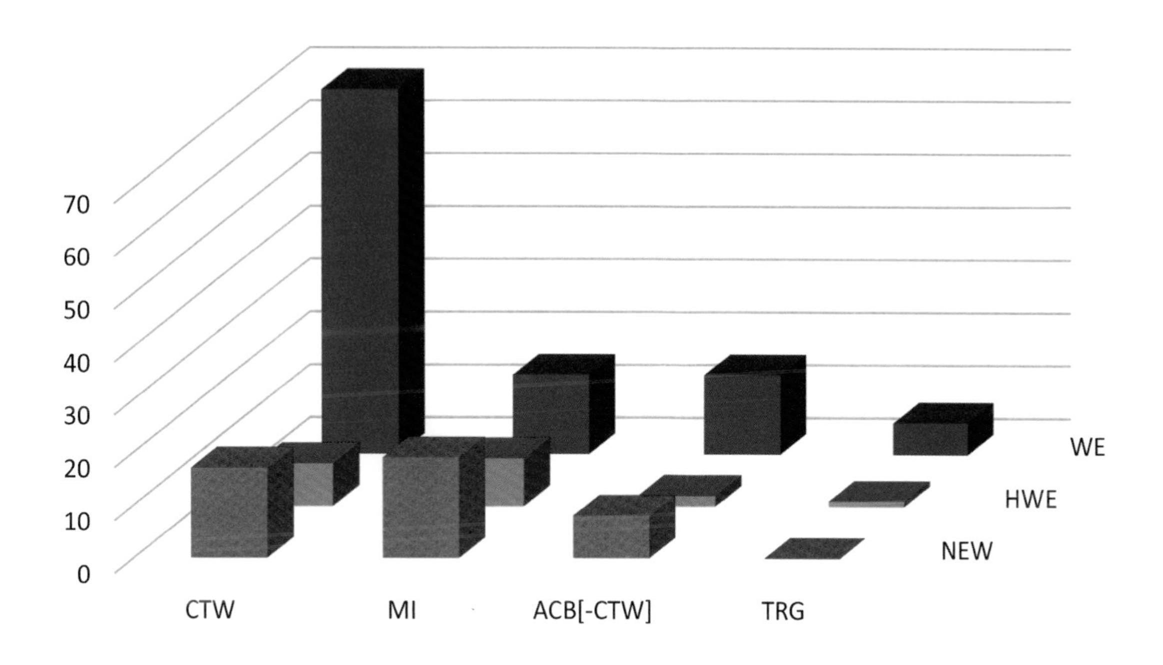
West-east distribution in Africa of local elements
70
60
50
40
30
20
10
0
CTW
MI
ACB[-CTW]
TRG
WE
HWE
NEW

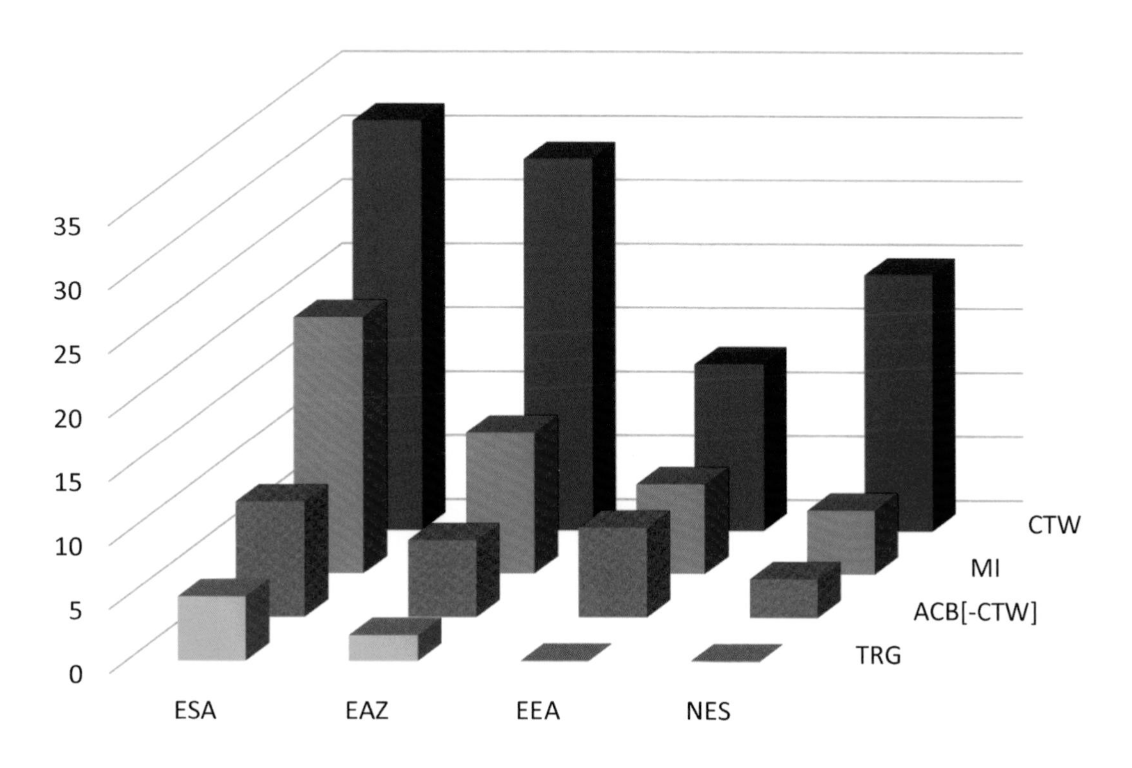
North-south distribution in Africa of local elements
35
30
25
20
15
10
5
0
ESA
EAZ
EEA
NES
CTW
MI
ACB[-CTW]
TRG

8
Environmental adaptations and ecological categories

In Section 8, we will summarize and discus the taxa recorded in the relevés in relation to various ecological adaptations. The data about the 169 species from the relevés are derived from all the information and literature cited in Section 6 and is summarized in Appendix 5 The size of the sample and the extensive area (Ethiopia) from which it is drawn should allow our results to be used for future comparisons with samples from other vegetation types, other parts of Africa or elsewhere in the tropics. Although several species in the relevés are relatively widespread from north to south, none of them occur in more than close to half of the relevés and many species occur in very few relevés. In fact, according to the data in Appendix 2 and the data extracted for Table 8-1, only about one third of the species in the total survey occur in more than ten relevés, and only ten of these species occur in one third of the relevés or more (*Anogeissus leiocarpa, Combretum collinum, Acacia seyal, Piliostigma thonningii, Terminalia schimperiana, Pterocarpus lucens, Ficus sycomorus, Ziziphus spina-christi, Lonchocarpus laxiflorus,* and *Stereospermum kunthianum*).

So instead of focussing entirely on taxonomy as done previously, we will focus in this Section on the variation in ecological adaptations, and analyse the relation between a number of ecological parameters assumed to be of importance for the woodland trees and shrubs in relation to their altitudinal range, their local phytochoria, their distribution on local geographical units and their African distribution. The idea to supplement the taxonomic division of plant species with division of them into groups based on their ecological functionality is well established, going back to the classification according to life-forms proposed by Humboldt (1806), and developed later by many botanists, for example in a sequence of papers by Raunkiær. The best known of these deals with his system of life-forms adapted to survive winter and drought (Raunkiær 1907), others deal with life-forms based on leaf size (Raunkiær 1916, 1934), and other adaptations. A sequence of papers by Warming on life-forms focussed more on growth form and branching systems, culminating with a review of his previous studies of the branching and growth of plants (Warming 1923). The range of models of life-forms of vascular plants was reviewed by Du Riets (1931). Recent data on life-forms in tropical Africa is still scarce, and the recent studies have focussed on the basic growth forms (tree, shrub, herb, liana, epiphyte; for example Dauby et al. 2016: 11-12; Sosef et al. 2017: 17). An exception is the studies of fire-resistance and resistance to browsing animals mentioned below. However, a few works (e.g., Grubb et al. 1963; Grubb & Whitmore 1966; Hamilton 1975; Friis 1992) have analysed the various vegetative features of forest taxa such as lamina size, leaf type (simple or compound), deciduousness, the possession of thorns and the possession of buttresses, and how these adaptations vary with altitude. Lind & Morrison (1974) has provided a very general overview for tropical East Africa, while only Hamilton (1975), dealing with Ugandan forests, and Friis (1992), dealing with forests on the Horn of Africa, have provided works specifically relevant to tropical East Africa, and these two authors deal with forest vegetation, not woodlands.

In the forests of Uganda, Hamilton (1975) has shown that lamina size and leaf-persistency, the possession of thorns or spines and the possession of buttresses all decline with increasing altitudes. In the forests of the Horn of Africa, the studies by Friis (1992) produced similar results. We will attempt to study if similar patterns exist in the woody taxa of the western Ethiopian woodlands, beginning with the maximum height of trees.

First, we will focus on plant height (Fig. 8-1), then on leaf composition (leaves simple or compound;

Fig. 8-2), leaf persistence during the dry season (Fig. 8-4), presence or absence of adaptations that can provide fire resistance (Fig. 8-5), and presence or absence of adaptations of adaptations that might protect against browsing animals (Fig. 8-6). According to the literature on life-forms, as in the review by Du Riets (1931), such characterisations of plant communities may be as useful as a floristic characterization. It should, however, be remembered that although the sample of species studied here is drawn from the relevés in the western Ethiopian woodlands, the ecological information here about each species represents the adaptations known for the whole distribution area of the species. We do not have information about locally acclimatized forms with adaptations only relating to the western Ethiopian woodlands.

8.1. Height of woody plants

Lind & Morrison (1974) suggest that the characteristic taxa in the *Combretum-Terminalia* woodlands, mainly the taxa of the genus *Combretum*, are stoutly branching trees that reach a height of about 12 m tall, and that very few taxa in these types of woodland would reach a height of more than 15 m tall. No source contradicts this generalisation but we will see what information our relevé-species may provide.

The graphs in Fig. 8-1 show that when the sample of relevé-species is distributed on a number of geographical and phytogeographical categories, the number of taxa of tall trees, i.e., trees reaching a maximum height of more than 10 m, is always lower than the number of trees and shrubs that reach a height of less than 10 m, resulting in woodlands of moderately tall trees and shrubs. The proportion between the two categories is also rather constant, both regarding altitudinal range and geographical distribution in Ethiopia. The notable exception from these rather constant figures is seen in the woodland taxa that also occur in ASO, among these species only about 10% have a maximum height of more than 10 m. This agrees with the general observation that the eastern lowland taxa in *Acacia-Commiphora bushland* (ACB) are generally more dominated by shrubs than the overall woody flora of the western Ethiopian woodlands. A frequently forwarded explanation is that droughts are more frequent and severe in the *Acacia-Commiphora bushland* (ACB) than in the western woodlands.

On the continental African scale, the groups with the highest proportion of taxa reaching a height of more than 10 m contain the most widespread taxa (ESA and WE). Apart from a simplistic suggestion, that tall species may have more effective dispersal than lower species, it is not clear why tall woodland species should be more widespread than lower ones.

Generally, we may conclude that the maximum hight of the trees is an evenly distributed parameter in the CTW-area.

Fig. 8-1 (figure on next page). Maximum height of woody taxa in the sample of woody species from the relevés (taxa reaching a maximum height of more than 10 m, left or below, *versus* taxa reaching a maximum height of less than 10 m, right or above). The graphs at the top show number and respective proportion of species on altitude with intervals of 500 m. The second row graphs show the distribution on local elements (summarized from Appendix 2 & 4: CTW (without sub-elements); ACB[-CTW]; MI; TRG). The third row graphs show the distribution on the geographical areas in Ethiopia, as shown in Fig. 7-2 and summarized from Appendix 4. The bottom row graphs show the distribution on geographical areas on a continental African scale (data in Table 7-2 & 7-3: NS-NAR: restricted to the Horn of Africa, Sudan and/or Yemen, south to northern D.R. Congo, Uganda and Kenya. EEA: Distributed southwards to Tanzania. EAZ: distributed southwards to Angola, Zimbabwe, Malawi and Mozambique. ESA: distributed southwards to South Africa and/or Namibia. EW-NAR: restricted to the Horn of Africa, Sudan and/ or Yemen, west to Sudan and South Sudan. HWE: distributed westwards to Cameroon, Nigeria and Niger. WE: distributed westwards to the Atlantic Ocean, Guinea, Senegal and/or Mauritania). All graphs show the number of taxa to the left; the proportional distribution of taxa to the right. Note that the data are based on the altitudinal distribution of the taxa in the whole of Ethiopia.

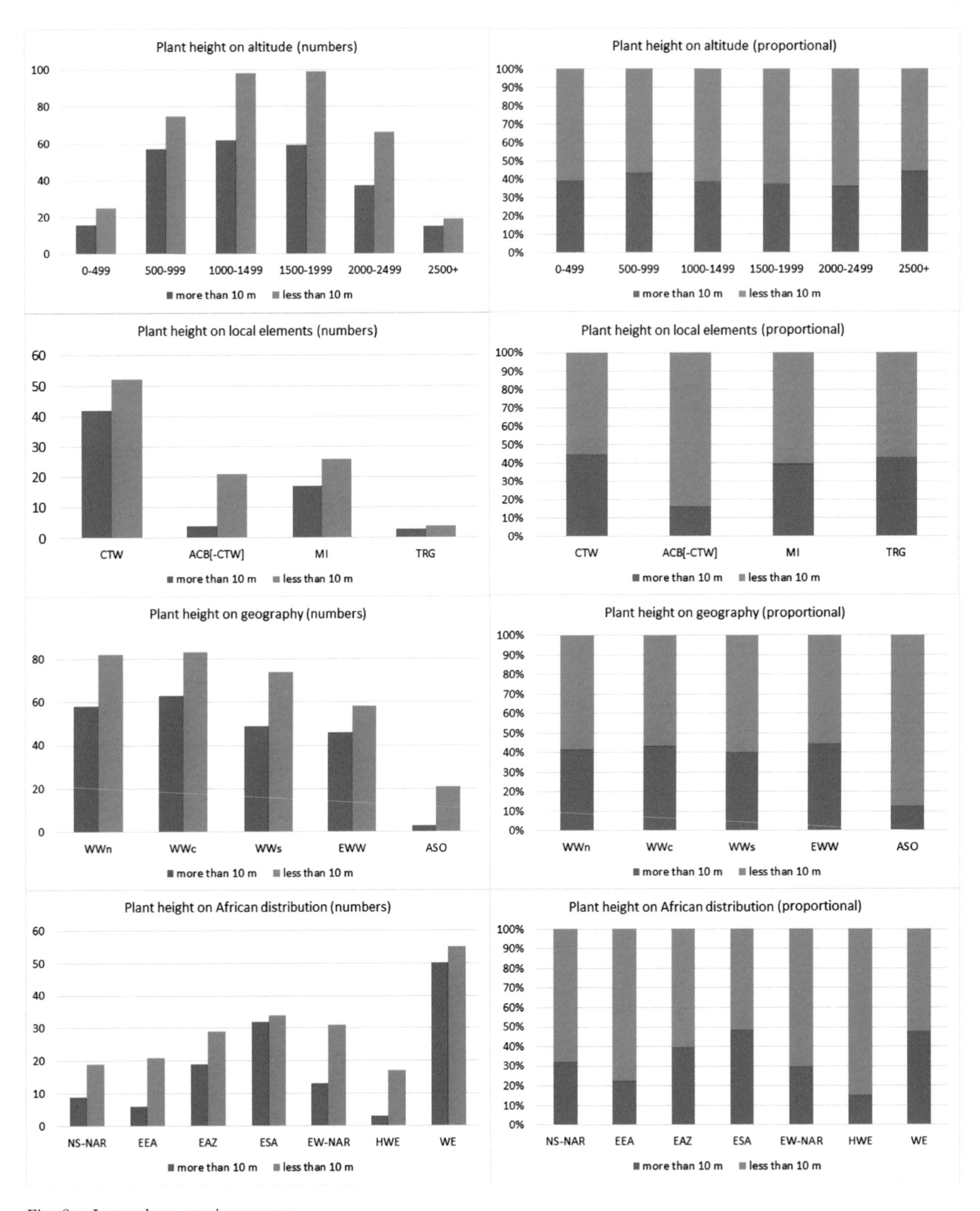

Fig. 8-1. Legend on previous page.

Table 8-1. Species occurring in more than 10 relevés.

Species	Number of relevés	% of all relevés
Anogeissus leiocarpa	78	46.2
Combretum collinum	71	42.0
Acacia seyal	59	34.9
Piliostigma thonningii	56	33.1
Terminalia schimperiana	55	32.5
Pterocarpus lucens	54	32.0
Ficus sycomorus	52	30.8
Ziziphus spina-christi	52	30.8
Lonchocarpus laxiflorus	50	29.6
Stereospermum kunthianum	49	29.0
Lannea fruticosa	47	27.8
Flueggea virosa	45	26.6
Sterculia africana	42	24.9
Balanites aegyptiaca	39	23.1
Dichrostachys cinerea	38	22.5
Gardenia ternifolia	38	22.5
Maytenus senegalensis	38	22.5
Terminalia laxiflora	38	22.5
Acacia hecatophylla	37	21.9
Combretum molle	36	21.3
Grewia mollis	36	21.3
Boswellia papyrifera	31	18.3
Cordia africana	31	18.3
Acacia polyacantha subsp. *campylacantha*	30	17.8
Annona senegalensis	28	16.6
Entada africana	26	15.4
Dalbergia melanoxylon	25	14.8
Vitex doniana	24	14.2
Bridelia scleroneura	23	13.6
Entada abyssinica	23	13.6
Ficus glumosa	23	13.6
Lannea schimperi	22	13.0
Strychnos innocua	22	13.0
Albizia malacophylla	20	11.8
Dombeya quinqueseta	20	11.8
Ziziphus abyssinica	20	11.8
Combretum hartmannianum	19	11.2
Ochna leucophloeos	17	10.1
Oxytenanthera abyssinica	16	9.5
Rhus vulgaris	16	9.5
Pseudocedrela kotschyi	15	8.9
Acacia senegal	13	7.7
Calotropis procera	13	7.7
Syzygium guineense subsp. *macrocarpum*	13	7.7
Boswellia pirottae	12	7.1
Croton macrostachyus	12	7.1
Terminalia brownii	12	7.1
Acacia mellifera	11	6.5
Ficus sur	11	6.5
Ozoroa insignis	11	6.5
Terminalia macroptera	11	6.5

8.2. Simple or compound leaves; total leaf-size

The ecological importance of a distinction between species with simple and species with compound leaves is not well documented. Compound leaves have sometimes been claimed to have reduced transpiration or to be less likely to be torn by strong wind, but the evidence for the evolutionary advantages of either leaf type seems insufficient (Warman et al. 2010). In their review of what they call 'rangelands' Lind & Morrison (1974) distinguish between two types of habitats suitable for the grazing of wild and domestic mammals. On one side they place grasslands, bushlands and thickets, woodlands and wooded grasslands with predominantly compound-leaved trees (to which belong the woodlands of *Brachystegia, Isoberlinia* and *Julbernardia*, which requires more humidity than the *Acacia* woodlands and wooded grasslands, and the more drought-tolerant *Acacia* woodlands and wooded grasslands). On the other side they place woodlands and wooded grasslands with predominantly simple-leaved trees, typically the *Combretum-Terminalia* woodlands. In the western woodlands of Ethiopia, we have no examples of woodlands of *Brachystegia, Isoberlinia* and *Julbernardia*, but apart from the many simple-leaved trees there is a substantial number of species with compound leaves and small leaflets, mostly species of *Acacia* in the wide sense and *Acacia*-like species, for example *Albizia* and *Entada*. We do not have enough data to compare the proportion between the number of species of *Acacia* (in the wide sense) and other woodland species in the western Ethiopian woodlands with the proportion further to the west in Africa, but it should be noted that there are as many as 18 species of *Acacia* (in the wide sense) recorded from the western woodlands. In their checklist, Brundu & Camarda (2013) recorded 15 species of *Acacia* (in the wide sense), of which 11 were also recorded in the relevés in the western Ethiopian woodlands. The importance of *Acacia* (in the wide sense) is increasing with latitude. Already White (1983: 203-208) noted the increasing importance of *Acacia*-dominated vegetation in the Sahel regional transition zone to the north of the Sudanian region; in the 18 vegetation types enumeration from the Sahel, nine are characterized by one or several species of *Acacia* (in the wide sense). As Ethiopia spans from 4° 25' N to 14° 50' N and Chad from 7° 20' N to 23° 10' N, and thus reaches further into the Sahel, one should expect a greater variety of *Acacia* species in Chad, but the majority of the Acacia species in Chad are the same as the species in the western Ethiopian woodlands. It should be noted that in the western Ethiopian woodlands there are not only species with small compound leaves, such as *Acacia*, *Albizia* and *Entada,* but also several compound-leaved species with large leaflets.

In our sample of woody plants from the relevés in the western woodlands, the number of taxa with simple leaves is consistently higher than the number with compound leaves (Fig. 8-2), with a percentage of simple-leaved taxa that varies between ca. 60 and 70% in all altitudinal zones, all local phytochoria and all local geographical areas in Ethiopia. This again we take to be an indication of the general homogeneity of the western woodlands. On an African scale, the highest number of species with simple leaves is found among the species that are widely distributed in a west-east direction, corresponding to the Sudanian region of White (1983).

Friis (1992; Fig. 11) also studied the percentage of taxa with compound leaves in Ethiopian forests on altitudinal zones. The highest frequency of compound-leaved species occurred between 1830 and 2135 m a.s.l., one altitudinal interval above the interval with the highest floristic diversity in the sample. This agrees with the large number of leguminous trees in the Ethiopian forests between 1800 and 2440 m a.s.l. However, this did not agree with Hamilton's results from Ugandan forests, where the highest percentage of taxa with compound leaves, about 35%, was found at the lowest altitude, about 1000 m a.s.l. An additional leaf-character that might have ecologically importance and be characteristic of plant communities is leaf-size. This was originally suggested by Raunkiær (1916, 1934), who tentatively characterized different plant communities with different leaf-size spectra, but the method seems to have been little used in studies of tropical vegetation.

Our results from the western woodlands show that the distribution of leaf size classes on altitudinal zone is rather uniform across the altitudes (Fig. 8-3), with a strong dominance of species with Mesophyll leaves. Very few taxa with very small leaves (Nanophyll) or very large leaves (Megaphyll) occur in the lower zones of the woodlands up to 1500 (-2000) m a.s.l. The number of Mesophyll taxa is always higher than the number of Macrophyll and Microphyll taxa. Except for the lowermost altitudinal zones, where two species of palms occur in the open woodlands, there are hardly any Megaphyll taxa in any zone. The proportional representation of leaf-size on altitudinal zones is also rather constant, with 40-45% Mesophyll taxa, 20-25% Macrophyll taxa and 15-22% Microphyll taxa. The distribution of leaf-size classes on geographical areas is again rather constant and always with strong dominance of Mesophyll species. However, the distribution of leaf-size classes in the local elements ACB[-CTW], MI and TRG differs notably from that of the CTW-species, with a higher percentage of Microphyll taxa and a lower percentage of Macrophyll taxa. This agrees with the general observation that the eastern lowland taxa in *Acacia-Commiphora* bushland have generally smaller leaves than the taxa in the western woodlands. A simplistic assumption about the economy of producing new leaves every year would suggest that plant communities with mainly deciduous leaves lasting for the wet months only would be dominated by species with relatively small leaves such as the Mesophyll; Megaphyll taxa would not seem economical. On the continental African scale, the group of Mesophyll taxa is prominent both in west-east direction across Africa (WE) and in the north-south direction (EEA, ESA)(Fig. 8-3).

Friis (1992: Fig. 15) found that in the Ethiopian forests the proportional distribution of leaf size classes in relation to altitude was rather uniform for the middle altitudinal zones. As in the woodlands and throughout the altitudinal range of forests, the dominance of the Mesophyll size-class was notable. On the other hand, Hamilton (1975) showed a clear decline in leaf-size with increasing altitude in Ugandan forests.

8.3. Ability to survive unfavourable conditions – leaf persistence

The relation between seasonality and leaf persistence has been difficult to study because of the poor information about the deciduousness of the individual woodland taxa. Hardly any literature exists for comparison with other tropical African woodlands. Without sufficient observations in the field it is also difficult to decide if a taxon is fully deciduous, only partly deciduous or almost evergreen, as is the case in species of *Ficus*. Some species, for example *F. sycomorus* and *F. sur*, can present leaves in some parts of the tree, while other parts are leafless (our personal observations, made repeatedly in the field). Information about the deciduousness of most taxa in the sample of woody species from our relevés in the western Ethiopian woodlands is lacking in the standard floristic works which have been used to support compilation of information for Section 6. The information provided in Section 6 is mainly based on personal observation or informed assumptions, sometimes from herbarium specimens.

Leaf persistence is here taken as the only ecological adaptation conferring the ability to survive dry periods. There are obviously others, such as ability to store water in trunks and roots, but it has not been possible systematically to gather enough information on this for all species in the sample. Such features would also seem related to the ability to survive fires, and a few examples of this are mentioned under that subject below.

The graphs in Fig. 8-4 show that the number of deciduous taxa is always higher than the number of evergreen taxa at all altitudes from 500 up to 2500 m a.s.l. and in all parts of the geographical area with woodlands. From zone to zone and from area to area, the proportion between deciduous and evergreen taxa remains relatively constant, with between 35 and 42% evergreen taxa, slightly increasing with altitude. Only above 2500 m a.s.l. does the proportion of evergreen taxa become higher than the proportion of deciduous taxa. The evergreen taxa represented above 2500 m a.s.l. do not occur in deciduous woodland at these

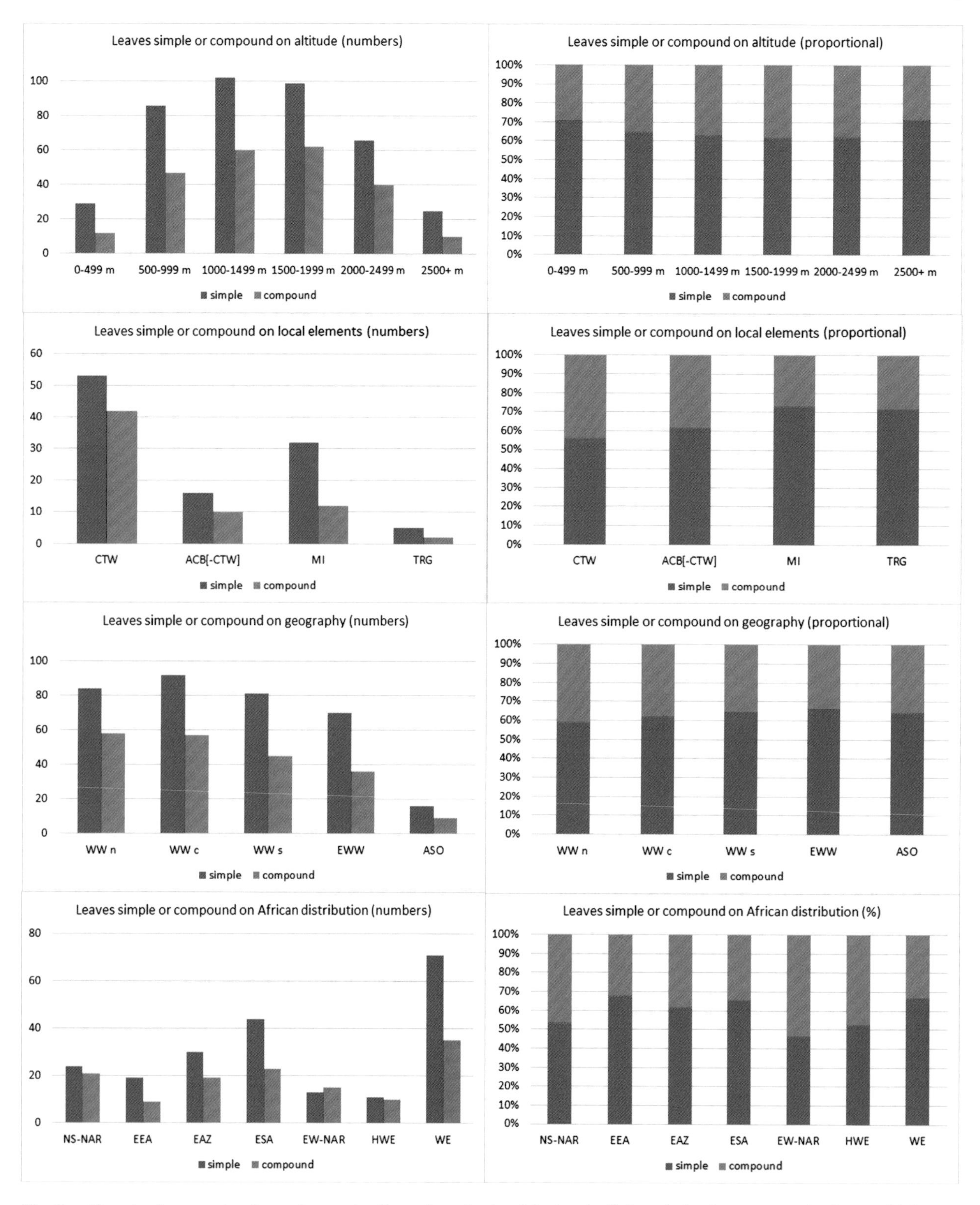

Fig. 8-2. Taxa in the sample of woody species from the relevés with simple (left or below) or composite leaves (right or above). The remaining part of the legend as for Fig. 8-1.

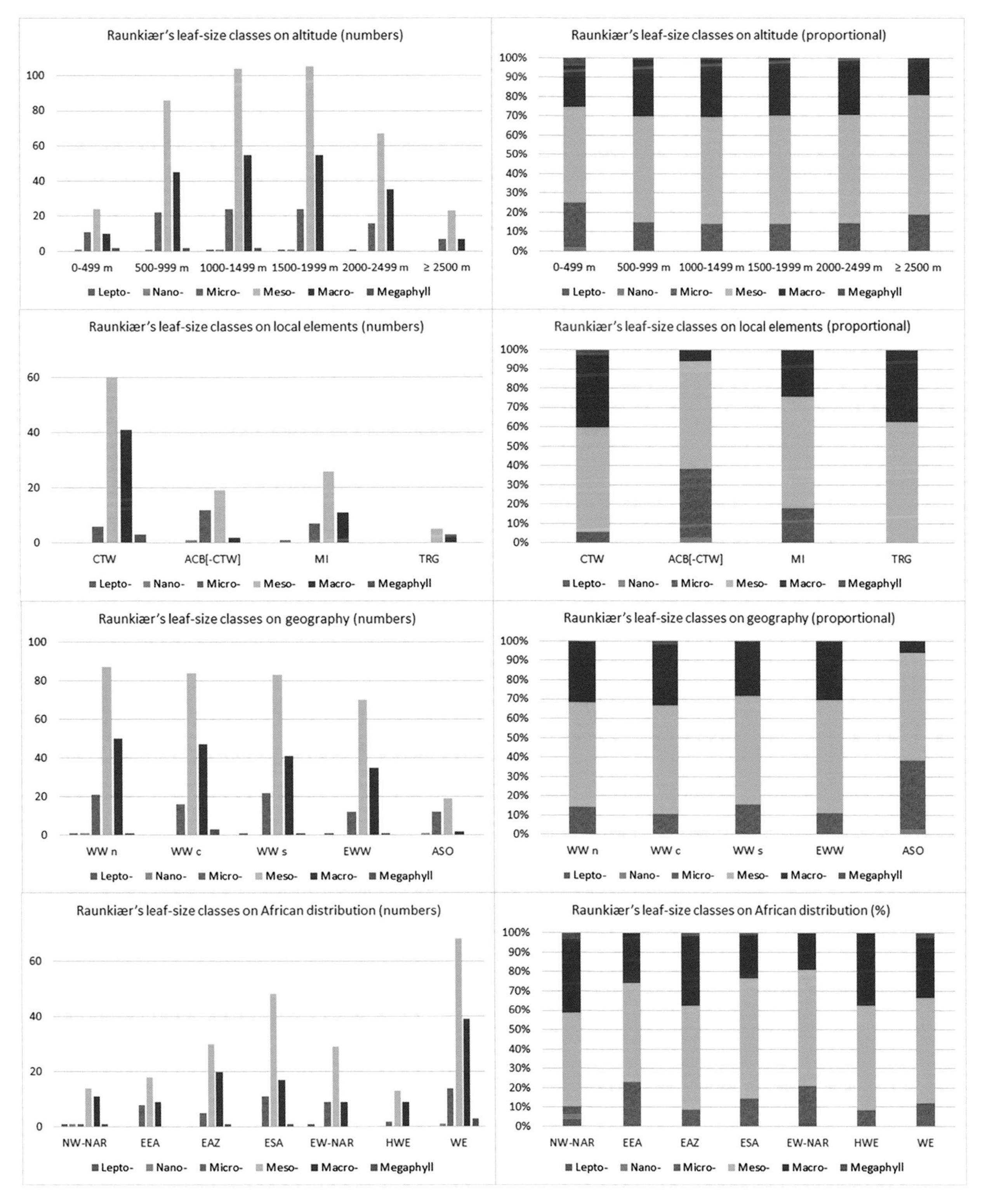

Fig. 8-3. Taxa in the sample of woody species from the relevés with leaves classified according to Raunkiær's leaf-size classes (left to right: Leptophyll taxa; Nanophyll taxa; Microphyll taxa; Mesophyll taxa; Macrophyll taxa; Megaphyll taxa). The remaining part of the legend as for Fig. 8-1.

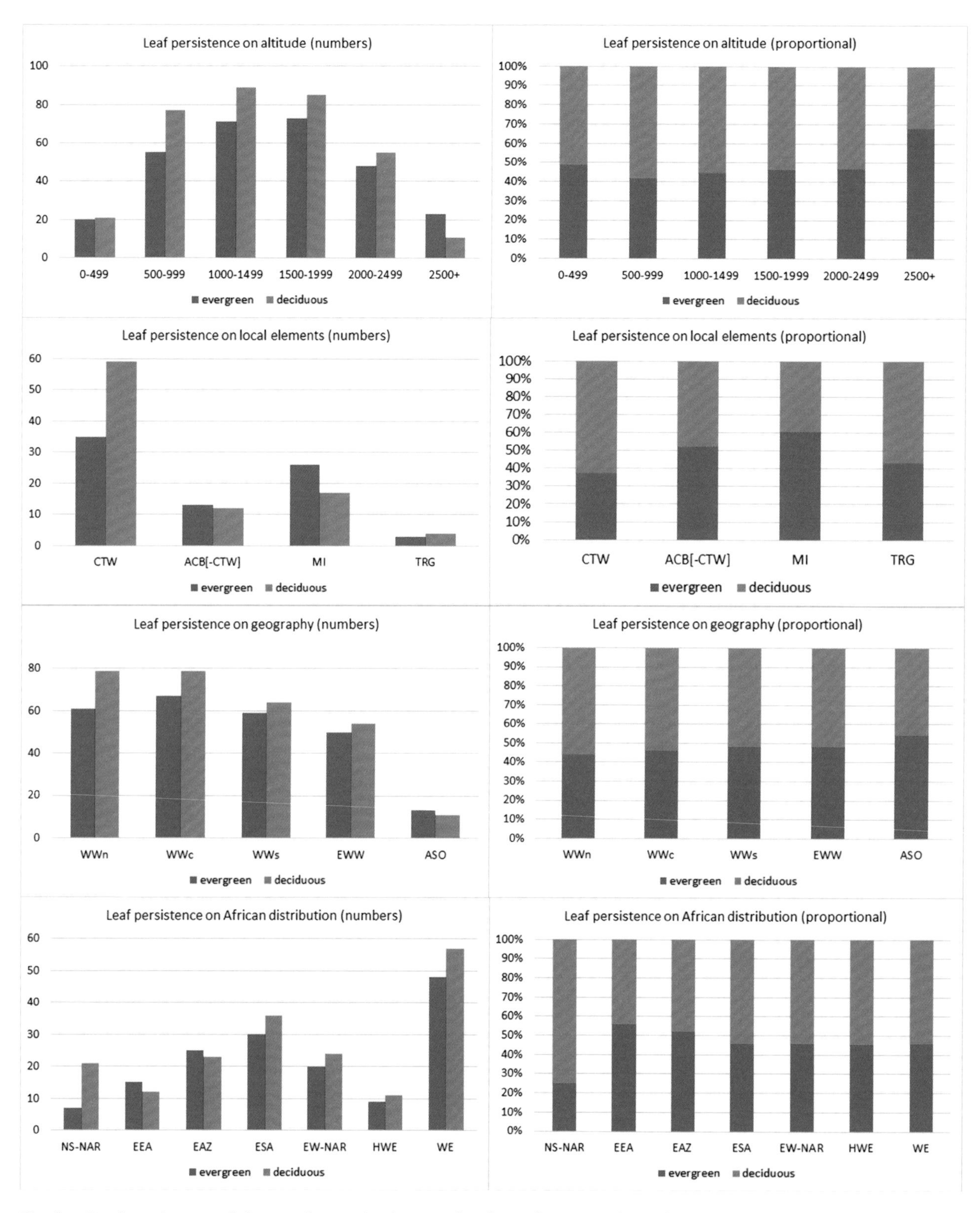

Fig. 8-4. Leaf persistence of the woody taxa in the sample of woody species from the relevés. Taxa with leaves evergreen (left or below) *versus* taxa with leaves deciduous (right or above). The remaining part of the legend as for Fig. 8-1.

altitudes and are marginal intruders (MI) that also occur above the woodlands.

Generally, the number and proportions of evergreen *versus* deciduous taxa vary slightly more between the local elements more than between altitudinal zones or geographical areas. Among the ACB[-CTW] and ASO-species, there is a slightly higher number and proportion of evergreen taxa than in the CTW-species, probably due to the number of taxa of the Capparidaceae in this group, in which semi-succulent, evergreen leaves are common.

On the continental African scale, the evergreen taxa are best represented among those reaching from Ethiopia into tropical eastern Africa (EEA, EAZ). Perhaps this is because of the number of ACB[-CTW]-species in this group. The WE-taxa, reaching in the western-eastern direction into or across the Sudanian zone, have more deciduous taxa than evergreen ones, deciduousness being characteristic of the Sudanian belt (White 1983). Also the taxa reaching to South Africa (ESA) have more deciduous taxa than evergreen ones.

Friis (1992) found amongst the forest trees a much higher proportion of deciduous taxa in the interval between 305 and 610 m a.s.l. than in the intervals below and above. In the forest flora of Ethiopia the interval between 305 and 610 m a.s.l. is particularly rich in species shared with the semi-deciduous Guineo-Congolian type forest (later renamed *Transitional Rain Forest* (TRF)), which occurs in the lowlands of western Ethiopia. A similar tendency to increased deciduousness is not seen in the woodland taxa. The highest percentage of deciduous taxa in the forest of the Horn of Africa was still much lower than in Hamilton's study from Uganda (Hamilton 1975), where the highest proportion of deciduous taxa was just under 30%. In Uganda, this figure was found at the lowest altitude, ca. 1000 m a.s.l., which could again be due to species shared with the semi-deciduous Guineo-Congolian type forest at low altitudes.

8.4. Ability to survive fires – adaptations

Woody plants in woodlands and wooded grasslands need to survive grass fires when the herbaceous ground-cover dries out and burns (Lind & Morrison 1974; Jensen & Friis 2001; Breugel et al. 2016a). As mentioned above, adaptations that help trees to survive fires are often also useful against drought. Several woody plants in woodlands develop a thick, corky or fibrous bark that both resists fire and prevents water from evaporating from the trunk. Other woody plants, like *Adansonia digitata* (the baobab) and large taxa of *Ficus*, have one or several adaptations that may help them survive the dry season, for example ability to store moisture in the trunk so they are both able to resist drought and fires (own field observations). Some woody plants grow other specialized organs that store water and thus sustain the plant against both droughts and fires. Groome (1955) studied how an important woodland genus, *Pterocarpus*, builds up water and nutrients in the roots, so that even if young plants are destroyed by fire, new saplings may sprout from the water-storing roots. A taxon in the Ethiopian woodlands with considerable capacity to store water underground is *Ozoroa pulcherrima,* which has corms from which new growth may sprout (own observations). Without more detailed field studies of all species in the sample it is difficult to discover the adaptation that may prevent destruction of trees by fire, so our results here represent only a first approximation.

The graphs in Fig. 8-5 show a slightly decreasing proportion of taxa with adaptation for fire resistance from the lowest interval to between 2000 and 2500 m a.s.l., and for altitudes above 2500 m a.s.l., well into the Afromontane vegetation, the number of taxa with adaptations to resist fires drops to below 50%. The illustration in Fig. 2-24, showing number of fire incidences between 2003 and 2013, outlines the border of the CTW-area but does not indicate altitudes; however, it is our impression from both comparison of Fig. 2-24 with maps with altitudes, and from observations from our field work that intense fires rarely occur at altitudes above 2000 m a.s.l. (apart from the fires in dried-out swamps and in Ericaceous bushland at

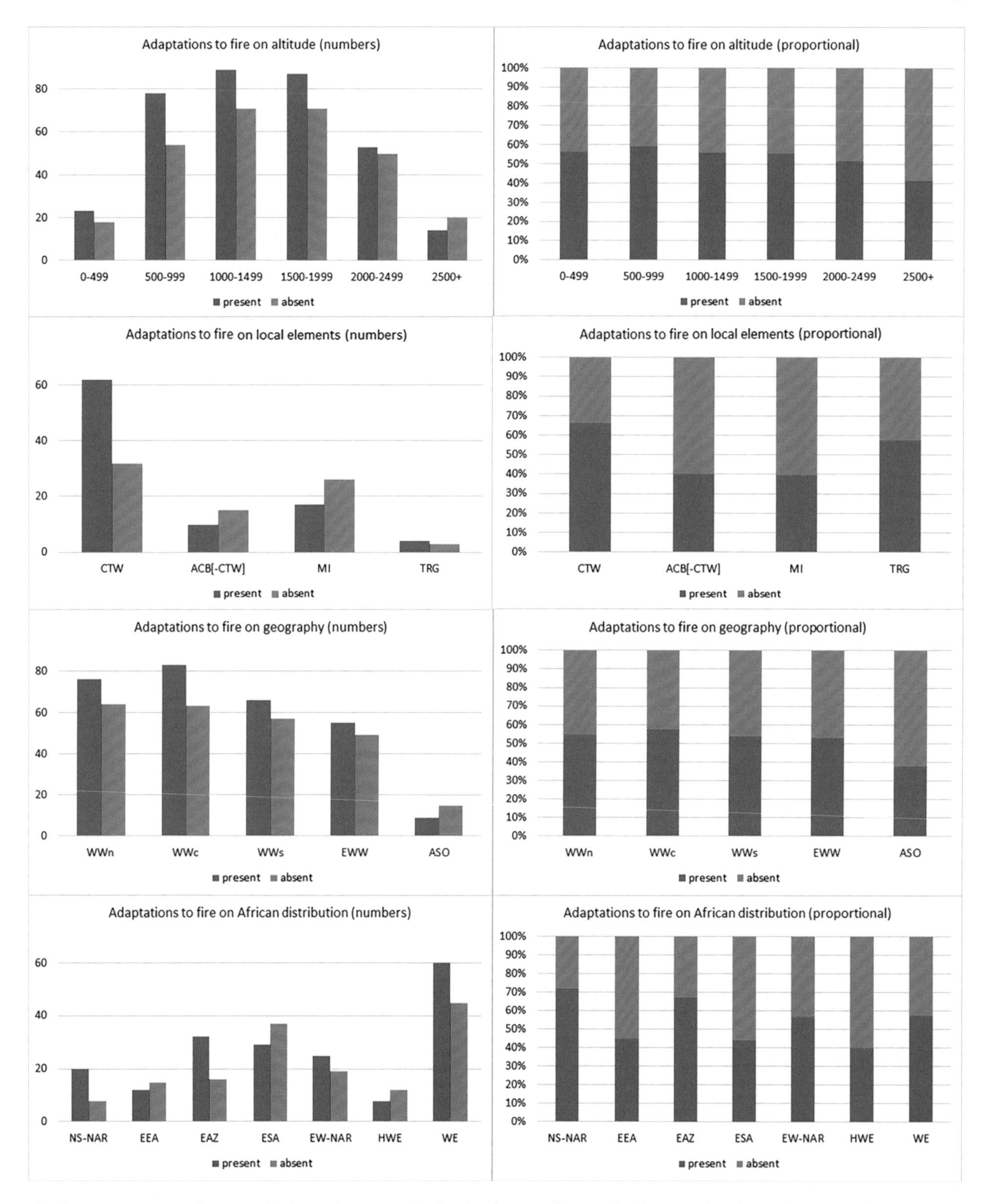

Fig. 8-5. Presence or absence of adaptations to resist fire in the woody taxa in the sample of woody species from the relevés. Taxa with adaptations present (left or below) *versus* taxa without adaptations (right or above). The remaining part of the legend as for Fig. 8-1.

much higher altitudes, as mentioned in Section 2.6). This is vaguely reflected in the proportion of taxa with adaptation for fire resistance in Fig. 8-5.

Concerning the geographical distribution of taxa with adaptation for fire resistance, one can see the highest number of taxa in the north, slightly fewer in the south (Fig. 8-5, right hand column). This does not agree well with the fire incidences between 2003 and 2013 shown in Fig. 2-24, where there are relatively fewer fires in most areas of the north compared to the south. However, it may be that the fluctuating rainfall from year to year in the north has the effect that insufficient biomass for fierce grass fires builds up every year. Only in years with relatively high rainfall do fires become fierce enough to influence survival of species with fire adaptations. Another possible reason for the difference between north and south could be that the high fire frequency in the south is a relatively new phenomenon. Jensen & Friis (2001) suggested that in the Gambela lowlands fire frequency increases with growing human population. The assumption of a growing population in the Gambela lowlands is supported by the historical information in Section 2.7. It is conceivable that the composition of the woody flora of the south-western woodlands has not yet adapted to the increasing fire frequency caused by the more numerous people living in the area.

As seen in Fig. 2-24 the number of fires is low in the ACB-area, and the number of adaptations to fire among the sub-element ACB[-CTW] is also lower than in CTW-species. The number of taxa with fire adaptations drop markedly in the local element MI, the species of which are Afromontane in their main distribution. The same can be observed among the taxa that penetrate to the east of the western woodlands (EWW); all these areas have much lower fire intensity than do the western woodlands (Fig. 2-24).

On the continental African scale, the species with adaptations to survive fires are best represented by WE-taxa, reaching in the western-eastern direction in the Sudanian zone all the way between the Atlantic Ocean and Ethiopia. In the north-south direction, the taxa reaching to southern tropical Africa (EAZ), with the large, frequently burning Zambezian woodlands, have the highest representation of species with adaptations to survive fires.

8.5. Ability to deter browsers – thorns and spines

In woodlands and wooded grasslands dominated by species of *Acacia*, a high proportion of the species in the woody flora is provided with different kinds of thorns to protect them against herbivores. In the African woodlands and wooded grasslands there are more than 40 different species of hooved mammals that browse woody plants, and many woody plants in Africa have thorns or spines that might deter browsers (Greve et al. 2012). The number of such browsers has been found to be correlated to the diversity of species of the umbrella-shaped species of *Acacia* (Greve et al. 2012: Fig. 1 & 2). The paper documents this not only by a detailed study of the thorny trees and shrubs, but also by documenting the diversity and the richness of large (> 10 kg) mammalian browser species across Africa. The two data sets show that there is a notable maximum of the number of both browsing mammals and the umbrella-shaped species of *Acacia* in the *Acacia* wooded grasslands in eastern Uganda, southern Kenya and northern and central Tanzania, much higher than in the areas in Ethiopia and western Africa we deal with in this work.

Trees characteristic of woodlands and wooded grasslands may have additional deterrents to grazing animals, such as chemicals that cause their leaves to taste unpleasantly (Cooper & Owen-Smith 1985), but the information of that kind of feature has not been available for all species in this study and is therefore not further considered here. Species of *Acacia* and other genera may even be awarded some protection by aggressive ants. Species of *Acacia* are known to have 'ant galls', which are inflated, hollow spines with a small opening that allows the ants to pass in and out of the interior of the gall (Lind & Morrison 1974). In our relevés, only *Acacia seyal* var. *fistula* is provided with such 'ant galls'.

In several other species of legumes, particularly in the genera *Cassia* and *Senna* and in many species of

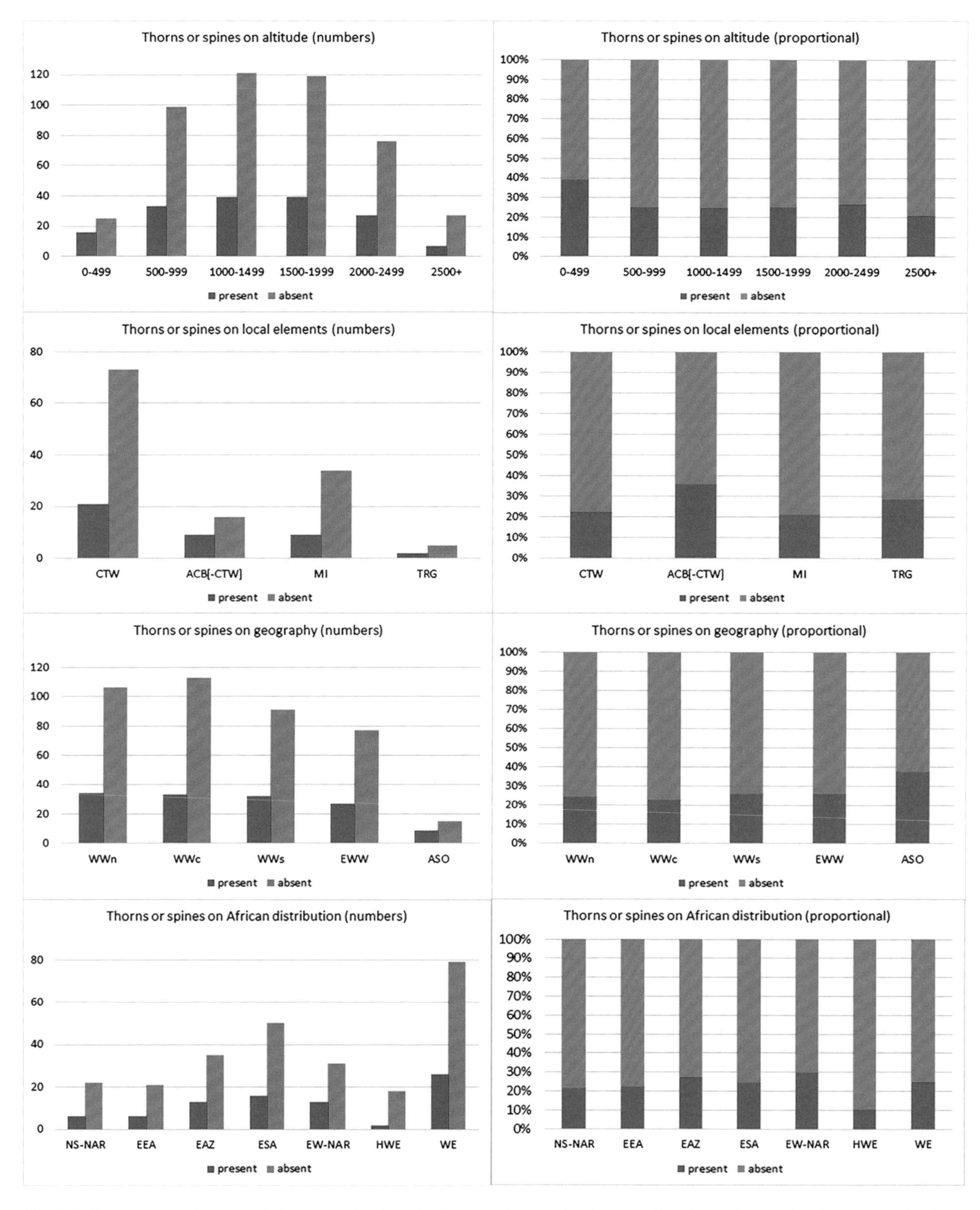

Fig. 8-6. Presence or absence of thorns and spines in the woody taxa in the sample of woody species from the relevés. Taxa with thorns or spines (left or below) *versus* taxa without thorns or spines (right or above). The remaining part of the legend as for Fig. 8-1.

Acacia, the rachis of the composite leaves is provided with glands that produce a nectar-like substance attractive to ants. This feature is also seen as a protective measure against browsers (Greve et al. 2012). However, due to lack of data on most species we have here restricted our study to analyse the presence and absence of thorns and spines.

The top graphs in Fig. 8-6 show that at all altitudes and in all local geographical areas in Ethiopia the number of species of woodland trees with thorns or spines is much lower than the number of species without thorns. The proportion of thorny or spiny taxa, only slightly above 20%, is fairly constant with increasing altitudes, except for the lowest altitudinal zone, below 500 m a.s.l., where the proportion of thorny or spiny taxa is ca. 40% (though this is based on a very low actual number of taxa).

The low level of protection against browsers in plants in the Sudanian woodland agrees well with the low number of wild species of browsers in the woodlands across Africa between Ethiopia and the Atlantic Ocean (Greve et al. 2012: Fig. 2(a)). Near the Ethiopian Highlands there are 2-5 browser species, occurring in a particularly wide zone in the south, while further to the west (and almost across Africa) and to the north, the richness of indigenous browsers drops to between zero and two. The number of domestic browsers in the Sudanian woodland may be increasing and is almost certainly far outnumbering indigenous browsers. It may therefore now play a role in depleting the vegetation. However, large areas of the Sudanian zone are infested with tsetse flies (*Glossina*), infecting both humans with sleeping sickness and cattle with trypanosomiasis. Although cattle has existed in West Africa for many thousands of years, it is unlikely that before the latest millennia it has been numerous enough to exert a pressure on the vegetation and has not influenced the development of protection against browsers in plants.

The number of species with thorns or spines is rather similar in all geographical areas within Ethiopia in the west and in the area to the east of the Rift Valley (Fig. 8-6, right hand side). Among the local elements and sub-elements, the proportion of taxa with spines or thorns is highest in the ACB[-CTW] sub-element, which has its main distribution in the landscapes to the east of the Ethiopian Highlands. In the ACB-area, the diversity of indigenous browsing animals is slightly higher than in the Sudanian woodlands (Greve et al. 2012). Similarly, the percentage of thorny species is slightly higher among the CTW-species that extend to the ASO-area, which agrees with the relatively higher number of thorny species in the eastern lowlands with *Acacia-Commiphora* bushland than in the western woodlands. This agrees with Greve et al. (2012: Fig. 1). To the number of spiny taxa of *Acacia* in Greve et al. (2012) should be added the considerable number of spiny species of *Commiphora*, which occur in the eastern lowlands.

Similarly, on the continental African scale, the proportion of taxa with spines and thorns is lower in Sudanian woodlands towards the Atlantic (WE) compared to the southern direction through East Africa towards South Africa (EAZ, ESA).

The even distribution on altitude of the number of woodland species with thorns and spines does not agree with the distribution of forest species with such adaptations (Friis 1992); in the Ethiopian forests there is a maximum of thorny and spiny taxa between 1220 and 2745 m a.s.l., which is also the zone with most domestic animals.

9

Modelled distributions of typical Ethiopian *Combretum-Terminalia* woodland species

There are several reasons why we modelled the distribution of twelve of the most characteristic species from the *Combretum-Terminalia* woodlands of the western Ethiopian lowlands. Five of the species modelled here occur in one third or more of the relevés, as seen in Table 8-1 (*Anogeissus leiocarpa*, *Combretum collinum*, *Terminalia schimperiana*, *Lonchocarpus laxiflorus*, *Pterocarpus lucens*), the other seven species belong to the genera *Combretum* and *Terminalia* and are elsewhere important in Sudanian vegetation across Africa from Ethiopia to the Atlantic Ocean.

The first reason for modelling the distributions of these species based on their total African distribution is to see how such models will represent the areas between the sampled areas inside Ethiopia. The best sampled areas in Ethiopia are those accessible by motorable roads, which do not cover more than a fraction of western Ethiopia. The second reason is to see how their distributions in Ethiopia agree with the distribution across Africa, and to examine the information about the species used in the analyses in Section 7.2, particularly to see if the local elements in Ethiopia are likely to represent similar groups of species with the same distribution in western Africa. A third reason is to document the general position of the western Ethiopian lowlands as part of the Sudanian region, in relation to the introductory remarks about this given in Section 2.1.

In modelling species-distributions the environmental parameters are most often climatic data (e.g., details about temperature and precipitation), but sometimes also other variables, such as soil type and land cover. We have considered that altitude is not a necessary environmental parameter in this model, as variation in altitude is reflected in variation in the annual mean temperature (Hahn & Knoch 1932; Liljequist 1986; Friis et al. 2010). The use of soil types based on the available databases, for example that of Africa Soil Information Service (http://africasoils.net/services/data/soil-databases/) was not possible to use here due to the size of the data involved in a model based on this information and the great difference in the reliability of data that is provided for the whole of Africa. The same would apply to inclusion of land cover data. The modelling is thus made entirely on climatic data from Bioclim. The choice of a limited number of climatic parameters was made to minimize multi-collinearity and to reduce the risk of overfitting, when the model begins to describe the random error in the data rather than the relationships between variables because it contains more parameters than are justified by the data. We selected a subset of the bioclimatic variables and have used Bio 1: Annual Mean Temperature (°C × 10); Bio 4: Temperature Seasonality (standard deviation ×100); Bio 7: Temperature Annual Range (BIO5 minus BIO6, from BioClim); Bio 12: Annual Precipitation (mm); Bio 15: Precipitation Seasonality (Coefficient of Variation).

We have previously worked with species distribution modelling across Africa from the Atlantic Ocean to Ethiopia (Weber et al. 2020), using both climatic envelop methods from DIVA-GIS and methods employing "machine learning," and we found that the "machine learning" method, MAXENT 3.4, was the best for the actual modelling. It was a significant advantage that it provided the necessary "random background" points of absence or "pseudo-absence" data to be fitted in. On each map, the scale from 0 to 1 indicates the modelled suitability for the species, from 0.0-0.2 (low suitability; green) to 0.8-1.0 (high suitability; red).

Additional georeferenced distributional data for the species of Combretaceae modelled in this Section was provided by Jan J. Wieringa, Naturalis Biodiversity Center, Leiden, the Netherlands, which includes records worked on by Carel C.H. Jonkind, Botanic Garden Meise, Belgium. Georeferenced distributional data for the species of *Lonchocarpus laxiflorus* and *Pterocarpus lucens* was derived from GBIF and our material from the Horn of Africa. The data sets are not reproduced in this work.

9.1. Anogeissus leiocarpa

Anogeissus leiocarpa is a frequent and widespread species in a rather broad zone across Africa from Senegal to Ethiopia and Eritrea, but it does not reach as far south as to Uganda or other countries in tropical East Africa (Fig. 9-1). Yet the zone across Africa of *Anogeissus leiocarpa* is relatively southern, avoiding much of the dry Sahel in the north, but towards the Gulf of Guinea reaching into the Guinea-Congolia/Sudania regional transition zone of White (1983). This is not in agreement with the distribution in Ethiopia, where the species is widespread from Eritrea to the Gambela lowlands and has not been recorded further to the south. The model predicts a large potential area in almost the whole of northern and western Ethiopia, including most of the Gambela lowlands. In northwestern Ethiopia, the notable gaps in the recorded distribution are filled with potential distribution by the model, which seems plausible from our field observations. There are recorded distributions from the deep river valleys in the Ethiopian Highlands, particularly in the Tacazze Valley, the Abay Gorge and the Didessa Valley, but not in the valleys further south, and there is no modelled distribution in the Omo or other southern river valleys.

9.2. Combretum adenogonium

Combretum adenogonium is a frequent and widespread species in a rather broad zone across Africa from Senegal to Ethiopia and Eritrea, reaching southwards to Uganda, western Kenya and large parts of Tanzania; in tropical East Africa it is frequently known as *Combretum fragrans* F. Hoffm. While *C. adenogonium* is more common in northern than in southern Ethiopia, its zone of distribution across Africa is relatively southern, avoiding much of the dry Sahel in the north; towards the Gulf of Guinea it reaches into the Guinea-Congolia/Sudania regional transition zone of White (1983) (Fig. 9-2). The species has a Sudano-Zambesian distribution, and numerable records have been made in the predicted distribution area in the Zambezian region; however, the species seem to be lacking from Angola. There are widespread records from Ethiopia, from the border with Eritrea to the Omo Valley; records have been made from the deep river valleys in the highlands, particularly in the Tacazze Valley, the Abay Gorge and the Gibe and Gojeb Valleys. The model predicts a large potential area in the whole of northern and western Ethiopia and more scattered in large parts of southern Ethiopia, where the predictions of the model seem to agree with the recorded distribution. However, a predicted distribution on the eastern escarpment of the Ethiopian Highlands does not appear to be filled by an actual distribution.

9.3. Combretum collinum

Combretum collinum is a frequent and widespread species in a rather broad zone across Africa from Senegal to Ethiopia, Uganda, Kenya and Tanzania; the species has been subdivided into a range of subspecies that partly overlap in distribution. The zone across Africa of *Combretum collinum* is relatively southern, avoiding much of the dry Sahel in the north, and towards the Gulf of Guinea it reaches into the Guinea-Congolia/Sudania regional transition zone of White (1983) (Fig. 9-3). The species has a Sudano-Zambesian distribution, and many records have been made in the predicted distribution area in the Zambezian region. There are quite a few actual records to the south of the predicted Zambezian zone, and we can see no explanation for this. It may be possible that the southern subspecies have environmental requirements that are not well predicted by the very numerous records in the Sudanian part of the distribution area. The spe-

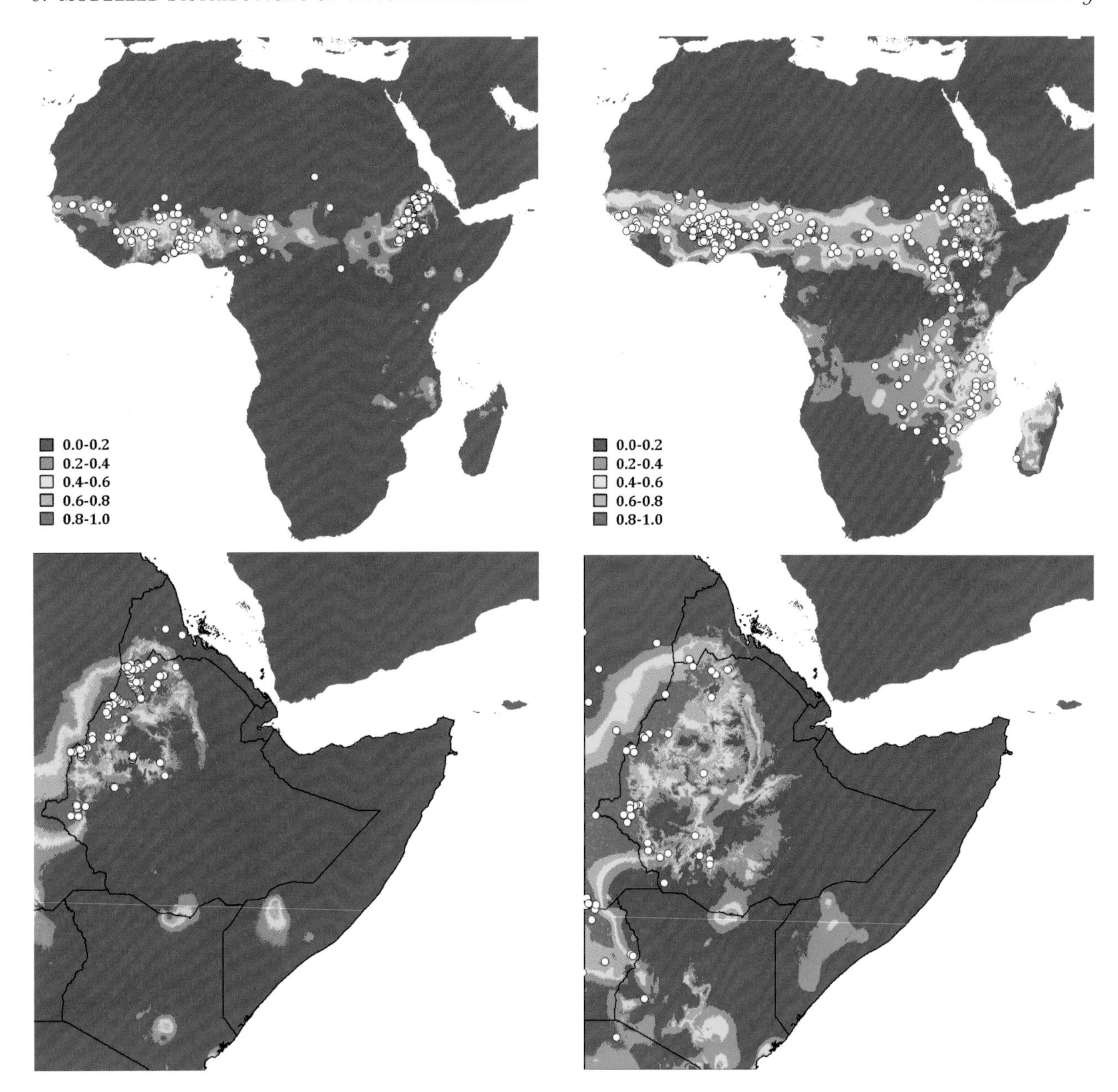

Fig. 9-1. *Anogeissus leiocarpa*. Modelled distribution in Africa (above) and on the Horn of Africa (below). Modelled suitability for the species varying from 0.0-0.2 (low suitability, indicated with green) to 0.8-1.0 (high suitability, indicated with red). White dots indicate observed records.

Fig. 9-2. *Combretum adenogonium*. Modelled distribution in Africa (above) and on the Horn of Africa (below). The remaining part of the legend as for Fig. 9-1.

cies occurs in Ethiopia in the entire CTW-area, from just south of the Eritrean border to the area south of Dima in south-western Ethiopia, and there are many records from the deep river valleys in the Ethiopian Highlands. The model predicts large potential areas in almost the whole of northern and western Ethiopia, in large parts of southern Ethiopia, in the deep river valleys in the Ethiopian Highlands, and in an extension to the east of the Rift Valley in the *Transitional semi-evergreen bushland* (TSEB). The species seems likely to occur in the whole or most of the predicted area, except that a modelled distribution on the eastern escarpment of the Ethiopian Highlands does not seem to be filled by an actual presence of the species.

9.4. Combretum hartmannianum

Combretum hartmannianum is a frequent and widespread species in the northern part of the western woodlands of Ethiopia and in a narrow zone of the adjacent part of the Sudan and in the Tacazze Valley (Fig. 9-4). Its southern limit in Ethiopia is approximately at the Abay River. Apart from small patches that may just represent accidental combination of the climatic parameters used for the prediction, there are no predicted areas outside the actual distribution, and both the recorded and the predicted area of distribution is fully inside the Sudanian regional centre of endemism of White (1983). The species seems likely to occur in the whole or most of the predicted area.

9.5. Combretum molle

Combretum molle is a frequent and widespread species in a broad zone across Africa from Senegal to Eritrea, Ethiopia, Somalia, and in nearly all low-lying parts of Uganda, Kenya and Tanzania in East Africa. The species is extremely variable in a range of characters that occur in various combinations throughout the extensive distribution area and it has not been possible to divide it into meaningful subspecies. The zone across Africa of *Combretum molle* is broad, but the species avoids the dry Sahel in the north; towards the Gulf of Guinea it reaches into the Guinea-Congolia/Sudania regional transition zone of White (1983) and actual records have been made near the Gulf of Guinea at the Dahomey Gap (Fig. 9-5). The species has a Sudano-Zambesian distribution with numerous records from the predicted Zambezian zone and even to the south of that zone. In Ethiopia the species has recorded distribution from the Eritrean border to localities with *Transitional semi-evergreen bushland* (TSEB) near the Kenyan border and from the Gambela lowlands to localities with *Transitional semi-evergreen bushland* (TSEB) near the border with Somalia east of Harar. The species reach altitudes above 2000 m a.s.l., but inside the area of the Ethiopian Highlands the documented records are usually associated with river valleys. The model predicts a very large potential area in almost the whole of Ethiopia apart from the south-eastern lowlands (and two records from that part of the country are from a small, isolated highland, the Gerire Hills, with *Transitional semi-evergreen bushland* (TSEB), as described by Friis (2019). The species is not likely to occur in the whole of the predicted area, and it is our impression from field work that despite its wide distribution and ecological range the potential distribution is here overpredicted.

9.6. Combretum rochetianum

Combretum rochetianum has often been accepted to be a near-endemic species, restricted to a small part of Eritrea, Ethiopia and the Sudan, as appears from the literature about the species cited in Section 6. However, when we modelled the African distribution from the data in Section 6, the resulting MaxEnt model suggested a narrow potential distribution across Africa to the Atlantic Ocean, a potential area in tropical East Africa and a relatively slender Zambezian extension. This agrees with the distributional data provided by Jan J. Wieringa and Carel C.H. Jonkind, which included records from Senegal and Ghana. When these additional records were added to the model they provided a slightly wider potential African distribution, as shown in Fig. 9-6. The taxonomy behind this distribution has not been finally sorted out and published, but in GBIF (GBIF.org

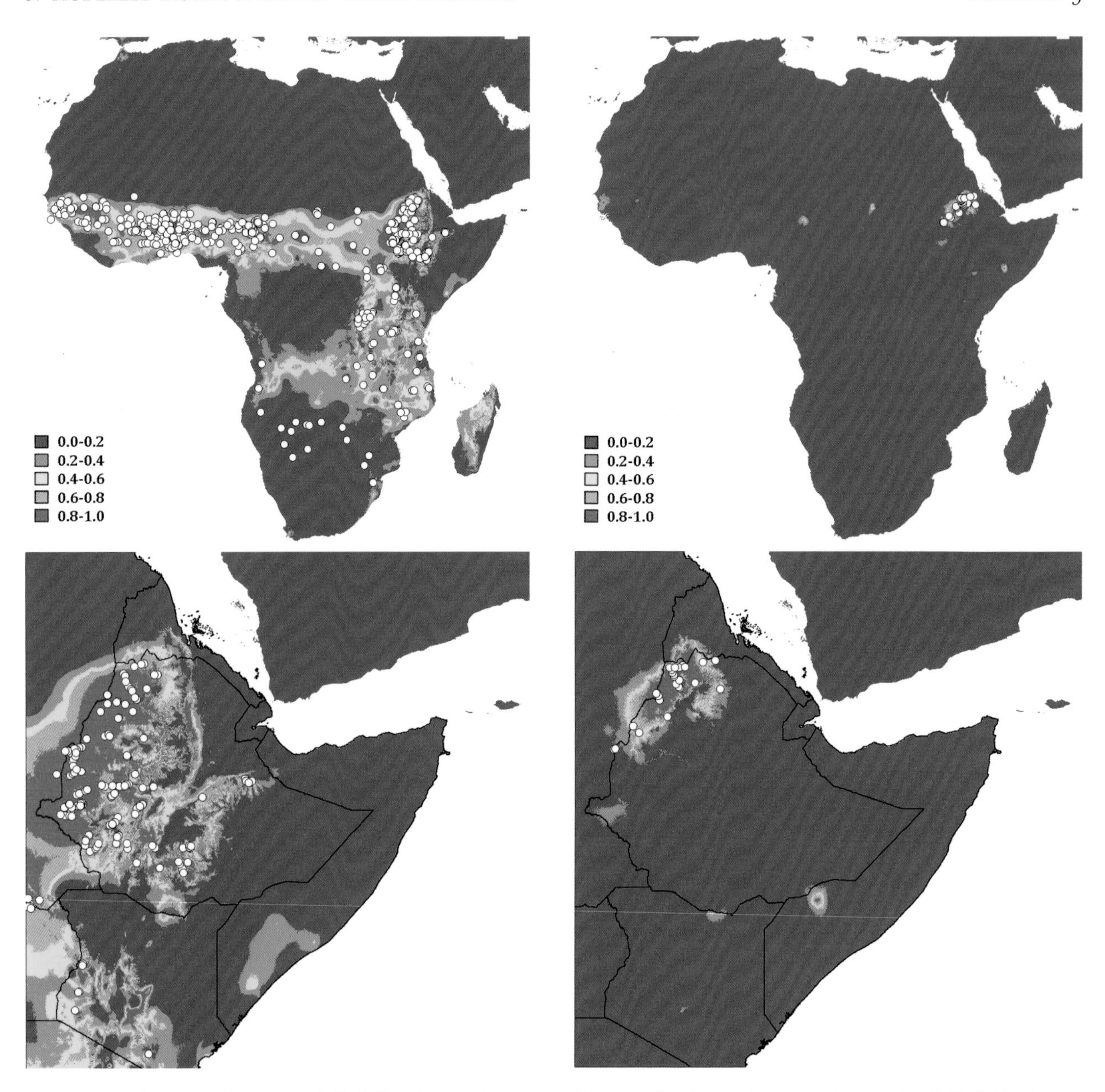

Fig. 9-3. *Combretum collinum*. Modelled distribution in Africa (above) and on the Horn of Africa (below). The remaining part of the legend as for Fig. 9-1.

Fig. 9-4. *Combretum hartmannianum*. Modelled distribution in Africa (above) and on the Horn of Africa (below).The remaining part of the legend as for Fig. 9-1.

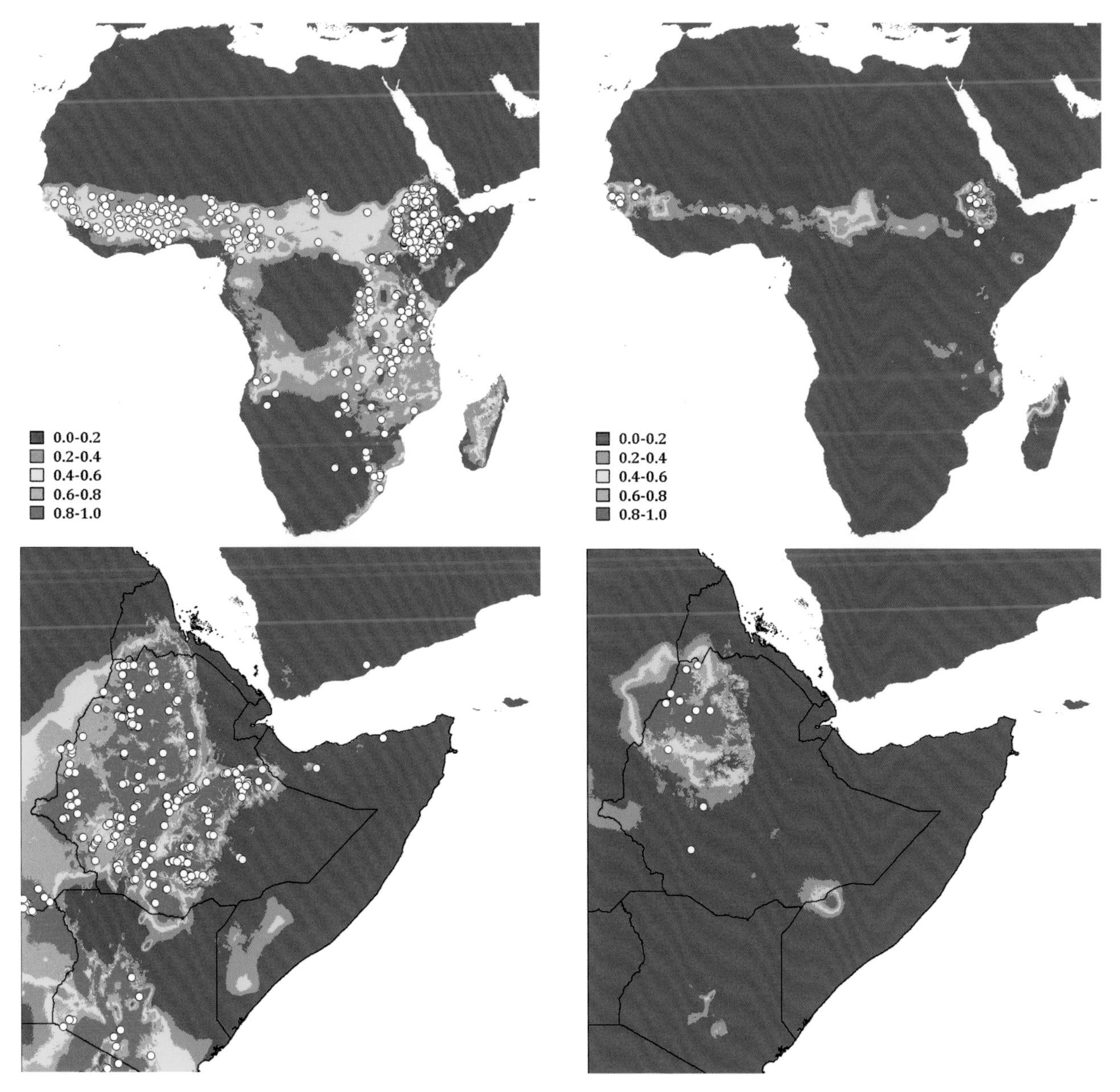

Fig. 9-5. *Combretum molle*. Modelled distribution in Africa (above) and on the Horn of Africa (below). The remaining part of the legend as for Fig. 9-1.

Fig. 9-6. *Combretum rochetianum*. Modelled distribution in Africa (above) and on the Horn of Africa (below). The remaining part of the legend as for Fig. 9-1.

2020) there are cited specimens of *C. rochetianum* from Senegal, Guinea-Bissau, Ghana, Central African Republic and Cameroon, and GBIF also cites *C. oubanguense* Exell (1937), with a type from the Central African Republic, as a synonym of *C. rochetianum*. In Ethiopia, *C. rochetianum* occurs in the north-western lowlands, in the woodlands around Lake Tana and in the GJ floristic region, but the predicted area in the north-west is wider than the area with documented records. On the other hand, two records from much further south (Gibe Gorge and from the GG floristic region do not agree well with the predicted area in the north-west and may be misidentified. Altogether it is difficult to comment on the agreement between the Ethiopian distribution and the central and western distribution across Africa.

9.7. Terminalia brownii

The African distribution of *Terminalia brownii* is different from the typical Sudanian distributions in the Combretaceae; the species occurs in Sudan, Ethiopia and Eritrea, where it has a more eastern distribution than is usually seen in CTW-species (see below). Moreover, it is a cultivated tree with the Konso people in the GG floristic region (see Section 6 and Engels & Goettsch 1991: 179). It is moderately common in the dry parts of Uganda and the D.R. Congo, in Kenya and some places in Tanzania (Fig. 9-7). The most surprising feature of its distribution is a disjunct population in northern Nigeria. In Ethiopia, it is most common in the dry *Transitional semi-evergreen bushland* (TSEB) and the nearby areas in the *Acacia-Commiphora bushland* (ACB), but it is restricted to zones that are marginal towards other vegetation types, and it does not occur in the core of the *Acacia-Commiphora bushland* (ACB); it penetrates into the northern and particularly into the southern parts of the CTW-area, for which reason its local distribution type has been classified as ACB[-CTW]. The model predicts a moderately large potential area in northern, eastern and southern Ethiopia, and the recorded distribution agrees with this, although the observed records are relatively scattered.

9.8. Terminalia laxiflora

Terminalia laxiflora is a frequent and widespread species in a rather broad zone across Africa from Senegal (but apparently more common from Mali and eastwards) to north-western Uganda, Ethiopia and Eritrea. In western Africa it avoids both the moist forest zone in the south and the dry Sahel in the north and does not reach the Gulf of Guinea in the Dahomey Gap. There are a few scattered potential areas in the Zambezian region, but they are not occupied by this species (Fig. 9-8). In Ethiopia it occurs rather widely in most of the CTW-area from the Eritrean border in the north to the area around Dima in south-western Ethiopia, but seems to thin out in the southern part of the CTW-area. It does enter the deep river valleys in the Ethiopian Highlands, particularly the Abay Valley. The model predicts a fairly narrow potential area along the western Ethiopian border, but not a wide distribution in southern Ethiopia, where a few documented records from the Gojeb and the Omo Valley are outside the modelled areas of distribution. The modelled distribution in Ethiopia does not agree with the wide range in the western part of the distribution area of *T. laxiflora*, while the documented distribution in Ethiopia agrees slightly better; difficulties mentioned in Section 4 with the distinction between some material of *T. laxiflora* and *T. schimperiana* may have causes misidentifications and incorrect areas of distribution.

9.9. Terminalia macroptera

Terminalia macroptera is a frequent and widespread species in a rather broad zone across Africa from Senegal to north-western Uganda, eastern D.R. Congo, Ethiopia and Eritrea. In West Africa it avoids the dry Sahel zone and approaches the moist forest zone in the Guinea-Congolia/Sudania regional transition zone of White (1983), and actual records have been made near the Gulf of Guinea at the Dahomey Gap (Fig. 9-9). The model predicts areas of potential distribution in the Zambezian region, but the species has not been recorded from the predicted areas. In Ethiopia it oc-

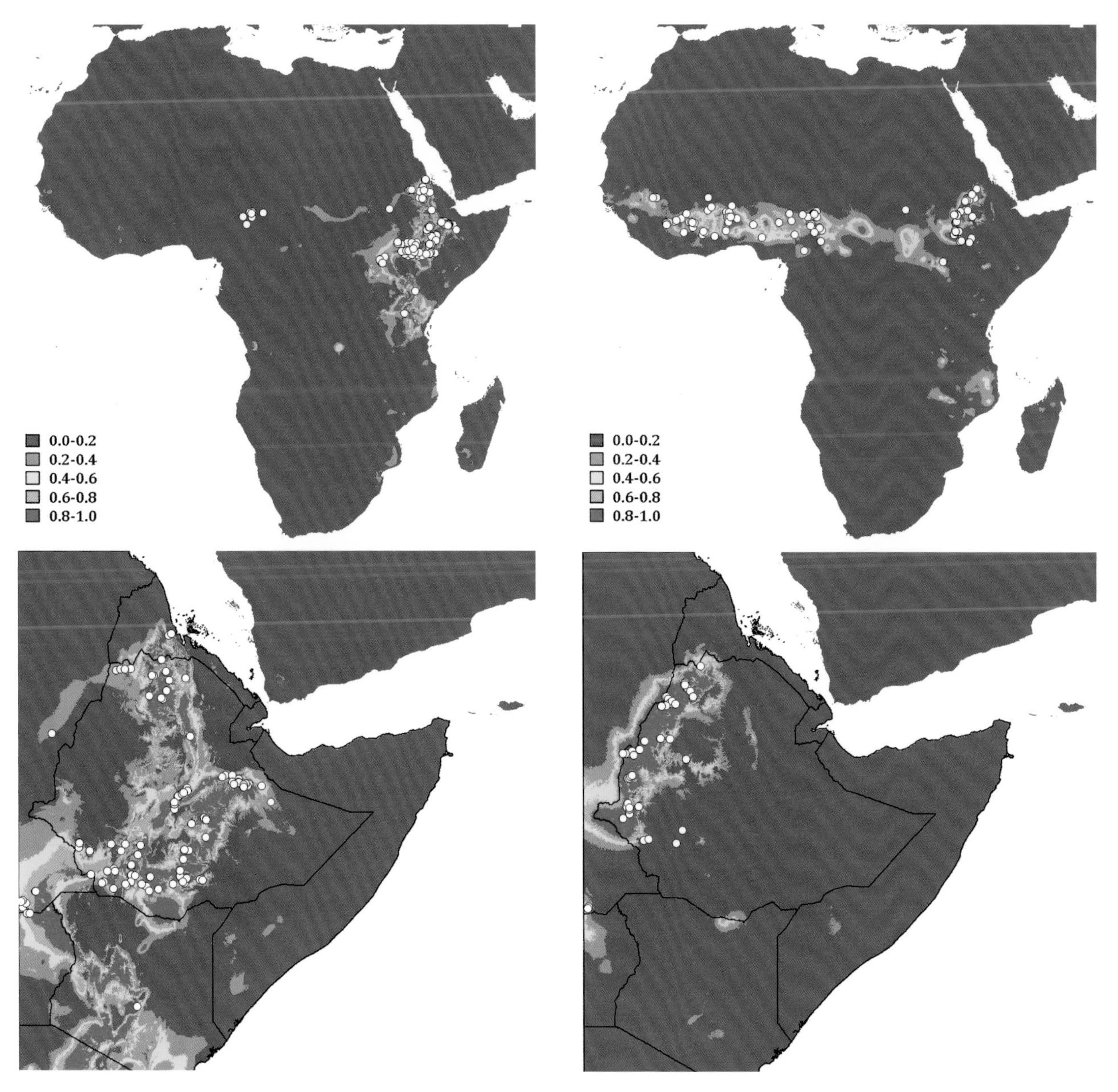

Fig. 9-7. *Terminalia brownii*. Modelled distribution in Africa (above) and on the Horn of Africa (below).The remaining part of the legend as for Fig. 9-1.

Fig. 9-8. *Terminalia laxiflora*. Modelled distribution in Africa (above) and on the Horn of Africa (below). The remaining part of the legend as for Fig. 9-1.

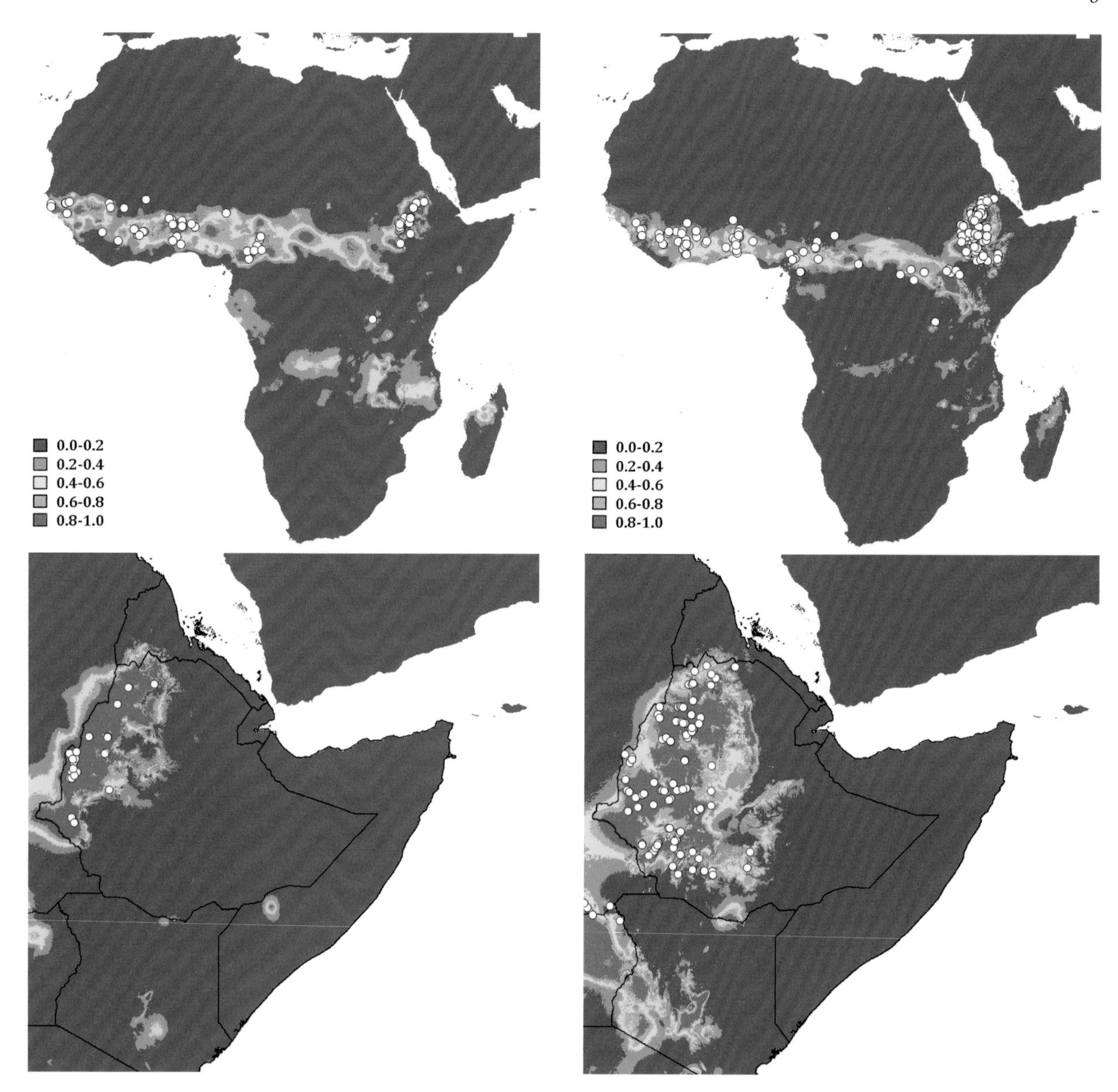

Fig. 9-9. *Terminalia macroptera*. Modelled distribution in Africa (above) and on the Horn of Africa (below). The remaining part of the legend as for Fig. 9-1.

Fig. 9-10. *Terminalia schimperiana*. Modelled distribution in Africa (above) and on the Horn of Africa (below). The remaining part of the legend as for Fig. 9-1.

curs from the GD floristic region southwards to the Gambela lowlands. It is recorded from the Didessa Valley but seems otherwise mostly restricted to the large, coherent areas of the western lowlands. The model predicts a fairly wide potential area in almost the whole of northern and western Ethiopia to the Gambela lowlands and in the Abay Valley, but it is uncertain if the species does really occupy this rather large area.

9.10. Terminalia schimperiana

Terminalia schimperiana is a frequent and widespread species in a rather narrow zone across Africa from Senegal to north-western Uganda, Eritrea and Ethiopia, avoiding the driest northern Sudanian zone and the Sahel. The species occurs in the Guinea-Congolia/Sudania regional transition zone of White (1983), does reach the Gulf of Guinea in the Dahomey Gap in the rain forest belt, and also gets close to the Gulf in the Ivory Coast (Fig. 9-10). There are minor areas that have been predicted as suitable in the Zambezian zone, but the species has not been recorded from these. However, the nomenclature may cause some confusion, as in western Africa the species has been and probably is still called *Terminalia glaucescens* (Friis & Sebsebe 2020). In Ethiopia the species is found from the Eritrean border to the area south of Dima and near Maji in south-western Ethiopia, and it occurs in the *Transitional Semi-Evergreen Bushland* (TSEB) to the east of the Rift Valley. The modelled distribution predicts a large potential area in almost the whole of northern Ethiopia and in large parts of southern Ethiopia, but the documented distribution is more restricted, and the eastern records from far into the Ethiopian Highlands seem to be from the deep river valleys in the highlands. As with *Combretum molle*, the species is not likely to occur in the whole of the predicted area.

9.11. Lonchocarpus laxiflorus

Lonchocarpus laxiflorus is a frequent and widespread species in a rather broad zone across Africa from Senegal to north-western Uganda, Ethiopia and Eritrea; it occurs in the Guinea-Congolia/Sudania regional transition zone of White (1983), and reaches the Gulf of Guinea in the Dahomey Gap in the rain forest belt, only avoiding the moist forest zone itself in the south and the dry Sahel in the north (Fig. 9-11). There is a related species, *Lonchocarpus bussei* Harms in the part of the Zambezian region that the model has predicted as suitable, and that taxon has sometimes been included in *Lonchocarpus laxiflorus*. The Sudanian species occurs in Ethiopia from the Eritrean border to the area south of Dima in south-western Ethiopia. The model predicts a large potential area in almost the whole of northern and western Ethiopia and in large parts of southern Ethiopia, but the actual distribution of the species agrees with the suggestion in Section 6 that the species is restricted to the lowlands, and the eastern records are from the deep river valleys that cut into the Ethiopian Highlands. The species is not likely to occur in the whole of the predicted area in the Gambela lowlands, as it is associated with thin, stony soil and rocky hills.

9.12. Pterocarpus lucens

Pterocarpus lucens is a frequent and widespread species in a rather broad zone across Africa from Senegal to north-western Uganda and Ethiopia; it locally occurs in the Guinea-Congolia/Sudania regional transition zone of White (1983), but does not reach the Gulf of Guinea in the Dahomey Gap. It avoids the dry Sahel in the north (Fig. 9-12). There is a closely related species, *Pterocarpus antunesii* (Taub.) Harms in the part of the Zambezian region that the model predicts as suitable for *P. lucens*, and that taxon is sometimes treated as a subspecies of the Sudanian species. That species occurs in Ethiopia from the border with Eritrea to the eastern part of the Gambela lowlands, and has been recorded in a few places in the deep river valleys in the highlands. In Ethiopia, the model predicts a large

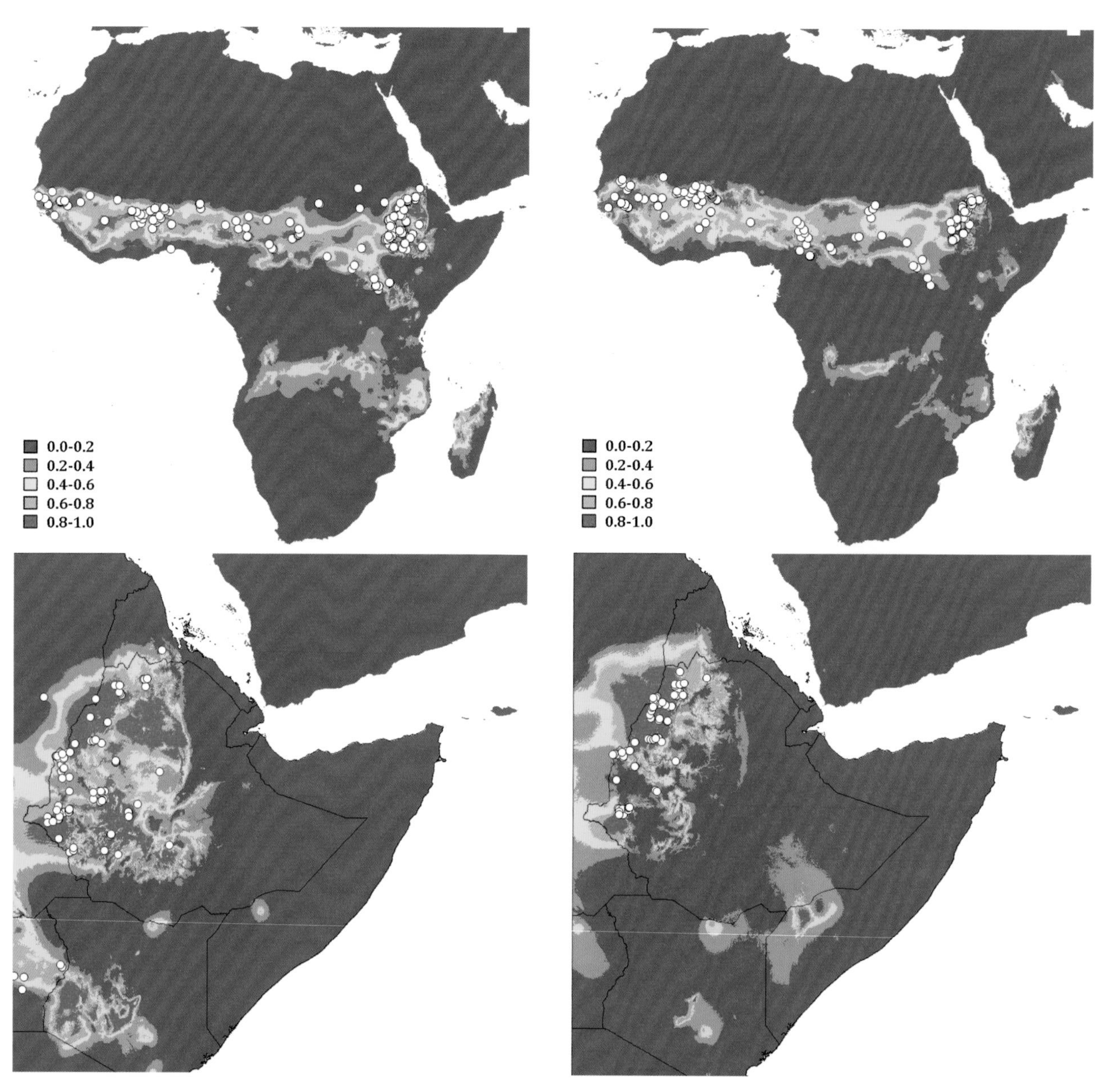

Fig. 9-11. *Lonchocarpus laxiflorus*. Modelled distribution in Africa (above) and on the Horn of Africa (below). The remaining part of the legend as for Fig. 9-1.

Fig. 9-12. *Pterocarpus lucens*. Modelled distribution in Africa (above) and on the Horn of Africa (below). The remaining part of the legend as for Fig. 9-1.

potential area in the GD, GJ and northernmost WG floristic regions, and a large, partly disjunct potential area in the Gambela lowlands, from where the species has not been recorded. It is not very likely to occur in the whole of the predicted area in the Gambela lowlands, as in the north it is associated with thin soil on rocks and rocky hills.

9.13. Conclusions about the models

For the western woodland species of Combretaceae and Legumes studied in this Section, most models predict a wider or narrower zone right across Africa from Ethiopia to the Atlantic Ocean. Although *Combretum hartmannianum* and *C. rochetianum* have both been considered near-endemics to a restricted area in Eritrea, Ethiopia and the Sudan, this seems to be true only for *C. hartmannianum*. One species, *Terminalia brownii*, has a deviating distribution pattern that does not include the westernmost Ethiopian lowlands and does not reach across Africa. The relation between the north-south width of the modelled zone across Africa and the real and predicted distribution in Ethiopia is complex. It is not necessarily always the case that a wide zone across Africa is reflected in a wide north-south distribution in Ethiopia, and similarly a relatively narrow northern or southern zone across Africa will not always be reflected in a similarly narrow distribution in Ethiopia. The wide zone of *Anogeissus leiocarpa* in West Africa ends in the western woodlands of Ethiopia in a relatively northern local distribution, coinciding with the Ethiopian distribution of *Terminalia laxiflora*, but *T. laxiflora* seems to occupy a narrower zone in the Sudanian region in West Africa than *Anogeissus leiocarpa*. *Pterocarpus lucens* occupies a very wide zone across Africa, but does not reach the southernmost parts of the CTW-area in western Ethiopia. Such observations, if confirmed by more examples of observed and modelled distributions, would confirm our idea that species may form part of different local phytochoria in different parts of the Sudanian zone.

The relation between the modelled and the recorded distribution inside Ethiopia is usually good. We assume only that the model has over-predicted the potential areas of distribution for *Combretum molle*, *Terminalia schimperiana* and *Lonchocarpus laxiflorus*. In all other species the recorded distributions along roads and other accessible areas are filled in with plausibly predicted areas.

The last mentioned reason for this Section, that it should document the western woodlands as part of the Sudanian region, is demonstrated by the fact that nine out of the twelve species dealt with in this Section, all occurring in more than 50% of our relevés in western Ethiopia, reach throughout the Sudanian region from Senegal to the western woodlands of Ethiopia. See also Section. 11.2.

10

Cluster analyses and ordinations; associations, indicator species and gradients

10.1. Materials and methods for clustering and ordinations

10.1.1 Data

During the field work described in Section 4 169 taxa were recorded in 151 relevés (numbered 1-148, of which three were each subdivided into two, and seven were supplemented with record by other authors from the literature). See the relevés and the recorded species in Appendix 1. For presentation of the data we have used the abbreviations for species names listed in Appendix 2 and on Fig. 10-1. Taxa (including species and subspecies) were recorded as present (1) or absent (0). For the analysis presented in this Section, taxa occurring in one sample site only were excluded. We will refer to the resulting dataset as 'all species.' A second data set was created in which we also excluded the marginal intruders (MI) and transgressors species (TRG). See Sections 6 and 7 for an explanation and justification of the elements MI and TRG. We will refer to this dataset as the 'CTW-species.' All analysis in this chapter were carried out using all species and using the 'CTW-species' only.

For each relevé location a number of topographic, edaphic and biomass characteristics were recorded to describe the environmental conditions in that relevé. In addition, a number of climatic and topographic variables were extracted from publicly available data sets. See Section 4 for the data gathering in the field. The variables and data sources are described and listed in Appendix 3.

The altitude and slope category of each relevé were recorded in the field. In addition, values were extracted from the 90-meter resolution CGIAR-CSI DEM (version 4, 2008). The correlation between the field-based observations of altitude and the DEM based altitude is high ($r = 0.981$; Fig. 10-2), providing confidence in the accuracy of this measure. The correlation between the slope estimates from the field and DEM is much weaker (Spearman rank order correlation = 0.6). This indicates that the DEM capture the more general patterns, but not all local variation (Fig. 10-2).

Two other relevé characteristics estimated in the field were the degree of burning (using 4 ordinal categories) and the biomass available for burning (using 3 ordinal categories). Comparing these with the fire frequencies for the period 2003-2013 shows that fire frequencies tend to be higher in locations with a higher degree of burning and with more available biomass (Fig. 10-2). The correlation is rather weak though, with a Spearman rank order correlation of 0.69 and 0.58 between the longer-term fire frequencies and the Degree of burnt area and Available biomass respectively. It should be noted that the field observations were done at a specific moment in time, sometimes during the height of the rainy season, sometimes during the height of the dry season, but mostly during the early dry season. The average frequency of fires, on the other hand, provides an estimate of the long-term pressure exerted by fire on the vegetation composition. This variable will therefore be used in further analysis.

10.1.2 Vegetation clusters

Earlier studies, as described in Section 3, and observations during the field surveys, as described in Section 5, suggest that different communities of woody species can be distinguished within the area referred to as the CTW-area (Fig. 4-1), also verified by comparison of

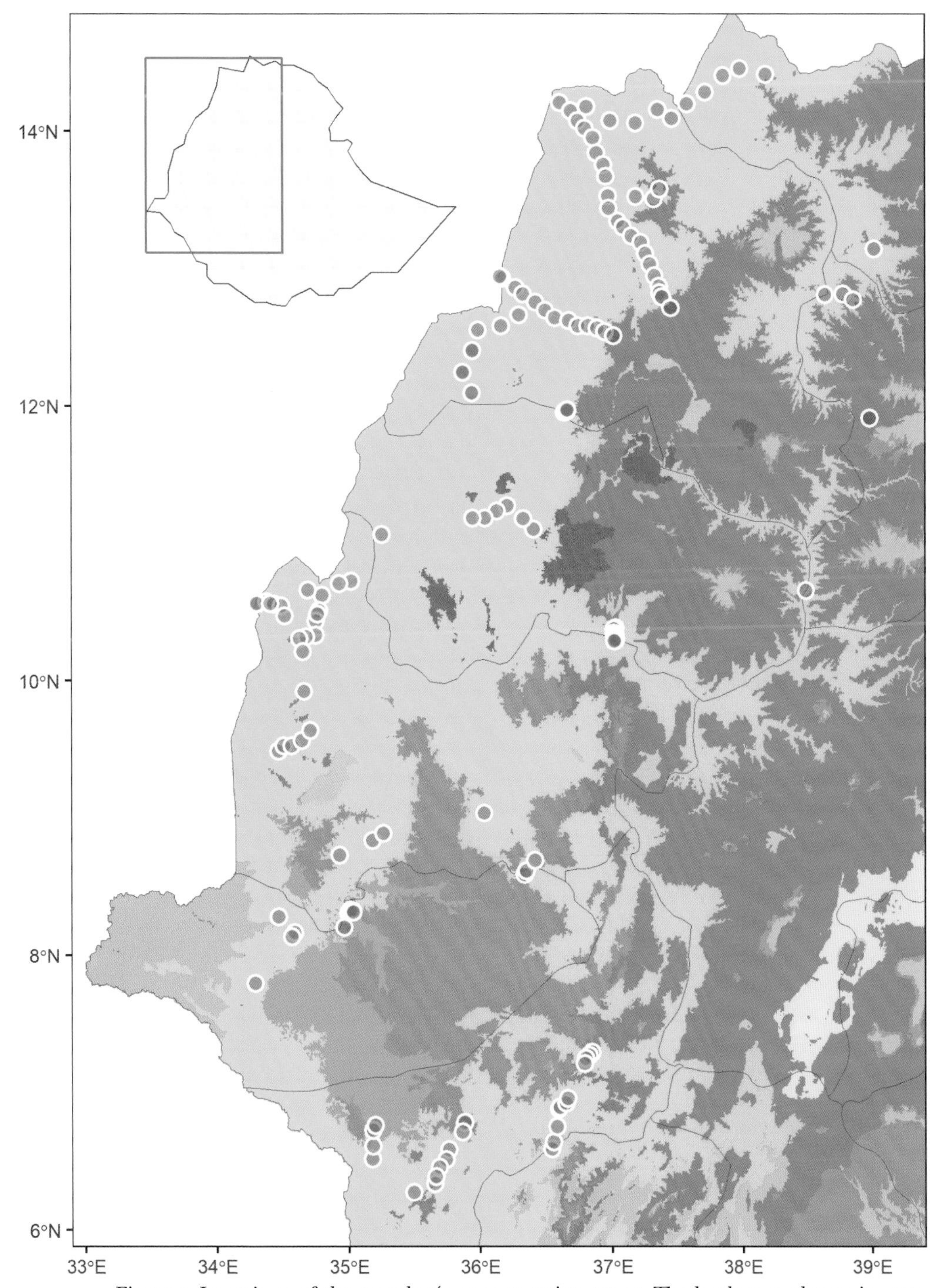

Fig. 10-1. Locations of the 151 relevés on vegetation types. The background map is part of the map in Fig. 4-1, showing the potential natural vegetation types; the floristic boundaries of the *Flora of Ethiopia and Eritrea* are inserted to help with geographical orientation. Inset maps shows Ethiopia with the extent of the studied region.

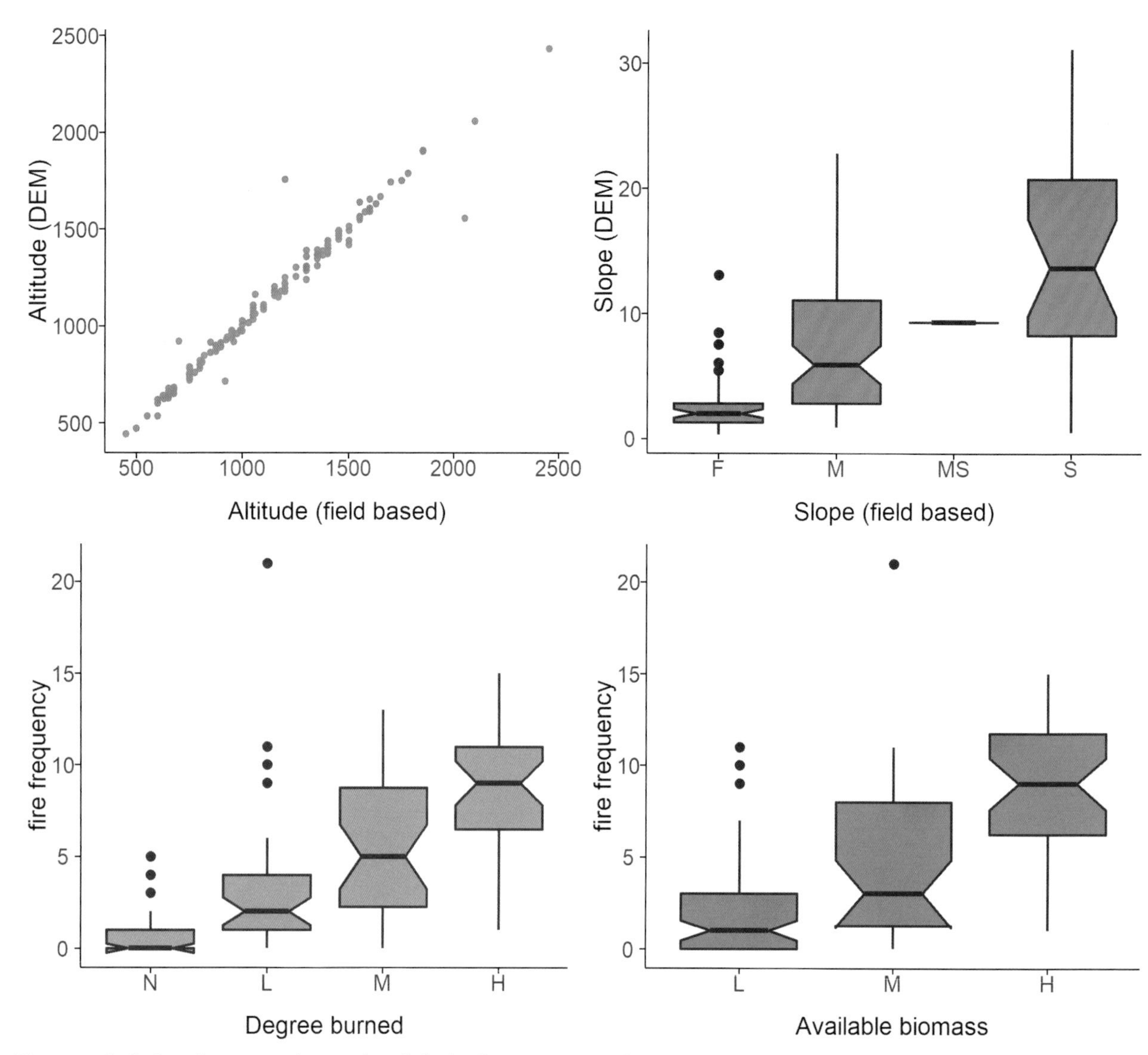

Fig. 10-2. Relations between observed and derived environmental parameters. In the upper left figure the altitudes measured in the field and those based on the 90 m DEM are shown, indicating good correlation. In the upper right figure, the slope categories recorded in the field are compared with the slope derived from the 90m SRTM DEM (CGIAR-CSI 2008). It shows that there is a positive correlation, but with the DEM based slope estimate varying considerable within each field-based slope category, indicating a less positive correlation and that not all variation in slope is expressed in the DEM. The two lower figures show the distribution of fire frequencies within sites classified in the field according to the estimated degree of vegetation burned (left) and according to the estimated biomass available (right), again also with positive correlation. Slope categories: F=flat, M=moderate, MS=moderately steep, S=steep. Codes lower graphs: N=none, L=low, M=moderate, H=high.

all distribution maps in Section 6). We used cluster analysis of the relevés to examine whether there is indeed a separation in distinct communities within the western woodlands of Ethiopia. Cluster analyses were carried out using the Simpson dissimilarity index (β_{sim}; Simpson 1943; Lennon et al. 2001). This measure has the distinct advantage that it efficiently discriminates turnover from nestedness (Baselga 2010). It furthermore takes into account the compositional differences, while ignoring differences in species richness (Koleff et al. 2003). Consequently, boundaries between biogeographical units reflect high species turnover rather than species richness gradients (Castro-Insua et al. 2018).

Different cluster methods were tested. Based on the results, we continued with the generalized *Unweighted Pair Group ArithMetic* Averages (aka flexible *UPGMA*) linkage method with β = -0.1 (Belbin et al. 1992). See Appendix 6.

To determine the optimum number of clusters in which to divide the relevés, we computed the maximum silhouette width (Rousseeuw 1987), the Cindex (Hubert & Levin 1976), the McClain and Rao index (McClain & Rao 1975), the Dunn index (Dunn 1974) and the Frey index (Frey & Groenewoud 1972). Based on the first two methods the 'optimum' number of clusters ranged from 51 to 83, depending on the cluster method and whether all species, or only the CTW-species were considered. These results come down to an average cluster size of 2 to 3 relevés, suggesting a lack of clearly defined vegetation communities at any but the most local scale. In contrast, based on the other three methods, the relevés could best be divided in 2 (McClain & Dunn methods) to 3 (Frey method) clusters (see Appendix 6). For this study, we were mostly interested in the larger scale patterns in vegetation communities across the western woodlands. We therefore looked further into this large-scale pattern based on two or three clusters.

To assess the fit of the individual relevés to the clusters as well as the quality of clusters and the entire classification, we computed the silhouette widths of the relevés (Lengyel & Botta-Dukát 2019) using the silhouette function in the cluster package for R (Rousseeuw 1987). High values for silhouette widths are close to +1, low values for silhouette widths are values below zero; a high value for a relevé indicate that it is well matched with its own cluster and poorly matched with the neighbouring clusters. If most relevés have high values for their silhouette widths and few low values, then the cluster is appropriate. In addition, we tested for homogeneity in the multivariate dispersion among the three clusters based on the Simpson dissimilarity in species composition (Anderson 2006). A permutation procedure was used to test for significance, using the *permutest* function in *vegan* (Oksanen et al. 2019).

10.1.3. Characteristic species

To determine which species can be considered as indicators for the different clusters, we used the indicator value (IndVal; Dufrêne & Legendre 1997). The IndVal index measures the association between a species and clusters (or sites), serving as a measure of how well a species predicts specific communities. It thus is particularly useful for the field determination of community types or for ecological monitoring purposes (De Cáceres et al. 2010; De Cáceres & Legendre 2009). Computations were done using the *multipatt* function in the R package *indicspecies* (De Cáceres & Jansen 2020). This function returns for each species the cluster with the highest association value (IndVal), the corresponding IndVal and the level of significance of the association, based on 999 random permutation tests.

The IndVal index is the product of two components, called A and B (Dufrêne & Legendre 1997; De Cáceres & Legendre 2009). Both components were computed separately as well. Component A is the groups equalized proportion of a species' occurrences that are within the cluster. This can also be interpreted as the sample estimate of the probability that a site belongs to the target cluster given the fact that the species has been found. It is referred to as specificity or positive predictive value of the species as indicator of the site group. The component B, also called the fidelity or sensitivity of the species indicator for the cluster, is the proportion of sites in the cluster that contain the species. This can be interpreted as the sample estimate of the probability of finding the species in sites belonging to the site group (De Cáceres & Legendre 2009).

10.1.4. Ordination

10.1.4.1 Unconstrained ordination

To further examine the differences in species compositions among the relevés, and to what extent these can be explained by the different environmental variables (see general description of the environment in Section 2 and enumeration of the specific environmental variables in Appendix 3), we carried out unconstrained and constrained ordination analyses. With the unconstrained ordination, the primary focus was on how the clusters identified in the previous part of the analysis were distributed in ordination space; either as distinct clusters, or as a more arbitrary division of relevés along a more or less continuous gradient. We furthermore examined to what extent the clusters were aligned along specific environmental gradients. Constrained ordination was used to further examine possible effects of major environmental variables on the species composition within the western woodlands, invisible in the unconstrained ordination (ter Braak & Šmilauer 2015).

For the unconstrained ordination, we used non-metric Multidimensional Scaling (NMDS). This method is commonly regarded as a robust unconstrained ordination method. Because it uses the rank orders of species in the community, it has fewer premises and it preserves non-linear patterns in the data (McCune et al. 2002). The latter may be of particular importance in large-scale studies like the present one where patterns are likely to be non-linear. Analysis were carried out with the metaMDS function from the R *vegan* package (Oksanen et al. 2019). We used the Simpson dissimilarity index (β_{sim}; Simpson 1943; Lennon et al. 2001) as it is not sensitive to species-richness, and thereby suitable to capture biogeographic patterns of species turnover. We also carried out the analysis using the Jaccard dissimilarity index, but this did not yield better results (in fact, results were fairly similar).

In order to obtain a global solution, rather than local optima, we used 200 iterations using different random starts, and repeated this once more, using the best result of the first run as starting point. We used the stress as a goodness of fit measure, i.e., a measure of how well the distances in ordination space represent the observed dissimilarities in species composition (Kindt & Coe 2005). Repeating this procedure 50 times showed that the procedure yielded stable results.

The NMDS was carried out using both all species and the CTW-species only. In addition, the analyses were carried out using two, three and four dimensions. Results for all species and CTW-species only were very similar. In both cases, the NMDS analysis with two dimensions resulted in a stress of 0.24, which is high enough to be considered suspect. Running the ordination with three and four dimensions reduced the stress in both cases to 0.18, indicating that the original dissimilarities are moderately well reflected in the ordination space.

The three main clusters, defined based on the f.UPGMA method (see Section 10.1.2), were fitted in the ordination space. To assess the goodness of fit, the r^2 was computed (Oksanen et al. 2019).

To assess the relation between species composition and climatic and edaphic conditions, topography and frequency of fire occurrences, we fitted isotropic smooth surfaces of a number of environmental variables (selected from those available in Appendix 3) using penalized thin plate regression splines (Wood 2003). The environmental variables we used were the altitude (*dem*), terrain wetness index (*twi*), slope (*slopedem*), the maximum fire frequency for the period 2003-2013 of all cells in a radius of 2 cells around the focal pixel (*firefreqmax*), the soil types recorded in the field, and the 19 bioclimatic variables listed in Appendix 3.

To assess the goodness of fit, we computed the deviance in ordination space explained by the variable. The significance of the fitted vectors or factors was assessed using permutation of the fitted variables. Computations were done with the *ordisurf* function in the R *vegan* package (Oksanen et al. 2019).

10.1.4.2 Constrained ordination

To focus on the variation in the vegetation composition explained by the different environmental variables, we carried out a distance-based redundancy analysis (Legendre & Anderson 1999) as implemented in the *capscale* function in the vegan package (Oksanen et al. 2019). Like for the unconstrained ordination, we used the Simpson dissimilarity index (β_{sim}). To assess the percentage of the variance in the community structures explained by the environmental variables, while adjusting for the number of terms, we computed the adjusted R^2 (Peres-Neto et al. 2006). A permutation test, using the *permutest* function in *vegan* with 1000 permutations, was carried out to assess the significance of the joint effect of the environmental constraints.

Running the analysis with *bioclim* variables showed many of the variables had a very similar direction of influence. Grouping these variables resulted in 4 groups representing climate variability (*CV*; bio2, bio4, bio7, bio15), rainfall in dry and warm periods (*PDW*; bio3, bio14, bio17, bio18), rainfall in cold or wet periods (*PCW*; bio13, bio16, bio19) and temperature (*T*; bio1, bio5, bio6, bio8, bio9, bio10, bio11). For each group, we carried out a principal component analysis (PCA) and took the first principal component (PC) as variable to represent that group. The first PC represented 78%, 77%, 90% and 94% of the overall variability of the *CV*, *RDW*, *RCW* and *TMP* variables respectively. Likewise, we computed the PCA of fire frequency, and median, maximum and mean fire frequency in a 200-meter radius, and used the first PC (accounting for 90% of the variability) as input in the capscale analysis. The other two variables we used were the altitude (*DEM*), the terrain wetness index (*TWI*), slope, and slope aspect.

10.1.5 Mapping the clusters

We used random forest (Breiman 2001) to model the distribution of the three clusters based on the CTW-species (Section 10.1.2). As dependent variables we used the relevés categorized into the three clusters. Only relevés with a positive silhouette width (Section 10.1.2) were included. As predictor variables we used the bioclim and fire frequency variables presented in Section 10.1.4.2. In addition, we used the 250 m resolution SoilGrid raster layers (Hengl et al. 2017; SoilGrids team 2020) representing the predicted pH, cation exchange capacity, percentage clay and sand, coarse fragments volumetric, and soil organic carbon stock at 0-5 cm depth. For the analysis, all layers were resampled to a resolution of 0.00208 degrees (approximately 250 m at the equator) and standardized. Analysis were carried out using the *r.learn.ml2 addon* (Pawley 2020) for GRASS GIS (GRASS Development Team 2020), which provides an interface to Scikit-learn (Pedregosa et al. 2011; Scikit-learn developers 2020). The number of trees (n_estimator) was set to 200. To balance the training data, the categories were assigned weights inversely proportional to category frequencies. Predictive performance of the model was evaluated using a four-fold cross-validation, using the accuracy (percentage of correctly predicted data points) and kappa statistics.

10.2. Results

10.2.1 General observations

In the 151 relevés a total of 169 species of woody plants were observed. In each relevé between 2 and 33 species and a few subspecies (in the following both are referred to as species) were recorded, with most relevés having 11 species and the whole data set having an average of 13 species per relevé (See Fig. 10-3 for the distribution). The majority of species occur in very few relevés, with 80 species that occur in less than 4 relevés, and 38 species occurring in one relevé only (the so-called singletons).

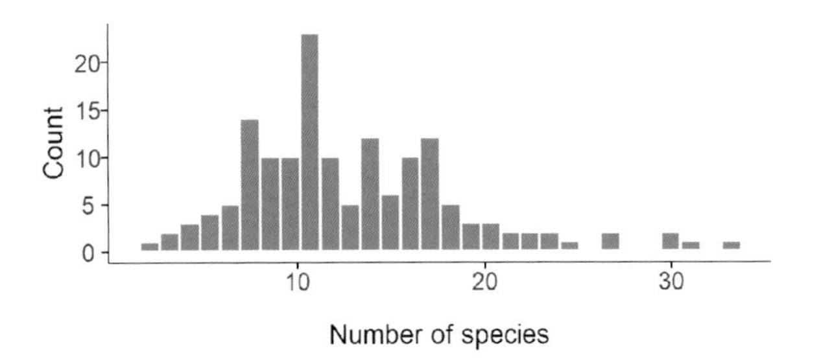

Fig. 10-3. Distribution of number of relevés (count) per number of species.

The relevés no. 80-84 with the highest number of species, 27 to 33, are all from the species-rich woodlands between the Abay River and Assosa (our profiles J.1 and J.3 in Section 5). The vegetation of these localities was also studied by us, but the lists of species quoted in Appendix 1 from the relevés 80-84 were derived from areas with more detailed ecological studies (Mindaye Teshome et al. 2017) than we normally were able to do, and the unspecified size of the relevés no. 80-84 must have been significantly larger than ours. The many relevés with low numbers of species and the many species that occur in one or few relevés only suggest a patchy, spatially heterogeneous distribution of species and species communities. Possible reasons have been suggested in the general Section 4.1 on Materials and Methods, including the fact that intensive cultivation may have limited the possibility for sufficiently large relevés with the full potential of species in some sites.

10.2.2. Vegetation clusters

In Fig. 10-4 we show how the different relevés are clustered based on the similarities in species composition based on all species and on the CTW-species. Relevés that are part of different clusters in the two dendrograms are linked with orange lines. The same relevés are marked as orange dots on the map. The results show that the two main clusters occur in geographically fairly distinct regions; one in the northern and north-western lowlands and one that includes most relevés in the south, but also the relevés higher up the slopes of the escarpments in the north (Fig. 10-4). Exceptions are the relevés at higher altitudes in the north, near the Afromontane Highlands, and in the relevés in eastern Gambella, between the *Wooded grasslands of the western Gambela region* (WGG) and the *Transitional rain forest* (TRF). When considering CTW-species only, these relevés are assigned to the other cluster, resulting in a more pronounced north-south division. This indicates that in these relevés the vegetation is to a larger extent determined by marginal intruders (MI) and transgressors species (TRG). See Section 7 for an introduction and justification of these floristic elements.

It is important to notice that overall, the clustering is not very strong. The average silhouette width of the relevés is 0.17. This includes 20 relevés with a negative silhouette width, indicating a weak to poor match to the cluster they are assigned to. This can be in part contributed to the marginal intruders and transgressors species. Removing these species results in a slightly higher average silhouette width (0.21) and less relevés with a negative silhouette width (16 relevés).

Dividing the relevés in three clusters results in two clusters similar to the ones above, and a third smaller cluster that includes many of the above-mentioned relevés in the north along the altitudinal margins of the CTW-area (Fig. 10-5). On average, clusters are slightly better defined (average cluster width = 0.18). However, the third cluster is very poorly defined, with an average silhouette width close to zero. This underlines the earlier observation that these relevés are less defined by their similarities, and more by the presence of marginal intruders (MI) and Transgressors species (TRG) from neighbouring vegetation types outside the *Combretum-Terminalia* woodland (CTW). The marginal intruders are most typical from the Afromontane Highlands, but in the case of the northern, third cluster, the intruding species are mainly distributed in relation to the *Acacia-Commiphora* bushland and have been designated to an element named ACB[-CTW], referring to the extensive distribution of species in this local element in the *Acacia-Commiphora* bushland. This is demonstrated by excluding the marginal intruders and transgressors species from the analysis.

This results in a third cluster with relevés limited to a few relevés in the north-eastern Tacazze Valley and the upper Abay Valley (Fig. 10-5).

In Fig. 10-6 we show the distribution of the three clusters based on CTW-species, based on a random forest model. The predictive accuracy and kappa statistics of the model are respectively 95% ± 0.035 and 0.89 ± 0.074. The map illustrates the broader geographic distribution of the two major clusters, a north-western and western (green), associated with the north-western lowland and the western escarpment and lowlands, and a more southern on the low southwestern plateaux and associated river valleys (blue). The third cluster (orange) includes the dry woodlands of the upper Tacazze and Mareb Valleys. Note that the woodlands in the Gibe and Omo River Valleys were not surveyed, which makes the designation of these areas to the northern cluster highly uncertain. From our field experience, we would assume that most of these river systems should be assigned to the southern (blue) cluster or, as suggested by Fig. 4-2 and the description 'R' of the Section 5 ('South of Maji and towards the lower Omo Valley'), the lowermost Omo Valley should be excluded from the CTW-area and referred to the western part of the *Acacia-Commiphora* bushland. The single relevé no. 143 in Profile D near Lalibela in the uppermost part of the Tacazze Valley has been classified as part of the northern (green) cluster, but the random forest model in Fig. 10-6 suggests that conditions are more similar to that in the northern third cluster (orange). This is in agreement with field experience which suggests that the vegetation around the river itself should indeed be classified with the northern, third cluster (orange).

Overall, the three clusters are slightly less well defined then when considering two clusters only, with an average silhouette width of 0.2. The larger northern (green) cluster is significantly more heterogeneous than the other two clusters ($F=6.6916$, $p<0.004$). The cluster is also less well defined, with an average silhouette width of 0.14, compared to an average silhouette width of 0.36 and 0.35 for the second (blue) and third (orange) clusters.

When considering more than three clusters, outcomes show patterns that are much less cohesive, differ more depending on the method, and are difficult to interpret. This reinforces the lack of clear vegetation patterns at any scale but the region-wide north-south and/or altitudinal divide.

10.2.3. Indicator species

Table 10-1 provides for each of the three clusters (see Section 10.2.2) the 4 species with the highest relative associative strength (*IndVal* index value). Included in the table is degree of significance of the association. The full list of species-cluster combinations, with their A, B, IndVal and p-value scores, are provided in Appendix 7. It shows that for respectively 11, 14, and 21 cluster-species combinations the association is significant. So, the majority of species is not strongly linked to one of the clusters.

When analysing both data sets with all species (Table 10-1) and CTW-species only (Table 10-2), the indicator species of the northern (first) cluster are the characteristic *Combretum-Terminalia* woodland species. *Anogeissus leiocarpa* is in Section 6 described as a characteristic and often dominant species at 500-1750 m a.s.l. in *Combretum-Terminalia* and *Anogeissus-Pterocarpus* woodland. *Pterocarpus lucens* is in Section 6 described as being typical at 600-1400 m a.s.l. in woodland on rocky hills. *Acacia hecatophylla* is in Section 6 described as typically western species at 600-1700 m a.s.l., restricted to areas north-west of the Abay River (apart from a doubtful record from the Gojeb River in the south), it is mostly associated with vertisol. None of these species reach to the east of the Rift Valley, while the fourth species, *Sterculia africana,* is more widespread, occurs at 550-1550 m a.s.l., very often in woodland on rocky ground. Other species in the northern (first) cluster and based on the data set of all species (Appendix 7) are *Ziziphus spina-christi* (600-1850 m a.s.l.; mainly in dry woodland, both on rocky ground and alluvial soil, extending far east of the Rift Valley), *Dalbergia melanoxylon* (650-1200 m a.s.l.; in woodland, often on rocky ground or alluvial ground, restricted to north-

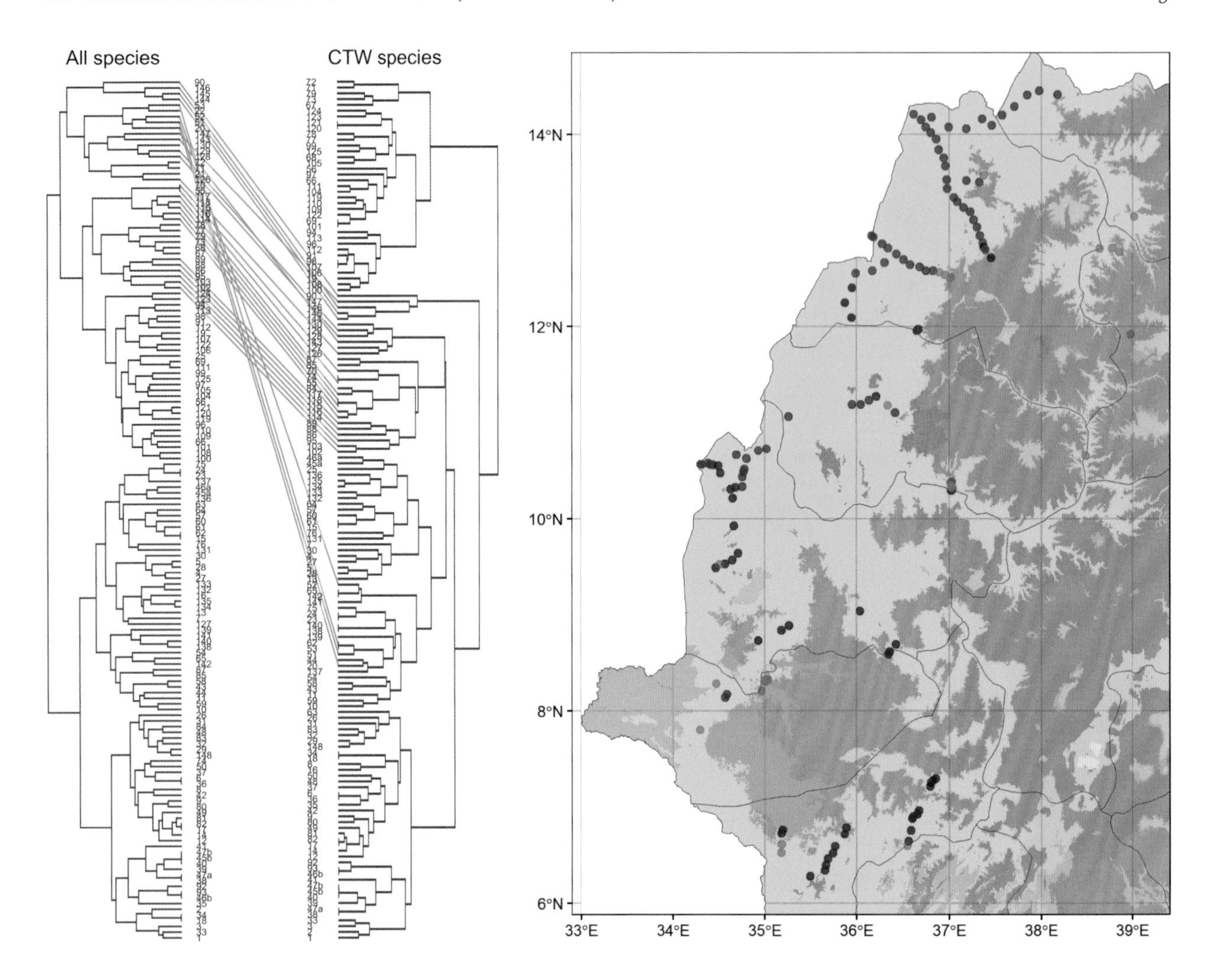

Fig. 10-4. Clustering with two clusters. Left: dendrograms showing the clustering of the relevés using the Simpson dissimilarity and the f.UPGMA based on all species and the CTW-species only. The relevés that are assigned to different clusters depending on the method are linked with orange lines. Right: considering all species, one cluster is formed by the blue and orange dots, the other by the green dots. Based on the CTW-species only, one cluster is formed by the blue dots, the other by the green and orange dots.

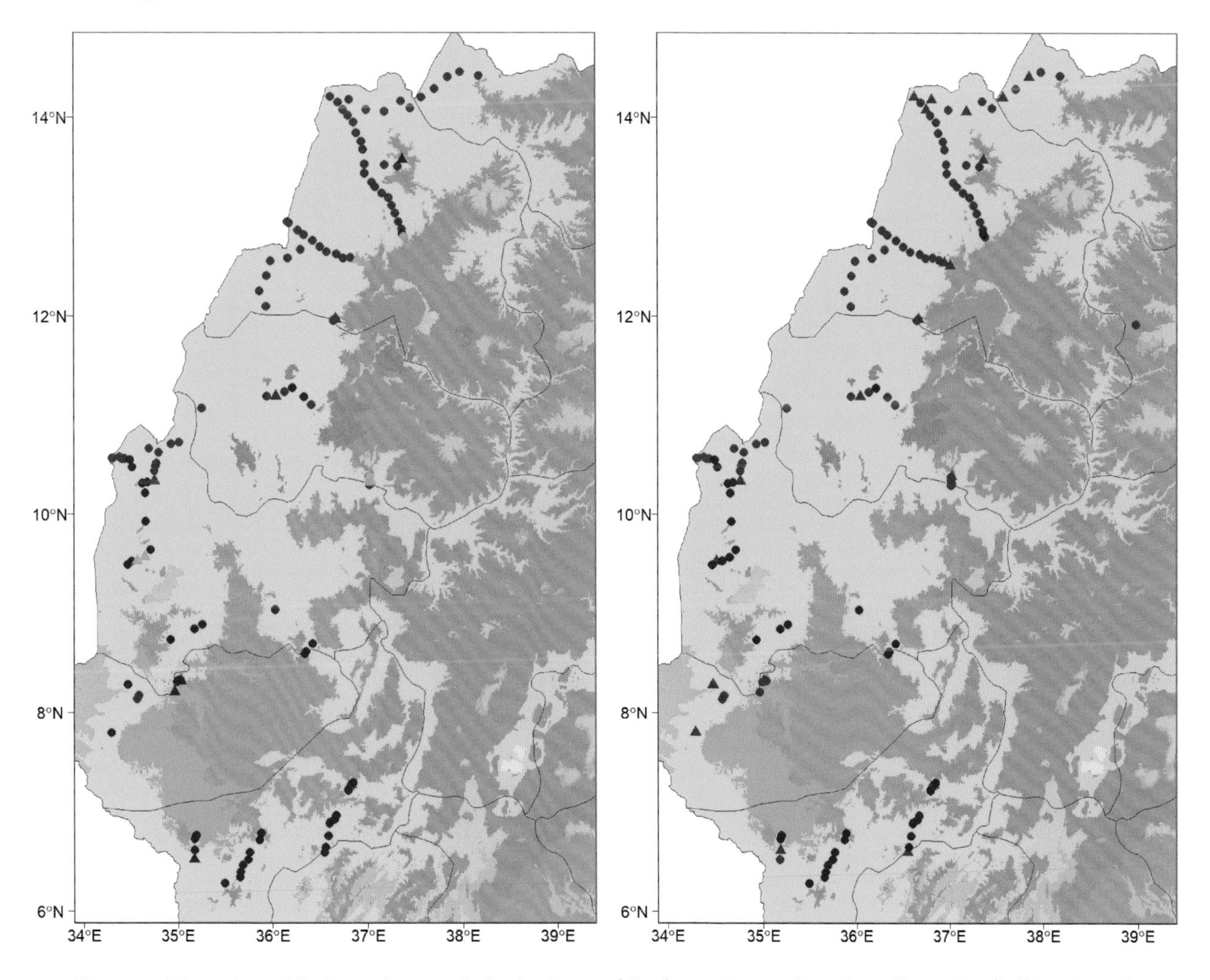

Fig. 10-5. Clustering with three clusters. Relevés clustered in three clusters, based on all species (left) and on the CTW-species only (right), both maps with green, blue and orange dots. Relevés with a negative silhouette width are indicated by a triangle. Clustering was done using the f.UPGMA method.

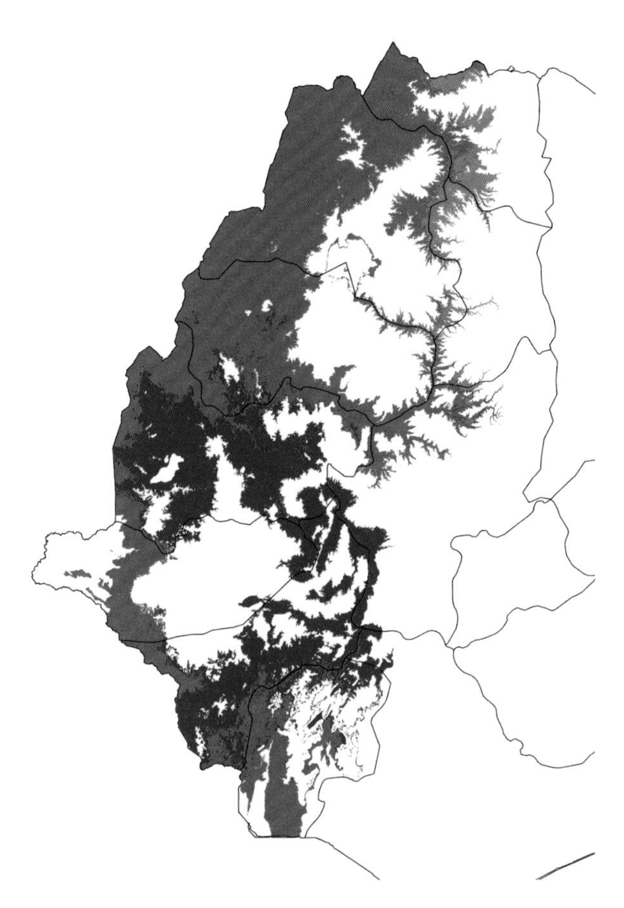

Fig. 10-6. Map of the western woodlands of Ethiopia which, as on the map in Fig. 4-1, are classified as CTW-area, subdivided according to the three clusters detected in this work. The clustering is based on a random forest model with as input categories the three clusters based on CTW-species only, and as predictors bioclimatic and soil variables. The colouring is as for the dots in Fig. 10-5 (green: northern cluster; blue: southern cluster; orange: third cluster). The species with the highest indicator values for the green cluster are: *Anogeissus leiocarpa*, *Pterocarpus lucens*, *Acacia hecatophylla* and *Sterculia africana*; for the blue cluster these are *Annona senegalensis* and *Terminalia schimperiana*, *Vitex doniana*, *Entada abyssinica* and *Combretum collinum*; and for the orange cluster these are *Capparis decidua*, *Acacia mellifera*, *Acacia asak* and *Adansonia digitata*.

western Ethiopia), *Boswellia papyrifera* (750-1600 m a.s.l.; dry *Combretum-Terminalia* woodland, *Pterocarpus* woodland or forming almost pure stands, often on rocky slopes, restricted to north-western Ethiopia), *Balanites aegyptiaca* (450-2000 m a.s.l.; dry woodland on vertisol, often with *Acacia seyal* and *A. hecatophylla*, less prominent in *Combretum-Terminalia* woodland, but also in other dry habitats, extending far east of the Rift Valley), *Combretum hartmannianum* (650-1200 m a.s.l.; in *Combretum-Terminalia* woodland on alluvial soil and vertisol, restricted to north-western Ethiopia), *Ficus glumosa* (550-1850 m a.s.l.; on rocks and rocky outcrops in woodland and bushland, widespread in the western woodlands of Ethiopia, extending far east of the Rift Valley), and *Boswellia pirottae* (750-1750 m a.s.l.; woodlands, mainly with *Combretum* and *Terminalia* on steep, rocky soil, restricted to north-western Ethiopia).

The indicator species of the southern (second) cluster in Table 10-1 (based on analyses with all species) are: *Combretum collinum* (600-1800 m a.s.l.; woodland with *Combretum*, *Terminalia*, *Stereospermum* and *Anogeissus*, extending far east of the Rift Valley), *Terminalia schimperiana* (700-2000 m a.s.l.; *Combretum-Terminalia* woodland in a wide range of habitats, extending to the east of the Rift Valley), *Bridelia scleroneura* (550-1700 m a.s.l.; *Combretum-Terminalia* woodland with well-developed shrub stratum, common in the southern part of the CTW-area, with a few records in the north, and extending to the east of the Rift Valley) and *Annona senegalensis* (500-1600 m a.s.l.; humid woodlands, mostly in the south-western part of the CTW-area, with one record north of the Abay River one record to the east of the Rift Valley). Based on analyses with the CTW-species the two first indicator species in Table 10-2 are the same as for the analyses based on all species: *Combretum collinum* and *Terminalia schimperiana,* the remaining two are *Vitex doniana* (500-1950 m a.s.l.; *Combretum-Terminalia* woodland, widespread in the western Ethiopian lowlands) and *Entada abyssinica* (950-2000 m a.s.l.; woodland and evergreen bushland, widespread in the western Ethiopian lowlands and at medium altitudes, extending to the east of the Rift Valley).

It is noticeable that, in the clustering with all species, most of the indicator species of the third cluster are marginal intruders (MI), transgressors (TRG) and ACB-species: *Croton macrostachyus, Maytenus arbutifolia*, *Senna petersiana*, and *Capparis decidua*), while in the clustering with CTW-species, the indicator species of the third cluster is to a large extent characterized by and ACB-species (*Capparis decidua, Acacia mellifera, Acacia asak,* and *Adansonia digitata*).

Excluding the marginal intruders and Transgressors from the analysis results in most relevés being assigned to the two main clusters (see presentation of clustering results in Section 10.2.2). Notably, in these two clusters, the association of respectively 4 out of 54 and 9 out of 31 cluster-species combinations were found to be significant. This indicates that there are only few species that can be considered as indicator species for these two main clusters.

The third cluster of the CTW-species consists of a few isolated relevés in the north-eastern Tacazze Valley and the upper Abay Valley (Fig. 10-5). The vegetation in these relevés consists mostly (72%) of species characteristic for the *Acacia-Commiphora* Bushland. In contrast to the two main clusters, a large proportion of the cluster-species combinations, 16 out of the 29, was found to be significant. These include three out of the four species with the highest IndVal values (Table 10-2), while the fourth is *Adansonia digitata*, which also reaches the upper Tacazze Valley. These are also species typical of dry, depauperated woodland, and that would seem the best characterisation of the third cluster.

Component B of the IndVal index, and by extension the IndVal index, are relative sensitive for common species. Yet rare species, if found, may be a good indicator of vegetation types. To identify such asymmetrical indicator species (Dufrêne & Legendre 1997), it is of interest to consider not only the IndVal index, but its components as well. For example, *Acacia hecatophylla* has a high predictive value (A=1) but low sensitivity (B=0.346) for cluster 1 (Table 10.2). In other words, one is not likely to encounter the species, but if encountered (within the western lowlands), it is indicative for the type of tree community in that location.

For most species, B values are low. This confirms that even the more characteristic species for a cluster were recorded in only a small part of the sites belonging to that cluster. In addition, note that the classification of the relevés into clusters is based on the species composition itself. In other words, they are not independent. This means one can expect more significantly associated species than expected by chance only. Furthermore, multiple testing has not been taken into account when computing the p-values. However, we did not use the p-values for formal testing, but rather as a measure of how significant indicator values of species are compared to others. We used this to filter out species with a high but not significant indicator value.

Table 10-1 (see next page). First four indicator species of cluster 1, 2 and 3, all species. Species with the highest significant IndVal value, as defined by the f.UPGMA linkage method. For the SpecCode, see Appendix 2. Phytogeographical categories, see Section 7. For A and B, see text.

Table 10-2 (see next page). First four indicator species of cluster 1, 2 and 3, CTW species only. Species with the highest significant IndVal value of three clusters, as defined by the f.UPGMA linkage method and based on the CTW species only. For the SpecCode, see Appendix 2. Phytogeographical categories, see Section 7. For A and B, see text.

Table 10-1

Species	SpecCode	Phytogeographical category	Cluster	IndVal	A	B
Anogeissus leiocarpa	anolei	CTW[n+c]	1	0.796	0.749	0.845
Pterocarpus lucens	pteluc	CTW[n+c]	1	0.632	0.745	0.536
Acacia hecatophylla	acahec	CTW[n+c]	1	0.616	0.885	0.429
Sterculia africana	steafr	CTW[all+ext]	1	0.600	0.795	0.452
Combretum collinum	comcol	CTW[all+ext]	2	0.644	0.581	0.714
Bridelia scleroneura	briscl	CTW[all+ext]	2	0.637	0.947	0.429
Terminalia schimperiana	tersch	CTW[all+ext]	2	0.631	0.629	0.633
Annona senegalensis	annsen	CTW[c+s+ext]	2	0.620	0.784	0.490
Croton macrostachyus	cromac	TRG	3	0.618	0.859	0.444
Maytenus arbutifolia	maytrb	MI	3	0.527	1.000	0.278
Senna petersiana	senpet	CTW[all+ext]	3	0.496	0.886	0.278
Capparis decidua	capdec	ACB[-CTW]	3	0.459	0.949	0.222

Table 10-2

Species	SpecCode	Phytogeographical category	Cluster	IndVal	A	B
Anogeissus leiocarpa	anolei	CTW[n+c]	1	0.833	0.964	0.720
Pterocarpus lucens	pteluc	CTW[n+c]	1	0.661	0.900	0.486
Sterculia africana	steafr	CTW[all+ext]	1	0.626	1.000	0.393
Acacia hecatophylla	acahec	CTW[n+c]	1	0.588	1.000	0.346
Terminalia schimperiana	tersch	CTW[all+ext]	2	0.773	0.763	0.784
Vitex doniana	vitdon	CTW[all]	2	0.735	0.953	0.568
Entada abyssinica	entaby	CTW[all+ext]	2	0.692	0.932	0.514
Combretum collinum	comcol	CTW[all+ext]	2	0.683	0.640	0.730
Capparis decidua	capdec	ACB[-CTW]	3	0.889	0.988	0.800
Acacia mellifera	acamel	ACB[-CTW]	3	0.860	0.924	0.800
Acacia asak	acaasa	ACB[-CTW]	3	0.775	1.000	0.600
Adansonia digitata	adadig	CTW[n]	3	0.746	0.928	0.600

Legends for Table 10-1 and Table 10-2 on previous page.

10.2.4. Vegetation composition and gradients

10.2.4.1 Unconstrained ordination

To further examine patterns in species compositions, and to what extent these can be explained by the different environmental variables, we used non-metric Multidimensional Scaling (NMDS). We used the ecological distances among the sites based on all species and on the CTW-species. Preliminary results of the NMDS based on all species showed relevé no. 22 to be a strong outlier, with a species composition very dissimilar to that in other relevés. The relevé is located at 2450 m a.s.l. at the watershed north-west of Gondar (see also the description of this environment by Pichi Sermolli in Section 3.1). Apart from some singletons, the vegetation in the relevé was composed of three marginal intruder species, viz., *Maytenus arbutifolia*, *Maesa lanceolata* and *Bersama abyssinica*. These species occur in other relevés as well, but always as part of a larger assemblage of species. The relevé was removed from further analysis.

As detailed in Section 10.1.4, running the NMDS ordination with three dimensions yielded a stress of 0.18, both when using all species and CTW-species. This indicates that the original dissimilarities are moderately well reflected in the ordination space. The distribution of the relevés, species and clusters based on all species and CTW-species in the ordination space formed by the first two axes are shown in Fig. 10-7. The corresponding goodness of fit statistics are given in Table 10-3, showing a better fit of clusters based on CTW-species only.

The two main clusters represented in the NMDS ordination in Fig. 10-7 (represented by green and blue dots) separate fairly well, whether using all species or CTW-species only. However, it should be noted that in the field the relevés of these two clusters occur along a more or less continuous gradient of species replacements. In the third cluster (orange), most relevés are differentiated from the other relevés based on the relative high abundances of marginal intruders and transgressors species, rather than on their similarity. An exception are the five relevés in the northeast (see Fig. 10-5) that are clearly separated from the relevés in the other clusters, even when clusters are defined based on CTW-species only. The same can be observed when plotting the clusters on the first and third NMDS axes (not shown here). These results confirm that one can distinguish two major clusters within the western lowlands. A third, more distinct, cluster can be distinguished in the north-eastern Tacazze Valley and the upper Abay Valley. However, note that this is based on a few relevés only. In the north, vegetation at the higher altitudinal limits of the CTW-area is characterized by marginal intruders and transgressor species. Curiously, we did not find the same in the south. The relevés in north-western Welega (WG floristic region), around and northwest of Assosa (Fig. 10-5), are more difficult to interpret, although it should be noted that this region includes some smaller mountains with potential vegetation classified as Afromontane (*Dry Afromontane Forest*, DAF).

The relevés of the northern cluster (green dots in Fig. 10-8) can be found along a wide range of environmental conditions, such as latitude, altitude, fire frequencies and Terrain Wetness Index (TWI), while the relevés of the southern cluster (blue dots in Fig. 10-8) have more limited distribution along these gradients.

One can distinguish a clear north-south gradient in the species compositions, with latitude explaining 68% of the variance in the ordination diagram. Fig. 10-8 also shows a division of the main clusters along a north-south gradient. The division between the northern (green) cluster and the southern (blue) cluster lies between 9° and 10° north, but perhaps closer to 11° north near the border with the Sudan.

Regarding altitude, the dividing line between the relevés of the northern (green) cluster and those of the southern (blue) cluster is at 1300 m a.s.l. or above; for one relevé even at 1600 m a.s.l. The unitless Terrain Wetness Index (TWI) seems also to separate the two main clusters. It is a factor that has elsewhere been found to control species composition and richness (Kopecký & Čížková (2010). However, in our study, compared to the other factors, the *TWI* only explains a small part of the variance in the ordination space, with a border between the northern cluster (green dots) and the southern (blue dots) at values between 7.5 and 8.

The variation in the Maximum fire frequency (Fig. 10-8, lower right figure) explains considerably more (40%) of the variance in the ordination space. There is considerable overlap between the two main clusters along this gradient, but with the clear difference that within the northern (green) cluster a much wider range of Fire Frequencies (FF) is found (1-12) than in the southern (blue) cluster (6-12).

The type of soil explains about 24% of the variance in ordination space. However, no clear relation with the clusters could be discerned (not shown). This is in line with the earlier observation that soil conditions can be very heterogeneous, varying over relative short distances and not closely related to the general soil patterns described in the introductory Section 2.5. In the field, we noticed that the soil types were highly dependent on the detailed topography of the often very rugged terrain, and that vertisols could occur in flat areas close to leptosols on rocky slopes. As such, soil conditions seem to contribute to the within-cluster heterogeneity, rather than explaining the differences between the clusters.

Climatic conditions are known to be important determinants of vegetation structure (Fayolle et al. 2014; Ganamé et al. 2019; Morin et al. 2018). The Bioclim variables listed in Appendix 3 provide a useful statistical summary with each variable describing different aspects of climate (Kriticos et al. 2014). We computed for each of the 19 variables the deviance in ordination space it explained, and selected the six variables that explained more than 60% of the deviance. Some of the resulting variables were highly mutually correlated, basically representing the same gradient. We therefore selected a subset of variables, such that the maximum variance inflation factor (VIF) was less than 10. The selected variables were plotted as isotropic smooth surfaces in the ordination diagram (Fig. 10-9).

Results show that vegetation patterns clearly follow a rainfall gradient, with the annual rainfall (bio12), precipitation of the driest month (bio14) and the precipitation seasonality (bio15) explaining respectively 65%, 66% and 69% of the deviance of the ordination space (Fig. 10-9), while the mean temperature of warmest quarter (bio10) explains the least, 62%. The vegetation in the northern cluster (green dots) generally occurs under drier conditions with a more variable rainfall. Average annual temperatures, but also temperature seasonality, are higher, adding to drier conditions under which water deficits are more likely. It is more difficult to see what environmental factors separate the cluster in the north-eastern Tacazze Valley and the upper Abay Valley from the other two clusters. The relevés in this cluster occur at the drier end of the annual rainfall gradient, with most noticeable the limited rainfall in the coldest quarter.

Table 10-3. Goodness of fit and permutation-based significance of the fit of clusters in ordination space

Cluster method	NMDS axes	r-sq	p-value
All species	1-2	0.395	0.001
	1-3	0.433	0.001
	2-3	0.055	0.002
CTW-species	1-2	0.457	0.001
	1-3	0.439	0.001
	2-3	0.064	0.002

10.2.4.2 Constrained ordination

A distance-based redundancy analysis (Legendre & Anderson 1999) was carried out to further examine the degree in which the different environmental factors relate to the major patterns in the species responses. This was done separately for the climate variables, the terrain wetness index (TWI), altitude (DEM) and fire frequencies (FF), and for the soil types. Initially, we included slope and slope aspect as well, but these were found to have no influence on species composition. Results are presented in Fig. 10-10. The ordination plots show for each combination of variables the relevé (left) and species (right) scores and variables used as constrained on the first two axes. Overall, climate variables represented 32% of the variance in the species compositions ($F=2.11$, $p<0.01$). There is a

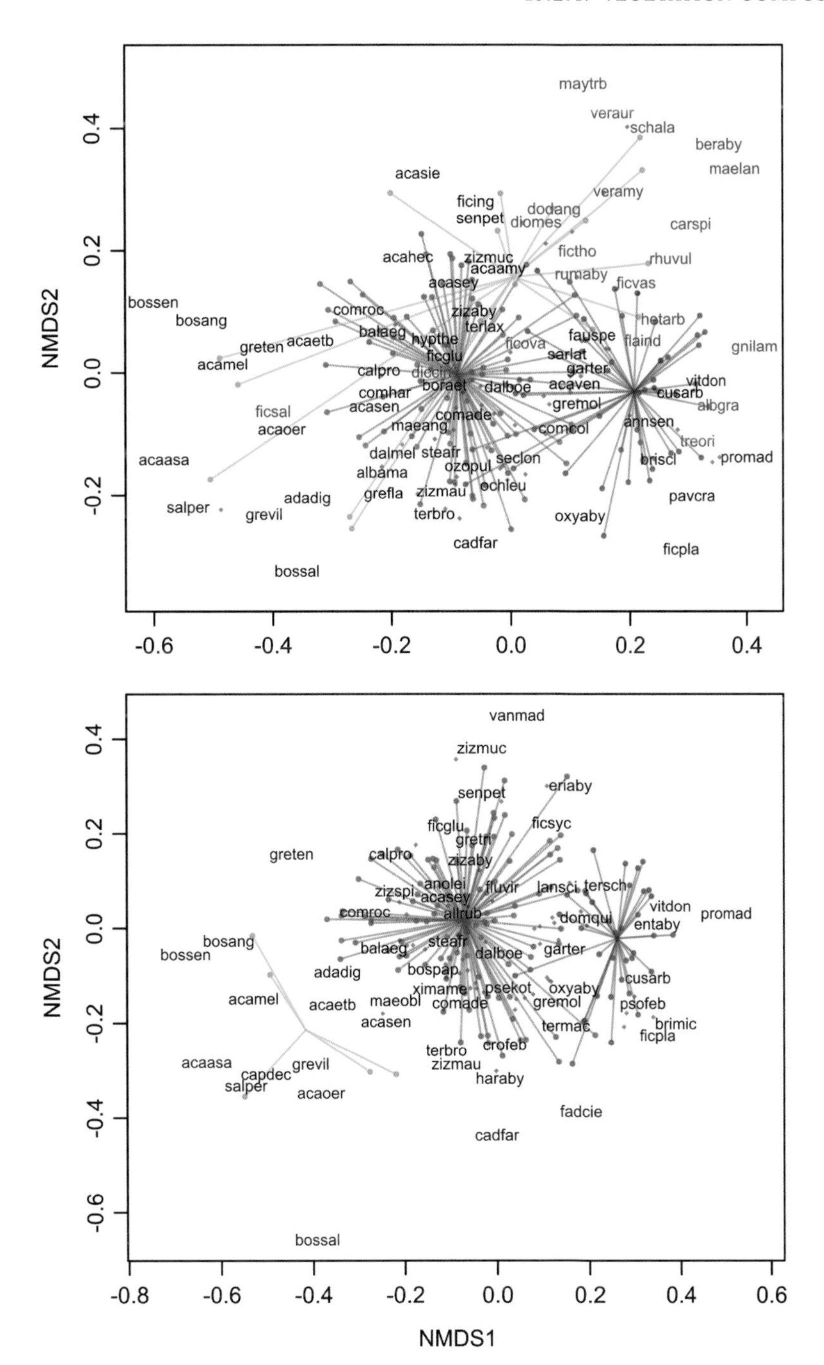

Fig. 10-7. Non-metric multidimensional scaling (NMDS); plots. NMDS-ordination of species assemblage structure among sites within the CTW-area. It provides the first two axes of a three-dimensional representation of Beta-dissimilarity distances. For the upper graph, the vegetation similarity (using the beta-dissimilarity index) was computed using all species. In the lower graph, only the CTW-species were used. Relevés are colour coded according to the clusters they were assigned to (see Section 10.2.2 and Fig. 10-5). They are furthermore linked to the medium of their respective clusters. The species are indicated by the first three letters of the genus and species names, as listed in Appendix 2, with the marginal intruders and transgressors species in red.

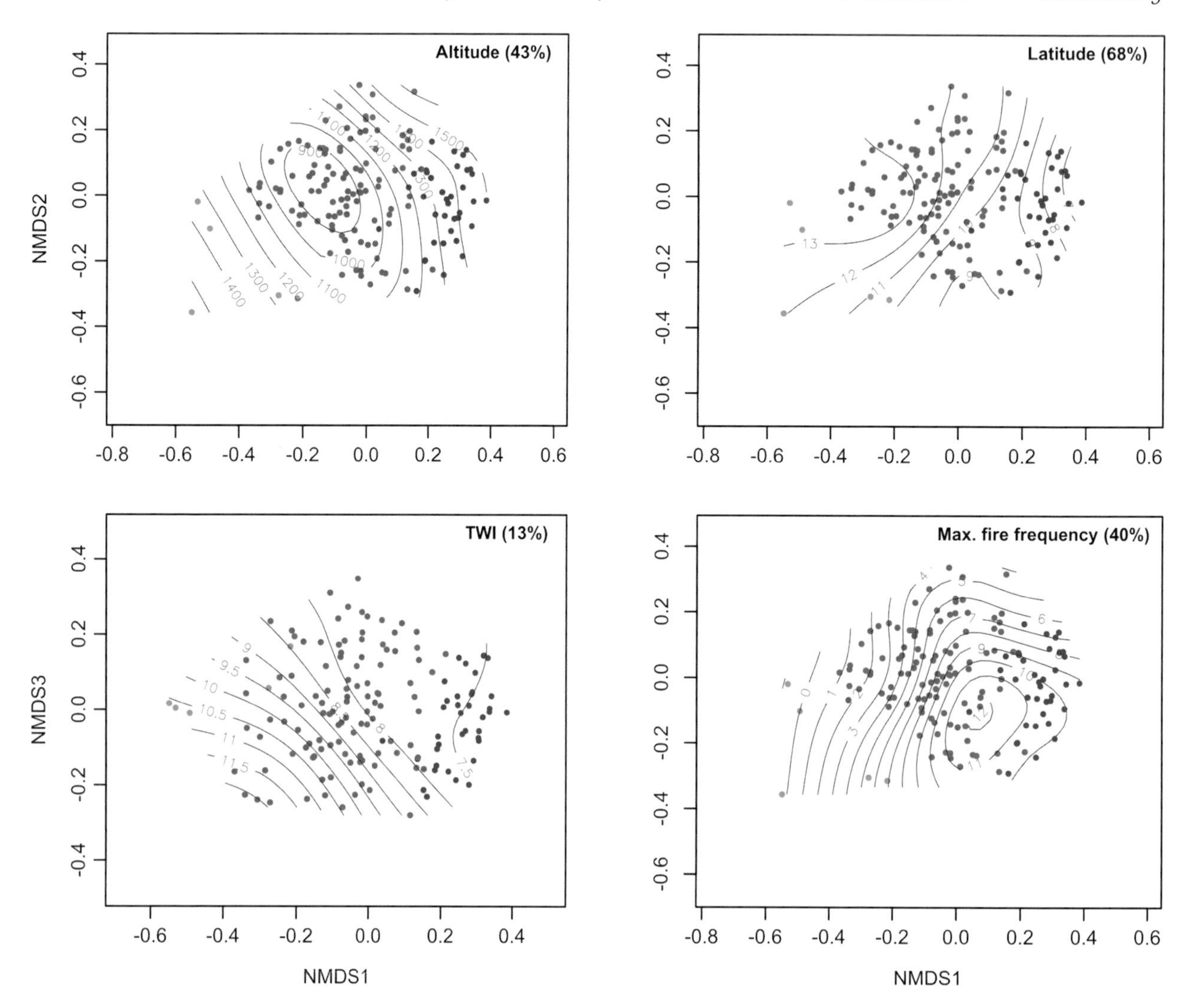

Fig. 10-8. Non-metric multidimensional scaling (NMDS); gradients. NMDS-ordination of species assemblage structures among sites within the CTW-area. Relevés are grouped in three major clusters as identified using the f.UPGMA method (see Section 10.1.2). Isolines represent the altitude, latitude, maximum fire frequencies and terrain wetness index, modelled in ordination space using a generalized additive modelling approach as implemented in ordisurf in the R vegan package. For altitude, latitude and maximum fires frequency, the surface is fitted onto the first two NMDS axes. For the terrain wetness index, this was done on the first and third NMDS axes, as this yielded the highest deviance explained (13% versus 7% on the first two axes). Between brackets the deviance explained by the corresponding variable is given.

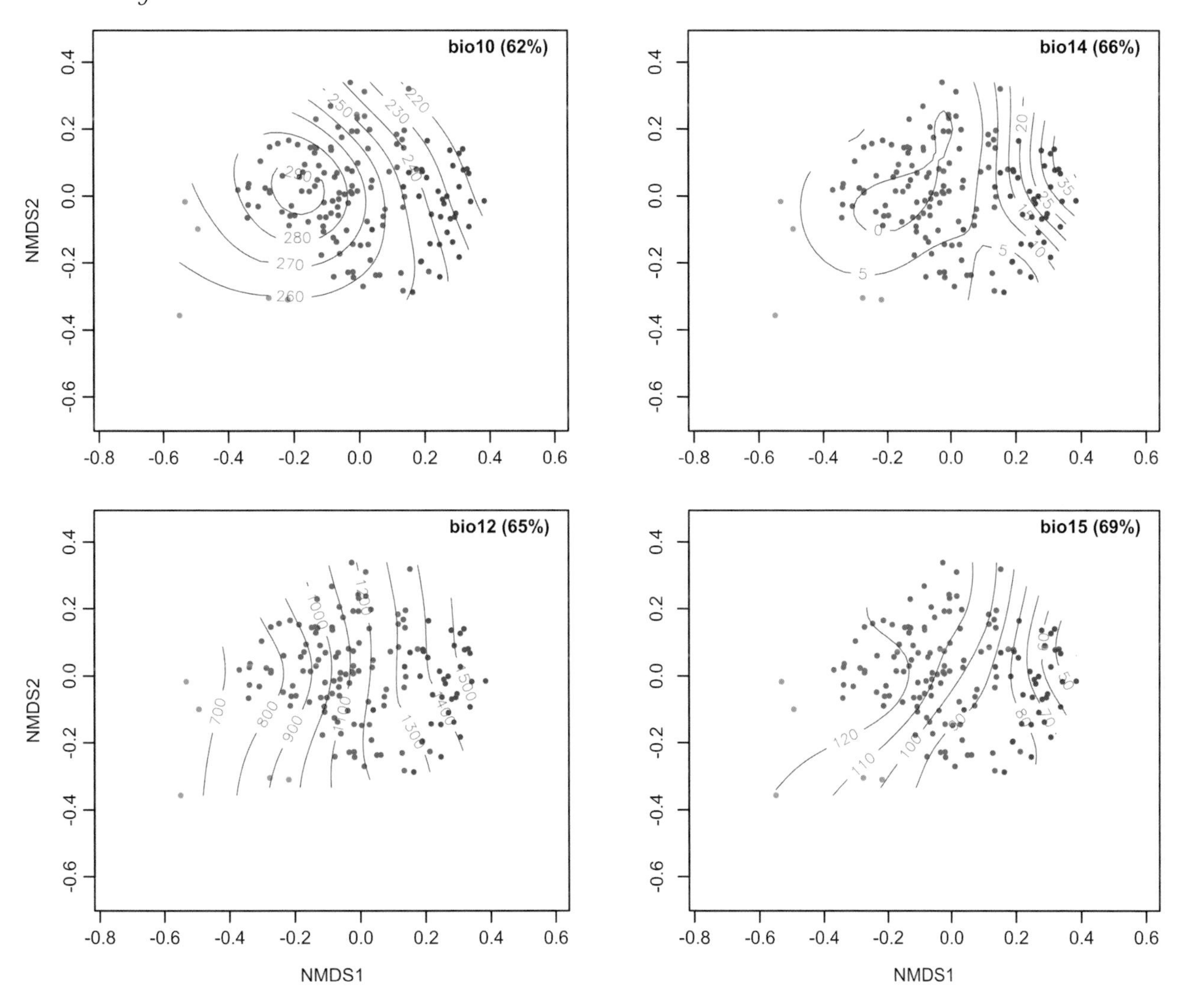

Fig. 10-9. Non-metric multidimensional scaling (NMDS); gradients. NMDS-ordination of species assemblage structures among sites within the CTW-area. Relevés are grouped in three major clusters as identified using the f.UPGMA method (see Section 10.1.2. Isolines represent the mean temperature of the warmest quarter (bio10), the annual precipitation (bio12), the precipitation of the driest month (bio14) and the precipitation seasonality (bio15). Between brackets the deviance explained by the corresponding variable is given.

relative clear separation of relevés according to the cluster they belong to. Similarly, TWI, DEM and FF together accounted for 19% (F=1.81, p==0.01) and the soil types for 20% (F=1.2, p< 0.01) of the total variance in species composition.

Relevés are coloured according to the cluster as defined in Section 10.2.2 (for the grey dots, see below). Overall, results presented in Fig. 10-9 are largely in agreement with those presented in Fig. 10-8 and Fig. 10-9. The relevés in the northern (green) cluster occur along very broad environmental gradients. Conditions in relevés of the southern cluster (blue) are generally wetter, especially in the dry and warm period, while temperatures, temperature variability and rainfall variabilities are lower. Fire Frequencies explain a small but significant part of the variation (marginal effects of terms; F=7.8, p < 0.01). Overall, Fire Frequencies (and relevés with observed high degree of burning) tend to be slightly higher for relevés of the southern (blue) cluster than for the northern (green) cluster. However, the considerable overlap shows that fire does not explain the separation in two clusters.

A number of relevés are marked separately (grey dots in Fig. 10-10). These are the northernmost sample sites along the road from Shire-Enda Selassie to Humera, passing Kafta-Shiraro National Park on the way. They are part of the 'green' cluster, but most have a low to negative silhouette width (Section 10.2.2) indicating a limited to poor fit to that cluster. Their position on the more extreme end of a number of environmental gradients may offer some explanation. Rainfall in the cold or wet periods (PCW) is considerably lower than in the rest of the sites of this cluster. As such, they have an intermediate position along this gradient between the 'green' and 'orange' clusters. The PCW is comparable with the sites of the southern (blue) cluster in the far south, but variability in rainfall and temperature are higher, and rainfall in the dry and warm season lower than in those sites (Fig. 10-10). These conditions explain the observed low availability of biomass for burning and consequently, the absence of fire in some parts of this region.

The soil types observed during the field work is rather well associated with the clusters. Vertisol ('black cotton soil') and similar alluvial soils (S1, S2) are clearly associated with the first, northern cluster (green dots). Humus-rich brown or reddish soils (S3-S5) occurs at the boundary between first, northern cluster (green dots) and the second, southern cluster (blue dots). Iron-rich red or sandy and grey soils (S6, S7) are associated with the first, northern cluster (green dots). Stony to rocky soil or bare rocks (S8, S10, S11) are associated the first, northern cluster (green dots), but also with the third cluster (orange dots). This is also the case with the stony, sandy soils of S9.

In general, one can conclude that the different environmental factors presented here explain small but significant part of the variation in species composition. The clusters as defined in Section 10.2.2 are not clearly separated, but rather represent a to some extent arbitrary split along the different environmental gradients. Note that if marginal intruders and transgressors are included, results are similar, except that the third (orange) cluster is less well defined, with relevés occurring along a much broader range of environmental conditions. Proximity to other vegetation types and land use history are factors that were not included, but are likely to be important determinants of the species composition.

10.2.5. Species and environment

The clusters of green (northern) and blue (southern) dots indicated on Fig. 10-7 are reflecting gradients in environmental parameters from species of dry and low habitats in the lower and left side of the plot towards higher and moister habitats in the upper right corner. A few examples of species that have not been discussed in Section 10.2.3 are discussed here. The lowland bamboo (*Oxytenanthera abyssinica*; oxyaby) is both in the plot for all species and in CTW-species located at the low and relatively moist habitats, but its geographical distribution would fit best in the area of the northern (green) cluster. *Ficus platyphylla* (ficpla) is also in both in the plot for all species and for CTW-species located at the low and relatively moist habitats, which agrees with our observations in Section 6, so both the environment and the distribution in the southern part of the western woodlands agree with a position in the blue (southern) cluster. *Cussonia arborea* (cusarb) is also in both in the plot for all species and for CTW-species located near the centre of the blue (southern) cluster; it occurs at medium altitudes in the central and southern part of the CTW-area, but it has a few records from the northern CTW-area. *Dalbergia boehmii* (dalboe) is an example of a species that both in the plot for all species and for CTW-species is located between the northern (green) and the southern (blue) cluster; it has a very narrow distribution in the in humid and dense woodland in the central part of the CTW-area. *Acacia seyal* (acasey) is in both the plot for all species and for CTW-species located above the centre of the green (northern) cluster, near where *Balanites aegyptiaca* (balaeg), *Anogeissus leiocarpa* (anolei) and *Calotropis procera* (calpro) are located. All four species may occur close to each other, as described in Section 5, profile A, although *Calotropis procera* is mostly associates with very dry and degraded ground. In the left part of the two plots in Fig. 10-7. Non-metric multidimensional scaling (NMDS); plots. In Fig. 10-7 there are a number of species associated with *Acacia-Commiphora* bushland and associated with the orange (third) cluster. They are mainly species of Capparidaceae (*Boscia senegalensis* (bossen), *Boscia angolensis* (bosang), *Boscia salicifolia* (bosal), *Capparis decidua* (capdes)), but also drought-tolerant species of *Acacia* (*Acacia etbaica* (acaetb), *Acacia mellifera* (acamel), *Acacia asak* (acaasa), and *Acacia oerfota* (acaoer)), as well as a number of other drought-tolerant species, for example *Salvadora persica* (salper), *Grewia tenax* (greten), *Grewia villosa* (grevil) and *Adansonia digitata* (adadig). In the plot based on all species there is a number of species in the upper right side of the plot which in Section 6 has been characterized as marginal intruders (MI). This group of species includes *Dodonaea angustifolia* (dodang), *Carissa spinarum* (carspi), *Rhus vulgaris* (rhuvul), *Ficus vasta* (ficvas) and a number of other species. They are located in a part of the plot where there are also a number of relevés near the upper part of the CTW-area, where the marginal intruders are particularly prominent.

In the upper right plot in Fig. 10-10, the horizontal axis indicates general climatic and temperature variation, whereas the vertical axis is related to the precipitation in the cold months, driest at the top of the plot, wettest at the bottom. *Ziziphus spina-christi* (zizspi), *Anogeissus leiocarpa* (anolei) and *Pteraocarpus lucens* (pteluc), *Balanitea aegyptiaca* (balaeg), *Boswellia papyrifera* (bospap) seem tolerant of high temperatures and significant variation, which agrees with their occurrence in lowland on rocky ground. On the vertical axis with precipitation in the cold and wet period axis the very drought tolerant *Acacia mellifera* (acamel) is at the highest point (lowest precipitation); *Terminalia brownii* (terbro), the most drought-tolerant species of Terminalia, is near the *Acacia mellifera*. At the other end of the vertical axis are *Terminalia laxiflora* (terlax), *Lonchocarpus laxiflorus* (lonlax) and *Flueggea virosa* (fluvir), typical species of the *Combretum-Terminalia* woodlands, with *Flueggea virosa* sometimes in riverine vegetation. Towards the right end of the horizontal axis are a number of typical species of moist *Combretum-Terminalia* woodlands, *Piliostigma thonningii* (pilsti), *Terminalia schimperiana* (tersch), *Vitex doniana* (vitdon), and *Bridelia scleroneura* (brisci).

In the plot in the middle of the right hand column of Fig. 10-10, the species associated with high fire frequencies are most notably *Lonchocarpus laxiflorus* (lonlax), *Grewia mollis* (gremol), *Combretum collinum* (comcol) and *Entada africana* (entafr). Other typical woodland species associated with medium to high fire frequencies are (in alphabetical order), *Acacia senegal* (acasen), *Flueggea virosa* (fluvir), *Lannea fruticosa* (lanfru), *Ochna leucophloeos* (ochleu), *Oxytenanthera abyssinica* (oxyaby), *Pseudocedrela kotschyi* (psekot), *Stereospermum kunthianum* (stekun) and *Terminalia schimperiana* (tersch). Species that can tolerate fire to some extent, but are less frequently found in areas with high fire frequencies are *Gardenia ternifolia* (garter), *Hyphaene thebaica* (hypthe), *Lannea barteri* (lanbar), *Pterocarpus lucens* (pterluc), *Sterculia africana* (sterafr), *Tarminalia laxiflora* (terlax), *Ziziphus abyssinica* (zizaby), and *Ziziphus spina-christi* (zizspi).

In the same plot, the opposite direction of the Terrain Wetness Index (TWI) suggests that species occurring in areas with a high TWI are less likely to be found in areas with high fire frequencies. Examples are *Acacia mellifera* (acamel), *Ziziphus spina-christi* (zizspi), *Balanites aegyptiaca* (balaeg) and *Anogeissus leiocarpa* (anolei). Note that the influence of TWI is limited, so other factors with the same direction as the TWI may in fact play a more important role. For example, *Anogeissus leiocarpa* typically occurs on stony or rocky soil or bare rocks (see below), which is therefore a more likely explanation for this species mostly occurring in sites with low fire frequencies than wet soil conditions and high TWI.

The plot in the lower right corner of Fig. 10-10 demonstrates relations between soil types and species. In the upper left corner are species associated with vertisols ('black cotton soils') and similar alluvial soils (S1, S2), for example *Acacia hecatophylla* (acahec), *Balanites aegyptiaca* (balaeg), *Terminalia laxiflora* (terlax), and *Acacia seyal* (acasey). Towards S7 is located *Combretum hartmannianum* (comhar), which is also usually found on vertisol. Humus-rich brown or reddish soils (S3-S5) are associated with *Entada africana* (entafr), *Lannea schimperi* (lansci), *Bridelia scleroneura* (briscl), *Acacia polyacantha* subsp. *campylacantha* (acacam), *Grewia mollis* (gremol), *Pterocarpus lucens* (pteluc), and *Piliostigma thonningii* (piltho). Iron-rich red or sandy and grey soils (S6, S7) is associated with *Allophylus rubifolius* (allrub), *Ficus ingens* (ficing), and *Fadogia cienkowskii* (fadcie). Stony to rocky soil or bare rocks (S8, S10, S11) are associated with *Anogeissus leiocarpa* (anolei), *Lannea fruticosa* (lanfru), *Boswellia papyrifera* (bospap), and *Sterculia africana* (sterafr). In the field between S4 and S11 there is a range of species located, and it is difficult to relate these to soil type.

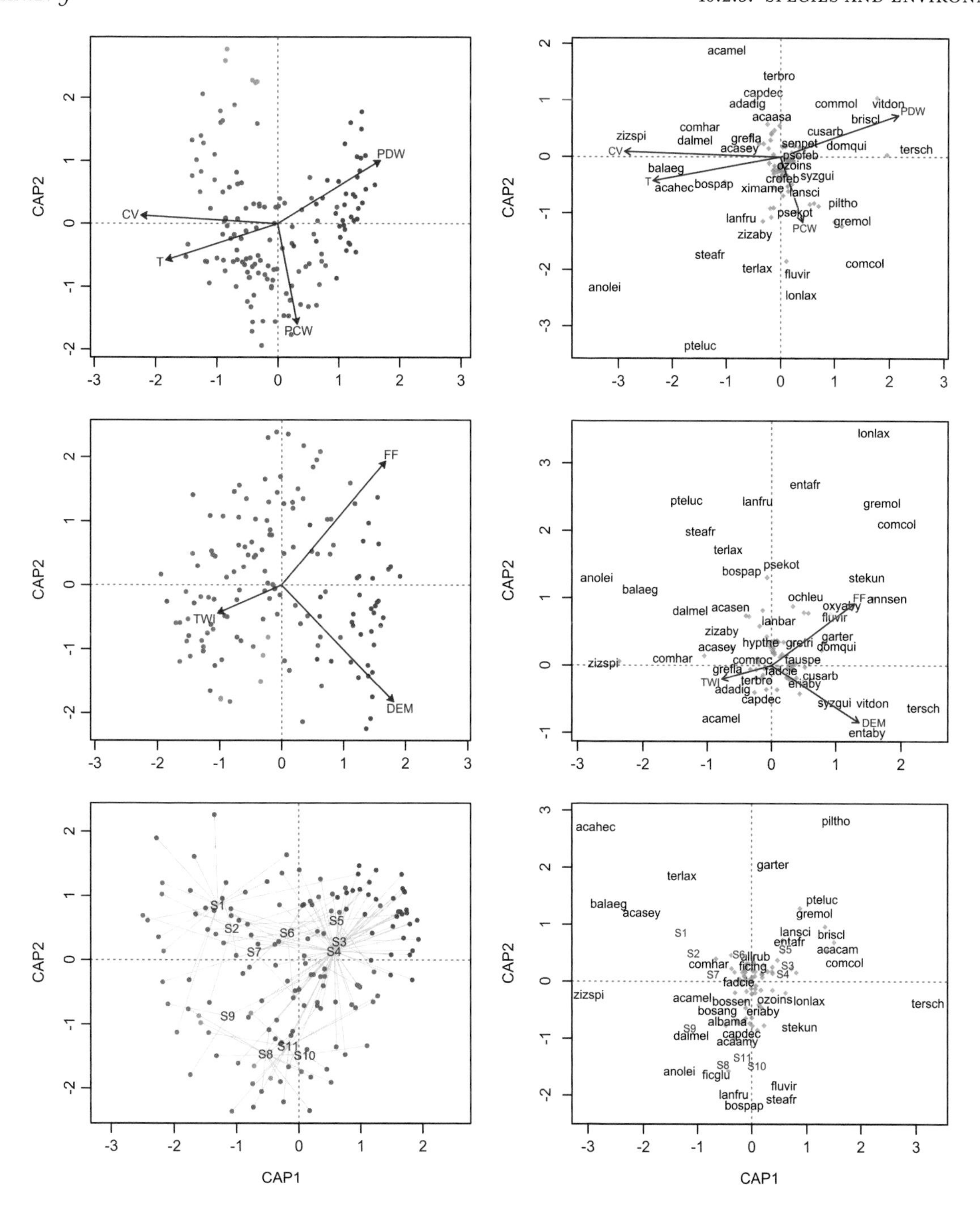

Fig. 10-10. Distribution of the relevés in ordination space, constrained. The plots are formed by the first two axes of a distance-based redundancy analysis (capscale). Of the variance explained by climatic variables, 81% is represented on the first two axes (upper row). For the variables Fire Frequency, altitude and Terrain Wetness Index (middle row) this is 90% and for the soil types (lower row) 63%. Arrows and red letter codes represent the constraining variables (see the text for details). Species are represented by the first three letters of the genus and species name, as indicated in Appendix 2. S1-S11 indicate the soil groups shown on Fig. 2-19 and listed in Appendix 3. Relevés are scaled for relevé scores (left column) or scaled symmetrically by square root of eigenvalues of both species and relevés (right column). In the left column, the green, blue and orange dots represent the three clusters identified earlier. The grey dots are part of the green cluster, but are singled out as they form a distinct group in the first graph and in geographic space (see text).

11
Conclusion: what are the answers to the scientific questions?

In this concluding Section 11 we will try to see how well the results of this work allow us to answer the research questions in Section 4. We will also discuss a few other issues touched upon in the previous Sections, particularly the methodology used for our studies and ideas about the future conservation of the western Ethiopian woodlands.

11.1. Can we confirm the extent of the *Combretum-Terminalia* woodland area? Was it defined correctly in Friis et al. (2010), including its northern and southern limits?

The data and basic discussions in relation to these questions are found in the Sections 6-10. Here we will mostly take the *Combretum-Terminalia woodland* (CTW) as the representative of the western Ethiopian woodlands, but also touch upon the question about inclusion of other types of woodland.

The north-western limit: We lack floristic data beyond the borders with Eritrea and Sudan, where other types of vegetation may exist. The most prominent soil type in north-western Ethiopia is vertisol, and it extends for a considerable distance north-west and west of the border between Ethiopia and the Sudan. It is probable that areas with this soil type outside Ethiopia will have vegetation types similar to those on Ethiopian vertisol, provided the rainfall is similar. As described in Profile A in Section 5, the vegetation on vertisol near the Sudan-Ethiopian border is dominated by *Balanites aegyptiaca*, associated with various species of *Acacia*, either shrubs or low trees associated with the *Acacia-Commiphora* bushland (*A. mellifera, A. oerfota, A. senegal*) or taller species of *Acacia* (*A. hecatophylla, A. seyal*). Associated with vertisol, the vegetation with these species forms a mosaic, where the typical *Combretum-Terminalia* woodland is found on sloping and more rocky ground. Species more frequent further south, *Anogeissus leiocarpa, Combretum hartmannianum Ziziphus-spina-christi*, do occur here, so the vegetation near Humera is similar to, but not absolutely identical with the Sudanian vegetation types of Harrison & Jackson (1958); '*Acacia mellifera* thornland (on dark cracking clays, alternating with grass areas)' and the '*Acacia-Balanites* savannah, alternating with grass areas.' White (1983) indicated a very small area in the extreme north-western part of Ethiopia as 'Edaphic grassland mosaic with *Acacia* wooded grassland' (Fig. 3-5), covering the area near Humera, but our observations could not confirm that there is a mosaic of edaphic grassland and *Acacia* wooded grassland, only that in that area there is *Acacia-Balanites* wooded grassland or various types of *Acacia* bushland. White's 'Edaphic grassland mosaic with *Acacia* wooded grassland' may have been his interpretation of Harrison & Jackson's '*Acacia-Balanites* savannah, alternating with grass areas' (Harrison & Jackson 1958: 12), combined with their observation that *Acacia mellifera* bushland formed a cycle with almost pure grassland (Harrison & Jackson 1958: 12), and how there was an almost imperceptible transition between '*Acacia-Balanites* savannah, alternating with grass areas' and *Acacia mellifera* bushland (Harrison & Jackson 1958: 13). According to our observations it will be difficult to distinguish between *Acacia-Balanites* wooded grassland and small areas inside this, dominated by grasses and *Acacia mellifera* bushland.

Changes in composition of the vegetation from site to site in this area are correlated with changes in soil, and already 50-60 km from Humera towards Gondar (in Profile A), at altitudes 750-800 m a.s.l. the flat areas with vertisol are generally replaced by sloping or undulating areas with luvisols, leptosols or other soil types (Fig. 2-18), and the *Balanites-Acacia* wooded grassland is replaced with a woody stratum that is dominated by *Anogeissus leiocarpa*, *Boswellia papyrifera*, *Dalbergia melanoxylon* and species of *Combretum*. Up to about 1000 m a.s.l., in relevé no. 16, ca. 150 km south of Humera, the most prominent woody species are still *Anogeissus leiocarpa*, *Boswellia papyrifera*, *Dalbergia melanoxylon*, with a combination of species that resembles the floristic composition in Harrison & Jackson's '*Anogeissus-Combretum hartmannianum* savanna woodland.' In this part of the profile we observed a number of species of woody Capparaceae (*Boscia senegalensis*, *Cadaba farinosa*, *Capparis decidua*), which are otherwise associated with the *Acacia-Commiphora* bushland and the dry Sahel zone across Africa (see the Sections 2.1, 3.12 and 6).

According to Harrison & Jackson (1958: 13-14) the two vegetation types '*Acacia mellifera* thornland (on dark cracking clays, alternating with grass areas)' and the '*Acacia-Balanites* savannah, alternating with grass areas' gradually change into 'hill catenas' in more hilly terrain with an annual rainfall above 1000 mm. At the lowest altitudes the 'hill catenas' have a woody flora that is very similar to that of lower altitudes, but richer in species. As can be seen in Fig. 2-7, the 1000 mm isohyet is near relevé no. 16, where we began to observe a more varied flora. We consider that there is at this point good agreement between Harrison & Jackson's observations and ours.

According to our cluster analyses in Section 10, we found that the relevés in Profile A nearest to Humera (Fig. 10-4 & 10-5) joined the first cluster (green dots), but also that the match between some of these relevés and the rest of the cluster was weak. In spite of the slightly different soil and the somewhat deviating floristic composition, we will consider the vegetation on the relevés near Humera in Profile A to be part of the Ethiopian *Combretum-Terminalia* woodland, with the note that it may have features that indicate transition to other vegetation types. A more precise delimitation here must include observations covering an area in three different countries, Eritrea, Sudan and Ethiopia, and that has been outside our reach. However, a northern limit of the *Combretum-Terminalia woodland* (CTW) located near the present border between Ethiopia and Eritrea will agree with a limit as predicted on climatic parameters, as shown in the widest and the narrowest model in Fig. 4-2.

In Section 9 we have looked at the northern and north-western limits of the *Combretum-Terminalia woodland* (CTW) in relation to the modelled distributions of typical CTW-species. In most cases there are very few records from western Eritrea and the adjacent parts of the Sudan. The models is Section 9 of species of *Combretum*, *Terminalia*, *Lonchocarpus laxiflorus* and *Pterocarpus lucens* show in most cases that very suitable environment for these species continues into the Sudan at least as far as the plains along the Atbara River. Most of the modelled distribution maps in Section 9 suggest that typical CTW-species also occur inside the south-western Eritrea, and that this area should probably also be part of the CTW-area, but the modelling is based on very few observed records from Eritrea and the Sudan, so we have to extrapolate from evidence from Ethiopia, and a more precise limit cannot currently be determined. According to the vegetation map by Harrison & Jackson (1958; Fig. 3-4) the *Balanites-Acacia seyal* wooded grassland continues to the town of Sennar on the Blue Nile. Further south along the Sudan-Ethiopian border the *Balanites-Acacia seyal* wooded grassland is replaced with woodland dominated by *Anogeissus leiocarpa* and *Combretum hartmannianum*, and this zone may probably mark the north-western limit of CTW-vegetation.

The northern limit at higher ground above the vertisol plains: This limit was studied along the Profile A.2 between Humera and the Shire Highlands, represented by the relevés 26-35. Near Humera, the vegetation was very similar to that of the first part of the main Profile A. The vegetation began changing from relevé 27, ca. 40 km from Humera, where the topography and soil type altered to an undulating landscape with brown soil and fragments of volcanic rock, and in places even to areas with exposed flat rock-faces. *Anogeissus leiocarpa*, *Boswellia papyrifera*, *Dalbergia melanoxylon*, *Adansonia digitata*, *Terminalia brownii* and species of *Combretum* became common and replacing wooded grassland with *Balanites* and species of *Acacia*. In some places extensive fire-scars were observed during our field work, but as it can be seen from the illustration in Fig. 2-24, fire frequencies for the period 2003-2013 were generally low in this region. Again, all relevés in the entire profile A2 are clustered with the northern cluster (green dots) in Fig. 10-4 & 10-5, in which both floristic and environmental data are clustered, and in spite of the slightly deviating flora we will include it in the western Ethiopian *Combretum-Terminalia* woodland. Yet two points are noteworthy with this. Firstly, the species compositions in the relevés are relatively weakly associated to the first (green) cluster. Secondly, the relevés are situated on the more extreme end of a number of environmental gradients (see section 10.2.4.2). Given all this, parts of the profile A2 is representative of a transition to the vegetation in the dry northern and eastern river valleys (further below). It is highly likely that it may also represent a transition to other types of woodland beyond the borders of Ethiopia. The information provided by Harrison & Jackson (1958) and White (1983) on the woodlands in the Sudan above the vertisol plains is not sufficient to allow a comparison between that vegetation and the vegetation in relevés at similar altitudes in Ethiopia, and there is no information about this in the Italian literature from western Eritrea.

The dry northern and eastern river valleys in the Ethiopian Highlands: In Friis et al. (2010) the vegetation in the northern and eastern river valleys in the Ethiopian Highlands (the upper Mareb valley on the border between Ethiopia and Eritrea, the upper Tacazze Valley and the valley with the eastern bend of the Abay River) were classified as western Ethiopian *Combretum-Terminalia* woodland. Our field studies of the Profiles C (relevés 144-147), D (relevés 143) and G (relevé 90) document that the vegetation in these localities is poor in species, particularly near the rivers. In the upper zones of these rivers there are records of *Boswellia papyrifera*, *Combretum molle*, *Terminalia brownii* and *Stereospermum kunthianum*. According to our cluster analyses (Section 10) we found that the relevés in the Profiles C, D and G in Fig. 10-4 all cluster as aberrant members of the first cluster (all species considered), or as a separate third cluster (orange) as shown in Fig. 10-5 (only CTW-species considered), with only relevé 143 in Profile D changing association from the first to the third cluster.

Again, because of the presence of species of *Boswellia*, *Combretum*, *Terminalia* and *Stereospermum* in these relevés we will consider them to be part of the western Ethiopian *Combretum-Terminalia* woodland, but as depauperate variants. The presence of some of the same species of small thorny *Acacia* (*A. mellifera*, *A. oerfota*) and of Capparaceae (*Boscia angustifolia*, *Boscia senegalensis*, *Capparis decidua*) suggests floristic links to the vegetation on vertisol near Humera. However, the soils in Profiles C, D and G are mostly hard or sandy soil derived from limestone. The presence of *Salvadora persica* (see distribution map in Fig. 6-106) is a floristic link with the driest parts of the *Acacia-Commiphora bushland* (ACB), or the *Semi-Desert Scrubland* (SDS) or *salt pans* (SLV/SSS).

The position of relevé 143, at an altitude of 1850 m a.s.l. along the uppermost reaches of the Tacazze River south of Lalibela, offers a particular problem. The relevé represents only the vegetation immediately next to the Tacazze River. Further away from the river the vegetation becomes very open scrub vegetation with evergreen shrubs such as *Euclea* and *Dodonaea*. *Acacia etbaica,* frequent in dry parts of *Dry Afromontane forest and woodland* (DAF), subtype *Afromontane woodland, etc.* (DAF/WG), see Friis et al. (2010: 88), is also very common on the slopes above the river. In many places between Lalibela and Sekota this semi-evergreen bushland has a very scattered tree-component of *Combretum molle* and (even less common) *Combretum collinum*, *Stereospermum kunthianum*, *Lannea fruticosa*, *Cordia africana* and *Croton macrostachyus*, which associates it with the main part of the northern (green) cluster.

The southern limit in the Omo Valley and between the Omo and South Sudan: Due to inaccessibility, the southern limit of the Ethiopian *Combretum-Terminalia* woodland has not been well defined in the previous literature. During our field work up to 2018 we could not study the vegetation in extreme southwestern Ethiopia because we were not able to cross the Omo

River, and during our latest field work in 2018 we were not able to progress further south than to Maji at 6° 10' N. Therefore, for this area we have to rely on information from other sources.

Pichi Sermolli (1957) marked the southern limit of his *Boscaglia caducifolio* (deciduous woodland) at ca. 6° 20' N (Fig. 3-3), which is north of Maji. This is too far to the north according to our field observations. Harrison & Jackson (1958) marked the northern limit of the Toposa area in the Sudan (now South Sudan), a vegetation type rather similar to that of the lower Omo Valley, see Section 3.3), at ca. 6° N, which is also too far to the north (Fig. 3-4). White (1983) moved the southern limit of his 'Undifferentiated [Sudanian] woodland – Ethiopian' further south, placing it at approximately 5° 50' N. (Fig. 3-5). Friis et al. (2010) and Lillesø et al. (2011) moved the southern limit of the *Combretum-Terminalia woodland and wooded grassland* (CTW) to the border with Kenya (Fig. 3-6, Fig. 3-7). However, throughout its range in western Ethiopia, with the exception of a few places in the extreme north-west and in the driest river valleys in the highlands, there are frequent grass fires in all areas of *Combretum-Terminalia* woodland. In the map showing the number of fires in Ethiopia and adjacent parts of the Sudan and South Sudan in Fig. 2-24, very few fires are indicated in the Toposa area, and in south-western Ethiopia south of the villages of Balala and Kibish, which are located on the border with South Sudan at ca. 5° 20' N. According to Breugel et al. (2016a, Fig. 1) the modelled fire frequencies drop to almost zero (less than one in 10 years) south of the same latitude. In the absence of direct observations, we will therefore argue that the boundary between frequent fires and few or no fires can be used to indicate the boundary of *Combretum-Terminalia* woodland, as approximated by the purple-blue line in Fig. 11-1.

In Section 4 (Fig. 4-2) we modelled the potential area of the *Combretum-Terminalia* woodland, using environmental parameters from the 151 relevés. We found that the northernmost position of the southern boundary of that vegetation was predicted to be approximately at the latitude of Maji (Fig. 4-2, lower map), and that the southernmost position of the southern boundary would be approximately where frequent fires change to very few or no fires, marked as a boundary by the purple-blue line in Fig. 11-1. This agrees with the conclusion in Section 5, in what we have called 'Profile R', which is not based on our relevés from 2014-2018, but on information by Carr (1998), Schloeder (1999), Jacobs & Schloeder (2002) and our older field observations on the eastern side of the Omo River. According to this information, it is impossible to draw a sharply marked line between the *Combretum-Terminalia* woodland and the *Acacia-Commiphora* bushlands. As an alternative to the above suggested line drawn on fire frequencies, the boundary might approximately follow a line from the South Sudan border south of Kibish and then follow a 1000 m contour in the Omo Valley towards the mountainous area near the town of Jinka, continuing south of these mountains and ending at the Afromontane vegetation in the mountains to the west of Arba Minch. This is also in agreement with the discussion in Breugel et al. (2016a; Supplementary material, Appendix 3). The 1000 m contour around the Omo Valley is approximately outlined with the red line in Fig. 11-1. At the moment, without detailed field observations, we think that a line between the positions of the purple-blue and the red lines in Fig. 11-1 may represent the best approximation to the southern border of the *Combretum-Terminalia* woodland in south-western Ethiopia. The boundary between the vertisols in the lower Omo Valley to the east and the hill slopes to the west, north and east may be close to such a line intermediate between the purple-blue and the red line in Fig. 11-1.

As a test of the suggested southern border for the *Combretum-Terminalia* woodland in south-western Ethiopia, we can compare that border with the modelled distributions presented in Section 9, which are based on the entire African data sets for the species, and the observed distributions in Section 6. For a number of species the southern limit of the modelled distribution is north of the southern border here suggested for the *Combretum-Terminalia woodland* (CTW). These include *Anogeissus leiocarpa* (Fig. 9-1), for which the southernmost observed records are in the Gambela lowlands (Fig. 6-22), *Terminalia laxiflora* (Fig. 9-8), for which the southernmost observed records are in woodlands some distance north of Maji (Fig. 6-29), and *Terminalia macroptera* (Fig. 9-9), for which the southernmost observed records are from the border zone between the *Combretum-Terminalia woodland* (CTW) and the *Transitional rain forest* (TRF) in the IL floristic region (Fig. 6-30). In addition, our results suggest that *Lonchocarpus laxiflorus* (Fig. 9-11) does not occur south of Maji, which is slightly north of the southern border here suggested for the *Combretum-Terminalia woodland* (CTW). The same seems to be true for *Terminalia schimperiana* (Fig. 9-10), although the species reaches further south within the *Transitional semi-evergreen bushland* (TSEB) in the GG floristic region (Fig. 6-31). Lastly, the modelled southern limit of *Pterocarpus lucens* (Fig. 9-12) is found near the southern limit of the IL floristic region, while its southernmost observed records are from near the northern limit of the IL floristic region (Fig. 6-91).

While the southern limit of the modelled distribution of the above mentioned species is north of the suggested southern border of the *Combretum-Terminalia woodland* (CTW), there are a number of species that actually do reach or almost reach those limits. The modelled southern limits of *Combretum collinum* (Fig. 9-3) and *Combretum molle* (Fig. 9-5) approach the two possible southern limits in Fig. 11-1, while the southernmost observed records are from around Maji for *Combretum collinum* and south of Maji for *Combretum molle*. And although the southernmost observed records of *Combretum adenogonium* (Fig. 9-2) are close to Maji, its modelled southern limit is close to the two possible southern limits in Fig. 11-1, or possibly even further south.

The modelled southern boundary of *Terminalia brownii* (Fig. 9-7) is south of the two possible southern limits in Fig. 11-1, and the southernmost observed records are just north of the estuary of the Omo River, near the northern shores of Lake Turkana, but the local sub-element of the latter species has been classified as ACB[-CTW], for which reason its distribution transgresses the border between the CTW- and ACB-areas.

The interpretation of this part of Ethiopia by Olson et al. (2001) does not agree with our observations and the discussion above. Near the Boma Plateau on the South Sudan side of the border there may be forest-savanna mosaics similar to vegetation in the Lake Victoria Basin, as is indicated in Fig. 3-10 from Olson et al. (2001), where the vegetation is marked as 'Vbfsm.' We have no evidence that a similar vegetation does extend into Ethiopia. The indication by Olson et al. of a forest-savanna mosaic near the southernmost border of the *Combretum-Terminalia* woodland would seem to be based on a misinterpretation of the forest-savanna mosaic further north in the IL floristic region, which Friis et al. (2010: 106-113) and Breugel et al. (2016a: Supplementary material, Appendix 1 & 2) have described as fragments of the *Transitional rain forest* (TRF).

We must therefore conclude that the reply to this research question is that we mainly confirm the extent of the *Combretum-Terminalia woodland and grassland* (CTW) as shown in Friis et al. (2010) as validly representing the extent of the western Ethiopian woodlands, with some modifications at the northern and southern limits. As appears from the above, the northern limit of the CTW-area is marked by a zone with gradually poorer flora and more intrusion of species from the *Acacia-Commiphora* bushland in the Somalia-Masai regional centre of endemism and from the Sahel regional transition zone (White 1983). Due to this it is difficult to draw a precise line, but since CTW-species keep occurring right up to the border between Ethiopia and Eritrea and to the upper reaches of the deep valleys of the Mareb, Tacazze and Abay Rivers, we suggest that these areas be included within the CTW-area for practical purposes, and that our vegetation map in Fig. 4-1 be accepted as actually indicating the northern limit of the CTW. A similarly clear conclusion is lacking with regard to the southern limit of the CTW-area, mainly due to inaccessibility and insufficient field work. Our evidence points towards a limit at the suggested lines in Fig. 11-1, with a gradual transition to *Acacia-Commiphora* bushland somewhere in the Omo National Park. Therefore, the limit on the map in Fig. 4-1 is probably too far to the

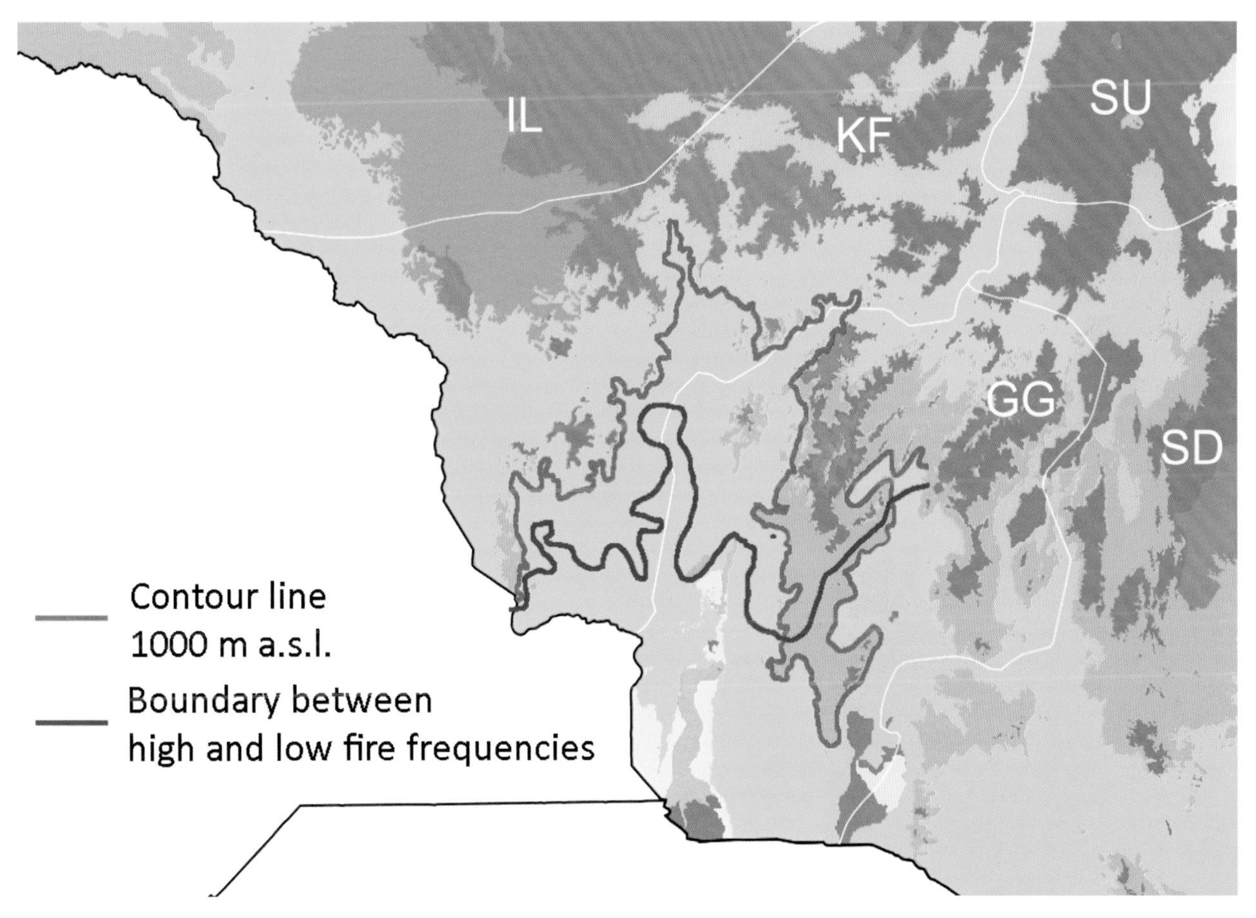

Fig. 11-1. Approximate southern boundary of the *Combretum-Terminalia* woodland and associated vegetation of the western woodlands in Ethiopia. Map of south-western Ethiopia with vegetation designated as in Fig. 4-1: Pale brown: *Combretum-Terminalia woodland* (CTB). Pale green: *Transitional semi-evergreen bushland* (TSEB). Pale purple: *Acacia-Commiphora bushland* (ACB). For legends to other vegetation types, see Fig. 4-1. The purple-blue line indicates the southern limit of areas with more than one fire in 10 years; according to that model, all areas to the north of this line should be indicated as *Combretum-Terminalia woodland* (CTB), *Transitional semi-evergreen bushland* (TSEB), or other vegetation types; all areas south of this line should be indicated as *Acacia-Commiphora bushland* (ACB), *Transitional semi-evergreen bushland* (TSEB), or highland vegetation types. The red line indicates the 1000 m contour, marking an alternative southern boundary of the *Combretum-Terminalia woodland* (CTB). The floristic regions (IL, KF, SU, GG, SD) are indicated.

south. However, in order to draw a more accurate limit more field data from west of the lower part of the Omo River is needed.

Finally, our studies suggest that the distribution of many CTW-species extends to the east of the CTW-area as indicated in this work, but there is no fixed limit to these extensions and according to our field observation the species do not grow together to make up a readily recognizable vegetation as in the western Ethiopian woodlands. Some species just reach the Rift Valley, others extend to or beyond the border with northern Somalia along the *Transitional Semi-Evergreen Bushland* (TSEB; Breugel et al. 2016b). Nowhere to the east of the Rift Valley the CTW-species are dominant in the vegetation. We will therefore suggest that the eastern limit of the CTW-area in southern Ethiopia is maintained.

11.2. How does the *Combretum-Terminalia* woodland in Ethiopia relate to the Sudanian woodlands between the western Ethiopian border and the Atlantic Ocean? Do the widely distributed species follow the same pattern across the continent? And how do the western Ethiopian woodlands relate to the Zambezian woodlands?

The relation between the western Ethiopian woodlands and the woodlands of the Sudanian region across Africa, with the corresponding data, is presented and discussed in Section 6, 7 and 9. In Section 3.12 we particularly discussed the results of a study by Fayolle et al. (2018). They compared the distribution of woodland and wooded grassland (savanna) species across Africa, based on flora lists for 298 samples all over the African continent. Based on this comparison, they suggested that there is a sharp discontinuity between the N&W savannas (comprising the Guinean and the Sudanian savannas) and the S&E savannas (Ethiopian and Ugandan savannas forming one group, the rest of the East and South African forming another; Fayolle et al. 2018: Fig. 1). In Section 3.12 we also discussed the findings of Marshall et al. (2020), in which similarity in flora between areas on the Horn of Africa, in particular in Ethiopia, and areas in western Africa is stressed.

In Section 3.2 we pointed out that all flora lists from Ethiopia used by Fayolle et al. were from the Ethiopian Highlands or from the eastern Somalia-Masai lowlands. Only one flora list from the Sudan represented a vegetation type near the vegetation we find in the western woodlands of Ethiopia. The distribution maps we have produced for Section 9 show that 7 out of 12 of the most characteristic trees of the western woodlands (mainly species of the genera *Combretum* and *Terminalia* and species selected from the list of widespread CTW-species in Table 8-1) are distributed from western Ethiopia to the Atlantic Ocean, and that for these species there is no evidence of a sharp discontinuity between western Ethiopia and the N&W savannas. The western woodlands of Ethiopia should therefore be floristically associated with the N&W savannas. Contrary to that, we are in full agreement with Fayolle et al. (2018) that there is a discontinuity between the N&W savannas and the Ethiopian Highlands (east of the western woodlands) and the eastern lowlands of Ethiopia, including the eastern slopes towards Somalia and north-eastern Kenya. As it appears from many distribution maps in Section 6 and as we have documented in Section 7, there is a transition, but no sharp discontinuity, between the woodland flora of the Ethiopian Highlands and the flora of the upper and intermediate slopes of the western escarpments with woodland, where many highland-species occur as marginal intruders (MI).

We therefore suggest that the sharp discontinuity between the N&W savannas and the S&E savannas claimed by Fayolle et al. (2018) is only clearly discernible in Ethiopia if samples from low altitudes along the border between Sudan and Ethiopia are compared with the sample from the Highlands and the eastern Ethiopian or Somali lowlands. The distribution maps of the Sudanian woodland species in Section 6 and 9 suggest an overlap or interdigitation between the highland flora and the flora of the western lowland of Ethiopia. We find that this is also a probable explanation why Marshall et al. (2020) have detected local and rather patchy similarities between areas inside the borders of Ethiopia and widely separated areas in the Sudanian Region to the west of Ethiopia.

In Section 9, we have documented how the *Combretum-Terminalia* woodland species of the N&W savannas, particularly species of *Combretum, Terminalia, Lonchocarpus* and *Pterocarpus,* occur in Ethiopia with a more or less notable overlap between the N&W savannas and the S&E savannas of Fayolle et al. (2018), and with deep penetration into the highlands along the deep river valleys. In the same Section, we have attempted to see how well the distribution models for these species will represent the observed distributions in Ethiopia as well as in the entire distribution area. We found that a number of the twelve modelled species have different latitudinal range in western Africa and in western Ethiopia, where some species are northern in Ethiopia, but with a broad latitudinal range in West Africa, particularly *Anogeissus leiocarpa*, *Terminalia macroptera* and *Pterocarpus lucens*, which are all mainly northern in Ethiopia, but with a wider latitudinal range in West Africa.

The floristic relation between the western woodlands of Ethiopia and the Zambezian woodlands (S&E savanna, according to Fayolle et al. 2018) is illustrated in Section 7, and more specifically documented in Fig. 7-8 and Fig. 7-9. The link between the Ethiopian Highlands, the eastern escarpment of Ethiopia and the lowlands of East Africa (Somalia, Kenya, Tanzania) is much stronger than the link to the N&W savannas.

11.3. How uniform is the *Combretum-Terminalia* woodland in Ethiopia? Does the vegetation and flora in the western lowlands and on the western Ethiopian escarpment represent one or several ecological or floristic units?

Data and discussion related to this question is presented and discussed in Section 7, 8 and 10. Throughout our studies for this work in Section 7-10 we found that the vegetation of the western woodlands is diverse on a small scale, but in spite of density of trees rather homogenous on a large scale, and that it generally looks uniform, at least when compared with other vegetation types in Ethiopia. This is in spite of our observation that both the physiognomy (as seen from illustrations in Section 5) and the floristic composition are variable on a local scale. Taken over larger areas, the species composition of the western woodlands is similar, or the turnover of the species is gradual when estimated from larger units than relevés.

The distribution of individual species can differ markedly, as appears from the distribution maps in Section 6. The distribution of some species is clearly limited to specific habitats. Other species are very widespread, but may occur with a frequency that varies from one geographical part of the CTW-vegetation to another. An example of a species that only occurs within a limited part of the western woodlands is the large palm *Borassus aethiopum*, which only occurs on relatively flat terrain from the GJ floristic region in the north to the IL floristic region in the south (Fig. 5-26, Fig. 5-34; map in Fig. 6-167). The changing local combination of species is demonstrated by the variation from one relevé to the next in Appendix 1. This is further established by the review of the species distributed on relevés in the introduction to Section 8, and by the observation that the majority of the species have been observed in less than 50% of all the relevés (Table 8-1).

The analyses in Section 10 show that the 151 relevés cluster in as many as 80 clusters. Our field observations attest to the same variability in both physiognomy and floristic composition. We consider this as sufficient evidence that the *Combretum-Terminalia* woodlands of western Ethiopia are not uniform on a local scale. It surprised us that in Profile A between Humera and Gondar we never observed precisely the same associations of species as those recorded by Pichi Sermolli (1940) along approximately the same route, and that our relevés were also very variable in species composition. In Section 2.1 we quoted the local variation in the floristic compositions of the woodlands of northern Nigeria as demonstrated by Kershaw (1968), while White (1983) rejected the recognition of vegetation types, even on a small scale.

Our observations from Ethiopia support the view of White; the floristic composition of the woodlands can indeed change within very small areas. However, the mutual coherence of these local associations of Sudanian trees is emphasized by a considerable floristic overlap in dominant species, such as appear from Section 10.2.3. In contrast to the floristic coherence on the highest regional level (the Sudanian region of White) and the somewhat flickering image of the plant associations on the local levels, one should also notice the limited variation in ecological adaptations to a range of characteristic Sudanian environmental parameters, which we have attempted to document in Section 8. As shown by the graphs that illustrate Section 8, overall patterns of ecological adaptation vary little from the lowest level of the woodlands around 500 m a.s.l. to the upper limits at approximately 1800 m a.s.l. Species with the same adaptations replace each other from the lowlands to ca. 1800 m and from south to north. Most trees are less than 10 m tall, have mostly simple leaves (apart from the legumes), most have Mesophyll leaves, deciduous and evergreen leaves occur fairly constantly, with a moderate overweight to deciduous leaves, most trees have a moderately strong tendency to adaptations against fires, and most trees have a rather low representation of adaptations to protect the trees against browsers.

So notwithstanding the apparent variation in physiognomy shown by the photographs illustrating Section 5 and the local variation in species associations, our findings, chiefly from Section 8 and 10, emphasize the overall uniformity of the western woodlands of Ethiopia only to be subdivided into two or three large floristically overlapping units: the northern *Anogeissus leiocarpa-Pterocarpus lucens-Acacia hecatophylla-Sterculia africana* woodland and the southern *Combretum collinum-Bridelia scleroneura-Terminalia schimperiana-Annona senegalensis* woodland, which we have mapped in and Fig. 10-6 together with a cluster of depauperate relevés from the upper reaches of the great rivers.

11.4. If the *Combretum-Terminalia* woodland is not uniform, is there a gradual transition from lowland to highland or from south to north? If the vegetation and flora represent distinct units, then what species or what environmental parameters can be used to define these?

The data and discussions of this question are mainly found in Section 7 and 10. Already in Section 11.3 we have in part answered these questions with the conclusion that there are many local associations in the western woodlands of Ethiopia. But as soon as we try to place these associations together in a hierarchical structure, we meet problems. In Section 10, we found three loosely defined clusters of relevés and species, changing along a gradient from south to north and from lowlands to highlands. We also found species that mainly occur in the north or the south. Typically northern are the two species of *Boswellia*, *B. papyrifera* and *B. pirottae*, that do not proceed much further south than Kurmuk, and *Combretum hartmannianum* that has a slightly more southern limit. Other northern species are *Adansonia digitata* and *Acacia amythethophylla*. Species of the central woodlands are *Psorospermum febrifugum*, *Dalbergia boehmii*, *Zanthoxylum gilletii*, *Commiphora pedunculata*, *Ozoroa pulcherrima*, *Crossopteryx febrifuga*, *Fadogia cienkowskii*. As an example of a southern species that does not occur north of Kurmuk is *Harrisonia abyssinica*. Other southern species are *Annona senegalensis*, *Faurea rochetiana*, *Protea madiensis*, *Grewia velutina*, *Ziziphus mauritiana*. However, there are no clearly defined and sharp environmental parameters that can be identified as delimiting more than individual species, not even the fire frequency.

From the above, we can probably answer the first part of the research question 10.4 with 'yes': there is gradual transition from lowland to highland and from south to north. On one hand, the vegetation consists of narrowly defined clusters or associations that represent local plant communities. On the other hand, there is a turnover between these clusters or associations that provides a gradient from lowland to highland and from north to south. The second part of the research question 10.4 should probably be answered with 'no', as we have observed no clusters we will define as completely distinct units, not even the green and the blue cluster in Section 10. In Section 10.2.3 we found only one species, *Anogeissus leiocarpa,* with an indicator value of ca. 0.8 or above, and the majority of the species in the relevés were not strongly linked to one of the two clusters. Regarding the final part of the question, what species or what environmental parameters can be used to define these, we have for the species decided to name the green, northern cluster the *Anogeissus leiocarpa-Pterocarpus lucens-Acacia hecatophylla-Sterculia africana* woodland and to name the blue, southern cluster the *Combretum collinum-Bridelia scleroneura-Terminalia schimperiana-Annona senegalensis* woodland. The indicator value for all other species than *Anogeissus leiocarpa* are around 0.6 or lower. Moreover, it is difficult in the field to refer a relevé from its species composition to one of the two clusters. Regarding environmental parameters, distinctions are not more clear-cut, but as seen in the discussion of the following question, we find environmental parameters (drivers) that may explain gradients without defining sharp ecological limits.

11.5. Can we point to important drivers with regard to the variation, continuous or discontinuous, along the increasing latitudes and/or altitude?

The opportunity to identify important drivers in the western Ethiopian woodlands is highly reliant on the parameters that were available for our studies. Here we have only used the directly observed and derived parameters listed in Appendix 3: geographical coordinates, altitude, slope, aspect, topographical wetness index, degree of burning and biomass available for burning, fire frequency, simplified soil types and a range of climatic parameters (Bio 1 to Bio 19).

Unfortunately this leaves out the influence of anthropogenic factors. In Section 2.8 and in Fig. 2-30, we tried to illustrate the human population density in the western woodlands and found that it was generally low, but apparently increasing and very variable from place to place. We have mentioned the location of large refugee camps established during the last 30 to 40 years in these areas, and we have occasionally mentioned the new infrastructure (roads, towns, bridges and recently also the large GERD Dam on the Abay River). All this points to a significant and growing influence of anthropogenic factors on the vegetation. This increase has lasted for at least a number of decades, perhaps even a century. However, due to lack of data, we have not been able to quantify the human influence in our analyses in Section 10.

Regarding our 151 relevés we studied by non-metric multidimensional scaling the importance of latitude, altitude, topographical wetness index, fire frequency and the climatic parameters temperature of warmest quarter of the year (bio10), annual precipitation (bio12), precipitation of the driest month (bio14) and precipitation seasonality (bio15) and the results are shown in Fig. 10-8 and Fig. 10-9. As pointed out in Section 10.2.5 the NMDS suggested that among the parameters studied, latitude could explain 68% of the variation, altitude 43%, and maximum fire frequency 40%, while among the climatic parameters various modifications of the precipitation were the most important factors. Precipitation seasonality explained 69% of the variation, precipitation of driest month 66%, and annual precipitation 65%, while the mean temperature of the warmest quarter explained 62% of the variation. Further studies of the importance of the climate variables, the topographical wetness index (TWI), altitude, fire frequencies and soil types were made via constrained ordination, more specifically a distance-based redundancy analysis, as shown in Fig. 10-10. The results show that the two large floristic units, the northern (green dots and areas in Fig. 10-5 and Fig. 10-6) and the southern (blue dots and areas in Fig. 10-5 and Fig. 10-6), roughly occupy the opposite ends of the gradients in temperature and rainfall during the dry and warm seasons, but overlap and are not easy to use for convenient climatic markers. The third smaller floristic unit in the north and in the northern and eastern valleys of the Ethiopian Highlands (orange dots and areas in Fig. 10-5 and Fig. 10-6) is characterized by low rainfall, high temperatures and high variability in temperature and rainfall. To sum up, the clusters are not well defined, with considerable overlap across species and environmental parameters.

11.6. Further conclusions

Two more conclusions seem relevant here, one general about the methodology for future studies of Ethiopian vegetation and vegetation in similar landscapes outside Ethiopia, and one specifically concerning the need for more conservation of the vegetation in the western woodlands.

Methodology: We have demonstrated that neither modelling of vegetation based entirely on environmental parameters (particularly Sayre et al. 2013), nor phytogeographical modelling based on the flora of previously defined areas large enough to be heterogenous (for example one degree squares in the examples discussed in Section 3.12) can predict the results we have found by a detailed study based on actual records from relevés in the field. Both these points are particularly valid for Ethiopia with very high topographic heterogeneity and points of low and high altitudes close together. In this work we have used statistically based adaptations of the methods by Dobremez (1976) in Nepal and Langdale-Brown et al. (1964), Trapnell (2001), White (1983) and many others in Eastern Africa. This has allowed us to employ objective, reproducible methods in Section 10 which largely agree with the conclusions reached by the more intuitive methods used in Section 5-9. Furthermore, we have tried to live up to the request of Chiarucci et al. (2010), claiming that conclusions about vegetation should be supported by detailed information about the distribution of the species that occur in the vegetation. We hope that some of the methods used here may be used by others, and may help the study of vegetation in parts of Africa of difficult access with extremely variable landscapes.

Conservation: Very little has been said in this work about conservation of the western woodlands and their species. It has been pointed out that the information about anthropogenic influence is limited, and in Section 2.9 we noticed that only few national parks provided protected areas for the western woodlands. In the north (the northern, green cluster of Section 10) there is the Kafta Shiraro National Park, and further south the Alatish National Park. The Dati Wolel National Park is in the southern, blue cluster of Section 10, but it is only poorly documented and without having seen it, except for the periphery, we are not sure how important it really is for plant conservation. The conservation areas in the Gambela Region seem to be those that best cover important areas of the western Ethiopian woodlands, but not the northern or the central areas.

To support identification of additional areas of importance for conservation we have only made a few estimations of species-richness, which seems to be highest in the area between Kurmuk, Gizen and the Abay River (see Section 4 and Fig. 4-5). This is an area on the border between our northern and southern cluster and observations by Mindaye Teshome et al. (2017) indicate that it is of importance for the conservation of naturally reproducing populations of *Boswellia papyrifera*. However, the area between Assosa and Kurmuk is also an area with high population density (see Fig. 2-30), many refugee camps and currently there is only one protected area, the Dabus Valley Controlled Hunting Area, which according to our field experience mainly covers wetlands. Slightly to the north of the Assosa-Kurmuk area, much of the area with relatively high species-richness (Fig. 4-4: 1/16 degree square) will be flooded by the reservoir behind the GERD Dam if the project develops as planned (see Section 2.4).

A detailed study based on data used in this work and along the lines of, for example Bachman et al. (2020) for the north-east African species of *Aloe*, would give more precise results, provided there is a high enough information density in this large area. However, the data assembled for this study is inadequate for a gap analysis.

With a detailed gap analysis not being possible, we hope that this work will at least increase interest in the western woodlands of Ethiopia and their conservation, and that, with its environmental, floristic and historical background information, it may provide a helpful framework or store of references for future work, not only inside Ethiopia, but also in the broad belt of Sudanian vegetation from the shores of the Atlantic Ocean to the Ethiopian Highlands. As pointed out by Friis & Sebsebe Demissew (2020) in their study of three species names of *Terminalia* (*Terminalia schimperiana*, *T. glaucescens*, and *T. avicennioides*) with an intermingled history, it is necessary to study the plants of the Sudanian woodlands in their distribution right across Africa, not only the Sudanian species, but also the Sudanian woodlands in the same west-eastern range.

Appendix 1: Field observations and species lists of observed woody species for each of the numbered relevés

This appendix contains the directly made field observations from each of the numbered relevés, including a designation of the geographical location (with the number of the relevé given to it in the field in brackets; this is the reference in our note books), the date of the observations, the directly observed coordinates and altitudes as recorded from our GPS used in the field, and a short description of the area with notes on plant density, dominant species, terrain, degree of cultivation, soil, ground-cover and signs of burning and grazing. These direct field observations of each relevé are followed by lists of the species of indigenous woody plants observed, either from our identifications in the field, from identifications finished at the National Herbarium of Ethiopia (ETH) or, if critical material was sent on loan, identifications made at the Herbarium of the Royal Botanic Gardens, Kew (K), in the United Kingdom. While the numbering of the relevés is mainly chronological (except for data from the literature), the sequence of the profiles from A to Q is arranged geographically from north to south and from west to east. The letters of the profiles (but not the branches of the profiles) are indicated in this list.

Profile A.

1 – (2014_01). On the Humera-Gondar road; after the road-forking Humera-Gondar-Shire, 9.5 km S of Humera. 20/10/2014
14° 12' 30" N, 36° 36' 54" E; 600 m a.s.l.
Open *Balanites aegyptiaca* woodland with undergrowth of small specimens of *Acacia* spp. (*Acacia mellifera*, *A. seyal*, *A. oerfota*); the ground now mostly cultivated with *Sesamum indicum*. Ground-cover, where there was no cultivation, consisted of short annual grasses (< 20 cm tall). No sign of burning. Vertisol (black cotton soil) over pale and porous basaltic rock, looking like limestone.

Acacia hecatophylla
Acacia mellifera
Acacia oerfota
Acacia seyal
Balanites aegyptiaca

2 – (2014_02). On the Humera-Gondar road; 10 km S of road-forking Humera-Gondar-Shire. 20/10/2014
14° 09' 00" N, 36° 41' 42" E. 630 m a.s.l.
Open *Balanites aegyptiaca* woodland with undergrowth of mostly sparse, but locally dense *Acacia* bushland, growing on sand drifts over vertisol (black cotton soil). Ground-cover of low (< 20 cm tall) annual herbs; hardly any grass. No sign of burning.

Acacia hecatophylla
Acacia mellifera
Acacia seyal
Anogeissus leiocarpa
Balanites aegyptiaca
Dalbergia melanoxylon
Dichrostachys cinerea
Maerua oblongifolia

3 – (2014_03). On the Humera-Gondar road; 20 km S of road-forking Humera-Gondar-Shire. 20/10/2014
14° 04' 30" N, 36° 44' 54" E; 675 m a.s.l.
Open *Balanites aegyptiaca* woodland, originally with undergrowth of locally dense *Acacia* bushland, but this has mostly been cleared for cultivation. The whole area with vertisol (black cotton soil). In a few places in between the large areas of intensive cultivation patches with ground-cover of tall grass (*Hyparrhenia*), but no sign of burning.

Acacia hecatophylla
Acacia mellifera
Acacia senegal
Acacia seyal
Balanites aegyptiaca
Boscia senegalensis
Combretum hartmannianum
Ziziphus spina-christi

4 – (2014_04). On the Humera-Gondar road; 30 km S of road-forking Humera-Gondar-Shire. 20/10/2014
14° 01' 12" N, 36° 48' 00" E; 750 m a.s.l.
Mixed open woodland on rocky ground (porous, probably basaltic rocks). Moderately dense ground-cover of annual and perennial herbs (including herbs with tuberous roots, e.g., *Chlorophytum*), no grass. No sign of burning.

Anogeissus leiocarpa
Boswellia papyrifera
Combretum molle
Dalbergia melanoxylon
Dichrostachys cinerea
Grewia flavescens
Ochna leucophloeos
Sclerocarya birrea
Sterculia africana
Strychnos innocua
Terminalia brownii
Ximenia americana

5 – (2014_05). On the Humera-Gondar road; 40 km S of road-forking Humera-Gondar-Shire. 20/10/2014
13° 57' 06" N, 36° 51' 36" E; 800 m a.s.l.
Open woodland on rocky ground; rocks basaltic. Hardly any cultivation; ground-cover of annual and low perennial grasses (< 20 cm tall). No sign of burning. [*Tamarindus indica* form riparian vegetation along small stream; this species not included as part of the relevé].

Anogeissus leiocarpa
Boswellia papyrifera
Combretum hartmannianum
Combretum rochetianum
Dalbergia melanoxylon
Dichrostachys cinerea
Grewia flavescens
Lannea fruticosa
Sclerocarya birrea
Sterculia africana
Ziziphus spina-christi

6 – (2014_06). On the Humera-Gondar road, 52 km S of road-forking Humera-Gondar-Shire (at turn to disused air strip). 21/10/2014
13° 50' 30" N, 36° 53' 12" E; 800 m a.s.l.
Partly cultivated woodland on vertisol (black cotton soil; according to local informant the area has been cultivated more or less intensively for about 10 years, mainly with *Sorghum*); uncultivated land with scattered annual grasses (some up to 2 m tall), but no sign of burning.

Acacia hecatophylla
Acacia seyal
Anogeissus leiocarpa
Balanites aegyptiaca
Combretum collinum
Combretum hartmannianum
Dichrostachys cinerea
Piliostigma thonningii
Ziziphus spina-christi

7 – (2014_07). On the Humera-Gondar road, 60 km S of road-forking Humera-Gondar-Shire. 21/10/2014
13° 45' 30" N, 36° 56' 18" E; 750 m a.s.l.
Open woodland on stony, red soil, presumably of volcanic origin. Ground-cover of short annual grasses (< 10 cm tall). No sign of burning. [*Tamarindus indica*, *Diospyros mespiliformis* and *Ficus sur* form scattered riparian vegetation along small stream, these species not included as part of the relevé].

Anogeissus leiocarpa

Combretum collinum
Combretum hartmannianum
Ficus glumosa
Ficus sycomorus
Sclerocarya birrea
Sterculia africana

8 – (2014_08). On the Humera-Gondar road; 70 km S of road-forking Humera-Gondar-Shire. 21/10/2014

13° 40' 24" N, 36° 57' 24" E; 800 m a.s.l.
Mixed, open woodland on a patchy mosaic of vertisol (black cotton soil) and rocky, grey to brown soil, presumably of volcanic origin. Ground-cover of annual herbs and grasses (< 20 cm tall). No sign of burning.

Acacia hecatophylla
Acacia seyal
Anogeissus leiocarpa
Combretum hartmannianum
Combretum molle
Dalbergia melanoxylon
Ficus glumosa
Ficus ingens
Flueggea virosa
Lannea fruticosa
Pterocarpus lucens
Sclerocarya birrea
Ziziphus abyssinica
Ziziphus spina-christi

9 – (2014_09). On the Humera-Gondar road, 85 km S of road-forking Humera-Gondar-Shire. 21/10/2014

13° 31' 42" N, 36° 58' 24" E; 810 m a.s.l.
Open woodland, rather heavily grazed, on brown clay soil and scattered dark grey stones of volcanic origin. Ground-cover of mainly perennial grasses (< 10 cm tall). No sign of burning.

Anogeissus leiocarpa
Combretum collinum
Dalbergia melanoxylon
Ficus glumosa
Grewia flavescens
Lannea fruticosa
Lonchocarpus laxiflorus
Maytenus senegalensis
Piliostigma thonningii
Pterocarpus lucens
Sterculia africana
Strychnos innocua
Terminalia laxiflora
Ziziphus spina-christi

10 – (2014_10). On the Humera-Gondar road; 95 km S of road-forking Humera-Gondar-Shire. 22/10/2014

13° 26' 12" N, 36° 58' 36" E; 850 m a.s.l.
Open woodland, dominated by *Anogeissus leiocarpa* and *Terminalia laxiflora,* on brown soil with numerous loose rocks and stones of volcanic origin. Ground cover of annual and perennial herbs (< 10 cm tall). No sign of burning. [*Diospyros mespiliformis* forms a very sparse riparian vegetation along small stream, this species not included in the relevé].

Anogeissus leiocarpa
Ficus glumosa
Ficus sycomorus
Flueggea virosa
Piliostigma thonningii
Pterocarpus lucens
Strychnos innocua
Terminalia laxiflora

11 – (2014_11): On the Humera-Gondar road; 105 km S of road-forking Humera-Gondar-Shire. 22/10/2014

13° 20' 36" N, 37° 02' 54" E; 875 m a.s.l.
Open woodland dominated by *Anogeissus leiocarpa*; *Combretum hartmannianum* also present, but not prominent, *Terminalia laxiflora* and *T. macroptera* seen, but not common. Vertisol (black cotton soil), rather damp and muddy, probably completely inundated during the rainy season. Ground-cover of *Aeschynomene americana* and *Hygrophila* sp.; hardly any grass. No sign of burning.

Acacia hecatophylla
Anogeissus leiocarpa
Combretum hartmannianum
Dichrostachys cinerea
Ficus sycomorus
Flueggea virosa
Gardenia ternifolia
Lannea schimperi
Pterocarpus lucens
Terminalia laxiflora
Terminalia macroptera

12 — (2014_12). On the Humera-Gondar road; 110 km S of road-forking Humera-Gondar-Shire. 22/10/2014

13° 18' 06" N, 37° 05' 12" E; 925 m a.s.l.
Open woodland, heavily grazed and without dominant woody species, on vertisol (black cotton soil), with rather extensive patches of cultivated land. Ground-cover of annual grasses (< 20 cm tall) and *Hygrophila*. No sign of burning.

Acacia hecatophylla
Acacia seyal
Albizia malacophylla
Anogeissus leiocarpa
Combretum collinum
Ficus glumosa
Ficus sycomorus
Piliostigma thonningii
Pterocarpus lucens
Terminalia laxiflora
Ziziphus spina-christi

13 — (2014_13). On the Humera-Gondar road; 120 km S of road-forking Humera-Gondar-Shire; near bridge on the Angereb River at Asheri village. 22/10/2014

13° 14' 12" N, 37° 09' 12" E; 850 m.
Dry and open woodland, sometimes with extensive bare patches, on sometimes steeply sloping ground; tree-cover of mixed floristic composition without dominant woody species. On brown soil, sometimes rocky, presumably of volcanic origin. Ground-cover scarce, with scattered annual and perennial grasses and other herbs (< 10 cm tall). No sign of recent burning.

Anogeissus leiocarpa
Boswellia papyrifera
Boswellia pirottae
Calotropis procera
Combretum hartmannianum
Dalbergia melanoxylon
Ficus glumosa
Flueggea virosa
Grewia flavescens
Lannea fruticosa
Lonchocarpus laxiflorus
Maytenus senegalensis
Pterocarpus lucens
Sterculia africana
Ziziphus spina-christi

14 — (2014_14). On the Humera-Gondar road; 130 km S of road-forking Humera-Gondar-Shire; ca. 12 km S of Asheri village. 30/10/2014

13° 11' 30" N, 37° 13' 24" E; 875 m a.s.l.
Mixed, rather dense woodland on deep vertisol (black cotton soil). No cultivation. Ground-cover of annual grasses (< 50 cm tall). No sign of burning.

Acacia hecatophylla
Acacia seyal
Anogeissus leiocarpa
Balanites aegyptiaca
Combretum collinum
Combretum hartmannianum
Combretum molle
Dalbergia melanoxylon
Dichrostachys cinerea
Gardenia ternifolia
Lonchocarpus laxiflorus
Piliostigma thonningii
Pterocarpus lucens
Terminalia laxiflora
Ziziphus spina-christi

15 — (2014_15). On the Humera-Gondar road; 140 km S of road-forking Humera-Gondar-Shire. 30/10/2014

13° 06' 30" N, 37° 15' 36" E; 950 m a.s.l.
Very open and heavily grazed woodland on steep slope with very stony soil (volcanic rocks and gravel); no cultivation. Ground-cover of mostly annual grasses, mostly dried out or eaten by grazing animals (< 30 cm tall). No sign of burning.

Anogeissus leiocarpa
Boswellia papyrifera
Combretum collinum
Lonchocarpus laxiflorus
Oxytenanthera abyssinica
Sterculia africana

16 — (2014_16). On the Humera-Gondar road; 150 km S of road-forking Humera-Gondar-Shire. 31/10/2014

13° 02' 00" N, 37° 17' 42" E; 1000 m a.s.l.
Moderately sloping, open woodland on very stony (stones of volcanic origin) and gravelly, grey soil. Ground-cover sparse, but heavily grazed (no cultivation), consisting mainly of annual grasses (*Aristida*, etc.; < 50 cm tall). [*Tamarindus indica* form a sparse riparian vegetation along small stream, the species not included in the relevé].

Acacia hecatophylla
Acacia hockii
Anogeissus leiocarpa
Dichrostachys cinerea
Ficus glumosa
Flueggea virosa
Lannea fruticosa
Sterculia africana
Ziziphus abyssinica
Ziziphus spina-christi

17 — (2014_17). On the Humera-Gondar road; 162 km S of road-forking Humera-Gondar-Shire. 31/10/2014

12° 56' 48" N, 37° 19' 42" E; 1100 m a.s.l.
Very flat and open area with widely scattered trees. Ground-cover consisting of almost pure *Pennisetum* grassland on vertisol (black cotton soil), presumably fallow grassland, with grasses up to ca. 75 cm high, following past cultivation that may have involved cutting trees. (In areas nearby similar *Pennisetum* grassland was dominated by more dense stands of *Acacia hecatophylla*). No sign of burning.

Acacia hecatophylla
Anogeissus leiocarpa
Terminalia laxiflora
Ziziphus spina-christi

18 — (2014_18). On the Humera-Gondar road; 170 km S of road-forking Humera-Gondar-Shire. 31/10/2014

12° 52' 18" N, 37° 21' 42" E; 1200 m a.s.l.
Flat, fallow grassland with a few species of scattered trees on vertisol (black cotton soil), degraded by cultivation and grazing and invaded by *Calotropis procera*, which normally is only found along roadsides and in other highly disturbed areas that are markedly different from the more natural vegetation. Ground-cover with patches of *Hygrophila*, *Aeschynomene americana* and perennial grasses (*Bothriochloa insculpta*), which seems to indicate temporary flooding during rainy season. [*Diospyros mespiliformis* form scattered riparian vegetation along small stream, species not included in the relevé]. No signs of burning.

Anogeissus leiocarpa
Acacia hecatophylla
Calotropis procera

19 — (2014_18*). Near Tucur Dinghia; woodland at edge of forest patch around the Beit Michael church at the foot of the otherwise very bare Kambanta Mountain. 31/10/2014

12° 49' 24" N, 37° 22' 06" E; 1400 m a.s.l.
Transition between protected and rather dense "church forest" and dense woodland on brown soil with rocks. Very sparse ground-cover due to shade. No sign of burning.

Acacia polyacantha subsp. *campylacantha*
Anogeissus leiocarpa
Bridelia scleroneura
Carissa spinarum
Combretum molle
Cordia africana
Croton macrostachyus
Diospyros mespiliformis
Ficus sycomorus
Flacourtia indica
Maytenus senegalensis
Oncoba spinosa
Schrebera alata
Terminalia schimperiana
Vitex doniana

20 — (2014_19). On the Humera-Gondar road; 180 km S of road-forking Humera-Gondar-Shire. 31/10/2014
12° 47' 48" N, 37° 23' 06" E; 1550 m a.s.l.
Transition between *Combretum-Terminalia* woodland and degraded Afromontane semi-evergreen bushland with trees. Small parch of disturbed scrub in rather intensively cultivated farmland on dark brown and stony soil, which is similar to the common highland soil. [*Syzygium guineense* subsp. *guineense* form scattered riparian vegetation along small stream, species not included in the relevé. Invasion of weedy species, *Caesalpinia decapetala*, *Ricinus communis* and others, which are not included in the sample in Section 6]. No signs of burning.

Anogeissus leiocarpa
Cordia africana
Croton macrostachyus
Diospyros mespiliformis
Ficus sycomorus
Maytenus arbutifolia
Senna singueana
Vernonia amygdalina
Ziziphus spina-christi

21 — (2014_20). On the Humera-Gondar road; 190 km S of road-forking Humera-Gondar-Shire. 31/10/2014
12° 47' 48" N, 37° 23' 06" E; 2050 m a.s.l.
Degraded Afromontane semi-evergreen bushland with scattered trees, of which most are not typical of *Combretum-Terminalia* woodland or associated *Acacia* woodland or wooded grassland. Extensive, heavily grazed areas. [*Syzygium guineense* subsp. *guineense* form riparian vegetation along small stream, species not included in the relevé]. No signs of burning.

Croton macrostachyus
Maesa lanceolata
Calpurnia aurea
Clutia abyssinica
Ficus thonningii
Maytenus senegalensis
Vernonia amygdalina

22 — (2014_21). On the Humera-Gondar road; near the pass of the road (watershed). 31/10/2014
12° 42' 48" N, 37° 26' 54" E; 2450 m a.s.l.
Degraded Afromontane semi-evergreen bushland with scattered trees, of which most are not typical of *Combretum-Terminalia* woodland or associated *Acacia* woodland or wooded grassland. Extensive, heavily grazed areas. [*Syzygium guineense* subsp. *guineense* form riparian vegetation along small stream, species not included in the relevé]. No sign of burning.

Apodytes dimidiata
Bersama abyssinica
Maesa lanceolata
Maytenus arbutifolia
Olea europaea subsp. *cuspidata*

23 — (2014_B1). On the Humera-Gondar road; side-road towards Shire branching off near Dancha, 20 km E of Dancha. 22/10/2014
13° 31' 24" N, 37° 11' 06" E; 900 m a.s.l.
Open woodland without dominant woody species on brown, stony soil (rocks, stones and gravel of volcanic origin). Little grazing and cultivation. Ground-cover

of mainly annual grasses (< 30 cm tall). No sign of burning.

Acacia polyacantha subsp. *campylacantha*
Anogeissus leiocarpa
Combretum hartmannianum
Lannea fruticosa
Lonchocarpus laxiflorus
Piliostigma thonningii
Pterocarpus lucens
Sterculia africana
Stereospermum kunthianum
Terminalia schimperiana
Ziziphus spina-christi

24 – (2014_B2). On the Humera-Gondar road; side-road towards Shire branching off near Dancha, 40 km E of Dancha. 22/10/2014
13° 30' 18" N, 37° 19' 30" E; 1150 m a.s.l.
Mixed, moderately dense woodland without dominant woody species on brown, stony soil (rocks, stones and gravel of volcanic origin). Ground-cover of mainly annual grasses (< 20 cm tall). No cultivation, no sign of grazing. No sign of burning.

Acacia polyacantha subsp. *campylacantha*
Anogeissus leiocarpa
Piliostigma thonningii
Pterocarpus lucens
Sterculia africana
Terminalia schimperiana
Ziziphus spina-christi

25 – (2014_B3). On the Humera-Gondar road; side-road towards Shire branching off near Dancha, 60 km E of Dancha. 22/10/2014
13° 34' 36" N, 37° 22' 00" E; 1300 m a.s.l.
Mixed, moderately dense woodland without dominant woody species on brown, stony soil (rocks, stones and gravel of volcanic origin). Ground-cover mainly of annual grasses (< 20 cm tall). No cultivation, no sign of grazing. No sign of burning.

Acacia polyacantha subsp. *campylacantha*
Anogeissus leiocarpa
Entada abyssinica
Erythrina abyssinica
Ficus glumosa
Lannea schimperi
Maytenus senegalensis
Pterocarpus lucens
Stereospermum kunthianum
Terminalia schimperiana
Vangueria madagascariensis

26 – (2014_A01). On the Humera-Shire (Enda Selassie) road; 20 km E of road-forking Humera-Gondar-Shire. 23/10/2014
14° 10' 42" N, 36° 48' 42" E; 650 m a.s.l.
Almost undisturbed *Balanites aegyptiaca* woodland, dominated by that species, with undergrowth of *Acacia* scrub. Black cotton soil, almost devoid of ground-cover under the scrub. No cultivation or sign of grazing. No sign of burning.

Acacia senegal
Acacia seyal
Adansonia digitata
Anogeissus leiocarpa
Balanites aegyptiaca
Calotropis procera
Dalbergia melanoxylon
Grewia tenax
Grewia villosa
Lannea fruticosa
Ziziphus spina-christi

27 – (2014_A02). On the Humera-Shire (Enda Selassie) road; 40 km E of road-forking Humera-Gondar-Shire. 23/10/2014
14° 04' 36" N, 36° 59' 36" E; 875 m a.s.l.
Mixed open woodland without dominant woody species on level terrain. Soil brown, covered with small brown lava blocks, 20-30 cm in diam. Ground-cover sparse with few annual grasses (< 20 cm tall) and annual perennial herbs (*Chamaecrista* common). No cultivation and no sign of grazing. No sign of burning.

Acacia mellifera
Acacia seyal
Anogeissus leiocarpa
Boswellia papyrifera
Combretum hartmannianum
Combretum molle
Dalbergia melanoxylon
Dichrostachys cinerea
Grewia flavescens
Ochna leucophloeos
Sclerocarya birrea
Sterculia africana
Terminalia brownii

28 – (2014_A03). On the Humera-Shire (Enda Selassie) road, 60 km E of road-forking Humera-Gondar-Shire. 23/10/2014

14° 03' 30" N, 37° 11' 06" E; 900 m a.s.l.
Very open mixed woodland without dominant woody species on flat, rocky ground with very thin, brown soil over volcanic rocks. Hardly any ground-cover, only consisting of a few scattered annuals. No cultivation or sign of grazing. No sign of burning.

Acacia polyacantha subsp. *campylacantha*
Acacia senegal
Adansonia digitata
Anogeissus leiocarpa
Boswellia papyrifera
Capparis decidua
Combretum hartmannianum
Combretum molle
Dalbergia melanoxylon
Dichrostachys cinerea
Grewia flavescens
Lannea fruticosa
Pterocarpus lucens
Sclerocarya birrea
Sterculia africana
Terminalia brownii
Ziziphus spina-christi

29 – (2014_A04). On the Humera-Shire (Enda Selassie) road; 85 km E of road-forking Humera-Gondar-Shire. 23/10/2014

14° 09' 30" N, 37° 21' 24" E; 1100 m a.s.l.
Open and almost pure *Boswellia papyrifera* woodland on rocky ground with very little soil, only covering the bare volcanic rocks in patches. Hardly any ground-cover. No cultivation (apart from tapping of frankincense from the *Boswellia* trees), no sign of grazing. It seems likely that human activity has amplified the dominance of *Boswellia papyrifera* in order to increase the output of frankincense. No burning. [*Diospyros mespiliformis* formed a sparse riparian vegetation along small stream; species not included in the].

Acacia seyal
Anogeissus leiocarpa
Boswellia papyrifera
Balanites aegyptiaca
Combretum collinum
Dalbergia melanoxylon
Ficus glumosa
Ozoroa insignis
Ziziphus spina-christi

30 – (2014_A05). On the Humera-Shire (Enda Selassie) road; 100 km E of road-forking Humera-Gondar-Shire. 24/10/2014

14° 05' 42" N, 37° 27' 30" E; 950 m a.s.l.
Open, mixed woodland without dominant woody species on mostly rocky ground with occasional patches of stony, dark brown soil. Ground-cover heavily grazed; what is left less than 10 cm high. No cultivation and no sign of burning.

Adansonia digitata
Anogeissus leiocarpa
Boswellia papyrifera
Combretum hartmannianum
Combretum molle
Dalbergia melanoxylon
Dichrostachys cinerea
Grewia bicolor

Terminalia brownii
Terminalia schimperiana

31 — (2014_A06). On the Humera-Shire (Enda Selassie) road; 120 km E of road-forking Humera-Gondar-Shire. 24/10/2014
14° 12' 06" N, 37° 34' 18" E; 900 m a.s.l.
Very open, degraded woodland on rocky ground, strongly dominated by *Anogeissus leiocarpa*. Soil gravelly, grey, derived from basement rocks (granites?). Ground-cover very dry and thin (< 20 cm tall). No cultivation or sign of grazing. No sign of burning.

Acacia mellifera
Anogeissus leiocarpa
Balanites aegyptiaca
Boscia angustifolia
Boswellia papyrifera
Dalbergia melanoxylon
Dichrostachys cinerea
Grewia bicolor
Lannea fruticosa
Ziziphus spina-christi

32 — (2014_A07). On the Humera-Shire (Enda Selassie) road; 140 km E of road-forking Humera-Gondar-Shire. 24/10/2014
14° 17' 12" N, 37° 42' 30" E; 1050 m a.s.l.
Very open, degraded wooded grassland on rocky ground, mostly dominated by *Balanites aegyptiaca and Acacia seyal,* in places dominated by *Anogeissus leiocarpa*. Soil gravelly, grey, derived from basement rocks (granites?). Ground-cover very dry and sparse (< 20 cm tall). No cultivation or sign of grazing. No sign of burning. [*Diospyros mespiliformis, Tamarindus indica* and *Vangueria madagascariensis* for a sparse riparian vegetation along small stream, which was dry at time of observation; species not included in the relevé].

Acacia seyal
Albizia amara
Anogeissus leiocarpa
Balanites aegyptiaca
Ziziphus spina-christi

33 — (2014_A08). On the Humera-Shire (Enda Selassie) road; 160 km E of road-forking Humera-Gondar-Shire. 24/10/2014
14° 24' 24" N, 37° 50' 54" E; 1050 m a.s.l.
Open *Balanites aegyptiaca* woodland with undergrowth of shrub dominated by *Acacia senegal* and *Acacia mellifera* (could be called "wooded bushland"). Soil a fine-grained brown clay verging towards vertisol (black cotton soil) in areas which are probably temporarily waterlogged during the rainy season. Ground-cover of annual grasses (*Chloris*, *Aristida*) (< 20 cm tall). No cultivation or sign of grazing. No sign of burning.

Acacia hecatophylla
Acacia mellifera
Acacia senegal
Acacia seyal
Balanites aegyptiaca
Boscia angustifolia
Dichrostachys cinerea

34 — (2014_A09). On the Humera-Shire (Enda Selassie) road; 178 km E of road-forking Humera-Gondar-Shire. 24/10/2014
14° 27' 18" N, 37° 58' 36" E; 1450 m a.s.l.
Dry and degraded open woodland dominated by *Balanites aegyptiaca* and *Acacia seyal*; *Calotropis procera* common in the most disturbed and open places. Soil a fine-grained grey soil, probably derived from schists. Ground-cover of short (< 20 cm tall) perennial grasses. Some terracing had been made to allow cultivation, but nothing seen to be cultivated at time of our observations; no sign of grazing. No sign of burning.

Acacia hecatophylla
Acacia mellifera
Acacia seyal
Anogeissus leiocarpa
Balanites aegyptiaca
Boscia angustifolia
Calotropis procera
Dichrostachys cinerea
Flueggea virosa
Ziziphus spina-christi

35 — (2014_A10). On the Humera-Shire (Enda Selassie) road; 200 km E of road-forking Humera-Gondar-Shire. 24/10/2014

14° 24' 54" N, 38° 10' 36" E; 1700 m a.s.l.
Rather intensively cultivated area with a few patches of remaining natural vegetation on slopes. Soil grey. Invasion of *Otostegia* and *Dodonaea angustifolia*, mainly in old cultivations, but also in natural vegetation. No or very low and scattered ground-cover. No sign of grazing. No sign of burning.

Acacia hecatophylla
Acacia seyal
Albizia amara
Balanites aegyptiaca
Cordia africana
Croton macrostachyus
Dodonaea angustifolia
Euphorbia abyssinica
Otostegia fruticosa
Ziziphus spina-christi

Profile B.

36 — (2015_01). On the Metema-Aykel road; 3 km from Metema, 28 km from Shehedi (Genda Wuha). 24/2/2015

12° 56' 00" N, 36° 10' 42" E; 750 m a.s.l.
Mixed open woodland without dominant woody species on undulating terrain. On hills very fine-grained sand and brown soil, in the flat terrain vertisol (black cotton soil), typically cracking when dry. Ground-cover of short perennial grasses (*Hyparrhenia*, < 1 m tall) or natural vegetation replaced by *Sesamum* fields; on the hills and slopes short annual grasses (< 20 cm tall). No sign of grazing. Burnt patches in the natural ground-cover on the vertisol.

Acacia hecatophylla
Acacia hockii
Acacia seyal
Anogeissus leiocarpa
Balanites aegyptiaca
Combretum collinum
Combretum hartmannianum
Dichrostachys cinerea
Entada africana
Ficus glumosa
Lannea fruticosa
Lannea schimperi
Ochna leucophloeos
Piliostigma thonningii
Pterocarpus lucens
Sterculia africana
Stereospermum kunthianum
Ximenia americana
Ziziphus abyssinica
Ziziphus spina-christi

37 — (2015_02). On the Metema-Aykel road; 13 km from Metema, 18 km from Shehedi (Genda Wuha). 24/2/2015

12° 51' 42" N, 36° 16' 30" E; 750 m a.s.l.
Open *Balanites aegyptiaca* woodland on level terrain with vertisol (black cotton soil), large parts of the natural ground-cover converted to cultivated fields, mainly with *Sorghum*. Fallow land being invaded by low, herbaceous dicots, mostly annuals (*Alyssicarpus, Hyptis*, etc.). No sign of grazing. No signs of burning. [*Diospyros mespiliformis* forming sparse riparian vegetation along small stream; species not included in the relevé].

Acacia hecatophylla
Anogeissus leiocarpa
Balanites aegyptiaca
Combretum hartmannianum
Ficus glumosa
Ficus sycomorus
Piliostigma thonningii
Stereospermum kunthianum
Terminalia laxiflora
Ziziphus abyssinica
Ziziphus spina-christi

38 — (2015_03). On the Metema-Aykel road; 22 km from Metema, 9 km from Shehedi (Genda Wuha). 24/2/2015

12° 48' 54" N, 36° 20' 06" E; 750 m a.s.l.
Open *Anogeissus leiocarpa-Balanites aegyptiaca-Pterocarpus lucens* woodland on dark brown, often cracking, but also stony soil with many patches of old cultivations. Although the soil is dark brown, it seems as fertile and as least as suited for cultivation as vertisol (black cotton soil). Ground-cover of old and dry perennial grasses (ca. 25 cm); the fallow farmland invaded by annual weeds. No sign of grazing. No sign of burning. [*Diospyros mespiliformis* and *Tamarindus indica* formed sparse riparian vegetation along small stream; species not included in the relevé].

Acacia hecatophylla
Acacia seyal
Anogeissus leiocarpa
Balanites aegyptiaca
Ficus glumosa
Ficus sycomorus
Flueggea virosa
Gardenia ternifolia
Pterocarpus lucens
Ziziphus abyssinica
Ziziphus spina-christi

39 — (2015_04). On the Metema-Aykel road; 3 km E of Shehedi (Genda Wuha) towards Aykel. 25/2/2015.

12° 45' 18" N, 36° 25' 42" E; 775 m a.s.l.
Woodland without dominant woody species on hilly ground. Hillsides with brown, stony soil, verging to vertisol (black cotton soil) at the base of the hill-slopes along a dry stream, where it was often cracking; on the slopes stony soil with many patches of old cultivations (soil seems as fertile as black cotton soil). Ground-cover of old and dry perennial grasses (ca. 25 cm) and old farmland invaded by annual weeds. No sign of grazing. No sign of burning. [*Diospyros mespiliformis*, *Tamarindus indica* and *Acacia nilotica* formed a sparse riparian vegetation along small stream; species not included in the relevé]. Ground-cover of dry, perennial grasses (< 20 cm tall). Areas with vertisol partly cultivated. No sign of grazing. No sign of burning.

Acacia hecatophylla
Acacia seyal
Acacia sieberiana
Anogeissus leiocarpa
Balanites aegyptiaca
Ficus ingens
Ficus sycomorus
Pterocarpus lucens
Sterculia africana
Stereospermum kunthianum
Strychnos innocua
Terminalia laxiflora
Ziziphus abyssinica
Ziziphus spina-christi

40 — (2015_05). On the Metema-Aykel road; 19 km E of Shehedi (Genda Wuha) towards Aykel. 25/2/2015

12° 41' 48" N, 36° 30' 06" E; 800 m a.s.l.
Woodland dominated by *Acacia hecatophylla* on pale, dark greyish fine-grained soil, drying and cracking as typical vertisol (black cotton soil), but paler. Very scarce natural ground-cover (< 20 cm tall) in patches under the dense stands of *Acacia*. Much cultivation. No signs of grazing. No signs of burning.

Acacia hecatophylla
Acacia seyal
Acacia sieberiana
Balanites aegyptiaca
Dombeya quinqueseta
Ficus sycomorus
Gardenia ternifolia
Piliostigma thonningii
Pseudocedrela kotschyi
Terminalia laxiflora
Ziziphus abyssinica
Ziziphus spina-christi

41 — (2015_06). On the Metema-Aykel road; 30 km E of Shehedi (Genda Wuha) towards Aykel. 25/2/2015

12° 38' 30" N, 36° 34' 36" E; 875 m a.s.l.

Open, mixed, but degraded woodland on slightly sloping stony, brown soil. Patches of fields with *Sorghum* cultivation, elsewhere natural ground-cover (< 80 cm tall). No sign of grazing. Here and there fire-scars in the natural ground-cover. [*Diospyros mespiliformis* forms sparse riparian vegetation along a small stream, not included].

Acacia hecatophylla
Acacia seyal
Anogeissus leiocarpa
Combretum collinum
Combretum rochetianum
Ficus sycomorus
Ficus vasta
Hymenodictyon floribundum
Ozoroa insignis
Pterocarpus lucens
Terminalia laxiflora
Terminalia macroptera
Ziziphus spina-christi

42 — (2015_07). On the Metema-Aykel road; 45 km E of Shehedi (Genda Wuha) towards Aykel. 25/2/2015

12° 37' 12" N, 36° 40' 48" E; 975 m a.s.l.

Open woodland dominated by *Pterocarpus lucens* on sloping terrain. Stony, brown soil, partly with *Sorghum* cultivations, partly with natural ground-cover of grasses (ca 0.8 m). No sign of grazing. Some sizeable fire-scars seen, particularly on the steeper slopes. [*Diospyros mespiliformis* forms sparse riparian vegetation along a small, dry stream; species not included in the relevé].

Acacia hecatophylla
Anogeissus leiocarpa
Combretum molle
Ficus glumosa
Ficus ingens
Gardenia ternifolia
Piliostigma thonningii
Pterocarpus lucens
Sterculia africana
Ziziphus spina-christi

43 — (2015_08). On the Metema-Aykel road; 57 km E of Shehedi (Genda Wuha) towards Aykel. 25/2/2015

12° 34' 48" N, 36° 44' 48" E; 1000 m a.s.l.

Relatively dense mixed woodland without dominant woody species on level, rocky terrain. Stony brown soil between the rocks. Hardly any ground-cover due to shade from the trees. No cultivation and no sign of grazing. No sign of burning. [*Carissa spinarum, Diospyros mespiliformis, Syzygium guineense* subsp. *guineense* and *Tamarindus indica* form sparse riparian vegetation along small stream; species not recorded in the relevé].

Acacia hecatophylla
Anogeissus leiocarpa
Piliostigma thonningii
Pterocarpus lucens
Stereospermum kunthianum
Terminalia schimperiana
Ziziphus abyssinica

44 — (2015_09). On the Metema-Aykel road; 64 km E of Shehedi (Genda Wuha) towards Aykel. 25/2/2015.

12° 35' 00" N, 36° 49' 18" E; 1200 m a.s.l.

Open, mixed woodland without dominant woody species on level terrain with much exposed rock. Rocky and stony brown soil between the rocks. Nearly all the area with soil had been cultivated with fields of *Sorghum*. No sign of grazing, No sign of burning. [*Tamarindus indica* growing on old termite mound; species not recorded in the relevé].

Acacia hecatophylla
Anogeissus leiocarpa
Dichrostachys cinerea
Ficus sycomorus
Flueggea virosa

Pterocarpus lucens
Strychnos innocua
Terminalia schimperiana
Ziziphus abyssinica
Ziziphus spina-christi

45a – (2015_A1). On the road from Shehedi (Genda Wuha) to Shahura (from Metema-Aykel road), 120 km S of Shehedi. 26/2/2015.
12° 05' 42" N, 35° 56' 30" E; 750 m a.s.l.
Open, mixed woodland without dominant woody species on mosaic of hills with rocky ground and stony soil (45a), where the ground-cover consists of perennial grasses (< 50 cm tall) and level ground with vertisol (black cotton soil) (45b) with slightly different vegetation, with perennial grasses (ca. 80 cm tall). No cultivation. No sign of grazing. Signs of burning only on the hilly ground.

Adansonia digitata
Anogeissus leiocarpa
Boswellia papyrifera
Entada africana
Lannea schimperi
Pterocarpus lucens
Sterculia africana
Terminalia schimperiana

45b – (2015_A1). On the road from Shehedi (Genda Wuha) to Shahura (from Metema-Aykel road), 120 km S of Shehedi. 26/2/2015.
12° 05' 42" N, 35° 56' 30" E; 750 m a.s.l.
Open, mixed woodland without dominant woody species on mosaic of hills with rocky ground and stony soil (45a), where the ground-cover consists of perennial grasses (< 50 cm tall) and level ground with vertisol (black cotton soil) (45b) with slightly different vegetation, with perennial grasses (ca. 80 cm). No cultivation. No sign of grazing. Signs of burning only on the hilly ground.

Acacia hecatophylla
Acacia nilotica
Acacia seyal
Ficus sycomorus
Ziziphus spina-christi

46a – (2015_A2). On the road from Shehedi (Genda Wuha) to Shahura (from Metema-Aykel road), 100 km S of Shehedi. 26/2/2015
12° 14' 48" N, 35° 52' 24" E; 675 m a.s.l.
Open, mixed woodland without dominant woody species on undulating terrain with brown, stony and rocky soil (46a) and small areas with vertisol (black cotton soil) (46b). Ground-cover of perennial grasses in both type ca. 80 cm. No cultivation. No sign of grazing. A few scars of burning. [*Diospyros mespiliformis* formed sparse riparian vegetation along small, dry stream; species not recorded in the relevé].

Adansonia digitata
Anogeissus leiocarpa
Combretum collinum
Combretum molle
Dalbergia melanoxylon
Ficus glumosa
Grewia flavescens
Lannea schimperi
Pterocarpus lucens
Sterculia africana
Stereospermum kunthianum
Strychnos innocua
Terminalia schimperiana

46b – (2015_A2). On the road from Shehedi (Genda Wuha) to Shahura (from Metema-Aykel road), 100 km S of Shehedi. 26/2/2015
12° 14' 48" N, 35° 52' 24" E; 675 m a.s.l.
Open, mixed woodland without dominant woody species on undulating terrain with brown, stony and rocky soil (46a) and small areas with vertisol (black cotton soil) (46b). Ground-cover of perennial grasses in both type ca. 80 cm tall. No cultivation. No sign of grazing. A few scars of burning. [*Diospyros mespiliformis* formed sparse riparian vegetation along small, dry stream; species not recorded in the relevé].

Acacia hecatophylla
Acacia seyal

47a – (2015_A3). On the road from Shehedi (Genda Wuha) to Shahura (from Metema-Aykel road), 80 km S of Shehedi. 26/2/2015

12° 24' 12" N, 35° 56' 54" E; 650 m a.s.l.
Open woodland on relatively flat terrain with mosaic of brown, stony and rocky soil (47a), dominated by *Anogeissus leiocarpa*, *Sterculia africana* and *Boswellia papyrifera* (the two latter not in the relevé, but nearby) and smaller areas with vertisol (black cotton soil) (see 47b), dominated by *Balanites aegyptiaca* and *Acacia hecatophylla*. Ground-cover grazed, and therefore low (< 20 cm tall) and with no signs of burning. No cultivation. No sign of grazing. Areas near this site ground-cover of tall perennial grasses (> 120 cm tall) and sign of burning.

Acacia seyal
Anogeissus leiocarpa
Ficus sycomorus
Gardenia ternifolia
Pterocarpus lucens
Ziziphus spina-christi

47b – (2015_A3). On the road from Shehedi (Genda Wuha) to Shahura (from Metema-Aykel road), 80 km S of Shehedi. 26/2/2015

12° 24' 12" N, 35° 56' 54" E; 650 m a.s.l.
Open woodland on relatively flat terrain with mosaic of brown, stony and rocky soil (47a), dominated by *Anogeissus leiocarpa*, *Sterculia africana* and *Boswellia papyrifera*, and smaller areas with vertisol (black cotton soil) (47b), dominated by *Balanites aegyptiaca* and *Acacia hecatophylla*. Ground-cover grazed, and therefore low (< 20 cm tall) and with no signs of burning. No cultivation. No sign of grazing. Areas near this site ground-cover of tall perennial grasses (> 120 cm tall) and sign of burning.

Acacia hecatophylla
Acacia seyal
Balanites aegyptiaca
Ficus sycomorus
Gardenia ternifolia
Ziziphus spina-christi

48 – (2015_A4). On the road from Shehedi (Genda Wuha) to Shahura (from Metema-Aykel road), 60 km S of Shehedi. 26/2/2015

12° 33' 12" N, 35° 59' 24" E; 650 m a.s.l.
Open, mixed woodland without dominant woody species on slightly undulating terrain with brown, stony and rocky soil. No cultivation. No sing of grazing. Ground-cover of perennial grasses (< 100 cm tall). A few scars from burning.

Acacia hecatophylla
Acacia seyal
Anogeissus leiocarpa
Balanites aegyptiaca
Combretum rochetianum
Dichrostachys cinerea
Entada africana
Lannea fruticosa
Pterocarpus lucens
Terminalia laxiflora

49 – (2015_A5). On the road from Shehedi (Genda Wuha) to Shahura (from Metema-Aykel road), 40 km S of Shehedi. 26/2/2015

12° 34' 48" N, 36° 10' 12" E; 600 m a.s.l.
Open, mixed woodland without dominant woody species on undulating terrain with dark brown, stony soil, cracking like black cotton soil. Very sparse and short ground-cover (< 20 cm tall). No cultivation. No sign of grazing. No sign of burning.

Acacia hecatophylla
Anogeissus leiocarpa
Balanites aegyptiaca
Combretum collinum
Ficus glumosa
Hyphaene thebaica
Piliostigma thonningii
Pterocarpus lucens
Terminalia laxiflora

Ziziphus abyssinica
Ziziphus spina-christi

50 – (2015_A6). On the road from Shehedi (Genda Wuha) to Shahura (from Metema-Aykel road), 20 km S of Shehedi. 26/2/2015
12° 39' 54" N, 36° 18' 06" E; 900 m a.s.l.
Mixed, in places moderately dense woodland without dominant woody species on slightly undulating terrain with dark brown, cracking soil (not black as in black cotton soil), but on hillsides a stony soil. Level terrain in part converted to farmland. Very sparse ground-cover outside cultivated areas. No sign of grazing. No sign of burning.

Acacia hecatophylla
Acacia seyal
Balanites aegyptiaca
Boswellia papyrifera
Combretum hartmannianum
Lannea fruticosa
Lannea schimperi
Piliostigma thonningii
Pterocarpus lucens
Sterculia africana
Stereospermum kunthianum
Terminalia laxiflora

51 – (2015_10). On the Metema-Aykel road; 75 km E of Shehedi (Genda Wuha) towards Aykel. 27/2/2015
12° 33' 54" N, 36° 53' 30" E; 1200 m a.s.l.
Open, mixed woodland without dominant woody species on sloping, rocky and stony ground. Soil brown, but rocky outcrops very prominent. All areas without rocks were partly cultivated. Ground-cover in uncultivated areas sparse to moderately dense and consisting of perennial grasses (< 50 cm tall). No sign of grazing. Scars of burning on the most sloping areas.

Acacia hecatophylla
Anogeissus leiocarpa
Diospyros mespiliformis
Ficus ingens
Ficus sycomorus
Flueggea virosa
Lannea fruticosa
Maytenus arbutifolia
Oncoba spinosa
Terminalia schimperiana
Vernonia auriculifera

52 – (2015_11). On the Metema-Aykel road; 85 km E of Shehedi (Genda Wuha) towards Aykel. 27/2/2015
12° 32' 18" N, 36° 56' 54" E; 1400 m a.s.l.
Mosaic of mixed, open woodland without dominant woody species and montane evergreen scrub on steep, rocky slopes. Soil brown and very stony. Much cultivation on the most level terrain. Ground-cover generally sparse and of very varying height (always < 50 cm tall), never forming a continuous layer, even between the rocks. Grazing likely, but not observed. Moderately large scars from burning on the steeper slopes.

Anogeissus leiocarpa
Combretum collinum
Croton macrostachyus
Dichrostachys cinerea
Diospyros mespiliformis
Ficus vasta
Ficus sycomorus
Flueggea virosa
Maytenus arbutifolia
Stereospermum kunthianum
Syzygium guineense subsp. *macrocarpum*
Terminalia schimperiana
Vernonia auriculifera
Vitex doniana

53 – (2015_12). On the Metema-Aykel road; 95 km E of Shehedi (Genda Wuha) towards Aykel. 27/2/2015
12° 30' 30" N, 37° 00' 42" E; 2100 m a.s.l.
Degraded Afromontane evergreen bushland without dominant woody species on steeply sloping ground. Soil brownish, very stony. Ground-cover heavily grazed, with a few ungrazed and higher patches be-

tween the rocks (< 20 cm tall). No cultivation. No sign of burning.

Acacia abyssinica
Bersama abyssinica
Brucea antidysenterica
Croton macrostachyus
Euphorbia ampliphylla
Ficus ingens
Ficus sycomorus
Maytenus arbutifolia
Vernonia auriculifera

Profile F.

54 — (2017_01). Near Manduria, just W of village. 22/06/2017.
11° 06' 12" N, 36° 24' 42" E; 1400 m a.s.l.
Open woodland with *Sarcocephalus latifolius* as the completely dominant woody species. Moderately steep rocky slope in most of the area. Soil brown, stony. Ground-cover of low perennial grasses and other annual and perennial herbs, in most areas scarce due to shading from the *Sarcocephalus* shrubs. No cultivation. No sign of grazing. No sign of burning.

Albizia amara
Anogeissus leiocarpa
Cordia africana
Ficus sur
Flueggea virosa
Lannea schimperi
Lonchocarpus laxiflorus
Oxytenanthera abyssinica
Piliostigma thonningii
Pterocarpus lucens
Sarcocephalus latifolius
Terminalia schimperiana

55 — (2017_02). Near Gilgil Beles, just W of river and small town. 22/06/2017
11° 10' 48" N, 36° 20' 00" E; 1050 m a.s.l.
Open woodland completely dominated by *Acacia polyacantha* subsp. *campylacantha* on level terrain not far from the Gilgil Beles river. Soil dark, almost like vertisol (black cotton soil). Ground-cover of short annual and perennial grasses (< 50 cm tall). No cultivation. Grazing likely, but not observed. No sign of burning.

Acacia polyacantha subsp. *campylacantha*
Cordia africana
Terminalia laxiflora

56 — (2017_03). Near Mambuk (sometimes spelt Mambruk), 7 km on the road W of the village towards Mankush. 22/06/2017
11° 16' 30" N, 36° 12' 42" E; 1150 m a.s.l.
Open, mixed woodland without dominant woody species on level terrain. Soil heavy and dark brown (not quite vertisol). Now heavily cultivated and has according to local informant been so for at least 5 years. Ground-cover almost totally removed by the cultivation. No sign of grazing. May have burnt before, but does not burn any longer due to the conversion to farmland.

Albizia malacophylla
Balanites aegyptiaca
Combretum collinum
Cordia africana
Ficus sycomorus
Gardenia ternifolia
Lannea schimperi
Terminalia schimperiana
Ziziphus abyssinica

57 — (2017_04). West of Mambuk (sometimes spelt Mambruk), on the road 17 km W of village towards Mankush. 22/06/2017
11° 14' 06" N, 36° 08' 06" E; 1000 m a.s.l.
Rather dense, mixed woodland without dominant woody species on moderately sloping rocky ground with pockets and thin layer of red soil. Ground-cover of very scarce stratum of perennial grasses (mostly < 50 cm tall). No cultivation. No sign of grazing. No sign of burning.

Boswellia pirottae
Combretum collinum
Entada africana
Flueggea virosa
Gardenia ternifolia
Lonchocarpus laxiflorus
Ochna leucophloeos
Oxytenanthera abyssinica
Pterocarpus lucens
Securidaca longipedunculata
Sterculia africana
Terminalia macroptera

58 – (2017_05). Near Mambuk (sometimes spelt Mambruk), 26 km W of village towards Mankush. 22/06/2017

11° 11' 12" N, 36° 02' 30" E; 950 m a.s.l.
Mixed, open woodland without dominant woody species on gently sloping terrain. Dark brown soil with many stones. Ground-cover sparse, with hardly any grass, but the climber *Tylosema fazogliensis* common in the most open parts of the woodland. No cultivation. No sign of grazing. Burning not likely.

Anogeissus leiocarpa
Annona senegalensis
Boswellia pirottae
Combretum collinum
Entada africana
Ficus sycomorus
Gardenia ternifolia
Grewia mollis
Lonchocarpus laxiflorus
Maytenus senegalensis
Piliostigma thonningii
Polyscias farinosa
Pseudocedrela kotschyi
Pterocarpus lucens
Stereospermum kunthianum
Terminalia schimperiana
Ziziphus abyssinica

59 – (2017_06). Near Mambuk (sometimes spelt Mambruk), 37 km W of village towards Mankush. 22/06/2017

11° 11' 12" N, 35° 56' 48"; 930 m a.s.l.
Mixed, open woodland without dominant woody species on gently slope sloping terrain. Dark brown soil with many stones. Natural vegetation deeply influenced by farming, which has almost completely disturbed the ground-cover. No sign of grazing. Signs of burning.

Acacia hecatophylla
Anogeissus leiocarpa
Balanites aegyptiaca
Cordia africana
Entada africana
Ficus sycomorus
Flueggea virosa
Grewia mollis
Lonchocarpus laxiflorus
Maytenus senegalensis
Oxytenanthera abyssinica
Piliostigma thonningii
Polyscias farinosa
Pseudocedrela kotschyi
Pterocarpus lucens
Stereospermum kunthianum
Terminalia laxiflora

Profile J.

60 – (2017_07). On the road 42 km from lower Abay Bridge towards Sherkole. 23/6/2017

10° 43' 36" N, 35° 01' 18" E; 750 m a.s.l.
Mixed, open woodland without dominant woody species on almost level terrain with scattered rocks. Soil dark brown, stony. Ground-cover uniform, consisting of perennial grasses (up to ca. 2 m high, according to local informant); almost continuously covering the soil. No cultivation. No sign of grazing. Although no sign of recent burning, it seems likely that this vegetation could burn fiercely.

Anogeissus leiocarpa
Balanites aegyptiaca
Boswellia papyrifera
Boswellia pirottae
Combretum collinum
Entada africana
Flueggea virosa
Grewia mollis
Lonchocarpus laxiflorus
Maytenus senegalensis
Ochna leucophloeos
Oxytenanthera abyssinica
Pterocarpus lucens
Steganotaenia araliacea
Sterculia africana
Strychnos innocua
Terminalia laxiflora

61 — (2017_08). On the road 52 km from lower Abay Bridge towards Sherkole. 23/6/2017
10° 42' 30" N, 34° 55' 48" E; 875 m a.s.l.
Mixed open woodland without dominant woody species on level terrain. Soil pale reddish and sandy. Ground-cover of perennial grasses growing to ca. 2 m tall. No cultivation. No sign of grazing. Signs of burning in patches.

Albizia amara
Anogeissus leiocarpa
Balanites aegyptiaca
Boswellia papyrifera
Boswellia pirottae
Cassia arereh
Combretum collinum
Dalbergia melanoxylon
Dichrostachys cinerea
Entada africana
Lonchocarpus laxiflorus
Oxytenanthera abyssinica
Piliostigma thonningii
Pterocarpus lucens
Sterculia africana
Stereospermum kunthianum

62 — (2017_09). On the road 6 km from Sherkole towards the lower Abay Bridge. 24/6/2017
10° 37' 30" N, 34° 47' 54" E; 750 m a.s.l.
Mixed, open woodland without dominant woody species on gently sloping terrain with scattered, large rocks. Soil pale reddish and sandy in upper part of slope, fine-grained, almost black cotton soil at base of slope. Ground-cover of perennial grasses growing to ca. 2 m tall. No cultivation. No sign of grazing. Signs of fierce burning in many places.

Acacia polyacantha subsp. *campylacantha*
Acacia seyal
Albizia amara
Albizia malacophylla
Anogeissus leiocarpa
Boswellia papyrifera
Combretum collinum
Entada africana
Ficus ingens
Ficus sycomorus
Flueggea virosa
Grewia mollis
Lannea fruticosa
Lannea schimperi
Lonchocarpus laxiflorus
Maytenus senegalensis
Oxytenanthera abyssinica
Ozoroa pulcherrima
Piliostigma thonningii
Pterocarpus lucens
Sterculia africana
Stereospermum kunthianum
Terminalia macroptera
Ziziphus abyssinica
Ziziphus spina-christi

63 — (2017_10). On the road 10 km S of Sherkole towards Mengi. 24/6/2017
10° 30' 48" N, 34° 46' 36" E; 880 m a.s.l.
Partly cleared, but regenerating mixed woodland, probably without dominant woody species, on level terrain. Dark, red soil with few stones. Ground-cover short (< 50 cm tall). The area has been cultivated, but

the farming seemed abandoned. No sign of grazing. Signs of burning in a few places.

Acacia polyacantha subsp. *campylacantha*
Albizia amara
Anogeissus leiocarpa
Balanites aegyptiaca
Boswellia papyrifera
Combretum molle
Commiphora pedunculata
Dalbergia melanoxylon
Entada africana
Lannea fruticosa
Lannea schimperi
Lonchocarpus laxiflorus
Maytenus senegalensis
Piliostigma thonningii
Steganotaenia araliacea
Stereospermum kunthianum

64 — (2017_11). On the road 10 km N of Mengi towards Sherkole. 24/6/2017
10° 26' 06" N, 34° 45' 12" E; 1030 m a.s.l.
Mixed, open woodland without dominant woody species on undulating terrain, very rocky. Soil pale rosy, sandy, but a few places with flat and darker red soil. Ground-cover of tall perennial grasses (ca 200 cm high). No cultivation. No sign of grazing. Signs of fierce burning; bark of trees all blackened.

Acacia polyacantha subsp. *campylacantha*
Albizia malacophylla
Anogeissus leiocarpa
Balanites aegyptiaca
Combretum collinum
Entada africana
Fadogia cienkowskii
Gardenia ternifolia
Grewia mollis
Lannea fruticosa
Lonchocarpus laxiflorus
Maytenus senegalensis
Ochna leucophloeos
Oxytenanthera abyssinica
Psorospermum febrifugum
Pterocarpus lucens
Securidaca longipedunculata
Sterculia africana
Stereospermum kunthianum
Terminalia schimperiana
Vernonia amygdalina

Ziziphus spina-christi

65 — (2017_12). On the road 4 km S of Mengi towards Homesha. 25/6/2017
10° 20' 00" N, 34° 45' 12" E; 1200 m a.s.l.
Mixed, open woodland without dominant woody species on undulating terrain, very rocky. Soil red, sandy in places. Ground-cover very sparse because of rocky ground, some tall perennial grasses (ca 200 cm high). No cultivation. No sign of grazing. No signs of burning.

Annona senegalensis
Anogeissus leiocarpa
Combretum collinum
Dalbergia boehmii
Dichrostachys cinerea
Ficus glumosa
Flacourtia indica
Flueggea virosa
Kigelia africana
Lannea fruticosa
Lonchocarpus laxiflorus
Maytenus senegalensis
Ochna leucophloeos
Ozoroa insignis
Pseudocedrela kotschyi
Rhus vulgaris
Sarcocephalus latifolius
Sterculia africana
Stereospermum kunthianum
Strychnos innocua
Syzygium guineense subsp. *macrocarpum*
Terminalia schimperiana
Terminalia macroptera
Vitex doniana

66 — (2017_13). On the road 10 km S of Mengi towards Homesha. 25/6/2017

10° 19' 18" N, 34° 40' 42" E; 1400 m a.s.l.

Patch of mixed, open woodland without dominant woody species in extensive farmland on level terrain. Red soil without rocks or gravel. Ground-cover in the uncultivated patch consisting of tall perennial grasses (> 100 cm tall). No sign of grazing. No sign of burning.

Acacia polyacantha subsp. *campylacantha*
Acacia seyal
Acacia venosa
Annona senegalensis
Combretum collinum
Dalbergia boehmii
Ficus sycomorus
Gardenia ternifolia
Grewia mollis
Ozoroa insignis
Piliostigma thonningii
Sclerocarya birrea
Stereospermum kunthianum
Strychnos innocua
Terminalia schimperiana
Vitex doniana
Ziziphus spina-christi

67 — (2017_14). On the road 1 km S of Homesha towards Assosa. 25/6/2017

10° 18' 24" N, 34° 37' 36" E; 1400 m

Mixed, open woodland without dominant woody species on sloping ground, rocky in places; the slopes form a broad gully with a small stream and a moderately well-developed riverine forest at the bottom. Dark red soil between the rocks. No cultivation. No sign of grazing. Signs of regular burning outside the riverine forest (which was not studied) at the bottom of the gully.

Annona senegalensis
Combretum collinum
Ficus sycomorus
Gardenia ternifolia
Maytenus senegalensis
Ochna leucophloeos
Piliostigma thonningii
Senna singueana
Syzygium guineense subsp. *macrocarpum*
Terminalia macroptera
Vitex doniana

68 — (2017_15). On the road 13 km S of Homesha towards Assosa. 25/6/2017

10° 12' 48" N, 34° 39' 06" E; 1600 m a.s.l.

Narrow strip of mixed, open woodland without dominant woody species in intensely cultivated, flat farmland. Soil red, with scattered rocks. Ground-cover moderately dense (perennial grass to ca. 50 cm tall). No sign of grazing. Hardly any sign of burning.

Grewia mollis
Syzygium guineense subsp. *macrocarpum*
Psorospermum febrifugum
Ochna leucophloeos
Securidaca longipedunculata
Gardenia ternifolia
Annona senegalensis
Maytenus senegalensis
Lonchocarpus laxiflorus
Stereospermum kunthianum
Terminalia schimperiana
Piliostigma thonningii
Combretum collinum
Vitex doniana
Bridelia micrantha
Cordia africana
Oxytenanthera abyssinica
Senna singueana

69 — (2017_16). On the road 15 km N of Tongo towards Bambasi. 26/6/2017

9° 29' 30" N, 34° 27' 54" E; 1375 m a.s.l.

Mixed, open woodland without dominant woody species, partly cultivated, on undulating terrain and rocks. Soil dark red, part of the site with rocky hillsides. Ground-cover moderately dense (ca 50 cm high), particularly in the rocky areas. No sign of grazing, but unlikely that it does not occur. Marked signs of burning on the hillsides, not in the cultivated terrain.

Acacia polyacantha subsp. *campylacantha*
Acacia seyal
Annona senegalensis
Cordia africana
Combretum collinum
Dombeya quinqueseta
Entada abyssinica
Faurea rochetiana
Ficus sycomorus
Lannea schimperi
Maytenus senegalensis
Oxytenanthera abyssinica
Piliostigma thonningii
Psorospermum febrifugum
Pterocarpus lucens
Securidaca longipedunculata
Steganotaenia araliacea
Stereospermum kunthianum
Terminalia schimperiana

70 — (2017_17). 25 km N of Tongo towards Bambasi. 26/6/2017
9° 31' 36" N, 34° 30' 24" E; 1450 m a.s.l.
Mixed, open woodland without dominant woody species on slightly undulating terrain, with some cultivation. Dark red soil with no rocks or gravel. Ground-cover very open, mostly exposing the bare soil, and with scattered, low annual grasses and herbs. No sign of grazing. No sign of burning.

Acacia polyacantha subsp. *campylacantha*
Annona senegalensis
Combretum collinum
Cordia africana
Croton macrostachyus
Dichrostachys cinerea
Dombeya quinqueseta
Strychnos innocua
Syzygium guineense subsp. *macrocarpum*
Terminalia laxiflora
Terminalia macroptera

71 — (2017_18). On the road 30 km N of Tongo towards Bambasi, just after turn-off towards Begi. 26/6/2017
9° 31' 48" N, 34° 34' 00" E; 1575 m a.s.l.
Open woodland on undulating terrain, dominated by *Syzygium guineense* subsp. *macrocarpum*. Soil dark red, without rocks and gravel. Ground-cover open, exposing the bare soil, and with scattered annual grasses and herbs up to ca. 30 cm high. No cultivation. No sign of grazing, but it probably occurs. No sign of burning.

Annona senegalensis
Combretum molle
Croton macrostachyus
Dombeya buettneri
Fadogia cienkowskii
Rhus vulgaris
Syzygium guineense subsp. *macrocarpum*
Terminalia laxiflora
Terminalia macroptera

72 — (2017_19). On the road 40 km N of Tongo towards Bambasi. 26/6/2017
9° 34' 06" N, 34° 38' 48" E; 1400 m a.s.l.
Mixed woodland without dominant woody species and almost completely closed canopy, shading the ground-cover, on level terrain. Dark red soil with no rocks or gravel. Ground-cover mostly low (< 50 cm tall), but rich in species of dicot herbs. No cultivation. No sign of grazing. No sign of burning.

Annona senegalensis
Carissa spinarum
Combretum collinum
Combretum molle
Croton macrostachyus
Entada abyssinica
Fadogia cienkowskii
Flacourtia indica
Flueggea virosa
Gardenia ternifolia
Grewia mollis
Maytenus senegalensis

Rhus vulgaris
Steganotaenia araliacea
Syzygium guineense subsp. *macrocarpum*
Terminalia laxiflora
Zanthoxylum gilletii

73 – (2017_20). On the road 50 km N of Tongo towards Bambasi. 26/6/2017
9° 38' 24" N, 34° 42' 36" E; 1375 m a.s.l.
Mixed, open woodland without dominant woody species on level terrain (probably liable to flooding during the rains). Soil dark brown to black, cracking, similar to vertisol (black cotton soil). Ground-cover of tall perennial grasses (ca. 2 m tall), liable to fierce burning. No cultivation. No sign of grazing. No direct sign of burning, but the rainy season was heavy at the time of observation.

Annona senegalensis
Combretum collinum
Flueggea virosa
Gardenia ternifolia
Lonchocarpus laxiflorus
Maytenus senegalensis
Piliostigma thonningii
Psorospermum febrifugum
Syzygium guineense subsp. *macrocarpum*
Terminalia laxiflora
Terminalia macroptera
Vitex doniana

74 – (2017_21). On the road 2 km E of Kurmuk towards Assosa. 27/6/2017
10° 33' 48" N, 34° 18' 42" E; 675 m a.s.l.
Mixed, open woodland without dominant woody species on hilly terrain. Deep red soil. Ground-cover of tall perennial grasses (ca. 2 m tall). No cultivation. No sign of grazing. Signs of fierce burning, which has blackened the bark of many trees.

Acacia hockii
Acacia polyacantha subsp. *campylacantha*
Acacia seyal
Anogeissus leiocarpa
Balanites aegyptiaca
Combretum hartmannianum
Entada africana
Lannea fruticosa
Lannea schimperi
Maytenus senegalensis
Piliostigma thonningii
Pseudocedrela kotschyi
Sclerocarya birrea
Sterculia africana
Terminalia laxiflora
Ziziphus spina-christi

75 – (2017_22). On the road 10 km E of Kurmuk towards Assosa. 27/6/2017
10° 34' 30" N, 34° 22' 48" E; 750 m a.s.l.
Mixed, open woodland without dominant woody species on hilly terrain. Deep red soil with much gravel and small stones. Ground-cover of tall perennial grasses (ca 2 m tall). No cultivation. No sign of grazing. Signs of fierce burning.

Acacia polyacantha subsp. *campylacantha*
Acacia senegal
Acacia seyal
Anogeissus leiocarpa
Balanites aegyptiaca
Combretum collinum
Combretum hartmannianum
Dombeya quinqueseta
Entada africana
Ficus sycomorus
Grewia mollis
Lannea fruticosa
Maytenus senegalensis
Ochna leucophloeos
Ozoroa insignis
Ozoroa pulcherrima
Piliostigma thonningii
Pseudocedrela kotschyi
Pterocarpus lucens
Sterculia africana
Stereospermum kunthianum
Ziziphus spina-christi

76 – (2017_23). On the road 15 km from Kurmuk towards Assosa. 27/6/2017

10° 33' 36" N, 34° 26' 00" E; 1050 m a.s.l.

Mixed, open woodland without dominant woody species on steep slopes with very stony, brown soil. Ground-cover of tall grasses (ca 2 m high). No cultivation. No sign of grazing. Liable to fierce burning and signs on the past fires on tree trunks.

Acacia hecatophylla
Anogeissus leiocarpa
Boswellia papyrifera
Combretum collinum
Entada abyssinica
Lannea fruticosa
Lonchocarpus laxiflorus
Ochna leucophloeos
Oxytenanthera abyssinica
Ozoroa pulcherrima
Strychnos innocua
Terminalia schimperiana

77 – (2017_24). On the road 22 km from Kurmuk towards Assosa. 27/6/2017

10° 32' 54" N, 34° 29' 24" E; 1250 m a.s.l.

Mixed, open woodland without dominant woody species on almost flat terrain, with (temporarily muddy because of the rainy season) dark red soil without stones. No cultivation. No sign of grazing. Ground-cover of tall perennial grasses (ca. 2 m tall). Liable to fierce burning and signs on the past fires on tree trunks.

Acacia hockii
Acacia seyal
Annona senegalensis
Boswellia papyrifera
Combretum collinum
Entada abyssinica
Gardenia ternifolia
Grewia mollis
Lannea fruticosa
Lonchocarpus laxiflorus
Maytenus senegalensis
Ochna leucophloeos
Oxytenanthera abyssinica
Ozoroa insignis
Piliostigma thonningii
Pterocarpus lucens
Stereospermum kunthianum
Terminalia macroptera
Terminalia laxiflora
Ziziphus abyssinica
Ziziphus spina-christi

78 – (2017_25). On the road 32 km from Kurmuk towards Assosa. 27/6/2017

10° 28' 30" N, 34° 31' 00" E; 1450 m a.s.l.

Mixed, open woodland without dominant woody species on moderately sloping ground; deep, dark brown soil with hardly any stones. No cultivation. No signs of grazing. Ground-cover of perennial grass (ca. 2 m high), liable to fierce burning and signs on the past fires on tree trunks.

Acacia hockii
Acacia venosa
Annona senegalensis
Combretum collinum
Cussonia arborea
Entada abyssinica
Grewia mollis
Lonchocarpus laxiflorus
Maytenus senegalensis
Ochna leucophloeos
Ozoroa insignis
Piliostigma thonningii
Protea gaguedi
Stereospermum kunthianum

79 – (2017_26). In the bamboo thicket called Anbessa Chaka, 20 km S of Assosa on the road to Bambasi. 28/6/2017

9° 55' 24" N, 34° 39' 42" E; 1450 m a.s.l.

Open woodland, here and there completely dominated by the lowland bamboo *Oxytenanthera abyssinica*, in other places mixed woodland, on shallow, brown soil over rocky ground, slightly sloping. Ground-cover sparse in patches dominated by dense bamboo, in open areas with tall grass (> 1.5 m tall), mixed with perennial herbs. No cultivation. No sign of grazing. Sign of fierce burning in the past fires on tree trunks.

Albizia malacophylla
Bridelia scleroneura
Combretum collinum
Dombeya quinqueseta
Gardenia ternifolia
Lonchocarpus laxiflorus
Oxytenanthera abyssinica
Ozoroa insignis
Piliostigma thonningii
Polyscias farinosa
Psorospermum febrifugum
Syzygium guineense subsp. *macrocarpum*
Terminalia macroptera
Vitex doniana

80 — (T_E&B_1). At Ashefabego, W of the road from Mengi to Sudan border at Gizen. Observations made ca. year 2015 (data from Mindaye Teshome, Abeje Eshete & F. Bongers (2017). Uniquely regenerating frankincense tree populations in western Ethiopia. Forest Ecology and Management 389: 127-135.)
10° 39' 48" N, 34° 41' 30" E; 960 m a.s.l.
This data is extracted from a publication, and the following environmental information is our interpretation: Woodland, apparently without dominant woody species, growing in shallow, red soil on rocky ground, slightly sloping. Ground-cover with tall grass (> 1.5 m tall). No cultivation. No sign of grazing. Sign of fierce burning.

Acacia hecatophylla
Albizia malacophylla
Allophylus rubifolius (in the publication of Mindaye Teshome et al. (2017) named as *A. abyssinicus*, which is a forest species not seen by us elsewhere in the western woodlands)
Anogeissus leiocarpa
Boswellia papyrifera
Cassia arereh
Combretum adenogonium
Combretum collinum
Combretum molle
Commiphora pedunculata
Crossopteryx febrifuga
Dalbergia melanoxylon
Entada africana
Ficus thonningii
Gardenia ternifolia
Grewia velutina
Lannea fruticosa
Lannea schweinfurthii (in the publication of Mindaye Teshome et al. (2017) named as *L. welwitschii*, but almost certainly *L. schweinfurthii*)
Lonchocarpus laxiflorus
Maytenus senegalensis
Ozoroa pulcherrima
Piliostigma thonningii
Pseudocedrela kotschyi
Pterocarpus lucens
Securidaca longipedunculata
Sterculia africana (in the publication of Mindaye Teshome et al. (2017) named as *S. setigera*; this may be possible, but we have nowhere else seen that species in the western woodlands)
Strychnos innocua
Terminalia laxiflora
Ximenia americana
Ziziphus spina-christi

81 — (T_E&B_2). At Arenja, on the new road Mengi to Guba (Mankush), across the Abay Valley. Observations made ca. year 2015 (data from Mindaye Teshome et al. 2017).
11° 04' 00" N, 35° 15' 24" E; 920 m a.s.l.
This data was taken from a publication, and the following environmental information is our interpretation: Mixed, open woodland, apparently without dominant woody species, growing on shallow, red soil on rocky ground, slightly sloping. Ground-cover with tall grass (> 1.5 m tall). No cultivation. No sign of burning. Sign of fierce burning. A record of the Afromontane forest tree *Prunus africana*, which nowhere

else has been recorded from the western Ethiopian woodlands, has been omitted as unlikely.

Acacia hecatophylla
Acacia senegal
Acacia seyal
*Allophylus rubifolius (*in the publication of Mindaye Teshome et al. (2017) named as *A. abyssinicus*, which is a forest species not seen by us elsewhere in the western woodlands)
Anogeissus leiocarpa
Balanites aegyptiaca
Boswellia papyrifera
Combretum adenogonium
Combretum collinum
Combretum molle
Crossopteryx febrifuga
Dalbergia boehmii
Dalbergia melanoxylon
Dichrostachys cinerea
Dombeya quinqueseta (in the publication of Mindaye Teshome et al. (2017) named as the Afromontane *D. torrida*, but almost certainly *D. quinqueseta*)
Entada africana
Ficus thonningii
Gardenia ternifolia
Grewia mollis
Hyphaene thebaica
Lannea fruticosa
Lonchocarpus laxiflorus
Maytenus senegalensis
Borassus aethiopum
Piliostigma thonningii
Pseudocedrela kotschyi
Pterocarpus lucens
Securidaca longipedunculata
Sterculia africana (in the publication of Mindaye Teshome et al. (2017) named as *S. setigera*; this may be possible, but we have nowhere else seen that species in the western woodlands)
Strychnos innocua
Terminalia laxiflora
Ximenia americana
Ziziphus spina-christi

82 — (T_E&B_3). At Baneshegol, on the road N of Mengi towards Gizen. Observations made ca. year 2015 (data from Mindaye Teshome et al. 2017).
11° 29' 12" N, 34° 45' 54" E; 700 m a.s.l.
This data was taken from a publication, and the following environmental information is our interpretation: Mixed, open woodland, apparently without dominant woody species, growing in deep, red soil, flat to slightly sloping. Ground-cover with tall grass (> 1.5 m tall). No cultivation. No signs of grazing. Sign of fierce burning.

Acacia hecatophylla
Acacia senegal
Acacia seyal
*Allophylus rubifolius (*in the publication of Mindaye Teshome et al. (2017) named as *A. abyssinicus*, which is a forest species not seen by us elsewhere in the western woodlands)
Anogeissus leiocarpa
Balanites aegyptiaca
Boswellia papyrifera
Boswellia pirottae
Combretum adenogonium
Combretum collinum
Crossopteryx febrifuga
Dalbergia boehmii
Dalbergia melanoxylon
Dombeya quinqueseta (in the publication of Mindaye Teshome et al. (2017) named as the Afromontane *D. torrida*, but almost certainly *D. quinqueseta*)
Entada africana
Gardenia ternifolia
Grewia mollis
Lannea fruticosa
Lannea schweinfurthii (in the publication of Mindaye Teshome et al. (2017) named as *L. welwitschii*, but almost certainly *L. schweinfurthii*)
Lonchocarpus laxiflorus
Maytenus senegalensis
Borassus aethiopum
Piliostigma thonningii
Pseudocedrela kotschyi
Pterocarpus lucens

Securidaca longipedunculata
Sterculia africana (in the publication of Mindaye Teshome et al. (2017) named as *S. setigera*; this may be possible, but we have nowhere else seen that species in the western woodlands)
Terminalia laxiflora
Ximenia americana
Ziziphus mauritiana
Ziziphus spina-christi

83 – (T_E&B_4). At Kurmuk, on the Sudan border NW of Assosa. Observations made ca. year 2015 (data from Mindaye Teshome et al. 2017).
10° 33' 48" N, 34° 18' 06" E; 675 m a.s.l.
This data has been adapted from a publication, and the following environmental information is our interpretation: Mixed, open woodland, apparently without dominant woody species, growing in deep red soil, flat to slightly sloping. Ground-cover with tall grass (> 1.5 m tall). No cultivation. No signs of grazing. Sign of fierce burning.

Acacia senegal
Acacia seyal
Allophylus rubifolius (in the publication of Mindaye Teshome et al. (2017) named as *A. abyssinicus*, which is a forest species not seen by us elsewhere in the western woodlands)
Anogeissus leiocarpa
Balanites aegyptiaca
Boswellia papyrifera
Combretum collinum
Combretum molle
Commiphora pedunculata
Cordia africana
Dalbergia boehmii
Dalbergia melanoxylon
Dichrostachys cinerea
Dombeya quinqueseta
Entada africana
Ficus thonningii
Gardenia ternifolia
Grewia mollis
Hyphaene thebaica
Lannea fruticosa
Maytenus senegalensis
Pterocarpus lucens
Sterculia africana (in the publication of Mindaye Teshome et al. (2017) named as *S. setigera*; this may be possible, but we have nowhere else seen that species in the western woodlands)
Stereospermum kunthianum
Terminalia laxiflora
Ximenia americana
Ziziphus spina-christi

84 – (T_E&B_5). At Gulashe, E of Kurmuk on the way to Assosa. Observations made ca. year 2015 (data from Mindaye Teshome et al. 2017).
10° 33' 42" N, 34° 24' 24" E; 820 m a.s.l.
This data was extracted from a publication, and the following environmental information is our interpretation: Woodland, apparently without dominant woody species, growing in deep red soil, flat to slightly sloping. Ground-cover with tall grass (> 1.5 m tall). No cultivation. No signs of grazing. Sign of fierce burning.

Acacia senegal
Acacia seyal
Albizia malacophylla
Allophylus rubifolius (in the publication of Mindaye Teshome et al. (2017) named as *A. abyssinicus*, which is a forest species not seen by us elsewhere in the western woodlands)
Anogeissus leiocarpa
Balanites aegyptiaca
Boswellia papyrifera
Boswellia pirottae
Combretum adenogonium
Combretum collinum
Dalbergia melanoxylon
Dichrostachys cinerea
Entada africana
Faurea rochetiana
Gardenia ternifolia
Grewia mollis
Lannea fruticosa
Lonchocarpus laxiflorus

Maytenus senegalensis
Pterocarpus lucens
Securidaca longipedunculata
Sterculia africana (in the publication of Mindaye Teshome et al. (2017) named as *S. setigera*; this may be possible, but we have nowhere else seen that species in the western woodlands)
Strychnos innocua
Terminalia laxiflora
Ximenia americana
Ziziphus mucronata
Ziziphus spina-christi

Profile N.

85 — (J&F_Site X). On the fenced land around Gambela Airport. April 1992-December 1993 (data from Jensen & Friis (2001).
8° 08' 24" N, 34° 34' 00" E; 600 m a.s.l.
Dense, mixed woodland with almost continuous canopy cover (ca. 90%) and without dominant woody species on flat terrain with deep brown to yellowish brown or red soil and few stones. Ground-cover of low herbs, but hardly any grasses. No cultivation. No signs of Grazing. According to local informants, the vegetation had not burnt for a significant length of time.

Allophylus rubifolius
Anogeissus leiocarpa
Diospyros mespiliformis
Erythrococca trichogyne
Erythroxylum fischeri
Flueggea virosa
Grewia tenax
Harrisonia abyssinica
Maerua triphylla
Meyna tetraphylla
Pterocarpus lucens
Strychnos innocua
Tamarindus indica
Vangueria madagascariensis
Ziziphus pubescens

86 — (J&F_Site A). At the foot of the Bure Escarpment, ca. 5 km towards Gambela from where the road from Bure to Gambela crosses the Baro River. April 1992-December 1993 (data from Jensen & Friis (2001)).
8° 12' 36" N, 34° 57' 54" E; 650 m a.s.l.

Mixed, relatively open woodland (ca 30% canopy cover) without dominant woody species on terrain gently sloping away from the escarpment, soil with many stones or loose gravel over a brown to reddish soil. Ground-cover of patchy areas of tall grasses (ca 2 m high). No cultivation. No signs of grazing. Burning regularly once or twice a year.

Acacia senegal
Balanites aegyptiaca
Cadaba farinosa
Combretum collinum
Entada africana
Flueggea virosa
Grewia mollis
Harrisonia abyssinica
Lannea barteri
Lannea fruticosa
Lonchocarpus laxiflorus
Pterocarpus lucens
Sterculia africana
Stereospermum kunthianum
Ximenia americana
Ziziphus abyssinica
Ziziphus mauritiana

87 — (J&F_Site B). Between 10 and 12 km on the road towards Akobo, S of the bridge on the Baro River next to the town of Gambela. April 1992-December 1993 (data from Jensen & Friis (2001)).
8° 10' 18" N, 34° 35' 12" E; 550 m a.s.l.
Mixed woodland with fairly dense canopy (ca. 60% cover) without dominant woody species on slightly undulating terrain, low ridges with coarser sand and lower ground with finer brown to red soil. Groundcover of tall grasses (ca 2 m high). No cultivation.

No sign of grazing. Burning regularly at least once a year.

Acacia senegal
Allophylus rubifolius
Anogeissus leiocarpa
Bridelia scleroneura
Cadaba farinosa
Combretum collinum
Combretum molle
Dichrostachys cinerea
Flueggea virosa
Gardenia ternifolia
Grewia mollis
Harrisonia abyssinica
Lonchocarpus laxiflorus
Maytenus senegalensis
Maerua oblongifolia
Pterocarpus lucens
Strychnos innocua
Tamarindus indica
Terminalia laxiflora
Ziziphus abyssinica

88 — (J&F_Site C). Between 80 and 85 km on the road to the S across the bridge on the Baro River next to the town of Gambela; between Akobo and Pugnido. April 1992-December 1993 (data from Jensen & Friis (2001)).
7° 48' 00" N, 34° 17' 30" E; 450 m a.s.l.
Mixed, relatively open woodland (canopy cover < 30%) without dominant woody species on slightly undulating terrain, low ridges with coarser and lower ground with finer yellowish-brown soil. Ground-cover of tall grasses (ca 2 m tall). No cultivation. No signs of grazing. Burning at least once or twice a year.

Acacia senegal
Annona senegalensis
Anogeissus leiocarpa
Bridelia scleroneura
Combretum adenogonium
Combretum collinum
Crossopteryx febrifuga
Ficus sycomorus
Grewia mollis
Grewia velutina
Harrisonia abyssinica
Lannea barteri
Lonchocarpus laxiflorus
Maytenus senegalensis
Ochna leucophloeos
Strychnos innocua
Terminalia laxiflora

89 — (J&F_Site D). 22 km W of Gambela toward Itang. April 1992-December 1993 (data from Jensen & Friis (2001)).
8° 17' 00" N, 34° 28' 06" E; 500 m a.s.l.
Very open mixed woodland or wooded grassland (canopy cover ca. 25%) without dominant woody species on flat terrain, with dark brown, almost black soil (black cotton soil). Ground-cover of tall annual and perennial grasses (> 2 m tall). No cultivation. No sign of grazing. Burning regularly. (unidentified records of *Cadaba* sp. and *Combretum* sp. and *Grewia* sp. have been omitted here from the relevé).

Annona senegalensis
Bridelia scleroneura
Ficus sycomorus
Flueggea virosa
Grewia mollis
Harrisonia abyssinica
Lannea fruticosa
Lonchocarpus laxiflorus
Maytenus senegalensis
Stereospermum kunthianum

Profile G.

90 — (2017_A1). In the Abay Gorge, on the road between Mekane Selam, Mertule Maryam and Mota. 9/06/2017.
10° 39' 20" N, 38° 29' 08" E; 1550 m a.s.l.
Very open woodland on rocky slopes, dominated by *Albizia amara*. Hardly any soil or ground-cover, except

for *Rumex nervosus*, and scattered perennial dicots. No cultivation. No sign of grazing. No signs of burning.

Acacia mellifera
Acacia senegal
Albizia amara
Boscia salicifolia

Profile Q.

91 – (2018_01). In the Gojeb Valley, south side of mountains south of Jimma, on the road from Jimma towards the Gojeb Valley. 9/10/2018
7° 17' 42" N, 36° 50' 54" E; 1500 m a.s.l.
At this site, on the side towards Jimma of the Gojeb Valley, the limit between the forest and the woodland is at ca. 1750 m a.s.l., but between 1750 and 1500 m the vegetation is disturbed by plantations of *Cupressus*, so ca. 1500 m is the highest area with relatively undisturbed mixed, open woodland without dominant woody species. Flat or slightly sloping ground, deep brown soil between stones. Ground-cover less than 1 m high, consisting of short *Hyparrhenia* and *Loudetia* species. No cultivation. No sign of grazing. No sign of recent burning.

Combretum molle
Entada abyssinica
Ficus sur
Ficus sycomorus
Grewia mollis
Heteromorpha arborescens
Lannea schimperi
Piliostigma thonningii
Rhus vulgaris
Terminalia schimperiana
Vernonia amygdalina
Vitex doniana

92 – (2018_02). In the Gojeb Valley, south side of mountains south of Jimma, on the road from Jimma towards the Gojeb Valley, lower than the previous relevé (no. 91). 9/10/2018
7° 16' 36" N, 36° 49' 30" E; 1170 m a.s.l.

Mixed, open woodland with two dominant woody species of *Acacia* on flat ground. Soil deep, dark brown between large stones. Ground-cover of rather tall, perennial grasses (ca. 1 m high). No cultivation. No sign of grazing. No sign of burning.

Acacia hecatophylla
Acacia seyal
Cordia africana
Dichrostachys cinerea
Entada abyssinica
Ficus sycomorus
Lonchocarpus laxiflorus
Piliostigma thonningii
Steganotaenia araliacea
Stereospermum kunthianum
Terminalia schimperiana

93 – (2018_03). In the Gojeb Valley, south side of mountains south of Jimma, on the road from Jimma towards the Gojeb Valley, just at the river and lower than the previous relevé (no. 92). 9/10/2018
7° 15' 18" N, 36° 48' 00" E; 1060 m a.s.l.
Mixed, open woodland without dominant woody species on slightly sloping ground. Soil dark brown, very deep, no stones. Undergrowth of tall grass up to c 1 m high, which seems to burn regularly, but also subshrubs like *Desmodium* sp., *Acalypha* sp. and *Hoslunda* sp. No cultivation. No sign of grazing.

Acacia seyal
Acacia hecatophylla
Bridelia scleroneura
Combretum molle
Cordia africana
Entada africana
Ficus sycomorus

Flueggea virosa
Grewia mollis
Lonchocarpus laxiflorus
Piliostigma thonningii
Stereospermum kunthianum
Terminalia schimperiana
Vernonia amygdalina

94 — (2018_04). In the Gojeb Valley, south of the Gojeb River on the road from the bridge towards Ameya. 9/10/2018
7° 13' 00" N, 36° 47' 36" E; 1350 m a.s.l.
Mixed, open woodland without dominant woody species on steep slope. Mostly deep, dark brown soil with few stones. Short undergrowth of grass (< 15 cm tall). No cultivation. No sign of grazing. No sign of burning.

Annona senegalensis
Bridelia scleroneura
Combretum molle
Cussonia arborea
Entada abyssinica
Gardenia ternifolia
Heteromorpha arborescens
Lannea schimperi
Piliostigma thonningii
Stereospermum kunthianum
Terminalia schimperiana

95 — (2018_05). In the Omo Valley, near the new bridge and dam, current end of road from Ameya. 9/10/2018
6° 35' 42" N, 36° 32' 36" E; 750 m a.s.l.
Mixed, very open woodland without dominant woody species on hilly terrain half way up the Omo Valley. Deep, red soil with few stones. Ground-cover 1-1.5 m tall grass in the undergrowth, dominated by *Loudetia arundinacea*. Many subshrubby legumes, *Desmodium*, etc. No cultivation. No sign of grazing. The vegetation seems prone to burning.

Annona senegalensis
Bridelia scleroneura
Combretum collinum
Crossopteryx febrifuga
Dombeya quinqueseta
Ficus platyphylla
Lannea fruticosa
Lonchocarpus laxiflorus
Pseudocedrela kotschyi
Pavetta crassipes
Terminalia schimperiana

96 — (2018_06). In the Omo Valley, S of the Chebera-Churchura National Park, ca. 6 km from new bridge and dam on the road from Ameya. 12/10/2018.
6° 38' 42" N, 36° 33' 30" E; 1050 m a.s.l.
Dense mixed woodland without dominant woody species on hilly terrain. Deep, but stony, brown soil. Shaded and weakly developed undergrowth, < 40 cm tall. No cultivation. No sign of grazing. Apparently little burning, apart from along the road, where there was tall *Hyparrhenia*.

Annona senegalensis
Bridelia scleroneura
Combretum molle
Entada africana
Ficus sycomorus
Lannea fruticosa
Oxytenanthera abyssinica
Piliostigma thonningii
Pseudocedrela kotschyi
Terminalia schimperiana
Vitex doniana

97 — (2018_07). In the Omo Valley, S of the Chebera-Churchura National Park, ca. 16 km from new bridge on the road from Ameya. 12/10/2018
6° 45' 24" N, 36° 35' 00" E; 950 m a.s.l.
Very dense mixed woodland without dominant woody species next to past and present farmland on undulating ground on the upper slopes of the Omo Valley. Deep, unusual dark brown soil with some stones.

Ground-cover of few and shaded herbs. No sign of grazing. Little burning because of dense *Vernonia* scrub over the ground-cover.

Annona senegalensis
Bridelia scleroneura
Combretum collinum
Cordia africana
Entada abyssinica
Ficus sur
Ficus sycomorus
Lannea schimperi
Stereospermum kunthianum
Terminalia laxiflora
Vitex doniana

98 – (2018_08). In the Omo Valley, S of the Chebera-Churchura National Park, ca. 35 km from new bridge on the road from Ameya. 12/10/2018
6° 52' 54" N, 36° 35' 54" E; 1450 m a.s.l.
Steeply sloping mixed woodland without dominant woody species at forest edge (forest with *Polyscias ferruginea, Vernonia amygdalina, Phoenix reclinata, Maesa lanceolate and Albizia grandibracteata*). Forest species not listed below. Deep brown soil with no stones. Ground-cover sparse due to shading. No cultivation. No sign of grazing. Little or no burning.

Combretum molle
Entada abyssinica
Ficus sycomorus
Gardenia ternifolia
Gnidia lamprantha
Rhus vulgaris
Terminalia schimperiana
Vitex doniana

99 – (2018_09). In the Omo Valley, near S boundary of Chebera-Churchura National Park, on the road from Ameya. 13/10/2018
6° 53' 54" N, 36° 36' 18" E; 1500 m a.s.l.
Mosaic of mixed, open woodland without dominant woody species and forest on sloping ground, marking the lower boundary of the forest zone. Soil dark brown. Ground-cover in the woodland tall *Hyparrhenia* (> 1.5 m tall). No cultivation. No sign of grazing. Vegetation prone to burning.

Bridelia scleroneura
Combretum collinum
Dombeya quinqueseta
Entada abyssinica
Gnidia lamprantha
Oxytenanthera abyssinica
Piliostigma thonningii
Protea madiensis
Terminalia schimperiana
Vitex doniana

100 – (2018_10). Inside of Chebera-Churchura National Park, on the road from Ameya. 13/10/2018
6° 55' 30" N, 36° 39' 06" E; 1600 m a.s.l.
Patch of mixed, open woodland without dominant woody species on steep slope inside Transitional Rainforest-type of forest (in which occurred *Pouteria altissima, Albizia grandibracteata, Malacantha alnifolia, Ficus vallis-choudae, Ficus lutea*, etc.) Deep brown soil with few stones. Groundcover in the woodland consisting of herbs and small shrubs, in part also tall grasses, mainly *Hyparrhenia*, up to 1.5 m tall. No cultivation. No sign of grazing. Vegetation prone to burning, but few signs of recent fires.

Bridelia scleroneura
Combretum collinum
Combretum molle
Dombeya quinqueseta
Entada abyssinica
Ficus sycomorus
Maesa lanceolata
Terminalia schimperiana
Vitex doniana

101 – (2018_11). Highest patch of woodland inside the Chebera-Churchura National Park, on the road from Ameya. 13/10/2018
6° 57' 42" N, 36° 40' 06" E; 1630 m a.s.l.

Small area with mixed open woodland without dominant woody species on flat terrain. Deep, dark brown soil with few stones. Ground-cover of short grass and little burning (the area included an old road building site), elsewhere slightly longer grass, up to 1.5 m tall. No cultivation. No sign of grazing. No sign of burning, but almost certainly possibility for fires.

Albizia grandibracteata
Combretum collinum
Ficus sycomorus
Terminalia schimperiana

Profile O.

102 — (2018_12). Just N of Dima, on the road from Bebeka to Dima via Gurefada. 16/10/2018
6° 31' 18" N, 35° 10' 42" E; 625 m a.s.l.
Mixed, open wooded grassland with much *Acacia seyal* on flat or slightly undulating terrain. Soil stony, brown. Ground-cover of tall grass (*Loudetia arundinacea*), up to 2 m tall. Some cultivation. No sign of grazing. No sign of recent burning, but the vegetation is liable to fire.

Acacia polyacantha subsp. *campylacantha*
Acacia seyal
Balanites aegyptiaca
Bridelia scleroneura
Combretum collinum
Harrisonia abyssinica
Lannea fruticosa
Lonchocarpus laxiflorus
Maytenus senegalensis
Pseudocedrela kotschyi
Sterculia africana
Terminalia brownii

103 — (2018_13). ca. 12 km N of Dima, on the road from Bebeka to Dima via Gurefada. 16/10/2018
6° 36' 48" N, 35° 10' 54" E; 750 m a.s.l.
Mixed, open woodland without dominant woody species on undulating terrain. Soil dark brown, sometimes stony. Ground-cover of tall grass up to 2 m tall. No cultivation. No sign of grazing. No sign of burning, but almost certainly possibility for fires.

Acacia polyacantha subsp. *campylacantha*
Acacia seyal
Annona senegalensis
Bridelia scleroneura
Boscia salicifolia
Combretum collinum
Entada africana
Grewia mollis
Lannea fruticosa
Lonchocarpus laxiflorus
Maytenus senegalensis
Pseudocedrela kotschyi
Sclerocarya birrea
Sterculia africana
Terminalia brownii

104 — (2018_14). Near the Gurefada village, on the road from Bebeka Forest to Dima via Gurefada. 16/10/2018
6° 43' 42" N, 35° 10' 54" E; 950 m a.s.l.
Mixed, open woodland without dominant woody species on undulating terrain, vegetation reinvading abandoned cultivation Soil dark brown, stony. Almost all the ground-cover consisted of low subshrubs of *Sida* sp., which may be slightly liable to burning. No sign of grazing. No sign of burning.

Acacia polyacantha subsp. *campylacantha*
Annona senegalensis
Combretum collinum
Dombeya quinqueseta
Erythrina abyssinica
Ficus sur
Ficus sycomorus
Ficus vasta
Grewia mollis
Lannea schimperi
Lonchocarpus laxiflorus
Maytenus senegalensis
Ozoroa insignis

Piliostigma thonningii
Stereospermum kunthianum
Terminalia brownii
Terminalia schimperiana
Vitex doniana

105 — (2018_15). Highest point with woodland along the road from Bebeka Forest to Dima via Gurefada. 16/10/2018

6° 45' 48" N, 35° 11' 48" E; 1300 m a.s.l.
Mixed, open woodland without dominant woody species on undulating terrain near the forest boundary, with scattered continued and abandoned cultivations. Soil brown with stones. Ground-cover with grasses and other herbs typical of forest, thin and < 40 cm tall. No sign of grazing. No sign of burning.

Annona senegalensis
Combretum collinum
Cordia africana
Entada abyssinica
Ficus sur
Grewia mollis
Terminalia schimperiana
Trema orientalis
Vitex doniana

Profile P.

106 — (2018_16). Highest point with woodland S of Bechuma, along the road from Shewa Gimira to Maji. 19/10/2018

6° 47' 12" N, 35° 53' 06" E; 1780 m a.s.l.
Open woodland on rather steep slopes, dominated by *Cordia africana*. Soil brown, stony. Slopes with moderately tall grass, up to 1 m tall in places, but usually much lower due to grazing. No cultivation. No sign of burning, but probably with some burning on the steepest slopes with high ground-cover.

Acacia polyacantha subsp. *campylacantha*
Combretum molle
Cordia africana
Entada abyssinica
Ficus vasta
Rhus vulgaris
Terminalia schimperiana
Vitex doniana

107 — (2018_17). Lowest point with woodland in the area S of Bechuma, along the road from Shewa Gimira to Maji. 19/10/2018

6° 43' 00" N, 35° 52' 00" E; 1350 m a.s.l.
This site is at the border between woodland and Transitional Rainforest in a valley near a river (tributary to Akobo) and the vegetation of the valley includes plants representing degraded Transitional Rainforest. The woodland-component is mixed, open woodland without dominant woody species on moderately steep sloping ground. Soil brown, stony. Ground-cover in the woodland short perennial grass, heavily grazed and 10-20 cm tall. Some cultivation. Burning not likely.

Albizia schimperiana
Combretum molle
Cordia africana
Ficus thonningii
Terminalia schimperiana
Vitex doniana

108 — (2018_18). Just S of the village of Djemu, along the road from Shewa Gimira to Maji. 19/10/2018

6° 35' 30" N, 35° 45' 48" E; 1350 m a.s.l.
Mixed, open and species-rich woodland without dominant woody species and with signs of scattered present and past cultivation. Soil dark with stones. Ground-cover very short with perennial grasses, 10-20 cm tall. No sign of grazing. No sign of burning.

Acacia polyacantha subsp. *campylacantha*
Annona senegalensis
Bridelia scleroneura
Combretum collinum
Combretum molle
Cussonia arborea

Dichrostachys cinerea
Dombeya quinqueseta
Ficus sycomorus
Gardenia ternifolia
Maytenus senegalensis
Piliostigma thonningii
Protea madiensis
Psorospermum febrifugum
Terminalia schimperiana
Vitex doniana

109 — (2018_19). Further S of the village of Djemu, along the road from Shewa Gimira to Maji. 19/10/2018

6° 31' 06" N, 35° 44' 30" E; 1060 m a.s.l.
Mixed, open woodland without dominant woody species on gently undulating ground. Soil dark brown, stony. Ground-cover moderately tall grass, up to 1 m tall. No cultivation. No sign of Grazing. Potential for burning, but no sign of fires.

Acacia polyacantha subsp. *campylacantha*
Annona senegalensis
Bridelia scleroneura
Combretum collinum
Combretum molle
Dombeya quinqueseta
Ficus platyphylla
Ficus sycomorus
Gardenia ternifolia
Grewia mollis
Ozoroa insignis
Piliostigma thonningii
Pavetta crassipes
Rhus vulgaris
Stereospermum kunthianum
Terminalia schimperiana

110 — (2018_20). Even further S of the village of Djemu (11 km S of previous relevé, no. 109), along the road from Shewa Gimira to Maji. 19/10/2018

6° 27' 36" N, 35° 41' 12" E; 1350 m a.s.l.
Patches of mixed, open woodland without dominant woody species on gentle slopes in farmland. Soil brown, stony. Ground-cover in places < 40 cm tall, in other places up to 1.5 m tall. No sign of grazing. The farmland borders grassland with moderate signs of burning (grass fires were actually observed nearby).

Acacia seyal
Annona senegalensis
Bridelia scleroneura
Combretum collinum
Cussonia arborea
Dombeya quinqueseta
Ficus sycomorus
Gardenia ternifolia
Grewia mollis
Maytenus senegalensis
Piliostigma thonningii
Stereospermum kunthianum
Terminalia schimperiana
Ziziphus abyssinica

111 — (2018_21). 5 km W of Tum along the road towards Dima. 20/10/2018

6° 16' 46" N, 35° 29' 38" E; 1180 m a.s.l.
Mixed, open woodland without dominant woody species on flat terrain, surrounded by farmland and surrounding a *Scleria* swamp. Soil deep brown. Ground-cover of tall, perennial grasses up to 1.5 m tall, but no sign of burning. No sign of grazing. It was noted that no woodland species of *Combretum* was observed. [Around the swamp*: Ficus vasta, Croton macrostachyus* and the climber *Combretum panniculatum;* species not included in the relevé].

Acacia polyacantha subsp. *campylacantha*
Cordia africana
Lannea schimperi
Entada abyssinica
Ficus sycomorus
Piliostigma thonningii
Stereospermum kunthianum
Terminalia schimperiana
Vitex doniana

112 – (2018_22). Along the road towards Djemu E of Tum, on the lowermost slope of the Maji mountains. 20/10/2018
6° 20' 36" N, 35° 39' 18" E; 1550 m a.s.l.
Mixed open woodland without dominant woody species on slightly undulating terrain, with marginal presence of Afromontane trees. Ground-cover rather short grassland, < 30 cm tall, with much *Piloselloides hirsuta* on black cotton soil. Occasional cultivation. No sign of grazing. Clear signs of burning of the ground-cover and on bark of trees.

Acacia seyal
Annona senegalensis
Combretum molle
Cordia africana
Cussonia arborea
Dombeya quinqueseta
Entada abyssinica
Faurea rochetiana
Gardenia ternifolia
Gnidia lamprantha
Grewia mollis
Lannea schimperi
Piliostigma thonningii
Protea gaguedi
Terminalia brownii
Terminalia schimperiana
Vitex doniana

113 – (2018_23). Along the road towards Djemu E of Tum and ca. 10 km from the previous relevé no. 112). 20/10/2018
6° 23' 36" N, 35° 39' 48" E; 1500 m a.s.l.
Mixed, moderately open woodland without dominant woody species on slightly undulating terrain. Soil brown with scattered termite mounds and no stones. Ground-cover of moderately tall perennial grasses, up to 1 m tall. Some cultivation. No sign of grazing. Clear signs of burning of the ground-cover and on bark of trees.

Acacia dolichocephala
Acacia seyal
Annona senegalensis
Bridelia scleroneura
Combretum collinum
Combretum molle
Cussonia arborea
Entada africana
Euclea racemosa subsp. *schimperi*
Ficus sur
Gardenia ternifolia
Heteromorpha arborescens
Piliostigma thonningii
Pavetta crassipes
Rhus vulgaris
Terminalia brownii
Terminalia schimperiana
Vitex doniana

Profile M.

114 – (2018_24). Mid-altitude slopes of the Bure Escarpment on the road from Gore and Metu to Gambela. 24/10/2018
8° 18' 30" N, 34° 59' 42" E; 800 m a.s.l.
Mixed, open woodland without dominant woody species on steep sloping ground. Soil brown, stony. Ground-cover dense and 2-2.5 m tall, dominated by the grass *Loudetia arundinacea*. No cultivation. No sign of grazing. Vegetation highly prone to burning.

Acacia polyacantha subsp. *campylacantha*
Acacia seyal
Albizia malacophylla
Annona senegalensis
Calotropis procera
Combretum collinum
Entada africana
Ficus ovata
Flueggea virosa
Grewia mollis
Harrisonia abyssinica
Lannea fruticosa
Lonchocarpus laxiflorus
Piliostigma thonningii

Pseudocedrela kotschyi
Pterocarpus lucens
Terminalia laxiflora

115 — (2018_25). Mid-altitude slopes of the Bure Escarpment on the road from Gore and Metu to Gambela, above, but near relevé no. 114. 24/10/2018
8° 19' 30" N, 35° 00' 12" E; 890 m a.s.l.

Mixed, open woodland without dominant woody species on steep sloping ground. Soil brown, stony. Ground-cover dense and 2-2.5 m tall, dominated by tall grasses (species of *Hyparrhenia* and *Loudetia arundinacea*). No cultivation. No sign of grazing. Vegetation highly prone to burning.

Acacia polyacantha subsp. *campylacantha*
Acacia seyal
Albizia malacophylla
Bridelia scleroneura
Combretum collinum
Combretum molle
Cordia africana
Dombeya quinqueseta
Entada africana
Ficus sycomorus
Flueggea virosa
Grewia mollis
Lannea fruticosa
Lonchocarpus laxiflorus
Piliostigma thonningii
Pseudocedrela kotschyi
Pterocarpus lucens
Terminalia laxiflora

116 — (2018_26). Mid-altitude slopes of the Bure Escarpment on the road from Gore and Metu to Gambela, above, but near previous relevé (no. 115). 24/10/2018
8° 19' 48" N, 35° 00' 48" E; 990 m a.s.l.
Mixed, open woodland without dominant woody species on steep sloping ground. Soil brown, stony. Ground-cover dense and 2-2.5 m tall, dominated by tall grasses (species of *Hyparrhenia* and *Loudetia arundinacea*). No cultivation. No sign of grazing. Vegetation highly prone to burning.

Acacia polyacantha subsp. *campylacantha*
Albizia malacophylla
Annona senegalensis
Balanites aegyptiaca
Cordia africana
Flueggea virosa
Grewia mollis
Harrisonia abyssinica
Lonchocarpus laxiflorus
Pterocarpus lucens
Terminalia laxiflora

117 — (2018_27). Mid-altitude slopes of the Bure Escarpment on the road from Gore and Metu to Gambela, above, but near the previous relevé (no. 116). 24/10/2018
8° 19' 36" N, 35° 01' 42" E; 1150 m a.s.l.
Mixed, open woodland without dominant woody species on steep sloping ground. Soil brown, stony. Ground-cover dense and 2-2.5 m tall, dominated by tall grasses (species of *Hyparrhenia* and *Loudetia arundinacea*). No cultivation. No sign of grazing. Vegetation highly prone to burning. [Along small stream *Trema orientalis, Sesbania melanocaulis, Ficus glumosa, Albizia grandibracteata* and *Ficus vallis-choudae*, not included in the list below].

Acacia seyal
Albizia malacophylla
Combretum collinum
Cordia africana
Cussonia arborea
Ficus ovata
Grewia mollis
Pterocarpus lucens
Terminalia laxiflora

118 – (2018_28). Mid-altitude slopes of the Bure Escarpment on the road from Gore and Metu to Gambela, above, but near the previous relevé (no. 117). 24/10/2018

8° 19' 12" N, 35° 02' 06" E; 1300 m a.s.l.
Mixed, open woodland without dominant woody species on steep sloping ground. Soil brown, stony. Ground-cover dense and 2-2.5 m tall, dominated by tall grasses (species of *Hyparrhenia* and *Loudetia arundinacea*). No cultivation. No sign of grazing. Vegetation highly prone to burning. [Along small stream *Ficus sur* and *Ficus vallis-choudae*, the latter not in the woodland and not included in the relevé].

Acacia seyal
Albizia malacophylla
Cordia africana
Flueggea virosa
Grewia mollis
Pterocarpus lucens
Rhus vulgaris
Terminalia laxiflora

Profile K.

119 – (2018_29). In the Didessa Valley, on the road between Bedele and Nekemt. 25/10/2018

8° 35' 30" N, 36° 20' 30" E; 1200 m a.s.l.
Mixed, open woodland without dominant woody species on undulating ground, almost completely converted to farmland. Soil brown, stony. Hardly any natural ground-cover between the stones. No sign of grazing. No sign of burning.

Albizia malacophylla
Bridelia scleroneura
Combretum collinum
Cordia africana
Cussonia arborea
Dodonaea angustifolia
Entada abyssinica
Ficus sycomorus
Gardenia ternifolia
Grewia mollis
Piliostigma thonningii
Rhus vulgaris
Stereospermum kunthianum
Terminalia schimperiana

120 – (2018_30). In the Didessa Valley on the road between Bedele and Nekemt, near bridge on Didessa River. 25/10/2018

8° 36' 54" N, 36° 21' 12" E; 1450 m a.s.l.
Mixed, open woodland without dominant woody species on undulating ground. Soil brown, stony. Ground-cover of grasses up to ca. 2 m tall. No cultivation. No sign of grazing. Vegetation liable to burning.

Albizia malacophylla
Bridelia scleroneura
Combretum collinum
Faurea rochetiana
Ficus sur
Gardenia ternifolia
Grewia mollis
Lannea schimperi
Lonchocarpus laxiflorus
Piliostigma thonningii
Stereospermum kunthianum
Syzygium Guineense subsp. *macrocarpum*
Terminalia schimperiana

121 – (2018_31). In the Didessa Valley on the road between Bedele and Nekemt, just N of bridge on Didessa River. 25/10/2018

8° 41' 30" N, 36° 25' 18" E; 1300 m a.s.l.
Mixed, open woodland without dominant woody species on flat valley bottom around a church with no cultivation attached. Soil brown, stony soil. Ground-cover in places of short, low grasses (cut or mowed) or elsewhere tall grass up to ca. 2 m tall. No sign of grazing. The tall grass liable to burning.

Albizia malacophylla
Bridelia scleroneura
Combretum collinum
Cordia africana

Ficus sur
Gardenia ternifolia
Grewia mollis
Lonchocarpus laxiflorus
Piliostigma thonningii
Stereospermum kunthianum
Terminalia schimperiana

Profile L.

122 – (2018_32). North-east of Dembidolo on the road from Dembidolo to Gimbi. 28/10/2018
8° 43' 54" N, 34° 55' 42" E; 1600 m a.s.l.
Mixed, open woodland without dominant woody species on undulating ground. Soil brown, stony. Ground-cover very low due to grazing. Abundant cultivation and grazing below the trees. No sign of burning.

Acacia abyssinica
Acacia seyal
Combretum collinum
Combretum molle
Cordia africana
Entada abyssinica
Ficus sycomorus
Ficus vasta
Albizia malacophylla
Maytenus senegalensis
Piliostigma thonningii
Rhus vulgaris
Stereospermum kunthianum
Terminalia schimperiana

123 – (2018_33). Further north-east of Dembidolo on the road from Dembidolo to Gimbi. 28/10/2018
8° 50' 24" N, 35° 10' 54" E; 1400 m a.s.l.
Mixed, open woodland without dominant woody species on undulating ground. Soil brown, stony. Ground-cover thin with very little grass and consisting mainly of weedy species of dicots with low potential for burning. Abundant cultivation and grazing below the trees. No sign of burning. Along the road between the relevés no. 122 and 123 there are many areas at lower altitudes, in which species more typical of Transitional Rainforest than woodlands, for example *Albizia grandibracteata, Cordia africana* and occasionally also *Trichilia dregeana* are dominant. This indicates that the area forms a transition to Transitional Rainforest. Nearly all these areas with forest trees have under-planting of *Coffea arabica*, and even the fern *Platycerium elephantotis* was growing on some of them, while the open, cultivated land have woodland trees left here and there.

Acacia abyssinica
Albizia grandibracteata
Bersama abyssinica
Bridelia micrantha
Clausena anisata
Combretum collinum
Croton macrostachyus
Entada abyssinica
Ficus sur
Flacourtia indica
Flueggea virosa
Gardenia ternifolia
Rhus vulgaris
Stereospermum kunthianum
Terminalia schimperiana
Trema orientalis

124 – (2018_34). Even further north-east of Dembidolo on the road from Dembidolo to Gimbi. 28/10/2018
8° 53' 18" N, 35° 15' 54" E; 1450 m a.s.l.
Mixed, open woodland without dominant woody species on undulating ground. Soil brown, stony. Ground-cover thin, with short grass, < 30 cm tall. Little or no sign of burning.

Albizia malacophylla
Combretum collinum
Entada abyssinica
Ficus lutea
Ficus sur
Ficus vasta

Flueggea virosa
Gardenia ternifolia
Syzygium guineense subsp. *macrocarpum*
Terminalia schimperiana
Trema orientalis

125 — (2018_35). In the Didessa Valley on the Gimbi side, along the road from Gimbi to Nekemt. 31/10/2018
9° 02' 12" N, 36° 02' 00" E; 1300 m a.s.l.
Undulating mixed, open woodland without dominant woody species on the valley bottom. Soil brown, stony. Ground-cover with patches of tall grass in the ground-cover, 1-1.5 m tall, where the ground is not cultivated. No sign of grazing. No sign of recent fires, but the vegetation away from the cultivations highly liable to burning.

Bridelia scleroneura
Combretum collinum
Cordia africana
Entada abyssinica
Erythrina abyssinica
Ficus sur
Lonchocarpus laxiflorus
Oxytenanthera abyssinica
Piliostigma thonningii
Stereospermum kunthianum
Terminalia schimperiana
Vitex doniana

Profile H.

126 — (2018_36). Uppermost part of the Abay Gorge south of Bure on the road between Bure and Nekemt, just below lower limit of typical highland vegetation. 10.11. 2018
10° 23' 06" N, 37° 01' 36" E; 1850 m a.s.l.
Mixed, open woodland without dominant woody species on gently sloping terrain with many rocks and areas of rocky outcrops. Soil red, stony. Ground-cover of very short grass < 20 cm high. No cultivation. Evidence of some grazing. Hardly any possibility for grass fires.

Acacia amythethophylla
Acacia persiciflora
Acacia seyal
Capparis tomentosa
Combretum molle
Croton macrostachyus
Cussonia ostinii
Dichrostachys cinerea
Dombeya quinqueseta
Erythrina abyssinica
Ficus thonningii
Ficus vasta
Flueggea virosa
Gardenia ternifolia
Grewia ferruginea
Lannea schimperi
Maytenus senegalensis
Pavetta oliveriana
Polyscias farinosa
Premna schimperi
Piliostigma thonningii
Rhus vulgaris
Schrebera alata
Senna petersiana
Vangueria madagascariensis

127 — (2018_37). Uppermost part of the Abay Gorge south of Bure, on the road between Bure and Nekemt, further below lower limit of typical highland vegetation, but not near the upper edge of the narrow part of the gorge. 10.11. 2018
10° 22' 30" N, 37° 00' 54" E; 1750 m a.s.l.
Open woodland without dominant woody species on moderately sloping terrain, almost totally dominated by *Anogeissus leiocarpa*, but with an admixture of *Acacia hecatophylla* and other woody species. Many rocks and hardly any soil. Ground-cover very thin and with very little and short grass. No cultivation. No sign of grazing. Hardly any opportunity for grass fires.

Acacia gerrardii
Acacia amythethophylla

Acacia hecatophylla
Acacia seyal
Albizia malacophylla
Anogeissus leiocarpa
Combretum collinum
Combretum molle
Dichrostachys cinerea
Entada abyssinica
Flueggea virosa
Heteromorpha arborescens
Hymenodictyon floribundum
Lonchocarpus laxiflorus
Ormocarpum pubescens
Piliostigma thonningii
Senna petersiana
Sterculia africana
Strychnos innocua
Vangueria madagascariensis

128 – (2018_38). Upper part of the Abay Gorge south of Bure, on the road between Bure and Nekemt, just above the upper edge of the inner narrow gorge; below a village. 11.11.2018
10° 20' 42" N, 37° 00' 54" E; 1650 m a.s.l.
Remains of mixed, open woodland without dominant woody species in a mosaic of farmland and heavily grazed woodland on sloping ground. Soil grey, stony. Ground-cover very low; some areas with prominent rocks (where *Strychnos innocua, Lannea fruticosa* and *Dombeya quinqueseta* occurred). No sign of burning.

Acacia amythethophylla
Acacia seyal
Cordia africana
Croton macrostachyus
Dombeya quinqueseta
Erythrina abyssinica
Ficus glumosa
Ficus sycomorus
Flueggea virosa
Grewia trichocarpa
Heteromorpha arborescens
Lannea fruticosa
Ozoroa insignis
Rhus vulgaris
Senna petersiana
Strychnos innocua
Vangueria madagascariensis

129 – (2018_39). Upper part of the Abay Gorge south of Bure, on the road between Bure and Nekemt, below the edge of the inner narrow gorge. 10.11.2018
10° 20' 00" N, 37° 01' 30" E; 1550 m a.s.l.
Mixed, open woodland without dominant woody species on steeply sloping ground. Soil grey, stony, consisting of fine gravel derived from crystalline, probably granitic rocks. Ground-cover of very short grass and, in places, the introduced *Chamaecrista rotundifolia* (*Friis et al.* 18519, a species native of Central America, new to Ethiopia), formed almost monospecific stands. No cultivation. No sign of grazing. No sign of burning.

Acacia seyal
Dichrostachys cinerea
Faurea rochetiana
Ficus glumosa
Ficus sycomorus
Flueggea virosa
Lannea fruticosa
Maerua angolensis
Rhus vulgaris
Senna petersiana
Vangueria madagascariensis
Ziziphus mucronata

130 – (2018_40). Ca. 1/3 down the inner, narrow part of the Abay Gorge, on the road between Bure and Nekemt. 11.11.2018
10° 19' 18" N, 37° 01' 30" E; 1450 m a.s.l.
Open, species-rich woodland without *Terminalia* and *Combretum* (*Anogeissus leiocarpa* present, but not dominant) on moderately sloping ground. Soil grey, stony, derived from crystalline rocks, e.g., granites. Ground-cover rather well-developed, partly consisting of annual or short-lived perennial grasses, e.g., *Aristida*, but no *Hyparrhenia* or *Loudetia*, ca. 1 m tall. No cultivation.

No sign of grazing. Signs of burning showing that grass fires may occur, but will be quickly passing.

Acacia amythethophylla
Acacia persiciflora
Acacia seyal
Acacia venosa
Anogeissus leiocarpa
Dichrostachys cinerea
Dombeya quinqueseta
Erythrina abyssinica
Ficus sycomorus
Flueggea virosa
Grewia flavescens
Grewia mollis
Lannea fruticosa
Maytenus senegalensis
Ochna leucophloeos
Senna petersiana
Sterculia africana
Stereospermum kunthianum
Vangueria madagascariensis
Ziziphus abyssinica

131 — (2018_41). Above 1/2 down the inner, narrow part of the Abay Gorge south of Bure, on the road between Bure and Nekemt. 11.11.2018
10° 18' 54" N, 37° 01' 30" E; 1350 m a.s.l.
Mixed, open woodland without dominant woody species on steep, rocky slope. Soil grey, stony, derived from crystalline rocks, e.g., granites; scarce among the rocks. Ground-cover of ca. 2 m high grasses, *Hyparrhenia* and *Loudetia*. No cultivation. No sign of grazing. Burning of this ground-cover is liable to produce fierce grass fires.

Acacia seyal
Anogeissus leiocarpa
Boswellia papyrifera
Boswellia pirottae
Combretum collinum
Dichrostachys cinerea
Entada africana
Flueggea virosa
Hymenodictyon floribundum
Lannea fruticosa
Ochna leucophloeos
Ozoroa pulcherrima
Senna petersiana
Stereospermum kunthianum
Strychnos innocua
Terminalia laxiflora

132 — (2018_42). Below 1/2 down the inner, narrow part of the Abay Gorge south of Bure, on the road between Bure and Nekemt. 12.11.2018
10° 18' 36" N, 37° 01' 42" E; 1250 m a.s.l.
Mixed, open woodland without dominant woody species on steep, rocky slope. Soil grey, stony, derived from crystalline rocks, e.g., granites; scarce among the rocks. Ground-cover of ca. 2 m high grasses, *Hyparrhenia* and *Loudetia*. No cultivation. No sign of grazing. Burning of this ground-cover is liable to produce fierce grass fires.

Acacia seyal
Anogeissus leiocarpa
Boswellia papyrifera
Boswellia pirottae
Combretum collinum
Combretum molle
Dalbergia melanoxylon
Dichrostachys cinerea
Erythrina abyssinica
Ficus glumosa
Flueggea virosa
Grewia trichocarpa
Lannea fruticosa
Lonchocarpus laxiflorus
Maerua angolensis
Ormocarpum pubescens
Senna petersiana
Stereospermum kunthianum
Terminalia schimperiana

133 — (2018_43). Nearly 2/3 down the inner, narrow part of the Abay Gorge south of Bure, on the road between Bure and Nekemt. 12.11.2018
10° 18' 06" N, 37° 01' 30" E; 1150 m a.s.l.
Mixed, open woodland without dominant woody species on steep, rocky slope. Soil grey, stony, derived from crystalline rocks, e.g., granites; scarce among the rocks. Ground-cover of ca. 2 m high grasses, *Hyparrhenia* and *Loudetia*. No cultivation. No sign of grazing. Burning of this ground-cover is liable to produce fierce grass fires.

Acacia seyal
Albizia amara
Anogeissus leiocarpa
Boswellia papyrifera
Boswellia pirottae
Combretum collinum
Dalbergia melanoxylon
Dichrostachys cinerea
Flueggea virosa
Grewia trichocarpa
Lannea fruticosa
Lonchocarpus laxiflorus
Maerua angolensis
Sterculia africana
Stereospermum kunthianum
Terminalia schimperiana
Ziziphus spina-christi

134 — (2018_44). 2-3 km from the bridge on the Abay River in the Abay Gorge south of Bure, on the road between Bure and Nekemt. 12.11.2018
10° 17' 54" N, 37° 01' 42" E; 1050 m a.s.l.
Mixed, open woodland without dominant woody species on steep, rocky slope. Soil grey, stony, derived from crystalline rocks, e.g., granites, scarce among the rocks. Ground-cover of ca. 2 m high grasses, *Hyparrhenia* and *Loudetia*. No cultivation. No sign of grazing. Burning of this ground-cover is liable to produce fierce grass fires (during our observations early in the fire season, the grass was already burnt in one spot).

Acacia amythethophylla
Anogeissus leiocarpa
Boswellia papyrifera
Calotropis procera
Cassia arereh
Combretum collinum
Combretum molle
Dalbergia melanoxylon
Dichrostachys cinerea
Ficus glumosa
Ficus cordata subsp. *salicifolia*
Flueggea virosa
Grewia trichocarpa
Lannea barteri
Lannea fruticosa
Lonchocarpus laxiflorus
Sterculia africana
Stereospermum kunthianum
Strychnos innocua
Terminalia schimperiana
Ziziphus spina-christi

135 — (2018_45). About 1 km from the bridge on the Abay River in the Abay Gorge south of Bure, on the road between Bure and Nekemt. 12.11.2018
10° 17' 36" N, 37° 01' 24" E; 950 m a.s.l.
Mixed, open woodland without dominant woody species on steep, rocky slope. Soil, grey, stony, derived from crystalline rocks, e.g., granites, scarce among the rocks. Ground-cover of ca. 2 m high grasses, *Hyparrhenia* and *Loudetia*. No cultivation. No sign of grazing. Burning of this ground-cover is liable to produce fierce grass fires (during our observations early in the fire season, the grass was already burnt in one spot). A number of tree-species, which seemed distinct from the ones listed here, were leafless and sterile and out of reach on the steep slopes.

Acacia amythethophylla
Anogeissus leiocarpa
Boswellia papyrifera
Calotropis procera
Combretum molle
Dichrostachys cinerea

Ficus glumosa
Flueggea virosa
Grewia trichocarpa
Lannea barteri
Lannea fruticosa
Lonchocarpus laxiflorus
Sterculia africana
Stereospermum kunthianum
Terminalia laxiflora
Ziziphus spina-christi

136 – (2018_46). Just above the bridge on the Abay River in the Abay Gorge south of Bure, on the road between Bure and Nekemt. 12.11.2018
10° 17' 30" N, 37° 01' 06" E; 850 m a.s.l.
Open woodland dominated by *Lannea barteri* on very steep, rocky slope. Soil grey, stony, derived from crystalline rocks, e.g., granites, scarce among the rocks. Ground-cover of up to ca. 2 m high grasses, *Hyparrhenia* and *Loudetia*, mostly on very steep slopes. No cultivation. No grazing. Burning of this ground-cover is liable to produce fierce grass fires (during our observations early in the fire season, the grass was already burnt in one spot). A number of tree-species, which seemed distinct from the ones listed here, were leafless and sterile and out of reach on the steep slopes.

Anogeissus leiocarpa
Boswellia papyrifera
Calotropis procera
Ficus glumosa
Grewia trichocarpa
Hymenodictyon floribundum
Lannea barteri
Lannea fruticosa
Lannea schimperi
Lonchocarpus laxiflorus
Pterocarpus lucens
Sterculia africana
Stereospermum kunthianum
Tamarindus indica
Terminalia schimperiana

Profile E.

137 – (2018_47). At the foot of the western escarpment west of Shahura (Alefa, west of Lake Tana), on the road from Shahura towards the western lowlands. 15.11.2018
11° 57' 18" N, 36° 38' 42" E; 1100 m a.s.l.
Mixed, open woodland without dominant woody species on hilly ground. Soil brown, stony. Ground-cover with many annual grasses and herbs, up to ca. 20 cm high. No cultivation, but ground-cover heavily grazed. No signs of burning and unlikely that this vegetation is prone for grass fires.

Acacia polyacantha subsp. *campylacantha*
Acacia seyal
Allophylus rubifolius
Anogeissus leiocarpa
Combretum collinum
Ficus sycomorus
Flacourtia indica
Flueggea virosa
Lannea schimperi
Lonchocarpus laxiflorus
Pterocarpus lucens
Sterculia africana
Terminalia schimperiana
Ziziphus spina-christi

138 – (2018_48). On the lower slopes of the western escarpment west of Shahura (Alefa, west of Lake Tana), on the road from Shahura towards the western lowlands. 15.11.2018
11° 57' 42" N, 36° 39' 24" E; 1200 m a.s.l.
Mixed, open woodland without dominant woody species on steeply sloping ground. Soil brown, stony. Ground-cover with more than 2 m tall grasses (*Loudetia*, *Hyparrhenia*) and herbs. No cultivation. No sign of grazing. Vegetation highly prone for fierce grass fires.

Acacia polyacantha subsp. *campylacantha*
Allophylus rubifolius

Anogeissus leiocarpa
Boswellia papyrifera
Boswellia pirottae
Calotropis procera
Combretum hartmannianum
Erythrina abyssinica
Ficus sycomorus
Flueggea virosa
Lannea fruticosa
Piliostigma thonningii
Pterocarpus lucens
Terminalia schimperiana
Trema orientalis
Ziziphus mucronata

139 – (2018_49). On the lower slopes of the western escarpment west of Shahura (Alefa, west of Lake Tana), on the road from Shahura towards the western lowlands. 15.11.2018

11° 57' 54" N, 36° 39' 42" E; 1300 m a.s.l.

Mixed, open woodland without dominant woody species on very steeply sloping ground. Soil brown, stony. Ground-cover with more than 2 m tall grasses (*Loudetia*, *Hyparrhenia*) and herbs. No cultivation. No sign of grazing. Vegetation highly prone for fierce grass fires.

Acacia polyacantha subsp. *campylacantha*
Acacia seyal
Albizia malacophylla
Anogeissus leiocarpa
Calotropis procera
Cordia africana
Ficus sycomorus
Lonchocarpus laxiflorus
Pterocarpus lucens
Rumex abyssinicus
Sarcocephalus latifolius
Sterculia africana
Vitex doniana
Woodfordia uniflora
Ziziphus abyssinica
Ziziphus mucronata

140 – (2018_50). On the middle slopes of the western escarpment west of Shahura (Alefa, west of Lake Tana), on the road from Shahura towards the western lowlands. 15.11.2018

11° 58' 06" N, 36° 39' 42" E; 1400 m a.s.l.

Mixed, open and degraded woodland without dominant woody species on very steeply sloping ground. Soil brown, stony. No cultivation. No sign of grazing. Ground-cover with more than 2 m tall grasses (*Loudetia, Hyparrhenia*) and herbs, highly prone for fierce grass fires.

Acacia polyacantha subsp. *campylacantha*
Ficus sycomorus
Flueggea virosa
Lannea fruticosa
Pterocarpus lucens
Rumex abyssinicus
Terminalia schimperiana

141 – (2018_51). On the upper slopes of the western escarpment west of Shahura (Alefa, west of Lake Tana), on the road from Shahura towards the western lowlands. 15.11.2018

11° 58' 06" N, 36° 40' 00" E; 1500 m a.s.l.

Mixed, open and heavily degraded woodland without dominant indigenous species, but with the invasive species *Lantana camara* occupying large areas on the steeply sloping ground (*Lantana camara* not recorded in the relevé). Soil brown, stony. Ground-cover with only very low grasses and herbs and much bare soil. No cultivation. No sign of grazing. Vegetation not prone for grass fires.

Acacia polyacantha subsp. *campylacantha*
Anogeissus leiocarpa
Calotropis procera
Combretum collinum
Ficus sycomorus
Flueggea virosa
Hymenodictyon floribundum
Rhus vulgaris
Rumex abyssinicus
Sarcocephalus latifolius

Syzygium guineense subsp. *macrocarpum*
Terminalia schimperiana
Woodfordia uniflora

142 — (2018_52). On the upper slopes of the western escarpment west of Shahura (Alefa, west of Lake Tana), on the road from Shahura towards the western lowlands. 15.11.2018
11° 58' 18" N, 36° 40' 00" E; 1600 m a.s.l.
Open, mixed and heavily degraded woodland and scrub. Soil, brown, stony. Natural ground-cover almost entirely replaced by cultivation. No sign of grazing. Due to the extensive cultivated areas, the vegetation is not prone for grass fires. At 1650 m the influence from a village has changed the vegetation entirely, and above that the vegetation is Afromontane woodland dominated by *Croton macrostachyus*, but *Anogeissus*, *Terminalia* and *Syzygium guineense* subsp. *macrocarpum* continue to be sporadically present in the Afromontane woodland up to 1800 m a.s.l.

Anogeissus leiocarpa
Cordia africana
Dichrostachys cinerea
Diospyros abyssinica
Embelia schimperi
Ficus vasta
Flueggea virosa
Sarcocephalus latifolius
Syzygium guineense subsp. *macrocarpum*

Profile D.

143 — (2018_53). Uppermost part of the Tacazze Valley, where the road from the main road (between Woldiya and Debre Tabor) to Lalibela crosses the valley. 18.11.2018
11° 54' 54" N, 38° 58' 48" E; 1850 m a.s.l.
Very open *Acacia-Ziziphus* woodland, devoid of all typical *Combretum-Terminalia* woodland species. Soil grey, stony and heavily overgrazed. Ground-cover with a few annual grasses and herbs, < 20 cm tall. No cultivation. No sign of grazing. No possibility for burning. The Tacazze Valley is here surrounded by very open scrub with evergreen shrubs like *Euclea* and *Dodonaea*; *Acacia etbaica* is also very common at higher altitudes. In many places between Lalibela and Sekota this semi-evergreen bushland has a very scattered tree-component of *Combretum molle* and (even less common) *Combretum collinum*, *Stereospermum kunthianum*, *Lannea fruticosa, Cordia africana* and *Croton macrostachyus.*

Acacia abyssinica
Acacia etbaica
Acacia seyal
Acacia tortilis
Senna petersiana
Ziziphus spina-christi

Profile C.

144 — (2018_54). Tacazze Valley north-west of Sekota, near the Tacazze bridge on the road from Sekota towards the Semien. 20.11.2018
12° 48' 42" N, 38° 38' 18" E; 1150 m a.s.l.
Very open woodland dominated by *Acacia asak* on moderately steep slopes, in many places the vegetation verging to subdesert scrubland with trees. Soil nearly white and very hard, with stones. No perennial ground-cover, at the time of our visit only seedlings of a number of dicot-species not yet identifiable. *Acacia asak* is also common in the Afar region on the east side of the Ethiopian plateau.

Acacia mellifera
Acacia asak
Adansonia digitata
Balanites aegyptiaca
Boscia angustifolia
Boscia senegalensis
Calotropis procera
Capparis decidua
Ficus cordata subsp. *salicifolia*
Salvadora persica
Ziziphus spina-christi

145 — (2018_55). Tacazze Valley north-west of Sekota, near the village of Zequala and the Tacazze bridge. 20.11.2018

12° 49' 00" N, 38° 46' 30" E; 1350 m a.s.l.

Open woodland dominated by *Terminalia brownii*. Soil pale and hard, very stony and with hardly any natural ground-cover, which has been replaced by cultivation, mainly consisting of drought-resistant races of *Sorghum*. Farming begins from ca. 1200 m a.s.l., from which altitude and upwards *Terminalia brownii* is common to ca. 1800 m a.s.l., in places mixed with *Acacia asak*.

Acacia mellifera
Acacia asak
Adansonia digitata
Balanites aegyptiaca
Boscia angustifolia
Boscia senegalensis
Calotropis procera
Capparis decidua
Grewia tenax
Terminalia brownii
Ziziphus spina-christi

146 — (2018_56). Tacazze Valley north-west of Sekota, ca. 10 km north-east of the village of Zequala on the road to Sekota. 20.11.2018

12° 46' 30" N, 38° 51' 12" E; 1550 m a.s.l.

Very open woodland without dominant woody species on flat terrain. Soil very pale grey and dry, stony. Much cultivation, natural ground-cover all replaced with drought-resistant races of *Sorghum*. No sign of grazing. No sign of fires.

Acacia asak
Adansonia digitata
Albizia amara
Capparis decidua
Stereospermum kunthianum
Terminalia brownii
Ziziphus spina-christi

147 — (AEW_1). Near Abergelle (data from Abeje Eshete Wassie (2011)). The Frankincense tree of Ethiopia – ecology, productivity and population dynamics. Ph.D.-Thesis, Wageningen – http://edepot.wur.nl/171534)

13° 08' 36" N, 39° 00' 48" E; 1600 m a.s.l.

Open woodland dominated by *Boswellia papyrifera*, used for tapping and production of frankincense. Ground-cover not described, but almost certainly the ground is very stony and with hardly any ground-cover. No grazing. Burning is very unlikely due to lack of biomass.

Acacia abyssinica
Acacia etbaica
Acacia mellifera
Acacia oerfota
Boswellia papyrifera
Capparis decidua
Combretum hartmannianum
Commiphora africana
Dichrostachys cinerea
Grewia erythraea
Grewia villosa
Lannea fruticosa
Lannea triphylla
Maerua angolensis
Salvadora persica
Senna singueana
Stereospermum kunthianum
Terminalia brownii

Profile B (part).

148 — (AEW_2). Near Metema (data from Abeje Eshete Wassie (2011)).

12° 56' 42" N, 36° 09' 42" E; 750 m a.s.l.

Mixed woodland. Habitat description very brief, but probably the terrain is flat, the soil is deep and brown, moderately stony, with moderately tall ground-cover, no cultivation and occasional burning. All information about this relevé is provided by Haile Adamu Wale et al. (2011).

Acacia polyacantha subsp. *campylacantha*
Acacia seyal
Albizia malacophylla
Anogeissus leiocarpa
Balanites aegyptiaca
Boswellia papyrifera
Boswellia pirottae
Combretum adenogonium
Combretum collinum
Combretum molle
Cordia africana
Dalbergia melanoxylon
Dichrostachys cinerea
Diospyros mespiliformis
Ficus glumosa
Flueggea virosa
Gardenia ternifolia
Grewia bicolor
Lannea fruticosa
Lonchocarpus laxiflorus
Maytenus senegalensis
Ochna leucophloeos
Pterocarpus lucens
Sterculia africana
Stereospermum kunthianum
Strychnos innocua
Terminalia laxiflora
Ximenia americana
Ziziphus abyssinica
Ziziphus spina-christi

Appendix 2: List of species codes, number of records per species and comparison with records in Friis et al. (2010)

The abbreviation of species names used in the analyses, diagrams and graphs in Section 10 are listed here with the full species names. Also indicated are the number of times the species have been recorded in the relevés and the number of collections (not the number of specimens) of the species that have been identified, georeferenced and used for the distribution maps in Section 6. The vegetation type from which the species were recorded in Friis et al. (2010) are extracted from Appendix 3 in that work ("Woody plants in the Flora of Ethiopia and Eritrea, assigned to vegetation types" on pp. 177-237), to be compared with the local elements, to which the species have been referred in Section 6 and 7. Here below are listed the taxa mentioned in Friis et al. (2010) as strict CTW-taxa or as taxa associated with CTW-area but not found in the relevés in the present study. See also Sections 3.6 and 4.3.

Strict CTW taxa listed in Friis et al. (2010) but not found in our relevés	*Catunaregam nilotica, Combretum nigricans, Dombeya longibracteolata, Hymenocardia acida, Strychnos henningsii, Vitellaria paradoxa, Euphorbia nigrispinoides, Euphorbia venefica, Turraea nilotica*
Taxa associated with CTW in Friis et al. (2010) but not found in our relevés	*Acacia drepanolobium* (ACB; CTW), *Albizia anthelmintica* (ACB; CTW), *Albizia coriaria* (CTW; TRF), *Allophylus africanus* (CTW; RV), *Commiphora schimperi* (ACB; CTW), *Euphorbia candelabrum* (ACB; CTW), *Ficus populifolia* (CTW; DAF), *Gardenia volkensii* var. *volkensii* (ACB; CTW; RV), *Pappea capensis* (CTW; DAF), *Pavetta gardeniifolia* var. *gardeniifolia* (CTW; DAF), *Psydrax schimperiana* subsp. *schimperiana* (ACB; CTW; DAF), *Rhus ruspolii* (CTW; DAF), *Sterculia cinerea* (?DSS;?ACB; CTW), *Sterculia setigera* (?ACB; CTW), *Strychnos spinosa* (CTW; RV), *Ximenia caffra* (ACB; CTW; RV).

Abbreviations used in analyses	Species	Records from relevés (numbers)	Records from specimens (numbers)	Recorded from literature (numbers)	Vegetation type in 2010 Atlas	Assigned to local element here
acaaby	*Acacia abyssinica*	5	79	1	DAF	MI
acaamy	*Acacia amythethophylla*	6	5		CTW	CTW[n]
acaasa	*Acacia asak*	3	20		ACB; RV	ACB[-CTW]
acadol	*Acacia dolichocephala*	1	47		ACB; CTW	CTW[all+ext]
acaetb	*Acacia etbaica*	2	86		ACB; DAF	ACB[-CTW]
acager	*Acacia gerrardii*	1	10		CTW	MI
acahec	*Acacia hecatophylla*	37	10		CTW	CTW[n+c]
acahoc	*Acacia hockii*	5	35		CTW	CTW[all+ext]
acamel	*Acacia mellifera*	11	57		ACB	ACB[-CTW]
acanil	*Acacia nilotica*	1	100		ACB; WGG	ACB[-CTW]

Abbreviations used in analyses	Species	Records from relevés (numbers)	Records from specimens (numbers)	Recorded from literature (numbers)	Vegetation type in 2010 Atlas	Assigned to local element here
acaoer	*Acacia oerfota*	2	42		ACB; DSS	ACB[-CTW]
acaper	*Acacia persiciflora*	1	22		?	MI
acacam	*Acacia polyacantha* subsp. *campylacantha*	30	46		CTW; RV	CTW[all+ext]
acasen	*Acacia senegal*	13	126		ACB; CTW	CTW[all+ext]
acasey	*Acacia seyal*	59	114	1	ACB; CTW; DAF; WGG	CTW[all+ext]
acasie	*Acacia sieberiana*	2	44	1	CTW; DAF	CTW[all+ext]
acator	*Acacia tortilis*	1	42		ACB; CTW; DSS	ACB[-CTW]
acaven	*Acacia venosa*	3	17		DAF	CTW[n+c]
adadig	*Adansonia digitata*	8	8		CTW	CTW[n]
albama	*Albizia amara*	9	29		DAF	CTW[all+ext]
albgra	*Albizia grandibracteata*	2	35		MAF; RV; TRF	MI
albmal	*Albizia malacophylla*	20	39		CTW; DAF; RV	CTW[n+c]
albsch	*Albizia schimperiana*	1	56		DAF; MAF	MI
allrub	*Allophylus rubifolius*	9	68		ACB; CTW; DAF	CTW[all+ext]
annsen	*Annona senegalensis*	28	19		CTW	CTW[c+s+ext]
anolei	*Anogeissus leiocarpa*	78	35		CTW	CTW[n+c]
apodim	*Apodytes dimidiata*	1	54		DAF; MAF; RV	MI
balaeg	*Balanites aegyptiaca*	39	77		ACB; CTW; WGG, DSS	CTW[all+ext]
beraby	*Bersama abyssinica*	3	95		DAF; MAF; RV	MI
boraet	*Borassus aethiopum*	2	2		CTW	CTW[n+c]
bosang	*Boscia angustifolia*	5	22		ACB; DSS	ACB[-CTW]
bossal	*Boscia salicifolia*	2	25		ACB; DAF	ACB[-CTW+ext]
bossen	*Boscia senegalensis*	3	12		ACB; RV	ACB[-CTW]
bospap	*Boswellia papyrifera*	31	26		CTW	CTW[n+c]
bospir	*Boswellia pirottae*	12	15		CTW	CTW[n+c]
brimic	*Bridelia micrantha*	2	54		DAF; RV	CTW[all+ext]
briscl	*Bridelia scleroneura*	23	48		CTW; RV	CTW[all+ext]
bruant	*Brucea antidysenterica*	1	85		DAF; MAF	MI
cadfar	*Cadaba farinosa*	2	121		not treated	ACB[-CTW]
calpro	*Calotropis procera*	13	40		ACB; DSS; RV	ACB[-CTW]

Abbreviations used in analyses	Species	Records from relevés (numbers)	Records from specimens (numbers)	Recorded from literature (numbers)	Vegetation type in 2010 Atlas	Assigned to local element here
calaur	*Calpurnia aurea*	1	93		DAF	MI
capdec	*Capparis decidua*	5	11		ACB; DSS	ACB[-CTW]
captom	*Capparis tomentosa*	1	150		ACB; DAF; RV	ACB[-CTW]
carspi	*Carissa spinarum*	2	169		ACB; DAF; RV	MI
casare	*Cassia arereh*	3	10		CTW; DAF	CTW[n+c]
claani	*Clausena anisata*	1	87		DAF; MAF	MI
cluaby	*Clutia abyssinica*	1	43		DAF	MI
comade	*Combretum adenogonium*	6	31		CTW; DAF/WG	CTW[all+ext]
comcol	*Combretum collinum*	71	90	1	CTW	CTW[all+ext]
comhar	*Combretum hartmannianum*	19	8		CTW	CTW[n+c]
commol	*Combretum molle*	36	173		CTW; DAF/WG	CTW[all+ext]
comroc	*Combretum rochetianum*	3	11		CTW	CTW[n]
comafr	*Commiphora africana*	1	87		ACB; DAF; DSS	ACB[-CTW]
comped	*Commiphora pedunculata*	3	3		CTW	CTW[c]
corafr	*Cordia africana*	31	94		CTW; DAF; MAF; TRF	TRG
crofeb	*Crossopteryx febrifuga*	5	16		CTW	CTW[c]
cromac	*Croton macrostachyus*	12	135		CTW; DAF/U-WG; MAF/P-BW	TRG
cusarb	*Cussonia arborea*	8	12		CTW; DAF	CTW[all+ext]
cusost	*Cussonia ostinii*	1	21		CTW; DAF	MI
dalboe	*Dalbergia boehmii*	5	1		not treated	CTW[c]
dalmel	*Dalbergia melanoxylon*	25	15		CTW	CTW[n+c]
diccin	*Dichrostachys cinerea*	38	78	6	ACB; CTW; DAF	TRG
dioaby	*Diospyros abyssinica*	1	41		DAF; MAF; RV; TRF	MI
diomes	*Diospyros mespiliformis*	6	44		CTW; RV	MI
dodang	*Dodonaea angustifolia*	2	118	1	DAF	MI
dombur	*Dombeya buettneri*	1	9		CTW	CTW[c+s]
domqui	*Dombeya quinqueseta*	20	36		CTW	CTW[all+ext]
embsch	*Embelia schimperi*	1	48		DAF; RV	MI
entaby	*Entada abyssinica*	23	73		CTW; DAF	CTW[all+ext]
entafr	*Entada africana*	26	14		CTW	CTW[all]
eriaby	*Erythrina abyssinica*	8	49		CTW; DAF	CTW[all+ext]

Abbreviations used in analyses	Species	Records from relevés (numbers)	Records from specimens (numbers)	Recorded from literature (numbers)	Vegetation type in 2010 Atlas	Assigned to local element here
eroaby	*Erythrococca trichogyne*	1	23		MAF	MI
eryfis	*Erythroxylum fischeri*	1	9		RV; TRF	MI
eucrac	*Euclea racemosa* subsp. *schimperi*	1	97		DAF	MI
eupaby	*Euphorbia abyssinica*	1	25		DAF	MI
eupamp	*Euphorbia ampliphylla*	1	20	1	DAF; MAF; TRF	MI
fadcie	*Fadogia cienkowskii*	3	6		not treated	CTW[c]
fauspe	*Faurea rochetiana*	5	19		not treated	CTW[c+s+ext]
ficglu	*Ficus glumosa*	23	42		ACB; CTW	CTW[all+ext]
ficing	*Ficus ingens*	6	45		CTW; DAF	CTW[all+ext]
ficlut	*Ficus lutea*	1	13		MAF; RV; TRF	MI
ficova	*Ficus ovata*	2	28		DAF; MAF; RV	MI
ficpla	*Ficus platyphylla*	2	10		TRF	CTW[c+s]
ficsal	*Ficus salicifolia*	2	25		CTW; DAF	MI
ficsur	*Ficus sur*	11	96		DAF; MAF; RV; TRF	TRG
ficsyc	*Ficus sycomorus*	52	106		DAF; RV	CTW[all+ext]
fictho	*Ficus thonningii*	6	77	10	DAF; MAF; RV	MI
ficvas	*Ficus vasta*	8	87		DAF; RV	MI
flaind	*Flacourtia indica*	5	49		ACB; DAF; RV	MI
fluvir	*Flueggea virosa*	45	73		ACB; CTW	CTW[all+ext]
garter	*Gardenia ternifolia*	38	116		CTW; DAF	CTW[all+ext]
gnilam	*Gnidia lamprantha*	3	24		CTW; DAF/WG	MI
grebic	*Grewia bicolor*	3	80		ACB; CTW	ACB[-CTW]
greery	*Grewia erythraea*	1	43		ACB	ACB[-CTW]
grefer	*Grewia ferruginea*	1	95		DAF	CTW[n+c+ext]
grefla	*Grewia flavescens*	8	33		ACB; DAF	CTW[n+c+ext]
gremol	*Grewia mollis*	36	77		CTW	CTW[all+ext]
greten	*Grewia tenax*	3	79		ACB	ACB[-CTW]
gretri	*Grewia trichocarpa*	6	65		ACB; RV	CTW[all+ext]
grevel	*Grewia velutina*	2	62		ACB; DAF	CTW[c+s+ext]
grevil	*Grewia villosa*	2	84		ACB	ACB[-CTW]

Abbreviations used in analyses	Species	Records from relevés (numbers)	Records from specimens (numbers)	Recorded from literature (numbers)	Vegetation type in 2010 Atlas	Assigned to local element here
haraby	*Harrisonia abyssinica*	8	48		ACB; CTW; WGG	CTW[c+s+ext]
hetarb	*Heteromorpha arborescens*	5	37	7	DAF	MI
hymflo	*Hymenodictyon floribundum*	5	40		CTW; DAF	CTW[all+ext]
hypthe	*Hyphaene thebaica*	3	13		ACB; CTW; DSS; RV	CTW[all+ext]
kigafr	*Kigelia africana*	1	21		RV	CTW[all+ext]
lanbar	*Lannea barteri*	5	11		CTW; RV	CTW[n+c]
lanfru	*Lannea fruticosa*	47	30		CTW; DAF	CTW[all]
lansci	*Lannea schimperi*	22	66	4	CTW; DAF	CTW[all+ext]
lanscw	*Lannea schweinfurthii*	2	21		CTW	CTW[c+s+ext]
lantri	*Lannea triphylla*	1	20		ACB; DSS	ACB[-CTW]
lonlax	*Lonchocarpus laxiflorus*	50	33		CTW	CTW[all]
maeang	*Maerua angolensis*	4	61		ACB; DAF	ACB[-CTW]
maeobl	*Maerua oblongifolia*	2	43		ACB; DSS	ACB[-CTW]
maetri	*Maerua triphylla*	1	71		RV	ACB[-CTW]
maelan	*Maesa lanceolata*	3	90		DAF; RV	MI
maytrb	*Maytenus arbutifolia*	5	229		DAF	MI
maysen	*Maytenus senegalensis*	38	144		ACB; CTW; DAF	TRG
meytet	*Meyna tetraphylla*	1	9		CTW; RV	CTW[c+s]
ochleu	*Ochna leucophloeos*	17	15		CTW	CTW[n+c]
oleeur	*Olea europaea* subsp. *cuspidata*	1	108	1	DAF	MI
oncspi	*Oncoba spinosa*	2	30		CTW; DAF/WG; RV	MI
ormpub	*Ormocarpum pubescens*	2	4		CTW	CTW[n]
otofru	*Otostegia fruticosa*	1	16	2	CTW; DAF; ACB	MI
oxyaby	*Oxytenanthera abyssinica*	16	17		CTW	CTW[all]
ozoins	*Ozoroa insignis*	11	84		ACB; CTW; DAF	CTW[all+ext]
ozopul	*Ozoroa pulcherrima*	5	5		CTW	CTW[c]
pavcra	*Pavetta crassipes*	3	9		?ACB;?CTW	CTW[c+s]
pavoli	*Pavetta oliveriana*	1	62		CTW; DAF; RV	MI
piltho	*Piliostigma thonningii*	56	70		CTW; WGG	CTW[all+ext]
polfar	*Polyscias farinosa*	4	11		CTW	CTW[all]

Abbreviations used in analyses	Species	Records from relevés (numbers)	Records from specimens (numbers)	Recorded from literature (numbers)	Vegetation type in 2010 Atlas	Assigned to local element here
presch	*Premna schimperi*	1	97		DAF; RV	MI
progag	*Protea gaguedi*	2	49		DAF	MI
promad	*Protea madiensis*	2	14		CTW	CTW[c+s+ext]
psekot	*Pseudocedrela kotschyi*	15	11		CTW	CTW[all]
psofeb	*Psorospermum febrifugum*	6	7		CTW	CTW[c]
pteluc	*Pterocarpus lucens*	54	17		CTW	CTW[n+c]
rhuvul	*Rhus vulgaris*	16	139	1	ACB; CTW; DAF	TRG
rumaby	*Rumex abyssinicus*	3	34	1	not treated	MI
salper	*Salvadora persica*	2	53		ACB; DSS; RV	ACB[-CTW]
sarlat	*Sarcocephalus latifolius*	5	14		CTW	CTW[n+c]
schala	*Schrebera alata*	2	63	1	DAF	MI
sclbir	*Sclerocarya birrea*	9	27		CTW	CTW[all+ext]
seclon	*Securidaca longipedunculata*	8	27		CTW	CTW[n+c]
senpet	*Senna petersiana*	8	34		DAF	CTW[all+ext]
sensin	*Senna singueana*	4	49		DAF	CTW[all+ext]
steara	*Steganotaenia araliacea*	5	73		ACB; CTW; DAF	CTW[all+ext]
steafr	*Sterculia africana*	42	43		?DSS; ACB; CTW	CTW[all+ext]
stekun	*Stereospermum kunthianum*	49	97		CTW; DAF	CTW[all+ext]
strinn	*Strychnos innocua*	22	23		CTW	CTW[all]
syzgui	*Syzygium guineense* subsp. *macrocarpum*	13	24		CTW; DAF/WG	CTW[c]
tamind	*Tamarindus indica*	3	49		ACB; DSS; RV	CTW[all+ext]
terbro	*Terminalia brownii*	12	96		ACB (incl. ACB/RV); CTW	ACB[-CTW]
terlax	*Terminalia laxiflora*	38	17		CTW	CTW[all]
termac	*Terminalia macroptera*	11	12		CTW	CTW[n+c]
tersch	*Terminalia schimperiana*	55	59		CTW	CTW[all+ext]
treori	*Trema orientalis*	4	40		DAF; RV	MI
vanmad	*Vangueria madagascariensis*	7	24		ACB; DAF	CTW[all+ext]
veramy	*Vernonia amygdalina*	5	70		DAF; MAF; RV	TRG
veraur	*Vernonia auriculifera*	3	25		DAF	MI
vitdon	*Vitex doniana*	24	21		CTW	CTW[all]

Abbreviations used in analyses	Species	Records from relevés (numbers)	Records from specimens (numbers)	Recorded from literature (numbers)	Vegetation type in 2010 Atlas	Assigned to local element here
woouni	*Woodfordia uniflora*	2	33		DAF; RV	MI
ximame	*Ximenia americana*	9	73	1	ACB; CTW	CTW[all+ext]
zangil	*Zanthoxylum gilletii*	1	2		RV	CTW[c]
zizaby	*Ziziphus abyssinica*	20	49		CTW; DAF	CTW[all+ext]
zizmau	*Ziziphus mauritiana*	2	30		ACB; RV	CTW[c+s+ext]
zizmuc	*Ziziphus mucronata*	4	69		ACB; CTW; RV	CTW[c+s+ext]
zizpub	*Ziziphus pubescens*	1	14		RV; TRF	CTW[c+s]
zizspi	*Ziziphus spina-christi*	52	38		ACB; DAF; DSS	CTW[all+ext]

Appendix 3: Observed and derived environmental data for each of the numbered relevés

The information presented in Appendix 3 is divided into two parts for suitable page width; Table Appendix 3-1 includes all parameters excluding the information from bio1-bio19, which are indicated in Table Appendix 3-2. For more information on the direct observations made during the field work, including geographical descriptions of the relevés, their coordinates in degrees, minutes and seconds, the dates of the observations, the descriptive field observations and the lists of observed species, see Appendix 1.

Legend to Table Appendix 3-1.

variables	description
ID	Relevé number (recorded in the field; refers to Appendix 1)
lat	Latitude (decimal degrees; recorded with GPS in the field)
long	Longitude (decimal degrees; recorded with GPS in the field)
Alt	Altitude (in m a.s.l.), based on field measurement with GPS.
Dem	Altitude (in m a.s.l.), based on data from 90 m SRTM DEM.
Slope	Slope, based on estimates in field (with categories flat, moderate and steep).
SlopeDem	Slope, measured in degrees, computed on data from 90 m SRTM DEM.
Aspect	Aspect, measured in degrees, clockwise, computed on data from 90 m SRTM DEM.
TWI	Topographical or Terrain Wetness Index (TWI). TWI = ln (a/tan β), where a is the specific catchment area (SCA): the local upslope area draining through a certain point per unit contour length, which is equal to a certain grid cell width, and β is the local slope. The TWI is computed on data from the 90 m SRTM DEM, using the GRASS GIS r.watershed function (GRASS Development Team 2009).
BurnDeg	Graded estimation of degree of burning, based on data from field observations. The categories are: none, low, moderate, high
BurnBiom	Biomass available for burning; the indications are based on field observations, with the categories: low, middle and high
Firefreq	Fire frequency; number of fires in the pixel, recorded between 2003-2013 by the MODIS programme on satellite data.
FirefreqAv	Average fire frequency for 2003-2013 of all cells in a radius of 2 cells around the focal pixel, calculated on data from the same source as Firefreq.
FirefreqMed	Median fire frequency for the period 2003-2013 of all cells in a radius of 2 cells around the focal pixel, calculated on data from the same source as Firefreq.
FirefreqMax	Maximum fire frequency for the period 2003-2013 of all cells in a radius of 2 cells around the focal pixel, calculated on data from the same source as Firefreq
SoilCategory	Soil types in descriptive terms, based on the field observation indicated in Appendix 1, are converted into 11 intuitively numbered categories, according the weathering of the rocks from which the soils are derived, the colour of the soil and the grain size. The sign => indicates range of variation within the categories:

1. Black cotton soil.
2. Brown soil => black cotton soil; brown soil => fine sand & black cotton soil; brown, stony soil => black cotton soil; dark brown, cracking soil (=> black cotton soil); stony, brown soil => black cotton soil; mosaic of stony soil and black cotton soil.
3. Brown soil.
4. Brown soil on rocks; brown, stony soil; stony brown soil; stony, brown (red) soil; stony, brown soil; stony, brown soil (sandy, pale reddish => almost black cotton soil); stony, brown soil (with lava blocks).
5. Stony, dark red soil; stony, red soil.
6. Dark red soil; red soil.
7. Grey soil; red soil => sandy soil; rocky, red soil; sandy, reddish soil.
8. Stony, grey (or nearly white) soil; stony, grey soil; grey, rocky soil.
9. Stony, sandy soil.
10. Rocky, and sandy soil; stony soil.
11. None [no soil]; rocks with little soil; rocky ground; steep rock).

Legend to Table Appendix 3.2

variables	description
bio1	Annual Mean Temperature (°C × 10), from BioClim, adapted from https://www.worldclim.org/
bio2	Mean Diurnal Range (Mean of monthly (max temp – min temp; °C × 10), from BioClim, adapted from https://www.worldclim.org/
bio3	Isothermality (BIO2/BIO7) (×100), from BioClim, adapted from https://www.worldclim.org/
bio4	Temperature Seasonality (standard deviation ×100), from BioClim, adapted from https://www.worldclim.org/
bio5	Max Temperature of Warmest Month (°C × 10), from BioClim, adapted from https://www.worldclim.org/
bio6	Min Temperature of Coldest Month (°C × 10), from BioClim, adapted from https://www.worldclim.org/
bio7	Temperature Annual Range (BIO5-BIO6), from BioClim, adapted from https://www.worldclim.org/
bio8	Mean Temperature of Wettest Quarter (°C × 10), from BioClim, adapted from https://www.worldclim.org/
bio9	Mean Temperature of Driest Quarter (°C × 10), from BioClim, adapted from https://www.worldclim.org/
bio10	Mean Temperature of Warmest Quarter (°C × 10), from BioClim, adapted from https://www.worldclim.org/
bio11	Mean Temperature of Coldest Quarter (°C × 10), from BioClim, adapted from https://www.worldclim.org/
bio12	Annual Precipitation, from BioClim (mm), adapted from https://www.worldclim.org/
bio13	Precipitation of Wettest Month (mm), from BioClim, adapted from https://www.worldclim.org/
bio14	Precipitation of Driest Month (mm), from BioClim, adapted from https://www.worldclim.org/
bio15	Precipitation Seasonality (Coefficient of Variation), from BioClim, adapted from https://www.worldclim.org/
bio16	Precipitation of Wettest Quarter (mm), from BioClim, adapted from https://www.worldclim.org/
bio17	Precipitation of Driest Quarter (mm), from BioClim, adapted from https://www.worldclim.org/
bio18	Precipitation of Warmest Quarter (mm), from BioClim, adapted from https://www.worldclim.org/
bio19	Precipitation of Coldest Quarter (mm), from BioClim, adapted from https://www.worldclim.org/

Table Appendix 3-1: Observed and calculated data

ID	Lat	Long	Alt	Dem	Slope	SlopeDem	Aspect	Twi	BurnDeg	BurnBiom	Firefreq	FirefreqAv	Firefreq-Med	Firefreq-Max	Soil category
1	14.208	36.615	600	620	flat	0.337624	225.0003	21.75472	none	low	0	0	0	0	1
2	14.150	36.695	630	624	flat	1.258069	71.56522	27.26171	none	low	0	0.153846	0	1	1
3	14.075	36.748	675	678	flat	0.450162	134.9997	8.060395	none	low	0	0.307692	0	2	1
4	14.020	36.800	750	749	flat	2.811304	151.26	7.703922	none	low	0	0	0	0	11
5	13.952	36.860	800	801	flat	2.63509	5.194478	9.960257	none	low	0	0.538462	0	2	11
6	13.842	36.887	800	796	flat	1.273061	180	8.001668	none	low	2	2.230769	2	5	1
7	13.758	36.938	750	752	flat	5.031382	353.6598	8.164697	none	low	3	2.230769	2	5	10
8	13.673	36.957	800	780	flat	2.024927	134.9997	8.826346	none	low	4	2.461538	3	4	2
9	13.528	36.973	810	812	flat	1.817568	156.8012	8.582442	none	low	0	0.538462	0	2	3

ID	Lat	Long	Alt	Dem	Slope	SlopeDem	Aspect	Twi	BurnDeg	BurnBiom	Firefreq	FirefreqAv	Firefreq-Med	Firefreq-Max	Soil category
10	13.437	36.977	850	862	flat	2.383356	205.7102	11.5259	none	low	0	0.230769	0	1	4
11	13.343	37.048	875	879	flat	6.038136	346.3286	8.556343	none	low	1	1.923077	1	6	1
12	13.302	37.087	925	927	flat	13.06584	191.041	5.902442	none	low	0	1.846154	2	5	1
13	13.237	37.153	850	862	moderately steep	9.312924	308.8111	4.971102	none	low	0	0.384615	0	2	3
14	13.192	37.223	875	868	flat	2.02804	205.5602	22.5292	none	low	3	3.769231	3	7	1
15	13.108	37.260	950	956	steep	2.024927	45.00027	8.544649	none	low	3	1.923077	2	4	10
16	13.033	37.295	1000	1004	moderate	3.710404	189.8659	9.164635	none	low	4	2	2	4	10
17	12.947	37.328	1100	1083	flat	1.610949	110.2247	7.623927	none	moderate	0	0.307692	0	2	1
18	12.872	37.362	1200	1195	flat	1.885885	242.3542	8.396596	none	moderate	1	0.230769	0	1	1
19	12.823	37.368	1400	1413	moderate	11.07366	186.5199	6.610367	none	low	0	0	0	0	3
20	12.797	37.385	1550	1557	moderate	16.9199	166.5347	5.553563	none	low	0	0	0	0	4
21	12.797	37.385	2050	1557	moderate	16.9199	166.5347	5.553563	none	low	0	0	0	0	4
22	12.713	37.448	2450	2431	moderate	7.977362	103.7607	7.433154	none	low	0	0	0	0	4
23	13.523	37.185	900	913	moderate	4.848621	140.3143	10.65596	none	low	3	2.230769	2	5	4
24	13.505	37.325	1150	1173	moderate	18.1484	180	5.800707	none	low	3	3.615385	3	7	4
25	13.577	37.367	1300	1359	moderate	3.767671	152.3538	11.04679	none	low	0	1.384615	1	4	4
26	14.178	36.812	650	679	flat	1.687653	134.9997	9.850363	none	low	0	0.307692	0	1	1
27	14.077	36.993	875	893	flat	2.820269	201.5016	5.878679	none	low	0	0.538462	0	2	4
28	14.058	37.185	900	891	flat	2.549849	273.5763	7.365272	none	low	0	1.076923	1	3	11
29	14.158	37.357	1100	1111	flat	7.504684	24.94411	9.032731	none	low	0	0	0	0	11
30	14.095	37.458	950	977	moderate	1.885885	242.3542	8.587892	none	low	0	0.615385	0	3	11
31	14.202	37.572	900	912	moderate	3.267929	175.815	7.755218	none	low	0	0.076923	0	1	9
32	14.287	37.708	1050	1056	flat	1.282977	277.1249	26.2303	none	low	0	0	0	0	9
33	14.407	37.848	1050	1070	flat	0.684537	215.5379	7.283257	none	low	1	0.846154	1	2	2
34	14.455	37.977	1450	1474	flat	8.445552	252.5822	7.862257	none	low	0	0	0	0	7
35	14.415	38.177	1700	1744	moderate	5.921237	20.37661	6.635883	none	low	0	0	0	0	7
36	12.933	36.178	750	761	moderate	2.567107	163.8107	7.533079	occasional	low	0	0.692308	1	2	2
37	12.862	36.275	750	720	flat	1.599111	185.7106	7.638064	none	low	0	1.307692	1	4	1
38	12.815	36.335	750	755	flat	1.17482	61.69947	11.00152	none	low	2	1	1	3	4

ID	Lat	Long	Alt	Dem	Slope	SlopeDem	Aspect	Twi	BurnDeg	BurnBiom	Firefreq	FirefreqAv	Firefreq-Med	Firefreq-Max	Soil category
39	12.755	36.428	775	760	steep	2.589154	10.61975	11.77978	none	low	1	0.846154	1	2	4
40	12.697	36.502	800	798	flat	0.927968	210.964	23.01934	none	low	0	0.846154	1	2	1
41	12.642	36.577	875	872	moderate	1.746588	149.9312	7.110481	moderate	low	0	0.461538	0	2	3
42	12.620	36.680	975	961	moderate	1.530376	171.0273	6.534355	moderate	moderate	0	0.692308	0	2	4
43	12.580	36.747	1000	972	flat	1.803603	228.5766	16.93975	low	low	2	1.692308	2	4	4
44	12.583	36.822	1200	1179	flat	0.811507	281.3098	5.371328	low	low	1	0.769231	0	3	4
45a	12.095	35.942	750	756	moderate	1.967929	255.9639	10.8305	moderate	moderate	3	3.307692	3	10	4
45b	12.095	35.942	750	756	flat	1.967929	255.9639	10.8305	moderate	moderate	3	3.307692	3	10	1
46a	12.247	35.873	675	650	moderate	0.900268	45.00027	24.63753	moderate	moderate	0	1.538462	1	6	4
46b	12.247	35.873	675	650	moderate	0.900268	45.00027	24.63753	moderate	moderate	0	1.538462	1	6	1
47a	12.403	35.948	650	637	flat	0.878922	84.80562	6.887398	low	low	1	1.538462	1	4	4
47b	12.403	35.948	650	637	flat	0.878922	84.80562	6.887398	low	low	1	1.538462	1	4	1
48	12.553	35.990	650	628	moderate	1.354941	40.23663	11.09897	moderate	moderate	7	5.153846	5	9	4
49	12.580	36.170	600	602	moderate	2.811298	351.8698	8.578118	low	low	1	1	0	5	2
50	12.665	36.302	900	897	moderate	3.273736	60.94563	6.630786	low	low	1	1.538462	1	5	3
51	12.565	36.892	1200	1215	moderate	9.233755	196.8869	9.379121	moderate	low	2	0.692308	1	2	4
52	12.538	36.948	1400	1441	1400 m	14.70894	58.4126	6.259945	moderate	low	0	0.384615	0	1	11
53	12.508	37.012	2100	2057	steep	8.198397	160.8662	9.874886	low	low	0	0	0	0	4
54	11.103	36.412	1400	1374	moderate	6.477365	165.1136	8.41051	moderate	low	3	5.153846	5	10	4
55	11.180	36.333	1050	1033	flat	2.527433	167.2755	8.660349	low	low	2	5.923077	6	9	1
56	11.275	36.212	1150	1168	flat	2.811304	331.26	7.63251	low	low	4	5	5	11	3
57	11.235	36.135	1000	1025	moderate	2.811311	314.9997	7.451927	low	low	11	8.846154	10	13	4
58	11.187	36.042	950	969	moderate	1.211946	246.8016	7.210046	low	low	9	11.23077	11	17	4
59	11.187	35.947	930	941	moderate	1.513749	3.012816	8.186582	moderate	low	7	8.846154	10	13	4
60	10.727	35.022	750	744	flat	2.012391	198.4351	6.405099	moderate	high	12	12.30769	13	17	4
61	10.708	34.930	875	899	flat	3.372368	351.8698	6.029866	moderate	high	13	10.92308	11	17	7
62	10.625	34.798	750	788	moderate	2.23279	265.9144	6.705621	high	high	10	6.166667	6	10	4
63	10.513	34.777	880	885	flat	1.479938	36.2541	6.769632	moderate	low	9	9.615385	10	14	6
64	10.435	34.753	1030	1016	moderate	2.703691	180	7.781796	high	high	11	10.76923	11	20	9
65	10.333	34.753	1200	1249	moderate	9.069992	79.47231	6.15132	none	low	5	7.153846	6	14	10

ID	Lat	Long	Alt	Dem	Slope	SlopeDem	Aspect	Twi	BurnDeg	BurnBiom	Firefreq	FirefreqAv	Firefreq-Med	Firefreq-Max	Soil category
66	10.322	34.678	1400	1382	flat	1.312228	165.9636	6.820069	low	moderate	3	5.153846	4	11	6
67	10.307	34.627	1400	1399	moderate	4.514792	39.28968	5.633783	moderate	moderate	8	6.769231	7	10	5
68	10.213	34.652	1600	1592	flat	1.368879	215.5379	6.968381	low	moderate	21	14.75	14.5	21	5
69	9.492	34.465	1375	1366	moderate	4.228748	151.9906	6.307099	moderate	moderate	9	10.76923	10	17	5
70	9.527	34.507	1450	1463	moderate	3.131966	113.9623	10.0737	low	low	4	2.153846	1	6	6
71	9.530	34.567	1575	1587	moderate	2.21859	284.5343	5.900411	low	low	0	0.615385	0	2	6
72	9.568	34.647	1400	1403	flat	5.416245	148.1723	7.634666	low	moderate	2	2.923077	3	7	6
73	9.640	34.710	1375	1387	flat	1.211946	246.8016	8.923581	high	high	11	8.538462	10	13	2
74	10.563	34.312	675	683	steep	0.503296	288.4348	7.085754	high	high	7	6.692308	7	12	6
75	10.575	34.380	750	728	steep	2.623081	165.9636	6.746033	high	high	1	4.461538	5	8	5
76	10.560	34.433	1050	1092	steep	13.10426	129.674	5.863021	high	high	4	6.384615	7	10	4
77	10.548	34.490	1250	1256	flat	2.589164	42.51072	7.621419	high	high	7	6.769231	6	11	6
78	10.475	34.517	1450	1467	moderate	5.206449	52.43167	6.724781	high	high	15	11.41667	12	16	3
79	9.923	34.662	1450	1485	moderate	7.674212	14.32285	7.162945	high	high	2	3.076923	3	5	4
80	10.663	34.692	960	917	moderate	17.6152	276.783	5.936615	high	high	10	8.461538	9	11	7
81	11.067	35.257	920	715	moderate	5.369829	325.8401	6.922809	high	high	9	9	9	15	5
82	10.487	34.765	700	920	moderate	2.386016	53.13036	9.549693	high	high	7	7.153846	7	11	6
83	10.563	34.302	675	669	moderate	1.282965	187.1251	7.187147	high	high	8	7.076923	8	12	6
84	10.562	34.407	820	847	moderate	19.95093	187.6961	4.825251	high	high	7	4.416667	4	9	6
85	8.140	34.567	600	536	flat	2.422793	23.19879	7.625009	moderate	moderate	9	9.076923	9	14	4
86	8.210	34.965	650	663	moderate	12.19668	316.562	7.290909	high	high	10	10.84615	10	19	4
87	8.172	34.587	550	534	moderate	3.814166	180	6.774483	high	high	9	7.076923	7	11	7
88	7.800	34.292	450	444	moderate	1.067554	243.4352	10.19283	high	high	9	8.846154	9	12	7
89	8.283	34.468	500	473	flat	0.355888	243.4352	6.468688	high	high	8	7.769231	7	12	1
90	10.656	38.486	1550	1637	steep	27.49812	3.671735	6.078545	low	low	0	0	0	0	11
91	7.295	36.848	1500	1419	moderate	10.01136	259.5743	7.109078	moderate	moderate	3	3.230769	3	7	4
92	7.277	36.825	1170	1150	flat	2.777436	256.7596	12.97002	moderate	moderate	8	6.923077	8	12	4
93	7.255	36.800	1060	1064	moderate	10.72055	292.4257	5.949269	high	high	10	7.692308	7	12	3
94	7.217	36.793	1350	1364	steep	17.34556	344.7894	7.254363	low	low	2	3.384615	3	9	3
95	6.595	36.543	750	758	moderate	19.81229	160.1677	6.763878	high	high	6	9.230769	9	13	5

ID	Lat	Long	Alt	Dem	Slope	SlopeDem	Aspect	Twi	BurnDeg	BurnBiom	Firefreq	FirefreqAv	Firefreq-Med	Firefreq-Max	Soil category
96	6.645	36.558	1050	1052	moderate	9.66289	331.7623	9.528012	moderate	moderate	11	7.615385	8	12	4
97	6.757	36.583	950	958	moderate	5.423165	339.4438	8.03763	low	low	5	5.923077	6	8	4
98	6.882	36.598	1450	1488	moderate	11.38645	335.5558	6.084891	low	low	5	7.076923	8	9	3
99	6.898	36.605	1500	1513	moderate	4.931962	176.3086	5.115691	moderate	moderate	9	10.46154	9	16	3
100	6.925	36.652	1600	1607	moderate	9.525633	109.3346	5.417812	moderate	moderate	1	2.615385	2	11	4
101	6.962	36.668	1630	1631	moderate	7.967272	130.1693	5.991709	low	moderate	1	2.153846	2	5	3
102	6.522	35.178	625	642	flat	3.06887	16.55722	8.411473	moderate	moderate	7	7.230769	7	10	4
103	6.613	35.182	750	781	moderate	3.070948	248.7497	6.857654	moderate	moderate	10	6.923077	7	11	4
104	6.728	35.182	950	960	moderate	1.721059	303.6898	6.961783	low	low	9	8.615385	8	13	4
105	6.763	35.197	1300	1307	moderate	9.617419	280.3888	7.067902	low	moderate	1	2.769231	3	5	4
106	6.787	35.885	1780	1787	steep	12.70413	247.5433	4.865855	moderate	moderate	3	1.307692	1	5	4
107	6.717	35.867	1350	1311	moderate	16.71113	337.7062	6.612638	moderate	low	2	1.538462	1	5	4
108	6.592	35.763	1350	1393	moderate	13.15098	217.7558	6.317966	low	low	10	6.769231	8	11	4
109	6.518	35.742	1060	1162	moderate	13.5201	242.103	6.43974	moderate	moderate	8	6.538462	7	10	4
110	6.460	35.687	1350	1369	moderate	4.73738	76.42969	7.61698	moderate	moderate	6	6.769231	6	10	4
111	6.279	35.494	1180	1180	flat	2.260906	39.28968	13.51458	low	high	3	3.615385	4	5	3
112	6.343	35.655	1550	1563	moderate	14.08852	141.2904	5.729534	moderate	moderate	5	8	8	11	1
113	6.393	35.663	1500	1494	moderate	9.285263	9.782498	6.703263	moderate	moderate	5	5.076923	5	10	3
114	8.308	34.995	800	819	steep	13.4556	122.675	8.629888	high	high	14	12.69231	13	17	4
115	8.325	35.003	890	892	steep	11.33148	136.123	7.767723	high	high	15	12.07692	13	17	4
116	8.330	35.013	990	975	steep	28.10433	121.168	6.215141	high	high	13	11	12	15	4
117	8.327	35.028	1150	1202	steep	22.51941	78.0126	4.914852	high	high	7	6.846154	7	13	4
118	8.320	35.035	1300	1389	steep	20.94634	37.03648	10.0476	high	high	3	3.615385	3	8	4
119	8.592	36.342	1200	1756	moderate	14.87636	41.60893	5.91916	low	low	2	2.461538	2	7	4
120	8.615	36.353	1450	1484	moderate	5.885322	255.9639	7.140434	moderate	moderate	11	9.384615	9	14	4
121	8.692	36.422	1300	1300	moderate	2.632713	244.9833	9.461106	moderate	moderate	4	6.615385	6	12	4
122	8.732	34.928	1600	1653	moderate	7.220317	37.87525	7.914795	low	moderate	3	5.307692	6	9	4
123	8.840	35.182	1400	1416	moderate	11.0388	199.9833	7.009875	moderate	moderate	2	1.846154	2	3	4
124	8.888	35.265	1450	1493	moderate	7.251159	323.8804	5.462435	low	low	4	1.923077	2	4	4
125	9.037	36.033	1300	1239	flat	3.431658	346.6074	17.02808	high	high	9	5.769231	5	15	4

ID	Lat	Long	Alt	Dem	Slope	SlopeDem	Aspect	Twi	BurnDeg	BurnBiom	Firefreq	FirefreqAv	Firefreq-Med	Firefreq-Max	Soil category
126	10.385	37.027	1850	1907	moderate	11.87385	136.6057	7.247695	low	low	1	1	1	2	5
127	10.375	37.015	1750	1752	moderate	12.80299	202.647	7.201738	low	low	2	1.230769	1	2	11
128	10.345	37.015	1650	1666	moderate	9.809341	43.69832	7.293997	low	low	2	3.384615	3	7	8
129	10.333	37.025	1550	1556	steep	19.0529	50.54873	5.843522	low	low	5	5.923077	6	12	8
130	10.322	37.025	1450	1447	steep	6.071132	319.2361	9.740927	low	low	6	8.076923	8	16	8
131	10.315	37.025	1350	1347	steep	8.330326	321.9341	6.676867	high	high	12	9.076923	10	13	8
132	10.310	37.028	1250	1302	moderate	22.77869	342.2841	5.728755	high	high	12	9.076923	10	13	8
133	10.302	37.025	1150	1182	steep	12.83998	313.5183	5.799532	high	high	11	8.846154	9	13	8
134	10.298	37.028	1050	1109	steep	14.59607	11.07031	6.863115	high	high	13	8.384615	9	13	8
135	10.293	37.023	950	940	steep	31.11371	250.6479	4.759078	high	high	13	8.384615	9	13	8
136	10.292	37.018	850	914	steep	26.87552	282.6649	4.637875	high	high	5	7.769231	8	13	8
137	11.955	36.645	1100	1097	steep	4.374889	105.8024	12.37689	low	low	2	2.769231	3	5	4
138	11.962	36.657	1200	1220	steep	15.08115	270.5906	4.333161	high	high	3	3.076923	3	6	4
139	11.965	36.662	1300	1287	steep	20.14742	212.0056	8.117733	high	high	5	3.692308	4	6	4
140	11.968	36.662	1400	1438	steep	13.86645	199.0468	4.263391	high	high	6	3.307692	3	6	11
141	11.968	36.667	1500	1441	steep	29.17206	261.9909	6.786856	moderate	moderate	3	3.461538	3	6	4
142	11.972	36.667	1600	1607	moderate	17.05235	222.431	4.99202	low	low	3	3.461538	3	6	4
143	11.915	38.980	1850	1901	moderate	3.619352	261.1583	9.977374	low	low	0	0	0	0	8
144	12.812	38.638	1150	1157	moderate	4.089678	82.18478	9.032138	low	low	0	0	0	0	8
145	12.817	38.775	1350	1366	flat	3.01706	71.56522	13.69308	low	low	0	0	0	0	8
146	12.775	38.853	1550	1547	flat	3.686565	187.4315	9.543931	low	low	0	0	0	0	8
147	13.143	39.013	1600	1591	moderate	NA	NA	NA	low	low	0	0	0	0	4
148	12.945	36.162	750	737	flat	1.467051	167.4711	8.97838	none	moderate	0	0.230769	0	1	4

Table Appendix 3-2: Bioclimatic data (bio1-bio19) from WorldClim

ID	bio1	bio2	bio3	bio4	bio5	bio6	bio7	bio8	bio9	bio10	bio11	bio12	bio13	bio14	bio15	bio16	bio17	bio18	bio19
1	280	170	69	1779	406	161	245	268	271	306	261	629	195	0	132	486	0	130	0
2	279	170	69	1741	405	161	244	267	271	305	262	644	197	0	132	494	0	136	0
3	278	170	69	1697	405	162	243	263	272	302	263	669	203	0	131	510	0	143	0
4	276	170	70	1728	403	161	242	261	262	301	261	686	206	0	129	519	0	149	519
5	272	169	70	1691	399	159	240	255	259	296	255	718	214	0	130	540	0	54	540
6	273	169	70	1672	400	161	239	256	262	296	256	734	217	0	128	547	0	56	547
7	274	169	71	1675	401	164	237	256	264	298	256	740	217	0	128	549	0	57	549
8	274	169	71	1678	401	165	236	255	265	298	255	754	220	0	128	555	0	60	555
9	272	168	72	1655	399	166	233	253	266	296	253	781	225	0	125	568	0	64	568
10	268	167	72	1616	394	163	231	258	263	291	250	810	231	0	125	585	0	68	585
11	269	167	72	1651	394	164	230	257	263	292	249	817	233	0	124	591	0	69	588
12	263	165	72	1621	387	160	227	252	259	287	244	842	239	0	124	609	1	73	602
13	269	167	72	1634	394	164	230	258	263	292	250	821	234	0	124	594	1	71	587
14	267	167	72	1616	391	162	229	256	262	290	248	828	236	0	123	600	2	73	591
15	262	165	73	1605	385	159	226	251	258	286	244	852	243	0	123	618	2	77	605
16	258	164	73	1589	380	157	223	247	254	281	240	872	248	1	124	633	3	80	617
17	254	162	73	1543	372	153	219	242	250	275	235	896	255	1	124	651	3	84	630
18	246	159	74	1506	362	148	214	234	243	268	229	926	265	1	122	673	4	89	648
19	232	154	75	1447	343	139	204	219	231	253	214	975	280	1	122	708	5	98	674
20	220	149	75	1407	327	129	198	207	219	240	203	1013	291	1	121	731	8	108	692
21	220	149	75	1407	327	129	198	207	219	240	203	1013	291	1	121	731	8	108	692
22	174	138	70	1328	278	81	197	166	170	195	162	1148	329	5	113	788	21	149	749
23	266	169	71	1741	394	158	236	256	259	292	247	795	231	0	126	581	1	66	578
24	253	166	71	1791	380	147	233	242	247	279	234	846	247	1	126	622	3	73	611
25	239	162	71	1814	363	135	228	227	233	264	218	892	264	1	126	658	3	81	643
26	275	169	68	1840	401	154	247	264	254	302	254	653	199	0	131	500	0	138	0
27	266	169	69	1750	393	151	242	251	252	291	251	721	216	0	130	545	0	156	545
28	266	170	69	1861	396	150	246	251	252	293	250	720	216	0	130	543	0	156	5
29	255	168	68	1946	384	140	244	237	242	283	237	768	233	0	131	579	0	165	579
30	264	172	69	1980	395	147	248	246	251	292	246	730	222	0	132	549	0	157	549
31	267	173	69	1991	399	149	250	251	253	297	251	692	212	1	131	520	3	145	9

ID	bio1	bio2	bio3	bio4	bio5	bio6	bio7	bio8	bio9	bio10	bio11	bio12	bio13	bio14	bio15	bio16	bio17	bio18	bio19
32	263	172	69	2099	394	145	249	257	249	294	245	717	223	1	132	539	3	149	539
33	263	172	69	2095	393	144	249	258	249	294	245	693	217	1	132	520	3	142	11
34	255	167	68	2397	381	136	245	251	241	289	234	778	245	0	133	589	2	163	585
35	225	163	70	2020	345	114	231	218	213	254	207	816	263	0	133	621	2	170	615
36	279	163	74	1336	402	182	220	262	281	297	262	907	232	0	117	618	0	8	618
37	280	163	75	1326	401	186	215	263	281	298	263	909	234	0	117	622	0	8	622
38	277	162	75	1329	398	184	214	260	278	295	260	923	238	0	118	633	0	8	633
39	275	162	75	1329	395	180	215	258	275	292	258	941	244	0	118	647	0	8	647
40	272	161	75	1320	390	177	213	260	272	289	255	958	249	0	119	660	0	9	660
41	266	160	75	1276	383	172	211	255	266	283	250	990	258	0	118	685	0	11	681
42	261	158	75	1283	377	167	210	249	261	278	245	1008	264	0	118	703	1	87	694
43	259	158	75	1283	374	165	209	248	259	277	243	1014	267	0	119	711	1	87	700
44	246	154	75	1312	359	154	205	234	246	264	230	1056	282	1	118	745	3	92	724
45a	267	157	72	1317	385	168	217	252	265	286	252	1107	282	0	116	752	0	94	752
45b	267	157	72	1317	385	168	217	252	265	286	252	1107	282	0	116	752	0	94	752
46a	276	158	72	1327	395	176	219	261	273	295	261	1013	260	0	117	693	0	81	693
46b	276	158	72	1327	395	176	219	261	273	295	261	1013	260	0	117	693	0	81	693
47a	278	160	73	1338	398	180	218	262	276	297	262	978	251	0	117	670	0	78	670
47b	278	160	73	1338	398	180	218	262	276	297	262	978	251	0	117	670	0	78	670
48	281	161	74	1335	401	184	217	265	279	299	265	940	242	0	117	645	0	74	645
49	283	162	74	1320	403	186	217	267	282	300	267	936	242	0	117	642	0	9	642
50	268	160	75	1289	386	175	211	251	268	285	251	997	253	0	116	680	1	8	680
51	241	152	74	1312	352	149	203	228	241	259	225	1062	288	1	120	755	3	93	729
52	226	146	74	1288	333	137	196	212	227	243	210	1103	302	1	119	786	5	101	748
53	183	128	72	1360	282	106	176	168	185	200	165	1204	329	3	113	837	15	129	787
54	217	149	68	1157	329	110	219	207	213	235	207	1833	448	2	108	1188	9	204	1188
55	241	151	68	1299	355	136	219	229	236	261	229	1609	401	1	110	1061	4	177	1061
56	234	150	68	1252	348	130	218	221	231	253	221	1650	410	1	110	1085	4	177	1085
57	242	150	68	1323	356	138	218	228	238	262	228	1569	389	1	110	1029	3	170	1029
58	245	150	68	1337	360	142	218	232	241	265	232	1523	376	1	110	994	3	167	994
59	247	150	69	1374	363	146	217	233	245	268	233	1476	363	0	109	960	2	162	960
60	264	141	68	1689	379	172	207	244	268	288	244	1122	260	0	101	678	1	38	678

ID	bio1	bio2	bio3	bio4	bio5	bio6	bio7	bio8	bio9	bio10	bio11	bio12	bio13	bio14	bio15	bio16	bio17	bio18	bio19
61	257	139	68	1731	371	168	203	236	263	282	236	1153	263	0	99	687	2	42	687
62	262	138	68	1760	377	176	201	241	270	287	241	1085	246	0	97	637	1	42	637
63	256	136	68	1767	370	172	198	236	265	282	236	1116	247	0	95	646	3	47	646
64	249	134	68	1755	361	166	195	229	259	274	229	1145	248	0	95	656	3	52	656
65	236	131	69	1679	344	156	188	217	245	260	217	1209	252	0	94	684	4	60	684
66	228	129	69	1669	335	150	185	209	237	252	209	1234	251	0	93	691	4	67	691
67	229	129	69	1695	336	151	185	210	238	253	210	1212	245	0	91	675	4	67	675
68	218	127	70	1620	322	142	180	200	227	242	200	1293	251	0	90	709	7	80	709
69	222	120	67	1467	326	148	178	208	232	244	208	1130	203	1	84	561	9	75	561
70	213	118	67	1451	315	140	175	199	222	234	199	1204	222	1	84	592	11	83	592
71	208	117	67	1376	308	134	174	195	216	227	195	1265	237	2	84	621	13	89	621
72	214	120	67	1423	316	138	178	203	222	233	200	1203	222	1	86	610	11	74	342
73	212	121	67	1465	314	135	179	202	221	232	198	1212	226	0	88	629	8	70	343
74	274	134	68	2058	388	191	197	248	287	302	248	952	214	0	93	537	2	44	537
75	270	134	67	2035	385	187	198	245	283	298	245	976	219	0	94	553	2	44	553
76	256	133	68	1893	368	174	194	232	267	282	232	1051	229	0	93	593	3	50	593
77	239	132	69	1776	348	159	189	217	248	263	217	1147	243	0	93	648	4	59	648
78	227	130	70	1695	334	149	185	207	236	251	207	1223	250	0	92	684	5	69	684
79	217	124	68	1507	320	139	181	204	226	237	202	1208	221	0	89	646	7	71	341
80	259	137	69	1810	372	174	198	237	267	285	237	1090	246	0	97	636	1	44	636
81	260	146	68	1575	375	162	213	242	260	283	242	1195	287	0	105	753	1	138	753
82	254	135	68	1794	367	171	196	233	264	280	233	1118	246	0	95	644	3	49	644
83	274	134	68	2064	388	192	196	248	287	302	248	945	212	0	93	533	2	44	533
84	265	134	68	1983	378	182	196	240	277	292	240	1008	223	0	93	569	2	47	569
85	269	149	72	1329	384	179	205	258	273	290	258	1180	239	6	78	609	28	112	609
86	252	146	73	1118	362	163	199	244	255	270	243	1267	238	8	78	646	36	121	334
87	270	148	71	1340	385	179	206	258	273	291	258	1178	236	6	78	608	28	109	608
88	263	150	73	1203	378	174	204	254	261	282	253	1267	271	4	77	657	34	149	527
89	276	146	70	1449	392	186	206	262	281	299	262	1055	213	4	80	543	19	85	543
90	223	149	75	1412	323	125	198	212	216	243	210	874	245	5	110	584	22	129	23
91	215	150	76	742	314	118	196	209	214	225	207	1453	199	33	51	593	114	293	554
92	223	149	76	748	323	128	195	217	223	234	216	1406	193	32	52	576	107	280	538

ID	bio1	bio2	bio3	bio4	bio5	bio6	bio7	bio8	bio9	bio10	bio11	bio12	bio13	bio14	bio15	bio16	bio17	bio18	bio19
93	228	149	77	739	328	135	193	222	229	239	221	1382	188	31	52	562	106	275	526
94	215	149	77	748	313	120	193	208	215	225	207	1487	198	36	50	586	121	310	550
95	249	134	77	849	343	171	172	248	255	261	240	1191	157	27	46	436	99	140	406
96	228	132	78	944	320	151	169	226	235	241	217	1499	201	37	44	552	135	189	489
97	237	137	78	850	331	156	175	236	243	249	227	1401	181	33	46	516	121	168	476
98	212	139	78	832	305	129	176	211	217	223	201	1670	216	42	43	611	156	220	551
99	209	139	78	845	303	127	176	209	215	220	198	1681	216	42	43	613	157	223	556
100	201	139	78	845	295	119	176	202	206	212	190	1689	211	43	43	607	157	401	566
101	203	141	78	817	297	118	179	204	207	214	192	1670	206	42	44	602	153	391	565
102	251	138	83	692	343	177	166	245	256	261	244	1049	179	14	62	473	63	107	365
103	241	139	83	666	332	166	166	235	245	251	234	1136	196	15	64	524	67	114	266
104	230	140	84	633	321	156	165	225	235	240	223	1210	213	15	68	577	68	117	280
105	213	137	86	622	298	139	159	206	219	222	206	1323	223	19	66	618	82	138	618
106	187	131	84	671	270	115	155	183	192	195	178	1790	232	46	41	615	185	262	608
107	213	137	83	760	303	138	165	208	219	224	204	1605	216	35	46	584	146	200	583
108	209	133	83	749	296	137	159	200	215	220	200	1551	210	35	45	562	144	204	562
109	223	134	82	715	311	149	162	214	229	233	214	1399	192	29	48	524	120	172	524
110	212	130	83	752	298	142	156	203	219	223	203	1452	198	33	46	529	133	193	529
111	221	128	83	750	306	152	154	213	228	232	213	1264	173	27	47	471	112	169	471
112	202	124	83	805	285	136	149	196	210	214	192	1493	200	38	42	514	151	219	513
113	207	127	83	754	291	138	153	198	214	218	198	1470	199	35	44	520	142	206	520
114	242	143	74	1118	349	157	192	232	247	259	232	1287	233	9	77	646	38	121	646
115	239	142	75	1115	344	155	189	228	243	255	228	1299	231	9	76	647	40	124	642
116	239	142	74	1127	345	154	191	229	244	257	229	1295	231	9	77	646	39	122	642
117	230	140	76	1083	333	150	183	219	236	246	219	1331	231	11	76	653	44	131	650
118	216	134	79	1051	312	144	168	204	223	230	203	1386	228	15	72	655	54	153	654
119	200	145	72	1201	303	103	200	191	198	217	188	1795	325	15	80	948	54	354	905
120	210	147	69	1341	316	104	212	203	207	228	199	1670	308	14	80	886	50	322	844
121	218	148	66	1440	326	105	221	210	214	236	206	1548	294	12	82	830	44	291	790
122	208	127	78	1146	302	140	162	198	215	224	194	1377	231	10	76	652	43	127	650
123	211	133	77	1148	309	137	172	199	219	227	197	1467	261	10	83	750	37	93	728
124	206	134	77	1185	304	132	172	195	214	223	192	1536	279	10	84	793	37	98	768

ID	bio1	bio2	bio3	bio4	bio5	bio6	bio7	bio8	bio9	bio10	bio11	bio12	bio13	bio14	bio15	bio16	bio17	bio18	bio19
125	222	142	60	1843	333	99	234	215	222	243	208	1502	280	5	87	807	20	72	778
126	180	144	72	1032	281	83	198	170	177	196	170	1563	343	8	94	934	34	225	934
127	194	146	71	1149	298	94	204	183	190	211	183	1484	333	5	98	904	23	206	904
128	197	146	71	1193	302	97	205	186	192	215	186	1460	331	5	99	892	21	203	892
129	202	147	71	1199	309	103	206	191	199	221	191	1420	327	4	100	875	17	197	875
130	210	148	71	1251	317	109	208	198	205	229	198	1380	324	3	101	857	15	189	857
131	219	149	71	1291	327	118	209	207	214	238	207	1329	318	3	103	833	12	180	833
132	215	148	70	1276	323	114	209	204	210	235	204	1343	320	3	103	840	13	182	840
133	226	150	70	1346	336	124	212	215	221	247	215	1277	311	2	104	808	9	171	808
134	232	150	70	1367	342	128	214	220	226	253	220	1241	305	2	105	788	9	166	788
135	233	150	70	1351	343	129	214	221	227	253	221	1240	305	2	105	788	9	165	788
136	233	150	70	1351	343	129	214	221	227	253	221	1240	305	2	105	788	9	165	788
137	243	153	73	1180	354	145	209	234	239	260	231	1268	337	1	118	879	3	108	869
138	240	152	72	1184	351	142	209	231	236	258	228	1275	339	1	117	884	3	108	874
139	232	150	73	1164	340	135	205	222	228	249	219	1312	349	1	117	909	4	111	895
140	226	149	73	1122	333	129	204	216	222	242	214	1326	353	1	117	920	4	110	904
141	215	146	73	1108	320	120	200	205	211	231	203	1353	360	1	116	937	6	112	918
142	215	146	73	1108	320	120	200	205	211	231	203	1353	360	1	116	937	6	112	918
143	193	161	77	1102	299	91	208	196	179	208	179	858	250	15	109	557	50	149	50
144	248	170	75	1566	360	136	224	246	236	270	232	683	218	7	124	483	23	126	27
145	239	169	77	1507	348	129	219	238	227	261	224	690	222	8	123	482	28	123	30
146	228	168	77	1394	336	119	217	230	216	248	213	705	227	8	123	487	28	123	30
147	229	167	76	1468	336	118	218	228	217	250	215	669	231	7	132	490	25	116	26
148	281	163	74	1339	403	184	219	264	282	299	264	904	234	0	116	615	0	11	615

Appendix 4: Distribution of taxa on geographical areas in Ethiopia and phytogeographical elements

This appendix contains first lists of taxa observed to occur in previously defined areas (WWn, WWc, WWs, EWW and ASO; see text and legend to Fig. 7-2) and second lists of taxa which have been referred to phytogeographical elements and sub-elements (CTW[all], CTW[all+ext], CTW[n], CTW[n+c], CTW[n+c+ext], CTW[c], CTW[c+s], CTW[c+s+ext], ACB[-CTW], MI, TRG; see text and legend to Fig. 7-1). See further discussion in Section 7.

Distributions on geographical areas

These lists include species observed to occur in the areas WWn, WWc, WWs, (northern, central and southern parts of the western woodlands), EWW (to the east of the western woodlands, the Rift Valley and east of the Rift) and ASO (*Acacia-Commiphora* bushland of the Somali region). Taxa may occur in one or several of these areas, and no attention is given to altitudinal range. For the geographical delimitation of the areas WWn, WWc, WWs, see Fig. 7-2. For each area or phytogeographical element, the sequence of taxa follows that of Section 6.

WWn

Total number of recorded taxa 140 [outliers 10]. *Boscia angustifolia*, *Boscia senegalensis*, *Cadaba farinosa*, *Capparis decidua*, *Capparis tomentosa*, *Maerua angolensis*, *Maerua oblongifolia*, *Securidaca longipedunculata*, *Rumex abyssinicus*, *Woodfordia uniflora*, *Protea gaguedi*, *Flacourtia indica*, *Oncoba spinosa*, *Ochna leucophloeos*, *Anogeissus leiocarpa*, *Combretum adenogonium*, *Combretum collinum*, *Combretum hartmannianum*, *Combretum molle*, *Combretum rochetianum*, *Terminalia brownii*, *Terminalia laxiflora*, *Terminalia macroptera*, *Terminalia schimperiana*, *Grewia bicolor*, *Grewia erythraea*, *Grewia ferruginea*, *Grewia flavescens*, *Grewia mollis*, *Grewia tenax*, *Grewia trichocarpa*, *Grewia villosa*, *Dombeya quinqueseta*, *Sterculia africana*, *Adansonia digitata*, *Bridelia micrantha*, *Bridelia scleroneura*, *Clutia abyssinica*, *Croton macrostachyus*, *Euphorbia abyssinica*, *Euphorbia ampliphylla*, *Flueggea virosa*, *Cassia arereh*, *Piliostigma thonningii*, *Senna petersiana*, *Senna singueana*, *Tamarindus indica*, *Acacia abyssinica*, *Acacia amythethophylla*, *Acacia asak*, *Acacia polyacantha* subsp. *campylacantha*, *Acacia dolichocephala*, *Acacia etbaica*, *Acacia hecatophylla*, *Acacia hockii*, *Acacia mellifera*, *Acacia nilotica*, *Acacia oerfota*, *Acacia persiciflora*, *Acacia senegal*, *Acacia seyal*, *Acacia sieberiana*, *Acacia tortilis*, *Acacia venosa*, *Albizia amara*, *Albizia malacophylla*, *Albizia schimperiana*, *Dichrostachys cinerea*, *Entada abyssinica*, *Entada africana*, *Calpurnia aurea*, *Dalbergia melanoxylon*, *Erythrina abyssinica*, *Lonchocarpus laxiflorus*, *Ormocarpum pubescens*, *Pterocarpus lucens*, *Trema orientalis*, *Ficus glumosa*, *Ficus ingens*, *Ficus lutea*, *Ficus ovata*, *Ficus sur*, *Ficus sycomorus*, *Ficus thonningii*, *Ficus vasta*, *Maytenus arbutifolia*, *Maytenus senegalensis*, *Apodytes dimidiata*, *Salvadora persica*, *Ximenia americana*, *Ziziphus abyssinica*, *Ziziphus spina-christi*, *Clausena anisata*, *Balanites aegyptiaca*, *Brucea antidysenterica*, *Boswellia papyrifera*, *Boswellia pirottae*, *Commiphora africana*, *Pseudocedrela kotschyi*, *Allophylus rubifolius*, *Dodonaea angustifolia*, *Bersama abyssinica*, *Lannea barteri*, *Lannea fruticosa*, *Lannea schimperi*, *Lannea triphylla*, *Ozoroa insignis*, *Rhus vulgaris*, *Sclerocarya birrea*, *Cussonia arborea*, *Cussonia ostinii*, *Polyscias farinosa*, *Heteromorpha arborescens*, *Steganotaenia araliacea*, *Diospyros abyssinica*, *Diospyros mespiliformis*, *Euclea racemosa* subsp. *schimperi*, *Embelia schimperi*, *Maesa lanceolata*, *Strychnos innocua*, *Olea europaea* subsp. *cuspidata*, *Schrebera alata*, *Carissa spinarum*, *Calotropis procera*, *Gardenia ternifolia*, *Hymenodictyon floribundum*, *Pavetta crassipes*, *Sarcocephalus latifolius*, *Vangueria madagascariensis*, *Vernonia amygdalina*, *Vernonia auriculifera*, *Cordia africana*, *Kigelia africana*, *Stereospermum kunthianum*, *Premna schimperi*, *Vitex doniana*, *Otostegia fruticosa*, *Borassus ae-*

thiopum, Hyphaene thebaica, Oxytenanthera abyssinica. **WWn [outliers].** *Annona senegalensis, Boscia salicifolia, Maerua triphylla, Faurea rochetiana, Erythrococca trichogyne, Acacia gerrardii, Ziziphus mucronata, Ozoroa pulcherrima, Crossopteryx febrifuga, Pavetta oliveriana.*

WWc

Total number of recorded taxa 146 [outliers 6]. *Annona senegalensis, Boscia salicifolia, Cadaba farinosa, Capparis tomentosa, Maerua angolensis, Maerua oblongifolia, Maerua triphylla, Securidaca longipedunculata, Rumex abyssinicus, Woodfordia uniflora, Gnidia lamprantha, Faurea rochetiana, Protea gaguedi, Protea madiensis, Flacourtia indica, Oncoba spinosa, Ochna leucophloeos, Syzygium guineense* subsp. *macrocarpum, Anogeissus leiocarpa, Combretum adenogonium, Combretum collinum, Combretum hartmannianum, Combretum molle, Terminalia laxiflora, Terminalia macroptera, Terminalia schimperiana, Psorospermum febrifugum, Grewia ferruginea, Grewia flavescens, Grewia mollis, Grewia trichocarpa, Grewia velutina, Dombeya buettneri, Dombeya quinqueseta, Sterculia africana, Erythroxylum fischeri, Bridelia micrantha, Bridelia scleroneura, Croton macrostachyus, Erythrococca trichogyne, Euphorbia ampliphylla, Flueggea virosa, Cassia arereh, Piliostigma thonningii, Senna petersiana, Senna singueana, Tamarindus indica, Acacia abyssinica, Acacia polyacantha* subsp. *campylacantha, Acacia dolichocephala, Acacia etbaica, Acacia gerrardii, Acacia hecatophylla, Acacia hockii, Acacia nilotica, Acacia persiciflora, Acacia senegal, Acacia seyal, Acacia sieberiana, Acacia venosa, Albizia amara, Albizia grandibracteata, Albizia malacophylla, Albizia schimperiana, Dichrostachys cinerea, Entada abyssinica, Entada africana, Calpurnia aurea, Dalbergia boehmii, Dalbergia melanoxylon, Erythrina abyssinica, Lonchocarpus laxiflorus, Pterocarpus lucens, Trema orientalis, Ficus glumosa, Ficus ingens, Ficus lutea, Ficus ovata, Ficus platyphylla, Ficus cordata* subsp. *salicifolia, Ficus sur, Ficus sycomorus, Ficus thonningii, Ficus vasta, Maytenus arbutifolia, Maytenus senegalensis, Apodytes dimidiata, Ximenia americana, Ziziphus abyssinica, Ziziphus mauritiana, Ziziphus mucronata, Ziziphus pubescens, Ziziphus spina-christi, Clausena anisata, Zanthoxylum gilletii, Balanites aegyptiaca, Brucea antidysenterica, Harrisonia abyssinica, Boswellia papyrifera, Boswellia pirottae, Commiphora pedunculata, Pseudocedrela kotschyi, Allophylus rubifolius, Dodonaea angustifolia, Bersama abyssinica, Lannea barteri, Lannea fruticosa, Lannea schimperi, Lannea schweinfurthii, Ozoroa insignis, Ozoroa pulcherrima, Rhus vulgaris, Sclerocarya birrea, Cussonia arborea, Cussonia ostinii, Polyscias farinosa, Heteromorpha arborescens, Steganotaenia araliacea, Diospyros abyssinica, Diospyros mespiliformis, Embelia schimperi, Maesa lanceolata, Strychnos innocua, Olea europaea* subsp. *cuspidata, Schrebera alata, Carissa spinarum, Calotropis procera, Crossopteryx febrifuga, Fadogia cienkowskii, Gardenia ternifolia, Hymenodictyon floribundum, Meyna tetraphylla, Pavetta crassipes, Pavetta oliveriana, Sarcocephalus latifolius, Vangueria madagascariensis, Vernonia amygdalina, Vernonia auriculifera, Cordia africana, Kigelia africana, Stereospermum kunthianum, Premna schimperi, Vitex doniana, Borassus aethiopum, Hyphaene thebaica, Oxytenanthera abyssinica.* **WWc [outliers].** *Combretum rochetianum, Grewia bicolor, Grewia tenax, Adansonia digitata, Clutia abyssinica, Euphorbia abyssinica.*

WWs

Total number of recorded taxa 123. [outliers 13]. *Annona senegalensis, Boscia angustifolia, Boscia salicifolia, Cadaba farinosa, Capparis tomentosa, Maerua angolensis, Maerua oblongifolia, Maerua triphylla, Rumex abyssinicus, Gnidia lamprantha, Faurea rochetiana, Protea gaguedi, Protea madiensis, Flacourtia indica, Oncoba spinosa, Combretum adenogonium, Combretum collinum, Combretum molle, Terminalia brownii, Terminalia laxiflora, Terminalia schimperiana, Grewia bicolor, Grewia erythraea, Grewia mollis, Grewia tenax, Grewia trichocarpa, Grewia velutina, Grewia villosa, Dombeya buettneri, Dombeya quinqueseta, Sterculia africana, Bridelia micrantha, Bridelia scleroneura, Clutia abyssinica, Erythrococca trichogyne, Euphorbia abyssinica, Flueggea virosa, Piliostigma thonningii, Senna petersiana, Senna singueana, Tamarindus indica, Acacia asak, Acacia polyacantha* subsp. *campylacantha, Acacia dolichocephala, Acacia etbaica, Acacia hockii, Acacia mellifera, Acacia nilotica, Acacia oerfota, Acacia senegal, Acacia seyal, Acacia sieberiana, Acacia tortilis, Albizia amara, Albizia grandibracteata, Albizia schimperiana, Di-*

chrostachys cinerea, *Entada abyssinica*, *Entada africana*, *Calpurnia aurea*, *Erythrina abyssinica*, *Lonchocarpus laxiflorus*, *Ficus glumosa*, *Ficus ingens*, *Ficus platyphylla*, *Ficus cordata* subsp. *salicifolia*, *Ficus sur*, *Ficus sycomorus*, *Ficus thonningii*, *Ficus vasta*, *Maytenus arbutifolia*, *Maytenus senegalensis*, *Apodytes dimidiata*, *Salvadora persica*, *Ximenia americana*, *Ziziphus abyssinica*, *Ziziphus mauritiana*, *Ziziphus mucronata*, *Ziziphus pubescens*, *Ziziphus spina-christi*, *Clausena anisata*, *Balanites aegyptiaca*, *Harrisonia abyssinica*, *Commiphora africana*, *Pseudocedrela kotschyi*, *Allophylus rubifolius*, *Dodonaea angustifolia*, *Lannea fruticosa*, *Lannea schimperi*, *Lannea triphylla*, *Lannea schweinfurthii*, *Ozoroa insignis*, *Rhus vulgaris*, *Sclerocarya birrea*, *Cussonia arborea*, *Polyscias farinosa*, *Heteromorpha arborescens*, *Steganotaenia araliacea*, *Diospyros abyssinica*, *Diospyros mespiliformis*, *Euclea racemosa* subsp. *schimperi*, *Embelia schimperi*, *Maesa lanceolata*, *Strychnos innocua*, *Olea europaea* subsp. *cuspidata*, *Schrebera alata*, *Carissa spinarum*, *Calotropis procera*, *Gardenia ternifolia*, *Hymenodictyon floribundum*, *Meyna tetraphylla*, *Pavetta crassipes*, *Pavetta oliveriana*, *Vangueria madagascariensis*, *Vernonia amygdalina*, *Vernonia auriculifera*, *Cordia africana*, *Kigelia africana*, *Stereospermum kunthianum*, *Premna schimperi*, *Vitex doniana*, *Hyphaene thebaica*, *Oxytenanthera abyssinica*. **WWs [outliers].** *Syzygium guineense* subsp. *macrocarpum*, *Combretum rochetianum*, *Psorospermum febrifugum*, *Grewia flavescens*, *Croton macrostachyus*, *Euphorbia ampliphylla*, *Acacia hecatophylla*, *Ficus ovata*, *Brucea antidysenterica*, *Bersama abyssinica*, *Lannea barteri*, *Cussonia ostinii*, *Sarcocephalus latifolius*.

EWW

Total number of recorded taxa 104 [outliers 1].
Annona senegalensis, *Boscia salicifolia*, *Rumex abyssinicus*, *Woodfordia uniflora*, *Gnidia lamprantha*, *Faurea rochetiana*, *Protea gaguedi*, *Protea madiensis*, *Flacourtia indica*, *Oncoba spinosa*, *Combretum adenogonium*, *Combretum collinum*, *Combretum molle*, *Terminalia schimperiana*, *Grewia ferruginea*, *Grewia flavescens*, *Grewia mollis*, *Grewia trichocarpa*, *Grewia velutina*, *Dombeya quinqueseta*, *Sterculia africana*, *Bridelia micrantha*, *Bridelia scleroneura*, *Clutia abyssinica*, *Croton macrostachyus*, *Erythrococca trichogyne*, *Euphorbia abyssinica*, *Euphorbia ampliphylla*, *Flueggea virosa*, *Piliostigma thonningii*, *Senna petersiana*, *Senna singueana*, *Tamarindus indica*, *Acacia abyssinica*, *Acacia polyacantha* subsp. *campylacantha*, *Acacia dolichocephala*, *Acacia gerrardii*, *Acacia hockii*, *Acacia persiciflora*, *Acacia senegal*, *Acacia seyal*, *Acacia sieberiana*, *Albizia amara*, *Albizia grandibracteata*, *Albizia schimperiana*, *Dichrostachys cinerea*, *Entada abyssinica*, *Calpurnia aurea*, *Erythrina abyssinica*, *Trema orientalis*, *Ficus glumosa*, *Ficus ingens*, *Ficus lutea*, *Ficus ovata*, *Ficus cordata* subsp. *salicifolia*, *Ficus sur*, *Ficus sycomorus*, *Ficus thonningii*, *Ficus vasta*, *Maytenus arbutifolia*, *Maytenus senegalensis*, *Apodytes dimidiata*, *Ximenia americana*, *Ziziphus abyssinica*, *Ziziphus mauritiana*, *Ziziphus mucronata*, *Ziziphus spina-christi*, *Clausena anisata*, *Balanites aegyptiaca*, *Brucea antidysenterica*, *Harrisonia abyssinica*, *Allophylus rubifolius*, *Dodonaea angustifolia*, *Bersama abyssinica*, *Lannea schimperi*, *Lannea schweinfurthii*, *Ozoroa insignis*, *Rhus vulgaris*, *Sclerocarya birrea*, *Cussonia arborea*, *Cussonia ostinii*, *Heteromorpha arborescens*, *Steganotaenia araliacea*, *Diospyros abyssinica*, *Diospyros mespiliformis*, *Euclea racemosa* subsp. *schimperi*, *Embelia schimperi*, *Maesa lanceolata*, *Olea europaea* subsp. *cuspidata*, *Schrebera alata*, *Carissa spinarum*, *Gardenia ternifolia*, *Hymenodictyon floribundum*, *Pavetta crassipes*, *Pavetta oliveriana*, *Vangueria madagascariensis*, *Vernonia amygdalina*, *Vernonia auriculifera*, *Cordia africana*, *Kigelia africana*, *Stereospermum kunthianum*, *Premna schimperi*, *Otostegia fruticosa*, *Hyphaene thebaica*. **EWW[outlier].** *Salvadora persica*.

ASO

Total number of recorded taxa 24 [outliers 3].
Boscia angustifolia, *Boscia salicifolia*, *Boscia senegalensis*, *Cadaba farinosa*, *Capparis decidua*, *Capparis tomentosa*, *Maerua angolensis*, *Maerua oblongifolia*, *Maerua triphylla*, *Terminalia brownii*, *Grewia bicolor*, *Grewia erythraea*, *Grewia tenax*, *Grewia villosa*, *Acacia asak*, *Acacia etbaica*, *Acacia mellifera*, *Acacia nilotica*, *Acacia oerfota*, *Acacia tortilis*, *Salvadora persica*, *Commiphora africana*, *Lannea triphylla*, *Calotropis procera*. **ASO[outliers].** *Ziziphus mauritiana*, *Balanites aegyptiaca*, *Hyphaene thebaica*.

Local phytogeographical elements

The taxa in these lists have only been referred to one phytogeographical element, based on a subjective overall evaluation of the general distribution in Ethiopia, the distribution elsewhere, on the available information on habitats in Section 6, supplemented with Ib Friis' and Sebsebe Demissew's specialist-knowledge of the taxa from field- and herbarium-work. Taxa that are mainly distributed in the CTW-area (the western *Combretum-Terminalia* woodlands), but only marginally extending above the CTW-area and into the Ethiopian Highlands, are referred to elements or sub-elements under CTW; taxa that occur mainly in the Ethiopian Highlands, but marginally intrude downwards into the lower altitudes of the CTW-area, are referred to the element MI. Taxa that are widespread in the CTW-area, the Ethiopian Highlands and in other vegetation types, are referred to the element TRG. Taxa distributed in the *Acacia-Commiphora* bushlands in the south-eastern Ethiopian lowlands, but marginally extending to the area of the CTW are referred to the element ACB[-CTW]. Taxa distributed in the *Acacia-Commiphora* bushlands in the south-eastern Ethiopian lowlands, but marginally extending to the area of the CTW and wider are referred to the element ACB[-CTW+ext].

The CTW-element is subdivided into CTW[all] (widespread in all parts of the CTW), CTW[all+ext] (widespread in all parts of the CTW and extending to the Rift Valley and to the east of the Rift), CTW[n] (restricted to the northern part of the CTW-area), CTW[n+c] (restricted to the northern and central part of the CTW), CTW[n+c+ext] (restricted to the northern and central part of the CTW-area and extending to the Rift Valley and to the east of the Rift), CTW[c] (restricted to the central part of the CTW), CTW[c+s] (restricted to the central and southern part of the CTW-area), CTW[c+s+ext] (restricted to the central and southern part of the CTW-area and extending to the Rift Valley and to the east of the Rift). Outliers with only one or few records outside the main area of the element have not been considered for this classification. The total number of taxa in the CTW-element is 95.

CTW[all]

Total number of recorded taxa 9.
Terminalia laxiflora, *Entada africana*, *Lonchocarpus laxiflorus*, *Pseudocedrela kotschyi*, *Lannea fruticosa*, *Polyscias farinosa*, *Strychnos innocua*, *Vitex doniana*, *Oxytenanthera abyssinica*.

CTW[all+ext]

Total number of recorded taxa 43.
Combretum adenogonium, *Combretum collinum*, *Combretum molle*, *Terminalia schimperiana*, *Grewia mollis*, *Grewia trichocarpa*, *Dombeya quinqueseta*, *Sterculia africana*, *Bridelia micrantha*, *Bridelia scleroneura*, *Flueggea virosa*, *Piliostigma thonningii*, *Senna petersiana*, *Senna singueana*, *Tamarindus indica*, *Acacia polyacantha* subsp. *campylacantha*, *Acacia dolichocephala*, *Acacia hockii*, *Acacia senegal*, *Acacia seyal*, *Acacia sieberiana*, *Albizia amara*, *Entada abyssinica*, *Erythrina abyssinica*, *Ficus glumosa*, *Ficus ingens*, *Ficus sycomorus*, *Ximenia americana*, *Ziziphus abyssinica*, *Ziziphus spina-christi*, *Balanites aegyptiaca*, *Allophylus rubifolius*, *Lannea schimperi*, *Ozoroa insignis*, *Sclerocarya birrea*, *Cussonia arborea*, *Steganotaenia araliacea*, *Gardenia ternifolia*, *Hymenodictyon floribundum*, *Vangueria madagascariensis*, *Kigelia africana*, *Stereospermum kunthianum*, *Hyphaene thebaica*.

CTW[n]

Total number of recorded taxa 4.
Combretum rochetianum, *Adansonia digitata*, *Acacia amythethophylla*, *Ormocarpum pubescens*.

CTW[n+c]

Total number of recorded taxa 16.
Securidaca longipedunculata, *Ochna leucophloeos*, *Anogeissus leiocarpa*, *Combretum hartmannianum*, *Terminalia macroptera*, *Cassia arereh*, *Acacia hecatophylla*, *Acacia*

venosa, Albizia malacophylla, Dalbergia melanoxylon, Pterocarpus lucens, Boswellia papyrifera, Boswellia pirottae, Lannea barteri, Sarcocephalus latifolius, Borassus aethiopum.

CTW[n+c+ext]

Total number of recorded taxa 2.
Grewia ferruginea, Grewia flavescens.

CTW[c]

Total number of recorded taxa 8.
Syzygium guineense subsp. *macrocarpum, Psorospermum febrifugum, Dalbergia boehmii, Zanthoxylum gilletii, Commiphora pedunculata, Ozoroa pulcherrima, Crossopteryx febrifuga, Fadogia cienkowskii.*

CTW[c+s]

Total number of recorded taxa 4.
Dombeya buettneri, Ficus platyphylla, Ziziphus pubescens, Meyna tetraphylla, Pavetta crassipes.

CTW[c+s+ext]

Total number of recorded taxa 9.
Annona senegalensis, Faurea rochetiana, Protea madiensis, Grewia velutina, Ziziphus mauritiana, Ziziphus mucronata, Harrisonia abyssinica, Lannea schweinfurthii.

ACB[-CTW]

Total number of recorded taxa 23.
Boscia angustifolia, Boscia senegalensis, Cadaba farinosa, Capparis decidua, Capparis tomentosa, Maerua angolensis, Maerua oblongifolia, Maerua triphylla, Terminalia brownii, Grewia bicolor, Grewia erythraea, Grewia tenax, Grewia villosa, Acacia asak, Acacia etbaica, Acacia mellifera, Acacia nilotica, Acacia oerfota, Acacia tortilis, Salvadora persica, Commiphora africana, Lannea triphylla, Calotropis procera.

ACB[-CTW+ext]

Total number of recorded taxa 1.
Boscia salicifolia

MI

Total number of recorded taxa 43.
Rumex abyssinicus, Woodfordia uniflora, Gnidia lamprantha, Protea gaguedi, Flacourtia indica, Oncoba spinosa, Erythroxylum fischeri, Clutia abyssinica, Erythrococca trichogyne, Euphorbia abyssinica, Euphorbia ampliphylla, Acacia abyssinica, Acacia gerrardii, Acacia persiciflora, Albizia grandibracteata, Albizia schimperiana, Calpurnia aurea, Trema orientalis, Ficus lutea, Ficus ovata, Ficus cordata subsp. *salicifolia, Ficus thonningii, Ficus vasta, Maytenus arbutifolia, Apodytes dimidiata, Clausena anisata, Brucea antidysenterica, Dodonaea angustifolia, Bersama abyssinica, Cussonia ostinii, Heteromorpha arborescens, Diospyros abyssinica, Diospyros mespiliformis, Euclea racemosa* subsp. *schimperi, Embelia schimperi, Maesa lanceolata, Olea europaea* subsp. *cuspidata, Schrebera alata, Carissa spinarum, Pavetta oliveriana, Vernonia auriculifera, Premna schimperi, Otostegia fruticosa.*

TRG

Total number of recorded taxa 7.
Croton macrostachyus, Dichrostachys cinerea, Ficus sur, Maytenus senegalensis, Rhus vulgaris, Vernonia amygdalina, Cordia africana.

Appendix 5: Ecological adaptations of species

Appendix 5 contains list of species sorted according to ecological adaptation as described in the field 'Description' in Section 6. 'A systematically arranged list ...' The information is drawn from material from the whole of Ethiopia, not only from the relevés. Under each set of adaptations, the sequence of taxa follows that of Section 6. See further discussions in Section 8 and graphical presentation on Fig. 8-1 to 8-6.

Shrubs, subshrubs or trees usually under or reaching 10 m

Annona senegalensis, Boscia angustifolia, Boscia salicifolia, Boscia senegalensis, Cadaba farinosa, Capparis decidua, Capparis tomentosa, Maerua angolensis, Maerua oblongifolia, Maerua triphylla, Securidaca longipedunculata, Rumex abyssinicus, Woodfordia uniflora, Gnidia lamprantha, Faurea rochetiana, Protea gaguedi, Protea madiensis, Oncoba spinosa, Ochna leucophloeos, Syzygium guineense subsp. *macrocarpum, Combretum rochetianum, Terminalia brownii, Terminalia schimperiana, Psorospermum febrifugum, Grewia bicolor, Grewia erythraea, Grewia ferruginea, Grewia flavescens, Grewia mollis, Grewia tenax, Grewia trichocarpa, Grewia velutina, Grewia villosa, Dombeya buettneri, Dombeya quinqueseta, Sterculia africana, Erythroxylum fischeri, Bridelia scleroneura, Clutia abyssinica, Erythrococca trichogyne, Euphorbia abyssinica, Flueggea virosa, Cassia arereh, Piliostigma thonningii, Senna petersiana, Acacia asak, Acacia dolichocephala, Acacia gerrardii, Acacia hecatophylla, Acacia hockii, Acacia mellifera, Acacia oerfota, Acacia persiciflora, Acacia seyal, Acacia venosa, Albizia malacophylla, Dichrostachys cinerea, Entada abyssinica, Entada africana, Calpurnia aurea, Dalbergia boehmii, Ormocarpum pubescens, Ficus glumosa, Ficus cordata* subsp. *salicifolia, Maytenus arbutifolia, Maytenus senegalensis, Salvadora persica, Ximenia americana, Ziziphus abyssinica, Clausena anisata, Zanthoxylum gilletii, Brucea antidysenterica, Boswellia papyrifera, Boswellia pirottae, Commiphora africana, Commiphora pedunculata, Allophylus rubifolius, Dodonaea angustifolia, Lannea fruticosa, Lannea schimperi, Lannea triphylla, Ozoroa pulcherrima, Rhus vulgaris, Cussonia ostinii, Polyscias farinosa, Heteromorpha arborescens, Steganotaenia araliacea, Euclea racemosa* subsp. *schimperi, Schrebera alata, Carissa spinarum, Calotropis procera, Fadogia cienkowskii, Gardenia ternifolia, Hymenodictyon floribundum, Meyna tetraphylla, Pavetta crassipes, Pavetta oliveriana, Sarcocephalus latifolius, Vernonia amygdalina, Vernonia auriculifera, Premna schimperi, Otostegia fruticosa, Oxytenanthera abyssinica.*

Trees usually or often reaching more than 10 m

Flacourtia indica, Anogeissus leiocarpa, Combretum adenogonium, Combretum collinum, Combretum hartmannianum, Combretum molle, Terminalia laxiflora, Terminalia macroptera, Adansonia digitata, Bridelia micrantha, Croton macrostachyus, Euphorbia ampliphylla, Senna singueana, Tamarindus indica, Acacia abyssinica, Acacia amythethophylla, Acacia polyacantha subsp. *campylacantha, Acacia etbaica, Acacia nilotica, Acacia senegal, Acacia sieberiana, Acacia tortilis, Albizia amara, Albizia grandibracteata, Albizia schimperiana, Dalbergia melanoxylon, Erythrina abyssinica, Lonchocarpus laxiflorus, Pterocarpus lucens, Trema orientalis, Ficus ingens, Ficus lutea, Ficus ovata, Ficus platyphylla, Ficus sur, Ficus sycomorus, Ficus thonningii, Ficus vasta, Apodytes dimidiata, Ziziphus mauritiana, Ziziphus mucronata, Ziziphus pubescens, Ziziphus spina-christi, Balanites aegyptiaca, Harrisonia abyssinica, Pseudocedrela kotschyi, Bersama abyssinica* Fresen., *Lannea barteri, Lannea schweinfurthii, Ozoroa insignis, Sclerocarya birrea, Cussonia arborea, Diospyros abyssinica, Diospyros mespiliformis, Embelia schimperi, Maesa lanceolata, Strychnos innocua, Olea europaea* subsp. *cuspidata* subsp. *cuspidata, Crossopteryx febrifuga, Vangueria madagascariensis, Cordia africana, Kigelia africana, Stereosper-*

mum kunthianum, Vitex doniana, Borassus aethiopum, Hyphaene thebaica.

Thorns or spines present

Capparis decidua, Capparis tomentosa, Flacourtia indica, Oncoba spinosa, Bridelia micrantha, Bridelia scleroneura, Euphorbia abyssinica, Euphorbia ampliphylla, Acacia abyssinica, Acacia amythethophylla, Acacia asak, Acacia polyacantha subsp. *campylacantha, Acacia dolichocephala, Acacia etbaica, Acacia gerrardii, Acacia hecatophylla, Acacia hockii, Acacia mellifera, Acacia nilotica, Acacia oerfota, Acacia persiciflora, Acacia senegal, Acacia seyal, Acacia sieberiana, Acacia tortilis, Acacia venosa, Dichrostachys cinerea, Erythrina abyssinica, Maytenus arbutifolia, Maytenus senegalensis, Ximenia americana, Ziziphus abyssinica, Ziziphus mauritiana, Ziziphus mucronata, Ziziphus spina-christi, Zanthoxylum gilletii, Balanites aegyptiaca, Harrisonia abyssinica, Commiphora africana, Commiphora pedunculata, Carissa spinarum, Meyna tetraphylla.*

Thorns or spines absent

Annona senegalensis, Boscia angustifolia, Boscia salicifolia, Boscia senegalensis, Cadaba farinosa, Maerua angolensis, Maerua oblongifolia, Maerua triphylla, Securidaca longipedunculata, Rumex abyssinicus, Woodfordia uniflora, Gnidia lamprantha, Faurea rochetiana, Protea gaguedi, Protea madiensis, Ochna leucophloeos, Syzygium guineense subsp. *macrocarpum, Anogeissus leiocarpa, Combretum adenogonium, Combretum collinum, Combretum hartmannianum, Combretum molle, Combretum rochetianum, Terminalia brownii, Terminalia laxiflora, Terminalia macroptera, Terminalia schimperiana, Psorospermum febrifugum, Grewia bicolor, Grewia erythraea, Grewia ferruginea, Grewia flavescens, Grewia mollis, Grewia tenax, Grewia trichocarpa, Grewia velutina, Grewia villosa, Dombeya buettneri, Dombeya quinqueseta, Sterculia africana, Adansonia digitata, Erythroxylum fischeri, Clutia abyssinica, Croton macrostachyus, Erythrococca trichogyne, Flueggea virosa, Cassia arereh, Piliostigma thonningii, Senna petersiana, Senna singueana, Tamarindus indica, Acacia hecatophylla, Albizia amara, Albizia grandibracteata, Albizia malacophylla, Albizia schimperiana, Entada abyssinica, Entada africana, Calpurnia aurea, Dalbergia boehmii, Dalbergia melanoxylon, Lonchocarpus laxiflorus, Ormocarpum pubescens, Pterocarpus lucens, Trema orientalis, Ficus glumosa, Ficus ingens, Ficus lutea, Ficus ovata, Ficus platyphylla, Ficus cordata* subsp. *salicifolia, Ficus sur, Ficus sycomorus, Ficus thonningii, Ficus vasta, Apodytes dimidiata, Salvadora persica, Ziziphus pubescens, Clausena anisata, Brucea antidysenterica, Boswellia papyrifera, Boswellia pirottae, Pseudocedrela kotschyi, Allophylus rubifolius, Dodonaea angustifolia, Bersama abyssinica, Lannea barteri, Lannea fruticosa, Lannea schimperi, Lannea triphylla, Lannea schweinfurthii, Ozoroa insignis, Ozoroa pulcherrima, Rhus vulgaris, Sclerocarya birrea, Cussonia arborea, Cussonia ostinii, Polyscias farinosa, Heteromorpha arborescens, Steganotaenia araliacea, Diospyros abyssinica, Diospyros mespiliformis, Euclea racemosa* subsp. *schimperi, Embelia schimperi, Maesa lanceolata, Strychnos innocua, Olea europaea* subsp. *cuspidata, Schrebera alata, Calotropis procera, Crossopteryx febrifuga, Fadogia cienkowskii, Gardenia ternifolia, Hymenodictyon floribundum, Pavetta crassipes, Pavetta oliveriana, Sarcocephalus latifolius, Vangueria madagascariensis, Vernonia amygdalina, Vernonia auriculifera, Cordia africana, Kigelia africana, Stereospermum kunthianum, Premna schimperi, Vitex doniana, Otostegia fruticosa, Borassus aethiopum, Hyphaene thebaica, Oxytenanthera abyssinica.*

Presumably or certainly evergreen

Boscia angustifolia, Boscia salicifolia, Boscia senegalensis, Cadaba farinosa, Capparis tomentosa, Maerua angolensis, Maerua oblongifolia, Maerua triphylla, Gnidia lamprantha, Faurea rochetiana, Protea gaguedi, Protea madiensis, Flacourtia indica, Oncoba spinosa, Ochna leucophloeos, Syzygium guineense subsp. *macrocarpum, Anogeissus leiocarpa, Combretum adenogonium, Combretum collinum, Combretum molle, Combretum rochetianum, Terminalia brownii, Terminalia laxiflora, Terminalia macroptera, Terminalia schimperiana, Psorospermum febrifugum, Grewia bicolor, Grewia erythraea, Grewia ferruginea, Grewia flavescens, Grewia mollis, Grewia trichocarpa, Grewia velutina, Dombeya buettneri, Dombeya quinqueseta, Erythroxylum fischeri, Bridelia micrantha, Bridelia scleroneura, Clutia abyssinica, Erythrococca trichogyne, Senna petersiana, Tamarindus indica, Acacia abyssinica, Albizia schimperiana, Calpurnia aurea, Trema orientalis, Maytenus arbutifolia, Maytenus senegalensis, Apodytes dimidiata, Salvadora persica, Brucea antidysenterica, Harrisonia abyssinica, Dodonaea angustifolia, Bersama abyssinica, Diospyros abyssinica, Diospyros mespiliformis, Euclea racemosa* subsp. *schimperi, Embelia schimperi, Maesa lanceolata, Strychnos innocua, Olea europaea* subsp. *cuspidata, Schrebera alata, Calotropis procera, Crossopteryx febrifuga, Fadogia cienkowskii, Gardenia ternifolia, Meyna tetraphylla, Pavetta crassipes, Pavetta oliveriana, Vernonia amygdalina, Vernonia auriculifera, Cordia africana, Kigelia africana, Otostegia fruticosa, Borassus aethiopum, Hyphaene thebaica, Oxytenanthera abyssinica.*

Presumably or certainly deciduous

Annona senegalensis, Capparis decidua, Securidaca longipedunculata, Rumex abyssinicus, Woodfordia uniflora, Combretum hartmannianum, Grewia tenax, Grewia villosa, Sterculia africana, Adansonia digitata, Croton macrostachyus, Euphorbia abyssinica, Euphorbia ampliphylla, Flueggea virosa, Cassia arereh, Piliostigma thonningii, Senna singueana, Acacia amythethophylla, Acacia asak, Acacia polyacantha subsp. *campylacantha, Acacia dolichocephala, Acacia etbaica, Acacia gerrardii, Acacia hecatophylla, Acacia hockii, Acacia mellifera, Acacia nilotica, Acacia oerfota, Acacia persiciflora, Acacia senegal, Acacia seyal, Acacia sieberiana, Acacia tortilis, Acacia venosa, Albizia amara, Albizia grandibracteata, Albizia malacophylla, Dichrostachys cinerea, Entada abyssinica, Entada africana, Dalbergia boehmii, Dalbergia melanoxylon, Erythrina abyssinica, Lonchocarpus laxiflorus, Ormocarpum pubescens, Pterocarpus lucens, Ficus glumosa, Ficus ingens, Ficus lutea, Ficus ovata, Ficus platyphylla, Ficus cordata* subsp. *salicifolia, Ficus sur, Ficus sycomorus, Ficus thonningii, Ficus vasta, Ximenia americana, Ziziphus abyssinica, Ziziphus mauritiana, Ziziphus mucronata, Ziziphus pubescens, Ziziphus spina-christi, Clausena anisata, Zanthoxylum gilletii, Balanites aegyptiaca, Boswellia papyrifera, Boswellia pirottae, Commiphora africana, Commiphora pedunculata, Pseudocedrela kotschyi, Allophylus rubifolius, Lannea barteri, Lannea fruticosa, Lannea schimperi, Lannea triphylla, Lannea schweinfurthii, Ozoroa insignis, Ozoroa pulcherrima, Rhus vulgaris, Sclerocarya birrea, Cussonia arborea, Cussonia ostinii, Polyscias farinosa, Heteromorpha arborescens, Steganotaenia araliacea, Carissa spinarum, Hymenodictyon floribundum, Sarcocephalus latifolius, Vangueria madagascariensis, Stereospermum kunthianum, Premna schimperi, Vitex doniana.*

Leptophyll (leaves or leaflets)

Euphorbia abyssinica, Acacia abyssinica (leaflet).

Nanophyll (leaves or leaflets)

Capparis decidua, Acacia amythethophylla (leaflet), *Acacia asak* (leaflet), *Acacia polyacantha* subsp. *campylacantha* (leaflet), *Acacia dolichocephala* (leaflet), *Acacia etbaica* (leaflet), *Acacia gerrardii* (leaflet), *Acacia hecatophylla* (leaflet), *Acacia hockii* (leaflet), *Acacia mellifera* (leaflet), *Acacia nilotica* (leaflet), *Acacia oerfota* (leaflet), *Acacia persiciflora* (leaflet), *Acacia senegal* (leaflet), *Acacia seyal* (leaflet), *Acacia sieberiana* (leaflet), *Acacia tortilis* (leaflet), *Acacia venosa* (leaflet), *Albizia amara* (leaflet), *Albizia malacophylla* (leaflet), *Albizia schimperiana* (leaflet), *Ormocarpum pubescens* (leaflet).

Microphyll (leaves or leaflets)

Boscia angustifolia, Boscia salicifolia, Boscia senegalensis, Cadaba farinosa, Capparis tomentosa, Maerua angolensis, Maerua oblongifolia, Securidaca longipedunculata, Woodfordia uniflora, Gnidia lamprantha, Grewia erythraea, Grewia tenax, Grewia velutina, Flueggea virosa, Cassia arereh (leaflet), *Senna petersiana* (leaflet), *Senna singueana* (leaflet), *Tamarindus indica* (leaflet), *Acacia etbaica* (leaf), *Acacia mellifera* (leaf), *Acacia oerfota* (leaf), *Albizia grandibracteata* (leaflet), *Dichrostachys cinerea* (leaflet), *Entada abyssinica* (leaflet), *Entada africana* (leaflet), *Calpurnia aurea* (leaflet), *Dalbergia boehmii* (leaflet), *Dalbergia melanoxylon* (leaflet), *Ormocarpum pubescens* (leaflet), *Maytenus arbutifolia, Ziziphus mucronata, Ziziphus pubescens, Ziziphus spina-christi, Clausena anisata* (leaflet), *Dodonaea angustifolia, Heteromorpha arborescens* (leaflet), *Diospyros abyssinica, Euclea racemosa* subsp. *schimperi, Olea europaea* subsp. *cuspidata.*

Mesophyll (leaves or leaflets)

Annona senegalensis, Boscia angustifolia, Boscia salicifolia, Boscia senegalensis, Capparis tomentosa, Maerua angolensis, Maerua triphylla, Rumex abyssinicus, Woodfordia uniflora, Faurea rochetiana, Protea gaguedi, Protea madiensis, Flacourtia indica, Oncoba spinosa, Ochna leucophloeos, Syzygium guineense subsp. *macrocarpum, Anogeissus leiocarpa, Combretum adenogonium, Combretum collinum, Combretum hartmannianum, Combretum molle, Combretum rochetianum, Terminalia brownii, Terminalia laxiflora, Terminalia schimperiana, Psorospermum febrifugum, Grewia bicolor, Grewia ferruginea, Grewia flavescens, Grewia mollis, Grewia tenax, Grewia trichocarpa, Grewia velutina, Grewia villosa, Dombeya buettneri, Dombeya quinqueseta, Sterculia africana, Adansonia digitata* (leaflet), *Erythroxylum fischeri, Bridelia micrantha, Bridelia scleroneura, Clutia abyssinica, Croton macrostachyus, Erythrococca trichogyne, Euphorbia ampliphylla, Cassia arereh* (leaflet), *Piliostigma thonningii* (leaf), *Senna petersiana* (leaf, leaflet), *Tamarindus indica* (leaf), *Acacia abyssinica* (leaf), *Acacia asak* (leaf), *Acacia polyacantha* subsp. *campylacantha* (leaf), *Acacia dolichocephala* (leaf), *Acacia etbaica* (leaf), *Acacia gerrardii* (leaf), *Acacia hecatophylla* (leaf), *Acacia hockii* (leaf), *Acacia nilotica* (leaf), *Acacia oerfota* (leaf), *Acacia persiciflora* (leaf), *Acacia senegal* (leaf), *Acacia seyal* (leaf), *Acacia sieberiana* (leaf), *Acacia tortilis* (leaf), *Acacia venosa* (leaf), *Albizia amara* (leaf), *Calpurnia aurea* (leaf), *Dalbergia boehmii* (leaf), *Dalbergia melanoxylon* (leaf), *Erythrina abyssinica* (leaf, leaflet), *Lonchocarpus laxiflorus* (leaflet), *Ormocarpum pubescens* (leaf), *Pterocarpus lucens* (leaflet), *Trema orientalis, Ficus glumosa, Ficus ingens, Ficus cordata* subsp. *salicifolia, Ficus sycomorus, Ficus thonningii, Maytenus senegalensis, Salvadora persica, Ximenia americana, Ziziphus abyssinica, Ziziphus mauritiana, Clausena anisata* (leaf, leaflet), *Balanites aegyptiaca* (leaf, leaflet), *Brucea antidysenterica* (leaflet), *Harrisonia abyssinica* (leaflet), *Boswellia papyrifera* (leaflet), *Boswellia pirottae* (leaflet), *Commiphora africana* (leaf, leaflet), *Commiphora pedunculata* (leaf, leaflet), *Pseudocedrela kotschyi* (leaflet), *Allophylus rubifolius* (leaflet), *Bersama abyssinica* (leaflet), *Lannea barteri* (leaflet), *Lannea fruticosa* (leaflet), *Lannea schimperi* (leaflet), *Lannea triphylla* (leaf, leaflet), *Lannea schweinfurthii* (leaf, leaflet), *Ozoroa insignis, Ozoroa pulcherrima, Rhus vulgaris* (leaf, leaflet), *Sclerocarya birrea* (leaflet), *Cussonia arborea* (leaflet), *Polyscias farinosa* (leaflet), *Heteromorpha arborescens* (leaf, leaflet), *Steganotaenia araliacea* (leaflet), *Diospyros mespiliformis, Embelia schimperi, Strychnos innocua, Schrebera alata* (leaf, leaflet), *Carissa spinarum, Crossopteryx febrifuga, Fadogia cienkowskii, Gardenia ternifolia, Hymenodictyon floribundum, Meyna tetraphylla, Pavetta crassipes, Pavetta oliveriana, Sarcocephalus latifolius* (in part), *Vernonia amygdalina, Cordia africana, Kigelia africana* (leaflet), *Stereospermum kunthianum* (leaflet), *Premna schimperi, Vitex doniana* (leaflet), *Otostegia fruticosa, Oxytenanthera abyssinica.*

Macrophyll (leaves, simple or compound leaves)

Annona senegalensis, Rumex abyssinicus, Protea madiensis, Combretum adenogonium, Combretum collinum, Terminalia laxiflora, Terminalia macroptera, Terminalia schimperiana, Grewia villosa, Dombeya buettneri, Dombeya quinqueseta, Sterculia africana, Adansonia digitata (leaf), *Croton macrostachyus, Cassia arereh* (leaf), *Piliostigma thonningii* (leaf), *Senna petersiana* (leaf), *Senna singueana* (leaf), *Acacia amythethophylla* (leaf), *Albizia grandibracteata* (leaf), *Albizia malacophylla* (leaf), *Albizia schimperiana* (leaf), *Dichrostachys cinerea* (leaf), *Entada abyssinica* (leaf), *Entada africana* (leaf), *Lonchocarpus laxiflorus* (leaf), *Pterocarpus lucens* (leaf), *Ficus lutea, Ficus ovata, Ficus platyphylla, Ficus sur, Ficus sycomorus, Ficus vasta, Zanthoxylum gilletii* (leaflet), *Brucea antidysenterica* (leaf), *Harrisonia abyssinica* (leaf), *Boswellia papyrifera* (leaf), *Boswellia pirottae* (leaf), *Pseudocedrela kotschyi* (leaf), *Allophylus rubifolius* (leaf), *Bersama abyssinica* (leaf), *Lannea barteri* (leaf), *Lannea fruticosa* (leaf), *Lannea schimperi* (leaf), *Lannea schweinfurthii* (leaf), *Sclerocarya birrea* (leaf), *Cussonia arborea* (leaf, leaflet), *Cussonia ostinii, Polyscias farinosa* (leaf), *Steganotaenia araliacea* (leaf), *Maesa lanceolata, Calotropis procera, Sarcocephalus latifolius, Vangueria madagascariensis, Vernonia auriculifera, Kigelia africana* (leaf), *Stereospermum kunthianum* (leaf), *Vitex doniana* (leaf, leaflet).

Megaphyll

Zanthoxylum gilletii (leaf), *Borassus aethiopum, Hyphaene thebaica.*

Obvious or likely adaptation for fire resistance present

Annona senegalensis, Maerua oblongifolia, Rumex abyssinicus, Faurea rochetiana, Protea gaguedi, Protea madiensis, Ochna leucophloeos, Anogeissus leiocarpa, Combretum adenogonium, Combretum collinum, Combretum molle, Terminalia brownii, Terminalia laxiflora, Terminalia macroptera, Terminalia schimperiana, Psorospermum febrifugum, Dombeya buettneri, Sterculia africana, Adansonia digitata, Bridelia scleroneura, Croton macrostachyus, Euphorbia abyssinica, Euphorbia ampliphylla, Cassia arereh, Piliostigma thonningii, Tamarindus indica, Acacia abyssinica, Acacia amythethophylla, Acacia asak, Acacia polyacantha subsp. *campylacantha, Acacia dolichocephala, Acacia etbaica, Acacia gerrardii, Acacia hecatophylla, Acacia hockii, Acacia mellifera, Acacia nilotica, Acacia persiciflora, Acacia senegal, Acacia seyal, Acacia sieberiana, Acacia tortilis, Acacia venosa, Albizia amara, Albizia grandibracteata, Albizia malacophylla, Albizia schimperiana, Dichrostachys cinerea, Entada abyssinica, Entada africana, Dalbergia boehmii, Dalbergia melanoxylon, Erythrina abyssinica, Lonchocarpus laxiflorus, Ormocarpum pubescens, Pterocarpus lucens, Ficus ovata, Ficus platyphylla, Ficus cordata* subsp. *salicifolia, Ficus sur, Ficus sycomorus, Ficus thonningii, Ficus vasta, Ziziphus abyssinica, Ziziphus mauritiana, Ziziphus pubescens, Balanites aegyptiaca, Boswellia pirottae, Commiphora africana, Commiphora pedunculata, Pseudocedrela kotschyi, Lannea fruticosa, Lannea schimperi, Lannea triphylla, Lannea schweinfurthii, Ozoroa insignis, Ozoroa pulcherrima, Sclerocarya birrea, Cussonia arborea, Cussonia ostinii, Steganotaenia araliacea, Diospyros abyssinica, Diospyros mespiliformis, Olea europaea* subsp. *cuspidata, Crossopteryx febrifuga, Fadogia cienkowskii, Hymenodictyon floribundum, Pavetta crassipes, Sarcocephalus latifolius, Cordia africana, Kigelia africana, Stereospermum kunthianum, Vitex doniana.*

No obvious or likely adaptation for fire resistance

Boscia angustifolia, Boscia salicifolia, Boscia senegalensis, Cadaba farinosa, Capparis decidua, Capparis tomentosa, Maerua angolensis, Maerua triphylla, Securidaca longipedunculata, Woodfordia uniflora, Gnidia lamprantha, Flacourtia indica, Oncoba spinosa, Syzygium guineense subsp. *macrocarpum, Combretum hartmannianum, Combretum rochetianum, Grewia bicolor, Grewia erythraea, Grewia ferruginea, Grewia flavescens, Grewia mollis, Grewia tenax, Grewia trichocarpa, Grewia velutina, Grewia villosa, Dombeya quinqueseta, Erythroxylum fischeri, Bridelia micrantha, Clutia abyssinica, Erythrococca trichogyne, Flueggea virosa, Senna petersiana, Senna singueana, Acacia oerfota, Calpurnia aurea, Trema orientalis, Ficus glumosa, Ficus ingens, Ficus lutea, Maytenus arbutifolia, Maytenus senegalensis, Apodytes dimidiata, Salvadora persica, Ximenia americana, Ziziphus mucronata, Ziziphus spina-christi, Clausena anisata, Zanthoxylum gilletii, Brucea antidysenterica, Harrisonia abyssinica, Boswellia papyrifera, Allophylus rubifolius, Dodonaea angustifolia, Bersama abyssinica, Lannea barteri, Rhus vulgaris, Polyscias farinosa, Heteromorpha arborescens, Euclea racemosa* subsp. *schimperi, Embelia schimperi, Maesa lanceolata, Strychnos innocua, Schrebera alata, Carissa spinarum, Calotropis procera, Gardenia ternifolia, Meyna tetraphylla, Pavetta oliveriana, Vangueria madagascariensis, Vernonia amygdalina, Vernonia auriculifera, Premna schimperi, Otostegia fruticosa, Borassus aethiopum, Hyphaene thebaica, Oxytenanthera abyssinica.*

Appendix 6: Selection of clustering method and number of clusters

We used cluster analysis to examine whether distinct tree communities were recognizable within the western woodlands of Ethiopia (Section 10.1.2). The analyses were carried out in R (R Core Team 2020), using the packages *vegan* (Oksanen et al. 2019) to compute the distance matrices, cluster (Maechler et al. 2019) and *dendextend* (Galili 2015) to carry out the cluster analysis, and *vegan* and *factoextra* (Kassambara & Mundt 2020) for the visualization of results. Below some more detailed results are provided that complement the results presented in Section 10.

We tested different clustering methods, including agglomerative hierarchical clustering, divisive hierarchical clustering and k-medoids clustering (Kaufman & Rousseeuw 2009). At a larger scale, all three yielded similar patterns, with vegetation relevés separating in two major clusters in the north and in the south respectively. However, the divisive hierarchical clustering and k-medoids clustering methods yielded similar but more tenuous patterns as compared to the agglomerative hierarchical clustering methods. We therefore continued with the latter.

We carried out agglomerative hierarchical clustering using five different linkage methods. These were the *Unweighted Pair Group ArithMetic Averages* (*UPGMA*), *Weighted Pair Group ArithMetic Averages* (*WPGMA*), complete linkage, generalized weighted average (aka flexible *WPGMA*) and the generalized UPGMA (aka flexible *UPGMA*). Following the recommendation in Belbin et al. (1992), β = -0.25 was used for the flexible WPGMA, and β = -0.1 for the flexible UPGMA.

To evaluate the clustering results, the agglomerative coefficient (Kaufman & Rousseeuw 2009) and the cophenetic correlation coefficient (Sokal & Rohlf 1962; Farris 1969) were computed. High values of the former indicate a high dissimilarity between the final clusters compared to the average dissimilarity between clusters (Odong et al. 2011). The cophenetic correlation is a measure of how well a dendrogram preserves the pairwise distances between the original data points (Saraçli et al. 2013).

Initial analyses showed that from the five linkage methods, UPGMA preserved the pairwise distances between the original data points best. However, the two β -flexible linkage methods yield higher agglomerative coefficients, indicating a stronger clustering structure, with better defined large clusters (Table Appendix 6-1).

The cophenetic correlation (Galili 2015) between the trees produced with the five linkage methods ranged from 0.48 to 0.78 for the cluster results based on all species (Table Appendix 6-2). The correlations between the flexible UPGMA and WPGMA, and between the UPGMA and flexible UPGMA were highest. This could be attributed to the fact that these methods all yielded more or less the same two or three major clusters. Differences deviated rapidly when increasing the number of clusters. Similar results were obtained for clusters based on all species and on CTW-species only.

Table Appendix 6-1: Evaluation of different linkage methods. Clustering was done based on the Simpson dissimilarity index

Statistics	species	Complete	UPGMA	WPGMA	f.WPGMA	f.UPGMA
agglomerative coefficient	All	0.7462	0.7302	0.7382	0.9059	0.8293
Cophenetic correlation	All	0.4343	0.6450	0.4589	0.5106	0.5746
agglomerative coefficient	CTW	0.7910	0.7813	0.7894	0.9320	0.8759
Cophenetic correlation	CTW	0.4326	0.6103	0.5856	0.4558	0.5400

Table Appendix 6-2: Cophenetic correlation between the dendrogram trees created using different linkage methods.

	Complete	UPGMA	WPGMA	f.WPGMA	f.UPGMA
Complete	1.00	0.56	0.56	0.48	0.53
UPGMA	0.56	1.00	0.55	0.64	0.79
WPGMA	0.56	0.55	1.00	0.53	0.50
f.WPGMA	0.48	0.64	0.53	1.00	0.78
f.UPGMA	0.53	0.79	0.50	0.78	1.00

Based on these preliminary results, the f.WPGMA and f.UPGMA linkage methods were considered the preferrable methods. Results of cluster analysis of the two methods were similar when considering two clusters. When dividing relevés in three main clusters, differences became more pronounced, especially for the northern relevés and along the margins of the study area. The f.UPGMA method yielded more cohesive and easier to interpret results, which are the ones presented in Section 10.2.2.

Number of clusters

We used five indices to determine the optimum number of clusters (k). These were the maximum silhouette width (Rousseeuw 1987), the Frey index (Frey & Groenewoud 1972), the McClain and Rao index (McClain & Rao 1975), the cindex (Hubert & Levin 1976) and the Dunn index (Dunn 1974). The maximum silhouette width was computed using the f.UPGMA and f.WPGMA methods. The other indices were computed using the NbClust package (Charrad et al. 2014). This package does not include the option to use the f.UPGMA and f.WPGMA linkage methods, so we used the UPGMA linkage method instead, as the results with this method correlated most with these linkage methods.

The 'optimum' number of clusters, as defined by the maximum silhouette width was 83 and 64 for the f.WPGMA and f.UPGMA methods respectively (Fig. Appendix 6-1). When only considering the CTW-species, these numbers were slightly lower, viz., 66 and 51 for the f.WPGMA and f.UPGMA methods respectively. The Cindex method gave a similar result, with an optimum number of clusters of 60. The resulting average cluster size would be 2 to 3, suggesting a lack of clearly defined vegetation communities at any but the most local scale. In contrast, based on the other methods the relevés could best be divided in 2 (McClain and Dunn methods) to 3 (Frey method) clusters.

When considering more than three clusters, geographic distribution patterns of these clusters were much less cohesive, differed more depending on the method, and were in general more difficult to interpret. These observations reinforces the lack of clear vegetation patterns at any scale but the region wide north-south and/or altitudinal divide. For this study, we were mostly interested in the larger scale patterns in vegetation communities across the western woodlands. We therefore looked further into this large scale pattern based on two or three clusters.

Results of the f.WPGMA and f.UPGMA methods yielded similar pattern when considering two clusters. Differences were more pronounced when considering three clusters. Differences were mostly related to how the two methods assign the more deviating relevés, which may be due to the higher sensitivity of the f.WPGMA method for outliers and peripheral objects (Belbin et al. 1992). With an average silhouette width of 0.15 (all species) and 0.14 (CTW-species), the f.WPGMA clusters were less well defined and cohesive then those defined by the f.UPGMA method. Furthermore, the geographic distribution of clusters based on the f.WPGMA was more difficult to interpret. Especially in the north, relevés at relatively close distances were grouped into alternating clusters.

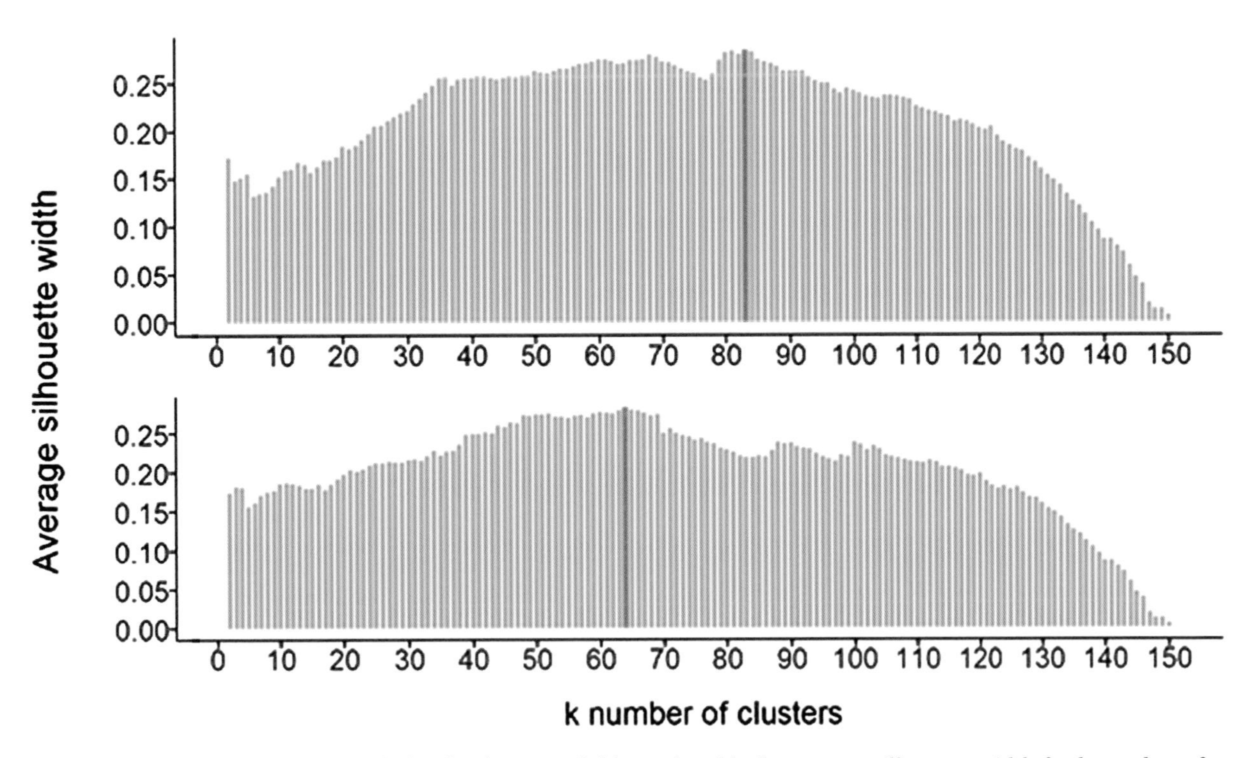

Fig. Appendix 6-1: Silhouette relevés for the data set of this work, with the average silhouette width for k number of clusters. The red line indicates the number of clusters that yields the maximum average silhouette width. The clustering is based on the Simpson dissimilarity, and the f.WPGMA (upper figure) and f.UPGMA (lower figure) methods (see main text for further explanation).

Appendix 7: All species and CTW-species distributed on three clusters according to *f.UPGMA* and *f.WPGMA*

All species. Flexible Unweighted Pair Group ArithMetic Averages (f.UPGMA)							
Species	SpecCode	Phytogeography	Cluster	IndicatorValue	p.value	A	B
Anogeissus leiocarpa	anolei	CTW[n+c]	1	0.7956	0.001	0.74886991	0.8452381
Pterocarpus lucens	pteluc	CTW[n+c]	1	0.6316	0.001	0.74468085	0.53571429
Acacia hecatophylla	acahec	CTW[n+c]	1	0.6159	0.001	0.8852459	0.42857143
Sterculia africana	steafr	CTW[all+ext]	1	0.5996	0.001	0.79482072	0.45238095
Ziziphus spina-christi	zizspi	CTW[all+ext]	1	0.5796	0.003	0.62707366	0.53571429
Dalbergia melanoxylon	dalmel	CTW[n+c]	1	0.5455	0.001	1	0.29761905
Boswellia papyrifera	bospap	CTW[n+c]	1	0.532	0.002	0.81965007	0.3452381
Balanites aegyptiaca	balaeg	CTW[all+ext]	1	0.5134	0.017	0.67086157	0.39285714
Combretum hartmannianum	comhar	CTW[n+c]	1	0.4125	0.019	0.79411765	0.21428571
Ficus glumosa	ficglu	CTW[all+ext]	1	0.3916	0.044	0.64417178	0.23809524
Boswellia pirottae	bospir	CTW[n+c]	1	0.378	0.03	1	0.14285714
Combretum collinum	comcol	CTW[all+ext]	2	0.644	0.001	0.58064516	0.71428571
Bridelia scleroneura	briscl	CTW[all+ext]	2	0.6372	0.001	0.94736842	0.42857143
Terminalia schimperiana	tersch	CTW[all+ext]	2	0.6309	0.001	0.62908681	0.63265306
Annona senegalensis	annsen	CTW[c+s+ext]	2	0.6197	0.001	0.78402904	0.48979592
Vitex doniana	vitdon	CTW[all]	2	0.6013	0.001	0.84375	0.42857143
Entada abyssinica	entaby	CTW[all+ext]	2	0.5602	0.001	0.80946746	0.3877551
Piliostigma thonningii	piltho	CTW[all+ext]	2	0.5498	0.011	0.56968959	0.53061224
Grewia mollis	gremol	CTW[all+ext]	2	0.5355	0.004	0.63870968	0.44897959
Cordia africana	corafr	TRG	2	0.5157	0.004	0.65158371	0.40816327
Gardenia ternifolia	garter	CTW[all+ext]	2	0.463	0.046	0.55295069	0.3877551
Acacia polyacantha subsp. *campylacantha*	acacam	CTW[all+ext]	2	0.4594	0.018	0.64646465	0.32653061
Ficus sur	ficsur	TRG	2	0.4391	0.005	0.94488189	0.20408163
Albizia malacophylla	albmal	CTW[n+c]	2	0.4199	0.015	0.72	0.24489796
Cussonia arborea	cusarb	CTW[all+ext]	2	0.4041	0.008	1	0.16326531
Psorospermum febrifugum	psofeb	CTW[c]	2	0.3023	0.029	0.89552239	0.10204082

All species. Flexible Unweighted Pair Group ArithMetic Averages (f.UPGMA)

Species	SpecCode	Phytogeography	Cluster	IndicatorValue	p.value	A	B
Croton macrostachyus	cromac	TRG	3	0.6178	0.001	0.85870756	0.44444444
Maytenus arbutifolia	maytrb	MI	3	0.527	0.001	1	0.27777778
Senna petersiana	senpet	CTW[all+ext]	3	0.4961	0.001	0.88607595	0.27777778
Capparis decidua	capdec	ACB[-CTW]	3	0.4593	0.001	0.94915254	0.22222222
Vangueria madagascariensis	vanmad	CTW[all+ext]	3	0.4305	0.004	0.83404255	0.22222222
Acacia asak	acaasa	ACB[-CTW]	3	0.4082	0.001	1	0.16666667
Vernonia auriculifera	veraur	MI	3	0.4082	0.002	1	0.16666667
Acacia mellifera	acamel	ACB[-CTW]	3	0.402	0.009	0.72727273	0.22222222
Rhus vulgaris	rhuvul	TRG	3	0.3988	0.028	0.57242991	0.27777778
Acacia amythethophylla	acaamy	CTW[n]	3	0.3705	0.009	0.82352941	0.16666667
Acacia abyssinica	acaaby	MI	3	0.3659	0.012	0.80327869	0.16666667
Diospyros mespiliformis	diomes	MI	3	0.3629	0.022	0.79032258	0.16666667
Adansonia digitata	adadig	CTW[n]	3	0.3504	0.012	0.73684211	0.16666667
Acacia etbaica	acaetb	ACB[-CTW]	3	0.3333	0.013	1	0.11111111
Salvadora persica	salper	ACB[-CTW]	3	0.3333	0.009	1	0.11111111
Boscia senegalensis	bossen	ACB[-CTW]	3	0.3168	0.026	0.90322581	0.11111111
Fadogia cienkowskii	fadcie	CTW[c]	3	0.3168	0.02	0.90322581	0.11111111
Bersama abyssinica	beraby	MI	3	0.3064	0.029	0.84482759	0.11111111
Maesa lanceolata	maelan	MI	3	0.3064	0.028	0.84482759	0.11111111
Maerua angolensis	maeang	ACB[-CTW]	3	0.3025	0.036	0.82352941	0.11111111
Boscia angustifolia	bosang	ACB[-CTW]	3	0.29	0.041	0.75675676	0.11111111

All species. Flexible Weighted Pair Group ArithMetic Averages (f.WPGMA)

Species	SpecCode	Phytogeography	Cluster	IndicatorValue	p.value	A	B
Acacia mellifera	acamel	ACB[-CTW]	1	0.6171	0.001	0.904293	0.421053
Acacia hecatophylla	acahec	CTW[n+c]	1	0.5955	0.001	0.612549	0.578947
Maytenus arbutifolia	maytrb	MI	1	0.513	0.001	1	0.263158
Balanites aegyptiaca	balaeg	CTW[all+ext]	1	0.4522	0.031	0.485695	0.421053
Boscia angustifolia	bosang	ACB[-CTW]	1	0.4442	0.002	0.937294	0.210526
Acacia asak	acaasa	ACB[-CTW]	1	0.3974	0.005	1	0.157895
Boscia senegalensis	bossen	ACB[-CTW]	1	0.3974	0.002	1	0.157895

All species. Flexible Weighted Pair Group ArithMetic Averages (f.WPGMA)

Species	SpecCode	Phytogeography	Cluster	IndicatorValue	p.value	A	B
Vernonia auriculifera	veraur	MI	1	0.3974	0.002	1	0.157895
Capparis decidua	capdec	ACB[-CTW]	1	0.3638	0.014	0.838204	0.157895
Calotropis procera	calpro	ACB[-CTW]	1	0.3613	0.01	0.619932	0.210526
Croton macrostachyus	cromac	TRG	1	0.3602	0.044	0.616162	0.210526
Diospyros mespiliformis	diomes	MI	1	0.3489	0.02	0.771098	0.157895
Adansonia digitata	adadig	CTW[n]	1	0.3304	0.018	0.691558	0.157895
Albizia amara	albama	CTW[all+ext]	1	0.3207	0.027	0.651376	0.157895
Bersama abyssinica	beraby	MI	1	0.3018	0.045	0.865248	0.105263
Anogeissus leiocarpa	anolei	CTW[n+c]	2	0.7891	0.001	0.68014	0.915493
Pterocarpus lucens	pteluc	CTW[n+c]	2	0.6831	0.001	0.770564	0.605634
Sterculia africana	steafr	CTW[all+ext]	2	0.671	0.001	0.864089	0.521127
Boswellia papyrifera	bospap	CTW[n+c]	2	0.6149	0.002	0.925693	0.408451
Ziziphus spina-christi	zizspi	CTW[all+ext]	2	0.5889	0.002	0.586185	0.591549
Lannea fruticosa	lanfru	CTW[all]	2	0.5438	0.009	0.636231	0.464789
Dalbergia melanoxylon	dalmel	CTW[n+c]	2	0.5408	0.001	0.865275	0.338028
Ficus glumosa	ficglu	CTW[all+ext]	2	0.4897	0.003	0.851361	0.28169
Combretum hartmannianum	comhar	CTW[n+c]	2	0.4311	0.016	0.776228	0.239437
Entada africana	entafr	CTW[all]	2	0.4136	0.021	0.63929	0.267606
Boswellia pirottae	bospir	CTW[n+c]	2	0.4111	0.007	1	0.169014
Annona senegalensis	annsen	CTW[c+s+ext]	3	0.6323	0.001	0.938008	0.42623
Combretum collinum	comcol	CTW[all+ext]	3	0.587	0.001	0.553102	0.622951
Bridelia scleroneura	briscl	CTW[all+ext]	3	0.561	0.001	0.872651	0.360656
Entada abyssinica	entaby	CTW[all+ext]	3	0.537	0.002	0.837665	0.344262
Vitex doniana	vitdon	CTW[all]	3	0.528	0.001	0.809909	0.344262
Grewia mollis	gremol	CTW[all+ext]	3	0.5276	0.004	0.67931	0.409836
Terminalia schimperiana	tersch	CTW[all+ext]	3	0.5081	0.043	0.507987	0.508197
Gardenia ternifolia	garter	CTW[all+ext]	3	0.4711	0.021	0.615445	0.360656
Rhus vulgaris	rhuvul	TRG	3	0.4521	0.009	0.890681	0.229508
Maytenus senegalensis	maysen	TRG	3	0.4506	0.031	0.589794	0.344262
Dombeya quinqueseta	domqui	CTW[all+ext]	3	0.4372	0.008	0.777372	0.245902
Cordia africana	corafr	TRG	3	0.4306	0.047	0.538502	0.344262

All species. Flexible Weighted Pair Group ArithMetic Averages (f.WPGMA)

Species	SpecCode	Phytogeography	Cluster	IndicatorValue	p.value	A	B
Ficus sur	ficsur	TRG	3	0.3885	0.005	0.920882	0.163934
Albizia malacophylla	albmal	CTW[n+c]	3	0.3817	0.028	0.683704	0.213115
Cussonia arborea	cusarb	CTW[all+ext]	3	0.3621	0.017	1	0.131148
Harrisonia abyssinica	haraby	CTW[c+s+ext]	3	0.3621	0.011	1	0.131148
Vangueria madagascariensis	vanmad	CTW[all+ext]	3	0.3388	0.018	1	0.114754
Heteromorpha arborescens	Hetarb	MI	3	0.2863	0.021	1	0.081967

CTW-species. Flexible Unweighted Pair Group ArithMetic Averages (f.UPGMA)

Species	SpecCode	Phytogeography	Cluster	IndicatorValue	p.value	A	B
Anogeissus leiocarpa	anolei	CTW[n+c]	1	0.8328	0.001	0.963802	0.719626
Pterocarpus lucens	pteluc	CTW[n+c]	1	0.6613	0.024	0.899906	0.485981
Sterculia africana	steafr	CTW[all+ext]	1	0.6265	0.03	1	0.392523
Acacia hecatophylla	acahec	CTW[n+c]	1	0.588	0.03	1	0.345794
Terminalia schimperiana	tersch	CTW[all+ext]	2	0.7735	0.004	0.763346	0.783784
Vitex doniana	vitdon	CTW[all]	2	0.7354	0.002	0.952926	0.567568
Entada abyssinica	entaby	CTW[all+ext]	2	0.6919	0.001	0.932141	0.513514
Combretum collinum	comcol	CTW[all+ext]	2	0.6832	0.035	0.639584	0.72973
Piliostigma thonningii	piltho	CTW[all+ext]	2	0.6663	0.012	0.684435	0.648649
Annona senegalensis	annsen	CTW[c+s+ext]	2	0.6643	0.009	0.859256	0.513514
Gardenia ternifolia	garter	CTW[all+ext]	2	0.6421	0.026	0.762651	0.540541
Bridelia scleroneura	briscl	CTW[all+ext]	2	0.585	0.028	0.844292	0.405405
Cussonia arborea	cusarb	CTW[all+ext]	2	0.4246	0.022	0.952926	0.189189
Capparis decidua	capdec	ACB[-CTW]	3	0.8892	0.001	0.988453	0.8
Acacia mellifera	acamel	ACB[-CTW]	3	0.86	0.001	0.924406	0.8
Acacia asak	acaasa	ACB[-CTW]	3	0.7746	0.001	1	0.6
Adansonia digitata	adadig	CTW[n]	3	0.7461	0.001	0.927746	0.6
Terminalia brownii	terbro	ACB[-CTW]	3	0.6988	0.003	0.813939	0.6
Salvadora persica	salper	ACB[-CTW]	3	0.6325	0.001	1	0.4
Boscia senegalensis	bossen	ACB[-CTW]	3	0.6252	0.004	0.977169	0.4
Boscia angustifolia	bosang	ACB[-CTW]	3	0.6114	0.003	0.934498	0.4
Albizia amara	albama	CTW[all+ext]	3	0.5863	0.005	0.859438	0.4

CTW-species. Flexible Unweighted Pair Group ArithMetic Averages (f.UPGMA)							
Species	SpecCode	Phytogeography	Cluster	IndicatorValue	p.value	A	B
Calotropis procera	calpro	ACB[-CTW]	3	0.5641	0.012	0.795539	0.4
Acacia etbaica	acaetb	ACB[-CTW]	3	0.4371	0.049	0.955357	0.2

CTW-species. Flexible Weighted Pair Group ArithMetic Averages (f.WPGMA)							
Species	SpecCode	Phytogeography	Cluster	IndicatorValue	p.value	A	B
Acacia hecatophylla	acahec	CTW[n+c]	1	0.7251	0.001	0.788732	0.666667
Acacia seyal	acasey	CTW[all+ext]	1	0.665	0.001	0.583759	0.757576
Balanites aegyptiaca	balaeg	CTW[all+ext]	1	0.6308	0.001	0.691021	0.575758
Ziziphus spina-christi	zizspi	CTW[all+ext]	1	0.5485	0.004	0.551573	0.545455
Acacia mellifera	acamel	ACB[-CTW]	1	0.4597	0.001	0.871595	0.242424
Senna petersiana	senpet	CTW[all+ext]	1	0.401	0.004	0.884211	0.181818
Vangueria madagascariensis	vanmad	CTW[all+ext]	1	0.3619	0.009	0.864198	0.151515
Capparis decidua	capdec	ACB[-CTW]	1	0.3322	0.014	0.910569	0.121212
Acacia amythethophylla	acaamy	CTW[n]	1	0.3183	0.027	0.835821	0.121212
Acacia asak	acaasa	ACB[-CTW]	1	0.3015	0.027	1	0.090909
Boscia senegalensis	bossen	ACB[-CTW]	1	0.3015	0.021	1	0.090909
Anogeissus leiocarpa	anolei	CTW[n+c]	2	0.6993	0.001	0.641823	0.761905
Pterocarpus lucens	pteluc	CTW[n+c]	2	0.5754	0.003	0.63206	0.52381
Flueggea virosa	fluvir	CTW[all+ext]	2	0.5323	0.004	0.643224	0.440476
Sterculia africana	steafr	CTW[all+ext]	2	0.5255	0.001	0.662651	0.416667
Lannea fruticosa	lanfru	CTW[all]	2	0.4904	0.016	0.561105	0.428571
Lonchocarpus laxiflorus	lonlax	CTW[all]	2	0.4803	0.036	0.523725	0.440476
Boswellia papyrifera	bospap	CTW[n+c]	2	0.4554	0.015	0.669985	0.309524
Dalbergia melanoxylon	dalmel	CTW[n+c]	2	0.4103	0.025	0.673469	0.25
Boswellia pirottae	bospir	CTW[n+c]	2	0.3261	0.044	0.812081	0.130952
Harrisonia abyssinica	haraby	CTW[c+s+ext]	2	0.3086	0.036	1	0.095238
Terminalia schimperiana	tersch	CTW[all+ext]	3	0.8097	0.001	0.723512	0.90625
Vitex doniana	vitdon	CTW[all]	3	0.7065	0.001	0.887324	0.5625
Entada abyssinica	entaby	CTW[all+ext]	3	0.693	0.001	0.853799	0.5625
Combretum collinum	comcol	CTW[all+ext]	3	0.6034	0.002	0.506531	0.71875
Annona senegalensis	annsen	CTW[c+s+ext]	3	0.5936	0.001	0.75179	0.46875

CTW-species. Flexible Weighted Pair Group ArithMetic Averages (f.WPGMA)							
Species	SpecCode	Phytogeography	Cluster	IndicatorValue	p.value	A	B
Bridelia scleroneura	briscl	CTW[all+ext]	3	0.5831	0.001	0.77703	0.4375
Piliostigma thonningii	piltho	CTW[all+ext]	3	0.5729	0.003	0.500155	0.65625
Gardenia ternifolia	garter	CTW[all+ext]	3	0.5354	0.002	0.573201	0.5
Stereospermum kunthianum	stekun	CTW[all+ext]	3	0.5117	0.008	0.492908	0.53125
Ficus sycomorus	ficsyc	CTW[all+ext]	3	0.4993	0.02	0.469344	0.53125
Grewia mollis	gremol	CTW[all+ext]	3	0.468	0.009	0.539041	0.40625
Combretum molle	commol	CTW[all+ext]	3	0.4623	0.013	0.526196	0.40625
Cussonia arborea	cusarb	CTW[all+ext]	3	0.4555	0.001	0.948387	0.21875
Lannea schimperi	lansci	CTW[all+ext]	3	0.4013	0.025	0.57257	0.28125
Protea madiensis	promad	CTW[c+s+ext]	3	0.25	0.049	1	0.0625

List of acronyms used in this work

This is a list of all the acronyms used in the work, including terms for vegetation types, geographical areas, herbaria, books, software, etc. The designations used for parameters in Appendix 3 have not been included.

AA: Afroalpine belt. Vegetation type in Friis et al. (2010) and in this work (Fig. 4-1).

ACB: *Acacia-Commiphora* woodland and bushland. Vegetation type in Friis et al. (2010) and in this work (Fig. 4-1).

ACB[-CTW]: Distribution type on the Horn of Africa based on generalized distribution pattern, with main distribution in the *Acacia-Commiphora bushland* (ACB), but marginally intruding in the *Combretum-Terminalia woodland and wooded grassland* (CTW) in the north or in the south, or both.

ACB/RV: *Acacia* wooded grassland in the Rift Valley. Vegetation type in Friis et al. (2010) and in this work (Fig. 4-1).

AF: Afar, area in Ethiopia, consisting of the lowland parts of the former Tigray, Welo, Shewa and Hararge regions, below and to the east of the 1000 m contour on the east side of the Ethiopian Highlands, extending to the Eritrean border in the east and the border with HA in the south, used to indicate distribution in the *Flora of Ethiopia* and Eritrea and in this work, see Fig. 2-28.

AR: The former Arsi region, area in Ethiopia used to indicate distribution in the *Flora of Ethiopia and Eritrea* and in this work, see Fig. 2-28.

a.s.l.: above sea level. Altitudes in the Afar below sea level are written 'below sea level.'

ASO: Distribution in Ethiopia based on direct observation: Distribution recorded from areas dominated by *Acacia-Commiphora bushland* (ACB), mainly in the Afar and Somalia regions.

BA: The former Bale region, area in Ethiopia used to indicate distribution in the *Flora of Ethiopia and Eritrea* and in this work, see Fig. 2-28.

BR: The herbarium of the Meise Botanic Garden, formerly the National Herbarium of Belgium, based in Meise north of Bruxelles.

BS 51: An annotated checklist in two volumes, entitled 'Flora of the Sudan-Uganda border area east of the Nile', written by Ib Friis & Kaj Vollesen, and published by the Royal Danish Academy of Sciences and Letters in 1998 and 2005 as volume 51(1) and 51(2) in the series *Biologiske Skrifter*.

C: Acronym for the largest herbarium in Denmark, based in Copenhagen, formerly the most important part of the Botanical Museum of the University of Copenhagen, but since 2004 incorporated in the Natural History Museum of Denmark, still under the University of Copenhagen and maintaining its former acronym, C.

CGIAR-CSI DEM: Digital elevation model provided by the Consortium for Spatial Information organised by CGIAR (Consultative Group on International Agricultural Research).

CTW: *Combretum-Terminalia* woodland and wooded grassland. Vegetation type in Friis et al. (2010) and in this work (Fig. 4-1). Also used in combinations: CTW-area, area covered by the CTW-vegetation and CTW-phytochorion; CTW-species: species in the CTW-element, that is species with local distribution in Ethiopia mainly in CTW-vegetation and in the CTW-phytochorion.

CTW[all]: Distribution type on the Horn of Africa with focus on the western woodlands and based on generalised distribution patterns: This type is distributed in the *Combretum-Terminalia* woodlands, from north to south or almost so, with no extension to the Rift Valley.

CTW[all+ext]: Distribution type on the Horn of Africa with focus on the western woodlands and based on generalized distribution patterns: This type is distributed in the *Combretum-Terminalia* woodlands, from north to south or almost so, with extension to the Rift Valley and east of the Rift.

CTW[c]: Distribution type on the Horn of Africa with focus on the western woodlands and based on generalized distribution patterns: This type is distributed in the central part of the *Combretum-Terminalia* woodlands, mainly south of the Abay River; no eastward extension.

CTW[c+s+ext]: Distribution type on the Horn of Africa with focus on the western woodlands and based on generalized distribution patterns: This type is distributed in the central and southern part of the *Combretum-Terminalia* woodlands, mainly south of the Abay River and with extension to the Rift Valley and east of the Rift.

CTW[n]: Distribution type on the Horn of Africa with focus on the western woodlands and based on generalized distribution patterns: This type is distributed in the northern part of the *Combretum-Terminalia* woodlands, mainly north of the Abay River; no eastward extension.

CTW[n+c]: Distribution type on the Horn of Africa with focus on the western woodlands and based on generalized distribution patterns: This type is distributed in the northern and central part of the *Combretum-Terminalia* woodlands, reaching the Gambela Region in the south; no eastward extension.

CTW[s]: Distribution type on the Horn of Africa with focus on the western woodlands and based on generalized distribution patterns: This type is distributed in the southern part of the *Combretum-Terminalia* woodlands, mainly south of the Abay River; no eastward extension.

CTW[s+ext]: Distribution type on the Horn of Africa with focus on the western woodlands and based on generalized distribution patterns: This type is distributed in the southern part of the *Combretum-Terminalia* woodlands, mainly south of the Abay River and with extension to the Rift Valley and east of the Rift.

CV: Acronym for Climate Variability.

DAF: Dry evergreen Afromontane forest and grassland complex. Vegetation type in Friis et al. (2010) and in this work (Fig. 4-1).

DANIDA: Stands for DANish International Development Agency, a section of the Ministry of Foreign Affairs of Denmark, which in the 1990s through its Council for Development Research supported studies of the effect of grass fires in the Sudanian vegetation zone from Senegal to Ethiopia, the FITES project.

DEM: Acronym for Digital Elevation Model.

DES: Desert. Vegetation type used in this work (Fig. 4-1).

DSS: Desert and semi-desert scrubland. Vegetation type in Friis et al. (2010).

EAZ: Category of distribution on the continental African scale and in South-Western Asia, including distributions from the Horn of Africa in the north to Angola, Zimbabwe and Mozambique in the south.

EB: Ericaceous belt. Vegetation type in Friis et al. (2010) and in this work (Fig. 4-1).

EE: The eastern lowlands of Eritrea, area in Eritrea below and to the east of the 1000 m contour on the east side of the Eritrean Highlands, used to indicate distribution in the *Flora of Ethiopia and Eritrea* and in this work.

EEA: Category of distribution on the continental African scale and in South-Western Asia, including distributions from the Horn of Africa in the north to East Tropical Africa (normally Tanzania) in the south.

ENRECA: Acronym standing for ENhanced REsearch CApacity; a former Danish bilateral programme for enhancement of research capacity in developing countries, sponsored by the Ministry of Foreign Affairs of Denmark.

ESA: Category of distribution on the continental African scale and in South-Western Asia, including distributions from the Horn of Africa in the north to South Africa and Namibia in the south.

ETH: The Ethiopian National Herbarium, part of the Addis Ababa University and located at its Arat Kilo Campus.

EW: The highlands and western lowlands of Eritrea, area in Eritrea above and to the west of the 1000 m contour on the east side of the Eritrean Highlands, used to indicate distribution in the *Flora of Ethiopia and Eritrea* and in this work, see Fig. 2-28.

EW-NAR: Category of distribution on the continental African scale, including distributions restricted to the Horn of Africa, Sudan and/or Yemen, west to Sudan and South Sudan.

EWW: Category of distribution in Ethiopia based on direct observation, including distributions recorded from areas in the western woodlands and to the east of the western woodland (mainly in the *Transitional semi-evergreen bushland* (TSEB).

f.UPGMA: Clustering method using Flexible Unweighted Pair Group ArithMetic Averages.

f.WPGMA: Clustering method using Flexible Weighted Pair Group ArithMetic Averages.

FEE: *Flora of Ethiopia and Eritrea*. A flora of the vascular plants of Ethiopia and Eritrea in ten volumes covering the vascular plants of Ethiopia and Eritrea, edited mainly by the staff of the Institute of Systematic Botany, Uppsala University, and the National Herbarium of Ethiopia, Addis Ababa University, supported by the Royal Botanic Gardens, Kew, and international team of botanists, published by the

Institute of Systematic Botany, Uppsala University, and the National Herbarium of Ethiopia, Addis Ababa University, from 1989 to 2012.

FF: Fire Frequency.

FITES: Stands for 'Fire in Tropical Ecosystems', project by the Centre for the Study of Fire in Tropical Ecosystems, a 'centre without walls' consisting of institutes at the University of Copenhagen and a range of African universities from Senegal to Ethiopia, supported by

FLW/MFS: Freshwater marches and swamps without or only very little open water. Vegetation type in Friis et al. (2010) and in this work (Fig. 4-1).

FLW/OW: Freshwater lakes with open water. Vegetation type in Friis et al. (2010) and in this work (Fig. 4-1).

FS: Flora of Somalia. A flora of the country of Somalia in four volumes, edited and largely written by Dr. M. Thulin, Uppsala, and published by the Royal Botanic Gardens, Kew, in the years 1993 to 2006.

FT: The acronym FT stands for the *Centro Studi Erbario Tropicale*, CSET, formerly the *Erbario Tropicale di Firenze* and before that the *Erbario e Museo Coloniale*. It was founded in Rome in 1904 and transferred to Florence in 1914. It is now a *Centro Studi* under the University of Florence, and has maintained the acronym FT.

FTEA: Flora of Tropical East Africa. A flora in numerous fascicles or volumes covering the vascular plants of Kenya, Uganda and Tanzania, edited and largely written by the staff at the Royal Botanic Gardens, Kew, and published by various publishers from 1948 to 2012.

FTNA: Forests and forest trees of Northeast Tropical Africa, their habitats and distribution patterns in Ethiopia, Djibouti, and Somalia. A monograph on the forests and forest trees of the Horn of Africa, written by Ib Friis and published in 1992 in *Kew Bulletin Additional Series* 15.

GD: The former Gondar region, area in Ethiopia, used to indicate distribution in the *Flora of Ethiopia and Eritrea* and in this work, see Fig. 2-28.

GG: The former Gamo Gofa region, area in Ethiopia, used to indicate distribution in the *Flora of Ethiopia and Eritrea* and in this work, see Fig. 2-28.

GJ: The former Gojam region, area in Ethiopia, used to indicate distribution in the *Flora of Ethiopia and Eritrea* and in this work, see Fig. 2-28.

GRASS GIS: GRASS is an acronym for Geographic Resources Analysis Support System and GIS an acronym for Geographical Information System.

HA: Hararge uplands and southern lowlands, area in Ethiopia, part of the former Hararge region, above and to the west of the 1000 m contour on the north side of the Ethiopian Highlands. Used to indicate distribution in the *Flora of Ethiopia and Eritrea* and in this work, see Fig. 2-28.

HWE: Category of distribution on the continental African scale and in South-Western Asia: Distributed from approximately halfway to Atlantic Ocean in the west (typically with western boundary at Niger, Nigeria or Cameroon) to Ethiopia in the east.

IAF: Intermediate Evergreen Afromontane Forest. Vegetation type used in this work (Fig. 4-1).

ICRAF: Acronym for the World Agroforestry Centre in Nairobi, Kenya; partner in the VECEA-project.

IL: The former Ilubabor region, area in Ethiopia used to indicate distribution in the *Flora of Ethiopia and Eritrea* and in this work, see Fig. 2-28.

ITCZ: Acronym for the Intertropical Convergence Zone, so named because it is the area where the northeast and the southeast trade winds converge. The position of ITCZ agrees with the zone where the sun is in zenith at that particular time of the year. The ITCZ encircles the Earth, but its specific position varies seasonally and between land and sea, being farthest to the north over land on the northern hemisphere by midsummer.

K: The very large herbarium of the Royal Botanic Gardens, Kew, outside London, UK.

KF: The former Kefa region, area in Ethiopia used to indicate distribution in the *Flora of Ethiopia and Eritrea* and in this work, see Fig. 2-28.

L: Formerly the acronym for the herbarium of the Leiden University, now incorporated in the National Herbarium of the Netherlands (Nationaal Herbarium Nederland), based in Naturalis in Leiden and also with the acronym L.

MAF: Moist Evergreen Afromontane Forest. Vegetation type in Friis et al. (2010) and in this work (Fig. 4-1).

MI: Distribution type based on generalized distribution pattern of the Horn of Africa: Marginally intruding species in the western woodlands, mainly with a main distribution in the highlands; usually with one or few records in our relevés.

MODIS: Acronym standing for 'Moderate Resolution Imaging Spectroradiometer', which is a key instrument aboard NASA's two satellites 'Terra' (originally known as EOS AM-1) and 'Aqua' (originally known as EOS PM-1) that record environmental data with the use of remote sensing to improve our understanding of processes occurring on land, in the oceans, and in the lower atmosphere.

NES: Category of distribution on the continental African scale and in South-Western Asia, including distributions restricted to the Horn of Africa or near in north-south direction.

NEW: Category of distribution on the continental African scale and in South-Western Asia, including distributions restricted to the Horn of Africa or near in the east-west direction.

NMDS: Acronym for Non-metric Multidimensional Scaling.

NS-NAR: Category of distribution on the continental African scale and in South-Western Asia: Restricted to the Horn of Africa, Sudan and/or Yemen, including distributions reaching southwards to northern D.R. Congo, Uganda and Kenya.

P: Acronym for the very large herbarium of the French Muséum National d'Histore Naturelle in Paris.

PCA: Acronym for Principal Component Analysis.

PCW: Acronym for Precipitation in cold and wet weather.

PDW: Acronym for Precipitation in dry and warm weather.

PSS: Abbreviation used in this work for *Plants of the Sudan and South Sudan*. An annotated checklist written by Iain Darbyshire, Maha Kordofani, Imadeldin Farag, Ruba Candia and Helen Pickering and published in 2015 by the Royal Botanic Gardens, Kew.

R package: A free software environment for statistical computing and graphics.

SD: The former Sidamo region. Used to indicate distribution in the *Flora of Ethiopia and Eritrea* and in this work, see Fig. 2-28.

SDS: Semi-Desert Scrubland. Vegetation type in this work (Fig. 4-1).

SIDA /SAREC: Acronym for a Swedish bilateral programme for enhancement of research capacity in developing countries, sponsored by the Swish Ministry of Foreign Affairs. The programme sponsored large parts of the Ethiopian Flora Project, and hence the *Flora of Ethiopia and Eritrea*.

SLV/OW: Salt lakes with open water. Vegetation type in Friis et al. (2010) and in this work (Fig. 4-1).

SLV/SSS: Salt pans, saline wetlands without or only very little open water. Vegetation type in Friis et al. (2010) and in this work (Fig. 4-1).

sqkm: Square kilometre.

SRTM DEM: Acronym standing for Shuttle Radar Topography Mission Digital Elevation Model. It was produced by NASA (National Aeronautics and Space Administration, part of the United States government.) and NGA (National Geospatial-Intelligence Agency, an agency under the United States Department of Defence) in cooperation with German and Italian space agencies, utilising instruments on the Endeavour space shuttle during its special mission in February 2000. The aim of this mission was to obtain precise elevation data on a near-global scale.

SU: Shewa uplands, area in Ethiopia, part of the former Shewa region, above and to the west of the 1000 m contour on the east side of the Ethiopian Highlands, area in Ethiopia used to indicate distribution in the *Flora of Ethiopia and Eritrea* and in this work, see Fig. 2-28.

TRF: Transitional Rain Forest. Vegetation type in Friis et al. (2010) and in this work (Fig. 4-1).

TRG: Distribution type based on generalized distribution pattern: Transgressing species, occurs similarly in both the *Combretum-Terminalia woodland and wooded grassland* (CTW) and in other vegetation types, particularly *Dry evergreen Afromontane forest* (DAF).

TSEB: Transitional Semi-Evergreen Bushland. Vegetation type in this work (Fig. 4-1).

TU: Tigray uplands, part of the former Tigray region, above and to the west of the 1000 m contour on the east side of the Ethiopian Highlands. Used to indicate distribution in the *Flora of Ethiopia and Eritrea* and in this work, see Fig. 2-28.

TWI: Topographical or Terrain Wetness Index.

UPGMA: Acronym for clustering method using Unweighted Pair Group ArithMetic Averages.

VECEA: Acronym for project supported by the Rockefeller Foundation to scientists in Denmark and at the World Agroforestry Centre (ICRAF) in Nairobi, Kenya, the first phase of which (2008-2010) partly supported the atlas of the potential vegetation of Ethiopia (Friis et al. 2010); The acronym stands for 'Vegetation and Climate Change in Eastern Africa.'

WAG: Acronym for the formerly independent herbarium of the Biosystematics Group of Wageningen University (WAG), focussing on African plants, now incorporated in the National Herbarium of the Netherlands (Nationaal Herbarium Nederland) and based in Naturalis in Leiden, but still referred to with the acronym WAG.

WE: Category of distribution on the continental African scale and in South-Western Asia, including distributions extending from the Horn of Africa westwards to the Atlantic Ocean, Guinea, Senegal and/or Mauritania).

WGG: Wooded grassland [moist] of the western Gambela Region. Vegetation type in Friis et al. (2010) and in this work (Fig. 4-1).

WPGMA: Acronym for clustering method using Weighted Pair Group ArithMetic Averages.

WU: Welo uplands, area, part of the former Welo region, above and to the west of the 1000 m contour on the east side of the Ethiopian Highlands, area in Ethiopia used to indicate distribution in the *Flora of Ethiopia and Eritrea* and in this work, see Fig. 2-28.

WWc: Distribution in Ethiopia based on direct observation: Recorded from the central part of the western woodlands (Fig. 7-2).

WWn: Distribution in Ethiopia based on direct observation: Recorded from the northern part of the western woodlands (Fig. 7-2).

WWs: Distribution in Ethiopia based on direct observation: Recorded from the southern part of the western woodland (Fig. 7-2).

List of references

The references to the floristic literature in the formal nomenclatural headings for each species in Section 6 are made with traditional standard abbreviations, which are accounted for in the beginning of Section 6, where bibliographic information for the fascicles of the *Flora of Tropical East Africa* is also provided (not repeated in this list of references). Ethiopian personal names are referred to in agreement with the Ethiopian tradition for personal names. According to this, the first name is the important one used for alphabetizing. The second name has also been written in full, in order to avoid confusion; this follows the authoritative reference work Brummitt & Powel (1992) and the author database in the International Plant Name Index (IPNI), https://www.ipni.org/.

Abbink, J. (2005). Gumuz ethnography. In: S. Uhlig (ed.), *Encyclopaedia Aethiopica*. Vol. 2. Harrassowitz Verlag, Wiesbaden. Pp. 916-917.

Abbink, J. (2010). Population history. Migrations in the south-west. In: S. Uhlig & A. Bausi (eds.), *Encyclopaedia Aethiopica.*, Vol. 4. Harrassowitz Verlag, Wiesbaden. Pp. 175-177.

Abeje Eshete Wassie (2011). *The Frankincense tree of Ethiopia – ecology, productivity and population dynamics.* Ph.D.-Thesis, Wageningen. http://edepot.wur.nl/171534)

Abiyot Berhanu, Sebsebe Dem[is]sew, Zerihun Woldu, Friis, I. & Breugel, P. van (2018). Intermediate evergreen Afromontane forest (IAF) in north-western Ethiopia: observations, description and modelling its potential distribution. *Phytocoenologia* 48(4): 351-367. https://doi.org/10.1127/phyto/2018/0207

Abrham Abiyu, Bongers, F., Abeje Eshete, Kindeya Gebrehiwot, Mengistie Kindu, Mulugeta Lemenih, Yitebitu Moges, Woldeselassie Ogbazghi & Sterck, F.J. (2010). Incense Woodlands in Ethiopia and Eritrea: Regeneration Problems and Restoration Possibilities. In: F. Bongers & T. Tenningkeit (eds.), *Degraded Forest in Eastern Africa: Management and Restoration.* Chapter 7. Routledge, London. https://www.taylorfrancis.com/books/e/9781849776400/chapters/10.4324/9781849776400-14

Addisalem Ayele Bekele, Bongers, F., Tilaye Kassahun & Smulders, M.J.M. (2016). Genetic diversity and differentiation of the frankincense tree (*Boswellia papyrifera* (Del.) Hochst.) across Ethiopia and implications for its conservation. *Forest Ecology and Management* 360: 253-260. https://doi.org/10.1016/j.foreco.2015.10.038

Aerts, R., Overtveld, K. van, November, E., Alemayehu Wassie, Abrham Abiyu, Sebsebe Demissew, Desalegn D Daye, Kidane Giday, Mitiku Haile, Sarah Tewolde Berhan, Demel Teketay, Zewge Teklehaimanot, Binggeli, P., Deckers, J., Friis, I., Gratzer, G., Hermy, M., Heyn, M., Honnay, O., Paris, M., Sterck, F.J., Muys, B., Bongers, F. & Healey, R. (2016). Conservation of the Ethiopian church forests: threats, opportunities and implications for their management. *Science of the Total Environment* 551: 404-414. http://dx.doi.org/10.1016/j.scitotenv.2016.02.034

Alelign, E., Demel Teketay, Yonas Yemshaw & Edwards, S. (2007). Diversity and status of regeneration of woody plants on the peninsula of Zegie, northwestern Ethiopia. *Tropical Ecology* 48: 37-50. http://www.tropecol.com/pdf/open/PDF_48_1/Alelign_et_al.pdf

Anderson, M.J. (2006). Distance-Based Tests for Homogeneity of Multivariate Dispersions. *Biometrics* 62(1): 245-253. On-line 2005: https://doi.org/10.1111/j.1541-0420.2005.00440.x.

APG III (2009). An update of the Angiosperm Phylogeny Group classification for the orders and families of flowering plants: APG III. *Botanical Journal of the Linnean Society* 161: 105-121. doi:10.1111/j.1095-8339.2009.00996.x

Areschough, F.W.C. (1867). *Bidrag til den skandinaviska vegetationens historia.* Lund, Berlingska Bocktryckeriet.

Arino, O., Ramos Perez, J.J., Kalogirou, V., Bontemps, S., Defourny, P. & Van Bogaert, E. (2012). *Global Land Cover Map for 2009 (GlobCover 2009).* European Space Agency (ESA) & Université catholique de Louvain (UCL). PANGAEA. https://doi.org/10.1594/PANGAEA.787668

Assédé, E.S.P., Azihou, A.F., Geldenhuys, C.J., Chirwa, P.W. & Biaou, S.S.H. (2020). Sudanian versus Zambezian woodlands of Africa: Composition, ecology, biogeography and use. *Acta Oecologica* 107: Article 103599, pp. 1-12. https://doi.org/10.1016/j.actao.2020.103599

Aubréville, A., Duvigneaud, P., Hoyle, A.C., Keay, R.W.J., Mendonca, F.A. & Pichi Sermolli, R.E.G. (1958). *Vegetation map of Africa south of the Tropic of Cancer*. 1:10,000,000. AETFAT, UNESCO, Paris. https://esdac.jrc.ec.europa.eu/images/Eudasm/Africa/images/maps/download/afr_veg.jpg

Bachman, S.P., Wilkin, P., Reader, T., Field, R., Weber, O., Nordal, I. & Sebsebe Demissew (2020). Extinction risk and conservation gaps for *Aloe* (Asphodelaceae) in the Horn of Africa. *Biodiversity and Conservation* 29: 77-98. https://doi.org/10.1007/s10531-019-01870-0

Bahru Zewde (1976). *Relations between Ethiopia and the Sudan on the Western Ethiopian frontier, 1898-1935*. Unpublished Ph.D. Thesis, University of London, London. https://eprints.soas.ac.uk/34127/1/11015954.pdf

Bahru Zewde (2003). [Ethiopian] Boundary with the Sudan. In: S. Uhlig (ed.), *Encyclopaedia Aethiopica*. Vol. 1. Harrassowitz Verlag, Wiesbaden. Pp. 612-614.

Barbosa P.M., Stroppiana, D., Grégoire, J.M. & Pereira, J.M.C. (1999). An assessment of vegetation fire in Africa (1981-1991): Burned areas, burned biomass, and atmospheric emissions. *Global Biogeochemical Cycles* 13(4): 933-950. https://agupubs.onlinelibrary.wiley.com/doi/abs/10.1029/1999GB900042

Baselga, A. (2010). Partitioning the Turnover and Nestedness Components of Beta Diversity. *Global Ecology and Biogeography* 19(1): 134-143. On-line 2009: https://doi.org/10.1111/j.1466-8238.2009.00490.x.

Beaumont, A.J., Beckett, R.P., Edwards, T.J. & Stirton, C.H. (*1999)*. Revision of the genus *Calpurnia* (Sophoreae: Leguminosae). *Bothalia* 29(1): https://doi.org/10.4102/abc.v29i1.568

Belbin, L., Faith, D.P. & Milligan, G.W. (1992). A Comparison of Two Approaches to Beta-Flexible Clustering. *Multivariate Behavioral Research* 27(3): 417-433. On-line 2010: https://doi.org/10.1207/s15327906mbr2703_6

Bender, M.L. (2005). Gumuz language. In: S. Uhlig (ed.), *Encyclopaedia Aethiopica*. Vol. 2. Harrassowitz Verlag, Wiesbaden. Pp. 914-916.

Blasco, F. (1988). The International Vegetation Map (Toulouse, France). In Küchler, A.W. & Zonneveld, I.S. (eds.), *Vegetation mapping*. Kluwer Academic Publishers, Dordrecht, The Netherlands. Pp. 443-460.

Braun-Blanquet, J. (1921). Prinzipien einer Systematik der Pflanzengesellschaften auf floristischer Grundlage. *Jahrbuch der St. Gallischen Naturwissenschaftlichen Gesellschaft* 57: 305-351.

Braun-Blanquet, J. (1964). *Pflanzensoziologie. Grundzüge der Vegetationskunde*. 3rd ed. Springer, Wien-New York.

Brenan, P.M. (1978). Some Aspects of the Phytogeography of Tropical Africa. *Annales of the Missouri Botanical Garden* 65(2): 437-478. https://doi.org/10.2307/2398859 & https://www.jstor.org/stable/2398859

Breugel, P. van, Friis, I., Sebsebe Demissew, Lillesø, J.-P.B. & Kindt, R. (2016a). Current and Future Fire Regimes and their Influence on Natural Vegetation in Ethiopia. *Ecosystems* 19: 369-386. On-line https://doi.org/10.1007/s10021-015-9938-x; the online version contains electronic supplementary material that is referred to and discussed in the present volume.

Breugel, P. van, Friis, I. & Sebsebe Demissew (2016b). The transitional semi-evergreen bushland in Ethiopia: characterization and mapping of its distribution using predictive modelling. *Applied Vegetation Science* 19: 355-367. https://doi.org/10.1111/avsc.12220

Breiman, L. (2001). Random Forests. *Machine Learning* 45(1): 5-32. https://doi.org/10.1023/A:1010933404324

Brownlie, I. & Burns, I.R. (1979). *African Boundaries. A Legal and Diplomatic Encyclopaedia*. C. Hurst & Co., London. [The Ethiopian-Sudan boundary: Pp. 852-887].

Bruce, J. (1790). *Travels to discover the source of the Nile in the years 1768, 1769, 1770, 1771, 1772, and 1773*. Vol. 5. Select specimens of Natural History collected in travels to discover the source of the Nile. J. Ruthen, Edinburgh; G.G.J. & J. Robinson, London.

Brummitt, R.K. & Powell, C.E. (1992). *Authors of Plant Names*. Royal Botanical Gardens, Kew.

Brundu, G. & Camarda, I. (2013). The Flora of Chad: a checklist and brief analysis. *PhytoKeys* 23: 1-17. https://phytokeys.pensoft.net/articles.php?id=1546

Bustorf, D. (2010). Slave raiding in the 19th century. In: S. Uhlig & A. Bausi (eds.), *Encyclopaedia Aethiopica*. Vol. 4. Harrassowitz Verlag, Wiesbaden. Pp. 676-678.

Cain, S.A. (1944). *Foundations of plant geography*. Harper and Brothers Publishers, New York.

Carr, C.J. (1998). Patterns of vegetation along the Omo River in southwest Ethiopia. *Plant Ecology* 135: 135-163. https://link.springer.com/article/10.1023/A:1009704427916

Castro-Insua, A., Gómez-Rodríguez, C. & Baselga, A. (2018). Dissimilarity Measures Affected by Richness Differences Yield Biased Delimitations of Biogeographic Realms. *Nature Communications* 9(1): 1-3. Online Nature Communications 9: Article 5084. https://doi.org/10.1038/s41467-018-06291-1.

Cei, G. & Pichi Sermolli, R.E.G. (1940). Felci raccolte da G. Cei nel territorio dei Galla e Sidama e cenni sulle loro stazioni. *Nuovo Giornale Botanico Italiana* 47(1): 1-23. https://bibdigital.rjb.csic.es/viewer/15068/?offset=#page=9&viewer=picture&o=bookmark&n=0&q=

CGIAR-CSI. (2008). *CGIAR-CSI SRTM 90m DEM Digital Elevation Database*. Version 4. Available on https://bigdata.cgiar.org/srtm-90m-digital-elevation-database/

Chaffey, D.R. (1978a). *Southwest Ethiopia forest inventory project. A glossary of vernacular names of plants in southwest Ethiopia with special reference to forest trees.* Ministry of Overseas Development. Land Resources Development Centre, London. Project Report 26: 1-75 & map.

Chaffey, D.R. (1978b). *Southwest Ethiopia forest inventory project. An inventory of Magada forest.* Ministry of Overseas Development. Land Resources Development Centre, London. Project Report 28: 1-52 & map.

Chaffey, D.R. (1978c). *Southwest Ethiopia forest inventory project. An inventory of forest at Munessa and Shashemane.* Ministry of Overseas Development. Land Resources Development Centre, London. Project Report 29: 1-97 & maps.

Chaffey, D.R. (1978d). *Southwest Ethiopia forest inventory project. An inventory of Tiro Forest.* Ministry of Overseas Development. Land Resources Development Centre, London. Project Report 30: 1-60 & maps.

Chaffey, D.R. (1979). *Southwest Ethiopia forest inventory project. A reconnaissance inventory of forest in southwest Ethiopia*. Ministry of Overseas Development. Land Resources Development Centre, London. Project Report 31: 1-316 & maps.

Chapman, J.D. & White, F. (1970). *The Evergreen Forests of Malawi*. Commonwealth Forestry Institute, University of Oxford, Oxford.

Charrad, M., Ghazzali, N., Boiteau, V. & Niknafs, A. (2014). NbClust: An R Package for Determining the Relevant Number of Clusters in a Data Set. *Journal of Statistical Software* 61(6): 1-36. On-line and software http://www.jstatsoft.org/v61/i06/.

Chiarucci, A., Araújo, M.B., Decocq, G., Beierkuhnlein, C. &Fernández-Palacios, J.M. 2010. The concept of potential natural vegetation: an epitaph? *Journal of Vegetation Science* 21: 1172-1178.

Chiovenda, E. (1912). Osservazioni Botaniche, Agrarie ed Industriali fatte nell'Abissinia Settentrionale nell'anno 1909. *Rapporti e Monografie Coloniali* n. 24: 1-132. Ministero delle Colonie, Roma.

Chiovenda, E. (1936). La Flora. *Le Vie d'Italia* 42(7): 457-465.

Christ, H. (1867). Ueber die Verbreitung der Pflanzen der alpinen Regionen der Europäischen Alpenkette. *Neue Denkschriften der allgemeinen Schweizerischen Gesellschaft für die Gesammten Naturwissenschaften* 22(7): 1-84. https://www.biodiversitylibrary.org/page/13890872

Calpham, C. (2007). Menelek II. In: S. Uhlig (ed.), *Encyclopaedia Aethiopica*. Vol. 3. Harrassowitz Verlag, Wiesbaden. Pp. 922-927.

Clayton, W.D. & Hepper, F.N. (1974). Computer-aided chorology of West African grasses. *Kew Bulletin* 29: 213-234. https://doi.org/10.2307/4108386 & https://www.jstor.org/stable/4108386

Cooper, S.M. & Owen-Smith, N. (1985). Condensed tannins deter feeding by browsing ruminants in a South African savanna. *Oecologia* 67: 142-146. https://link.springer.com/article/10.1007/BF00378466

Darbyshire, I., Kordofani, M., Farag, I., Candiga, R. & Pickering, H. (2015). *The Plants of Sudan and South Sudan. An annotated Checklist.* Royal Botanic Gardens, Kew.

Dauby, G., Zaiss, R., Blach-Overgaard, A., Catarino, L., Damen, T., Deblauwe, V., Dessin, S., Dransfield, J., Droissart, V., Duarte, M.C., Engledow, H., Fadeur, G., Figueira, R., Gereau, R.E., Hardy, O.J., Harris, D.J., de Heij, J., Janssens, S.B., Klomberg, Y., Ley, A.C., Mackinder, B.A., Meerts, P., van de Poel, J.L., Sonké, B., Sosef, M.S.M., Stévart, T., Stoffelen, P., Svenning, J.-C., Sepulchre, P., van der Burgt, X.M., Wieringa, J.J. & Couvreur, T.L.P. (2016). RAINBIO: a mega-database of tropical African vascular plants distributions. *PhytoKeys* 74: 1-18. https://doi.org/10.3897/phytokeys.74.9723

De Cáceres, M. & Jansen, F. (2020). *R Package Indicspecies*. Article and software https://cran.r-project.org/web/packages/indicspecies/indicspecies.pdf

De Cáceres, M. & Legendre, P. (2009). Associations Between Species and Groups of Sites: Indices and

Statistical Inference. *Ecology* 90: 3566-3574. http://sites.google.com/site/miqueldecaceres/.

De Cáceres, M., Legendre, P. & Moretti, M. (2010). Improving Indicator Species Analysis by Combining Groups of Sites. *Oikos* 119(10): 1674-1684. https://doi.org/10.1111/j.1600-0706.2010.18334.x

De Candolle, A.P. (1820). Geographie botanique. In: M.F. Cuvier (ed.), *Dictionaire des Sciences Naturelles dans lequel on traite methodiquement des differents etres de la nature*. Ed. 2. Vol. 18. Strassbourg, Paris. Pp. 359-422.

Dei Gaslini, M. (1940). *Le ricchezze del Galla-Sidamo*. Tipografia del Popolo d'Italia, Milano.

Denys, E. (1980). A Tentative Phytogeographical Division of Tropical Africa Based on a Mathematical Analysis of Distribution Maps. *Bulletin du Jardin botanique National de Belgique / Bulletin van de Nationale Plantentuin van België* 50(3/4): 465-504. https://doi.org/10.2307/3667842 & https://www.jstor.org/stable/3667842

Dewitte, O., Jones, A., Spaargaren, O., Breuning-Madsen, H., Brossard, M., Dampha, A., Deckers, J., Gallali, T., Hallett, S., Jones, R., Kilasara, M., Le Roux, P., Michaeli, E., Montanarella, L., Thiombiano, L., Van Ranst, E., Yemefack, M. & Zougmore, R. (2013). Harmonisation of the soil map of Africa at the continental scale. *Geoderma* 211-212: 138-153. https://doi.org/10.1016/j.geoderma.2013.07.007

Dobremez, J.F. (1976). *Le Nepal. Ecologie et Biogeography*. Editions du Centre National de la Recherche Scientifique, Paris. 355 pp.

Dowsett-Lemaire, F. (2001). 2. Environment, 3. The chorology of the forest flora, 4. A synopsis of the vegetation of Malawi.& Appendix 1. Membership of the major chorological elements in the evergreen forest flora of Malawi. In: F. White, F. Dowsett-Lemaire & J.D. Chapman, *Evergreen Forest Flora of Malawi*. Royal Botanic Gardens, Kew. Pp. 9-70, 603-607.

Droissart, V., Hardy, O.J., Sonké, B., Dahdouh-Guebas, F., Stévart, T. (2012). Subsampling herbarium collections to assess geographic diversity gradients: a case study with endemic Orchidaceae and Rubiaceae in Cameroon. *Biotropica* 44: 44-52. https://onlinelibrary.wiley.com/doi/10.1111/j.1744-7429.2011.00777.x

Droissart, V., Dauby, G., Hardy, O.J., Deblauwe, V., Harris, D.J., Janssens, S., Mackinder, B.A., Blach-Overgaard, A., Sonké, B, Sosef, M.S.M., Stévart, T., Svenning, J-C., Wieringa, J.J. & Couvreur, T.L.P. (2018). Beyond trees: Biogeographical regionalization of tropical Africa. *Journal of Biogeography* 45: 1153-1167. https://doi.org/10.1111/jbi.13190

Dufrêne, M. & Legendre, P. (1997). Species Assemblages and Indicator Species: the Need for a Flexible Asymmetrical Approach. *Ecological Monographs* 67(3): 345-366. https://doi.org/10.1890/0012-9615(1997)067[0345:SAAIST]2.0.CO;2.

Dunn, J.C. (1974). Well-Separated Clusters and Optimal Fuzzy Partitions. *Journal of Cybernetics* 4(1): 95-104. On-line 2008: https://doi.org/10.1080/01969727408546059.

Du Rietz, G.E. (1931). Life-forms of terrestrial flowering plants. *Acta Phytogeographica Suecica* 3(1): 1-95.

Eig, A. (1931a). Les elements et les groupes phytogeographiques auxillaires dans la flore palestinienne. *Feddes Repertorium specierum novarum regni vegetabilis*. Beiheft 63: 1-201.

Eig, A. (1931b). Quelques faits de la phytogeographie palestinienne precedes par des remarques sur les notions phytogeographiques. *Bulletin de la Société botanique de France* 78: 297-305. https://bibdigital.rjb.csic.es/viewer/14588/?offset=#page=351&viewer=picture&o=bookmark&n=0&q=

Elagib, N.A. & Basheer, M. (2021). Would Africa's largest hydropower dam have profound environmental impacts? *Environmental Science and Pollution Research* 28: 8936-8944. https://doi.org/10.1007/s11356-020-11746-4

Engels, J.M.M. & Goettsch, E. (1991). Konso agriculture and its plant genetic resources. In: J.M.M. Engels, J.G. Hawkes & Melaku Worede (eds), *Plant genetic resources of Ethiopia*. Cambridge University Press, Cambridge. Pp. 169-186.

Engler, A. (1882). *Versuch einer Entwichlungsgeschichte der Pflanzenwelt, inbesondere der Florengebiete seit der Tertiärperiode*. 2. Theil. Die extratropischen Gebiete der südlichen Hemisphare und die tropischen Gebiete. W. Engelmann, Leipzig.

ESA [European Space Agency] (2017). *Land cover CCI product user guide*. Version 2.0. Available at http://maps.elie.ucl.ac.be/CCI/viewer/download/ESACCI-LC-PUG-v2.5.pdf

Falge, C. (2007). Nuer ethnography. In: S. Uhlig (ed.), *Encyclopaedia Aethiopica*. Vol. 3. Harrassowitz Verlag, Wiesbaden. Pp. 1199-1200.

Farris, J.S. (1969). On the Cophenetic Correlation Coefficient. *Systematic Biology* 18(3): 279-285. https://doi.org/10.2307/2412324

Fattorini, S. (2015). On the concept of chorotype. *Journal of Biogeography* 42(11): 2246-2251. https://doi.org/10.1111/jbi.12589

Fattorini, S. (2016). A history of chorological categories. *History and Philosophy of the Life Sciences* 38(3): Art. 12, pp. 1-21. https://www.jstor.org/stable/44752350

Fayolle, A., Swaine, M.D., Aleman, J., Azihou, A.F., Bauman, D., te Beest, M., Chidumayo, E.N., Cromsigt, J.P.G.M., Dessard, H., Finckh, M., Gonçalves, F.M.P., Gillet, J.-F., Gorel, A., Hick, A., Holdo, R., Kirunda, B., Mahy, G., McNicol, I., Ryan, C.M., Revermann, R., Plumptre, A., Pritchard, R., Nieto-Quintano, P., Schmitt, C.B., Seghieri, J., Swemmer, A., Talila, H. & Woollen, E. (2018). A sharp floristic discontinuity revealed by the biogeographic regionalization of African savannas. *Journal of Biogeography* 46(2): 454-465. https://doi.org/10.1111/jbi.13475. Digital edition in 2018, printed edition 2019.

Fayolle, A., Swaine, M.D., Bastin, J.-F., Bourland, N., Comiskey, J.A., Dauby, G., Doucet, J.-L., Gillet, J.F., Gourlet-Fleury, S., Hardy, O.J., Kirunda, B., Kouamé, F.N. & Plumptre, A.J. (2014). Patterns of Tree Species Composition Across Tropical African Forests. *Journal of Biogeography* 41(12): 2320-2331. https://doi.org/10.1111/jbi.12382

Fiori, A. (1909-1912). Boschi e Piante Legnose dell'Eritrea. *L'Agricultura Coloniale* 3(6): 369-391 (1909); 4(1): 2-23, (2):73-98, (3):171-184, (6):285-302, (8):365-386 (1910); 5(2): 41-61, (3):81-100, (4-5):182-206, (6):266-296, (Append.): 1-173. (1912).

Frey, T. & Groenewoud, H. van (1972). A Cluster Analysis of the D^2 Matrix of White Spruce Stands in Saskatchewan Based on the Maximum-Minimum Principle. *Journal of Ecology* 60(3): 873-886. https://doi.org/10.2307/2258571 & https://www.jstor.org/stable/2258571

Friis, I. (1992). Forests and forest trees of Northeast Tropical Africa, their habitats and distribution patterns in Ethiopia, Djibouti and Somalia. *Kew Bulletin*. Additional Series 15: i-iv, 1-396.

Friis, I. (1994). Some general features of the Afromontane and Afrotemperate floras of the Sudan, Ethiopia and Somalia. In: J.H. Seyani & A.C. Chikuni (eds.), *Proceedings of the XIIIth plenary meeting of AETFAT, Zomba, Malawi, 2-11 April 1991*. National Herbarium and Botanic Gardens of Malawi, Zomba, Malawi. Pp. 953-968.

Friis, I. (1998). Frank White and the Development of African Chorology. In: C.R. Huxley, J.M. Lock & D.F. Cutler (eds), *Chorology, Taxonomy and Ecology of the Floras of Africa and Madagascar*. Royal Botanic Gardens, Kew, England, UK. Pp. 25-51.

Friis, I. (2006). Distribution Patterns of Solanum in the Horn of Africa. In: S.A. Ghazanfar & h. Beentje (eds.), *Taxonomy and Ecology of African Plants, their Conservation and Sustainable Use. Proceedings of the 17th AETFAT Congress, Addis Abeba*. Royal Botanic Gardens, Kew. Pp. 279-289.

Friis, I. (2009a). The scientific study of the Flora of Ethiopia and Eritrea up to the beginning of the Ethiopian Flora Project (1980). In: I. Hedberg, I. Friis & E. Persson (eds.), *Flora of Ethiopia and Eritrea*. Vol. 8. Addis Ababa University, Addis Ababa, Ethiopia; University of Uppsala, Uppsala, Sweden. Pp. 5-25.

Friis, I. (2009b). Collectors of botanical specimens from the Flora area mentioned in the Flora of Ethiopia and Eritrea. In: I. Hedberg, I. Friis & E. Persson (eds.), *Flora of Ethiopia and Eritrea*. Vol. 8. Addis Ababa University, Addis Ababa, Ethiopia; University of Uppsala, Uppsala, Sweden. Pp. 97-133.

Friis, I. (2013). Travelling among fellow Christians (1768-1833): James Bruce, Henry Salt and Eduard Rüppell in Abyssinia. *Scientia Danica. Series H, Humanistica, 4to.* Vol. 2: 161-194.

Friis, I. (2019). Cenni monografici sul paese dei Gherire (Mogadiscio, 1938)–a pioneer work on the Gerire Hills in western Ogaden, south-eastern Ethiopia-introduced, translated and provided with annotations and a revision of the botanical collections. *Webbia* 74 (supplement): 1-108. https://doi.org/10.1080/00837792.2018.1553281

Friis, I., Gilbert, M.G., Paton, A.J., Weber, O., Breugel, P. van & Sebsebe Demissew (2018). The Gerire Hills, a SE Ethiopian outpost of the Transitional semi-evergreen bushland: vegetation, endemism and three new species, *Croton elkerensis* (Euphorbiaceae), *Gnidia elkerensis* (Thymelaeaceae), and *Plectranthus spananthus* (Lamiaceae). *Webbia* 72(2): 203-223. https://doi.org/10.1080/00837792.2018.1505379

Friis, I., Gilbert, M.G., Breugel, P. van, Weber, O. & Sebsebe Demissew (2017). Kalanchoe hypseloleuce (Crassulaceae), a new species from eastern Ethiopia, with notes on its habitat. *Kew Bulletin* 72: Article 30. Pp. 1-11. https://doi.org/10.1007/S12225-017-9704-7

Friis, I., Gilbert, M.G., Weber, O. & Sebsebe Demissew (2016). Two distinctive new species of *Commicarpus* (Nyctaginaceae) from gypsum outcrops in eastern

Ethiopia. *Kew Bulletin* 71: Article 34. Pp. 1-19. https://doi.org/10.1007/s12225-016-9648-3

Friis, I., Gilbert, M.G., Weber, O., van Breugel, P. van & Sebsebe Demissew (2019). The Gerire Hills, SE Ethiopia: ecology and phytogeographical position of an additional local endemic, *Anacampseros specksii* (Anacampserotaceae). *Webbia* 74(2): 185-192. https://doi.org/10.1080/00837792.2019.1670020

Friis, I. & Ryding, O. [eds.] (2001). Biodiversity Research in the Horn of Africa Region. Proceedings of the Third International Symposium on the Flora of Ethiopia and Eritrea at the Carlsberg Academy, Copenhagen, August 25-27, 1999. *Biologiske Skrifter* 54.

Friis, I. & Sebsebe Demissew (2001). Vegetation maps of Ethiopia and Eritrea. A review of existing maps and the need for a new map for the Flora of Ethiopia and the need for a new map for the Flora of Ethiopia and Eritrea. In: I. Friis & O. Ryding (eds). Biodiversity Research in the Horn of Africa Region, Proceedings of the 3rd International Symposium on the Flora of Ethiopia and Eritrea. *Biologiske Skrifter* 54: 399-439.

Friis, I., Sebsebe Demissew & Breugel, P. van (2010). Atlas of the Potential Vegetation of Ethiopia. *Biologiske Skrifter* 58. http://www.royalacademy.dk/Publications/Low/3607_Friis,%20Ib,%20Demissew,%20Sebsebe%20and%20van%20Breugel,%20Paulo.pdf

Friis, I. & Sebsebe Demissew (2020). *Terminalia* (Combretaceae) in northern tropical Africa: Priority and typification of *T. schimperiana* and *T. glaucescens*; typification of other synonyms of *T. schimperiana* and of *T. avicennioides*. *Taxon* 69(2): 372-380. https://doi.org/10.1002/tax.12181

Friis, I. & Vollesen, K. (1998). Flora of the Sudan-Uganda border area east of the Nile, 1. Catalogue of vascular plants, 1st part. *Biologiske Skrifter* 51(1): 1-389. https://www.researchgate.net/publication/324341622_Flora_of_the_Sudan-Uganda_Border_Area_East_of_the_Nile_I_Catalogue_of_Vascular_Plants_1st_Part

Friis, I. & Vollesen, K. (2005). Flora of the Sudan-Uganda border area east of the Nile, 2. Catalogue of vascular plants, 2nd part, vegetation and phytogeography. *Biologiske Skrifter* 51(2): 390-855. https://www.researchgate.net/publication/260050805_Flora_of_the_Sudan-Uganda_border_area_east_of_the_Nile_II_Catalogue_of_Vascular_plants_2nd_part_vegetation_and_phytogeography

Galili, T. (2015). Dendextend: An r Package for Visualizing, Adjusting, and Comparing Trees of Hierarchical Clustering. *Bioinformatics* 31(22): 3718-3720. https://doi.org/10.1093/bioinformatics/btv428, software http://cran.r-project.org/package=dendextend

Ganamé, M., Bayen, P., Ouédraogo, I., Dimobe, K. & Thiombiano, A. (2019). Woody Species Composition, Diversity and Vegetation Structure of Two Protected Areas Along a Climatic Gradient in Burkina Faso (West Africa). *Folia Geobotanica* 54(3): 163-175. https://doi.org/10.1007/s12224-019-09340-9. On paper 2020.

Gascon, A. & Abbink, J.G. (2010). Parks, National. In: S. Uhlig & A. Bausi (eds.), *Encyclopaedia Aethiopica*. Vol. 4. Harrassowitz Verlag, Wiesbaden. Pp. 114-117.

Gascon, A. (2010). Population History. Introduction. In: S. Uhlig & A. Bausi (eds.), *Encyclopaedia Aethiopica*. Vol. 4. Harrassowitz Verlag, Wiesbaden. Pp. 170-172.

GBIF.org (2020). *GBIF Occurrence Download*. Https://doi.org/10.15468/dl.gw2u7q (filter applied 10 November 2020).

Giordano, G. (1940). Il problema forestale dell'Impero. *Quaderni Italiani,* Ser. 2, 9: 1-30.

GRASS Development Team (2020). *Geographic Resources Analysis Support System (GRASS) software* (7.8.4) [Computer software]. Open Source Geospatial Foundation. http://grass.osgeo.org

Greve, M., Lykke, A.M., Fagg, C.W., Bogaert, J., Friis, I., Marchant, R., Marshall, A.R., Ndayishimiye, J.I., Sandel, B.S., Sandom, C., Schmidt, M., Timberlake, J.R., Wieringa, J.J., Zizka, G. & Svenning, J.-C. (2012). Continental-scale variability in browser diversity is a major driver of diversity patterns in acacias across Africa. *Journal of Ecology* 100: 1093-1104. https://doi.org/10.1111/j.1365-2745.2012.01994.x

Groome, J.S. (1955). Muninga (*Pterocarpus angolensis* D.C.) in the western province of Tanganyika. I. Description, distribution and silvicultural characters. *East African Agricultural Journal* 21: 130-137. On-line 2015 https://doi.org/10.1080/03670074.1955.11665021

Grubb, P.J., Lloyd, J.R., Pennington, J.D. & Whitmore, T.C.A. (1963). A comparison of montane and lowland rain forest in Ecuador, I. The forest structure, physiognomy, and floristics. *Journal of Ecology* 51: 567-601. https://doi.org/10.2307/2257748 & https://www.jstor.org/stable/2257748

Grubb, P.J. & Whitmore, T.C.A. (1966). A comparison of montane and lowland rain forest in Ecuador, II. The climate and its effects on the distribution and physiognomy of the forests. *Journal of Ecology* 54: 303-333.

https://doi.org/10.2307/2257951 & https://www.jstor.org/stable/2257951

Guo Ruo, Brhane Weldegebrial, Genet Yohannes and Gebremedhin Yohannes (2018). The Impact of Fencing on Regeneration, Tree Growth and Carbon Stock in Desa Forest, Tigray, Ethiopia. *Biomedica Joutnal of Science & Technical Research* (ISSN: 2574-1241) 5(4): DOI: 10.26717/BJSTR.2018.12.002183.

Gwynn, C. (1937). The Frontiers of Abyssinia: A Retrospect. *Journal of the Royal African Society* 36(143): 150-161. https://www.jstor.org/stable/717627

Hahn, J. von & Knoch, K. (1932). *Handbuch der Klimatologie*. 1. Band. Allgemeine Klimalehre. Vierte, wesentlich umarbeitete und vermehrte Auflage. J. Engelhorns Nachfolger, Stuttgart.

Hahn, J. (1908). *Handbuch der Klimatologie*. 1. Band. Allgemeine Klimalehre. Dritte, wesentlich umarbeitete und vermehrte Auflage. J. Engelhorn, Stuttgart.

Haile Adamu Wale, Tamrat Bekele & Gemedo Dalle (2012). Plant community and ecological analysis of woodland vegetation in Metema Area, Amhara National Regional State, Northwestern Ethiopia. *Journal of Forestry Research* 23(4): 599-607. https://doi.org/10.1007/s11676-012-0300-2

Hamilton, A.C. (1975). A quantitative analysis of altitudinal zonation in Uganda forests. *Vegetatio* 30: 99-106. https://doi.org/10.1007/BF02389611

Hargreaves, G.H. & Allen, R.G. (2003). History and Evaluation of Hargreaves Evapotranspiration Equation. *Journal of Irrigation and Drainage Engineering* 129(1): 53-63. https://ascelibrary.org/doi/10.1061/%28ASCE%290733-9437%282003%29129%3A1%2853%29

Harmonized World Soil Database (2008). *Harmonized World Soil Database* (version 1.2). FAO, Rome. http://www.fao.org/soils-portal/data-hub/soil-maps-and-databases/harmonized-world-soil-database-v12/en/

Harrison, M.N. & Jackson, J.K. (1958). Ecological classification of the vegetation of the Sudan. *Forest Bulletin for the Sudan*. N.S., 2: 1-45, and coloured and folded vegetation map.

Hedberg, I. (2009). The Ethiopian Flora Project-an Overview. In: I. Hedberg, I. Friis & E. Persson (eds.), *Flora of Ethiopia and Eritrea*. Vol. 8. National Herbarium of Ethiopia, Addis Ababa University, Addis Ababa & Institute of Systematic Botany, University of Uppsala, Uppsala. Pp. 1-4.

Hedberg, O. (1951). Vegetation belts on the East African mountains. *Svensk Botanisk Tidskrift* 45: 140-202.

Heide, F. zur (2012). *Feasibility Study for a Lake Tana Biosphere Reserve, Ethiopia*. Federal Agency for Nature Conservation, Bonn. BfN Skripten 317. https://www.nabu.de/imperia/md/content/nabude/international/machbarkeitsstudie_lake_tana.pdf

Hengl, T., Mendes de Jesus, J., Heuvelink, G.B.M., Gonzalez, M.R., Kilibarda, M., Blagotić, A., Shangguan, W., Wright, M.N., Geng, X., Bauer-Marschallinger, B., Guevara, M.A., Vargas, R., MacMillan, R.A., Batjes, N.H., Leenaars, J.G.B., Ribeiro, E., Wheeler, I., Mantel, S. & Kempen, B. (2017). SoilGrids250m: Global gridded soil information based on machine learning. *PLOS ONE* 12(2): e0169748. https://doi.org/10.1371/journal.pone.0169748

Hijmans, R.J., Cameron, S.E., Parra, J.L., Jones, P.G. & Jarvis, A. (2005a) Very high-resolution interpolated climate surfaces for global land areas. *International Journal of Climatology* 25: 1965-1978. https://doi.org/10.1002/joc.1276

Hijmans, R.J., Guarino, L., Jarvis, A., O'Brien, R. Mathus, P. Bussink, C., Cruz, M., Barrantes, I. & Rojas, E. (2005b). *DIVA-GIS*. Version 5.2. Manual. Available on http://www.diva-gis.org/docs/DIVA-GIS_manual_7.pdf

Hubert, L.J. & Levin, J.R. (1976). A General Statistical Framework for Assessing Categorical Clustering in Free Recall. *Psychological Bulletin* 83(6): 1072-1080. https://doi.org/10.1037/0033-2909.83.6.1072.

Hulton, P., Hepper, F.N. & Friis, I. (1991). *Luigi Balugani's drawings of African plants. From the collections made by James Bruce of Kinnaird on his travels to discover the source of the Nile 1767-1773*. Yale Center for British Art, Yale & A.A. Balkema, Rotterdam.

Humboldt, A. von (1806). *Ideen zu einer Physiognomik der Gewachse*. Cotta, Tübingen

Humphries, C.J. (2001). Vicariance biogeography. In: S.A. Levin et al. (eds.), *Encyclopedia of Biodiversity*. Vol. 5. Pp. 767-779. Academic Press, San Diego. https://www.sciencedirect.com/science/article/pii/B0122268652002819?via%3Dihub

IUCN Red List (2020a). *IUCN Red List of Threatened species. Boswellia papyrifera*. Tantani. https://dx.doi.org/10.2305/IUCN.UK.2018-2.RLTS.T34394A128137387.en

IUCN Red List (2020b). *IUCN Red List of Threatened species. Combretum hartmannianum*. https://dx.doi.org/10.2305/IUCN.UK.1998.RLTS.T36464A9997265.en

Jackson, J.K. (1956). The vegetation of the Imatong Mountains, Sudan. *Journal of Ecology* 44: 341-374. https://doi.org/10.2307/2256827 & https://www.jstor.org/stable/2256827

Jacobs, M.J. & Schloeder, C.A. (2002). Fire frequency and species associations in perennial grasslands of south-west Ethiopia. *African Journal of Ecology* 40: 1-9. https://doi.org/10.1046/j.0141-6707.2001.00347.x

Jardine, N. (1972). Computational methods in the study of plant distributions. In: D.H. Valentine (ed.), *Taxonomy, Phytogeography and Evolution*. Academic Press, London. Pp. 381-393.

Jensen, M. & Friis, I. (2001). Fire regimes, floristics, diversity, life forms and biomass in wooded grassland, woodland and dry forest at Gambella, western Ethiopia. *Biologiske Skrifter* 54: 349-387.

Jensen, M., Michelsen, A. & Minassie Gashaw (2001). Responses in plant, soil inorganic and microbial nutrient pools to experimental fire, ash and biomass addition in a woodland savanna. *Oecologia* 128: 85-93. https://www.researchgate.net/publication/225534443_Responses_in_plant_soil_inorganic_and_microbial_nutrient_pools_to_experimental_fire_ash_and_biomass_addition_in_a_woodland_savanna

Jones, A., Breuning-Madsen, H., Brossard, M., Dampha, A., Deckers, J., Dewitte, O., Gallali, T., Hallett, S., Jones, R., Kilasara, M., Le Roux, P., Micheli, E., Montanarella, L., Spaargaren, O., Thiombiano, L., Van Ranst, E., Yemefack, M. & Zougmore, R. (eds.) (2013). *Soil Atlas of Africa*. European Commission, Publications Office of the European Union, Luxembourg. https://esdac.jrc.ec.europa.eu/content/soil-map-soil-atlas-africa

Jordaan, M. & Wyk, A.E. van (2006). Sectional classification of *Gymnosporia* (Celastraceae), with notes on the nomenclatural and taxonomic history of the genus. *Taxon* 55(2): 515-525.

Kassahun Embaye Yikuno, Demelash Alem Ayana & Abdella Gure (2015). Flowering and Causes of Seed Defects in Lowland Bamboo (*Oxytenanthera abyssinica*): A Case Study in Benishangul Gumuz Regional State, North-western Ethiopia. *International Journal of Life Sciences* 4(4): 251-259. https://www.researchgate.net/publication/284186180_Flowering_and_Causes_of_Seed_Defects_in_Lowland_Bamboo_Oxytenanthera_abyssinica_A_Case_Study_in_Benishangul_Gumuz_Regional_State_Northwestern_Ethiopia

Kassambara, A. & Mundt, F. (2020). *Factoextra: Extract and Visualize the Results of Multivariate Data Analyses*. Software https://CRAN.R-project.org/package=factoextra.

Kaufman, L. & Rousseeuw, P.J. (2009). *Finding Groups in Data: An Introduction to Cluster Analysis*. John Wiley & Sons, Hoboken, New Jersey. https://onlinelibrary.wiley.com/doi/book/10.1002/9780470316801

Kershaw, K.A. (1968). A Survey of the Vegetation in Zaria Province, N. Nigeria. *Vegetatio* 15(4): 244-268. https://www.jstor.org/stable/20035382

Kindt, R. & Coe, R. (2005). *Tree Diversity Analysis. A Manual and Software for Common Statistical Methods for Ecological and Biodiversity Studies*. Nairobi, Kenya: World Agroforestry Centre (ICRAF). http://apps.worldagroforestry.org/downloads/Publications/PDFS/MN08242.pdf

Koleff, P., Gaston, K.J. & Lennon, J.J. (2003). Measuring Beta Diversity for Presence-Absence Data. *Journal of Animal Ecology* 72(3): 367-382. https://doi.org/10.1046/j.1365-2656.2003.00710.x.

Kopecký, M. & Čížková, Š. (2010). Using Topographic Wetness Index in Vegetation Ecology: Does the Algorithm Matter? *Applied Vegetation Science* 13(4): 450-459. https://doi.org/10.1111/j.1654-109X.2010.01083.x

Kriticos, D.J., Jarošik, V. & Ota, N. (2014). Extending the Suite of Bioclim Variables: A Proposed Registry System and Case Study Using Principal Components Analysis. *Methods in Ecology and Evolution* 5(9): 956-960. https://doi.org/10.1111/2041-210X.12244.

Kurimoto, E. (2003). Anwaa ethnography. In: S. Uhlig (ed.), *Encyclopaedia Aethiopica*. Vol. 1. Harrassowitz Verlag, Wiesbaden. Pp. 97-103.

Kurimoto, E. (2005). Gambella. In: S. Uhlig (ed.), *Encyclopaedia Aethiopica*. Vol. 2. Harrassowitz Verlag, Wiesbaden. Pp. 668-669.

Kyalangalilwa, B., Boatwright, J.S., Daru, B.H., Maurin, O. & van der Bank, M. 2013. Phylogenetic position and revised classification of *Acacia* s.l. (Fabaceae: Mimosoideae) in Africa, including new combinations in *Vachellia* and *Senegalia*. *Botanical Journal of the Linnean Society* 172: 500-523.

Langdale-Brown, I., Omaston, H.A. & Wilson, J.G. (1964). *The vegetation of Uganda and its bearing on land-use*. Entebbe, Government. Printer, Uganda.

Last, G. (2009). The geology and soils of Ethiopia and Eritrea. Pp. 25-26 in: Ash, J. & Atkins, J., *Birds of Ethiopia and Eritrea-an atlas of distribution*. Christopher Helm, London.

Lebrun, J. (1947). *Exploration du Parc National Albert. Mission J. Lebrun (1937-1938).* Fascicule 1. La végétation de la plaine alluviale au sud du lac Edouard. Institut des Parcs Nationaux du Congo Belge, Brussels. http://cd.chm-cbd.net/archives_rdc/archives/publications/exploration-national-park-albert/exploration-national-park-albert-first-series/mission-j-lebrun-1937-1938/1947-fascicule-1-la-vegetation-de-la-plaine-07698/lebrun1947chap1c.pdf

Legendre, P. & Anderson, M.J. (1999). Distance-Based Redundancy Analysis: Testing Multispecies Responses in Multifactorial Ecological Experiments. *Ecological Monographs* 69(1): 1-24. https://doi.org/10.1890/0012-9615(1999)069[0001:DBRATM]2.0.CO;2.

Lengyel, A. & Botta-Dukát, Z. (2019). Silhouette Width Using Generalized Mean. A Flexible Method for Assessing Clustering Efficiency. *Ecology and Evolution* 9(23): 13231-13243. https://doi.org/10.1002/ece3.5774.

Lennon, J.J., Koleff, P., Greenwood, J.J.D. & Gaston, K.J. (2001). The Geographical Structure of British Bird Distributions: Diversity, Spatial Turnover and Scale. *Journal of Animal Ecology* 70(6): 966-979. On-line 2002 https://doi.org/10.1046/j.0021-8790.2001.00563.x

Liljequist, (1986). Some aspects of the climate of Ethiopia. *Symbolae Botanicae Upsaliensis* 26(2): 19-30.

Lillesø, J.-P.B., Shrestha, T.B., Dhakal, L.P., Nayaju, R.P. & Shrestha, R. (2005). *The map of potential vegetation of Nepal: a forestry/agro-ecological/biodiversity classification system.* Center for Skov, Landskab og Planlægning/Københavns Universitet. Development and Environment No. 2/2005. https://curis.ku.dk/ws/files/20497354/de2_001.pdf

Lillesø, J.-P.B., Breugel, P. van, Kindt, R., Bingham, M., Sebsebe Demissew, Dudley, C., Friis, I., Gachathi, F., Kalema, J., Mbago, F., Minani, V., Moshi, H.N., Mulumba, J., Namaganda, M., Ndangalasi, H.J., Ruffo, C., Jamnadass, R., Graudal, L.O.V. (2011). *Potential natural vegetation of Eastern Africa (Ethiopia, Kenya, Malawi, Rwanda, Tanzania, Uganda and Zambia).* Volume 1: The atlas. Forest & Landscape, University of Copenhagen. Forest & Landscape Working Papers, No. 61. https://www.worldagroforestry.org/publication/potential-natural-vegetation-eastern-africa-ethiopia-kenya-malawi-rwanda-tanzania-8 & https://www.researchgate.net/publication/275523909_Potential_natural_vegetation_of_Eastern_Africa_Ethiopia_Kenya_Malawi_Rwanda_Tanzania_Uganda_and_Zambia_Volume_1_The_atlas

Lind, E.M. & Morrison, M.E. (1974). *East African vegetation.* Longman, London.

Linder, H.P., de Klerk, H.M., Born, J., Burgess, N.D., Fjeldså, J. & Rahbek, C. (2012). The partitioning of Africa: Statistically defined biogeographical regions in sub-Saharan Africa. *Journal of Biogeography* 39: 1189-1205. https://doi.org/10.1111/j.1365-2699.2012.02728.x

Linder, H.P., Lovett, J.C., Mutke, J., Barthlott, W., Jürgens, N., Rebelo, A.G. & Küper, W. (2005) A numerical re-evaluation of the sub-Saharan phytochoria of mainland Africa. *Biologiske Skrifter* 55: 229-252. https://www.nees.uni-bonn.de/research-/systematics-evolution-ecology/biogeography-and-macroecology-biomaps/biomaps-pdf/linder-lovett-mutke-et-al.-2005-a-numerical-re-evaluation-of-the-sub-saharan-phytochoria.-biologiske-skrifter-55-229-252

Maechler, M., Rousseeuw, P., Struyf, A., Hubert, M. & Hornik, K. (2019). *Cluster: Cluster Analysis Basics and Extensions.* R package, available on CRAN. Article https://www.researchgate.net/publication/272176869_Cluster_Cluster_Analysis_Basics_and_Extensions – software https://cran.r-project.org/web/packages/cluster/cluster.pdf

Mantel-Niećko, J. (2003). Administrative divisions. Historical concepts of administrative division; Modern concepts of administrative division. In: S. Uhlig (ed.), *Encyclopaedia Aethiopica.* Vol. 1. Harrassowitz Verlag, Wiesbaden. Pp. 97-103.

Marcus, H.G. (1963). Ethio-British Negotiations concerning the Western Border with Sudan, 1896-1902. *The Journal of African History* 4(1): 81-94. https://www.jstor.org/stable/179614

Marshall, C.A.M., Wieringa, J.J. & Hawthorne, W.D. (2020). An interpolated biogeographical framework for tropical Africa using plant species distributions and the physical environment. *Journal of Biogeography* (preprint 2020): 1-14. https://doi.org/10.1111/jbi.13976

Mattei, G.E. (1909). Il Bambù dell'Eritrea (*Oxytenanthera Borzii* nov. sp.). *Bolletino Reale Orto Bot. Palermo* 8: 29-39 & one table. https://www.biodiversitylibrary.org/page/40172609

McClain, J.O. & Rao, V.R. (1975). CLUSTISZ: A Program to Test for the Quality of Clustering of a Set of Objects. *Journal of Marketing Research* 12(4): 456-460. https://www.jstor.org/stable/3151097.

McCune, B., Grace, J.B. & Urban, D.L. (2002). *Analysis of Ecological Communities*. Vol. 28. MjM software design, Gleneden Beach, Oregon.

Mehari Girmay, Tamrat Bekele, Sebsebe Demissew & Ermias Lulekal (2020). Ecological and floristic study of Hirmi woodland vegetation in Tigray Region, Northern Ethiopia. *Ecological Processes* (2020) 9: Article 53, pp. 1-19. https://doi.org/10.1186/s13717-020-00257-2

Mekete Belachew & Metz, M. (2003). Agriculture. In: S. Uhlig (ed.), *Encyclopaedia Aethiopica*. Vol. 1. Harrassowitz Verlag, Wiesbaden. Pp. 148-153.

Mengesha Asefa, Cao, M., He, Y., Ewuketu Mekonnen, Song, X. & Jie, Y. (2020). Ethiopian vegetation types, climate and topography. *Plant Diversity* 42(4): 302-311. https://doi.org/10.1016/j.pld.2020.04.004

Mengesha Tefera, Tadiwos Chernet & Workineh Haro (1996). *Geological map of Ethiopia*. 2. Edition. Ethiopian Ministry of Mines and the Geological Survey of Ethiopia, Addis Ababa.

Mengistu, T., Demel Teketay, Hulten, H. & Yonas Yemshaw (2005). The role of enclosures in the recovery of woody vegetation in degraded dryland hillsides of central and northern Ethiopia. *Journal of Arid Environments* 2: 259-281. https://doi.org/10.1016/j.jaridenv.2004.03.014

Minassie Gashaw & Michelsen, A. (2001). Soil seed bank dynamics and above-ground-cover of a dominant grass, Hyparrhenia confinis, in regularly burned savanna types in Gambella, western Ethiopia. *Biologiske Skrifter* 54: 389-397.

Minassie Gashaw (2002). Influence of heat shock on seed germination of plants from regularly burnt savanna woodlands and grasslands in Ethiopia. *Plant Ecology* 159: 83-93. https://doi.org/10.1023/A:1015536900330

Minassie Gashaw, Michelsen, A., Friis, I., Jensen, M., Sebsebe Demissew & Zerihun Woldu (2002a). Post-fire regeneration strategies and tree bark resistance to heating in frequently burning tropical savanna woodlands and grasslands in Ethiopia. *Nordic Journal of Botany* 22: 19-33. https://doi.org/10.1111/j.1756-1051.2002.tb01615.x

Minassie Gashaw, Michelsen, A., Jensen, M. & Friis, I. (2002b). Soil seed bank dynamics of fire-prone wooded grassland, woodland and dry forest ecosystems in Ethiopia. *Nordic Journal of Botany* 22: 5-17. https://doi.org/10.1111/j.1756-1051.2002.tb01614.x

Mindaye Teshome (2013). *Structure and Composition of Woody Plants in Boswellia Dominated Woodland of Western Ethiopia*. MSc Thesis, Wageningen.

Mindaye Teshome, Abeje Eshete & Bongers, F. (2017). Uniquely regenerating frankincense tree populations in western Ethiopia. *Forest Ecology and Management* 389: 127-135. https://doi.org/10.1016/j.foreco.2016.12.033

Miran, J. (2010). Red Sea slave trade in the 19th century. In: S. Uhlig & A. Bausi (eds.), *Encyclopaedia Aethiopica*. Vol. 4. Harrassowitz Verlag, Wiesbaden. Pp. 674-676.

Mohamed, M.M. & Elmahdy, S.I. (2917). Remote sensing of the Grand Ethiopian Renaissance Dam: a hazard and environmental impacts assessment. *Geomatics, Natural Hazards and Risk* 8(2): 1225-1240. DOI: http://dx.doi.org/10.1080/19475705.2017.1309463

Monod, T. (1957). Les Grandes Divisions Chorologiques de l'Afrique: rapport présenté à la réunion de spécialistes sur la phytogéographie (Yangambi, 29 juillet – 8 août 1956). *Publications of the Commission for Technical Cooperation in Africa South of the Sahara / Scientific Council for Africa South of the Sahara (C.C.T.A./C.S.A) No. 24*. London.

Morin, X., Fahse, L., Jactel, H., Scherer-Lorenzen, M., García-Valdés, R. & Bugmann, H. (2018). Long-Term response of forest productivity to climate change is mostly driven by change in tree species composition. *Scientific Reports* 8(1): 5627. https://doi.org/10.1038/s41598-018-23763-y.

Motyka J. & Pichi Sermolli, R.E.G. (1952). Usneae in Missione ad Lacum Tana et Semien a R. Pichi Sermolli anno 1937 lectae. *Webbia* 8(2): 383-404. 1952. https://doi.org/10.1080/00837792.1952.10669605

Mucina, L. & Rutherford, M.C. (2006). *The vegetation of South Africa, Lesotho and Swaziland*. Pretoria, South African National Biodiversity Institute (SANBI).

Mulatu Wubneh (2015). This land is my land: the Ethio-Sudan boundary and the need to rectify arbitrary colonial boundaries. *Journal of Contemporary African Studies* 33(4): 441-466. https://doi.org/10.1080/02589001.2016.1143602

Negri, G. (1940). Per uno schema cartografico della vegetazione dell'Africa Orientale Italiana. *Rivista Geografica Italiana* 47: 2-16, plus one folded map.

Odong, T.L., Heerwaarden, J. van, Jansen, J., Hintum, T.J.L. van & Eeuwijk, F.A. van (2011). Determination of Genetic Structure of Germplasm Collections: Are Traditional Hierarchical Clustering Methods Appropriate for Molecular Marker Data? *TAG. Theoretical and Applied Genetics / Theoretische und Angewandte Genetik* 123(2): 195-205. https://doi.org/10.1007/s00122-011-1576-x.

Oksanen, J., Guillaume Blanchet, F., Friendly, M., Kindt, R., Legendre, P., McGlinn, D., Minchin, P.R., O'Hara, B., Simpson, G.L., Stevens, H. & Wagner, H.H. (2019). *Vegan: Community Ecology Package.* https://CRAN.R-project.org/package=vegan.

Olson, D.M., Dinerstein, E., Wikramanayake, E.D., Burgess, N.D., Powell, G.V., Underwood, E.C., D'Amico, J.A., Itoua, I., Strand, H.E., Morrison, J.C., Loucks, C.J., Allnutt, T.F., Ricketts, T.H., Yumiko Kura, Lamoreux, J.F., Wettengel, W.W., Prashant Hedao, P. & Kassem, K.R. (2001). Terrestrial ecoregions of the world: A new map of life on earth: A new global map of terrestrial ecoregions provides an innovative tool for conserving biodiversity. *BioScience* 51: 933-938. https://doi.org/10.1641/0006-3568(2001)051[0933:TEOTWA]2.0.CO;2 With GIS map https://www.worldwildlife.org/publications/terrestrial-ecoregions-of-the-world

Pankhurst, R. (1961). *An introduction to the economic history of Ethiopia*. Lalibela House, Woodford Green, London.

Pankhurst, R. (1997). *The Ethiopian Borderlands: Essays in Regional History from Ancient Times to the End of the 18th Century*. Red Sea Press, Lawrenceville, New Jersey.

Pankhurst, R. (2010a). Slave trade from ancient times to the 19th century. In: S. Uhlig & A. Bausi (eds.), *Encyclopaedia Aethiopica*. Vol. 4. Harrassowitz Verlag, Wiesbaden. Pp. 673-674.

Pankhurst, R. (2010b). Trade in the middle and early modern ages. In: S. Uhlig & A. Bausi (eds.), *Encyclopaedia Aethiopica*. Vol. 4. Harrassowitz Verlag, Wiesbaden. Pp. 976-978.

Passalacqua, N. (2015). On the definition of element, chorotype and component in biogeography. *Journal of Biogeography* 42: 611-618. doi:10.1111/jbi.12473

Paton, A.J., Friis, I. & Sebsebe Demissew (2018). A new species of *Leucas*, *L. gypsicola* (Lamiaceae), from gypsum outcrops in eastern Ethiopia. *Kew Bulletin* 73: Article 59. Pp. 1-6. https://doi.org/10.1007/s12225-018-9781-2

Pawley, S. (2020). *R.learn.ml2 addon for GRASS GIS*. Open Source Geospatial Foundation. [Computer software]. https://grass.osgeo.org/grass78/manuals/addons/r.learn.ml2.html

Pedregosa, F., Varoquaux, G., Gramfort, A., Michel, V., Thirion, B., Grisel, O., Blondel, M., Prettenhofer, P., Weiss, R., Dubourg, V., Vanderplas, J., Passos, A., Cournapeau, D., Brucher, M., Perrot, M. & Duchesnay, É. (2011). Scikit-learn: Machine Learning in Python. *Journal of Machine Learning Research* 12(85): 2825-2830. https://www.jmlr.org/papers/volume12/pedregosa11a/pedregosa11a.pdf

Peres-Neto, P.R., Legendre, P., Dray, S. & Borcard, D. (2006). Variation Partitioning of Species Data Matrices: Estimation and Comparison of Fractions. *Ecology* 87(10): 2614-2625. https://doi.org/10.1890/0012-9658(2006)87[2614:VPOSDM]2.0.CO;2.

Phillips, S. (1995). Poaceae (Gramineae). In: I. Hedberg & S. Edwards (eds), *Flora of Ethiopia and Eritrea*. Vol. 7. National Herbarium, Addis Ababa University, Addis Ababa; Uppsala University, Uppsala.

Pichi Sermolli, R.E.G. (1938). Ricerche botaniche nella regione del Lago Tana e nel Semien. In: Anonymous (ed.), *Missione di Studio al Lago Tana*. Volume Primo: Relazioni preliminari. Reale Accademia d'Italia, Roma. Pp. 77-103.

Pichi Sermolli, R.E.G. (1940). Osservazioni sulla vegetazione del versante occidentale dell'altipiano etiopico. *Nuovo Giornale Botanico Italiano*. N.S., 42(3): 609-623. https://doi.org/10.1080/11263504009440591

Pichi Sermolli, R.E.G. (1951). Ricerche botaniche. Parte 1. Fanerogame raccolte nel Bacino Idrografico del Lago Tana, nel Semièn, nella Regione di Tucùr-Dinghià ed in Eritrea. In: [Anonymous ed.], *Missione di Studio al Lago Tana*. Vol. 7. Pp. 1-320 & Tab. I-LX. Accademia nazionale del Lincei, Roma.

Pichi Sermolli, R.E.G. (1955). Tropical East Africa (Ethiopia, Somaliland, Kenya, Tanganyika). Pp. 202-260 in UNESCO, *Arid Zone Research*. VI. Plant Ecology. Reviews of Research.

Pichi Sermolli, R.E.G. (1957). Una carta geobotanica dell'Africa Orientale (Eritrea, Ethiopia, Somalia). *Webbia* 13: 15-132 & 1 map. https://doi.org/10.1080/00837792.1957.10669673

Platts, P. J, Omeny, P. & Marchant, R. (2014). AFRICLIM: high-resolution climate projections for ecological applications in Africa. *African Journal of Ecology* 53(1): 103-108. https://doi.org/10.1111/aje.12180

Poncet, C.J. (1709). *A voyage to Æthiopia, made in the Years 1698, 1699, and 1700*. Describing Particularly that Famous Empire; as also the Kingdoms of Dongola, Sennar, part of Egypt, &c, with The Natural History of those Parts. By Monsieur Poncet, M.D.

Faithfully Translated from the French Original. London, Printed for W.Lewis at the Dolphin, next Tom's Coffee-House in Russel Street, Govent-Garden. https://books.googleusercontent.com/books/content?req=AKW5QafGU6chD8Qoy-v5NCjok-BtWkrSbpvRfnfJBfHcC-YAJ6_zofwkhnwq9MEi-LJBT-RaNGTc5caA6TeGnRb8woEcYWoi1Xhpx-chxGiqEnEdzTg6Y22AfunSKFURIQXUccq3WOjo-qSNbxk56veWrY1NEBJQrwnvrVbmWoFdbhAfCY-WQmhoYtqQf3GUtTdo_uBItFM2vco_I6z60pw_eBJRDzXi1707IfLxb21KbK40W7k33a3hnOR-JQ42gTloZgIoXYPDiYet-l3whCa94mliiSKGI1culyH-Dx6hoyjDMuGQLwt_aP3jo

R Core Team. (2020). *R: A Language and Environment for Statistical Computing*. Vienna, Austria: R Foundation for Statistical Computing. https://www.R-project.org/.

Raunkiær, C. (1907). *Planterigets Livsformer og deres betydning for Geografien*. Gyldendahl, København.

Raunkiær, C. (1916). Om Bladstørrelsens anvendelse i den Biologiske Plantegeografi. *Botanisk Tidsskrift* 34: 1-13.

Raunkiær, C. (1934). *The Life Forms of Plants and Statistical Plant Geography*. Oxford University Press, London. https://archive.org/details/in.ernet.dli.2015.271790

Rousseeuw, P.J. (1987). Silhouettes: A Graphical Aid to the Interpretation and Validation of Cluster Analysis. *Journal of Computational and Applied Mathematics* 20 (November): 53-65. https://doi.org/10.1016/0377-0427(87)90125-7.

Saraçli, S., Doğan, N. & Doğan, I. (2013). Comparison of Hierarchical Cluster Analysis Methods by Cophenetic Correlation. *Journal of Inequalities and Applications* 2013(1): 203. https://doi.org/10.1186/1029-242X-2013-203.

Sayre, R., Comer, P., Hak, J., Josse, C., Bow, J., Warner, H., Larwanou, M., Ensermu Kelbessa, Tamrat Bekele, Kehl, H., Amena, R., Andriamasimanana, R., Ba, T., Benson, L., Boucher, T., Brown, M., Cress, J., Dassering, O., Friesen, B., Gachathi, F., Houcine, S., Keita, M., Khamala, E., Marangu, D., Mokua, F., Morou, B., Mucina, L., Mugisha, S., Mwavu, E., Rutherford, M., Sanou, P., Syampungani, S., Tomor, B., Vall, A., Vande Weghe, J., Wangui, E. & Waruingi, L. (2013). *A New Map of Standardized Terrestrial Ecosystems of Africa*. Washington, DC: Association of American Geographers (pp. 1-24). http://www.aag.org/galleries/publications-files/Africa_Ecosystems_Booklet.pdf

Sayre, R., Dangermond, J., Frye, C., Vaughan, R., Aniello, P., Breyer, S., Cribbs, D., Hopkins, D., Naumann, R., Derrenbacher, W., Wright, D., Brown, C., Convis, C., Smith, J., Benson, L., Paco VanSistine, D., Warner, H., Cress, J., Danielson, J., Hamann, S., Cecere, T., Reddy, A., Burton, D., Grosse, A., True, D., Metzger, M.J., Hartmann, J., Moosdorf, N., Dürr, H.H., Paganini, M., De Fourny, P., Arino, O., Maynard, S., Anderson, M. & Comer, P. (2014). *A New Map of Global Ecological Land Units e an Ecophysiographic Stratification Approach*. Association of American Geographers, Washington, DC.

Sayre, R., Karagulle, D., Frye, C., Boucher, T., Wolff, N.H., Breyer, S., Wright, D., Martin, M., Butler, K., Graafeiland, K. van, Touval, J., Sotomayor, L., McGowan, J., Game, E.T. & Possingham, H. (2020). An assessment of the representation of ecosystems in global protected areas using new maps of World Climate Regions and World Ecosystems. *Global Ecology and Conservation* 21 (2020): e00860

Schloeder, C.A. (1999). *Investigation of the Determinants of African Savanna Vegetation Distribution: A Case Study from the Lower Omo Basin, Ethiopia*. All Graduate Theses and Dissertations. 6574. https://digitalcommons.usu.edu/etd/6574

Schouw, J.F. (1822). *Grundtræk til en almindelig Plantegeographie*. Gyldendal, København.

Schweinfurth, G. (1905). Vegetationstypen aus der Kolonie Eritrea. In: Karsten, G. & Schenck, H. (eds.), *Vegetationsbilder* 2(8): Tab. 55-60. https://archive.org/details/VegetationstypenAusDerKolonieEritrea1905

Scikit-learn developers (2020). *Scikit-learn* (0.23.2) [Computer software]. https://scikit-learn.org/stable/modules/generated/sklearn.ensemble.RandomForestClassifier.html

Scott, H. (1955). Journey to the high Simien district, Northern Ethiopia, 1952-1953. *Webbia* 11(1): 425-450. https://doi.org/10.1080/00837792.1956.10669641

Sebsebe Demissew (1988). The Floristic Composition of the Menagesha State Forest and the Need to Conserve Such Forests in Ethiopia. *Mountain Research and Development* 8(2/3): 243-247. https://doi.org/10.2307/3673454

Sebsebe Demissew & Friis, I. (2009). The vegetation types in Ethiopia. In: I. Hedberg, I. Friis & E. Persson (eds.). *Flora of Ethiopia and Eritrea*. Vol. 8. National Herbarium, Addis Abeba University, Addis Abeba & Uppsala University, Uppsala. P.p. 27-32.

Sebsebe Demissew, Brochmann, C., Ensermu Kelbessa, Friis, I., Nordal, I., Sileshi Nemomissa, Tadesse Woldemariam, Tamrat Bekele, Zemede Asfaw & Zerihun Woldu · (2011). The Ethiopian Flora Project: A springboard to other projects. *Acta Universitatis Upsaliensis. Symbolae Botanicae Upsaliensis* 35(2): 189-215. https://www.researchgate.net/publication/260050501_The_Ethiopian_Flora_Project_A_springboard_to_other_projects

Sebsebe Demissew, Cribb, P. & Rasmussen, F. (2004). *Field guide to Ethiopian Orchids*. Royal Botanic Gardens, Kew.

Sebsebe Demissew, Mengistu Wondafrash & Yilma Dellellegn (1996). Ethiopia's natural resource base. In Sue Edwards (ed)., *Important Bird Areas of Ethiopia. A First Inventory*. Ethiopian Wildlife and Natural History Society, Addis Abeba. Pp. 36-53.

Sebsebe Demissew, Nordal, I. & Stabbetorp, O. (2003). *Flowers of Ethiopia and Eritrea: aloes and other lilies*. Shama Books, Addis Abeba.

Senni, L. (1938). Problema forestale e selvicoltura nell' Africa Orientale Italiana. *L'Alpe* 25: 1-16.

Simpson, G.G. (1943). Mammals and the Nature of Continents. *American Journal of Science* 241(1): 1-31. https://doi.org/10.2475/ajs.241.1.1.

SoilGrids team (2020). *Soilgrids250m*. Vers. 2.0. ISRIC. https://soilgrids.org/

Sokal, R.R. & Rohlf, F.J. (1962). The Comparison of Dendrograms by Objective Methods. *Taxon* 11(2): 33-40. https://doi.org/10.2307/1217208.

Solomon Tilahun, Edwards, S. & Tewolde Berhan Gebre Egziabher (1996). *Important bird areas of Ethiopia. A first inventory*. Ethiopian Wildlife and Natural History Society, Addis Ababa.

Song, C & Cao, M. (2017). Relationships Between Plant Species Richness and Terrain in Middle Sub-Tropical Eastern China. *Forests* 8(9): 344. https://doi.org/10.3390/f8090344.

Sosef, M.S.M., Dauby, G., Blach-Overgaard, A., van der Burgt, X., Catarino, L., Damen, T., Deblauwe, V., Dessein, S., Dransfield, J., Droissart, V., Duarte, M.C., Engledow, H., Fadeur, G., Rui Figueira, R., Gereau, R.E., Hardy, O.J., Harris, D.J., de Heij, J., Janssens, S., Klomberg, Y., Ley, A.C., Mackinder, B.A., Meerts, P., van de Poel, J.L., Sonké, B., Stévart, T., Stoffelen, P., Svenning, J.-C., Sepulchre, P., Zaiss, R., Wieringa, J.J. & Couvreur, T.L.P. (2017). Exploring the floristic diversity of tropical Africa. *BMC Biology* 15: Article 15. P.p. 1-23. https://doi.org/10.1186/s12915-017-0356-8

Spaulding, J.L. (2003). Fung. In: S. Uhlig (ed.), *Encyclopaedia Aethiopica*. Vol. 2. Harrassowitz Verlag, Wiesbaden. Pp. 591-592.

Stropp, J., Ladle, R.J., Malhado, A.C.M. Hortal, J. Gaffuri, J., Temperley, W.H., Skøien, J.O. & Mayaux, P. (2016). Mapping ignorance: 300 years of collecting flowering plants in Africa. *Global Ecology and Biogeography* 25: 1085-1096. https://onlinelibrary.wiley.com/doi/10.1111/geb.12468

Sulieman, H.M. & Ahmed, A.G.M. (2013). Monitoring changes in pastoral resources in eastern Sudan: A synthesis of remote sensing and local knowledge. *Pastoralism: Research, Policy and Practice* 3: Article 22. P.p. 1-16. https://link.springer.com/article/10.1186/2041-7136-3-22

Tadesse Alemu (2015). *Geology and tectonics of Ethiopia. Basic Geoscience Mapping Core Process*. Geological Survey of Ethiopia, Ministry of Mines and Energy, Addis Ababa.

Tarekegn Tadesse (2005). Geology. In: S. Uhlig (ed.), *Encyclopaedia Aethiopica*. Vol. 2. Harrassowitz Verlag, Wiesbaden. Pp. 757-760.

Tefera Darge Delbiso, Rodriguez-Llanes, J.M., Altare, C., Masquelier, B. & Debarati Guha-Sapir (2016). Health at the borders: Bayesian multilevel analysis of women's malnutrition determinants in Ethiopia. *Global Health Action* (internet publication). https://doi.org/10.3402/gha.v9.30204

ter Braak, C.J.F. & Šmilauer, P. (2015). Topics in Constrained and Unconstrained Ordination. *Plant Ecology* 216(5): 683-696. https://doi.org/10.1007/s11258-014-0356-5.

Thryambakam, P. & Saini, S.K. (2014). Wetland and Eco Tourism-A Case Study of Lake Ashenge, Dess'aa National Forest & Hermi Natural Forest, Tigray Region Of Northern Ethiopia. *Journal of Agroecology and Natural Resource Management* 1(3): 193-198.

Thulin, M. (2020). The genus *Boswellia* (Burseraceae). The Frankincense Trees. *Acta Universitatis Upsaliensis. Symbolae Botanicae Upsaliensis* 39: 1-142.

Thulin, M. (ed.) (1993-2006). *Flora of Somalia*. Published in four volumes with separate pagination. Royal Botanic Gardens, Kew.

Trapnell, C.G. 2001. *Ecological survey of Zambia. The traverse records of C.G. Trapnell 1932-43*. Vol. 1-3. Royal Botanic Gardens Kew, Kew.

Triulzi, A. (2003). Beni Šangul. In: S. Uhlig (ed.), *Encyclopaedia Aethiopica*. Vol. 1. Harrassowitz Verlag, Wiesbaden. P. 529.

Ullendorff, E. (1967). The Anglo-Ethiopian Treaty of 1902. *Bulletin of the School of Oriental and African Studies, University of London* 30(3): 641-654. http://www.jstor.com/stable/612393

UNEP (1997). *World Atlas of Desertification*. United Nations Environment Programme, London. https://wedocs.unep.org/handle/20.500.11822/30300

UNEP (2008). *Africa: Atlas of Our Changing Environment*. Division of Early Warning and Assessment (DEWA) United Nations Environment Programme (UNEP), Nairobi, Kenya. http://wedocs.unep.org/bitstream/handle/20.500.11822/7717/817.pdf?sequence=3&isAllowed=y

UNEP-WCMC & IUCN (2020). *Protected Planet: The World Database on Protected Areas (WDPA)* [On-line publication]. Accessed September 2020. Cambridge, UK: UNEP-WCMC and IUCN. Available at: www.protectedplanet.net, Ethiopia. https://www.protectedplanet.net/country/ET

Various editors (1952-2012). *Flora of Tropical East Africa*. Published in one or more parts for each family, 263 parts in all. Separate pagination. Various publishers. See introduction to Section 6.

Various editors (1989-2009). *Flora of Ethiopia and Eritrea* (Vol. 3 as *Flora of Ethiopia*). Published in eight volumes with separate pagination. National Herbarium of Ethiopia, Addis Ababa University, and Institute of Systematic Botany, Uppsala University. See introduction to Section 6.

Vilhena, D.A. & Antonelli, A. (2015). A network approach for identifying and delimiting biogeographical regions. *Nature communications* 6: 6848. https://doi.org/10.1038/ncomms7848

Vollesen, K. (1995a). Combretaceae. In: S. Edwards, Mesfin Tadesse & I. Hedberg (eds.), *Flora of Ethiopia and Eritrea*. Vol. 2(2). National Herbarium, Addis Ababa University, Addis Ababa; Uppsala University, Uppsala. Pp. 115-132.

Vollesen, K. (1995b). Sterculiaceae. In: S. Edwards, Mesfin Tadesse & I. Hedberg (eds.), *Flora of Ethiopia and Eritrea*. Vol. 2(2). National Herbarium, Addis Ababa University, Addis Ababa; Uppsala University, Uppsala. Pp. 165-185.

Wallmark, P. (1986). *I höglandets skugga: economi, social organisation och etnisk identitet hos begafolket i norra Wollegas lågland, Etiopien*. Uppsala Universitet, Kulturantropologiska Institutionen, Uppsala.

Walter, H. & Lieth, H. (1960-1967). *Klimadiagramm-Weltatlas*. Fischer Verlag, Jena.

Warman, L., Moles, A.T. & Edwards, W. (2010). Not so simple after all: searching for ecological advantages of compound leaves. *Oikos* 120(6): 813-821. https://doi.org/10.1111/j.1600-0706.2010.19344.x

Warming, E. (1923). Økologiens Grundformer. Udkast til en systematisk Ordning. *Det Kongelige Danske Videnskabernes Selskabs Skrifter, Naturvidenskabelig og Mathematisk Afdeling* 8. Række, Vol. IV(2): 118-187.

Weber, O., Ergua Atinafe, Tesfaye Awas & Friis, I. (2020). *Euphorbia venefica* Trémaux ex Kotschy (Euphorbiaceae) and other shrub-like cylindrically stemmed *Euphorbia* with spirally arranged single spines. *Bulletin de la Société des naturalistes luxembourgeois* 122: 57-82. https://www.snl.lu/publications/bulletin/SNL_2020_122_057_082.pdf

White, F. (1965). The savanna woodlands of the Zambezian and Sudanian Domains. *Webbia* 19(2): 651-68.

White, F. (1970). Classification. In: J.D. Chapman & F. White (eds.): *The evergreen forests of Malawi*. Commonwealth Forestry Institute, University of Oxford. Pp. 78-112.

White, F. (1976). The vegetation map of Africa. The history of a completed project. *Boissiera* 24: 659-666.

White, F. (1980). Notes on the Ebenaceae VIII. The African sections of *Diospyros*. *Bulletin du Jardin Botanique National de Belgique / Bulletin van de Nationale Plantentuin van België* 50(3/4): 445-460. https://doi.org/10.2307/3667840 & https://www.jstor.org/stable/3667840

White, F. (1983). *The vegetation of Africa. A descriptive memoir to accompany the UNESCO/AETFAT/ UNSO vegetation map of Africa*. With 4 coloured vegetation maps (1:5,000,000). Natural Resources Research 20: 1-356.

White, F. (1993). Chorological Classification of Africa: History, Methods and Applications. *Bulletin du Jardin botanique national de Belgique / Bulletin van de National Plantentuin van België* 62(1/4): 225-281. https://doi.org/10.2307/3668279 & https://www.jstor.org/stable/3668279

Wickens, G.E. (1976). Flora of Jebel Marra (Sudan Republic) and its Geographical Affinities. *Kew Bulletin*. Additional Series, Vol. 5: 1-368.

Wild, H. & Grandvaux Barbosa, L.A. (1967). *Vegetation map (1:2,500,000 in colour) of the Flora Zambesiaca*

Area. Map & Descriptive memoir. Flora Zambesiaca Supplement. Collins, Salisbury.

Wolde-Selassie Abute (2010). Resettlement. In: S. Uhlig & A. Bausi (eds.), *Encyclopaedia Aethiopica*. Vol. 4. Harrassowitz Verlag, Wiesbaden. Pp. 376-379.

Wood, S.N. (2003). Thin Plate Regression Splines. *Journal of the Royal Statistical Society*. Series B (Statistical Methodology) 65 (1): 95-114. https://doi.org/10.1111/1467-9868.00374

Wossenu Abtew & Shimelis Behailu Dess (2019). *The Grand Ethiopian Renaissance Dam on the Blue Nile*. Springer Nature, Cham. https://link.springer.com/book/10.1007/978-3-319-97094-3

Zach, M.H. (2007). Meroë. In: S. Uhlig (ed.), *Encyclopaedia Aethiopica*. Vol. 3. Harrassowitz Verlag, Wiesbaden. Pp. 936-938.

Zerihun Woldu & E. Feoli (2001). The shrubland vegetation of Adwa, northern Ethiopia. *Biologiske Skrifter* 54: 319-333.

Index to plant names

This index includes page references to plant names, providing references to all mentioning of names in the main text and in the appendices, as well as names in the legends to illustrations. Names listed in the tables are not indexed.

Index to geographical names

The index includes page references to local Ethiopian names for places of our relevés, for topographical features that form limits of vegetation, etc., as well as to names of oceans and tropical countries outside Ethiopia, used in the phytogeographical notes and discussions. A few relevant institutions outside Ethiopia are also indexed. The name 'Ethiopia', which is mentioned nearly 1800 times, has not been indexed, but the names of the neighbouring countries, 'Djibouti', 'Eritrea', 'Kenya', 'Somalia' 'South Sudan' and 'Sudan' are in the index.

About authors and assistants

Authors (left to right): Ib Friis (taking notes in Transitional semi-evergreen bushland, Dolo Mena, Ethiopia, 2013; photo O. Weber). Paulo van Breugel (at computer in Nairobi, 2009). Odile Weber (in boat on the lakes at Silkeborg, Denmark, 2015; photo: Ib Friis). Sebsebe Demissew (at the Tyrrhenian Sea during visit to the *Centro Studi Erbario Tropicale*, Florence, 2012; photo: Ib Friis).

Ib Friis (OrcID: 0000-0002-2438-1528) is Danish. Educated in biology at the University of Copenhagen and in tropical botany at the University of Uppsala, Sweden, he received a degree as Fil. dr. from Uppsala and as Dr. scient. from Copenhagen. He has worked as a lecturer, associate professor, curator of collections and full professor of botany at the University of Copenhagen, and is now professor emeritus at the Natural History Museum of Denmark. Nearly every year since 1970 he has carried out field work in remote parts of Ethiopia and has written, co-authored or edited ca. 340 papers and books, mainly on floristics, ecology, phytogeography and the history of botanical exploration of the Horn of Africa, systematics and evolution of the plant family Urticaceae, botanical nomenclature and the general history of botany. Ib Friis is member of the Royal Danish Academy of Sciences and Letters, foreign member of the Ethiopian Academy of Sciences, foreign member of the Royal Physiographic Society of Lund, Sweden, honorary foreign member of the Linnean Society of London, and honorary research associate at the Royal Botanic Gardens, Kew, UK.

Paulo van Breugel (OrcID: 0000-0001-9579-0831) is Dutch. Educated at the Wageningen Agricultural University in tropical forestry, ecology and silviculture, and at the University of Copenhagen, where he obtained his Ph.D. entitled 'The potential natural vegetation of eastern Africa – distribution, conservation and future changes.' He has worked on forest genetic resources, macro-ecology, eco- and biogeography and conservation and sustainable management in Syria, Lebanon and Brazil for Biodiversity International, in Kenya for the World Agroforestry Centre, and on the VECEA project, an international project coordinated from the University of Copenhagen, focusing on the study of vegetation and effects of climate change in Eastern Africa. He is now lecturer at Applied Geo-information Science programme of the HAS University of Applied Sciences, 's-Hertogenbosch, the Netherlands (the acronym 'HAS' commemorates the *Hogere Agrarische School*, from which the university developed). Paulo van Breugel has done field work in East Africa and in other parts of the tropics and written or co-authored ca. 65 publications, and written 13 'addons' for the computer software GRASS GIS, all mostly related to tropical vegetation, the production of vegetation maps and the modelling of potential vegetation, using GIS and multivariate statistics.

Odile Weber (OrcID: 0000-0002-0861-2752) is Luxembourgish. Educated in ecology, conservation and the environment at the University of York, she has worked since 2006 as a student, assistant botanist and research team member at the herbarium of the Royal Botanic Gardens, Kew, UK, where she mainly focused on Monocotyledons. She is now a scientific assistant and research fellow at the herbarium of the Musée national d'histoire naturelle in Luxemburg. Odile Weber has done field work in various African countries, including two periods in the field in remote eastern and western parts of Ethiopia. She has written or co-authored 28 papers, mainly on the taxonomy and conservation of Monocotyledons, especially on species of the genus *Aloe* occurring in the Horn of Africa, but also on other African plants, including little known species in the Ethiopian flora, on the flora and vegetation of Luxemburg and on the curation of the national herbarium of Luxemburg.

Sebsebe Demissew (OrcID: 0000-0002-0123-9596) is Ethiopian. He obtained his first degrees in biology at the University of Addis Ababa, followed by a degree of fil. dr. in botany at the University of Uppsala, Sweden. He has worked as a lecturer, associate professor and full professor at the University of Addis Ababa, has been leader of the Ethiopian Flora Project (international project funded from Sweden), of the National Herbarium of Ethiopia and of the Gullele Botanic Garden in Addis Ababa, as well as General Secretary of *L'Association pour l'Étude Taxonomique de la Flora d'Afrique Tropicale* (AETFAT). He has carried out field work in most parts of Ethiopia and elsewhere in Africa, and has written or co-authored ca. 250 scientific papers and books on African plants and on ecology, vegetation and ethnobotany of Ethiopia. He was Ethiopian representative at the VECEA project, is a founding member of the Ethiopian Academy of Sciences, Co-Chair of the Multidisciplinary Expert Panel for the Intergovernmental Platform for Biodiversity and Ecosystem Services (IPBES), foreign member of the Royal Danish Academy of Sciences and Letters and of the Royal Society of London, winner of the Kew International Medal, the "Prof. Luigi Tartufari" International Prize for Biological Science of the *Accademia Nazionale dei Lincei*, and the Jose Cuatrecasas Medal for Excellence in Tropical Botany of the Smithsonian Institution, and he is a honorary research associate at the Royal Botanic Gardens, Kew, UK.

This work would not have been possible without the highly competent and always friendly assistance from a number of assistants. The primary of these are Wege Abebe, herbarium assistant at the National Herbarium of Ethiopia, and Ermias Getachew, driver at the Addis Ababa University, during our field work in Ethiopia and in Addis Ababa. Wege and Ermias worked with the authors on nearly every field trip during this project, ranging from our trips to the relevés near the border with Eritrea in the north to our trips to the relevés in the Omo Valley in the south.

The two principal assistants during the field work (left to right): Wege Abebe (in Tigray, 2009; photo: Ib Friis). Ermias Getachew (in Harar, 2006; photo: Ib Friis).

DET KONGELIGE DANSKE VIDENSKABERNES SELSKAB
udgiver følgende publikationsrækker:
THE ROYAL DANISH ACADEMY OF SCIENCES AND LETTERS
issues the following series of publications:

	AUTHORIZED ABREVIATIONS
Scientia Danica. Series B, Biologica *Formerly: Biologiske Skrifter,* 4° (Botany, Zoology, Palaeontology, general Biology)	Sci.Dan.B
Scientia Danica. Series H, Humanistica, 4 *Formerly: Historisk-filosofiske Skrifter,* 4° (History, Philosophy, Philology, Archaeology, Art History)	Sci.Dan.H.4
Scientia Danica. Series H, Humanistica, 8 *Formerly: Historisk-filosofiske Meddelelser,* 8° (History, Philosophy, Philology, Archaeology, Art History)	Sci.Dan.H.8
Scientia Danica. Series M, Mathematica et physica *Formerly: Matematisk-fysiske Meddelelser,* 8° (Mathematics, Physics, Chemistry, Astronomy, Geology)	Sci.Dan.M
Oversigt, Annual Report, 8°	Overs.Dan.Vid.Selsk.

Correspondence
Manuscripts are to be sent to

The Editor
Det Kongelige Danske Videnskabernes Selskab
H.C. Andersens Boulevard 35
DK-1553 Copenhagen V, Denmark
Tel: +45 33 43 53 00
E-mail: kdvs@royalacademy.dk
royalacademy.dk

Questions concerning subscription to the series should be directed to the Academy

Editor Marianne Pade

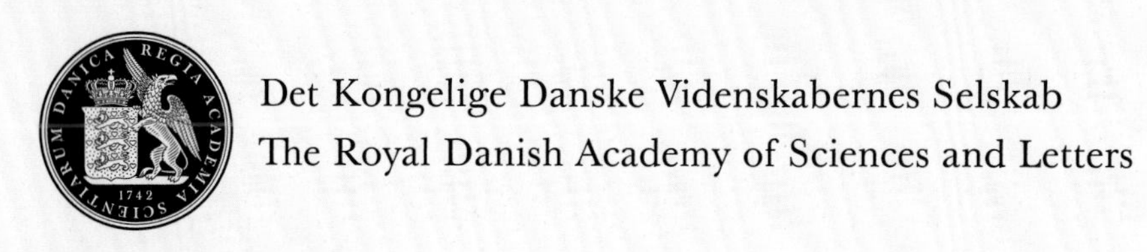

Biologiske Skrifter, BS

Biol.Skr.Dan.Vid.Selsk.

Scientia Danica. Series B, Biologica

Sci.Dan.B

VOL. DKK

1 Inge Bødker Enghoff:
Regionality and biotope exploitation in Danish Ertebølle and adjoining periods. 2011. 394 pp. Lavishly illustrated. 320.-

2 Jesper Guldberg Hansen, Reinhardt Møbjerg Kristensen and Aslak Jørgensen:
The armoured marine tardigrades (Arthrotardigrada, Tardigrada). 2012. 92 pp. Lavishly illustrated. 150.-

3 Sachin Rustgi, Nuan Wen, Claudia Osorio, Rhoda A.T. Brew-Appiah, Shanshan Wen, Richa Gemini, Jaime H. Mejias, Nii Ankrah, Charles P. Moehs and Diter von Wettstein:
Natural dietary therapies for the 'gluten syndrome'. 2014. 87 pp. Lavishly illustrated. 150.-

4 Elvira Cotton, Finn Borchsenius and Henrik Balslev:
A revision of Axinaea (Melastomataceae). 2014. 120 pp. Lavishly illustrated. 160.-

5 Pétur M. Jónasson, Kirsten Hamburger, Claus Lindegaard, Ebbe Lastein and Peter Dall †:
Respiratory adaptation of zoobenthos to access to oxygen and habitat structure in dimictic eutrophic lakes affecting vertical distribution. 2015. 38 pp. Illustrated. 80.-

6 *Tropical Plant Collections: Legacies from the Past? Essential Tools for the Future?* Proceedings of an international symposium held by The Royal Danish Academy of Sciences and Letters in Copenhagen, 19th–21st of May, 2015.
Edited by Ib Friis and Henrik Balslev. 2017. 320 pp. Lavishly illustrated. 300.-

7 Ruth Nielsen og Steffen Lundsteen:
Danmarks Havalger bind 1. Rødalger (Rhodophyta). 2019. 398 s. Rigt illustreret. Sælges kun sammen med bind 2 for 500.-

8 Ruth Nielsen og Steffen Lundsteen:
Danmarks Havalger bind 2. Brunalger (Phaeophyceae) og Grønalger (Chlorophyta). 2019. 476 s. Rigt illustreret. Sælges kun sammen med bind 1 for 500.-

9 Ib Friis, Paulo van Breugel, Odile Weber and Sebsebe Demissew:
The Western Woodlands of Ethiopia. 2021. 524 pp. Lavishly illustrated. 450.-

Priser ekskl. moms / Prices excl. VAT

General guidelines

The Academy invites original papers that contribute significantly to research carried on in Denmark. Foreign contributions are accepted from temporary residents in Denmark, participants in a joint project involving Danish researchers, or those in discussion with Danish contributors.

Instructions to authors

Please make sure that you use the stylesheet on our homepage www.royalacademy.dk. All manuscripts will be refereed. Authors of papers accepted for publication will receive digital proofs; these should be returned promptly to the editor. Corrections other than of printer's errors will be charged to the author(s) insofar as the costs exceed 15% of the cost of typesetting.

Authors receive a total of 50 free copies. Authors are invited to provide addresses of up to 20 journals to which review copies could profitably be sent.

Manuscripts can be returned, but only upon request made before publication of the paper. Original photos and artwork are returned upon request.

Manuscript

General
Book manuscripts and illustrations must comply with the guidelines given below. The digital manuscript and illustrations plus one clear printed copy of both should be sent to the editor of the series. Digital manuscripts should be submitted in a commonly used document format (contact the editor if you are in doubt), and the illustrations should be sent as separate files. Please do not embed illustrations within text files.

A manuscript should not contain less than 48 printed pages. This also applies to the Sci.Dan.M. where contributions to the history of science are welcome.

Language
Manuscripts in Danish, English, German and French are accepted; in special cases other languages too. Linguistic revision may be made a condition of final acceptance.

Title
Titles should be kept as short as possible, preferring words useful for indexing and information retrieval.

Abstract, Summary
An abstract in English is required. It should be of 10 to 15 lines, outline main features, stress novel information and conclusions, and end with the author's name, title, and institutional and/or private postal address. – Papers in Danish must be provided with a summary in another language as agreed between author and editor.

Manuscript
Page 1 should contain title, author's name and the name of the Academy. Page 2: Abstract, author's name and address. Page 3: Table of contents if necessary. Consult a recent issue of the series for general layout. Indicate the position of illustrations and tables. A printout must accompany manuscripts submitted electronically.

Figures
All illustrations submitted must be marked with the author's name. It is important that the illustrations are of the highest possible quality. Foldout figures and tables should be avoided.

References
In general, the editor expects all references to be formally consistent and in accordance with accepted practice within the particular field of research. Bibliographical references should be given in a way that avoids ambiguity.